Arnold's Problems

Springer
*Berlin*
*Heidelberg*
*New York*
*Hong Kong*
*London*
*Milan*
*Paris*
*Tokyo*

ℝ
ℂ
ℍ
$E_6$
$E_7$
$E_8$
$A_3$
$B_3$

04 Math

# Arnold's Problems

Edited by Vladimir I. Arnold

Springer

PHASIS

Vladimir I. Arnold
Steklov Mathematical Institute
ul. Gubkina 8
117966 Moscow, Russia
*e-mail:* arnold@genesis.mi.ras.ru
and
CEREMADE
Université de Paris-Dauphine
Place du Maréchal de Lattre de Tassigny
75775 Paris Cedex 16, France
*e-mail:* arnold@ceremade.dauphine.fr

Originally published in Russian as "Zadachi Arnolda"
by PHASIS, Moscow, Russia, 2000 (ISBN 5-7036-0060-X)

Library of Congress Control Number: 2004105411

Mathematics Subject Classification (2000): 00-02

ISBN 3-540-20614-0 Springer-Verlag Berlin Heidelberg New York



Springer-Verlag is a part of Springer Science+Business Media

springeronline.com


Printed in Germany



Edited by PHASIS using a special LaTeX macro package
Cover design: Erich Kirchner, Heidelberg

Printed on acid-free paper 46/3142/LK - 5 4 3 2 1 0

## *Author's Preface to the Second Edition*

For a working mathematician, it is much more important to know what questions are not answered so far and failed to be solved by the methods already available, than all lists of numbers already multiplied, and than the erudition in the ocean of literature that has been created by previous generations of researchers over twenty thousand years.

To tell the students about the most (in the author's opinion) interesting unsolved problems—this is the purpose of the present book which is composed of problems formulated at seminars in Moscow and Paris starting from 1958. The main body of the book is formed by comments of my former students about the current progress in the problems solution (featuring bibliography inspired by them).

The observed half-life of the problem (of its more or less complete solution) is about seven years on average. Thus, many problems are still open, and even those that are mainly solved keep stimulating new research appearing every year in journals of various countries of the World.

The invariable peculiarity of these problems was that Mathematics was considered there not as a game with deductive reasonings and symbols, but as a part of natural science (especially of Physics), that is, as an experimental science (which is distinguished among other experimental sciences primarily by the low costs of its experiments).

Problems of binary type admitting a "yes–no" answer (like the Fermat problem) are of little value here. One should rather speak of wide-scope programs of explorations of new mathematical (and not only mathematical) continents, where reaching new peaks reveals new perspectives, and where a preconceived formulation of problems would substantially restrict the field of investigations that have been caused by these perspectives. It is not sufficient to know whether there is a river beyond the mountain; it does remain to cross this river! Evolution is more important than achieving records.

In the raw cases where the imperatives of simplicity and beauty contradicted each other, the author usually has chosen the latter, having in mind that it was the beauty rather than the utility of science (including Mathematics) that historically played the role of the main engine leading researchers to the discoveries proved to be most useful nowadays (such as the conic sections for space navigation, or the Maxwell equations for television and radar).

I would wish the reader not to be held back by the fact that such applications are not evident at the beginning: if a result is truly beautiful then it will certainly be of use in due course!

*V. I. Arnold*

Moscow, 2003

*Le monde est soutenu*
*par les enfants*
*qui étudient.*

Roger Peyrefitte

Les juifs. Paris: Flammarion, 1965, p. 281

# *Author's Preface to the First Edition*

Moscow has a long-standing fame for its mathematical seminars. At the beginning of each academic term I formulate problems, usually a dozen or two. The future analysis shows that the average half-life of a problem (after which it would be more or less solved) is about *seven years*.

Poincaré used to say that precise formulation, as a question admitting a "*yes* or *no*" answer, is possible only for problems of little interest. Questions that are really interesting would not be settled this way: they yield *gradual* forward motion and *permanent* development.

In Poincaré's opinion, the main essence of any problem is to understand what is definitive in its formulation, and what can be varied (like boundary conditions in an elliptic problem).

I. G. Petrovskiĭ, who was one of my teachers in Mathematics, taught me that the most important thing that a student should learn from his supervisor is that some *question is still open*. Further choice of the problem from the set of unsolved ones is made by the student himself. To select a problem for him is the same as to choose a bride for one's son.

Mainly, I did not write my problems down, especially in the sixties; therefore most of them are probably lost. Some problems are included in my papers and books. Sometimes I reconstructed my problems to the seminar from conversations with my colleagues and friends. I hope that below the authors are quoted in most of such situations.

There are two principal ways to formulate mathematical assertions (problems, conjectures, theorems, ...): Russian and French. The *Russian way* is to choose *the most simple and specific* case (so that nobody could simplify the formulation preserving the main point). The *French way* is *to generalize the statement as far as nobody could generalize it further*.

I assume that this division more or less coincides with the division of people into the right-hemisphere resolvers of posed problems, and the left-hemisphere authors of research programs.

Once, when I was a younger student, I asked R. L. Dobrushin (who was a graduate student) a question. "A fool can ask so many questions that a hundred of intellectuals could not answer them," Dobrushin said. As for me, questions should nevertheless be published. By the way, it turned out that the question that I had asked Dobrushin that time—*whether the perimeter of a rectangle can increase as the result of a sequence of foldings and unfoldings*—remains open and is treated as folklore (although, seems to me, I published it, say, 40 years ago).

Ya. B. Zeldovich thought that posing a problem is a much finer art than its solution. "*Once you formulate a precise question*, he said, *there already appears a mathematician able to solve it.* In fact, *mathematicians are like flies*, fit to walk on the ceiling!"

This had led him to a well-known struggle, where Pontryagin and Logunov tried to criticize the mathematical rigor of his theories. It resulted in the following phrase in Pontryagin's book: "Some physicists think that one can make a correct use of the mathematical analysis without full knowledge of its foundation. And I do agree with them."

Zeldovich was offended by this phrase. "Why hasn't he named me?" Yakov Borisovich said to me then.

I am deeply indebted to a large number of my former and present students who have written this book. I tried to quote them appropriately.

Mathematical training in Moscow usually begins before the school age. Here is a couple of excercises (children 4–5 years old would have solved them in half an hour):

1) From a barrel of wine, a spoon was poured into a cup of tea, and then the same spoon of the obtained (nonhomogeneous!) mixture was poured from the cup back into the barrel. Where did the amount of the foreign beverage become greater?

2) On a chess board, two opposite angle squares (a1, h8) are cut off. Can the 62 remaining squares be covered by 31 domino pieces (without overlaps), every piece covering two (neighboring) squares?

Leibnitz thought that a curve intersects its curvature circle at four coincident points and that $d(ab) = (da)(db)$.

Hilbert argued that a really interesting work in mathematics rarely happens to be correct. For example, in his survey of relativity theory, he affirmed that "*simultaneity* exists by itself." His description of geometry of numbers from the article dedicated to Minkowski is beyond any critics at all.

A. Weil wrote that his famous dissertation had been read only by two opponents; but even they understood too little because of their lack of proficiency (the work was erroneous). And this is yet one of the most important works of our century (1926–1928) in number theory.

Errors committed by Poincaré himself are too widely known to be recalled here: he confused homology with homotopy and missed the 3-manifold of dodecahedral lens type which is now named after him. Many questions in the theory of differential equations, dynamical systems, and celestial mechanics "solved" by him still remain open.

Descartes wrote to Huygens: "If I see any vacuum in Nature, it is only in Pascal's head."

Mathematician N refused to correct misprints while re-editing his book, in order not to rob the reader of pleasure in finding errors.

It seems, Napoleon said that a person, who is unable to *think*, cannot also be taught anything.

I hope that the present book will teach at least anybody to think (by the above problems 1 and 2 though).

*V. I. Arnold*

Garches (France), 1999

# *Editorial to the Second Edition*

You are looking at the second edition of the title "Arnold's Problems," which is now in English. Its size has noticeably grown compared with the first Russian edition of 2000—by more than a one third; for new problems and comments have appeared, and some old comments have been supplemented. The number of authors of comments has doubled, from 29 to 59.

The format of the comments has also been modified. The name of the comment's author is now shown at the beginning of the comment (beside the problem's number), no longer at its end. If there are several comments to a problem, then the problem number in every comment is preceded by a symbol indicating if this comment is the first one ($\triangledown$), an intermediate one ($\frac{\vartriangle}{\triangledown}$) or the last one ($\vartriangle$). Each comment is opened by a notation indicating its nature: the letter $\mathcal{H}$ means that the comment is *historic*, and $\mathfrak{R}$ means that the comment is devoted to the *results* of the research on the problem.

Just as in the first (Russian) edition of this book, twin problems appear here (see the explanation on page XIII).

For the problems appearing in the first edition, the numbers have been preserved. In cases when problems of the preceding years forgotten in the former edition have since been discovered, they are appended at the end of the list of problems of the corresponding year.

We also point out a feature of the bibliography. If an article was published in a journal in Russian that is translated into English on a regular basis (cover-to-cover), then its bibliographical description includes only the translation of the article (since the original is easily found in this case). In the cases when it might be difficult to find respectively the English translation or the Russian original of an article, the references to both of them are provided.

We acknowledge our pleasant duty to thank Professors M. S. P. Eastham, A. G. Khovanskiĭ, L. P. Kotova, M. B. Sevryuk, and O. V. Sipacheva who have contributed to this edition by improving the English text.

All formulations of the problems and all the comments have been checked by Vladimir Igorevich Arnold. Some comments, in comparison with the first edition, have been reduced by excluding the descriptions of unpublished and unverified results. Unfortunately, not all potential authors of comments accepted our suggestion to write comments to the problems they had studied. Now we keep on inviting all the colleagues to participate in commenting Arnold's Problems. For more information, see the Internet site `http://www.phasis.ru`.

In order to make the author's famous Russian original edition accessible to readers worldwide, PHASIS and Springer-Verlag have collaborated in the publication of this enlarged and updated English edition using the know-how, experience and abilities of both publishers.

*V. Philippov*
*A. Yakivchik*

*M. Peters*

Heidelberg, 2004

Moscow, 2004

*To ask the right question*
*is harder than to answer it.*

Georg Cantor

# *Editorial to the First Edition*

The present title represents the problems that have been posed by Vladimir Igorevich Arnold during a period of over 40 years.

This is principally a fairly complete list of problems presented by him at his seminar on the theory of singularities of differentiable mappings, twice a year at the beginning of each academic term. (This famous seminar has been working at the Department of Mechanics and Mathematics of Moscow State University for over 30 years and deserves the title of one of the leading World centers of mathematical science.) In addition, there are problems published by Vladimir Igorevich in his numerous papers and books. It is clear, however, that not all Arnold's problems have been collected so far, and we would be grateful to those readers who will report to us any problems not appearing in the present volume.

The book consists of two parts. The first part comprises the formulations of the problems; brief explanations that are italicized there are due to the author. The second part is a collection of comments including a survey of results on the given problem or, in some cases, a historic reference. Almost all the comments are signed by their authors (which are mostly the former students of Vladimir Igorevich); the brief unsigned comments belong either to the author or to the editor. In a few cases, the authors include a description of their unpublished and unverified results in their comments, sometimes even those on classical problems; such assertions should be regarded as conjectures. The bibliography to all comments has been carefully checked by the editor.

For the sake of historic certainty, we preserve the so-called twin problems, i. e., the problems that date back to different years but are almost identical in their essence. Only one of these problems (and not always the earliest) is commented on in such a case, the other twin problems being supplied with a reference: "See the comment to problem ⟨*number*⟩." Such references are used in some other cases when information from the comment to one problem applies to another problem as well.

All mathematical notations appearing in the book are commonly used. However, the notations for spheres and balls of various dimensions must be clarified. Non-parallelizable $n$-dimensional spheres (i. e., for $n \notin \{0,1,3,7\}$) are always denoted by $S^n$. The spheres of dimensions $n = 0,1,3,7$ are generally denoted by $\mathbb{S}^n$, but in some exceptional cases (either pointed out by V. I. Arnold or, for example, in dealing with the bouquet of spheres $S^2 \vee S^1$) also by $S^n$. The closed ball of dimension $n \geq 3$ is denoted by $B^n$. For the two-dimensional ball (disk) the notation $D^2$ is mainly used. Finally, the one-dimensional ball (line segment) $\{x \in \mathbb{R} \mid a \leq x \leq b\}$ is denoted by $[a;b]$.

We hope the reader appreciates the tough work that we had to perform while preparing this title, and we would like to thank all participants in this project, especially the authors of comments. We are not entirely satisfied by the quality of our own efforts, but our main desire was the early appearance of the book. Many problems have been left without comments; with several exceptions this means only that nobody has undertaken the task to write such a comment so far.

At the same time, we believe that the work on this project is still only at its first steps, and we would be indebted to everybody who will contribute to the next edition of this title with remarks, suggestions, corrections, new comments or historic references.

*V. B. Philippov*
*M. B. Sevryuk*

Moscow, 1999

# *Contents*

*Et à quoi bien exécuter des projets,*
*puisque le projet est en lui-même*
*une jouissance suffisante?*

Charles Baudelaire

Le spleen de Paris, XXIV (Les projets)

# The Problems

# 1956

**1956-1.** "The rumpled dollar problem": is it possible to increase the perimeter of a rectangle by a sequence of foldings and unfoldings?

# 1958

**1958-1.** Let us consider a partition of the closed interval $[0;1]$ into three intervals $\Delta_1$, $\Delta_2$, $\Delta_3$ and rearrange them in the order $\Delta_3$, $\Delta_2$, $\Delta_1$. Explore the resulting dynamical system $[0;1] \to [0;1]$: is it true that the mixing rate and similar ergodic characteristics are the same for almost all lengths $(\Delta_1, \Delta_2, \Delta_3)$ of the partition intervals?

An analogous question may be asked for $n$ intervals and for arbitrary permutations as well (changing the orientation of some intervals also being allowed).

**1958-2.** Let all four faces of a tetrahedron have equal areas. Prove that the lengths of opposite edges are equal (and all faces are congruent!). *The idea is quite simple: cut along three edges from a vertex and develop.*

**1958-3.** Find a multidimensional version of the Hilbert conjecture on the number of limit cycles of a polynomial vector field. *For instance, one is interested in the number of integral curves connecting two algebraic or invariant manifolds and sufficiently "monotone."*

# 1959

**1959-1.** Let the biholomorphic mapping $z \mapsto z + a + b\sin z \mod 2\pi$ of the circle $\operatorname{Im} z = 0$ onto itself be not conjugate analytically to a rotation but have an irrational rotation number. Is it true that in any neighborhood of the circle, there is a periodic orbit?

# 1963

**1963-1.** Is there true instability in multidimensional problems of perturbation theory where the invariant tori do not divide the phase space?

**1963-2.** Prove the presence of nondegenerate hyperbolic points (and separatrix splitting) in any neighborhood of an elliptic fixed point 0 of a generic analytic area-preserving mapping $(\mathbb{R}^2, 0) \hookleftarrow$.

**1963-3.** Are there bounded motions filling up a set of positive measure in the three (and $n$) body problem, for any values of the masses and for the distances comparable with each other? Does there exist a critical value of the perturbation parameter $\mu$ at which the invariant torus with given Diophantine frequency vector breaks up?

**1963-4.** Let $T$ be an orientation-preserving analytic diffeomorphism of a circle onto itself with Diophantine rotation number $\omega$. Can one always turn $T$ into the rotation $T_0$ through the angle $2\pi\omega$ via an analytic change of variables $S$: $STS^{-1} = T_0$?

**1963-5.** Consider a system of linear differential equations with quasi-periodic coefficients

$$\dot{q} = \omega, \quad \dot{x} = A(q)x; \qquad q \in \mathbb{T}^k = \mathbb{R}^k/2\pi\mathbb{Z}^k, \quad x \in \mathbb{R}^n,$$

where $\omega \in \mathbb{R}^k$ is a constant vector with Diophantine components while $A : \mathbb{T}^k \to \mathrm{gl}(n, \mathbb{R})$ is an analytic function. Is such a system always reducible for $k > 1, n > 1$?

**1963-6.** Let $\Gamma$ be a (generally noncommutative) group with finitely many generators $a_1, \ldots, a_s$. By a dynamical system with the "time" $\Gamma$ we shall mean an *action* of the group $\Gamma$ on a space with measure $\Omega$ by measure-preserving transformations $A_\gamma$ ($\gamma \in \Gamma$). For such a system, time averages may be defined as follows. Let us consider the set $\Gamma_n$ of elements of $\Gamma$ that can be obtained by $n$ (but not less than $n$) multiplications from $a_1, a_1^{-1}, \ldots, a_s, a_s^{-1}$, and let $N(n)$ be the number of such elements. Then define the "time average" $f_n$ of a function $f$ as

$$f_n(x) = \frac{1}{N(n)} \sum_{\gamma \in \Gamma_n} f(A_\gamma x), \quad x \in \Omega.$$

Now let $\Omega$ be a homogeneous space, with a transitive action of a compact Lie group $G$ on it; and let the transformations $A_\gamma$ ($\gamma \in \Gamma$) belong to $G$.

Are the ergodic theorems of Birkhoff and von Neumann true for such dynamical systems with a noncommutative time?

The next three problems also concern dynamical systems $(\Omega, G, \Gamma)$ with a noncommutative time $\Gamma$.

**1963-7.** For some groups $\Gamma$ the sequence of points $A_\gamma x$ is uniformly distributed in its closure, if the closure is connected. In other words, the time averages $f_n(x)$ of a continuous function converge to the space average over the closure $\overline{\Gamma(x)}$ of a trajectory $A_\gamma x$ ($\gamma \in \Gamma$):

$$\lim_{n\to\infty} f_n(x) = \frac{1}{\operatorname{mes}\overline{\Gamma(x)}} \int_{\overline{\Gamma(x)}} f(y)\, d\mu(y).$$

Examples are given by the free group $\Gamma$ with two generators $a$, $b$ and the group $\Gamma$ with generators $a$, $b$, $c$ and the relation $abc = e$.

Does this result extend to arbitrary groups $\Gamma$ with finitely many generators?

**1963-8.** Does the result mentioned in the previous problem extend to the noncompact case? (For instance, let $\Omega$ be the Euclidean plane or the Lobachevskian plane.)

**1963-9.** What is the generalization of the result mentioned in problem *1963-7* to the case where a Lie group, e. g., the isometry group of the Lobachevskian plane, is considered as time?

**1963-10.** In what cases is the monodromy group of the system $dx = [A(z)\,dz]x$ of linear differential equations on a Riemann surface $M$ bounded? Here $z \in M$, $x \in \mathbb{C}^n$, and $A(z)\,dz$ is a matrix of differentials which are analytic in $z$ except for a finite set of singular points.

**1963-11.** Consider a system of linear differential equations $dx/dz = A(z)x$, where $z \in \mathbb{C}P^1$, $x \in \mathbb{C}^n$, and $A$ is a matrix which depends on $z$ analytically, except for three singular points $z_1$, $z_2$, $z_3$ on the Riemann sphere $\mathbb{C}P^1$. Denote $\mathbb{C}P^1 \setminus \{z_1, z_2, z_3\}$

by $Z$. If the monodromy group of the system $dx/dz = A(z)x$ is bounded, then this system has a single-valued first integral $(B(z)x, \bar{x}) = \text{const}$, where $B(z)$ is a positive definite self-adjoint matrix, single-valued for $z \in Z$.

Is it true that the surface depicting the solutions of this system in the $(2n+1)$-dimensional manifold $M_c$: $(Bx, \bar{x}) = c$, is uniformly distributed with respect to the following metric: on $Z$, we introduce a metric of constant negative curvature, and on $\mathbb{C}^n(z)$ the metric is defined by the scalar product $(B(z)x, y)$?

**1963-12.** The system $dx/dz = A(z)x$ from the previous problem can be considered as a dynamical system where the role of the time is played by the universal covering of $Z$, i. e., by the Lobachevskian plane. But an ordinary dynamical system with continuous time can also be related to this system. In order to do so, consider a new phase space whose points are the points $(z, x) \in M_c$ together with the direction $\xi$ of a vector tangent to $Z$ at $z$. The motion is defined in the following way: the point $z$ is moving uniformly along the geodesic in the direction of $\xi$, and $x$ over $z$ is moving according to the equations $dx/dz = A(z)x$. The metric and the invariant measure are defined as in the previous problem.

This construction allows us to "multiply" the flow defined on a manifold by a group of automorphisms (which is a representation of the fundamental group of the manifold). The problem is in the study of the resulting "products."

# 1965

**1965-1.** Let $A: \Omega \to \Omega$ be a globally canonical homeomorphism of the $2n$-dimensional toroidal annulus $\Omega = \mathbb{T}^n \times B^n$, where $\mathbb{T}^n = \mathbb{R}^n / 2\pi\mathbb{Z}^n$ denotes the $n$-torus while $B^n \subset \mathbb{R}^n$ is a domain in $\mathbb{R}^n$ homeomorphic to a closed $n$-dimensional ball. Let $p_0$ be an interior point of $B^n$, and $T \subset \Omega$ be the torus $\mathbb{T}^n \times \{p_0\}$. Do $T$ and $AT$ always intersect at not less than $n+1$ (geometrically distinct) points?

In this problem and the subsequent two problems, a mapping $A: \Omega \to \Omega$, where $\Omega = \mathbb{T}^n \times B^n$,

$$\mathbb{T}^n = \{q = (q_1, \ldots, q_n) \operatorname{modd} 2\pi\}, \quad B^n \subset \mathbb{R}^n = \{p = (p_1, \ldots, p_n)\},$$

is said to be *globally canonical* if it is homotopic to the identity transformation and

$$\oint_{\gamma} p\, dq = \oint_{A\gamma} p\, dq$$

($p\,dq = p_1\,dq_1 + \cdots + p_n\,dq_n$) for any closed curve $\gamma \subset \Omega$ (not necessarily homologous to zero).

**1965-2.** Let $A : \Omega \to \Omega$ be a globally canonical diffeomorphism of the $2n$-dimensional toroidal annulus $\Omega = \mathbb{T}^n \times B^n$, where $\mathbb{T}^n = \mathbb{R}^n/2\pi\mathbb{Z}^n$ denotes the $n$-torus while $B^n \subset \mathbb{R}^n$ is a domain in $\mathbb{R}^n$ homeomorphic to a closed $n$-dimensional ball. Let $p_0$ be an interior point of the domain $B^n$ and let $T \subset \Omega$, the torus $\mathbb{T}^n \times \{p_0\}$. Do $T$ and $AT$ always intersect at not less than $2^n$ points (counting multiplicities)?

**1965-3.** Let $A : \Omega \to \Omega$ be a globally canonical diffeomorphism of the $2n$-dimensional toroidal annulus $\Omega = B^n \times \mathbb{T}^n$, where $B^n \subset \mathbb{R}^n$ is a domain in $\mathbb{R}^n$ homeomorphic to a closed $n$-dimensional ball while $\mathbb{T}^n = \mathbb{R}^n/2\pi\mathbb{Z}^n$ denotes the $n$-torus. Suppose that, for any $q \in \mathbb{T}^n$, the spheres $S^{n-1}(q) = \partial B^n \times \{q\}$ and $AS^{n-1}(q)$ are linked in $\partial B^n \times \mathbb{R}^n$ where $\mathbb{R}^n \to \mathbb{T}^n$ is the universal covering. Is it true that, in this set-up, the diffeomorphism $A$ possesses at least $2^n$ fixed points in the annulus $\Omega$ (counting multiplicities)?

# 1966

**1966-1.** What is the connection between the $I$-component $I(t)$ of the solution of the system

$$d\varphi/dt = \omega(I) + \varepsilon f(I,\varphi), \quad dI/dt = \varepsilon F(I,\varphi)$$

($\varphi \in \mathbb{T}^k$, $I \in \mathbb{R}^l$, $0 < \varepsilon \ll 1$) and the solution $J(t)$ of the "evolution equation"

$$dJ/dt = \varepsilon \bar{F}(J), \quad \bar{F}(J) := \frac{1}{(2\pi)^k} \oint_{\mathbb{T}^k} F(J,\varphi)\,d\varphi$$

with the same initial data on the interval $0 < t < 1/\varepsilon$?

**1966-2.** What is the behavior of orbits in the complement to the union of the invariant tori of a nearly integrable Hamiltonian system? Is it true, in particular, that these orbits exhibit no evolution in the $s$-th approximation, i. e., $|I(t) - J(t)| \ll 1$ for $0 < t < 1/\varepsilon^s$? Here $I$ denotes the vector of the action variables, $J(t)$ is the solution of the $s$-th order "evolution equation" with the initial conditions $J(0) = I(0)$, while $0 < \varepsilon \ll 1$ is the perturbation parameter.

**1966-3.** Prove or disprove the following conjecture. Consider a nearly integrable Hamiltonian system with $k \geq 3$ degrees of freedom and with the Hamilton function $H_0(I) + \varepsilon H_1(I, \varphi)$, where $(I, \varphi)$ are the action–angle variables. Then "generically" for every pair of neighborhoods of the tori $I = I'$, $I = I''$ with $H_0(I') = H_0(I'')$, there is an orbit passing through both neighborhoods provided that $\varepsilon$ is sufficiently small.

**1966-4.** Let a diffeomorphism $A\colon q \mapsto q + f(q)$ of the torus $\mathbb{T}^2 = \{(q_1, q_2) \operatorname{modd} 2\pi\}$ preserve the measure $dq_1 \wedge dq_2$ and the center-of-mass:

$$\oint_{\mathbb{T}^2} f(q)\, dq_1 dq_2 = 0.$$

Prove that $A$ has at least 4 fixed points counting multiplicities and at least 3 geometrically distinct fixed points.

**1966-5.** Let $\Omega = \mathbb{T}^k \times B^k$ ($\mathbb{T}^k = \{q \operatorname{modd} 2\pi\}$, $B^k = \{p \in \mathbb{R}^k,\ |p| \leq 1\}$) be the toroidal annulus equipped with the canonical structure $\omega^1 = p\, dq$, and let $A\colon \Omega \to \Omega$ be a canonical diffeomorphism homotopic to the identity transformation and such that each sphere $\{q\} \times \partial B^k$ is linked with its image on the covering of the boundary $\mathbb{T}^k \times \partial B^k$. Then $A$ possesses at least $2^k$ fixed points counting multiplicities and at least $k+1$ geometrically distinct fixed points.

**1966-6.** Investigate the ergodic properties of motions in the complement of the union of the invariant tori of a nearly integrable Hamiltonian system. In particular, is the entropy of such a system positive?

# 1968

**1968-1.** What collections of numbers $B_0$, $B_1$, $B_2$, ... can be realized as collections of Morse numbers $B_0 = M_0$, $B_1 = M_1 - M_0$, $B_2 = M_2 - M_1 + M_0$, ... for a polynomial in $n$ variables of degree $d$?

**1968-2.** What topological characteristics of a real (complex) polynomial are computable from the Newton diagram (and the signs of the coefficients)?

# 1969

**1969-1.** An embedding of a torus into $\mathbb{R}^3$ is given. Can it have nontrivial (at least infinitesimal?) isometric deformations? *The question is connected with small denominators, taking into account the dynamical system defined by the asymptotic lines on the parabolic curve. This system itself is worth examining.*

**1969-2.** Given a function in the plane (a germ at 0), is it possible to find a function, that is smoothly equivalent to the given function, and is the Gaussian curvature function of (a germ of) a surface $z = f(x,y)$ in $\mathbb{R}^3$? *Can merely the original function in the plane be itself realized in this form? The answer may depend on the singularity at* 0: *for example, it may happen that finite multiplicity,* $\mu < \infty$, *is required.*

# 1970

**1970-1.** Construct versal unfoldings of endomorphisms (of vector spaces and groups).

**1970-2.** Is the problem of distinguishing a center from a focus algebraically trivial? What about the general problem of the algebraic classification of the equilibrium points of a system of ordinary differential equations $\dot{x} = v(x)$ in $\mathbb{R}^n$?

**1970-3.** Investigate the connection between the rotation numbers of a Hamiltonian system and the property that the Hamiltonian is single-valued.

**1970-4.** Carry over Poincaré's Last Theorem about an annulus (and its conjectural generalizations) to the case of multi-valued Hamiltonians.

**1970-5.** Study the Diophantine approximations on generic submanifolds (and the bifurcations in $k$-parameter families).

**1970-6.** Explore the equations in variations along a stationary solution of the Euler hydrodynamic equation (for example, the existence of conjugate points), in particular, for the Kolmogorov flow and for the flow on the torus with the stream function $\sin y$.

**1970-7.** Compute the curvatures of the groups $\mathrm{SDiff}(S^2)$ and $\mathrm{SDiff}(\mathbb{T}^3)$.

**1970-8.** Investigate the birth of discrete spectrum at the point of maximum speed, from the viewpoint of genericity: non-degenerate case, bifurcations, etc. (in particular, for flows on the torus with the stream function $f$ at the critical points of the function $v = f'$).

**1970-9.** Investigate the inertia indices of the stationary points of the kinetic energy on an orbit of the co-adjoint representation (from the viewpoint of bifurcations and genericity!).

**1970-10.** Prove that a divergence-free vector field on $S^2$ has at least two zeros. Prove an analogous statement for the mappings $S^2 \to S^2$ preserving oriented area (verify beforehand that the index of a fixed point of an area- and orientation preserving diffeomorphism of a plane does not exceed 1).

**1970-11.** What can one say about $\pi_2(\mathbb{C}\mathrm{P}^n \setminus V)$, where $V$ is a generic hypersurface of degree $m$?

**1970-12.** Evaluate the fundamental groups and the homologies of the spaces of curves with the simplest singularities that split completely into lines in $\mathbb{C}\mathrm{P}^2$ (the spaces of surfaces that split into planes in $\mathbb{C}\mathrm{P}^3$, etc.).

**1970-13.** Evaluate the topological invariants of the manifold of nonsingular cubic curves in $\mathbb{C}\mathrm{P}^2$.

**1970-14.** Evaluate the fundamental group of the space of embeddings of a circle into a solid torus (the answer is a knot invariant!).

**1970-15.** Investigate topological properties of the stratification of the space of meromorphic functions on a Riemann surface (rational functions in the case of $S^2$).

**1970-16.** Is the problem of Lyapunov stability of an equilibrium of the system $\dot{x} = v(x)$, $x \in \mathbb{R}^n$ algebraically trivial? What about the problem of asymptotic stability? Does there exist an analytic Lyapunov function for this system?

# 1971

**1971-1.** Let $A$ be a germ of a diffeomorphism, $A: (\mathbb{R}^n, 0) \hookleftarrow$, or $A: (\mathbb{C}^n, 0) \hookleftarrow$. Let $A = B^k$. Does this imply that $A$ commutes with some diffeomorphism $C$ such that $C^k = \mathrm{id}$? This is true for the formal power series. Is this true for the diffeomorphisms of the circle?

**1971-2.** Bifurcations of invariant manifolds in neighborhoods of singular points: see the conjecture on page 3 in the paper: ARNOLD V. I. Remarks on singularities of finite codimension in complex dynamical systems. *Funct. Anal. Appl.*, 1969, **3**(1), 1–5 [*the Russian original is reprinted in*: Vladimir Igorevich Arnold. Selecta–60. Moscow: PHASIS, 1997, 129–137].

**1971-3.** The algebraic unsolvability of the problem of stability of the equilibrium and of the problem of topological classification of dynamical systems in a neighborhood of a fixed point. See the papers: ARNOLD V. I. Local problems of analysis. *Moscow Univ. Math. Bull.*, 1970, **25**(2), 77–80; ARNOLD V. I. Algebraic unsolvability of the problem of stability and the problem of topological classification of singular points of analytic systems of differential equations. *Uspekhi Mat. Nauk*, 1970, **25**(2), 265–266 (in Russian); ARNOLD V. I. Algebraic unsolvability of the problem of Lyapunov stability and the problem of topological classification of singular points of an analytic system of differential equations. *Funct. Anal. Appl.*, 1970, **4**(3), 173–180.

**1971-4.** Prove the instability of the equilibrium 0 of an analytic system $\ddot{x} = -\partial U/\partial x$ in the case where the isolated (in $\mathbb{C}^n$?) critical point 0 of the potential $U$ is not a minimum.

**1971-5.** A smooth map $A: M \to M$ is called *coarse* if any map $B$ that is close to $A$ (with derivatives) is topologically equivalent to $A$ (that is, $B = CAC^{-1}$). Are the coarse maps dense in the space of all smooth maps $\mathbb{S}^1 \hookleftarrow$?

**1971-6.** Do there exist singular points of a vector field of finite codimension that do not allow a topologically versal unfolding with the number of parameters equal to the codimension (or with a finite number of parameters)? The conjectural example in dimension 3: two pairs of imaginary roots with ratio 3 (thesis of R. J. Sacker).

**1971-7.** Is it true that the set of germs of vector fields at a singular point, whose topological type cannot be determined by any jet of finite order, has infinite codimension? The same question—for Lyapunov stability and asymptotic stability.

**1971-8.** Investigate the pathology of the decomposition of the space of finite order jets of diffeomorphisms at a singular point, into topological equivalence classes. *Conjecturally, if the dimension and the codimension are large enough, then*:

1) *the set of the equivalence classes is infinite and even continual*;

2) *there exists a manifold in the space of jets such that each jet from this manifold defines the topological type of its germs, but this type changes along the manifold so that for any point in the manifold there are points of another topological type in its neighborhood.*

Investigate analogous questions for the decomposition into Lyapunov (asymptotically) stable and unstable jets. Is the number of connected components of the sets of stability and unstability in the space of jets infinite?

**1971-9.** Generalize the Hilbert problem on limit cycles to systems with discrete time.

**1971-10.** Explore the system of biocenos evolution without predators: $\dot{x}_i = x_i\bigl(A_i\bigl[\exp\bigl(\sum_k[-\lambda_{ik}x_k]\bigr) - 1\bigr]\bigr)$.

**1971-11.** Find (upper and lower?) estimates for the Hausdorff dimension of Navier–Stokes attractors in terms of the Reynolds number.

# 1972

**1972-1.** Investigate the topology of the complement of the caustic $\Sigma^2_\pm$ in $\mathbb{C}^3$: is it true that this complement is a $K(\pi,1)$ space?

**1972-2.** Investigate the monodromy group of the singularity $x^3+y^3+z^3$ (and also the topology of the complement of the discriminant).

**1972-3.** Is it true that $\min_y F(x,y)$ is topologically equivalent to a smooth function: a) for a generic $F$, b) always?

**1972-4.** Investigate the local convexity of the boundary of the stability domain (in the families of matrices and polynomials).

**1972-5.** Prove the uniform estimate for an oscillatory integral: how can one calculate the uniform index for a neighborhood in terms of the phase at the degenerate point?

**1972-6.** Is it true that the only singularities whose intersection form is positive or negative definite are $A$, $D$, $E$?

**1972-7.** Is the following conjecture on transversality of the stratification of a space of quadratic forms true: the manifold of quadratic forms in a Hilbert space that are determined by oscillations of arbitrary membranes is transversal to the stratified manifold of quadratic forms with multiple eigenvalues?

**1972-8.** Find "the most probable" representations of symmetry groups.

**1972-9.** Investigate the error of the method of averaging in the case of two frequencies, when in average the ratio of the frequencies changes with nonzero rate in the averaged motion (although the instantaneous rate of change in some fast phases changes its sign).

**1972-10.** Investigate the error of the method of averaging in generic multi-frequency systems under the assumption of passing through a resonance.

**1972-11.** Evaluate the cohomology of the braid groups of the series $D$ and $E$.

**1972-12.** Classify the singularities of convex hulls of generic submanifolds in a vector space.

**1972-13.** Find the number of moduli for the Brieskorn singularities $\sum_i x_i^{a_i}$.

**1972-14.** Is it true that the complement of a bifurcation diagram is always a $K(\pi, 1)$ space?

**1972-15.** Prove that simple orbits coincide with orbits that are adherent only to orbits of smaller codimension (but not to unions of orbits of greater codimension).

**1972-16.** Find all the self-consistent gravitational potentials on the straight line (the stationary points, possibly generalized, of the Poisson–Vlasov equation).

**1972-17.** Prove that a diffeomorphism of the two-dimensional torus homotopical to the identity has at least four fixed points (counting multiplicities) and at least three of them are geometrically distinct, whenever this diffeomorphism preserves areas and leaves the center-of-mass invariant.

**1972-18.** Show that any orientation- and area-preserving diffeomorphism of the two-dimensional sphere onto itself has at least two geometrically distinct fixed points.

**1972-19.** Are the structurally stable maps of $\mathbb{S}^1$ into itself dense?

**1972-20.** Straightening the circle diffeomorphisms (by a smooth change of variables) for almost all the rotation numbers (*solved by M. R. Herman*) and the topological obstacle to analytic straightening: the existence of periodic orbits arbitrarily close to the real circle (*maybe, even in a neighborhood of any point of the* *circle*?). The similar obstacle to prolonging the reducibility annulus to a rotation by a holomorphic change of variables or the reducibility disk in Siegel's problem.

**1972-21.** The Floquet theory over the torus.

**1972-22.** A sufficiently curved submanifold is extremal in the Diophantine sense (with probability 1, the Diophantine exponent is the same as in the ambient space).

**1972-23** (R. Thom)**.** A gradient vector field with a singular point has a trajectory entering the singular point with tangency to some straight line.

**1972-24.** Investigate the connections between the invariants of a singularity of a plane complex curve and the local fundamental group of its complement.

**1972-25.** Action of the monodromy $M$ on the homology of the Milnor fiber. Decompose the singularity having included it in the family $f(z) - pz$ where $p \in \mathbb{C}^n$ is a parameter. Examine the bifurcation manifold $\Sigma = \{p, \varepsilon : \varepsilon$ is a critical value of the function $z \mapsto f(z) - pz\} \subset \mathbb{C}^{n+1}$. (This manifold is determined by the equation $\varepsilon = H(p)$, where $H$ is the Legendre transform of $f$.) Consider $\pi_1(\mathbb{C}^{n+1} \setminus \Sigma)$ (germs at 0). Conjecturally, the properties of $M$ (nilpotency, etc.) reflect the properties of $\pi_1$. For example, if a path $\varepsilon_0 e^{i\varphi}$, $0 \leq \varphi \leq 2\pi N$, commutes with all the generators of $\pi_1$, then is it true that it does not shift vanishing cycles (so that $M^N = 1$)?

**1972-26.** What are the restrictions imposed on the topology of a manifold by the hypothesis that the manifold is a degree $n$ algebraic hypersurface in $\mathbb{R}^m$ (in $\mathbb{R}\mathrm{P}^m$)?

**1972-27.** Is it possible to represent an algebraic function $z(a,b,c)$, $z^7 + az^3 + bz^2 + cz + 1 = 0$, as one of the components of a superposition of algebraic functions in two variables? Find the conditions on the fundamental group, the adjacency of the strata, monodromy, and other topological invariants under which the algebraic function is not representable as a component of a superposition (conjecturally, these topological invariants are more complicated for the functions that are not representable in such a form).

*In this problem algebraic functions can be replaced by "pseudoalgebraic" functions, which are topologically (or combinatorially) equivalent to them—conjecturally the nonrepresentability persists even for superpositions of such pseudoalgebraic maps.*

**1972-28.** Find the three-dimensional characteristic class of the foliation of either $P(x,y) = C$ or $P\,dx + Q\,dy = 0$ in $\mathbb{C}\mathrm{P}^2 \setminus$ (singular points). (Here $P$ and $Q$ are polynomials.)

The following three problems are related to this class.

**1972-29.** Determine if this class is integral (for example, in the real case).

**1972-30.** Determine the conditions on the deformations of the coefficients or on the cobordisms that preserve this cocycle.

**1972-31.** Try to relate this class to limit cycles (not simply-connected fibers).

**1972-32.** Are the Boardman classes $\Sigma^I$ *topologically* invariant?

**1972-33.** Prove that a symplectic diffeomorphism of a compact symplectic manifold $M$ onto itself possesses at least as many fixed points as a smooth function on $M$ has critical points, whenever this diffeomorphism is homologous to the identity.

# 1973

**1973-1.** Describe the typical singularities appearing in the problems on differential games.

**1973-2.** Find the typical singularities of convex hulls.

**1973-3** (S. Smale – J. Debreux). **Apply the singularity theory to economic models.**

**1973-4.** Prove that the equilibria points stability problems and the problems about limit cycles are algorithmically unsolvable.

**1973-5.** Explore the normal forms of implicit differential equations unresolved with respect to derivatives, and their bifurcations.

**1973-6.** Investigate three-parameter bifurcations of the topological type of the dynamics in a neighborhood of a singular point of a vector field (the zero and an imaginary pair, etc.).

**1973-7.** The problem of smoothness of the stratum $\mu = \text{const}$.

**1973-8.** The problem of semicontinuity of the modality (the number of moduli).

**1973-9.** Investigate the lower deformations of the critical points of functions (a generalization of the theory of algebraic hypersurfaces!): the structure of discriminants, fundamental groups, vanishing cycles, etc.

**1973-10.** Prove the "$(2,2)$" formula for the number of moduli of a $\Gamma$-nondegenerate function in two variables, and deduce analogous "stereometric" formulae for the other invariants ($\mu$, etc.).

**1973-11.** Generalize the classification of the admissible types of quasihomogeneity of nondegenerately-quasihomogeneous critical points (which is known only in the case of two and three variables). The question is related to the theory of cyclotomic polynomials.

**1973-12.** Is it true that the complement of the discriminant of a function's singularity of finite multiplicity is $K(\pi, 1)$?

**1973-13.** Investigate the topological invariants of bifurcation diagrams of functions (at least within the scope of tables, in order to work out general conjectures!) in the real and the complex case.

**1973-14.** What restrictions on the coexistence of singularities (on the same fiber, on different fibers) are imposed by the condition that the singularities belong to a versal unfolding of a given singularity of finite multiplicity (the problem is related to the 16th Hilbert problem)?

*This is the problem that formed the basis of the semicontinuity of the spectrum of a singularity, estimates for the number of Morse points on a hypersurface, etc.*

**1973-15.** Develop the theory of cobordisms of the critical points of functions.

**1973-16.** Carry over the achievements of the theory of critical points of functions to the study of smooth complex maps into spaces of greater dimension.

**1973-17.** Describe completely the stratification of the space of functions in two variables.

**1973-18.** Is there any relation between the Minakshisundaran–Pleijel coefficients and the coefficients of the polynomial whose value is the volume of the $\varepsilon$-neighborhood (e. g., for an isoperimetric embedding into $\mathbb{R}^N$)?

**1973-19.** Does each function have Morsifications with any number of critical values, from 1 to $\mu$? How many distinct critical values are necessary in the real case?

**1973-20.** Find the transformation group preserving the ratio of the forms $\int u^2\,dx$ and $\int (u')^2\,dx$ in the space of functions $u$.

**1973-21.** Construct Dynkin diagrams for simple singularities as the quivers of some subspaces of local rings (derive the quivers from the structure of ideals?). *A. N. Shoshitaĭshvili suggested a construction that solves this problem for all cases except $E_7$, which is, therefore, unsatisfactory.*

**1973-22.** The Jacobian of the map $\binom{x}{y} \mapsto \binom{x}{xy}\begin{smallmatrix}=u\\=v\end{smallmatrix}$ is degenerate on the line $x=0$, and the line $u=0$ is not covered by this map (with the exception of the point 0).

The Lyashko–Looijenga map for (unimodal) parabolic singularities has an analogous property. What is the general formulation of the corresponding conservation law: the more degeneracy in the domain, the more is uncovered in the range (or: the less is covered in the range, the more singularities are in the domain)?

**1973-23.** Is the asymptotic Hopf invariant (or helicity) of a divergence-free vector field in $\mathbb{S}^3$ invariant under volume-preserving homeomorphisms?

**1973-24.** Study the relation between the asymptotic Hopf invariant and the Reidemeister (*Ray–Singer*) torsion.

**1973-25.** A. D. Sakharov's conjecture: if a frozen-in vector field has linked or knotted trajectories, it cannot relax to arbitrarily small energies by the action of $\mathrm{SDiff}(B^3)$.

**1973-26.** The relaxation paradox: one cannot believe that formerly non-integrable fields have to relax to the eigenfields of the operator rot. What happens to them? Does the limit field encounter singularities? Or there is no limit field at all?

**1973-27.** Consider the mapping $\mathbb{C}^k \to \mathbb{C}^k$ associating to a point in a versal deformation the polynomial whose roots are the critical values of the corresponding function. For the versal deformation of the singularity $A_k$ the multiplicity of this ramified covering, $(k+1)^{k-1}$, is equal to the number of trees with $k+1$ numbered vertices. Give a similar interpretation to the multiplicity of this mapping for other simple singularities (which is, according to O. V. Lyashko, $k!\,h^k/|W|$, where $h$ is the Coxeter number, and $|W|$ is the order of the Weyl group of the singularity).

**1973-28.** Consider a random set of points in $\mathbb{R}^n$ with density $\rho$. Let $V(d)$ be the $d$-neighborhood of this set. Consider the averaged Betti numbers

$$\lim_{R\to\infty} \frac{b_i(V(d)\cap(\text{ball of radius } R))}{R^n} = \beta_i(d,\rho).$$

Investigate these functions.

# 1974

**1974-1.** The reconstruction of a quasihomogeneous Lie algebra from its root system. Consider a collection of positive exponents (weights) of quasihomogeneity $w_i$ of the coordinates $x_i$ in $\mathbb{C}^n$ $(i = 1, \dots, n)$. A *generator* of a quasihomogeneous Lie algebra is a monomial $x^m \partial/\partial x_i$ of weight zero ($m \in \mathbb{Z}^n$, $m_j \geq 0$, $\sum w_j m_j = w_i$). A *root* of this generator is a vector $\widetilde{m} = m - 1_i \in \mathbb{Z}^{n-1} = \{\widetilde{m} : \sum w_j \widetilde{m}_j = 0\}$. Is it possible to reconstruct the Lie algebra generated by these generators (up to an isomorphism of Lie algebras) from the system of its roots, considered up to a linear transformation of the hyperplane $\mathbb{R}^{n-1}$ that does not necessarily preserve the coordinate hyperplanes $m_i = 0$ in $\mathbb{R}^{n-1}$?

*The system of weights cannot be reconstructed, but the algebra is almost reconstructible (modulo the signs of some structural constants). In all the examples ever considered, different choices of these signs result in isomorphic algebras. But it is unclear whether this is always the case.*

**1974-2.** In the theory of the duality of convex polyhedra there appears a Lagrangian or Legendrian manifold with singularities. In the same way, in optimal control theory there appear generalizations of Hamiltonian systems with non-smooth Hamiltonians (a manifold of phase curves can pass through one point, as in the case of the Hamiltonian $H = |p_1| + |p_2|$). Nevertheless, their "flows" in some sense satisfy the Liouville theorem and should be considered as generalized symplectomorphisms (which are, most probably, not maps but Lagrangian submanifolds with singularities in the product space).

Develop a theory of Lagrangian manifolds with singularities, and generalized symplectomorphisms applying to such situations (and even obtain estimates from below for the number of intersection points of exact Lagrangian manifolds, and for the number of fixed points of exact symplectomorphisms, generalizing the Poincaré "geometric theorem").

**1974-3.** Find all singular values of the moduli of parabolic singularities (that change the topological or the combinatorial type of the projection of the manifold of the discriminant's singularities onto the bifurcation diagram of functions, i. e., the set of clauses for a decomposition of a critical point into several clusters of simpler critical points on (generally) several critical levels, realized by small deformations of the function). What are the elliptic curves corresponding to these values of moduli celebrated for?

**1974-4.** Find the classification problem of the theory of Lagrangian (Legendrian?) singularities, the answer to which would be in natural bijection with the list of Coxeter reflection groups.

**1974-5.** Find applications of the (Shephard–Todd) complex reflection groups to singularity theory.

**1974-6.** Symplectize the topology: Poincaré's index theory of singular points, apparently, turns into the theory of fixed points of symplectomorphisms and generalizations of Poincaré's last geometric theorem (i. e., to a generalization of the Morse theory). Do other topological theories have symplectizations? *Similarly to a noticeable difference between* $\mathbb{Z}_2$ *and its complexification* $\mathbb{Z}$, *the symplectization can also be as far from the initial object as the Coxeter group* $C_k$ *is from* $A_k$.

**1974-7.** Classify the simple singularities of functions on a manifold with an action of a group (for example, finite) up to equivariant diffeomorphisms (commuting with the group action).

**1974-8.** Investigate the typical perestroikas of a wave front moving with time (and of the corresponding Legendrian map).

**1974-9.** Give a topological classification of the Legendrian singularities corresponding to the parabolic critical points of functions.

**1974-10.** A conic singularity over a given base carries topological invariants of the base into the singular point. For non-conic singularities (e. g., quasihomogeneous?) one may try to find traces of the discrete invariants of the base (e. g., the rank and the signature of the Milnor fiber?) in the local algebra of the singularity.

What algebraic objects are encountered in this way? What happens to the characteristic classes and numbers?

# 1975

**1975-1.** Every interesting discrete invariant of a generic singularity with Newton polyhedron $\Gamma$ is an interesting function of the polyhedron. Study: the signature, the number of moduli, the singularity index, the integral monodromy, the variation, the Bernstein polynomial, and $\mu_i$ (for generic sections).

**1975-2.** Is it possible to reconstruct the Newton polyhedron $\Gamma$ from a $\Gamma$-nondegenerate function $f \in \mathfrak{m}^3$? In the quasihomogeneous case, is it possible to reconstruct the exponents? Is the main term reconstructible (or are those on the faces)?

**1975-3.** Let $f$ be a quasihomogeneous but degenerate function. Is it possible to make the Newton polyhedron of $f$ smaller by a quasihomogeneous coordinate transform? This is a particular case of the question of whether any function with $\mu < \infty$ is stably equivalent to a $\Gamma$-nondegenerate one.

**1975-4.** Let a function $f$ be $\Gamma$-nondegenerate. Is it true that there exists a correct upper basis $\{e_k\}$ such that $f \sim f_0 + \sum c_k e_k$? Does there exist a correct upper basis serving for all sums $f_0$ with upper summands? (If yes, then the answer to the first question is positive.)

**1975-5.** Let $(\alpha_1, 1)$ and $(\alpha_2, 1)$ be two types of quasihomogeneity with affinely equivalent patterns (i. e., sets of integers $m \geq 0$ of the hyperplane $\{m : (m, \alpha) = 1\}$). Is it true that the upper patterns $\{m > 0 : (m, \alpha) = 1 + \beta\}$ are mutually equivalent (with a non-monotone re-enumeration $\beta_1 \mapsto \beta_2$), and that the upper basis of the first singularity is mapped to the upper basis of the other one?

**1975-6.** The stratum $\mu = \text{const}$ of a quasihomogeneous function in the standard versal deformation is linear and generated by weakly upper monomials. Does this hold for a $\Gamma$-nondegenerate function? (Generally—is the stratum $\mu = \text{const}$ smooth?)

**1975-7.** Can the complex singularities belonging to distinct strata $\mu = \text{const}$ be topologically equivalent?

**1975-8.** Is the singularity index semicontinuous?

**1975-9.** Is it true that the number $s(\mu)$ of the strata $\mu = \text{const}$ with $\mu = 32$ is a power of 2? For $\mu = 1, 2, 4, 8, 16$ we have $s(\mu) = 1, 1, 2, 4, 32$, respectively. Is there a logical pattern in the sequence $s(\mu) =$ **1**, **1**, 1, **2**, 2, 3, 3, **4**, 4, 7, 11, 15, 14, 17, 22, **32**, where the boldface numbers are the values of $s(\mu)$ that correspond to $\mu = 1, 2, 4, 8, 16$?

**1975-10.** Is the set of non-equivalent quasihomogeneous patterns with a given number $n$ of variables finite? The equivalence is the combinatorial (or affine?) type of the convex hull of the pattern $\{m \in \mathbb{Z}^n : m \geq 0, (m, \alpha) = 1\}$.

**1975-11.** Is it true that in the complex case the complement of the bifurcation diagram of a function is always a $K(\pi, 1)$ space? Are the components contractible in the real case? *Conjecturally no, although R. Thom had thought that yes*!

**1975-12.** Does every real-valued function have a real Morsification (with $\mu$ real critical points)?

**1975-13.** What is the minimal number of critical values obtained by a perturbation of a critical point of multiplicity $\mu$ with $\mu$ Morse critical points? Conjecturally it is $n+1$, where $n$ is the number of variables (or corank).

**1975-14.** Is the corank a topological invariant?

**1975-15.** What singularities can absorb $A_1$? split $A_1$ off? Why is every stratum $\mu = \text{const}$ connected to the stratum $A_1$ by a chain of strata of all codimensions?

**1975-16.** Suppose $f \oplus g \sim f \oplus h$ ($f$, $g$, $h$ are isolated singularities). Is it true then that $g \sim h$?

**1975-17.** Give an "objective" definition of a series of singularities.

**1975-18.** List all decompositions of simple singularities.

**1975-19.** Calculate the stable cohomology ring of the complement of bifurcation diagrams: a) of functions of $n$ variables, b) stable over $n \to \infty$.

**1975-20.** Compose a list of simple singularities of maps from $m$-dimensional manifolds to $n$-dimensional ones.

How does the *A–D–E* classification show up in this list?

**1975-21.** Express the main numerical invariants of a typical singularity with a given Newton diagram (e. g., the signature, the genus of the 1-dimensional Milnor fiber) in terms of the diagram.

**1975-22.** The problem of stabilization of invariants: investigate the behavior of the main invariants of a singularity when adding squares of new variables.

**1975-23.** Compare the stratifications of real and complex singularities of functions. Distinguish $M$-singularities among real forms. Compare real and complex modalities. Is a complex stratification always the complexification of a real one?

**1975-24.** Investigate the stratum $\mu = \text{const}$ (defined by the condition that the codimension of the orbit is constant). Is the stratum smooth (for algebraic group actions, for natural problems of the singularity theory, e. g., for the classification of singularities of caustics and wave fronts)?

Is it true that every such stratum becomes irreducible in the base of the complex versal deformation of some suitable "deeper" singularity?

Does the cohomology ring of the complement of the stratum stabilize in this "growing" base?

**1975-25.** Investigate the Lagrangian singularities of bifurcating caustics from the cosmological "pancake theory" of Zeldovich (in particular, taking into account the gravitation and particle fusion, and for nonpotential flows).

**1975-26.** Evaluate the normal forms of versal deformations of matrices of various types (symmetric, unitary, etc.), and investigate the corresponding bifurcation diagrams and cohomology rings.

**1975-27.** Explore the asymptotics of oscillatory integrals (in particular, find uniform estimates near singularities of caustics and calculate the highest individual singularity indices appearing unremovable in typical families with a given number of parameters).

Carry over these estimates to integrals of the saddle-point method.

**1975-28.** Investigate the singularities of envelopes of typical families of submanifolds from the viewpoint of the symplectic and contact theory of Lagrangian and Legendrian maps.

**1975-29.** Explore the singularities of solutions of generic variational problems (as well as those appearing in typical families with prescribed or not prescribed finite number of parameters).

**1975-30.** Investigate the singularities of implicit differential equations (both ordinary and partial).

# 1976

**1976-1.** Given a system of Newton polyhedra, is there a system of *real* polynomials with these polyhedra which has the correct number of real roots (i. e., the same as for a system with generic complex coefficients)?

**1976-2.** Consider two plane polynomial vector fields of degrees $m$ and $n$, respectively. Is it possible to estimate the number of intersection points of their limit cycles in terms of $n$ and $m$ (find a sharp attainable estimate)?

**1976-3.** Investigate the convergence of the normal forms of equations of the form $y'' = f(x,y,y')$.

**1976-4.** Build a theory of the "non-Desargues curvature form" (that measures local non-equivalence to a linear equation) for $y'' = f(x,y,y')$.

**1976-5.** Construct a symplectic (or contact) version of the asymptotic Hopf invariant: $H: M^{2n} \to \mathbb{R}$, $\omega$ is a symplectic structure, $\omega|_{H=h} = d\alpha$, $\beta = \alpha \wedge \omega^{n-1} \in \Omega^{2n-1}$, $\int_{H=h} \beta = \text{const}$ is a symplectic analog of the Hopf invariant. Given a Hamiltonian vector field, study how to measure the average rate of evolution of a Lagrangian subspace of the tangent space under the flow.

**1976-6.** Elaborate a theory of $\mathbb{C}P^1$-neighborhoods in complex manifolds (similar to the theory of neighborhoods of elliptic curves already constructed, and preceding the theory of neighborhoods of higher genera curves).

**1976-7** (A. Tresse)**.** Justify the finiteness theorems of differential invariant theory.

**1976-8.** Consider a function with Newton diagram $\Gamma$. Is it true that each of its singularities of finite multiplicity is stably equivalent to a $\Gamma$-nondegenerate one?

**1976-9.** Classify the typical singularities of synthesis in a generic problem of optimal control given by a typical indicatrix field—a generic family of mappings of a fixed manifold into all tangent spaces to the base manifold (with the point of the base as a parameter).

**1976-10.** Investigate the asymptotic behavior of the measure of deviated trajectories in the problem of a generic perturbation of a generic $k$-frequency conditionally-periodic system with $m$ slow variables.

**1976-11.** For a given plane vector field with a singular point, construct an algebraic complex whose homology describes the limit cycles vanishing at the singular point.

**1976-12** (A. G. Kushnirenko)**.** The Descartes rule implies that the number of real roots of a polynomial has the number of its monomials as an upper bound. Extend this observation to polynomials of several variables: the simplicity of a formula implies bounds on the topology of the variety defined by it. *A theory of "fewnomials" has been elaborated by K. A. Sevast'yanov and A. G. Khovanskiĭ, but the estimates obtained in the multidimensional case probably are strongly nonsharp.*

**1976-13.** Is the stratum $\mu = \text{const}$ smooth? *The smoothness of the stratum would follow from an affirmative solution to the following question.*

*Given an algebraic action of a complex algebraic group on a finite-dimensional affine space (e. g., a linear representation), consider the set of all points whose stationary subgroups have a fixed dimension. Is this set a smooth manifold?*

**1976-14.** Does the real modality of a real-valued function coincide with the complex one?

**1976-15.** For which weights $\alpha_s = A_s/N$ does there exist a nondegenerate quasi-homogeneous function of degree 1?

**1976-16.** Evaluate the modality of a $\Gamma$-nondegenerate function in terms of its Newton diagram $\Gamma$. In particular, prove that for semi-quasihomogeneous functions the modality is equal to the number of monomials in a basis of the local ring on the diagram and above it.

**1976-17.** Evaluate the signature of the quadratic form defined by the intersection index in the middle-dimensional homology of a local nonsingular level set of a $\Gamma$-nondegenerate function of $n$ variables when $n \equiv 3 \pmod 4$.

**1976-18.** Find a normal form for all the $\Gamma$-nondegenerate functions with a given Newton diagram $\Gamma$.

**1976-19.** Find the Jordan canonical form of the monodromy operator of a $\Gamma$-nondegenerate function with a given Newton diagram $\Gamma$.

**1976-20.** Let $f$ be the Morsification of a real-valued function with a given singularity (say, $\Gamma$-nondegenerate). What is the maximum number of components a local real level set of $f$ can have?

**1976-21.** Chebyshev polynomials of several variables. With every critical point of a function of finite multiplicity, one can associate a "Chebyshev polynomial," which is the Morsification of this function with the least possible number of critical values. (The usual Chebyshev polynomials come from one-variable functions of the form $z^n$.) Which of the nice properties of Chebyshev polynomials in one variable hold for the above defined polynomials of several variables?

**1976-22.** Uniform estimates for oscillatory integrals. An *oscillatory integral* has the form

$$I(h,\lambda) = \int_{\mathbb{R}^n} e^{iF(x,\lambda)/h}\varphi(x)\,dx, \quad h \to 0, \tag{1}$$

where $\varphi$ is a smooth function concentrated at a sufficiently small neighborhood of the origin; $F$ is a real-valued deformation of the function $f = F(\cdot,0)$ depending smoothly on the parameter $\lambda$; $h$ is a small parameter; $F(0,0) = 0$. The *uniform index* $\beta$ of the singularity of the function $f$ in the point 0 is the infimum of $\gamma$ such that for any deformation $F$

$$|I(h,\lambda)| \leq C(\varphi)|h|^{\frac{1}{2}n-\gamma} \tag{2}$$

for all sufficiently small $|\lambda|$.

The content of the problem is to evaluate the index $\beta$ (say, for $\Gamma$-nondegenerate functions $f$).

For every pair of integers $n$ and $l$ the *universal uniform index* $\beta(n,l)$ can be defined as the infimum of $\gamma$ such that the oscillatory integral (1) has an estimate (2) which is uniform in $\lambda$ for all families $F$ of functions of $n$ variables $x$ and $l$ parameters $\lambda$ except a meager subset in the function space.

The problem of evaluating rational numbers $\beta(n,l)$ appears to be very difficult, because it seems to be almost equivalent to the problem of the full classification of all singularities.

For a fixed $l$ and $n \to \infty$, the numbers $\beta(n,l)$ stabilize:

$$\beta(n,l) = \beta(\infty,l) = \beta(l), \quad \text{if } n \text{ is large enough.}$$

The rational number $\beta(l)$ is the greatest singularity index among singularities of codimension $l$.

A problem for optimists: find all $\beta(l)$. A problem for pessimists: find $\beta(1000)$.

**1976-23.** Uniform estimates of preimage variations. Let $g$ be the germ of a diffeomorphism $(\mathbb{R}^n, O_1) \to (\mathbb{R}^n, O_2)$ or $(\mathbb{C}^n, O_1) \to (\mathbb{C}^n, O_2)$, and let $\mu$ be the multiplicity of the point $g(O_1) = O_2$. The preimages of $O_2$ can merge in different ways depending on the type of the singularity of $g$ at $O_1$. Investigate the asymptotic behavior of various geometric characteristics of the preimage of a small ball of radius $\delta$ centered at $O_2$ as $\delta \downarrow 0$. Thus, describe the different ways in which the preimages of $O_2$ can merge.

An example of such a characteristic could be the so-called *variations* of all dimensions.

The variation $\sigma_k(D)$ of a sufficiently good set $D \subset \mathbb{R}^n$ is the mean value of the $k$-dimensional volume of the orthogonal projection of $D$ onto a $k$-dimensional subspace $L$, over all subspaces $L \subset \mathbb{R}^n$. The volume is counted with multiplicities, i. e., the number of connected components that are projected into one point. *In particular, $\sigma_n(D)$ is the volume of Dm and $\sigma_0(D)$ is the number of its connected components.*

**1976-24.** The $A$, $D$, $E$ problem. Surprisingly, the Dynkin diagrams

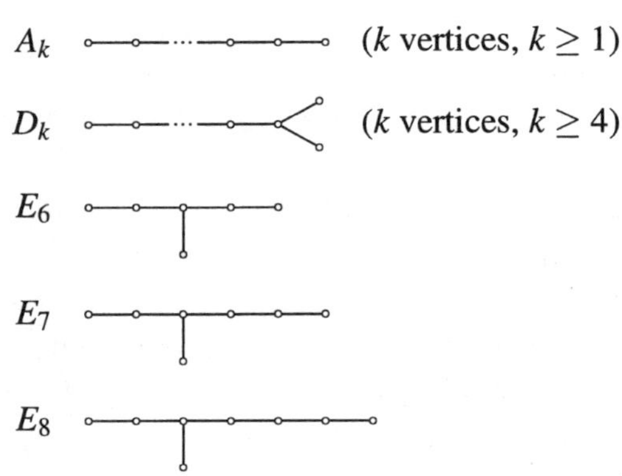


appear while solving various classification problems, such as the classification of:

1) critical points of a function;

2) regular polytopes (or finite orthogonal groups) in $\mathbb{R}^3$;

3) categories of vector spaces and linear maps;

4) caustics;

5) wave fronts;

6) groups generated by reflections (or Weyl groups with roots of equal norm);

7) simple Lie groups;

8) singularities of algebraic hypersurfaces with positive or negative definite intersection form of a neighboring smooth fiber.

Some connections between these objects are known. However, in most cases, no explanation of the same answer for different problems has been given.

The *A*, *D*, *E* problem: Find a *general* classification theorem from which the solutions to all of the above problems would follow.

**1976-25.** The $K(\pi, 1)$ problem. In the case of simple singularities *A*, *D*, *E* the complement of the bifurcation manifolds of a function and the complement of the bifurcation varieties of its level set are the Eilenberg–MacLane $K(\pi, 1)$ spaces. Can this result be generalized to the case of nonsimple singularities?

More general problem: investigate topological properties of the complements of the bifurcation subsets of differentiable maps.

**1976-26.** Complete to a commutative diagram:

| | | |
|---|---|---|
| Constructing simple singularities from regular polyhedra | $\longrightarrow$ | Constructing quasihomogeneous two-dimensional singularities from automorphic forms |
| $\downarrow$ | | $\downarrow$ |
| Constructing bifurcation diagrams of simple singularities | $\longrightarrow$ | ? |

**1976-27.** Complete to a commutative diagram:

| | | |
|---|---|---|
| Morse Theory | $\longrightarrow$ | Generalized Whitehead groups |
| $\downarrow$ | | $\downarrow$ |
| Picard–Lefschetz Theory | $\longrightarrow$ | ? |

**1976-28.** The stable cohomology ring. One can associate with a critical point of a holomorphic function $f$ the cohomology ring $H^*(f)$ of the complement of the bifurcation diagram of the level sets in the base of a versal deformation.

Let $f_2$ be the germ of a function from a versal deformation of $f_1$. Then the transversal to the stratum corresponding to $f_2$ in the base of the versal deformation defines an inclusion of the complements, and hence a homomorphism of the cohomology rings $H^*(f_1) \to H^*(f_2)$. For example, if $f_1 = x^n$ and $f_2 = x^{n-1}$ then $H^*$ are the cohomology rings of the braid groups with $n$ and $n-1$ threads, respectively.

Moreover, the homomorphism induces the stabilization of the cohomology rings of the braid groups as $n \to \infty$.

Does a similar situation appear in general? If yes, what would be the stable cohomology ring?

Similar questions arise for the complements of the bifurcation diagrams of functions.

**1976-29.** The converse of the Lagrange–Dirichlet theorem. Prove that the equilibrium 0 of a Newton system $\ddot{x} = -\operatorname{grad} U$ is unstable whenever the critical point 0 of the polynomial (or arbitrary analytic function) $U(x_1, \ldots, x_n)$ is not a point of local minimum.

**1976-30** (R. Thom)**.** Let $U : \mathbb{R}^n \to \mathbb{R}$ be a polynomial. Prove that at least one phase curve of the system $\dot{x} = \operatorname{grad} U$ meets the critical point 0 with tangency to some straight line.

**1976-31.** Algorithmic insolvability of the problem of stability. Is the problem of stability of the equilibrium 0 of a system $\dot{x}_k = P_k(x)$ algorithmically unsolvable? Here $P_k(x)$, $k = 1, \ldots, n$, are polynomials with rational coefficients.

There are closely related problems whose algorithmic insolvability might imply the algorithmic insolvability of the previous one:

1) The problem of existence of a limit cycle for the system $\dot{x} = P(x,y)$, $\dot{y} = Q(x,y)$, where $P$, $Q$ are polynomials with rational coefficients.

2) The problem of positiveness of a real Abelian integral $\oint R(x,y)\,dx$ along an oval $P(x,y) = 0$, where $P$, $R$ are polynomials with rational coefficients.

**1976-32.** Typical singularities of solutions of variational problems. It is known that variational problems lead to discontinuities and singularities even when everything is smooth in the setting of the problem. The singularities that appear can be pathologically complex because of infinite degeneracies. Is it possible to avoid the pathology by considering generic problems?

Examples: the problem of bypassing an obstacle; the problem of the quickest path with bounded velocity $\dot{x} \in F_x \subset T_x\mathbb{R}^n$, $\forall x \in \mathbb{R}^n$; the problem of attainable points. If we replace the indicatrix $F_x$ with its convex hull, then we get the following problem: describe the singularities of the convex hull of a generic $k$-dimensional submanifold in $\mathbb{R}^n$.

**1976-33.** Singularities in the theory of partial differential equations. Consider a generic partial differential equation with smooth initial data and smooth boundary conditions. Describe the nature of singularities of solutions of the system on surfaces, curves, and at points in space, that are responsible for the situation when a solution is not smooth but belongs to some functional space.

**1976-34.** Compositions of algebraic functions. Consider an algebraically closed field $k$ and $n$ independent variables $x_1,\dots,x_n$ over $k$. Let $K$ be the algebraic closure of the field $k(x_1,\dots,x_n)$. For every natural number $r \le n$ define the subfield $M_r \subset K$ of all elements of $K$ that can be obtained by a composition with not more than $s-1$ iterations of algebraic functions of $r$ variables. Clearly, $M_1 \subset M_2 \subset \dots \subset M_n = K$.

Can it happen that $M_r = K$ for some $r < n$? More generally, how many distinct fields are there among $M_1,\dots,M_n$? How many distinct fields are there among $M_r^s$, where $M_r^s \subset M_r \subset K$ is the subfield of all elements of $K$ that can be obtained by a composition of not more than $s-1$ algebraic functions of $r$ variables. For example,

$$M_n^1 = k(x_1,\dots,x_n), \quad \bigcup_{s=1}^{\infty} M_r^s = M_r.$$

Find the minimal $M_r$ (or $M_r^s$) which contains an element $f$ satisfying

$$f^n + x_1 f^{n-1} + \dots + x_{n-1} f + x_n = 0.$$

Similar questions make sense for the field $k(x_1,x_2,\dots)$ with an infinite number of variables $x_i$.

**1976-35.** How many connected components can the complement of a degree $n$ algebraic hypersurface in $\mathbb{R}P^k$ have? *This is unknown already for* $k = 3$.

**1976-36.** What are the possible arrangements of ovals of a plane projective curve of degree $d$ such that the number of ovals is maximal possible, i. e., equal to $1 + \frac{1}{2}(d-1)(d-2)$?

**1976-37.** Can a planar vector field defined by two quadratic polynomials have more than 3 limit cycles?

**1976-38.** Determine the singularities and other analytic properties of thermodynamic functions when the interaction potential is known.

**1976-39.** Does a symplectic diffeomorphism of the two-dimensional torus have a fixed point whenever this diffeomorphism is homologous to the identity?

**1976-40.** What could be the mathematical equivalent of the physical notion of turbulence? One of the aspects of this question: find "good" theorems of existence and uniqueness for the 3-dimensional Navier–Stokes equations.

**1976-41.** Find mechanical (physical, chemical, etc.) phenomena which can be described by systems with exponential repulsion of trajectories and with internally unstable attracting modes.

**1976-42.** The numerical qualitative or ergodic investigation of multidimensional dynamical systems (and, in particular, of limit modes in these systems) relies on posing questions that are realistic, rather than those that usually appear in abstract classification theorems. The high-priority problems here are:

1) teach a computer how to determine whether a trajectory enters a neighborhood of an attracting invariant set;

2) if it does enter, teach a computer how to determine the dimension of this set and, if possible, its topology;

3) teach a computer how to find the ergodic characteristics of motion on this set; first of all how to determine whether the trajectories have exponential instability on this set (i. e., whether the entropy is positive).

# 1977

**1977-1.** Investigate the connection between the spectral sequence of the Newton filtration and the mixed Hodge structure of a $\Gamma$-nondegenerate singularity.

**1977-2.** Deduce generalizations of Petrovskiĭ's inequalities for curves with singularities from the mixed Hodge structures (hypersurfaces, etc.).

**1977-3.** Give a classification of unimodal boundary singularities.

**1977-4.** Give a classification of the simple singularities in the presence of a fixed singular hypersuface (or another algebraic subvariety).

**1977-5.** Explore the discriminant of $H_3$.

**1977-6.** Axiomatize the theory of complete and linearized invariants convolutions.

**1977-7.** Determine and investigate the indices of singular points of 1-forms on singular varieties.

**1977-8.** How do the Reidemeister and Ray–Singer torsion appear in singularity theory?

**1977-9.** Give a classification of nondegenerate quasihomogeneous maps $\mathbb{C}^2 \to \mathbb{C}^3$ and $\mathbb{C}^3 \to \mathbb{C}^3$ (similar to the decomposition of the space of maps $\mathbb{C}^2 \to \mathbb{C}^3$ into three types, and of the space of maps $\mathbb{C}^3 \to \mathbb{C}^3$ into seven types).

**1977-10.** Prove Lyashko's statements about the Poincaré polynomial of a quasihomogeneous map $f : \mathbb{C}^m \to \mathbb{C}^n$ with weights $A_i$ in the domain and $D_i$ in the range:

$$p(t) = (-1)^{m-n} t^{\Sigma D - \Sigma A} \left[ -1 + \operatorname*{res}_{s=0} \prod \frac{1 - st^{A_i}}{s - st^{A_i}} \prod \frac{s - st^{D_i}}{1 - st^{D_i}} \frac{s\,ds}{1-s} \right] + \\ + \frac{\prod(1 - t^{D_i})}{\prod(1 - t^{A_i})} \left[ \sum t^{-D_i} - \sum t^{-A_i} + 1 \right].$$

For $m - n = 1$,

$$p(t) = \frac{\prod(1 - t^{D_i})}{\prod(1 - t^{A_i})} \left[ \sum t^{-D_i} - \sum t^{-A_i} + 1 \right] + t^{\Sigma D - \Sigma A}.$$

For $m - n = 2$,

$$p(t) = \frac{\prod(1 - t^{D_i})}{\prod(1 - t^{A_i})} \left[ \sum t^{-D_i} - \sum t^{-A_i} + 1 + t^{\Sigma D_i - \Sigma A_i} \right] - t^{\Sigma D - \Sigma A}.$$

If $m-n=1$, then $p(t)=t^{\Sigma D-\Sigma A}h(t)$, where $h$ is the Poincaré polynomial of the Hamm–Gruel filtration:

$$h(t)=(-1)^{m-n}\left[-1+\operatorname*{res}_{s=0}\prod\frac{st^{A_i}+1}{t^{A_i}-1}\prod\frac{st^{D_i}-s}{st^{D_i}+1}\frac{ds}{s(s+1)}\right],$$

$$\Omega_{\mathrm{rel}}=\Omega^{m-n}/d\Omega^{m-n-1}+df\wedge\Omega^{m-n-1}$$

(the slash here means "modulo").

For $m-n=2$ this fails: $D_1=D_2=2$, $A_1=\cdots=A_4=1$, $h(t)=3t^4+4t^3$, $p(t)=2t^{-2}+4t^{-1}+1$. If $m-n=2$, then $h(1)=\mu(1)$. It is not known whether this is the case if $m-n>2$.

**1977-11.** Investigate the mapping that associates with each (unordered) set of critical points (of a function from a versal deformation) the (unordered) set of critical values. Investigate also the corresponding maps: (ordered sets of critical points) $\rightsquigarrow$ (unordered sets of critical values). Find the discriminants, fundamental groups and other invariants of branched coverings. How do the critical points rearrange after a circuit around a caustic?

**1977-12.** Investigate the bifurcations (with the parameters $\mathrm{Re}\,\varepsilon$, $\mathrm{Im}\,\varepsilon$) of the family of vector fields on the plane $\dot{z}=\varepsilon z+Az|z|^2+\bar{z}^3$ if the values of $A$ are generic. (Conjecturally, they are the topologically versal deformations of $\mathbb{Z}_4$-symmetric fields for each of the 48 domains in the $A$-plane.)

# 1978

**1978-1.** Investigate the topological properties of functions $f(x)=\max_y F(x,y)$.

**1978-2.** Explore the singularities of the boundary of the attainability manifold in a typical controlled system.

**1978-3.** Explore the singularities of the Nekhoroshev steepness indices (the stratification of the variety of Hamilton functions with respect to the indices). Calculate the indices of a typical system with 1, 2, 3 degrees of freedom at all the points.

**1978-4.** Let $\{I_\alpha\}$ be a collection of first integrals of a Hamiltonian system. Assume that $\{I_\alpha\}$ is closed under taking Poisson brackets, so that $(I_\alpha, I_\beta) = \mathcal{F}_{\alpha\beta}(I)$. Is it possible to replace $\{I_\alpha\}$ with $\{J_\alpha\}$ such that $(J_\alpha, J_\beta) = \sum_\gamma C^\gamma_{\alpha\beta} J_\gamma$? (If not, what deformations of the initial terms Lie algebra are not equivalent to each other?)

**1978-5.** Formalize the principle: whatever is good, is also delicate.

**1978-6.** Relaxed Hilbert 16th problem.

**1978-7.** What resonances in a three-frequency Hamiltonian system are strong (the strong resonances in a two-frequency system are $|\omega_1| : |\omega_2| = 1:1,\ 1:2,\ 1:3,\ 2:3,\ 1:4,\ 3:4,\ 2:5,\ 4:5$)?

**1978-8.** Describe the boundary singularities $B_\mu$ and $C_\mu$ that appear in the problem of bypassing an obstacle.

**1978-9.** How many cycles emerge from generic two-parameter bifurcations when eigenvalues pass through $\pm i\omega_1$, $\pm i\omega_2$ (in the corresponding slow system, i. e., for a vector field in the plane that is tangent to the sides of an angle, from a bifurcation with nonzero eigenvalues) in a two-parameter family of such fields, or—which is the same—for $(\mathbb{Z}_2 \times \mathbb{Z}_2)$-equivariant fields in the plane?

**1978-10.** Investigate the stratification of the manifold of linear elements of generic surfaces near each of the 10 strata.

**1978-11** (V. L. Popov)**.** Find a connection between the theory of singularities and the quotients of $\mathbb{C}^2$ by finite subgroups of U(2) (not SU(2)!).

**1978-12.** Investigate geometric (topological?) properties of $p$-real submanifolds in $\mathbb{C}^N$ or in a Kähler manifold (with a restriction on the dimensions of the intersections of the tangent planes with the tangent planes multiplied by $i$).

**1978-13.** Investigate the relations between Smith's theory of complex conjugations and the mixed Hodge structure on a manifold.

**1978-14.** Investigate Lie subsemigroups (e. g., of $\mathrm{SL}(2,\mathbb{R})$) and their tangent cones at 1.

**1978-15.** How many limit cycles can emerge from a zero of a $\Gamma$-nondegenerate vector field with a given Newton diagram $\Gamma$? Is it true that their number is less than some constant $N(\Gamma)$?

**1978-16.** Investigate the singularities of Gaussian maps globally.

**1978-17.** Investigate the theory of symmetric hyperbolic systems of partial differential equations in the framework of singularities.

**1978-18.** Construct explicitly the local topological classification of Lagrangian and Legendrian maps in the cases where the smooth classification has modules, or even functional modules. *The smooth classification has been described by V. M. Zakalyukin (it contains functional parameters) up to dimension* 10*, inclusive of the mapped Lagrangian or Legendrian manifold.*

*But the following is not clear:*

a) *Does the Zakalyukin class define the topological type of the Lagrangian (Legendrian) map? That is, is this type constant along each class?*

b) *Does this class define the topology of the decomposition into simpler classes of a neighborhood in the space of jets? That is, are the bifurcation diagrams locally diffeomorphic or at least homeomorphic?*

c) *Here the bifurcation diagram can be interpreted as:*

— *A: discriminant (a bifurcation diagram of zeros);*

— *B: bifurcation diagram of functions (in the truncated base);*

— *C: the projection of A onto B;*

— *D: the decomposition into Lagrange classes in the space of jets;*

— *E: the decomposition into Legendre classes.*

d) *Similar questions for multi-jets.*

e) *In order to apply transversality arguments to these stratifications of universal objects we need to know whether Whitney's A and B conditions are satisfied. (Conjecturally no, hence Zakalyukin's "stratification" is subject to a refinement!).*

**1978-19.** Investigate the bifurcations of type $D_5$ in the 3-space topologically (*the problem has been studied by V. I. Bakhtin*).

**1978-20.** Investigate the singularities of bicaustics of type $D_5$, up to diffeomorphisms.

**1978-21.** Investigate the process of sweeping the bicaustic $D_4$, up to equivalence (strong equivalence): given three smooth curves $\varphi_i : (\mathbb{R},0) \to (\mathbb{R}^2,0)$ starting from 0 with the same velocity $v \neq 0$, in all other generic. The equivalence is provided by the diagrams

$$\begin{array}{ccc} (\mathbb{R},0) & \xrightarrow{\varphi_i} & (\mathbb{R}^2,0) \\ \downarrow{\scriptstyle \tau} & & \downarrow{\scriptstyle h} \\ (\mathbb{R},0) & \xrightarrow{\psi_i} & (\mathbb{R}^2,0) \end{array}$$

(where the diffeomorphisms $\tau$ and $h$ are independent of $i$); the strong equivalence is: $\tau(t) = t + \text{const}$.

**1978-22.** What is the behavior of the mixed Hodge structure of a singularity under the action of a complete monodromy group? (This can distinguish subgroups in $\pi_1$?)

# 1979

**1979-1.** How can one construct the quivers $A$, $D$, $E$ from the singularities $A$, $D$, $E$ (and their local rings)?

**1979-2.** Show that, for a generic function $F$, the function $\min_y F(x,y)$ is a topologically Morse function.

**1979-3.** Prove the semicontinuity of the spectrum of a singularity. Is it the spectrum of an oscillating system with $\mu$ degrees of freedom? In this case its interlacing by the spectrum of a close system with $\mu - 1$ degrees of freedom would follow from the Rayleigh–Courant–Fisher theory.

**1979-4.** Construct a "complexification" of the homology theory (replacing a boundary with a two-sheet branched covering). What is the complexification of orientation? (Apparently, it assigns an element of $\mathbb{Z} = \pi_1(\mathrm{U}(n))$ to a loop?)

**1979-5.** Construct the characteristic classes of Lagrangian singularities from the stable cohomology ring of complements of caustics.

**1979-6.** Do the real and the complex modality for a critical point of a function always coincide?

**1979-7.** Analyze the theory of envelopes in the framework of the theory of singularities. Find versal unfoldings, bifurcation diagrams, and the connection with symplectic and contact geometry.

**1979-8.** Why are caustics irreducible? How many irreducible components does the singularity manifold of a caustic have?

**1979-9.** Investigate the properties of the discriminants of non-quasihomogeneous Legendrian singularities. No topological classification has been found, even in the cases where a smooth classification (with moduli) is available.

**1979-10.** Describe the mixed Hodge structures of superpositions of functions.

**1979-11.** Investigate typical singularities of the boundary of the time-like attainability domain.

**1979-12.** Investigate the singularities of the time of shortest bypass of an obstacle.

**1979-13.** Is it true that the singularities of the boundary of the attainability domain in a generic controlled system are the same as those of a generic projection of a manifold with boundary? More generally, a "parameter" in optimization problems is a choice of control from a function space (which can have a boundary or other singularities). Are the singularities of the boundary of attainability in this case the same as those of a generic projection of finite-dimensional boundary manifolds with the same singularities?

**1979-14.** Is it true that the function of the shortest time within the attainable set has the same type of singularities as the minimum $\min_y F(x,y)$ of a generic family of functions?

**1979-15.** Investigate the bifurcations of the phase portrait in two-parameter generic systems of vector fields in the plane for the fields which are tangent to: a) a line, b) a pair of intersecting lines. (The normal forms for the eigenvalues are $0, \pm i\omega$ and $\pm i\omega_1, \pm i\omega_2$.)

**1979-16.** Study the number of zeros of the integral $I(h) = \oint_{\gamma_h}(P\,dx + Q\,dy)$, where $\gamma_h$ is a closed curve from the (continuous) family of periodic orbits of a polynomial vector field [e. g., $\gamma_h = \{x,y : H(x,y) = h\}$, say, for $H = y^2 + x^3 - x$]—an infinitesimal version of the Hilbert 16th problem on cycles. What can be the maximal number of roots of $I(h)$ when $I(h)$ is not identically zero?

**1979-17.** Give an asymptotically sharp bound for the number of connected components of the space of nonsingular real algebraic hypersurfaces of degree $d$.

**1979-18.** Is the equality in the Petrovskiĭ–Oleĭnik inequality attainable?

**1979-19.** Does the Ragsdale conjecture hold? One may reformulate this conjecture as follows. Let $f(x,y,z)$ be a homogeneous polynomial of an even degree, $F_\pm = f \pm t^2$ and $\mathbb{R}V_\pm$ is the local level surface $F_\pm = \pm\varepsilon$. The Ragsdale conjecture is in the estimates of the number of components

$$b_0(\mathbb{R}V_+) \le h_1^{2,2}(F_+), \quad b_0(\mathbb{R}V_-) \le h_1^{2,2}(F_-) + 1$$

in terms of the mixed Hodge structure (if $f$ has appropriate sign).

**1979-20.** Give the best possible estimates (through the degree or a Hodge number) for the individual Betti numbers of real algebraic hypersurfaces, in particular, for the number of components $b_0$. Probably, it is easier to estimate the numbers $b_0$, $b_0 - b_1$, $b_0 - b_1 + b_2$, ..., etc. and the combinations of the local type Morse numbers $M_0$, $M_0 - M_1$, $M_0 - M_1 + M_2$, ... ($M_i$ is the number of critical points of index $i$ merging at zero for some Morsification of the homogeneous equation of a hypersurface).

**1979-21.** What is the maximal number of handles that a component of an algebraic surface of degree $n$ in $\mathbb{R}P^3$ can have?

**1979-22.** Estimate the number of ovals of a curve with a fewnomial equation, through the number of its terms.

**1979-23.** How many nonconvex ovals can a plane algebraic curve of degree $n$ have?

**1979-24.** Does the isotopy type of the pair (plane $M$-curve, its complex orientation) determine a connected component in the space of nonsingular projective real curves of a given degree?

**1979-25.** Explore the fundamental group $\pi_1$ of the complement of the set of singular hypersurfaces in the complex projective space of all hypersurfaces of a fixed degree in $\mathbb{C}P^m$. Find the corresponding monodromy group (a representation of $\pi_1$ by automorphisms of the homology group of a hypersurface).

**1979-26.** Consider the system $\dot{x} = P(x,y)$, $\dot{y} = Q(x,y)$ where $P$, $Q$ are polynomials of the second degree, and let $H(x,y)$ be a first integral of this system (not necessarily a polynomial one). How many limit cycles can emerge from components of level curves of $H$ by small variations of $P$, $Q$ leaving them quadratic polynomials?

**1979-27.** In the system of differential equations $\dot{x} = P(x,y)$, $\dot{y} = Q(x,y)$, let $P$, $Q$ be power series starting with homogeneous polynomials $P_n$, $Q_n$ of degree $n$. Is it true then that for almost all pairs $(P_n, Q_n)$ the number of limit cycles emerging from the origin by a small perturbation of the system, is bounded by a constant depending only on $n$?

# 1980

**1980-1.** $I(h) = \oint_{H=h}(P\,dx + Q\,dy)$. Find an upper bound for the number of zeros of the function $I$.

**1980-2.** The boundary value problem for $\dot{x} = P(x, e^{it})$, $x(2\pi) = x(0)$: the number of solutions.

**1980-3.** The number of limit cycles emerging in the "Lotka–Volterra" system

$$\begin{cases} \dot{x} = x\,(\alpha + \beta x + \gamma y + \cdots) \\ \dot{y} = y\,(\delta + \varepsilon x + \zeta y + \cdots) \end{cases}$$

near $\alpha = \delta = 0$. In particular, integrals along $x^p y^q z^r = h$, $z = 1 - x - y$.

**1980-4** (E. A. Demëkhin)**.** Explain the strange bifurcations of $2\pi$-periodic solutions of the equation $k^3 x^{\mathrm{IV}} + k\ddot{x} + \dot{x}^2 = 0$ as the parameter $k$ varies.

**1980-5.** Investigate the structural stability of contact fields in $\mathbb{R}^3$.

**1980-6.** Apply the mixed Hodge structures to the Jacobian problem (in both cases analyticity differs from algebraicity!).

**1980-7.** Construct a theory of caustic cobordisms (different from that of Lagrangian cobordisms).

**1980-8.** In the theory of singularities (e. g., critical points of functions), why is the codimension in the real case the same as in the complex case? *Compare with the* $\mathbb{R}$- *and* $\mathbb{C}$*-modality and with the (co)dimension of the prolonged self-intersection line of the swallowtail or the umbrella.*

**1980-9.** Apply mixed Hodge structures to real algebraic geometry. For example, for estimation of topological invariants of real Morsifications, and for investigation of the topology of discriminants.

**1980-10.** Apply mixed Hodge structures to problems concerning superpositions—for they "remember" the dimension of the smooth algebraic cycle from which a given (co)cycle originates (say, on the graph, or on the discriminant, or on the complement).

**1980-11.** Prove the semicontinuity of the spectrum of singularity. If the singularity $S$ is adjacent to a simpler singularity $S'$ with $\mu' < \mu$, then $l_k \leq l'_k$ for $k = 1, \ldots, \mu'$.

**1980-12.** Complexify the homology theory.

**1980-13.** Do there exist any formulae for the complete invariants convolution in terms of the linearized convolution (similar to the Campbell–Hausdorff formula representing multiplication in a Lie group via the commutator of its Lie algebra)?

**1980-14.** What is the complex analog of the generalized Whitehead groups in algebraic $K$-theory? *One of the candidates is the "quasiresolvent" of the fundamental group of the complement of the bifurcation diagram of a singularity.*

**1980-15.** The embedding of the base $\mathbb{C}^{\mu'}$ of a versal unfolding of a simpler singularity $S'$ into the base $\mathbb{C}^{\mu}$ of a versal unfolding of a more complicated singularity $S$ ($\mu > \mu'$) induces a homomorphism

$$H^*(\mathbb{C}^{\mu} \setminus \Sigma) \to H^*(\mathbb{C}^{\mu'} \setminus \Sigma')$$

of the cohomology rings of complements of the corresponding bifurcation diagrams.

Are these homomorphisms canonical? Is it possible to define the stable cohomology ring?

**1980-16.** Does the real modality of a real singularity with finite multiplicity $f\colon (\mathbb{R}^n, 0) \to (\mathbb{R}, 0)$ always coincide with the complex modality?

**1980-17.** Show that the function $F = \max_x f(x, \cdot)$ is topologically equivalent to a Morse function for a generic family $f$.

# 1981

**1981-1.** Lannér schemes (Coxeter schemes for the groups generated by reflections in the walls of simplices in the Lobachevskian space).

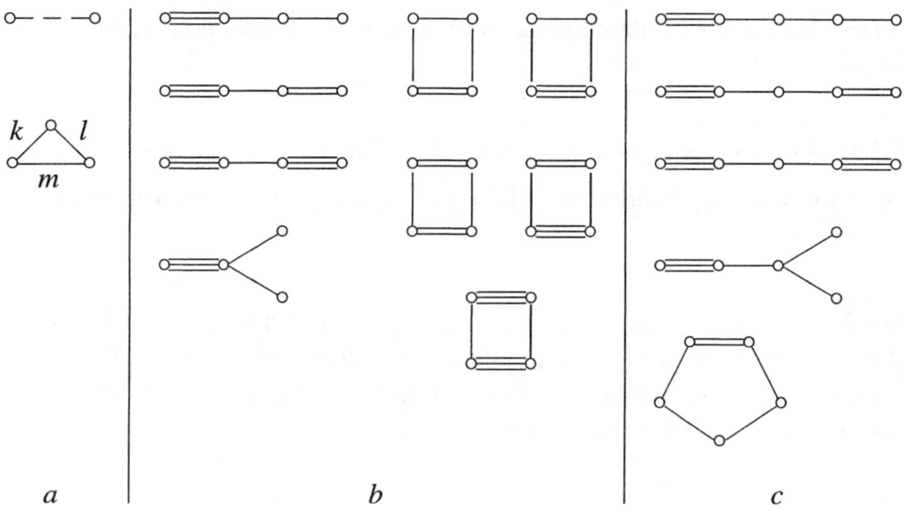


Simplices in the Lobachevskian space: a) the series on the plane, b) 9 simplices in the three-dimensional space, c) 5 simplices in the four-dimensional space.

Find applications of these schemes in singularity theory.

**1981-2.** Calculate the worst Nekhoroshev indices for generic Hamiltonians with $n$ degrees of freedom (or at least the asymptotics of these indices for $n$ large).

**1981-3.** Let

$$I_h(\lambda) = \int\limits_{x \in \mathbb{R}^k} e^{iS(x,\lambda)/h} a(x,\lambda)\, dx,$$

where $S$ is a generic function. Prove that there is the following bound for $\lambda$ such that $S(\cdot,\lambda)$ is a Morse function:

$$|I_h(\lambda)| \le Ch^{k/2} \sum_{x \in \operatorname{crit} S(\cdot,\lambda) \cap \operatorname{supp} a(\cdot,\lambda)} \left| \det \frac{\partial^2 S}{\partial x^2} \right|^{-1/2},$$

where crit is the set of critical points of a function, and supp is its support. *Y. Colin de Verdière proved this for simple or parabolic singularities.*

**1981-4.** Does there exist an exact Lagrangian embedding of $\mathbb{T}^2$ into the standard symplectic space $\mathbb{R}^4$?

**1981-5.** Will a nonstandard contact structure of $\mathbb{R}^3$ remain nonstandard after an arbitrary complexification?

**1981-6.** Evaluate the cohomology rings $L_k = \lim_{n\to\infty} \pi_{n+k} T\lambda_n$, where $\lambda_n$ are the tautological Grassmann bundles over $\mathrm{U}(n)/\mathrm{O}(n)$ or over $\mathrm{U}(n)/\mathrm{SO}(n)$, and $T$ is the Thom space.

**1981-7.** A quasifunction is an exact Lagrangian embedded submanifold of $T^*V$ that is isotopic to the zero section in the class of such embeddings. Critical points are intersections with the zero section. Conjecture: the number of critical points for a quasifunction is not less than for a function.

**1981-8.** What function on the collar can be extended over the ball without critical points?

**1981-9.** Consider closed contractible (bounding a disk on the universal covering) curves of constant geodesic curvature $K \neq 0$ on a surface $M^2$. There are at least as many such curves as critical points of a function on $M^2$. *Counterexample: horocycles on a surface of constant negative curvature. However, for* $\mathbb{T}^2$ *and* $S^2$ *this conjecture has not been disproved.*

**1981-10.** Construct a bifurcation theory for optical caustics, in particular, prove that "flying saucers" caustics do not exist.

**1981-11.** Find a Lagrangian singularity related to the hypericosahedron group $H_4$.

**1981-12.** Find the (Zariski) relations between the (Zariski) relations of swallowtails (and, in general, explore "syzygies," or "noncommutative resolvents" of the fundamental groups of complements of algebraic hypersurfaces, associated with the sequence of complete flags of generic projections, and with generators and relations of the sequence of the fundamental groups of complements of discriminants of those projections).

**1981-13.** Evaluate the fundamental group of the space of nondegenerate plane curves of fixed degree $d$.

**1981-14.** Investigate the singularities of the density of a gravitationally evolving dust-like medium, if the initial potential field of velocities is generic (even on the line!).

**1981-15.** Can the barycenter of a convex part of a closed convex surface coincide with the barycenter of the surface?

**1981-16.** Is it true that a polynomial vector field on the plane has only finitely many limit cycles? *H. Dulac committed an error proving it.*

**1981-17.** Investigate the winding number of an analytic diffeomorphism of $\mathbb{S}^1$ ($x \mapsto x + a + b \sin x$, etc.) as the limit (as $\operatorname{Im} a \to 0$) of the modules of elliptic curves formed by the orbit space for $\operatorname{Im} a \neq 0$. What are the singularities of the analytic extension of the rotation number as a function of $a$?

**1981-18.** Is there a kinematic magnetic dynamo in the topology of the three-dimensional ball $B^3$?

**1981-19.** Give a contact version of the problem of bypassing an obstacle.

**1981-20.** Is it true that the singularities of the increment of a generic family of matrices (polynomials) are topologically equivalent to convex polyhedral or at least Morse functions (possibly, polyhedrally convex, Morse modified along the parameters, on which everything depends smoothly)?

**1981-21.** Explore singularities in typical controlled systems.

**1981-22.** Develop the theory of versal unfoldings of differential forms $f(x)(dx)^{\alpha}$.

**1981-23.** Evaluate the number of different "inflections" of algebraic surfaces of degree $d$ in $\mathbb{C}P^3$.

**1981-24.** Investigate what the mixed structures and the spectra can provide for the Bruce problem about the maximum number of Morse points on a hypersurface of degree $d$.

**1981-25.** Work out a monodromy theory (of a representation of $\pi_1$ of the complement to a bifurcation diagram) for complete intersections (not only for hypersurfaces): one should consider flags of hypersurfaces and sequences of Dynkin diagrams.

**1981-26.** Explore the effects of singularities (inflections of various types) on the asymptotics of the numbers of integer points on submanifolds of the Euclidean space and inside its domains (as well as the effects on the Diophantine approximations).

**1981-27.** Construct a theory of self-intersections of Lagrangian and Legendrian manifolds. To what extent are the self-intersections topologically inevitable (locally and globally)?

**1981-28.** Investigate the singularities of the convex hulls of $M^3$ in $\mathbb{R}^4$ (especially their modules).

**1981-29.** Elliptic coordinates in $\mathbb{R}^n$:

a) a "magnet" generalization of the Ivory theorem (to forms);

b) infinite-dimensional versions (with either discrete or continuous spectrum): what happens to the Jacobi formulae? in particular, to the surprising duality between the expression of impulses in the elliptic coordinates and the inversion formula of the coordinates;

c) elliptic coordinates and the Hilbert transformation;

d) equations of mathematical physics, integrable with the help of b).

# 1982

**1982-1.** Is the symplectic structure of a neighborhood of the Lagrangian opening of the swallowtail standard?

**1982-2.** The Morse–Darboux super-lemma.

**1982-3.** Give a description of the liftable diffeomorphisms and fields in terms of their behavior on the singularities in the base.

**1982-4.** In the theory of integrable systems, Coxeter groups $A$, $D$, $E$ appear (A. M. Perelomov and others). In the theory of integrable systems with a boundary (E. K. Sklyanin), do $H_{3,4}$ also appear?

**1982-5.** Describe the shapes of the resonance zones for torus mappings defined by trigonometric polynomials which perturb a translation (Mathieu type systems).

**1982-6.** Study the asymptotics of solutions of the thermoconductivity equation on differential forms with transfer ("dynamo"): uniqueness of the stationary solution in a given homology class.

**1982-7.** Investigate the singularities of the boundaries of the manifolds of elliptic and hyperbolic polynomials.

**1982-8.** It is known that the first sheet of a hyperbolic surface is convex. What can be said in this vein about the second, third etc. sheet?

**1982-9.** What happens to Legendre transforms (fronts) if the initial functions (hypersurfaces) depend on parameters and become singular for some parameter values? What perestroikas of dual objects take place there?

**1982-10.** Supplement the formal analysis of normal forms performed in the paper ARNOLD V. I. Reconstructions of singularities of potential flows in a collision-free medium and caustic metamorphoses in three-dimensional space. *Trudy Semin. Petrovskogo*, 1982, **8**, 21–57 (in Russian); *the English translation*: *J. Sov. Math.*, 1986, **32**(3), 229–257, by a study of smooth and analytic normal forms.

**1982-11.** Prove that taking gravitation into account in a dust-like medium does not affect the topological features of caustic perestroikas (with typical initial potential flow).

**1982-12.** Given a stratum $\mu = \text{const}$, what maximal value of $\mu$ do the adherent singularities have? For instance, the adjacency $P_8 \to E_6$ exists whereas $P_8 \to A_7$ and $P_8 \to D_7$ do not.

**1982-13.** Find normal forms for a typical contact structure in a neighborhood of the swallowtail (and investigate the hierarchies arising from the constraints on the ranks along submanifolds or on their tangent planes at the singularity).

**1982-14.** Develop the algebraic (analytic?) symplectic (contact) geometry that treats all the things in terms of ideals. Example: replace $df \neq 0$ with $\exists h$ : the Poisson bracket of $f$ and $h$ is 1. Some theorems known in the nonsingular situation may happen to be more general (say, for isolated singularities?).

**1982-15.** Let

$$\prod_{k=1}^{3} \frac{z^N - z^{A_k}}{z^{A_k} - 1} = \sum_r p_r z^r$$

($A_k$ and $N$ are natural numbers) be a polynomial with nonnegative coefficients $p_r$. Consider the number $B(a) = \sum(p_r : aN < r < (a+1)N)$.

Increase the fractions $N/A_k$ (so that the coefficients of the polynomial remain nonnegative). Prove that the number $B(a)$ will then also increase (possibly nonstrictly).

*In the n-dimensional case, $A_k/N$ are the weights of a quasihomogeneous function with an isolated singularity at* 0.

**1982-16.** Consider a Newton polyhedron $\Delta$ in $\mathbb{R}^n$ and the number $\mu(\Delta) = n!\,V - \sum(n-1)!\,V_i + \sum(n-2)!\,V_{ij} - \cdots$, where $V$ is the volume under $\Delta$, $V_i$ is the volume under $\Delta$ on the hyperplane $x_i = 0$, $V_{ij}$ is the volume under $\Delta$ on the hyperplane $x_i = x_j = 0$, and so on.

Then $\mu(\Delta)$ grows (non strictly monotonically) as $\Delta$ grows (whenever $\Delta$ remains coconvex *and integer*?). *There is no elementary proof even for* $n = 2$.

**1982-17.** Consider the boundary value problem $\Delta u = 0$ in the domain bounded by a quadric (say, a hyperbola in the plane, with the boundary value 1 on one component and 0 on the other). Then there exists a "natural" solution (moreover, there is a natural condition at the infinity which selects it).

Does there exist any reasonable filtration for harmonic functions and forms in the case of generic (hyperbolic?) algebraic hypersurfaces that yields a one-to-one correspondence between (relative) homology classes and harmonic representatives (for quadrics, the answers of Vainshtein and Shapiro would appear)? Is there a real version of the mixed Hodge structure?

**1982-18.** Develop the singularity theory for mappings between symplectic (contact) manifolds (a singularity is a violation of symplecticity).

**1982-19.** Explore symplectic correspondences, i. e., multivalued symplectomorphisms

$$X^{2n} \subset (A^{2n} \times B^{2n}), \quad (\pi_A^* \omega_A + \pi_B^* \omega_B)|_{X^{2n}} \text{ is symplectic.}$$

Find the hierarchy of the germs of such correspondences.

**1982-20.** Study the rationality of Poincaré series in natural analytic classification problems, e. g., for the germs of typical mappings in the worst dimensions where functional moduli are inevitable (virtually excluding the germs from a set of infinite codimension). Another example: apply this to the classification of the equations $y'' = F(x, y, y')$.

**1982-21.** What happens to the Givental triads when the quadraticity condition is violated (generically)?

**1982-22.** Can a divergence-free vector field tangent to the layers of Rieb's foliation have an exponential repulsion of trajectories?

**1982-23.** Investigate the singularities in the problem of bypassing an obstacle when the latter is not a hypersurface in the ambient space (e. g., for curves in $\mathbb{R}^3$).

**1982-24.** Can the center of mass of a convex domain in a homogeneous sphere coincide with the center of the sphere? Since it cannot, it makes sense to try to prove the existence of two closed curves (magnetic trajectories) of constant positive geodesic curvature on the sphere as follows. Fiber the space of convex disks over $S^2$ by associating to a disk its "center" point on the sphere. Find constrained critical points along fibers using variational methods, and then apply Morse theory techniques to look for critical points along the base.

# 1983

**1983-1.** How many points (curves, ...) of inflections of various types are merged at a singular point of a hypersurface (subjected to a generic diffeomorphism)? *J. Plücker*: 6 *inflections meet in* $A_1$, *and* 8 *in* $A_2$.

**1983-2.** Courant's theorem says that the zeros of the $n$-th eigenfunction of the Dirichlet problem for the Laplace equation divide the domain into at most $n$ parts. Carry over Courant's theorem to the case of systems (when the zeros form a set of codimension greater than 1).

**1983-3.** Can one carry over the Conley–Zehnder theory to reversible systems (*the latter resemble Hamiltonian systems so much that one would like to treat the property of being Hamiltonian as a variety of "superreversibility"*)?

**1983-4.** Let $N$ lines be given in the real plane, and their complement be chess-like painted black and white. What is the greatest difference between the number of black and white regions?

**1983-5.** How many points of maximum can a polynomial of degree $d$ in two ($n$) variables have? In particular, what would it be if all $(d-1)^2$ critical points are real?

**1983-6.** Find local contact classification of pairs of surfaces in $J^1(\mathbb{R},\mathbb{R})$ (in $C^\infty$).

**1983-7.** How many nondegenerate periodic orbits can a diffeomorphism of $\mathbb{S}^1$ have if it is given by a trigonometric polynomial of degree $n$? The same question for a smooth map which is onto, or for a diffeomorphism which is given trigonometrically-rationally.

**1983-8.** Investigate real forms of reflection groups.

**1983-9.** Is it true that the number of periodic trajectories of a diffeomorphism of $\mathbb{S}^1$ is bounded by the integers which are the invariants (e. g., genera, bidegrees) of the algebraic self-correspondence that is the complexification of the diffeomorphism?

**1983-10.** Consider a projection $\mathbb{R}^{(n=)3} \to \mathbb{R}^{(k=)2}$ and the preimage of the integral points $\mathbb{Z}^{(k=)2}$, that are parallel lines (subspaces of dimension $n-k$) in $\mathbb{R}^{(n=)3}$. Consider a generic curve (manifold of dimension $k-1$) $\gamma$ in $\mathbb{R}^{(n=)3}$ and its linking number with all the parallel lines (subspaces of dimension $n-k$). Investigate the behavior of the linking number under dilations of $\gamma$ in terms of inflection points of $\gamma$. (If $n=k$ then this is a question about the number of integral points in a domain!)

**1983-11.** Is it true that the integrals $I(h) = \oint_{H=h}(P\,dx + Q\,dy)$ with varying polynomials $P$, $Q$ form a Chebyshev system (or, at worst, the number of zeros is not too much greater)? Here, for instance, $H$ is a cubic polynomial $y^2+x^3-x$. A similar question is also interesting about perturbations of other integrable polynomial systems of the Lotka–Volterra type [where $H = x^\alpha y^\beta z^\gamma$, $z = 1-x-y$, with the corresponding (non-polynomial) $P$, $Q$].

**1983-12.** Carry over the relation of indeterminacies (which connects projections of a Lagrangian manifold onto $p$- and $q$-subspaces) to Lagrangian manifolds with singularities and to the duality of convex polytopes. *For example, the stronger is a singularity of an oscillatory integral (as the wave length $h \to 0$), the less is the number of points (in the $\lambda$-space) with this asymptotic (since $S(x,\lambda)-\lambda x$ is a Morse function in $(x,\lambda)$). But one can probably say more!*

**1983-13.** De Rham mixed structure theory: Define filtrations in a neighborhood of a singularity of a form in the *real* case in terms of the type of the singularity.

**1983-14.** Describe the Gibbs distribution of the density evolution under the action of a small diffusion $\varepsilon$ and a flow $v$ with multivalued potential $U$ on a non-simply connected manifold as $\varepsilon \to 0$: $u_t + (uv)_x = \varepsilon \Delta u$, $v = -\operatorname{grad} U$ (*e. g., $v$ is a pseudoperiodic function on $\mathbb{R}^2$, $v = ax + by +$ periodic part, and $u$ is a function on a torus $\mathbb{R}^2/$periods*). Describe how the limit is being approached (for a generic $U$).

**1983-15.** Is it true that the singularities of the ellipticity and hyperbolicity boundaries in generic families are the same (topologically, smoothly) as the singularities of graphs of functions $\max_y F(x,y)$ for generic families $F$?

**1983-16.** Is it true that the number of limit cycles emerging from a singular point of an analytic system, is bounded (except for systems forming a set of codimension infinity, or possibly except for the *integrable* ones only)?

# 1984

**1984-1.** Examine the singularities of the boundary of the space of Chebyshev systems of functions.

**1984-2.** Construct a Morse theory with nonholonomic constraints, say, for higher derivatives.

**1984-3.** Investigate global topological restrictions on caustics implied by the condition that the eiconal is positive definite.

**1984-4.** Prove that on $\mathbb{T}^2$ there are (generically) at least four closed (on the universal covering) curves of constant geodesic curvature $K > 0$.

**1984-5.** Consider the circle $x^2 + y^2 = 1$ and a quadratic function with parabolic level lines intersecting the circle not more than twice [e. g., $y + (x - a)^2$, $|a| > a_0$, where the value $a_0 > 0$ is determined by the condition that the parabola $y + (x - a_0)^2 = c_0$ has a point of *cubic* tangency with the circle under a suitable choice of $c_0$]. Consider the correspondence permuting the intersection points. The product of two such correspondences changing orientation (the second, for example, changes the sign of $y$ or of $x$) determines an orientation-preserving diffeomorphism of the circle onto itself. Is the number of cycles (periodic trajectories) of this diffeomorphism bounded by a constant independent of $a$?

**1984-6.** Classify the germs of "generic" Poisson structures in $\mathbb{R}^3$. *The term "generic" needs to be defined. The situation is the same as in classifying Lie algebras or commuting pairs of functions on the symplectic plane and in similar problems: the initial infinite-dimensional space is not smooth and, generally, may have components of "different dimensions."*

**1984-7.** Build the theory of versal deformations of Fuchsian systems. Is it true that regular singularities are isomonodromic limits of (confluent) Fuchsian points? Which matrices from the monodromy group converge to the Stokes matrices in the irregular case?

**1984-8.** Give an axiomatic definition of skew-symmetric versions of the monodromy groups of simple singularities (which would lead to their classification, similar to the classification of reflection groups or Weyl groups in the symmetric case). Apply this definition to complete intersections (considering a flag of embedded hypersurfaces and sequences of root systems).

**1984-9.** Is the number of Dynkin diagrams (of strongly distinguished bases) of a fixed singularity finite?

**1984-10.** Describe variational and symplectic properties of Picard–Fuchs equations (the Gauss–Manin connection). Are they not the Euler equations for an appropriate group?

**1984-11.** Translate the relative Morse theory into the symplectic language of the theory of Lagrangian intersections or Legendrian links.

**1984-12.** Carry over the asymptotic ergodic definition of the Hopf invariant of a divergence-free vector field to S. P. Novikov's theory generalizing the Whitehead product in homotopy groups.

**1984-13.** Does there exist a mapping $\mathbb{R}P^2 \to \mathbb{R}^2$ with only one Whitney cusped singularity? *Yes; solved by Yu. V. Chekanov on October* 23, 1984.

**1984-14.** What is known about $\mathbb{C}$-contact structures in $\mathbb{C}^3$?

**1984-15.** How can we extract information independent of the choice of generating loops in the successive fundamental groups of complements of points on the $\mathbb{C}$ fibers of successive bundles from the "resolutions" of the fundamental groups of complements of algebraic hypersurfaces?

**1984-16.** Study the equation $dy/dx = f(x,y)$ where $x$ and $y$ are angular coordinates on the circle while $f$ is a trigonometric polynomial: How many limit cycles can occur for a given Newton polygon?

**1984-17.** Prove that the standard symplectic space $\mathbb{R}^4$ contains no exact embedded Lagrangian torus.

**1984-18.** Complexify the Rolle theorem: if the image of the boundary of a disk equals 0 modulo 2, then the disk contains a critical point inside.

**1984-19.** Classify the umbrellas in a contact space (that is, germs at the vertex up to contactomorphisms).

**1984-20.** Calculate the number of vanishing inflections (of type $A_n$) at a singular point of a hypersurface $A_2$ in $\mathbb{C}^3$ (in $\mathbb{C}^n$) subjected to a generic diffeomorphism (if $n = 2$ then there are 8 inflection points of type $A_2$—Plücker's formula).

**1984-21.** Consider a "generalized Bernoulli scheme"—a network of identical automata with finite radius of action (and memory) in $\mathbb{Z}^n$ ($n = 1$?). Can one derive from their work a difference approximation to something non-Gaussian (i. e., not to the equation of heat conductivity)? Just to what?

**1984-22.** Does there exist a finite number (as R. Thom assumed) of various (germs of) bifurcations of the phase portrait of a gradient system, generically depending on 4 parameters? *R. Thom stated that there are* 7 *types of such systems. According to B. A. Khesin, there are at least* 13 *of them, but probably their number is infinite*?

**1984-23.** Develop a "supertheory" whose even component corresponds to reversible systems, and odd one to Hamiltonian systems.

# 1985

**1985-1.** Examine the singularities on the boundary of the space of fundamental systems of solutions to $n$-th order linear ordinary differential equations.

**1985-2.** Examine the singularities on the boundary of the space of Chebyshev systems of functions.

**1985-3.** Study the topological properties of the stability boundary of $n$-th order linear ordinary differential equations and of the graph of increment.

**1985-4.** Given the equation

$$u_t + (uv)_x = \varepsilon u_{xx}, \qquad v \text{ is a potential field,}$$

on the circle $x \bmod 2\pi$, investigate the eigenfunctions of this equation with the eigenvalues close to zero, as $\varepsilon \to 0$ (also study the case of a multivalued potential).

**1985-5.** Given a contact structure (say, the standard one) on $\mathbb{S}^3$ and a curve being a Legendrian knot of a certain type. How many characteristic chords of the knot are ensured (for an arbitrary contact form)?

**1985-6.** Transfer the Ragsdale conjecture to singularity theory (express the right-hand sides of the Ragsdale-type inequalities for Morsifications of a singularity in terms of the invariants of the singularity rather than in terms of degree). *Even for* $x^n + y^n$, *a new theory is obtained because of upper deformations.*

**1985-7.** Prove theorems on the stabilization of various objects: the cohomology rings of complements of bifurcation diagrams (in $\mathbb{C}$ and $\mathbb{R}$?), the multiplicities of strata adjacency, the increment, the boundary of hyperbolicity, Vassiliev's complex of strata, etc.

**1985-8.** Develop $\mathbb{R}$- and $\mathbb{C}$-theories of vanishing inflections (and flattenings).

**1985-9.** Give an axiomatic description of the Poisson structures arising from mappings of periods of general forms (even for $A_\mu$): a) determine the ranks (e. g., the Lagrangian property) on tangent spaces of various strata of discriminants, b) classify all Poisson structures with given ranks. *The example of a usual swallowtail in $\mathbb{C}^3$ has been cleared up.*

**1985-10.** Is it true that the singularities of the hyperbolicity boundary include the singularities of the ellipticity boundary (at least stably)?

**1985-11.** How is the informal complexification of the notion of orientation related to the spinor structures?

**1985-12.** Are the Picard–Fuchs equations Hamiltonian with respect to some natural symplectic structure, and do they possess a positive Lagrangian responsible for some kind of non-oscillatory behavior?

**1985-13.** Can the awful formulae of representation theory (Klebsch–Gordan coefficients, etc.) be simplified by the aid of the theory of convex polyhedra? *Volumes of sections and numbers of lattice points in them are expressed in an equally complicated way via, say, equations of faces or coordinates of vertices of a polyhedron, but conceptually these are simple objects. Maybe one will feel easier if awful formulae are replaced with these simple geometric constructs. In particular—what is the geometry of the* $6j$*-symbol (it is nonzero if a tetrahedron can be formed with* 6 *lengths*)*: won't integer volumes appear there?*

**1985-14.** Develop the theory of uniform estimates for both oscillatory and exponential multidimensional integrals (Laplace's method) depending generically on parameters.

**1985-15.** Create either a symplectic or a contact version of Shcherbak's theory of $H_3$ and $H_4$, bypassing an obstacle being replaced in it with a general symplectic construction (similar to the way R. B. Melrose interpreted the billiard problem).

**1985-16.** Rewrite the Jacobi formulae of the theory of elliptic coordinates for the infinite-dimensional case (assuming that the spectrum is discrete and the axes lengths have the asymptotics required for the series to converge).

**1985-17.** Is the preservation of the intersection form of a singularity of a function under the stabilizing addition of four squares related to the Bott 8-periodicity? (*Under the stabilization, an* 8*-fold suspension of the Milnor fiber occurs.*)

**1985-18.** Study the behavior of the mixed Hodge structures under superpositions of algebraic functions.

**1985-19.** Is the moment map which sends an $n$-tuple of points $x_1 \leq x_2 \leq \cdots \leq x_n$ with given masses $m_i > 0$ into the $n$-tuple of momenta $M_k = \sum_i m_i x_i^k$ $(k = 1, 2, \ldots, n)$ a homeomorphism of a convex polyhedron onto its image?

**1985-20.** Homotopy classification of nondegenerate homogeneous vector fields of fixed degree: how many connected components does this space have? *For example, cubic fields in* $\mathbb{R}^3$: *What is the maximal index of such vector fields*?

**1985-21.** Does the Courant theorem on the zeros of the $n$-th eigenfunction of the Laplace operator admit a complexification (provided that the values are complex and the zeros do not divide the space)?

**1985-22.** Investigate the topology of the Maxwell set of simple real and complex singularities; is there a stabilization of cohomology rings of complements?

**1985-23.** How many Whitney cusped singularities does a generic mapping $S^2 \to S^2$ of degree $n$ necessarily have?

**1985-24.** Let an open swallowtail lying in a discriminant (either as a multiple self-intersection or as an $A_{\approx n/2}$ stratum) be Lagrangian in some symplectic structure. Classify the extensions of these structures over the entire discriminant.

**1985-25.** How is the stratification of the univalence boundary in the space of holomorphic mappings of the disk to the plane organized? Have the strata of small codimensions and the bifurcation diagrams been described?

**1985-26.** J. M. Ball's conjecture: Consider the pyramid inside the swallowtail,

$$\left\{x^{n+1}+a_1x^{n-1}+\cdots+a_n=\prod_{i=1}^{n+1}(x-x_i),\ x_i\in\mathbb{R}\right\}\subset\mathbb{R}^n.$$

Restrict it by the condition $|a_1|\leq 1$. Then for any two points of the bounded domain obtained, there is a curve of length less than $Cd$ ($d$ being the distance between points in $\mathbb{R}^n$) connecting these points inside the domain, where the constant $C$ is independent of the points.

More generally: How can one describe the semialgebraic sets possessing such property of pseudoconvexity (called the *Whitney property*)?

# 1986

**1986-1.** Consider the space of Lagrangian tori in $T^*\mathbb{T}^2$ that are isotopic to the zero section among all the tori. How many connected components does it have?

**1986-2.** Consider a Hamiltonian in $T^*\mathbb{T}^2$ quadratically convex with respect to momenta. Suppose that the tori mentioned in the preceding problem lie on its level-1 hypersurface (deformations are then applied to pairs torus–Hamiltonian). How many connected components are there in the space of such pairs (topologically trivial)?

**1986-3.** A rigid body is controlled by a momentum of a given intensity; the orientation of the momentum with respect to the body (satellite) can be taken as a controlling parameter. It is required to turn the body from one state to another (perform a rotation in $SO(3)$) as fast as possible, say, at zero initial and final angular velocity.

Describe the optimal control (first and foremost, the topology of the manifold of the discontinuity of this control on $SO(3)$).

**1986-4.** To a purely imaginary pair of a vectorfield's eigenvalues there corresponds, generally speaking, a Lyapunov invariant surface. Explore the perestroikas (bifurcations) of these surfaces at resonances.

**1986-5.** Transfer the Smale–Hirsch theory to the Lagrangian and Legendrian bands (germs of Lagrangian and Legendrian manifolds along curves belonging to these manifolds) or to the corresponding framed curves.

**1986-6.** Is the diameter of the symplectomorphism group of the ball $B^{2n}$ bounded? Conjecturally, no. (*In the two-dimensional case this was proved by A. I. Schnirelmann. In the higher-dimensional case, thanks to non-simple-connectedness of the group of symplectic matrices* ($\pi_1 = \mathbb{Z}$), *one can strongly twist a central ball, and the corresponding diffeomorphism is conjecturally rather far from the identity.*)

**1986-7.** Find the asymptotic form of the number of meanders with $n \to \infty$ bridges.

**1986-8.** Study the singularities of the apparent contours of convex bodies.

**1986-9.** In optimization theory, there occur situations where a nonconstant (say, periodic) control gives better (on average over a long time) results than any fixed parameter.

Study these situations from the viewpoint of genericity and bifurcations. *The situation resembles a phase transition. Generally, the regime optimal on the average may be more complex than a periodic one*!

**1986-10.** Reformulate the theorem about three inflection points of a projective curve and about four vertices of a Euclidean curve in terms of symplectic or contact topology.

**1986-11.** In addition to models with internal degrees of freedom along a small fiber of a bundle over space-time, models of the surface tension type are conceivable, where the fundamental laws of hydrodynamics act in a larger space but an observer on the surface only sees their manifestations in a smaller space (the difference in the dimensions can even be greater than 1). What common features do the models of this type have—what is the structure of their equations of motion?

**1986-12.** Study the singularities of the level $u = 0$ for a function $u$ of two variables satisfying the (Euler stationary) equation: there exists a function $f$ such that $\Delta u = f(u)$. Investigate the typical cases and bifurcations of codimensions 1 and 2.

# 1987

**1987-1.** Carry over the theory of the Gibbs distribution (for the one-dimensional evolution $u_t + (uv)_x = \varepsilon u_{xx}$) to the case of a discrete time (a map $\mathbb{S}^1 \to \mathbb{S}^1$, close to the identity, is being perturbed by a small diffusion). What is an analog of the theory of eigenvalues close to zero that correspond to the point attractors of the field $v$?

**1987-2.** Symplectize the nonoscillation theory (including the Pólya theorem concerning factorization on an interval).

**1987-3.** The transformation $z \mapsto z^2$ sends the trajectories of small oscillations to the Newton ellipses. And what about the transformation $z \mapsto z^\alpha$?

**1987-4.** Consider hypersurfaces in $\mathbb{RP}^n$ of constant signature, e. g., of signature $(1,1)$ in $\mathbb{RP}^3$ (a compactification of the Hilbert problem on embedding a surface of everywhere negative curvature into $\mathbb{R}^3$).

a) Is it true that the space of such hypersurfaces is connected?

b) Is it true that any such surface separates two lines (in the case of a hypersurface of signature $(k,l)$, separates $\mathbb{RP}^k$ and $\mathbb{RP}^l$ in $\mathbb{RP}^{k+l+1}$)? *The answer is positive for $k = 0$: any convex hypersurface is affine.*

c) Is it true that any such hypersurface is a two-fold covering of $\mathbb{RP}^k \times \mathbb{RP}^l$ (or, better, that the subspaces $\mathbb{RP}^k$ and $\mathbb{RP}^l$ separated by the hypersurface can be chosen in such a way that any line joining them intersects the hypersurface at exactly two points)? *This is true for $k = 0$: a convex hypersurface is star-like with respect to any point of the domain bounded by this hypersurface.*

d) The "maximum principle": consider a hyperbolic surface lying in the strip $|z| \leq 1$ in $\mathbb{R}^3$ and tangent to the cone $x^2 + y^2 = z^2$ along the circles $z = \pm 1$. Prove that the surface does not intersect the interior $z^2 > x^2 + y^2$ of the cone. Generalize to other boundary conditions.

**1987-5.** Examine the global topological properties of the caustics and fronts of Legendrian manifolds (and, separately, of optical manifolds; their properties may be different!).

**1987-6.** Evaluate $\pi_3(\mathbb{C}^n \setminus \Sigma^{n-2})$, where $\Sigma^{n-2}$ is the cuspidal edge of the swallowtail. *Of course, here also the similar questions in $\mathbb{R}^n$ and for strata of greater codimension and higher $\pi_i$ are assumed.*

**1987-7.** How many connected components does the complement of the trail of a complete flag in a neighborhood of this flag in $\mathbb{R}^n$ have? *There are two for $n = 2$ and six for $n = 3$.*

**1987-8.** How many connected components are possessed by the complements of (i) bifurcation diagrams of functions and (ii) discriminants of (at least) simple singularities in spaces of real versal deformations?

**1987-9.** Is M. E. Kazarian's list of the Young diagrams of simple singularities a solution to some other classification problem?

**1987-10.** How does the number of critical points of the $N$-th eigenfunction of the Laplacian in an $n$-dimensional domain increase as $N \to \infty$? Like $N^{1/n}$?

**1987-11.** What singularities are encountered in solutions of the variational problem to minimize the Dirichlet integral $\int (\nabla u)^2 \, dx$ over all functions $u$ obtained from a given one by the action of area-preserving diffeomorphisms of the domain (say, of the disk)? *If a given function vanishes on the boundary of the disk and has one maximum inside, then the extremum in the above problem is a centrally-symmetric function on the disk such that the area of the set where the latter function is less than a number equals the area of the set where the former given function is less than the same number. If the initial function has two maxima, similar to the two summits of Elbrus, and a saddle, then physicists observed in numerical experiments that the extremal function has singularities along the segment replacing the saddle.*

**1987-12.** Study the decomposition of the space of linear complex equations with singularities into isomonodromy classes (of special interest are the limits of isomonodromic systems with merged singular points, namely, their versal deformations, bifurcation diagrams, etc.).

**1987-13.** Study the degeneracies of symplectic structures in the space of closed 2-forms, namely, the stratification of the boundary of the manifold of symplectic structures, the bifurcation diagrams at the points of finite-codimensional strata on the boundary, ...

**1987-14.** Do there exist smooth hypersurfaces in $\mathbb{R}^n$ (other than the quadrics in odd-dimensional spaces), for which the volume of the segment cut by any hyperplane from the body bounded by them is an algebraic function of the hyperplane? *For these quadrics the volume is an algebraic function (Archimedes), and the area of segments of plane curves is never algebraic (Newton).*

**1987-15.** Define the "asymptotic Sturm invariant" describing the mean Lagrangian oscillations of variational equations for a Hamiltonian system (in the same sense as the asymptotic Hopf invariant counts the mean number of zeros (with signs) for solutions to normal variational equations; the latter assertion also needs be formalized).

**1987-16.** Study the boundary of the set of second order linear equations with alternating roots of solutions (and carry over the results to the Lagrangian alternation in Hamiltonian systems with $n$ degrees of freedom). Roots alternation property of a second order equation means: in the interval between any two roots of any solution there exists a root of any other solution.

**1987-17.** Are the transformations of phase flows of contact fields in $\mathbb{S}^3$ dense among the contactomorphisms from the identity component?

# 1988

**1988-1.** Classify the singularities of contact-Poisson structures.

**1988-2.** What is the maximum difference between the number of maxima and the number of minima for an $n$-th degree polynomial in $\mathbb{R}^2$? The same question for the $\mathbb{R}$-Morsification of singularities.

**1988-3.** Investigate normal forms of a quadratic cone in the contact space $\mathbb{R}^3$ ($\mathbb{R}^5$) with respect to $C^\infty$ and analytic germs of contactomorphisms at the vertex.

*The question is related to the theory of wave transfomation and relaxation oscillations* (*see the paper*: Arnold V. I. Surfaces defined by hyperbolic equations. *Math. Notes*, 1988, **44**(1), 489–497; *the Russian original is reprinted in*: Vladimir Igorevich Arnold. Selecta–60. M.: PHASIS, 1997, 397–412).

**1988-4.** What is the maximum number of periodic orbits for the diffeomorphism of $\mathbb{S}^1$ which is determined by elliptic functions similar to $x \mapsto x + a + \varepsilon \operatorname{sn} x$?

**1988-5.** Find the upper bound for the Hölder exponent of a continuous ("Peano") mapping of the square to the cube (is $2/3$ attained?). *Solved by E. V. Shchepin.*

**1988-6.** Can the number of intersection points of the image of a circle—under the $n$-th iteration of an analytic diffeomorphism of a surface—with another (fixed) circle grow faster than any exponent of $n$? *Solved by O. S. Kozlovskiĭ.*

**1988-7.** Can the number of periodic trajectories of a real analytic mapping of a surface to itself grow faster than any exponent of the period?

**1988-8.** Can various topological invariants of the intersection $(A^nX^k) \cap Y^l$ as well as the Milnor numbers and other local characteristics of the tangency of germs $(A^nX^k, 0)$ and $(Y^l, 0)$, given that $A(0) = 0$, grow faster than any exponent of $n$?

**1988-9.** Prove the stabilization for $\mu \to \infty$ of the homotopy type of the complement $\mathbb{R}^\mu \setminus A_k$ where $A_k$ is the corresponding stratum of the discriminant $A_\mu$ (of codimension $k$).

**1988-10.** Prove the analogous stabilization for complexifications, $\mathbb{C}^\mu \setminus {}^{\mathbb{C}}A_k$.

**1988-11.** Carry over the four-vertex theorem from planar curves to curves on the sphere $S^2$.

**1988-12.** If the Jacobian of a germ of a mapping $\mathbb{R}^2 \to \mathbb{R}^2$ is identically zero, then the mapping can be factored through a curve as $\mathbb{R}^2 \to K^1 \to \mathbb{R}^2$. Give a precise meaning to this assertion (algebraize it); for instance, begin with formal series and end by $C^\infty$.

**1988-13.** Prove that, in $\mathbb{R}^{2n}$, there are no (nonconvex) algebraically integrable hypersurfaces (i. e., the volume of the part cut off by a hyperplane cannot be an algebraic function of the hyperplane). *Proved by Newton for* $n = 1$.

**1988-14.** Give a formal definition of integrability of a differential equation determined by a vector field on a manifold (the definition must be independent of the algebraic and similar structures on the manifold, i. e., the integrability property must be invariant under diffeomorphisms of the manifold). Prove the nonintegrability in this sense for, e. g., typical Hamiltonian systems close to generic integrable systems.

**1988-15.** Transfer the four-umbilical-point theorem from surfaces to the symplectic or contact topology of Lagrangian or Legendrian singularities (prove the inevitability of $D^4$).

**1988-16.** The theory of second braids: Consider a hypersurface $\Gamma_0 : z^{\mu+1} + \lambda_1 z^{\mu-1} + \cdots + \lambda_\mu = 0$ in $\mathbb{C}^{\mu+1}$ and a sequence of projections $\mathbb{C}^{\mu+1} \to \mathbb{C}^\mu \to$

$\mathbb{C}^{\mu-1} \to \cdots$ (along the axes $z, \lambda_\mu, \lambda_{\mu-1}, \dots$). The discriminant of the projection of the hypersurface $\Gamma_0$ onto $\mathbb{C}^\mu$ is a hypersurface $\Gamma_1$ in $\mathbb{C}^\mu$. The fundamental group of the complement of $\Gamma_1$ is the braid group.

Let us define recursively hypersurfaces $\Gamma_k$ in $\mathbb{C}^{\mu+1-k}$ as the discriminants of the projections of $\Gamma_{k-1}$ from $\mathbb{C}^{\mu+2-k}$ onto $\mathbb{C}^{\mu+1-k}$.

Study these hypersurfaces. Are their complements $K(\pi,1)$ spaces? What are their fundamental groups? Can we describe the fundamental group of such a complement as the group of "Zariski relations" between Zariski relations in the preceding fundamental group?

*The cases $k = 1$ (where the braid group is described as a subgroup of the automorphism group of a free group) and $k = 2$ (where the fundamental group of the complement of the bifurcation diagram is described) have been examined, but the case $k = 3$ still remains to be studied, even for small $\mu$. Though, perhaps, it would be more in spirit of the description of fundamental groups of the complement by Zariski relations to replace the given flag of projections by a generic flag (for our flag, some strata are projected on the same submanifold).*

**1988-17.** Consider the "stochastic web"

$$\left\{x \in \mathbb{R}^2 : \sum_{i=1}^{5} \cos(x, v_i) = c\right\}$$

(where the vectors $v_i$ form a regular pentagon). Is it true that the diameters of this curve's closed components with interior point 0 are bounded above?

**1988-18.** Consider the mapping $T = AB$ of the plane to itself, where $B(x,y) = (x, y + \varepsilon \sin x)$ and $A$ is the rotation through the angle $2\pi/5$. Consider the closed invariant curves of $T$ bounding a domain with interior point 0. Are their diameters bounded above?

**1988-19.** Parametric Morse inequalities for $A_3$ and other singularities. Consider a generic smooth function on the space of a smooth bundle (for instance, with fiber the circle and with two-dimensional base). Over certain points of the base, the restrictions of the function to the fiber have non-Morse singularities, such as $A_2$ on some hypersurface in the base (on a caustic), $A_3$ on a stratum of codimension 2 in the base (at certain points of the base in the case of two-dimensional base, namely, at cusps of a caustic).

Study the relations between the nontriviality of a bundle (e. g., the differentials in its spectral sequence) and the inevitable singularity strata on the base (for instance, the minimum number of cusps of caustics when the base is two-dimensional).

**1988-20.** Consider a diffeomorphism of the boundary of a manifold to itself, which extends to a diffeomorphism of the manifold. Can this diffeomorphism be always extended as a volume-preserving diffeomorphism? What properties of the diffeomorphism of the boundary guarantee the existence of fixed points of the volume-preserving extension? *Example*: $\mathbb{S}^1 \times D^2$.

**1988-21.** Consider a field of directions on $\mathbb{S}^3$. Can these directions be included in planes in such a way that the obtained distribution of planes be invariant with respect to the flow of a vector field $v$ of given direction? ($\exists \alpha, \beta : \alpha|v = 0$, $d\alpha = \alpha \wedge \beta$.)

**1988-22.** Consider a field of divergence 0 on $\mathbb{S}^3$. Does there exist a contact structure in which this field is Legendrian? Or such a structure diffeomorphic to standard?

**1988-23.** Transfer the construction of Pontryagin and Thom from cobordism theory to real algebraic functions. The Serre property for bundles corresponds to the possibility of covering a typical deformation of the set of real roots of a polynomial (which can vanish in pairs) by a deformation of the polynomial itself. The Pontryagin isomorphism between the homotopy groups of spheres and the cobordism groups of framed manifolds corresponds to the isomorphism between the homotopy groups of the space of functions with moderate singularities and the cobordism groups of plane curves without horizontal inflectional tangent lines in the theory of real algebraic functions in one variable (see ARNOLD V. I. Spaces of functions with moderate singularities. *Funct. Anal. Appl.*, 1989, **23**(3), 169–177; *the Russian original is reprinted in*: Vladimir Igorevich Arnold. Selecta–60. Moscow: PHASIS, 1997, 455–469). This example suggests that the similarity extends much farther and can be formalized as the corresponding calculus of singularities. This similarity had first been explicitly mentioned and used in ARNOLD V. I. Braids of algebraic functions and the cohomology of swallowtails. *Uspekhi Mat. Nauk*, 1968, **23**(4), 247–248 (in Russian); *reprinted in*: Vladimir

Igorevich Arnold. Selecta–60. Moscow: PHASIS, 1997, 125–127, and especially in ARNOLD V. I. Cohomology classes of algebraic functions invariant under Tschirnhausen transformations. *Funct. Anal. Appl.*, 1970, **4**(1), 74–75; *the Russian original is reprinted in*: Vladimir Igorevich Arnold. Selecta–60. Moscow: PHASIS, 1997, 151–154.

**1988-24.** Quadratic forms in the Euclidean space $\mathbb{R}^n$ having a multiple eigenvalue constitute a variety of codimension two in the space of all the forms. Can one represent the corresponding discriminant as the sum of squares of two functions (polynomials, power series)? *This is so for* $n = 2$.

For the Hermitian case, the codimension (and the number of squares?) is three. For the hyper-Hermitian case of SU(2)-invariant quadratic forms in $\mathbb{R}^{4n}$, the codimension is five.

**1988-25.** Consider a (possibly, anti-) commutative graded ring (or, better, an $\mathbb{R}$- or $\mathbb{C}$-algebra) with Poincaré series $1+t+t^2+\cdots$ (having one additive generator of each degree). Classify such rings (algebras) with given degrees of multiplicative generators.

*In the simplest nontrivial case of a commutative ring with three multiplicative generators of degrees* 1, 2, *and* 3, *the number of such algebras is* 5. *In the general case, it is not clear for what sets of degrees the object is simple* (*admits no moduli*): *presumably, this is always so for three multiplicative generators.*

**1988-26.** The eccentricity of a Hilbert space. Let $R(N)$ be the minimum number such that $N$ balls of radius $R(N)$ can cover the unit ball in $\mathbb{R}^n$, and let $r(N)$ be the maximum number such that $N$ balls of radius $r(N)$ contained in the unit ball in $\mathbb{R}^n$ can be disjoint. As $N$ increases, the ratio $R(N)/r(N) = \rho(N)$ tends to a limit $\rho$ called the *eccentricity* of the space $\mathbb{R}^n$. Examine the asymptotic behavior of the eccentricity as the dimension $n$ increases. *Possibly,* $\lim\limits_{n\to\infty} \rho = \sqrt{2}$.

**1988-27.** Let $K : \mathbb{T}^2 \to \mathbb{R}_+$ be an arbitrary smooth positive-valued function on a Riemannian torus. Consider the motion of a charged particle on this torus in the presence of a magnetic field $K$ normal to the torus, i. e., its motion along curves on the torus such that their geodesic curvatures at each point are a prescribed (for this point of the torus) positive number $K$.

Suppose that the metric on the torus is flat. The motion of the particle (at velocity 1) is described by a curve in $\mathbb{T}^3 = T_1\mathbb{T}^2$. The standard metric determines a parallelization, namely, the decomposition $\mathbb{T}^3 = \mathbb{S}^1 \times \mathbb{T}^2$. The positivity of the curvature $K$ implies that the phase curves on $\mathbb{T}^3$ are transversal to the fibers $\{\varphi\} \times \mathbb{T}^2$. Thus, we obtain a Poincaré mapping of the fiber $\{0\} \times \mathbb{T}^2$ to itself. This mapping is a symplectomorphism homologous to the identity symplectomorphism for a suitable symplectic structure on $\{0\} \times \mathbb{T}^2$.

Prove that such a Poincaré mapping is homologous to the identity mapping also in the case of motion on a torus with an arbitrary Riemannian metric close to flat.

**1988-28.** Prove that, in the situation considered in the preceding problem, the Poincaré mapping is homologous to the identity mapping for a motion on the torus $\mathbb{T}^2$ with an arbitrary Riemannian metric provided that the geodesic curvature $K$ is sufficiently large.

**1988-29.** Consider the torus $\mathbb{T}^2$ with an arbitrary metric and an arbitrary positive-valued function $K$ on $\mathbb{T}^2$. Does there exist a Poincaré mapping or even a surface transversal to the vector field of the motion of a charged particle on $\mathbb{T}^2$ in the magnetic field $K$ and isotopic to a section of the bundle $T_1\mathbb{T}^2 \to \mathbb{T}^2$?

**1988-30.** Prove the existence of the expected number of closed trajectories of the motion of a charged particle in a magnetic field on an arbitrary surface, at least in the cases where the field $K$ is sufficiently strong or where the metric is close to that of constant curvature. *I believe that it is expedient to directly apply the "hyperbolic Morse theory" rather than to reduce the problem to examining fixed points of a symplectomorphism. In the case of a sufficiently strong magnetic field K, this conjecture is proved: the number of closed orbits is not less than* $2g+2$ *on surfaces of genus g; cf. problem 1994-14.*

**1988-31.** Generalization of the preceding problem: Consider a nontrivial bundle $M^3 \to N^2$ with fiber $\mathbb{S}^1$ endowed with a connection (specified by a field of two-dimensional planes transversal to the fibers). Let $\tau$ denote some volume element on $M^3$, and let $v$ be a divergence-free (with respect to $\tau$) vector field transversal to the plane of the connection. Is the number of closed orbits of such a vector field bounded below by the minimum number of critical points of functions on the surface $N^2$ (supposed to be an oriented surface without boundary)?

**1988-32.** A special case of the preceding problem: Is it true that an arbitrary divergence-free vector field on $\mathbb{S}^3$ making an acute angle with the Hopf field at each point has at least two geometrically different closed orbits?

# 1989

**1989-1.** Classify the simple singularities of functions on supermanifolds.

**1989-2.** Can the number of fixed points of the $n$-th iteration of an infinitely smooth mapping of a compact manifold to itself grow, as $n$ increases, faster than any prescribed sequence $a_n$ (for some subsequence of time values $n_i$)?

**1989-3.** Calculate $\pi_2$(the complement of the stratum $A_3$ of the swallowtail in $\mathbb{R}^n$) for non-stable dimensions $n$.

**1989-4.** Study the cohomology rings of the complements of bifurcation diagrams of functions $A_k$ in $\mathbb{C}^{k-1}$ (including the stabilization as $k \to \infty$, the behavior under the Lyashko–Looijenga mapping, and the relation to stratum diagrams). *This is the cohomology of the "second braid group," because the complement of a bifurcation diagram in* $\mathbb{C}^{k-1}$ *is* $K(\pi, 1)$.

**1989-5.** What functions on manifolds can serve as Jacobians?

**1989-6.** Give a relative version of the Moser theorem on symplectic structures (fix a submanifold and a 2-form on it).

**1989-7.** Carry over the inequalities of Harnack, of Petrovskiĭ, etc. to the pseudoperiodic hypersurfaces determined by sums of (incommensurable) harmonics of the form $A\cos((k,x)+a)$ in $\mathbb{R}^n$ (study the densities of topological objects in unit volumes). *For instance, we can divide by* $R^n$ *the number of maxima, or the Betti numbers, or the Euler characteristic of the domain* $f \leq c$ *in a large ball of radius* $R$ *and send* $R$ *to infinity; it is required to estimate the limit "density of* *maxima," or "density of Betti numbers," or "density of the Euler characteristic" from above in terms of the number of harmonics (or, if possible, of the Newton polyhedron).*

**1989-8.** Nonconvex Minkowski problem. Given a generic mapping $S^2 \to S^2$ of degree 1, consider its Jacobian as a (set-valued) function on the image sphere. What conditions on this function ensure the existence of a Gauss mapping (of a sphere immersed in $\mathbb{R}^3$) with such a Jacobian?

*In the absence of singularities, the only condition is that the center of gravity of the corresponding mass distribution on the sphere should be at zero (the Minkowski theorem).*

**1989-9.** Classify the flags in a symplectic space and simple symplectic quivers.

**1989-10.** Study the systems of fronts and of rays defined by hyperbolic variational principles near typical singularities of the surface of zeros of the symbol (for two- and three-dinensional physical spaces).

**1989-11.** Classify the neighborhoods of Riemann curves of genus $g$ on complex surfaces. *The case of an elliptic curve, $g = 1$, is studied in detail, e. g., in the following book*: ARNOLD V. I. Geometrical Methods in the Theory of Ordinary Differential Equations, 2nd edition. New York: Springer, 1988 (Grundlehren der Mathematischen Wissenschaften, 250); *the Russian original* 1978.

**1989-12.** The infinitesimal version of the problem about periodic orbits of correspondences: Let $A: \mathbb{S}^1 \to \mathbb{S}^1$ be a diffeomorphism of a real oval for an algebraic curve such that its analytic continuation is a correspondence on a Riemannian surface and $A^k = \mathrm{id}$. How many periodic orbits (of period $n$) can arise under a small perturbation of this diffeomorphism (in the class of real algebraic self-correspondences of the same bigenus and bidegree)? Is this number bounded by a function of $n$ of by a constant independent of $n$ (uniformly over perturbations or at least in the first approximation of perturbation theory)?

**1989-13.** In the problem of bypassing an obstacle, examine the asymptotics as the obstacle diffuses and turns into a steep potential.

**1989-14.** In the space of polynomials $\mathbb{R}^n = \{x^{n+1} + a_1 x^{n-1} + \cdots + a_n\}$, consider the subvariety $A_3$ (of codimension 2) consisting of the polynomials with threefold roots. The fundamental group of the complement of this subvariety is $\mathbb{Z}$. The polynomial in two variables $x^{n+1} + a_1(y)x^{n-1} + \cdots + a_n(y)$ naturally defines a curve in $\mathbb{R}^n$. A generic curve does not intersect the subvariety $A_3$. Fixing the boundary conditions for $y \to \pm\infty$, we can associate with such a curve an integer [an element of $\pi_1(\mathbb{R}^n \setminus A_3) \approx \mathbb{Z}$] called the *index* and counting the number of rotations of the curve around $A_3$.

Find the minimal degree of the polynomial in two variables (or of polynomials $a_j$ in $y$) for which a given value $i$ of this index is realized.

*Investigation of this question led V.A. Vassiliev to the problem on the minimal degree of a polynomial mapping $\mathbb{R} \to \mathbb{R}^3$ realizing a fixed knot. The investigation of the arising knot invariant led him to the theory of invariants of finite order.*

**1989-15.** What is the maximum number of parts into which the sphere can be divided by the zeros of a spherical function being a polynomial of degree $n$?

*The well-known Courant theorem gives the upper bound of $n^2/2 + O(n)$ (for the 2-sphere), and examples of V.N. Karpushkin give the lower bound of $n^2/4 + O(n)$.*

What is the largest number of maxima for such a function?

**1989-16.** Find the number of components in the space of nondegenerate homogeneous equations $\dot{x} = P(x)$, where $x \in \mathbb{R}^n$ and the components of $P$ are second-degree homogeneous polynomials having no common zeros but the origin.

*The geometric problem (for $n = 4$) reduces to studying deformations of quadruples of quadrics (ellipsoids) in the projective space. The quadrics are allowed to degenerate and even vanish, but they are forbidden to have a point common to all of them. The question is, how many quadruples are there that cannot be so deformed into each other? (For $n = 3$, triples of ellipses should be considered; in this case, the answer is 2: the ellipses from one triple are disjoint, and in the other triple, each ellipse separates the two intersection points of the two other ellipses.)*

**1989-17.** How many limit cycles can arise under a small polynomial (of degree $n$) perturbation of an integrable polynomial system of degree $n$?

*The question reduces to exploring the number of zeros of the integral*

$$I(h) = \oint \frac{P\,dx + Q\,dy}{M}$$

*along ovals $H = h$ of the system $\dot{x} = X(x,y)$, $\dot{y} = Y(x,y)$ with integrating factor $M$, where $X$, $Y$, $P$, $Q$ are polynomials of degree $n$. It is unsolved even for $n = 2$ and even in the case $M = 1$ where $H$ is a polynomial. In the case where $M = 1$ and $H$, $P$, $Q$ are polynomials of a fixed degree, there is a uniform upper bound for the number of zeros (A. N. Varchenko, A. G. Khovanskiĭ) but it is ineffective.*

**1989-18.** The sequence of meandric numbers 1, 1, 2, 3, 8, 14, 42, 81, ... is defined as follows. Suppose an infinite river running from south-west to north-east intersects an infinite straight road going from the west to the east under $n$ bridges numbered $1, \ldots, n$ in the order from west to east. The order of the bridges along the river determines a *meandric permutation* of the numbers $1, \ldots, n$. The *meandric number $M_n$* is the number of meandric permutations on $n$ elements.

*Meandric numbers possess many remarkable properties; for example, $M_n$ is odd iff $n$ is a power of 2 (S. K. Lando).* Find the asymptotics of $M_n$ as $n \to \infty$. *It is known that $c\,4^n < M_n < C\,16^n$ for some constants $c$, $C$.*

**1989-19.** Is it true that the minimum Hausdorff dimension of a minimal attractor of the Navier–Stokes equation (on the 2-torus, say) increases with the Reynolds number?

*Even the existence of some minimal attractors of dimensions growing with the Reynolds number is not proved; only upper estimates for the dimensions of all attractors by powers of the Reynolds number (obtained by Yu. S. Il'yashenko, M. I. Vishik, and A. V. Babin) are known.*

**1989-20** (V. P. Kostov)**.** Describe the singularities of the pseudo-Stokes hypersurface of a typical family of polynomials. *The pseudo-Stokes hypersurface of the family of polynomials $x^n + a_1 x^{n-2} + \cdots + a_{n-1}$ ($x$, $a_i \in \mathbb{C}$) is the set of values of the coefficients $a_i$ for which two of the roots have the same real part.*

# 1990

**1990-1.** Let $A\colon (\mathbb{C}^2,0) \to (\mathbb{C}^2,0)$ be a germ of a mapping of finite multiplicity holomorphic in a neighborhood of 0. Let also $X$ and $Y$ be complex straight lines (or holomorphic curves) passing through 0. Let $\mu$ denote the intersection multiplicity. Is the multiplicity $\mu(A^nX,Y)$ majorized by an exponent of $n$? *All the multiplicities* $\mu(A^nX,Y)$ *are assumed to be finite.*

**1990-2.** Translate the classification of umbilical points into the language of symplectic toplogy of Lagrangian singularities (possibly optical) and at least formulate conjectures on their topological necessity.

**1990-3.** The caustic of a point on the convex sphere $S^2$ (the manifold of points conjugate to the initial point along the geodesics from this point) is naturally partitioned into connected components (of its preimage in the tangent space at the initial point under the geodesic exponent mapping). We can partition it into the first caustic (generated by the first conjugate points), the second, and so on.

Can we divide the caustic of a point on $S^3$ (or $S^n$) into infinitely many connected components? For example, under a sufficiently small perturbation of the standard metric of the sphere $S^3$, the first $N$ components having the form of a double sphere $S^2$ apparently give precisely $N$ connected components, each consisting of two copies of $S^2$ attached to each other at several (how many?) conic points (of type $D_4$). But it is not improbable that, starting with some (very large) $N$, these two-sphere components begin to merge (I know no examples!) or even form infinite chains (all the more, no examples!), even if the perturbation is very small. Maybe, it is easier to obtain examples on $S^n$ rather than on $S^3$ [when each pair of spheres $S^2$ is replaced by $n-1$ copies of $S^{n-1}$; by the way, the precise arrangement of the $D_4$-point bridges connecting these copies (note that the $D_4$ points form a set of codimension 2 on $S^{n-1}$) is not calculated even in the framework of the first approximation of perturbation theory; this question is apparently related to caustics (focal sets) of ellipsoids in $\mathbb{R}^4$].

**1990-4.** A hypersurface in $\mathbb{RP}^n$ is $k$-quasiconvex if, at each point, its second quadratic form has constant signature $\{k,l\}$, where $k+l=n-1$ (the set $\{k,l\}$ is not ordered: the hypersurfaces are not co-oriented and may be non-co-orientable!). For $k = 0$, this is the usual convexity.

Is it true that any (connected) $k$-quasiconvex closed hypersurface embedded in $\mathbb{RP}^n$ is disjoint from certain subspaces in $\mathbb{RP}^k$ and $\mathbb{RP}^l$ (this is so for $k = 0$)?

**1990-5.** What topological invariants of a submanifold in a Euclidean space do admit an upper bound in terms of the complete absolute curvature (the volume of the manifold of tangent planes of this submanifold in the Grassmannian bundle over the Euclidean space)?

*The sum of Betti numbers can be estimated, and so can the Morse number, while the lengths of relations in the fundamental group, apparently, cannot! It seems that the set of admissible homotopy types of submanifolds whose complete curvatures fall in a fixed range is infinite (at what (co)dimensions of the submanifold and the space?).*

**1990-6.** Prove that a typical Hamiltonian system on the torus with pseudoperiodic Hamiltonian $ap + bq +$ (periodic function) having critical points involves mixing. *Solved by K. M. Khanin and Ya. G. Sinaĭ.*

**1990-7.** Consider a family of analytic diffeomorphisms $x \mapsto x + a + bf(x)$ of the circle, where $f$ is a periodic function. Is the multiplicity of periodic points arising at infinitely small $b$ bounded (uniformly with respect to $a$)?

**1990-8.** Two conducting ($k$-dimensional) surfaces with potential difference 1 move toward each other (in $\mathbb{R}^n$) until the distance between them becomes $\varepsilon$ (the charge distribution is electrostatic). Determine the asymptotic behavior of the force of attraction between the surfaces in terms of the singularities of their tangency at $\varepsilon = 0$ (*for a pair of cylinders in* $\mathbb{R}^3$, *this is a problem of A. D. Sakharov*).

**1990-9.** Give a precise meaning to the assertion (of M. Berry) that the asymptotics of an oscillatory integral, after all terms polynomial in the wave length are subtracted, exhibits exponentially small "jumps" of the universal form erf.

**1990-10.** Make a precise sense of the statement (due to V. V. Fock) that the asymptotics of slowly decaying eigenfunctions in the problem on small diffusion in a potential dynamical system with several attractors $[u_t + (uv)_x = \varepsilon\Delta u,$

$v = -\operatorname{grad} U$] have "jumps" of the universal form erf at the borders of the attractor basins.

**1990-11.** Give the exact meaning of the statement (of A. D. Sakharov) affirming that the average number of vertices of the polygonal pieces, into which a planar domain is divided by many lines, is equal to 4. Generalize it to the multidimensional case. *According to F. Aicardi, the mean number of faces of any dimension of the pieces in $\mathbb{R}^n$ is the same as for the n-dimensional cube. But a rigorous probabilistic proof seems to be lacking.*

**1990-12.** Consider the manifold of non-negative functions on a manifold $M$. Study the singularities of this manifold (stratification, stabilization, bifurcation diagrams, homological properties of the stratification, reconstruction of $M$). *A generic point of the boundary is a function with a single Morse minimum. The manifold of such functions is fibered over M with a contractible fiber.*

**1990-13.** Study the singularities of the boundary of the manifold of contact structures on a (three-dimensional?) manifold and of the boundary of the manifold of contact forms for a given structure.

**1990-14.** The "Hopf invariant" $\int \alpha \wedge d\alpha$ or $\int \alpha \wedge (d\alpha)^n$ on a contact manifold does not require the condition $H^2 = 0$ or $\pi_2 = 0$. Therefore, on a contact manifold, one can try to define a Morse–Floer type complex in a non-simply-connected and/or higher-dimensional case, hoping to get an invariant of the contact structure.

**1990-15.** Does the signature of the Milnor fiber of a function in $\mathbb{C}^3$ has an expression in the form of an integral over the 3-knot of a singularity? Can we "drag over" $p_1$ to this 3-manifold (possibly, with the use of its contact structure)?

**1990-16.** Which of the knot invariants can be "diffused" to invariants of divergence-free vector fields (and, apparently, of Legendrian fields on a contact manifold)? Can one calculate the "linking" of diffused Legendrian submanifolds in higher dimensions?

**1990-17.** Let $f : M^n \to S^n$ be a smooth mapping of a closed manifold to the unit sphere in $\mathbb{R}^{n+1}$, and let $\tau$ be the volume element on $M$. Under what conditions does there exist an immersion $i\colon M^n \to \mathbb{R}^{n+1}$ such that $f = g \circ i$, where $g$ is the Gauss mapping, and $\tau$ coincides with the volume element of the Riemannian metric on $M$ induced (via the immersion $i$) by the Euclidean metric on $\mathbb{R}^{n+1}$?

**1990-18.** Find the group $\pi_2(G_6) = \pi_2(G_8)$, where $G_n$ is the space of real polynomials $x^n + a_1x^{n-2} + \cdots + a_{n-1}$ having no real roots of multiplicity higher than 2.

**1990-19.** Let $X$ be one of the types of critical points of holomorphic functions which forms a set of codimension $k$ in the space of functions in $k$ variables. By an *inflection point of type* $X$ of a hypersurface in a projective space we shall mean a point at which the pair (hypersurface, its tangent hyperplane) is diffeomorphic to the pair (graph of the function, its tangent hyperplane) at a critical point of type $X$. Let also $Y$ be a type of critical points of functions in $k+1$ variables. Find the number of inflections of type $X$ on a level hypersurface of a generic function of type $Y$ (in $k+1$ variables) "vanishing" (i. e., merged) at the critical point.

**1990-20.** Let $f$ be a germ of a $C^\infty$-mapping of a real space onto itself at a fixed point of finite multiplicity. Assume that this point is a fixed point of finite multiplicity for all the iterations $f^n$ of the mapping $f$. Is it true that the multiplicity of this fixed point for the iteration $f^n$ is majorized by some exponential function $ae^{cn}$?

**1990-21.** Is it true that the number of isolated cycles of periods $< T$ of an analytic vector field on a compact manifold is majorized by some exponential function $ae^{cT}$?

**1990-22.** Describe the neighborhoods of Riemannian spheres in holomorphic surfaces with positive self-intersection numbers.

**1990-23.** An algebraic *correspondence* of an algebraic curve to itself is an algebraic curve in the Cartesian product of the initial curve with itself. The *discrete invariants* of such a correspondence are the genus of the initial curve, the genus of the correspondence, and the "bidegree" of the correspondence (i. e., the intersection numbers of the correspondence and the factors). Suppose that a correspondence is the graph of a diffeomorphism of the circle in a real domain. Is it true that the number of isolated cycles of this diffeomorphism is bounded above by a constant depending only on the aforesaid discrete invariants?

**1990-24.** How large can the number of isolated zeros of the complete Abelian integral

$$I(h) = \oint_{\gamma_h} (P\,dx + Q\,dy)$$

be, where $\gamma_h$ is a closed component of the level curve $\{(x,y) : H(x,y) = h\}$, if $P$, $Q$, $H$ are polynomials of given degrees?

**1990-25.** Let $g$ be a natural number $\geq 2$ and $U(x)$ a fixed polynomial of degree $2g+2$. Consider the family of hyperelliptic integrals of the first kind,

$$I(h) = \oint_{\gamma_h} \frac{P(x)}{y}\,dx,$$

where $\gamma_h$ is a closed component of the level curve $\{(x,y) : y^2 + U(x) = h\}$, and $P(x)$ an arbitrary polynomial of degree $\leq g$. Is this family of integrals a Chebyshev one (i. e., is it true that for any $P$ the number of isolated zeros of the function $I$ is at most $g-1$)?

**1990-26.** A *full flag* in $\mathbb{R}^n$ consists of vector subspaces

$$\{0\} = V_0 \subset V_1 \subset \cdots \subset V_n = \mathbb{R}^n$$

of all dimensions. Two flags are called *transversal* if their constituent subspaces of complementary dimensions are transversal. The set of flags not transversal to a given flag is called the *trail* of this flag. Find the number of connected components into which the trail of a flag divides a neighborhood of this flag.

**1990-27.** An ovaloid in $\mathbb{R}^n$ (that is, a closed hypersurface bounding a convex body) is said to be *algebraically integrable* if the volume cut off by a hyperplane from this ovaloid is an algebraic function of the hyperplane. Do there exist algebraically integrable smooth ovaloids different from ellipsoids in $\mathbb{R}^n$ with *odd* $n$?

**1990-28.** Since Poincaré, a "nonintegrable dynamical system" is usually understood to be a dynamical system having no analytic first integrals. However, we can suggest a number of other meanings of the term non-integrability, such as

1) the absence of invariant hypersurfaces (principal ideals),
2) the absence of invariant closed 1-forms (multivalued first integrals),

3) the absence of invariant distributions of tangent subspaces (invariant Pfaffian modules),

4) the absence of invariant foliations (invariant completely integrable Pfaffian systems).

Consider a dynamical system with discrete time (a diffeomorphism on a compact manifold) and an object of one of these types (function, ideal, closed 1-form, ...). The images of this object under the iterations of the diffeomorphism can form a finite set (if they are periodically repeated) or an infinite set, and they can generate a finite- or infinite-dimensional space. These properties reflect the "degree of chaos" in the dynamical system. Prove the non-integrability (in the sense of each of the four definitions given above) of all dynamical systems from some open set in the space of dynamical systems on manifolds of sufficiently high dimension.

# 1991

**1991-1.** Consider the rotation field of a three-dimensional ball around an axis. Is it possible to decrease its energy to arbitrarily small values by acting on this field by volume-preserving diffeomorphisms? *Sakharov's conjecture* (1973): *it is possible for this field, but not for a field with at least one knotted trajectory or with at least one pair of linked trajectories.*

**1991-2.** The Bernoulli–Euler sequence $(1, 1, 1, 2, 5, 16, 61, 272, 1385, \dots)$ gives the numbers of topologically different Morsifications of the singularities $A_n$ (i. e., the numbers of connected components in the complements of their bifurcation diagrams). What is the nonformal complexification of this theory? *The nonformal complexification of* $\pi_0$ *is* $\pi_1$. *Therefore, the answer is apparently the Lyashko–Looijenga covering.*

**1991-3.** Consider the recurrent sequence of degree $n$ (say, 3)

$$x_{m+n} = a_1 x_{m+n-1} + \cdots + a_n x_m \quad (m = 0, 1, 2, \dots).$$

Suppose that the number of zeros among the $x_i$ is finite (the sequence is then said to be nonresonant). How many zeros can there be? Is their number bounded for a given $n$?

**1991-4.** Study the singularities of the manifold of normal operators.

**1991-5.** Study the singularities of the exponential mappings of Lie algebras (at least, of the matrix algebra) to groups (including the stratifications of singularity manifolds and uncovered parts of the groups, stabilization, local and global homotopy and homology groups of the complements of uncovered sets).

**1991-6.** Do the open umbrellas possess the Petrovskiĭ $M$-property (do the sums of Betti numbers of their complements in the real case equal those in the complex case)?

**1991-7.** Does the manifold of singular $n$-th degree polynomials in two variables possess the Petrovskiĭ $M$-property? *Singular = having less than $(n-1)^2$ different critical values.*

**1991-8.** Consider a linear operator $A : \mathbb{C}^m \hookleftarrow$ and two planes $X$ and $Y$ of complementary dimensions. Describe explicitly the conditions guaranteeing the existence of infinitely many integers $n$ such that the space $(A^n X) \cap Y$ is of positive dimension.

**1991-9.** Construct a theory of connections with singularities. *Deform (in the sense of some equivalence) a given connection into a connection which is flat almost everywhere and its all the curvature is concentrated on a certain special submanifold. Then extract the invariants from the combinatorics of these singularities (and, possibly, from the "residues" of the connection at the singular points).*

**1991-10.** Is it true that a (smooth) pseudoperiodic curve in $\mathbb{R}^3$ has only one unbounded connected component? *Negatively solved by D. A. Panov.*

*A pseudoperiodic curve is defined as the preimage of a point under a pseudoperiodic mapping $f : \mathbb{R}^3 \to \mathbb{R}^2$, where $f = (linear) + (\mathbb{Z}^3\text{-}periodic)$ and the incommensurability conditions $\ker(linear) = \mathbb{R}$, $\ker(linear) \cap \mathbb{Z}^3 = \{0\}$ hold (they almost always hold).*

**1991-11.** Consider the convex hull of the set of integer points in the pyramid $z > ax + by$, $x > 0$, $y > 0$ ($a$, $b$ are arbitrary positive numbers). Examine the asymptotics of the polyhedral surface bounding this convex hull (for example, the mean number of edges in a vertex or on a face, the mean number of integer points on an edge; the probability that a random face is a triangle, a quadrangle, ...).

*Generalize Gaussian distribution of continued fraction elements by transferring it to trihedral (general?) pyramids in the space* $\mathbb{R}^3$ *containing* $\mathbb{Z}^3$. *In this situation, prove the multidimensional generalization of Lagrange's theorem on the periodicity of continued fractions: topological periodicity is present if and only if planes are eigenplanes of a lattice-preserving operator. The two-dimensional case shows that the boundary of the convex hull should be colored (where colors correspond to affine* $\mathrm{SL}(2,\mathbb{Z})$*-types of stars of vertices or generalized r-stars containing vertices connected to a given one by a path of at most r edges). In the two-dimensional case,* 1*-stars determine integer-valued angles of the boundary polygonal line; these, together with integer-valued edge lengths, are the elements of the continued fraction. The generalization of Lagrange's theorem to dimension* 3 *states that the topological periodicity of the coloring implies the pyramid's provenance from the eigenplanes of an* $\mathrm{SL}(3,\mathbb{Z})$ *operator.*

**1991-12.** Consider bundles whose fibers are surfaces, namely, the Milnor bundle for the $A_n$ singularities of a function in two variables or the tautological bundle over the moduli space of curves of given topological type. The fundamental group of the base is represented by automorphisms of homology groups of the fiber (by means of the monodromy). Can it be represented directly into the group of diffeomorphisms (rather than of their isotopy classes)? A similar question can be asked for higher dimensions and symplectomorphisms.

*In the case of* $A_1$ *and symplectomorphisms, the answer is affirmative for all dimensions: there are the symplectic Dehn twists. In the case of* $A_2$ *and curves, there also exists an explicit construction. According to V. V. Fock, there is no representation into the homeomorphisms of a fiber for the* $A_{\geq 4}$ *curve singularities.*

**1991-13.** Examine the topological properties of the manifold of the Legendrian curves (immersed or embedded) disjoint from a given Legendrian knot (find its fundamental group and cohomology).

*According to A. B. Givental, the space of all Legendrian submanifolds is similar to a Lagrangian Grassmannian, and the submanifold of those intersecting a given submanifold is similar to the trail of a Lagrangian plane (formed by the Lagrangian planes intersecting it nontransversally).*

**1991-14** (S. P. Novikov). A submanifold of the Euclidean space $\mathbb{R}^n$ is called *$\mathbb{Z}^r$-periodic* if it is invariant under translations by the vectors of some integral sublattice $\mathbb{Z}^r \subset \mathbb{R}^n$. Consider a generic irrational (affine) planar section of a $\mathbb{Z}^3$-periodic surface (*Fermi-surface*) in $\mathbb{R}^3$. Is it true then that every unbounded component of this curve lies in the $R$-neighborhood (with a finite $R > 0$) of some straight line?

# 1992

**1992-1** (B. Teissier). Consider a function $f$ in $\mathbb{R}^n$ with a critical point of index zero. Is it possible to change $f$ in an arbitrarily small neighborhood of this point so that the critical point disappears?

Suppose that the critical point of index zero has finite multiplicity. Is it true that there is a function in the class of versal deformations of $f$ without critical points?

**1992-2.** Study the natural action of the braid group on the manifold of full flags and on the spaces of their cotangent bundles, which arises from the coadjoint representation of the group $\mathrm{SL}(n,\mathbb{C})$. Construct a monodromy and variation theory for non-isolated singularities which takes into account the tower of boundary conditions near strata of various dimensions intersecting the boundary of a ball centered at the point under consideration (instead of the condition that the monodromy is fixed on the boundary of the Milnor fiber).

*The obtained theory must apply to the mapping that assigns characteristic polynomials to matrices. It must generalize the Brieskorn–Grothendieck description of simple singularities for the A, D, E surfaces over nonquasiregular elements of Lie algebras (thus, the theory must apply to the family of four-dimensional intersections of general orbits with the local transversal to the corresponding non-general orbit).*

**1992-3.** Study the analytic continuation of elliptic curves embedded in the orbit space of a holomorphic mapping of a complex curve to itself. What singularities does the continuability boundary have? How large is this boundary (and the corresponding Riemannian surface)? *For instance, for the mappings* $z \mapsto z + \omega + \varepsilon \sin z$

*and* $z \mapsto z + \omega + \varepsilon e^{iz}$*, where* $z \in \mathbb{C} \bmod 2\pi$*, we should consider continuation with respect to the parameters* $\omega$ *and* $\varepsilon$ *such that* $\varepsilon = 0$ *and* $\operatorname{Im}\omega \neq 0$ *[in this case, the curve is* $\mathbb{C}/(2\pi\mathbb{Z} + \omega\mathbb{Z})$*].*

**1992-4.** Draw the discriminant of the family of odd polynomials $x^7 + ax^5 + bx^3 + cx$. Study the topological properties of this discriminant (such as the fundamental group, the number of components in the complement, stabilization, cohomology, etc.) in the $\mathbb{R}$ and $\mathbb{C}$ cases.

**1992-5.** Consider several elements $a_1, \ldots, a_k$ (e. g., with $k = 2$) in the semigroup of germs of holomorphic mappings (or formal series) from $(\mathbb{C}, 0)$ to itself [from $(\mathbb{C}^N, 0)$ to itself]. Compose all words $s$ of length $n$ from the letters $a_i$. Suppose that every equation $s(x) = x$ has a root of finite multiplicity $\mu(s)$ at zero. Consider the maximum $M(n)$ of this multiplicity over all (nontrivial) words $s$ of length $n$. Can the function $M(n)$ grow faster than an arbitrary prescribed function $A(n)$ as $n \to \infty$ (at least, on some subsequence of the $n$ values) under a suitable choice of $a_i$'s?

*Or, maybe, we always have* $M(n) < Cn$ *or* $M(n) < Ce^{\lambda n}$ *with a constant* $C$ *(depending on* $a_i$*) for analytic* $a_i$*? It is natural to ask similar questions about infinitesimal maps, that is, germs of vector fields at a point. It is then reasonable to compose words from multiple Poisson brackets of given fields (or even from their sums and differences) and estimate the order of zero in the field obtained.*

*All these problems, which are nontrivial even on the straight line* ($N = 1$)*, arise in studying bifurcations of limit cycles in relation to the Hilbert* 16*th problem, which is, in particular, concerned with estimating their number.*

**1992-6.** The Milnor fiber (say, of a simple singularity) has a natural symplectic structure (originating from the coadjoint representation). The vanishing cycles $c_i$ are Lagrangian. Is it possible to represent the covanishing cycles ($\operatorname{var}^{-1} c_i \in H^*$) as Lagrangian submanifolds of the Milnor fiber such that their boundaries are Legendrian in the natural contact structure of the singularity knot? Can the intersection matrix (or even variation) of the Milnor fiber be described in terms of the obtained Legendrian link?

**1992-7.** Write explicitly an analytic (polynomial? trigonometric polynomial?) vector field without singular points in $\mathbb{R}^5$ such that its smooth manifold of trajectories is homeomorphic but not diffeomorphic to the standard 4-space $\mathbb{R}^4$. *Such a manifold is called a "fake" space* $\mathbb{R}^4$.

**1992-8.** The double factorial appears as an answer in the classification problem of symplectic flags and in the list of Vassiliev diagram in knot theory. Is there a direct map, associating a symplectic flag to a diagram or a knot? *A natural symplectic structure on the space of knots (and even on the space of immersions with double points) does exist (J.-L. Brylinski): a curve representing a knot can be regarded as a point of a degenerate coadjoint orbit of the hydrodynamical group* $\mathrm{SDiff}(\mathbb{R}^3)$.

**1992-9.** Does the Goresky–MacPherson theory of perverse sheaves have a distributed version where the constraints are imposed on ranks of chains with respect to the symplectic (contact) structure (or merely to the distribution) everywhere rather than only at the points of stratification?

*It is easy to invent a lot of definitions for, say, curves or surfaces in a symplectic or contact space (they are more likely to lead to cobordisms rather than to homologies, but we might try to set conditions at singular points too). The versions are so numerous that we need a rule for selecting suitable definitions.*

**1992-10.** Calculate the moduli spaces of germs for hyper-Kähler structures; are their Poincaré series almost always (except for spaces of germs of infinite codimension) rational functions?

**1992-11.** Consider the Navier–Stokes equation (say, on the 2-torus) with external force proportional to the viscosity (Kolmogorov's model). Is it true that, as the viscosity tends to 0 (i. e., the Reynolds number grows), there appear attractors of dimensions increasing with the Reynolds number (and containing no smaller attractors)?

Is it true that, moreover, the minimum dimension of all attractors unboundedly increases with the Reynolds number?

*A. N. Kolmogorov suggested* (*in* 1958) *that the answer to the first question was affirmative, but he doubted that so was the answer to the second because of the experiments on delaying loss of stability.*

**1992-12.** Prove exponential upper estimates (with probability 1) for topological invariants of the intersections $(A^nX)\cap Y$ in the case where $A$ is not a diffeomorphism, as in the note ARNOLD V. I. Dynamics of complexity of intersections. *Bol. Soc. Brasil. Mat. (N. S.)*, 1990, **21**(1), 1–10; *the Russian translation in*: Vladimir Igorevich Arnold. Selecta–60. Moscow: PHASIS, 1997, 489–499, but an arbitrary smooth mapping.

**1992-13.** Prove exponential upper estimates (with probability 1) for the number of periodic trajectories of period $n$ of a typical diffeomorphism (or a smooth mapping of a manifold to itself).

**1992-14.** Does there exist, for any increasing sequence of positive integers $a_n \to \infty$, a polynomial vector field on $\mathbb{R}^m$ (entering the spheres $\{|\mathbf{x}| = r > 1\}$), which has more than $a_{n_i}$ periodic orbits of periods $\leq n_i$ for some increasing sequence $n_i \to \infty$ (provided that all the periodic orbits are nondegenerate)?

**1992-15.** How large can the set of elliptic curves in the space of orbits of a polynomial vector field (or of a polynomial mapping, or of an algebraic correspondence with fixed discrete invariants, such as genera or degrees) be?

# 1993

**1993-1.** Carry over the theory of neighborhoods of elliptic curves in holomorphic surfaces to pseudoholomorphic surfaces (develop theories of normal forms, resonances, bifurcations, series divergence, ...).

**1993-2** (G. Moore)**.** Is there a relation between the invariants $J^{\pm}$, St of plane curves and polynomials in the areas of the components of curves' complements and their exponents arising in the theory of dual asymptotics of multiplicative integrals over Wilson loops (V. A. Kazakov, Yu. M. Makeenko, ...)? *Solved by M. B. Polyak in* 1997.

**1993-3.** Study the surface of changing four vertices for six in general families of curves $f_{a,b}(x,y) = c$ on the Euclidean plane with parameters $a$, $b$, $c$ such that, at $a = b = 0$, the function $f$ has critical point of minimum $c$ with symmetric second differential $K(dx^2 + dy^2)$.

*The supposed answer: a "dish" whose horizontal section has the form of a six-vertex hypocycloid and vertical sections through the axis are parabolic. But, more likely, there are functional moduli.*

**1993-4.** Study vanishing flattenings in general families of curves in $\mathbb{C}^n$ given as preimages of a general mapping $f\colon \mathbb{C}^n \to \mathbb{C}^{n-1}$. Find the numbers of vanishing flattenings, bifurcation diagrams, and so on.

*Certainly, the normal form of the mapping $f$ does not give an answer by itself: the preimage should be subjected to a general diffeomorphism. For instance, at $n = 2$, the normal form $f = xy$ gives no planar points on the curve $xy = c \neq 0$. The correct (Plücker) answer—6 vanishing inflection points—is only given by the equivalent mapping $f = x^2 - y^2 - y^3$.*

**1993-5.** Find all weight systems of multiplicative generators of a commutative $\mathbb{N}$-graded $\mathbb{C}$-algebra with the simplest Poincaré series $1+t+t^2+\cdots$, for which the classification of such algebras with respect to a) isomorphism of algebras, b) isomorphism of graded algebras is simple (has no modules).

*For three generators, and any weights $1 < u < v$, the number of algebras is $2(a_1 + a_2 + \cdots) + 1$ where $v/u = a_0 + 1/a_1 + \cdots$ is a continued fraction. In the case of four generators, there are Sturmfels' examples of a nonsimple weight system, for instance, $(1,3,4,7)$, $(1,3,4,9)$, $(1,4,5,6)$. Unfortunately, the complete list of weights for which there are no modules is unknown even for 4 generators.*

**1993-6.** Describe the Fintushel–Stern numbers related to the Floer numbers of quasihomogeneous knots of homology 3-spheres in terms of Newton polyhedra (admitting a multidimensional generalization).

*According to* FINTUSHEL R., STERN R. J. Integer graded instanton homology groups for homology three-spheres. *Topology*, 1992, **31**(3), 589–604, *the Poincaré polynomials of the Floer homology of the manifolds $x^a + y^b + y^c = 0$, $|x|^2 + |y|^2 + |z|^2 = 1$ have the form*

| | | | |
|---|---|---|---|
| 2 3 5 | $t+t^5$ | 2 3 7 | $t^{-1}+t^3$ |
| 2 3 11 | $t+t^3+t^5+t^7$ | 2 3 13 | $t^{-1}+t+t^3+t^5$ |
| 2 3 17 | $t+t^3+2t^5+t^7+t^9$ | 2 3 19 | $2t^{-1}+t+2t^3+t^5$ |
| 2 3 23 | $t+2t^3+2t^5+2t^7+t^{11}$ | 2 3 25 | $2t^{-1}+2t+2t^3+2t^5$ |
| 2 3 29 | $t+t^3+3t^5+2t^7+2t^9+t^{11}$ | 2 3 31 | $2t^{-1}+2t+3t^3+2t^5+t^7$ |

**1993-7.** If the class of a plane curve (the orbit of a typical immersion $\mathbb{S}^1 \to \mathbb{R}^2$ under the action of the groups of orientation-preserving diffeomorphisms of $\mathbb{S}^1$ and $\mathbb{R}^2$) is symmetric (invariant) with respect to a symmetry (reflection of $\mathbb{S}^1$ or $\mathbb{R}^2$, or both), then this class has a representative which is a symmetric curve (instead of the diffeomorphisms, we can take the second-order isometries; they transform the immersion into itself).

*A similar assertion is valid for atypical curves, that is, for various other orbits or strata of the manifold of immersions. But in different cases (it seems, even for curves in* $\mathbb{R}^3$*), there occur symmetric classes without symmetric representatives. Is there a simple criterion for the existence of such symmetric representatives (in the classification problem for maps, immersions, embeddings, ...)?*

**1993-8.** Vanishing Chern classes. In addition to vanishing inflections, we can consider other, non-point, strata of singularities on the dual hypersurface. They correspond to (singular) submanifolds of the initial hypersurface with various dimensions enumerated by the types of critical points of functions. The germs of these submanifolds at a singular point determine their infinitesimal analogues in the local ring of the singularity. The problem is to give an algebraic description (e. g., in the form of a flag or quiver of ideals in the local ring) and calculate the discrete invariants of the obtained algebraic objects for each singularity of the initial hypersurface.

**1993-9.** Can we join the curves  and  in the class of fronts of the Legendrian immersions in $ST^*\mathbb{R}^2$ having two (or having at most two) cusps?

**1993-10.** Consider two plane immersed curves in the same $J^+$ class. Join them by a generic path in the space of immersions, along which no perestroikas $J^+$ happen (i. e., no equally oriented self-tangencies). Consider the minimal number of the (other) perestroikas on such a path. Is this number bounded by a constant depending only on the complexity (the number $n$ of double points) of the initial curves? If yes, how does this function grow with $n$? May be, it is not computable because of its growing faster than any computable function?

Is the problem of determining whether two curves belong to the same $J^+$-component algorithmically solvable (*probably, not*)?

*Similar questions arise for all classification problems considered at the seminar; for example, for* St*-classes, for fronts, for fronts with a fixed or an upper bounded number of cusps, etc.*

**1993-11.** Do the periodic continued fractions satisfy the Gauss statistics for the elements? For instance:

A) One can consider random matrices in $\mathrm{SL}(2,\mathbb{Z})$ or in $\mathrm{GL}(2,\mathbb{Z})$ in a large ball of radius $R$, expand them in continued fractions, and explore

a) the statistic of the elements of these periodic fractions;

b) the statistic of the period length.

Does the limit of the distribution as $R \to \infty$ exist, and does it coincide with the Gauss distribution? Is this limit the same for any homothetically widening domains in place of balls?

B) One can also consider random trinomials $\lambda^2 + a\lambda + b$ (with real roots) in the domain $a^2 + b^2 \le R^2$ in $\mathbb{Z}^2$ and explore the statistics over these trinomials (one may also use other domains, e. g., $|a| \le R$, $|b| \le R$).

C) One can even start with rational fractions $p/q$, expand them in continued fractions, and try to calculate the limit of the statistics for $p^2 + q^2 \le R^2$, $R \to \infty$; again, one may replace the disks with other domains. *Conjecturally, the answer is independent of the shape of the domains and, in all the cases, it is the same, as the Gauss invariant measure of the endomorphism* $x \mapsto 1/x - [1/x]$ *of the interval* $(0;1)$ *into itself indicates.*

**1993-12.** Describe the action of the braid group (and of its subgroups corresponding to various non-isolated singularities of fibers) on the homology of generic orbits, i. e., of the manifold $T^*F_{n+1}$ ($F_{n+1}$ is the space of full flags in $\mathbb{C}^{n+1}$), specified by the coadjoint representation $A_n = \mathrm{SL}(n+1,\mathbb{C})$:

| | | |
|---|---|---|
| manifold $\mathbb{C}^{n^2+2n}$ of $(n+1)$-matrices with trace 0 | $\supset$ | nonsingular fiber $\simeq T^*F_{n+1}$ = nonsingular orbit |
| mapping $\det(A-\lambda E)$ $\downarrow$ | | $\downarrow$ fibration |
| characteristic polynomials | $\supset$ | noncritical values = complement to the swallowtail |

**1993-13.** Does there exist any planar not necessarily symplectic connection in the Milnor stratification at least for $A_2$? In other words, can one choose Dehn twists along a parallel and a meridian on a torus with hole $V$ so that they satisfy the relation $aba = bab$ in the group $\mathrm{Diff}\,V$ [or better in the group $\mathrm{Diff}(V,\partial V)$ leaving all points of the boundary stationary], but not in $\pi_0(\mathrm{Diff}\,V)$?

**1993-14.** In COHEN P. -B., WOLFART J. Dessins de Grothendieck et variétés de Shimura. *C. R. Acad. Sci. Paris, Sér.* I *Math.*, 1992, **315**(10), 1025–1028, the Lobachevskiĭ triangles with angles $\frac{\pi}{p}, \frac{\pi}{q}, and \frac{\pi}{r}$ and the groups generated by these triangles are considered. We know that among these triangles, there are 14 especially remarkable ones (physicists' "mirror symmetry" = the "strange duality" between the Gabrielov–Dolgachev numbers). The question is, whether or not these 14 triangles are somehow distinguished in the arithmetic-topological theory of Galois–Grothendieck–Shabat, too.

**1993-15.** In a paper by Mourtada [MOURTADA F. -Z. Familles génériques à quatre paramètres de champs de vecteurs quadratiques dans le plan. Singularité à partie linéaire nulle. *C. R. Acad. Sci. Paris, Sér.* I *Math.*, 1993, **316**(7), 673–678] (presented by R. Thom to the PDE section for some reason), bifurcations of phase portraits are given, and the abstract of the paper claims that all portraits in domains contained in the complement of the bifurcation diagram are considered, while in the text (at the end), it is mentioned that the limit cycles are not studied. It is necessary to finish the study of limit cycles in the context of this paper (at least, determine their number!) and describe their bifurcations.

**1993-16.** Pierre-Louis Lions has recently been awarded a prize for a study of the influence of small viscosity on the Hamilton–Jacobi–Bellman equation; the prize announcement says that he invented viscous solutions and proved their convergence to shock waves in an appropriate sense.

As far as I remember, some work in this direction has been done before Lions (in particular, by S. N. Kruzhkov). How is this work related to Lions' results? What new contribution has Lions made?

**1993-17.** Is there the following fact in popular literature: The binomial coefficient $C_i^x$ coincides modulo $p^p$ ($p$ is an odd prime) with the value of a degree $x$ polynomial in $i$ having *integer* coefficients if $x < p$?

**1993-18.** In *C. R. Acad. Sci. Paris, Sér.* I *Math.*, 1993, **316**(5), 513–518, a weird paper [FLIESS M., LÉVINE J., MARTIN PH., ROUCHON P. Défaut d'un système non linéaire et commande haute fréquence] about employing rapidly oscillating actions in control is published. The authors criticize the notions of complete controllability etc. and suggest something instead. This paper needs be thoroughly investigated, because the authors appeal to differential algebra, which by no means can be relevant. Have the authors obtained new results concerning the considered problems about inverted ordinary and double pendulums with rapidly oscillating suspension points?

**1993-19.** In *C. R. Acad. Sci. Paris, Sér.* I *Math.*, 1993, **316**(6), 573–577 there is the paper POLLACK R., ROY M.-F. On the number of cells defined by a set of polynomials, where for $n$ variables and $s$ equations of degree $d$ in $\mathbb{R}^n$, the number of components of sets determined by $s$ equations *or* inequalities for any sign choice is estimated: $\leq O\big((sd/n)^n\big)$. The only reference is WARREN H. E. Lower bounds for approximation by nonlinear manifolds. *Trans. Amer. Math. Soc.*, 1968, **133**, 167–178.

Does this result follow from Petrovskiĭ–Oleĭnik theory? What is known in the case of full intersection: how many components are there if no inequalities are present? Or—for the complement of the union of $s$ hypersurfaces—what is a hypersurface of degree $sd$? What is the reason for $(sd/n)^n$ here? In standard inequalities for a hypersurface of degree $sd = D$, one may rather expect $D^n/n$. For example, if $d = 1$ and $n = 2$ then the number of domains $\sim s^2/2$ but not $s^2/4$; by integration, it seems, in $\mathbb{R}^n$ for $d = 1$ we get: roughly speaking, $s^n/n!$, more precisely, $e^n(s/n)^n + \cdots \gg s^n/n^n$ which contradicts the result of the paper. Maybe the authors mean $O$ for $n$ fixed? Why do they then take all of $n^n$?

**1993-20.** Is it possible to evaluate the Casson invariant of knots of singularities (at least, for the Brieskorn singularities $x^a + y^b + z^c$, whose associated knots are not homology spheres) [the definition can be found in the paper LESCOP C. Sur l'invariant de Casson–Walker: formule de chirurgie globale et généralisation aux variétés de dimension 3 fermées orientées. *C. R. Acad. Sci. Paris, Sér.* I *Math.*, 1992, **315**(4), 437–440] by analogy with the evaluations performed by R. Fintushel and R. Stern for homology spheres? Can we obtain the signature of the Milnor fiber?

**1993-21.** In DAX J.-P. Points singuliers normaux, points singuliers normaux simples et modèles d'élimination. *C. R. Acad. Sci. Paris, Sér.* I *Math.*, 1992, **315**(3), 315–319, a classification of maps $X \to Y$ taking $A \subset X$ inside $B \subset Y$ is given. What is this, mapping diagrams or quite a new problem?

**1993-22.** In PECKER D. Courbes gauches ayant beaucoup de points multiples réels. *C. R. Acad. Sci. Paris, Sér.* I *Math.*, 1992, **315**(5), 561–565, unicursal space curves with maximum number of double points are constructed; they all turn out to be *real*. Thus, in the problem about *space* curves, as opposed to plane curves, *everything* can be realized in a real domain (including any sets of double points and cusps?). Is there a general phenomenon, namely, that the singularities of mappings to multidimensional spaces can be "driven" into real domains (i. e., realized at $\mathbb{R}$-points for $\mathbb{R}$-mappings)?

**1993-23.** I. Ekeland et al. have recently proved that each *centrally symmetric* (quadratically) convex closed surface in $\mathbb{R}^n$ has an *elliptic* (probably, nonhyperbolic and non-Jordan?) closed geodesic [DELL'ANTONIO G., D'ONOFRIO B., EKELAND I. Les systèmes hamiltoniens convexes et pairs ne sont pas ergodiques en général. *C. R. Acad. Sci. Paris, Sér.* I *Math.*, 1992, **315**(13), 1413–1415].

Is there an example of a *nonsymmetric* surface without *elliptic* (in the same sense) closed geodesics? In particular, is it true that any closed surface close to a sphere has an *elliptic* closed geodesic? If a surface is close to a triaxial ellipsoid, then this, *seemingly*, follows from the Poincaré–Birkhoff theorem (but I have *not* verified whether this does indeed—points with negative eigenvalues also have positive indices).

What can we do when the surface is close to a sphere? Probably, we could perform averaging over great disks and again apply the Poincaré theorem—has anybody done this? It is convenient to define the metric by a function of the form $f \cdot$ the standard metric.

The question is, how does the center of the instantaneous great disk approximating the trajectory move in this averaged motion? Probably, there arises a Hamiltonian system on the sphere specified in terms of $f$, and the Hamiltonian function is related to the integrals of $f$ over the great disks; what functions are obtained under such an integration?

**1993-24.** Study the "caustic–Maxwell stratum" duality.

**1993-25.** Jürgen Moser has recently found a new version of KAM-type theorems: Consider the complex torus $\mathbb{C}^n/(\Gamma \approx \mathbb{Z}^{2n})$ with, say, $n = 2$ and the foliation $\omega_1\,dz_1 + \omega_2\,dz_2 = 0$, generally nonresonant. The complex structure is perturbed into an *almost complex* one.

**Question:** What becomes of holomorphic foliations? The answer is as follows: For the directions of complete Lebesgue measure, they survive (in higher dimensions of leaves, which are left unstudied, the foliation is also quasiholomorphic, i. e., the tangent plane is invariant with respect to the almost complex structure $J: T_xM \hookleftarrow$, $J^2 = -E$; but the one-dimensional leaves are *complex* rather than only almost complex).

**Question:** What becomes of my theory of bifurcations of elliptic curves on complex surfaces when the surfaces are *almost complex*? This topic is quite extensive, because everything needs to be explored from the very beginning, including normal forms of normal bundles, neighborhoods with positive, zero, and negative self-intersection numbers, resonances, their realization, the Grauert theorem on negative neighborhoods, and so on.

**1993-26.** Study the singularities of the manifold of normal matrices.

There is yet another excellent unexplored manifold, the Taylor series of one-to-one mappings of the disk $|z| < 1$ to the plane ("the coefficient problem for univalent functions"): stratify the boundary and investigate the singularities of small codimensions in the space of series.

**1993-27.** Second, third, and succeeding braid groups: noncommutative resolvents. This is a very old problem, and it is time to clear it up.

Consider a general projection of a hypersurface $\Sigma_0$ in $\mathbb{C}^n$ onto the hyperplane $\mathbb{C}^{n-1}$ (germ at zero). The *discriminant* is a hypersurface $\Sigma_1$ in $\mathbb{C}^{n-1}$ over which the number of preimages is less than the degree of $\Sigma_0$ in 0. We obtain the chain of the discriminants $\Sigma_i \subset \mathbb{C}^{n-i}$ of the projections $p_i: \mathbb{C}^{n-i+1} \to \mathbb{C}^{n-i}$ and the chain of the groups $\Gamma_i = \pi_1(\mathbb{C}^{n-i+1} \setminus \Sigma_{i-1}, b_i)$ (near 0), where $b_i \in \mathbb{C}^{n-i+1}$. We have $\Gamma_i = F_i/R_i$, where $F_i$ is the group generated by loops around $\Sigma_{i-1}$ in the fiber $p_i^{-1}(p_i b_i)$ and $R_i$ is the normal closure in $F_i$ of the subgroup generated by the products $(A_\varphi f)f^{-1}$; here $f \in F_i$ and $A_\varphi$ is the action on $F_i$ of the groups $F_{i+1}$ by the braids $(\varphi \in F_{i+1})$.

Clearly, the generators of $\Gamma_{i+1}$ (denoted by $\varphi$ above) correspond to relations in $\Gamma_i$. Moreover, the generators $\alpha$ of the group $\Gamma_{i+2}$ correspond to relations from $R_{i+1}$ in $\Gamma_{i+1}$ and, therefore, to "relations between relations" in $\Gamma_i$.

Thus, the relations (elements of $R_{i+1}$) in $\Gamma_{i+1}$ correspond to the generators (elements of $F_{i+2}$) of the group $\Gamma_{i+2}$. If $\alpha \in F_{i+2}$ is such a generator, then $A_\alpha$ acts as a braid on $F_{i+1}$ and takes $\varphi$ to $A_\alpha\varphi$; the element $\xi = (A_\alpha\varphi)\varphi^{-1} \in F_{i+1}$ acts now on $F_i$. We obviously have the following "Poincaré's $d^2 = 0$ lemma":

$$A_\xi f \equiv f \quad \forall \xi = (A_\alpha\varphi)\varphi^{-1} \quad \forall f \in F_i.$$

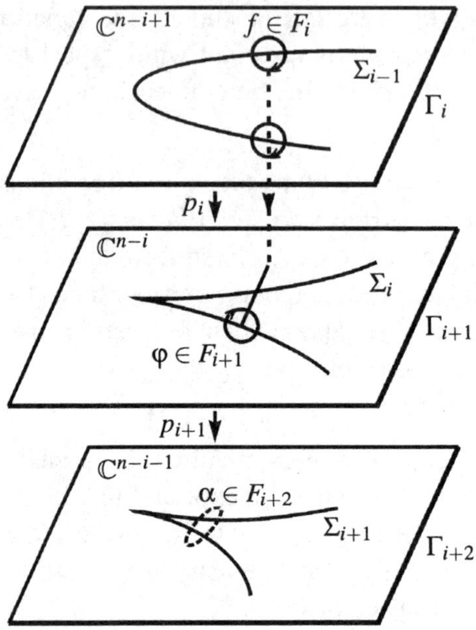


**Fig. 1:** The sequence of discriminants $\Sigma_i$ and groups $\Gamma_i$

**Question:** To what degree is $R_{i+1}$ less than the subgroup $\widehat{R}_{i+1}$ of the braid group acting on $F_{i+1}$ that is defined by

$$\widehat{R}_{i+1} = \left\{\sigma \in \mathrm{Br}(F_{i+1}) \mid \forall \xi = (\sigma\varphi)\varphi^{-1} \in F_{i+1} \ \forall f \in F_i \ A_\xi f \equiv f\right\}?$$

This is a general question referring to any germ of a hypersurface at a point.

*Now, let* $\Sigma_0$ *be a swallowtail* in $\mathbb{C}^{\mu=n}$ (it can be obtained from $\Sigma_{-1} = \{x, \lambda \mid x^{\mu+1} + \lambda_1 x^{\mu-1} + \cdots + \lambda_\mu = 0\}$ by projecting along the $x$ axis in $\mathbb{C}^{\mu+1}$). Then $\Gamma_0 = \mathbb{Z}$, $\Gamma_1 = \mathrm{Br}(\mu+1)$, and $\Gamma_2 = F_2/R_2 \equiv \widehat{R}_2$.

**Urgent question:** Is it true that $\Gamma_3 = F_3/R_3 \equiv \widehat{R}_3$? or $\widehat{R}_3 \supset R_3$? How can we describe $\Gamma_3$? Should we take quasihomogeneous, rather than general, projections $p_i$?

**1993-28.** Yet another old topic which it is timely to recall is singularities in "Cartan's geometric theory of PDE." The subject matter is systems of differential equations, that is, submanifolds in finite-order jet spaces, or, which is the same, modules of consequences.

**1993-29.** Suppose that $v$ is a vector field in $\mathbb{R}^n$ which has a singular point and the real parts of the eigenvalues of its Jacobi matrix are negative (*everywhere*, rather than only at the given singular point). Is it true that the basin of attraction of this singular point is the entire space $\mathbb{R}^n$?

*Perhaps, the condition will look less embarrassing if we consider the control system* $\dot{x} = v(x) + u$*; for an arbitrary* $u$*, the fixed point of this system is an attractor (with negative Lyapunov exponent).*

**1993-30.** Compare the studies of the normal forms of Stokes surfaces performed by A. I. Neĭshtadt and S. K. Lando.

**1993-31.** M. R. Herman has presented a nice construction of an area-preserving diffeomorphism of a disk with positive Lyapunov exponents in the whole domain (see below). Is it possible to adapt this construction for solution of Sakharov's problem on fast ideal dynamo?

Recall that the collection of objects $\{A: B^3 \to B^3$ satisfying $\det A_* = 1$ and a divergence-free field $v$ on the ball $B^3\}$ is called a *fast ideal dynamo* if $\iiint_B |A_*^n v|^2\,dx \geq C \exp \lambda n$, $\lambda > 0$.

**The construction communicated by Herman:** Let $A: \mathbb{T}^2 \to \mathbb{T}^2$ be an Anosov map, say, $\left(\begin{smallmatrix} 2 & 1 \\ 1 & 1 \end{smallmatrix}\right)$, and $\sigma: \mathbb{T}^2 \to \mathbb{T}^2$ be a holomorphic involution with 4 fixed points [e. g., the covering $w^2 = P_4(z)$ of an elliptic curve over $S^2$]. In $\mathbb{R}^2/\mathbb{Z}^2$, 4 points $(0,0)$, $(0,1/2)$, $(1/2,0)$, $(1/2,1/2)$ remain fixed under the action of $A^6$ (since $(0,0)$ is fixed under $A$, and the other points permute). Therefore, $A^6$ acts on the sphere (with 4 fixed points) as an Anosov system. Now it remains to resolve these 4 points.

**1993-32.** Multidimensional continued fractions and $A$-algebras.

D. Eisenbud has recently constructed an example (see below) of an $A$-algebra over $\mathbb{C}$ with moduli (not "simple"). Recall that an $A$-algebra is graded and has a Poincaré series $1 + t + t^2 + \cdots$ of polynomials in one variable. The degrees of multiplicative generators are determined uniquely: $1 = u_0 < u_1 < u_2 < \cdots$ ($u_i$ fills the first lacuna in the degrees of monomials in lower-degree generators).

Thus, we can compose a Young diagram; for example, the anomaly $\alpha_i = u_i - i$ gives $1 = \alpha_0 \leq \alpha_1 \leq \alpha_2 \leq \cdots$. *Is there a relation* between the presence of moduli in $A$-algebras with given anomaly and in flattenings?

Eisenbud hopes to *prove* transversality at the Weierstrass points.

*Certainly, most likely, there is no relation, but nevertheless, the simplicity (the absence of moduli) selects simple ones among all Young diagrams, and of interest is the list of Young diagrams simple in this sense. The bifurcation diagrams also might deserve attention (though I do not know what they are, for the space of A-algebras is not linear). Maybe, we should consider one-dimensional extensions of the ready $A_\mu$-algebra (of dimension $\mu$ over $\mathbb{C}$), because if an algebra has moduli, then they manifest themselves for the first time somewhere in the chain of extensions, and the space of extensions is less singular (it is not improbable that it is even not singular at all for one-dimensional extensions).*

**Eisenbud's example:** The generators are $x_1, x_2, x_3, x_4, x_5, y_6, y_7, y_8$, and $x_{17}$ (the subscripts indicate their degrees); the relations are $x_i x_j = 0$, $x_i y_j y_k = 0$, and $x_1 y_i = x_2 y_i = 0$; the relations between $y_i$ are the same as between $y^i$ (e. g., $y_6 y_8 = y_7^2$; the multiplication by $y_i$ acts on $x_3$, $x_4$, and $x_5$ as

$$x_3y_7 = x_4y_6, \qquad x_4y_8 = x_5y_7 = x_5y_8 = 0, \qquad x_{17}y_i = 0,$$
$$x_3y_8 = x_5y_6, \qquad x_4y_7 = ax_3y_8 = ax_5y_6.$$

We claim that $a$ is a modulus. Indeed, multiplying the generator of degree $i$ by $\lambda_i$ for various $i$, we obtain that

$$\left.\begin{array}{l} x_3y_7 = x_4y_6 \text{ implies } \lambda_3\lambda_7 = \lambda_4\lambda_6 \\ y_6y_8 = y_7^2 \quad \text{implies } \lambda_6\lambda_8 = \lambda_7^2 \end{array}\right\} \Longrightarrow (\lambda_4\lambda_7 = \lambda_3\lambda_8) \Longrightarrow a \text{ is a modulus!}$$

**1993-33.** Explore the asymptotic properties of random integer planes: Is their statistics similar to the Gauss statistics for continued fractions?

**1993-34.** Model the spectral sequence of a bundle by singularities in the same fashion as the homology complex is modeled by the Morse complex; namely, put geometric objects in correspondence with differentials and obtain "Morse inequalities," i. e., express the existence of some singularities (and bound characteristics of these singularities from below) in terms of differentials from the spectral sequence.

A concrete question: For the bundle

$$\begin{array}{c} S^{2n+1} \\ {\scriptstyle p}\Big\downarrow{\scriptstyle \mathbb{S}^1} \\ \mathbb{C}\mathrm{P}^n \end{array}$$

and a generic function $f$ on $S^{2n+1}$, specify the necessary multiplicity $\mu$ of a critical point of the restriction of $f$ to a fiber $p^{-1}(b)$ for the worst fiber; the word "necessary" means "minimum over all generic $f$."

**1993-35** (S. P. Novikov). Consider a cyclic covering of a compact manifold and a general pseudoperiodic smooth Morse function $f$ on the covering space (the differential of $f$ is lifted from the initial compact manifold). Let $+1$ denote the action of the group $\mathbb{Z}$ on the covering space, and let $f(x+1) \equiv f(x)+1$. Suppose that $f$ has critical points $p$ and $q$ of indices $i$ and $i-1$, respectively. Consider the "instantons" (trajectories of the field grad $f$) joining the points $p$ and $q-n$. Is the number of such instantons bounded by the exponent of $n$?

**1993-36.** Take a neighborhood $U$ of a hyperbolic fixed point 0 of a diffeomorphism of the plane $A$. The *order* of a homoclinic point $p$ (i. e., such that $A^m p \to 0$ as $m \to \pm\infty$) is the number of the points on the orbit of $p$ that fall outside $U$:

$$\operatorname{ord}(p) := \#\{m \in \mathbb{Z} \mid A^m p \notin U\}.$$

Is the number of homoclinic points of given order $n$ bounded by the exponent of $n$?

**1993-37.** A connected smooth hypersurface in the real projective space is said to be *locally hyperbolic* if its second quadratic form is everywhere nondegenerate. Is it true that all closed connected locally hyperbolic nonconvex surfaces in $\mathbb{RP}^3$ are quasiconvex and separate pairs of projective subspaces, having just two intersection points with every straight line connecting these subspaces (see problem *1990-4*?

**1993-38.** Is the set of closed connected locally hyperbolic nonconvex surfaces in $\mathbb{RP}^3$ connected? Is it true that any such surface has a convex plane section?

**1993-39.** Is it true that the generic caustic formed by the $r$-th conjugate points along the geodesics from a given point on the sphere $S^2$ has at least four cusps for any Riemannian metric on $S^2$?

**1993-40.** Is it true that the generic caustic formed by the $r$-th conjugate points along the geodesics from a given point on the sphere $\mathbb{S}^3$ has at least four $D_4$-type singularities for any Riemannian metric on $\mathbb{S}^3$?

**1993-41.** This problem and the six subsequent ones are devoted to critical points and Lagrangian singularities.

Let us consider a generic convex smooth closed curve $\gamma$ on $\mathbb{R}^2$ and its normal lines. The unit vectors on these lines determining the same orientation as internal normals to $\gamma$ form a Lagrangian submanifold $M$ of the space $T^*\mathbb{R}^2$ (we identify tangent and cotangent vectors using the Euclidean metric of the plane). Whitney cusped singularities of the projection of this Lagrangian submanifold onto the plane—they are the curvature centers of $\gamma$ for its vertices—are singular points of the caustic $\Gamma$ consisting of the curvature centers of $\gamma$ for all its points.

The manifold $M$ is diffeomorphic to a cylinder. Unit vectors applied outside a large disk containing the caustic form two "collars" (semicylinders) on $M$. These collars are projected into the plane diffeomorphically, and the middle part of the cylinder—with singularities (the set of critical values is the caustic $\Gamma$).

Can the middle part of the cylinder $M$ be replaced with another Lagrangian embedding, so that the resulting projection of the embedded Lagrangian cylinder into the plane has no Whitney cusped singularities (and coincides with the original projection on the collars)?

**1993-42.** A relaxed vesrion of the previous problem: can the middle part of the cylinder $M$ be replaced with a Lagrangian *immersion*, so that the resulting projection of the *immersed* Lagrangian cylinder into the plane has no Whitney cusped singularities (and coincides with the original projection on the collar)?

**1993-43.** The cylinder $M$ mentioned in problem *1993-41* is *optical*, i. e., it lies in the hypersurface $p^2 = 1$.

Can we replace this cylinder (the boundary collars being left intact) with an *optical* immersed (or embedded) Lagrangian cylinder whose projection on the plane has no Whitney cusped singularities?

**1993-44.** The topological invariants of the space of Morse functions on a given compact manifold (or of the space of functions whose critical points are not more complex than singularities from a given class) are interesting invariants of smooth manifolds; cf. ARNOLD V. I. Spaces of functions with moderate singularities. *Funct. Anal. Appl.*, 1989, **23**(3), 169–177; *the Russian original is reprinted in*: Vladimir Igorevich Arnold. Selecta–60. Moscow: PHASIS, 1997, 455–469.

Are the homotopy types of these function spaces determined by the topological type of the initial manifold, or do they indeed depend on the smooth structure?

**1993-45.** Consider a Morse function on a connected compact manifold. A suitable diffeomorphism sends all critical points of this function into a small ball in the manifold. The restriction of our function to a neighborhood of the boundary of this ball determines a Lagrangian (or Legendrian) collar, that is, the set of the first differentials of the function or its 1-jet at the points of a spherical annulus.

Is it possible to reconstruct a manifold from its Lagrangian collar? For what pairs of manifolds $M_1^n$ and $M_2^n$ do there exist functions $f_1 : M_1^n \to \mathbb{R}$ and $f_2 : M_2^n \to \mathbb{R}$ that coincide on balls containing all critical points?

**1993-46.** Consider a family of smooth functions as a function on the space of a smooth bundle (with compact base and fibers). Can the numbers of degenerate critical points (of different types) of the restrictions of this function to the fibers be estimated from below in terms of the topology of the bundle?

**1993-47.** Consider a smooth function in a neighborhood of a critical point 0 of finite multiplicity. Suppose that the index of the corresponding gradient vector field at 0 is zero. Consider the Lagrangian collar determined by the restriction of this function to a neighborhood of a sphere $\partial B$ centered at 0. Does this collar bound a Lagrangian disk (or other Lagrangian manifold embedded in the cotangent bundle of the ball $B$) disjoint from the zero section?

**1993-48** (M. B. Sevryuk)**.** Let a smooth involution $G\colon M \to M$ of an $N$-dimensional manifold $M$ possess an invariant $n$-torus $L \subset M$, $L \cong \mathbb{T}^n$, the restriction $G|_L$ being conjugate to the transformation $\varphi \mapsto -\varphi$ ($\varphi$ denotes the angular coordinate on $\mathbb{T}^n$) and therefore having $2^n$ isolated fixed points. What types of involution $G$ can be at these points?

*If $a \in M$ is a fixed point of the involution $G$ then by definition the type of involution $G$ at this point is $(p, N-p)$, whenever the linear part of $G$ at the point $a$ is a reflection in an $(N-p)$-dimensional plane. If*

$$(p_1, N-p_1), \ldots, (p_{2^n}, N-p_{2^n})$$

*are the types of involution G at fixed points $a_1, \dots, a_{2^n}$ on the torus L, then $n \le p_i \le N$ for all i. Do all the collections of numbers $p_i$ meeting these inequalities indeed occur?*

# 1994

**1994-1.** We use the term *pseudofunction* for an immersion $\mathbb{S}^1 \to S^2$ bounding half the sphere area and homotopic to an embedding of the equator in the class of immersions such that no subloop smaller than the entire curve bounds half the sphere.

Prove that a pseudofunction intersects any equator. *Proved by A. B. Given-tal′ even for Lagrangian $\mathbb{R}\mathrm{P}^n$ in the symplectic $\mathbb{C}\mathrm{P}^n$.*

**1994-2.** Prove that the number of inflections of a pseudofunction is at least four.

**1994-3.** Consider the cylinder $\mathbb{S}^1 \times I$. An immersion of $\mathbb{S}^1$ into this cylinder is called a 0-*pseudofunction* if it bounds half the cylinder area and is homotopic to an embedding of the boundary of an embedded disk in the class of immersed curves bounding half the cylinder area and containing no subloops bounding such an area.

Prove that a 0-pseudofunction intersects the equator. Study the existence of four inflection points for a 0-pseudofunction. *A curve on the cylinder $x^2+y^2=1$, $|z|<1$ can be projected onto the sphere $x^2+y^2+z^2=1$ either from the center or by the horizontal radii from the points on the vertical axis of the cylinder (i. e., by means of the Archimedean symplectomorphism). The former projection transforms the inflection points on the cylinder into inflection points on the sphere. The latter transforms the inflection points into points of double tangency with projections of great circles. The perturbations of the cylinder equator that, together with the equator, bound zero area have four inflection points in both senses.*

**1994-4.** If a curve embedded in $S^2$ meets the great circle $2k$ times, then it has at least $2k$ inflection points. Find the symplectic (or contact) setting of this geometric theorem and transfer it to general Chebyshev systems.

**1994-5.** A curve (immersed circle $\mathbb{S}^1$) in $\mathbb{R}^{2n}$ is called *convex* if no hyperplane intersects it in more than $2n$ points (counting multiplicity). Is it true that any convex curve in $\mathbb{R}^{2n}$ has a convex projection on $\mathbb{R}^{2n-2}$? or is a projection of a convex curve in $\mathbb{R}^{2n+2}$? *A similar question for projective convex curves in* $\mathbb{R}\mathrm{P}^m$ *with not necessarily even* $m$ *is also interesting.*

**1994-6.** Smooth curves in $\mathbb{R}^3$ close to plane convex curves have at least four flattening points. To give a contact formulation of this assertion (in the spirit of the Morse–Chekanov Legendrian theory), it would be useful to understand how large can a deformation be while still preserving the lower bound of four flattening points. Is it sufficient to assume that the initial curve as well as the dual curve remain trivial (embedded and unknotted) in a deformation?

**1994-7.** Consider the Legendrian self-linking numbers $L_i$ of a Legendrian curve in the solid torus $ST^*\mathbb{R}^2$. Are they contact-invariant (i. e., are they preserved by the contactomorphisms of the solid torus onto itself that preserve the orientation of the basis circle and the co-orientations of the contact planes)?

*Solved affirmatively by E. Giroux. A positive answer would follow from the connectedness of the contactomorphism group described above, but this connectedness is not proved. It is only proved that the contactomorphisms of the above-mentioned type cannot change the type of trivialization of a torus bundle "at infinity"* ($x^2+y^2 \gg 1$) *and over the basis circle.*

**1994-8.** What is the analog of the Bennequin inequality for Legendrian curves in $ST^*M^2$?

**1994-9.** Does the universal Milnor fibration of surfaces for $A_2$ in $\mathbb{C}^3$ ($x^3+\lambda_1 x+\lambda_2+y^2+z^2=0$) have a symplectic flat connection?

*For curves in* $\mathbb{C}^2$, *such a connection is constructed as follows: an elliptic curve with a marked point is identified with a neighboring elliptic curve with a marked point by a real linear realification transformation of the covering plane which maps the basis of the initial period lattice to the basis of a close lattice.*

**1994-10.** How does the number of isotopy classes of plane (or spherical) curves with $n$ double points grow? What is the distribution of these curves in the index (whether the limit distribution is the Gaussian one)?

*Here is the empirical distribution for the plane curves having* $n = 5$: 26, 133, 290, 364, 290, 133, 26.

**1994-11.** Examine the singularities of the curvature form of the natural (adiabatic) connection of the bundle of the Hermitian matrix eigenvalue manifold near the discriminant of multiple eigenvalues.

**1994-12.** Compare the versal deformation's curves of the mappings $(\mathbb{R},0) \to (\mathbb{R}^2,0)$ with the classification of long curve immersions on the plane: What classes are realized; what are the bifurcation diagrams (remember stabilization!); how many connected components does the complement of a bifurcation diagram have; does the smooth type of a long curve determine the connected component of the complement; what are the expressions of the values that the invariants ($J^+$, $J^-$, St, and others) take on in terms of the local algebra of the singularity; what becomes of all this theory under complexifications, that is, for the mappings $(\mathbb{C},0) \to (\mathbb{C}^2,0)$?

**1994-13.** Consider a particle in a magnetic field on a surface $M^2$. Study Legendrian divergence-free vector fields on $ST^*M^2$ and, in particular, their closed orbits. More generally, consider divergence-free Legendrian vector fields on $\mathbb{S}^3$ for some (standard?) contact structure. Does there exist a counterexample to the Seifert conjecture (that a divergence-free field without singular points has at least two closed trajectories) in this class of vector fields?

**1994-14.** Consider a particle in a magnetic field on a Riemannian manifold of an arbitrary dimension. The magnetic field is given by a closed two-form on the manifold, twisting the symplectic form of the phase space. In the case of a strong magnetic field (large curvature trajectories) apply the averaging method and, at least, formulate conjectures on topological lower bounds for the number of periodic orbits. These conjectures should generalize the theorem on the existence of $2g+2$ curves of large geodesic curvature on a surface of genus $g$.

**1994-15.** Is it true that a projective curve which does not intersect any more with its osculating hyperplanes is convex (that is, the number of intersection points of this curve with any hyperplane, counted with their multiplicities, does not exceed the dimension of the ambient space)? Investigate the number of connected components and the boundary of the manifold of convex curves in $\mathbb{R}P^n$ (stratification, bifurcation diagrams, stabilization, ...). *There is a single connected component if the orientation is not taken into account* (*S. S. Anisov*).

**1994-16.** Prove that a curve in $\mathbb{R}P^n$ that is projectively dual to a convex one is convex itself. *Proved by B. A. Khesin and V. Yu. Ovsienko, a simpler proof was given by M. E. Kazarian.*

**1994-17.** Find all projective curves projectively equivalent to their duals. *The answer seems to be unknown even in* $\mathbb{R}P^2$.

**1994-18.** Examine the boundary of the manifold of Möbius curves in $\mathbb{R}P^2$ (the Möbius curves are those from the connected component of the space of curves having at least three inflection points, that contains all the curves close to $\mathbb{R}P^1$).

**1994-19.** Examine the boundary of the manifold of tennis immersions $\mathbb{S}^1 \to S^2$ (a *tennis immersion* is an immersion from the connected component of the space of immersions that halve the area and have at least four inflection points, that contains all curves halving the area and close to equator in the space of curves in $S^2$).

**1994-20.** Explore the singularities of the caustic of an ellipsoid in $\mathbb{R}^4$ (or in $\mathbb{R}^n$, $n > 4$). *Conjecturally these singularities are topologically inevitable*: *caustics of other* (*convex*?) *surfaces have not less singularities, and this is true even for the Lagrangian collapse on* $\mathbb{R}^n$ (*V. M. Zakalyukin's conjecture*).

**1994-21.** Is it true that any knot in $ST^*\mathbb{R}^2 = \mathbb{S}^1 \times \mathbb{R}^2$ can be realized as a Legendrian knot of an immersion $\mathbb{S}^1 \to \mathbb{R}^2$? *Yes*; *solved by A. Shumakovich.*

**1994-22.** Prove that a convex curve in $\mathbb{R}P^{2n}$ is affine (does not intersect a hyperplane). *Proved by S. S. Anisov* (*and others*).

**1994-23.** Consider the front of a convex curve in $\mathbb{R}P^n$ (its points are the hyperplanes tangent to the curve). Are the fronts of different convex curves homeomorphic? diffeomorphic? Describe the topology (combinatorics) of a front: find the number of connected components in the complement, and so on. This is interesting even for the simplest curve $x_k = \cos kx$, $y_k = \sin kx$ ($k = 1, \dots, n$) in $\mathbb{R}^{2n}$ (and even if the answer for other curves is different). *This problem has given rise to studies of complex and real trigonometric polynomials, the Lyashko–Looijenga–Laurent mapping, and graph combinatorics, but it is still unsolved itself.*

**1994-24.** Are the Poincaré series of numbers of moduli in jet spaces rational functions in the majority of local problems in analysis? *For instance, is it true for almost all $f$ (that is, for all $f$ not belonging to some subset of infinite codimension in the space of Taylor series) in the following classification problems:*

*— classification of the Riemannian (or Einsteinian) metrics $f$ in a neighborhood of a point in a space modulo local diffeomorphisms of this space that leave this point fixed;*

*— classification of the vector fields $f$ on a manifold in a neighborhood of a singular point of a field modulo local diffeomorphisms of this manifold that leave this point fixed;*

*— classification of the smooth mappings $f: M^m \to N^n$ in a neighborhood of a point $x \in M$ modulo local diffeomorphisms of $M$ and $N$ that leave $x$ and $f(x)$ fixed;*

*— classification of the Hamiltonian vector fields $f$ in a neighborhood of a singular point of a field modulo local symplectomorphisms that leave this point fixed;*

*— local classification of the second order differential equations $y'' = f(x, y, y')$;*

*— classification of the germs $f$ of hyper-Kähler structures on a $4n$-manifold modulo local diffeomorphisms?*

*Recall that the Poincaré series of numbers of moduli for a given (local) object is the series $M(t) = \sum_{k=0}^{\infty} m(k)t^k$, where $m(k)$ is the number of moduli of the $k$-jet of this object (i. e., the dimension of the moduli space).*

**1994-25.** Is it possible to construct a theory of sufficient jets for expansions with logarithmic terms?

**1994-26.** Does there exist a minimal attractor for a system of Navier–Stokes equations whose dimension unboundedly increases as the viscosity diminishes ($\dim \to \infty$ as $\nu \to 0$)?

**1994-27.** Is it true that the minimum dimension of an attractor of a Navier–Stokes system unboundedly increases as the viscosity diminishes?

**1994-28** (Ya. B. Zeldovich)**.** Does there exist a divergence-free field $\mathbf{v}$ on a three-dimensional torus $\mathbb{T}^3$ such that a magnetic field $\mathbf{B}$ satisfying the system

$$\frac{\partial \mathbf{B}}{\partial t} + \{\mathbf{v}, \mathbf{B}\} = \mu\, \Delta \mathbf{B}, \quad \operatorname{div} \mathbf{B} = 0,$$

grows exponentially as $t$ increases for some initial field $\mathbf{B}_0$? Is there a divergence-free vector field $\mathbf{v}$ on $\mathbb{T}^3$ which is a fast kinematic dynamo?

**1994-29** (Ya. B. Zeldovich – A. D. Sakharov)**.** Does there exist a volume-preserving diffeomorphism of the three-dimensional ball $B^3$, whose iterations make the energy of some initial divergence-free vector field grow exponentially with the number of iterations?

**1994-30.** Consider a smooth function $u_0$ defined on the disk $x^2 + y^2 \leq 1$. Find the infimum of the Dirichlet integral

$$I[u] = \iint \left( \left( \frac{\partial u}{\partial x} \right)^2 + \left( \frac{\partial u}{\partial y} \right)^2 \right) dx\, dy$$

over the set of all smooth functions $u$ obtained from $u_0$ by an area-preserving diffeomorphism of the disk.

**1994-31.** Consider a dust-like gravitating medium in the standard Euclidean 3-space. Describe the singularities of the caustic hypersurfaces and the particle density in the physical space after the formation of the first caustics. Is it true that the singularities of the solution to the Vlasov–Poisson equations for generic initial distributions concentrated along generic smooth Lagrangian sections of the cotangent bundle have the same topological structure as for the Vlasov equation (where the gravitational interaction is not taken into account)? Do the density singularities in neighborhoods of points on caustics and of caustic singularities have the same orders as those for non-interacting particles?

**1994-32.** Calculate the asymptotic behavior of the maximum oscillation indices $\beta(p)$ and $\beta_n(p)$ encountered in general $p$-parameter families of oscillatory integrals of functions in $n$ variables.

**1994-33.** Consider a generic analytic nearly integrable Hamiltonian system: $H = H_0(p) + \varepsilon H_1(p,q,\varepsilon)$, where the perturbation $H_1$ is $2\pi$-periodic in the angle variables $(q_1,\dots,q_n)$ and where the unperturbed Hamilton function $H_0$ depends on the action variables $(p_1,\dots,p_n)$ generically. Let $n$ be greater than two.

Prove or disprove the following conjecture. For any two points $p'$, $p''$ on the same connected component of a level hypersurface of function $H_0$ in the action space, there exist orbits connecting an arbitrarily small neighborhood of the torus $p = p'$ with an arbitrarily small neighborhood of the torus $p = p''$, provided that $\varepsilon \neq 0$ is sufficiently small and $H_1$ is generic.

**1994-34.** Prove or disprove the following conjecture: An equilibrium point 0 of a general analytic Hamiltonian system is Lyapunov unstable if the quadratic part of the Hamiltonian function at 0 is neither positive nor negative definite.

**1994-35.** Find lower bounds for the number of periodic orbits of a charge in a magnetic field, where the motion of the charge is confined to a surface and the magnetic field is orthogonal to the surface. *Conjecturally, on a surface of genus g, a charge should generically have at least* $2g+2$ *periodic orbits. From a mathematical perspective, this is a problem about closed curves with given positive geodesic curvature on the surface. When the magnetic field is sufficiently strong, the conjecture is proved, cf. problem 1994-14.*

**1994-36.** Consider $q$ vectors $(\mathbf{k}_1,\dots,\mathbf{k}_q)$ applied to the origin in the Euclidean plane such that their endpoints are the vertices of a regular $q$-gon. Consider the sum of $q$ equal intensity harmonic waves with these wave vectors. If $q \neq 1, 2, 3, 4, 6$ (say, if $q = 5$), then this sum is not a periodic function (though it is quasiperiodic). *Example*: $q = 5$, $H(\mathbf{r}) = \sum_{j=1}^{5} \cos(\mathbf{k}_j, \mathbf{r})$.

Is it true that all closed components of the level lines $H = h$ that bound regions containing the origin lie in a bounded neighborhood of the origin? Does a Hamiltonian system with Hamiltonian function $H$ have an unbounded phase curve?

**1994-37.** Is the problem of the stability of an equilibrium point for a vector field whose components are polynomials with integer coefficients algorithmically solvable?

**1994-38.** This and the following four problems are concerned with the analytical (and geometric) solvability of analytical problems.

Let us introduce sets of "feasible manifolds" and "feasible mappings" with the following properties:

— the arithmetical spaces $\mathbb{R}^n$ and $\mathbb{C}^n$ are feasible for any $n$;

— any rational mapping is feasible;

— the image and preimage of a feasible manifold under a feasible mapping are feasible manifolds;

— the intersection, union, and mutual complements of two feasible manifolds is a feasible manifold;

— the superposition of two feasible mappings is a feasible mapping;

— if $f(x,y)$ is a feasible function, then its derivative with respect to $x$ and its primitive function determined by its value at some feasible point are feasible.

Now, consider an analytical problem specified by some choice of functions (components of vector fields, or Hamiltonian functions, etc.), which may depend on parameters. These functions are the *data* of the problem. A *feasible set* of the problem is a minimal feasible set containing the problem data. A problem is called *analytically solvable* if its solution is a feasible function of parameters.

Prove or disprove the following conjecture: There exist a number $M$ and two functions $N$ and $D$ such that the problem of the stability of an equilibrium point 0 for a vector field in $\mathbb{R}^n$ whose components are $n$-th degree polynomials is not analytically solvable

a) if $n$ and $d$ are greater than $M$,

b) if $d > 1$ and $n$ is greater than $N(d)$,

c) if $n > 2$ and $d$ is greater than $D(n)$.

**1994-39.** Prove or disprove the following conjecture: The problem of the integrability of a differential equation specified by a vector field in a space of dimension $n > 1$ whose components are polynomials of degree $d > 1$ is not analytically solvable.

**1994-40.** Prove or disprove the following conjecture: The problem of the complete integrability of a canonical Hamiltonian system specified by a polynomial Hamiltonian of degree $d > 2$ in a space of dimension $2n > 2$ is not solvable analytically.

**1994-41.** Definition: A problem is *geometrically unsolvable* if there are no analytically solvable problems among the problems obtained from the given one by diffeomorphic changes in the parameter space. Conjecture: The problems mentioned in *1994-38–1994-40* are geometrically unsolvable.

**1994-42.** Definition: A problem involving a function as a parameter is *almost solvable* if the function space contains a decreasing sequence of exceptional submanifolds of increasing codimensions such that the problem is solvable outside each of these submanifolds. Conjecture: There are no almost solvable problems among those mentioned in *1994-38–1994-40*.

**1994-43.** Consider a vector field in the Euclidean space $\mathbb{R}^5$. The manifold of orbits of such a field (suitably chosen) can be made diffeomorphic to an arbitrary *fake manifold* $\mathbb{R}^4$ (that is, a differentiable manifold homeomorphic but not diffeomorphic to the vector space $\mathbb{R}^4$).

Can we obtain a fake $\mathbb{R}^4$ from a vector field with polynomial components? trigonometric? analytic? elementary? Can we explicitly write at least one such vector field?

**1994-44.** A *pseudoperiodic mapping* is the sum of two mappings, a linear and a periodic one. A *pseudoperiodic manifold* is a point's inverse under a pseudoperiodic mapping. Consider a pseudoperiodic (but not periodic) curve in $\mathbb{R}^n$ (with respect to the fixed period lattice $\mathbb{Z}^n$). Suppose that the rank of the linear part of the corresponding mapping is maximal (i. e., equals $n-1$). In that case, evidently, the curve contains an infinite branch (finitely distant from some straight line).

Is it true that a noncompact component of such a pseudoperiodic curve is always unique? *Solved in the negative by D. A. Panov.*

**1994-45.** Let $A: M \to M$ be an analytic diffeomorphism of a compact analytic manifold (e. g., of the torus $\mathbb{T}^2$). Is it true that the number of periodic points of period $n$ of such a diffeomorphism is majorized by an exponential function of $n$?

It is assumed here that periodic points $x$ are nondegenerate (i. e., that 1 is not an eigenvalue of the derivative of the mapping $A^n$ at $x$). Generic diffeomorphisms $A$ have no degenerate periodic points.

**1994-46.** Is it true that the number of periodic orbits of periods at most $T$ of a polynomial vector field on a compact ball in $\mathbb{R}^m$ is majorized by an exponential function of $T$?

**1994-47.** Conjecture: The number of periodic points of a mapping of class $C^\infty$ grows almost always not faster than some exponential function of the period.

Here "almost always" means "for almost all (in the sense of the Lebesgue measure) the parameter values in each typical family of mappings depending on sufficiently many parameters."

**1994-48.** Consider two compact submanifolds $X^k$ and $Y^l$ in a compact manifold $M^m$. Let $A: M \to M$ be a differentiable mapping. Consider the successive images of the manifold $X$ under the iterations $A^n$ of the mapping $A$. To measure their complexity (which grows as $n$ increases), one studies their intersections $Z(n) = (A^n X) \cap Y$ with a fixed manifold $Y$. These intersections $Z(n)$ are, as a rule, smooth manifolds of dimension $s = k + l - m$.

Explore the asymptotic behavior of topological complexity $|Z(n)|$ of the manifold $Z(n)$ as a function in time $n$.

In particular, is it true that for manifolds and mappings of class $C^\infty$, the topological complexity of the intersection $Z(n)$ is almost always majorized by some exponential function of time $n$? *As the topological complexity measure one might consider the sum of the Betti numbers, the characteristic numbers, the Morse and Ljusternik–Schnirelmann numbers, the numbers of the generators and of the relations of the fundamental group, and so on.*

**1994-49.** Consider two germs of holomorphic curves passing through the origin of the plane $\mathbb{C}^2$:

$$(X,0) \hookrightarrow (\mathbb{C}^2,0) \hookleftarrow (Y,0),$$

and a germ of a holomorphic mapping leaving the origin invariant:

$$A: (\mathbb{C}^2,0) \to (\mathbb{C}^2,0).$$

We shall apply the iterations of $A$ to $X$ and study the intersections of $A^n X$ with $Y$. The *Milnor number* $\mu(n)$ is by definition the multiplicity of the intersection of curves $A^n X$ and $Y$ at the origin.

Do the Milnor numbers $\mu(n)$ admit an upper bound exponential in time $n$?

It is assumed here that $A$ is a mapping of finite multiplicity and that, for each $n$, the curves $A^n X$ and $Y$ do not coincide.

**1994-50.** Consider an algebraic filtration

$$V_1 \supset V_2 \supset V_3 \supset \cdots$$

of the space $V_1 = J^\infty$ of infinite jets of pairs of holomorphic mappings

$$f\colon (\mathbb{C}^k,0) \to (\mathbb{C}^m,0), \qquad g\colon (\mathbb{C}^l,0) \to (\mathbb{C}^m,0)$$

at the origin. The varieties $V_i$ are algebraic subvarieties in $J^\infty$, i. e., each of these varieties is defined by polynomial equations on a finite number of Taylor coefficients. This finite number however depends on $i$. The *generalized Milnor number* $\mu(f,g)$ is by definition the maximum over numbers $i$ for which the pair $(f,g)$ belongs to $V_i$.

Now consider holomorphic embeddings

$$(X^k,0) \hookrightarrow (\mathbb{C}^m,0) \hookleftarrow (Y^l,0)$$

and a germ of a holomorphic mapping

$$A\colon (\mathbb{C}^m,0) \to (\mathbb{C}^m,0).$$

Conjecture: The generalized Milnor numbers $\mu(n)$ of the pairs $(A^nX,Y)$ admit an upper estimate exponential in $n$ (provided that $A$ is a mapping of finite multiplicity, and that all its Milnor numbers are finite).

**1994-51.** Infinitesimal version of the Hilbert 16th problem. Assume that a polynomial vector field on the plane admits a first integral whose level curves are cycles (filling at least some annulus in the plane). Consider small polynomial perturbations (of prescribed degree) of this vector field. The location of the limit cycles appearing in this perturbation is given in the first approximation by zeros of a certain integral (found by Poincaré) along nonperturbed closed curves (which are the level curves of the first integral).

Is the number of zeros of the Poincaré integral bounded (by a constant depending only on the degree of the perturbation)?

**1994-52.** A partial case of the previous problem: consider the full Abelian integral

$$I(h) = \oint (P\,dx + Q\,dy)$$

along an oval of an algebraic curve $H(x,y) = h$. The polynomials $P(x,y)$ and $Q(x,y)$ represent an infinitesimal variation of the Hamiltonian vector field, and $I(h)$ is the Poincaré integral. Find an upper bound for the number of real zeros of the function $I$ for all polynomials $(P,Q)$ of a fixed degree.

**1994-53.** Materialization of resonances in holomorphic dynamics. Consider a holomorphic mapping of a neighborhood $G$ of the circle $\mathbb{S}^1$ (in the complex plane $\mathbb{C}$) onto another neighborhood of the same circle:

$$A : (G, \mathbb{S}^1) \to (G', \mathbb{S}^1).$$

Suppose that $A$ induces a diffeomorphism of the circle $\mathbb{S}^1$ conjugate to rotation $R_\lambda$ through the angle $2\pi\lambda$, the conjugating diffeomorphism $B$ being holomorphic in some neighborhood of the circle: $A = BR_\lambda B^{-1}$. Assume that the Poincaré rotation number $\lambda$ is irrational.

Suppose that the maximal disc $M$ (diffeomorphic to $\mathbb{S}^1 \times \mathbb{R}$), where the mapping $A$ is conjugate to the rotation, is contained in the neighborhood $G$ of the circle $\mathbb{S}^1$ together with its boundary $\partial M$.

Is it true that any neighborhood of each point of the boundary $\partial M$ contains a point in a periodic orbit of the mapping $A$, this orbit lying in an arbitrarily small neighborhood of the boundary? Is this true at least generically?

# 1995

**1995-1.** Explore the topology of the stratification of the space of trigonometric (real and complex) polynomials modulo topological equivalence.

**1995-2.** Investigate mappings of Lyashko–Looijenga type for rational functions, especially in the cases of two poles (Laurent polynomials) and three poles ("modular polynomials"), when the set of poles has no moduli and the answer does not depend on the location of poles.

Evaluate the multiplicities of these mappings on various strata of the discriminant (generalizing Cayley's formula for the number of trees).

**1995-3.** Prove that a surface dual to a small perturbation of the projective plane in $\mathbb{R}P^3$ has at least four connected cuspidal edges (Aicardi's conjecture), even at the level of infinitesimal perturbations.

*B. Segre proved that this is true for a cubic surface, and attempts to find counterexamples by the aid of higher order spherical harmonic functions were unsuccessful. The number of swallowtails on the dual surface is found to be not less than* 6. *If the decomposition of the perturbation into spherical harmonic functions does not contain cubic harmonics and starts with fifth order harmonics then, according to Aicardi's examples, one obtains at least* 8 *connected cuspidal edges and at least* 14 *swallowtails.*

*Counterexample*: *D. Panov,* 1997 (*published in*: PANOV D.A. Parabolic curves and gradient mappings. *Proc. Steklov Inst. Math.*, 1998, **221**, 261–278): *there exist smooth perturbations of the projective plane in* $\mathbb{R}P^3$ *having only one parabolic line.*

**1995-4.** A point on a smooth plane curve is called an $n$-inflection point if the order of tangency with a suitable algebraic curve of degree $n$ at this point is higher than usually. For example, the 1-inflection points are the ordinary inflection points (where the multiplicities of the intersections of the curve with its tangents are at least 3). The multiplicity of the intersection with the nearest curve of degree $n$ usually equals $(n^2+3n)/2$.

How many 4-inflection points does a plane curve carry if it is sufficiently smoothly close to a) a circle, b) a cubic oval, c) an oval of a fourth-degree curve? Similar questions can be asked for any $n$.

*Any convex curve carries at least six* 2-*inflection points* (*the intersections at these points have multiplicity* 6; *for this reason, such points are called* sextactic). *A curve smoothly close to a circle has at least eight* 3-*inflection points* (*and there exist such curves with precisely eight points of nondegenerate* 3-*inflection*). *But a curve smoothly close to an oval of a cubic curve has not less than ten* 3-*inflection points* (*the intersections with suitable cubics are of multiplicity* 10 *at these points*). *It is interesting to determine where the boundary between the "closeness to an oval of a cubic" and the "closeness to a circle" passes, and what happens on this boundary. Possibly, when the higher derivatives are taken into account, the circle becomes an insufficiently convex curve, and there exists an interesting class of n-convex plane curves with specially good properties for each n.*

**1995-5.** The caustic of a general Lagrange collapse over $\mathbb{R}^3$ has at least three cusp edges (a conjecture of V. M. Zakalyukin). *Three edges are realized in an ellipsoid's caustic*; *thus, the conjecture asserts that the case of an ellipsoid is minimally complicated*: *the encountered singularities are topologically necessary.*

**1995-6.** Construct a parametric Morse theory that substantiates the topological necessity of the presence of complex critical points of functions on the fiber under certain parameter values, in terms of the topological complexity of the bundle on the total space of which the initial smooth function is defined. Carry over this theory to set-valued functions (that is, Lagrangian intersections).

**1995-7.** Study the singularities of the manifold of real projective curves completely decomposable into real lines.

*The Maxwell–Sylvester theory of spherical harmonics asserts that this strange submanifold of the projective space of n-th degree curves is "linked" with the complementary projective space of curves containing the imaginary circle* $x^2+y^2+z^2=0$ *as a component in a surprising way (namely, through each point of the complement of both spaces, there passes precisely one straight line joining them and intersecting each of them at one point). Do there occur other such "links"?*

**1995-8.** Find the simplest (i. e., with the minimal number of singularities) pairs of positively co-oriented curves immersed in the plane having equal Legendrian knots in $ST^*\mathbb{R}^2$, for which no regular homotopy without equally directed self-tangencies has been constructed (and try to prove that the latter does not exist).

**1995-9.** Find the simplest pairs of positively co-oriented curves (or fronts) immersed in the plane for which equipped knots coincide but Legendrian equivalence of knots in $ST^*\mathbb{R}^2$ has not been proved (and try to prove Legendrian non-equivalence).

**1995-10.** Find the simplest front with zero Maslov index whose Legendrian knot in $ST^*\mathbb{R}^2$ has not been realized by a Legendrian curve with smooth front (and try to prove that such a realization does not exist).

**1995-11.** How can we evaluate the minimum number of inflection points on realizations of a given class for immersions of the circle into the plane (sphere, projective plane, surface of genus $g$ with the Lobachevskian metric)? *For example, the figure eight has not less than two inflection points on the plane, and it can have none on the sphere.*

*There is a paper by B. Z. Shapiro on this topic; cf. the dissertation of E. Ferrand (who proved the symplectic or contact equivalence of the family of curves of an Hadamard manifold to a standard one; in particular, for the Lobachevskian plane, this gives all four-vertex-type results).*

**1995-12.** Transfer the theory of completely integrable Hamiltonian systems from symplectic geometry to contact geometry (where, e. g., the Lagrangian invariant manifolds with their natural affine structures determined by Lagrangian fibrations must be substituted by Legendrian invariant manifolds with their natural projective structures determined by Legendrian fibrations). Carry over the Liouville theorem to this context and find applications to the infinite-dimensional case (where the equations of characteristics are partial differential).

**1995-13.** Is the "last geometric theorem" of Jacobi valid, according to which the first caustic (the set of first conjugate points to an arbitrary "pole" along all geodesics starting from it) of a typical ellipsoid has exactly four cusps?

# 1996

**1996-1.** The Eisenbud–Levin formula for the index of a vector field singularity "drives" a global topological invariant (mapping degree) into the local algebra of the singularity. What becomes of the other global invariants, such as characteristic classes and numbers, under a similar localization (both in the complex and, especially, in the real case)?

**1996-2.** Calculate the cohomology and fundamental groups of complements of strata of codimension 2 (and higher) in the space of immersed plane curves. *In the case of higher codimensions of strata the homotopy (and hence homology) groups probably are trivial. It is interesting to compare the results with those for analogous problems concerning the spaces of versal deformations of germs of maps* $(\mathbb{R},0) \to (\mathbb{R}^2,0)$ *(stabilization over the growing complexity of singularities).*

**1996-3.** Prove that the $n$-th symmetric power of $\mathbb{R}\mathrm{P}^2$ is $\mathbb{R}\mathrm{P}^{2n}$.

**1996-4.** Prove that the caustic is diffeomorphic to the Maxwell stratum for the singularity $B_4$ and transfer this result to higher singularities $B_k$ (taking into account the symplectic or contact structures). *The symplectic version was constructed by F. Napolitano in* NAPOLITANO F. Duality between the generalized caustic and Maxwell stratum for the singularities $B_{2k}$ and $C_{2k}$. *C. R. Acad. Sci. Paris, Sér.* I *Math.*, 1997, **325**(3), 313–317.

**1996-5** (P. G. Grinevich). Let $f(x)$ be a real Fourier integral,

$$f(x) = \int_{-\infty}^{\infty} F(k)e^{ikx}\,dk, \quad F(-k) = \overline{F(k)},$$

with vanishing low-frequency harmonics [$F(k) = 0$ for $|k| \leq \omega$]. Then the limiting averaged number of zeros of $f$ on long intervals is not less than the averaged number of zeros of the function $\cos \omega x$ (i. e., the limiting density of zeros is not less than $\omega/\pi$).

*For a Fourier series the number of its sign changes on the circle is not less than the number of zeros for the lowest Fourier harmonic that has a non-zero coefficient in the series.*

**1996-6** (F. Aicardi). Compare the following one-parameter families of hypersurfaces in the Euclidean space $\mathbb{R}^3$ given by a positive definite quadratic form $f$: a) the family of equidistants from the ellipsoid $f = 1$; b) the family of "quadraticoids" defined by the support functions $f + t$ on the unitary sphere.

Calculations show that in these families, when $t$ varies, the perestroikas are topologically equivalent for corresponding (different) forms (and, moreover, the bifurcation diagrams in the spaces of the parameters defining the forms are diffeomorphic).

Explain this equivalence of families, by constructing the natural mapping between them. Does it hold in $\mathbb{R}^n$?

*A quadraticoid and an equidistant, for two chosen corresponding forms, define the same fields of crosses on the Gauss sphere (images under the Gauss map of the fields of principal directions).*

*Question: Is the entire set of perestroikas occurring in these families topologically necessary for the eversion of a front realized by the Legendrian collapse (or even by any Legendrian isotopy)?*

**1996-7.** Consider a typical function $z = f(x,y)$ of two variables. The asymptotic directions ($d^2 f = 0$) on its graph determine a field of crosses in the hyperbolic domain of negative Hessian determinant (with standard singularities on the boundary, which consists of the parabolic points where the Hessian determinant vanishes and the crosses degenerate into straight lines). What restriction on the topology of the field of crosses (i. e., on the section of the corresponding bundle over the hyperbolic set) are imposed by the hypothesis that this field arises from a function as the field of asymptotic directions?

**1996-8.** Investigate the multiplicities and the transversal multiplicities of the Lyashko–Looijenga mapping for polynomials, Laurent polynomials, modular polynomials on various strata and pairs of strata. *For polynomials the solution has been given by D. Zvonkine; the transversal multiplicities are the same in all cases.*

**1996-9.** M. Barner defines a *strongly convex* curve in $\mathbb{RP}^n$ as a curve such that for every $n-1$ of its points there is a hyperplane passing through them and not intersecting the curve elsewhere. For example, a curve whose projection from a point to a hyperplane is convex in $\mathbb{RP}^{n-1}$ is strongly convex in Barner's sense in $\mathbb{RP}^n$.

Investigate the manifold of strongly convex curves: the number of its connected components, singularities of the boundary, properties of dual curves, the existence of strongly convex projections and suspensions.

**1996-10.** Let us say that a plane of codimension 2 in the projective space $\mathbb{RP}^{2n}$ is *interior* with respect to a convex curve if each hyperplane containing this plane intersects the curve at $2n$ points. Do there exist interior planes? What are the topological invariants of the manifold of such planes? *For $n = 1$, the interior planes are the points in the region bounded by the curve. The problem has been solved (affirmatively) by S. S. Anisov and S. M. Gusein-Zade.*

**1996-11.** Let us say that a straight line in the projective space $\mathbb{RP}^{2n}$ is *exterior* with respect to a convex curve if, through every point of this line, $2n$ tangent hyperplanes pass. Do there exist exterior lines? What are the topological invariants of the manifold of such lines? *For $n = 1$, the exterior lines are those disjoint from the curve. The problem has been solved (affirmatively) by S. S. Anisov and S. M. Gusein-Zade.*

**1996-12.** Evaluate the cohomologies of the subgroups of the braid group corresponding to the coverings L2 (Lyashko–Looijenga) and L3 (Lyashko–Looijenga–Laurent) of the complement of the swallowtail.

*The classifying $K(\pi, 1)$ spaces of these groups are known: they are the complements of the bifurcation diagram of the spaces of ordinary and Laurent polynomials, respectively.*

**1996-13.** Investigate the variety of rational functions with three poles and the mapping L4 on it (taking a function to the set of its critical values).

**1996-14.** Define and explore the Morse complex of a solenoidal vector field in $\mathbb{S}^3$ (determined by the function on the space of closed curves whose value on a curve equals the field's flow through a surface bounded by this curve).

*The extremals of this functional are closed trajectories of the field. The second differential has infinitely many both positive and negative squares, but one may try to examine "index difference" for a pair of closed trajectories with the help of bifurcation theory. If, moreover, the field is Legendrian with respect to some contact structure, then one may try to calculate such difference of indices of two closed trajectories using the geometry of the restriction of the contact* 1*-form to a surface whose boundary is the difference of these trajectories.*

**1996-15.** Consider a discrete subgroup of the isometry group of the Lobachevskian plane [for example, the modular group $\mathrm{SL}(2, \mathbb{Z})$]. This group acts not only on the Lobachevskian plane but also in the de Sitter world (represented by the hyperboloid $x^2 + y^2 - z^2 = 1$ of one sheet in the Klein model, where the Lobachevskian plane is modeled by a sheet of the two-sheeted hyperboloid $x^2 + y^2 - z^2 = -1$).

To the metric of the Lobachevskian plane, there corresponds an invariant Lorentzian metric on the de Sitter hyperboloid. In the projective model, the Lobachevskian plane corresponds to the interior of the unit disk, and the de Sitter world, to its exterior; in both cases, the geodesics are the straight lines and the isometries are the projective transformations of the plane that leave the separating circle invariant.

How is the dense orbit of a point in the de Sitter world under the action of the discrete group under consideration (e. g., of the modular group) distributed? Is it possible to define pseudo-fundamental domains, replacing the Voronoĭ polygonal domains on the Lobachevskian plane, for this world? *The question is provoked by works of E. Brieskorn and his successors on monodromy groups.*

**1996-16** (Generalization of the Chevalley theorem?). The Coxeter group $D(n)$ acts on the space $(\mathbb{C}P^1)^n$ as follows: to a permutation of coordinates in $\mathbb{R}^n$, there corresponds a permutation of factors, and to the change of sign of a coordinate, there corresponds the antipodal involution of a factor. The manifold of orbits of this action is diffeomorphic to $S^{2n}$ (this is the Maxwell–Sylvester theorem of the theory of spherical functions). We obtain a real linear action of a $(2n-1)$-dimensional Lie group in $\mathbb{C}^{2n}$ with smooth orbit manifold $\mathbb{R}^{2n+1}$. How can we describe all such actions?

**1996-17.** Consider a sign-changing generic smooth function $F$ on the plane 2-torus. Study the motion of a charged particle with small energy in such a magnetic field (that is, the curves of geodesic curvature $F/\varepsilon$ with $\varepsilon \to 0$ at each point).

*In the region where $F \neq 0$, the particle experiences a Larmor rotation along a circle of small radius $\varepsilon/F$ the center of which slowly drifts along a level line of the function $F$. The trajectories intersecting the line $F=0$ consist of loops with alternating orientation joined by segments of a trajectory whose inflection points lie on the curve $F=0$. It is required to write the corresponding asymptotic formulae in a neighborhood of the curve $F=0$ (where the assumptions of the standard averaging method are violated) and, in particular, evaluate the drift direction.*

*Would these evaluations lead to counterexamples for the problem about four closed phase trajectories homotopic to a fiber of the sphericized (co)tangent bundle of the torus in the case where the magnetic field $F$ changes its sign?*

**1996-18.** Consider a generic positive smooth function $F$ on the standard sphere $S^2$. Study the motion of a charged particle at velocity 1 in such a magnetic field (i. e., examine the curves of geodesic curvature $F$ at every point). Do there exist (two?) closed trajectories whose phase curves are homotopic to a fiber of the sphericized (co)tangent bundle of the sphere?

*Such trajectories exist if the function $F$ is sufficiently large. Is it true that they always exist for a zero-divergence Legendrian vector field of the natural contact structure in $ST^*S^2$ without singular points? Our phase velocity field does have these properties, and, in addition, it is transversal to the field of planes in $ST^*S^2$ determined by the Riemannian connection. The situation seems to be similar to that in the conjecture of A. Weinstein, which was proved by C. Viterbo, and can be modeled with the use of fields on $\mathbb{S}^3$ instead of $ST^*S^2$.*

**1996-19.** Study the asymptotic curves on cubic surfaces in $\mathbb{RP}^3$ (for example, on those close to the plane $\mathbb{RP}^2$). Is this dynamical system integrable or chaotic? What is the design of the first return function on a parabolic curve? (To each point on the parabolic curve, this function assigns the next point where the asymptotic line returns to the parabolic curve.)

*The 27 (complex) lines on a cubic surface are asymptotic lines, so we can learn something by applying the theory of normal forms nearby.*

**1996-20** (M. B. Sevryuk). Introduce the following definition. A symplectic structure is said to be *r-exact* if its $r$-th exterior power is exact whereas the $(r-1)$-th power is not ($r \in \mathbb{N}$). In particular, 1-exact structures are just exact ones.

Given a fixed number $r$, do systems Hamiltonian with respect to $r$-exact symplectic structures possess any special properties?

**1996-21** (M. B. Sevryuk). Does there exist a smooth vector field on $\mathbb{R}^n$ irreversible with respect to any phase space involution but such that its time 1 flow map is reversible?

If the answer to this question is affirmative then: Does there exist a smooth vector field $V$ on $\mathbb{R}^n$ possessing the following properties: 1) the field $V$ is irreversible with respect to any phase space involution, 2) for each $\tau_0 > 0$, there is $\tau \in (0;\tau_0)$ such that the time $\tau$ flow map of the field $V$ is reversible?

# 1997

**1997-1.** Study the combinatorics of the bifurcation diagram of the space of real trigonometric polynomials outside the set of $M$-polynomials (all critical points of which are real). *For the M-polynomials of degree n, there is an explicit polyhedral model. For example, at $n = 2$, the bifurcation diagram reduces to an astroid with diagonals, and the model is a square with diagonals.*

**1997-2.** We define a *selector* to be a piecewise linear function in $\mathbb{R}^n$ with coordinates $(x_1,\ldots,x_n)$, which coincides, in every region where all the coordinates are different, with one of the coordinates. Examples are given by *Matov selectors*, defined by expressions like $\max(x_1, x_2, \min(x_3, \max(x_4, x_5, x_6)), \ldots)$ (each argument enters once).

How many selectors exist in all, and how many Matov selectors? How can we recognize whether a selector is a Matov selector?

*V. I. Matov proved that, if $f_1, \ldots, f_n$ are generic smooth functions on a manifold $M$, and $S$ is a Matov selector, then the function $S(f_1, \ldots, f_n) : M \to \mathbb{R}$ is topologically equivalent to a Morse function (and described the possible indices in terms of the selector). Does any other selector satisfy this property?*

*According to calculations of F. Aicardi, the numbers of Matov selectors for* $n = 1, 2, \ldots$ *are equal to* 1, 2, 8, 52, 472, 5 504, 78 416, 1 320 064, 25 637 824, 564 275 712, ...

**1997-3** (A. A. Agrachev – M. Ya. Zhitomirskiĭ)**.** Let $\alpha$ be a 1-form nondegenerate on the boundary of a disk and vanishing at its tangent vectors, and let $d\alpha = \alpha \wedge \beta$. Then $d\beta$ necessarily vanishes somewhere. *The authors claim that this is not so for surfaces with boundary different from disks.*

**1997-4.** In the theory of wave front propagation, all deformations of a Legendrian manifold under which it remains non-self-intersecting are usually considered admissible. In real-life problems on the propagation of a co-oriented front, the front can only move forward (in the direction determined by its co-orientation) and cannot move backward. The introduction of this constraint changes the problem setting both in the theory of wave fronts and in immersion theory. For instance, we can consider the oriented graph where the vertices are classes of curves and two classes $A$ and $B$ are joined by an arrow from $A$ to $B$ if $A$ has a representative (together with its motion forward) such that, moving forward, this representative arrives at the class $B$.

Calculate the part of this graph that corresponds to immersions (fronts) with small numbers of self-intersections (and, for fronts, cusps). Does there exist a perestroika of 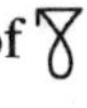 into  (with a different co-orientation) in the class of fronts with two cusps?

**1997-5.** Is the problem of the possibility of connecting two immersions of the circle into the plane by a path in the space of immersions without direct self-tangencies algorithmically solvable? *The conjecture is that it is not solvable, because in its framework, the problem of knot equivalence can (?) be modeled.*

**1997-6** (D. A. Panov). Does a generic function exist on the plane, whose Hessian is positive in a region, bordered by a smooth connected curve, and the field of asymptotic directions $d^2 f = 0$ on this parabolic curve has only one special elliptic point? Is it true that the number of hyperbolic special points on such a curve is not less than the number of elliptic ones?

*I recall the definition of elliptic (and hyperbolic) special points on a parabolic curve.*

*The special points are the points of tangency of the asymptotic direction of the graph with the parabolic curve. Over the hyperbolic region, the field of asymptotic directions defines a two-sheeted covering surface in the manifold of the non-oriented tangent elements (each point of the hyperbolic domain of the plane is lifted to the two asymptotic directions at that point).*

*For generic functions, this surface is smoothly continued by the asymptotic directions at the parabolic points. The critical line of the projection of this surface to the plane lies above the parabolic curve.*

*The asymptotic directions at the hyperbolic points are lifted to a field of directions on the surface constructed above. This field of directions on the surface is smoothly continued to the critical line, except for those "special" points of the parabolic curve, where the asymptotic direction is tangent to this curve.*

*For generic functions a special point is a singular point (a zero) of a smooth generic vector field (in a neighborhood of the point in question on the surface constructed above). A special point can be a saddle (index $-1$), and in this case is called* hyperbolic, *either a node or a focus (index $+1$), and in this case is called* elliptic.

**1997-7** (D. A. Panov). Consider a generic smooth function $F$ on the 2-torus. Let us construct the mapping of the torus to $\mathbb{RP}^2$ which takes each point of the torus to the point with homogeneous coordinates $[F_{xx} : F_{xy} : F_{yy}]$. Is it true that every point of the projective plane has no less than four preimages under this mapping $\mathbb{T}^2 \to \mathbb{RP}^2$?

*For a generic function, all the three derivatives cannot vanish simultaneously. The parabolic points are mapped to points on the zero Hessian circle $AC = B^2$ ($A = F_{xx}$, $B = F_{xy}$, $C = F_{yy}$). Each point on this circle indeed has not less than four preimages. This follows from the Morse inequality for functions on a circle: For each translation-invariant vector field $a\partial/\partial x + b\partial/\partial y$ on the torus, consider the derivative of $F$ along this field. This derivative has four critical points, which give four preimages of the point on the circle of parabolic points that corresponds to the direction of the field.*

**1997-8.** Stability of pyramids. Solutions to many problems of singularity theory (such as bifurcation diagrams or caustics) have the form of a pyramid in the 3-space whose horizontal section is (more or less) akin to a hypocycloid on the plane contracting to a point as the section plane approaches the critical "zero" position.

**Example 1.** Consider a general one-parameter family of surfaces in the Euclidean space $\mathbb{R}^3$ that passes through the "North pole" $N$ and contains the usual sphere (corresponding to the zero parameter value). Let us mark out the first caustic of the North pole $N$ on each surface. On the sphere, this is the South pole $S$. On nearby surfaces, these caustics are small curves with four (for generic families) cusps. Together, all such caustics sweep out a surface. It has the shape of the pyramid described above.

**Example 2.** Consider a generic positive function $F$ (magnetic field) on the plane. Let charged particles move from the point 0 in all possible directions on the plane at a small initial velocity $v$. If the function $F$ were constant, the trajectories of the particles would be Larmor circles of small radius $v/F$. The corresponding phase curves would form an exact Lagrangian torus in the phase space such that its projection into the plane would have two envelopes, a degenerate inner point caustic at the initial point 0 and an outer caustic being a circle of radius twice the Larmor radius. The entire picture depends on the parameter $v$.

If $F$ is not constant, then the interior caustic is no longer a point. It turns into a small closed envelope of the perturbed trajectories of the particles moving from 0 at an initial velocity of given magnitude $v$.

This envelope has (for a generic $F$) four cusps and is small together with the initial velocity $v$. Let us place each envelope in the separate plane $v = \text{const}$ in the 3-space. All these envelopes sweep out a pyramid-shaped surface.

A similar pyramid was obtained by A. A. Agrachev as the caustic of a simplest system with nonholonomic constraint in control theory (the example of magnetic field fits in this scheme).

**Example 3.** Consider the four-parameter family of trigonometric polynomials

$$F_{A,a,b,c}(t) = A\cos 2t + a\cos t + b\sin t + c.$$

The caustic of this family consists of the parameters values $(A, a, b)$ such that the corresponding function has a degenerate critical point. This surface in the 3-space has the form of a pyramid whose horizontal sections ($A = \text{const}$) are hypocycloids with four cusps, being small for small $A$.

**Example 4.** Consider a typical two-parameter family of functions for which 0 is a point of zero minimum [e. g., $H_{a,b}(x,y) = x^2 + y^2 + a(x^2 - y^2) + 2bxy + Ax^3 + Bx^2y + Cxy^2 + Dy^3$].

Consider the three-parameter family of vanishing cycles

$$\gamma_{a,b,c} = \{x, y : H_{a,b}(x,y) = c\}.$$

The number of vertices (extrema of the curvature) of a curve $\gamma$ with a very small $c$ is almost always four. However, at the point $a = b = c = 0$ of the parameter space, a narrow tongue of the locus of curves with six vertices reaches generically the plane $c = 0$.

This set of curves intersects the plane $c = \text{const} > 0$ in a small plane region bounded by a curve with six cusps, similar to a hypocycloid. As $c$ approaches zero, this "hypocycloid" contracts to a point. The entire boundary of the tongue of the locus of curves with six vertices in the parameter space has the shape of a pyramid near the point $a = b = c = 0$.

The problem is to determine the stability of the pyramid singularities mentioned above. In all cases, the question reduces to examining families of functions on the circle.

The conjectured answers are: the caustic (and the corresponding family of functions on the circle) is stable (with respect to the analytic or smooth deformations of the condition of the problem and, respectively, to the analytic or smooth normalizing diffeomorphisms) in a "conic neighborhood" of the corresponding pyramid in the parameter space. This "conic neighborhood" of the pyramid is itself bounded by a larger pyramid with the same vertex. Such a "neighborhood" contracts to one point at the vertex of the pyramid (and, therefore, the diffeomorphism reducing the caustic to normal form becomes only a homeomorphism at the vertex).

**1997-9.** The mathematical trinities. In addition to the pairs (an object, its complexification) in various mathematical theories, one often encounters triples of objects. The conjecture is that it is not a coincidence, and all the triples are related by commutative diagrams. The arrows joining two such triples usually form a natural triple themselves. The problem is to verify this conjecture and to study such triples systematically.

Here are several examples of such triples:

| | | |
|---|---|---|
| $\mathbb{R}$ | $\mathbb{C}$ | $\mathbb{H}$ |
| $E_6$ | $E_7$ | $E_8$ |
| $P_8$ | $X_9$ | $J_{10}$ |
| $A_3$ | $B_3$ | $H_3$ |
| $D_4$ | $F_4$ | $H_4$ |
| tetrahedron | octahedron | icosahedron |
| $6 = 2 \cdot 3$ | $12 = 3 \cdot 4$ | $30 = 5 \cdot 6$ |
| $60°, 60°, 60°$ | $45°, 45°, 90°$ | $30°, 60°, 90°$ |
| $\mathbb{S}^1 \xrightarrow{\mathbb{S}^0} S^1$ | $\mathbb{S}^3 \xrightarrow{\mathbb{S}^1} S^2$ | $\mathbb{S}^7 \xrightarrow{\mathbb{S}^3} S^4$ |
| coverings | connections | ? |
| monodromy | curvature | ?? |
| $w_i$ | $c_i$ | $p_i$ |
| usual polynomials | trigonometric polynomials | modular polynomials |
| usual numbers | trigonometric numbers | elliptic numbers |
| cohomology | $K$-theory | elliptic cohomology |

The symbol ? denotes a conjectural "hyperconnection"; probably the latter is some quaternionic thing turning into the connection of a fibering over complex curves in the base which, however, has many complex structures over whose curves these connections have some discordance.

The symbol ?? should denote a conjectural hypercurvature 4-form (which most probably measures the extent of violation of some generalization of the Bianchi identity by a hyperconnection).

# 1998

**1998-1.** Combine É. Cartan's theory of differential systems with singularity theory. *One should distinguish two higher (and infinite) codimension exceptions from the generic cases studies, while in the present form the Cartan theory, like the* *algebraic geometry, is more interested in "general statements" (like the Hilbert finiteness theological theorem) admitting no exceptions (in the analytic category), while to extend these to the smooth categories one should distinguish different degrees of degeneration, whose representatives might behave quite differently.*

**1998-2.** Consider a generic smooth surface in the three-dimensional real projective space. Can this surface have less than six special parabolic points (where the asymptotic direction is tangent to the parabolic curve, and the dual surface has a swallowtail)? If the number of special points is less than six, can the number of connected parabolic curves be less than four? *The number of special points in Panov's example with only one parabolic line is equal to* 12.

**1998-3.** Consider a smooth parabolic curve of constant multiplicity on a surface in the projective 3-space. Into how many parabolic curves of multiplicity one does it decompose under a small generic perturbation?

*The question is open even for the surfaces whose equations in affine coordinates have the form $z = f(x,y)$ and whose parabolic curves are the infinitely distant straight line. In this case, the multiplicity is even (and equals two in the simplest situation), and the conjectured number of multiplicity-one parabolic curves of the perturbed surfaces is three.*

Is it true that a general surface close to the surface specified by the equation $z = 1/(x^2+y^2)$ $[z = x/(x^2+y^2)]$ in affine coordinates has not less than three [respectively, two] parabolic curves in a neighborhood of infinity?

**1998-4.** The *spherical second differential* of a function on the sphere is the quadratic form on the tangent space that measures the difference between the given function and the nearest restriction to the sphere of a function linear in the ambient space.

A function is called *hyperbolic* if its spherical second differential is hyperbolic everywhere except at finitely many points (where the function can have singularities).

Can an odd hyperbolic function on the 2-sphere have less than six logarithmic poles?

Can an odd function obtained from an odd hyperbolic function on the 2-sphere by a generic smoothing have less than eight parabolic curves (along which the spherical second differential degenerates)?

**1998-5.** Does there exist a surface $z = f(x,y)$ whose Gaussian curvature [at every point $(x,y,f(x,y))$] is the given function $g(x,y)$? Here, $f$ and $g$ are functions smooth in a neighborhood of the given point.

Answering a similar question for the Hessian requires solving the equation $f_{xx}f_{yy} - f_{xy}^2 = h(x,y)$, where $h$ is a function defined in a neighborhood of the given point.

What singularities can a parabolic curve ($h = 0$) of a smooth surface have in a neighborhood of its flattening point (where $df = 0$ and $d^2 f = 0$)?

**1998-6.** Let us consider a curve specified by the equations $x = \cos t$, $y = \sin t$, $z = \cos 3t$. This curve has six flattening points (of zero torsion). Is it possible to annihilate all these flattening points by an admissible regular homotopy of a curve?

A regular homotopy is called *admissible* if, in course of the deformation, there are no events of

a) self-intersections of the curve (changes of the knot type);

b) self-intersections of the dual curve (formed by osculating planes of the initial curve in the dual space).

c) inflection points (zero-curvature points);

d) tangencies of the dual curve with the surface of the front (formed by tangent planes to the initial curve).

**1998-7.** The curves admitting a convex projection in the projective space form a domain in the space of curves. Examine the boundary of this domain, namely, its stratification, the singularities of the intersection of the boundary with the transversals to its strata, and the complex of its strata.

Similar questions for convex curves themselves and for "strongly convex" Barner curves (see problem *1996-9*) are also interesting.

**1998-8.** Study the cohomology rings of the complements of the bifurcation diagrams of holomorphic functions: Is it true that these complements are the Eilenberg–MacLane $K(\pi,1)$ spaces? What are their stable Betti numbers (and cohomology rings)?

**1998-9.** To an entire algebraic function, in addition to its braid group, there are related a series of the groups of its second, third, etc. braids. These groups are defined as the fundamental groups of local complements of successive discriminants, each being the set of atypical values of the projection of the preceding discriminant along the fibers of a generic one-dimensional bundle. As the initial "discriminant," we take the graph of the function treated as a hypersurface in the product of the domain of the function by its range fibered over the domain.

These groups, discriminants, their complements, the cohomologies of these complements, and the corresponding monodromies remain absolutely unexplored even for the simplest algebraic function $z(a)$ specified by the equation $z^n + a_1 z^{n-1} + \cdots + a_n = 0$.

Even a description of generators and relations in these groups is of interest. In addition to generic projections, it is interesting to consider the sequence of projections successively forgetting $a_n, a_{n-1}, \ldots$

**1998-10.** How to complexify braid theory? the Maslov index? the theory of Vassiliev's knot invariants?

**1998-11.** When the number $m$ is large, what is the behavior of the greatest multiplicity (Milnor number) of a critical point of a holomorphic function in two variables depending generically on $m$ parameters?

**1998-12.** Can an asymptotic line on the surface $z = f(x,y)$ all of whose points are hyperbolic be closed?

**1998-13.** Does the Euler equation for an ideal fluid have new conservation laws in addition to the classical ones? Are there such conservation laws along coadjoint orbits of the group of volume-preserving diffeomorphisms of the domain?

**1998-14.** How can we complexify the ring $\mathbb{Z}$? On the set of homotopy classes of the mappings of a Lie group into itself that leave the identity fixed, two generally noncommutative operations act: "addition," defined by $(a+b)(g) = a(g)b(g)$, and "multiplication," defined by $(ab)(g) = a(b(g))$.

For instance, we obtain the ring $\mathbb{Z}$ from the group $\mathrm{U}(1) = \mathrm{SO}(2)$ and the field $\mathbb{Z}_2$ from the group $\mathrm{O}(1)$. What is obtained from $\mathrm{SO}(3)$? from $\mathrm{Spin}(4)$? from other groups?

**1998-15.** What is the quaternionic analogue of the determinant?

**1998-16.** How can we complexify the notions of one-dimensional holomorphic bundle and of its connection and curvature? What becomes of the theories of quantum Hall effect and of Berry phase under such a complexification?

**1998-17.** Contactize the symplectic Liouville theorem on completely integrable Hamiltonian systems.

**1998-18.** A vector field $v$ of divergence zero on a 3-manifold is called *Hofer* if it is the field of kernels of a 2-form having contact potential [this means that the field $v$ is specified by the condition $i_v(d\alpha) = 0$, where $\alpha \wedge d\alpha$ nowhere vanishes].

Consider the motion of a charged point on a surface under the action of a magnetic field orthogonal to the surface. Under what conditions is the corresponding vector field on the 3-manifold of unit tangent vectors to the surface Hofer? Even the case of a nonvanishing field on the sphere with standard metric is interesting.

**1998-19.** The Heisenberg indeterminacy relation leads to the following conjecture. Let $\Gamma$ be a closed subgroup of the commutative group of the Euclidean space $\mathbb{R}^n$ such that the quotient space by $\Gamma$ is compact (e. g., a lattice).

Suppose that a ball of radius $r$ is contained in the complement of $\Gamma$. Then in the dual Euclidean space there is a nonzero "wave vector" $k$ of length not exceeding $c/r$ such that the scalar product $(k,x)$ takes only integer values when $x$ is in $\Gamma$.

Here $c$ is a constant depending only on $n$.

**1998-20.** Classify the simple curve singularities in a contact space.

**1998-21.** The following problem about Legendrian links was communicated to me by R. Penrose.

Consider the space-time $\mathbb{R}^{2+1}$ (with pseudo-Riemannian metric of signature $++-$ positive definite on the isochrones $t = \text{const}$).

The manifold of light rays in such a space has a natural contact structure. The rays from one point of the space-time form a Legendrian submanifold in this manifold.

The problem is to study the relation between the causality (the possibility of joining two points in the space-time by a time-like curve) and the linking of the corresponding Legendrian manifolds of dimension $n-1$ in the $(2n+1)$-dimensional space of rays in the $(n+1)$-dimensional space-time.

**1998-22.** Consider an $n$-edge polygonal knot in $\mathbb{R}^3$ or in $\mathbb{S}^3$. How does the minimum number of simplices in a triangulation of this space whose 1-skeleton contains the most complex $n$-edge knot grow with increasing $n$?

**1998-23** (N. A. Nekrasov)**.** Consider the quotient space of the space of (germs of) pairs of functions with zero Poisson bracket modulo the group of (germs of) symplectomorphisms. We claim that this "manifold" has a natural symplectic structure and is endowed with a natural discriminant of complex codimension one.

The problem is to study the fundamental group of the complement of this discriminant. Is this complement an Eilenberg–MacLane space? What is its cohomology ring?

**1998-24** (A. N. Varchenko)**.** The equation $u_{xx}u_t^2 + u_{tt}u_x^2 - 2u_xu_tu_{xt} = 0$ has the property that, if $u$ is its solution, then so is $f(u)$. What other operators have similar invariance properties (and how can they be used to construct hydrodynamical analogues, topological invariants, topological variational principles, etc.)?

**1998-25.** The problem on Jordan matrices by M. L. Kontsevich. Consider the space of square complex matrices of a fixed size. Can one choose one representative of each class of conjugate matrices so that all these representatives form a collection of affine subspaces of the matrix space?

# 1999

**1999-1.** Compile a *complete* list of the adjacencies of simple curve singularities in $\mathbb{C}^N$.

This and the following six problems are concerned with complex curves, that is, germs $(\mathbb{C},0) \to (\mathbb{C}^N,0)$; however, the same questions for real curves, that is, germs $(\mathbb{R},0) \to (\mathbb{R}^N,0)$, also make sense.

**1999-2.** Compile a list of the *semigroups* of simple curve singularities in $\mathbb{C}^N$.

a) Does a semigroup determine the type of a (simple) singularity?

b) What pairs of semigroups exclude the adjacency of the corresponding singularities (probably, simplicity is not essential here)?

c) Are the remaining adjacencies realized for some pair of singularities (simple? not simple?) with given semigroups?

**1999-3.** Is it true that the *simple* curve singularities in $\mathbb{C}^N$ are precisely those *stably simple* singularities that can be realized in $\mathbb{C}^N$?

**1999-4.** Compile a list of the filtered *Artin algebras* of simple singularities for the curves $f\colon (\mathbb{C},0) \to (\mathbb{C}^N,0)$.

a) Does such a filtered algebra (or its action on $\mathfrak{m}^1/\mathcal{A}_f$ by operators) determine the type of a simple singularity (or its semigroup)? Here, $\mathfrak{m}^1$ is the maximal ideal in the space of germs of functions $(\mathbb{C},0) \to (\mathbb{C},0)$ and $\mathcal{A}_f$ is the ideal generated by the components of the mapping $f$.

b) Does the semigroup of a singularity determine its Artin algebra or the filtration?

**1999-5.** *Resolution of singularities* of simple curves in $\mathbb{C}^N$.

a) Compile a list of resolution graphs. How are they related to question a) in problem *1999-2*?

b) Is it true that moduli of curves arise precisely when moduli of resolutions do (in the case of 4 points on $\mathrm{P}^1$, etc.)?

**1999-6.** *Stabilization of curves.* Consider the base $\mathbb{C}^{M(N)}$ of versal deformation of a more complex singularity containing the stratum $\Sigma$ of a simpler singularity.

a) How many (locally) irreducible components does the stratum $\Sigma$ have?

b) In what sense do the topological (homological? homotopy?) properties of the complement $\mathbb{C}^{M(N)} \setminus \Sigma$ stabilize as $N \to \infty$?

c) In what sense do these properties of the complement (whether or not it stabilizes as $N \to \infty$) stabilize when the type of the simpler singularity is fixed and the type of the initial more complex singularity (simple? any?) becomes more complicated?

**1999-7.** *The stratum* $\mu = \text{const}$ *for curves.* Consider the "manifold" of singularities of given codimension $\mu$ of the orbit in the function space as a submanifold in the base $\mathbb{C}^\mu$ of its versal deformation.

a) Is this "manifold" smooth? irreducible?

b) How can its dimension $m$ in $\mathbb{C}^\mu$ ("number of internal moduli") be evaluated (with the use of the semigroup? algebra? resolution?) or at least estimated?

c) Is it true that $m$ is semicontinuous with respect to the choice of the initial singularity, i. e., coincides with its usual modality?

**1999-8.** Fix a positive integer $n \geq 3$ and consider $n$ positive integers $a_1, a_2, \ldots, a_n$. Their sums (linear combinations with integer non-negative coefficients) constitute the semigroup $S(a)$ of positive integers:

$$S(a) := \{\langle k, a\rangle \mid k \in \mathbb{Z}_+^n\}$$

$(\mathbb{Z}_+ = \{0;1;2;\ldots\})$. Suppose that $\gcd(a_1, a_2, \ldots, a_n) = 1$. Then, starting from certain $K(a) \in \mathbb{Z}_+$, all the non-negative integers lie in $S(a)$. *For instance,* $K(a) = (a_1-1)(a_2-1)$ *for* $n = 2$. *Note that this value of* $K(a)$ *is always even (since the numbers* $a_1$ *and* $a_2$ *are relatively prime they cannot be even simultaneously). The problem of calculating* $K(a)$ *for* $n$ *large is called the* Frobenius problem.

Explore the statistics of $K(a)$ for typical large vectors $a$. *Conjecturally,*

$$K(a) \approx c \sqrt[n-1]{a_1 a_2 \cdots a_n}, \quad c = \sqrt[n-1]{(n-1)!}\,.$$

The subsequent three problems are devoted to the statistics of the semigroups of positive integers $S(a)$ for relatively prime $a_1, a_2, \ldots, a_n$ as well. *All these problems are intended mainly for a computer experiment—with the prospects of concluding with proofs. The case* $n = 3$ *is already interesting.*

**1999-9.** *For* $n = 2$, *a number* $N \in \mathbb{Z}$ *belongs to the semigroup* $S(a)$ *if and only if the number* $K(a) - 1 - N$ *does not (J. J. Sylvester). Thus, for* $n = 2$ *the semigroup* $S(a)$ *occupies precisely one half of the segment* $[0; K(a)-1]$ *(recall that, for* $n = 2$, *the number* $K(a) - 1$ *is always odd).*

Determine what fraction of the segment $[0; K(a)-1]$ is occupied by the semigroup $S(a)$ for $n \geq 3$ and for large vectors $a$. *Conjecturally, this fraction is asymptotically equal to* $1/n$ *(with overwhelming probability for large* $a$*).*

**1999-10.** *Examples show that $S(a)$ fills the right half of the segment $[0;K(a)-1]$ more densely.*

Find the typical density of filling the segment $[0;K(a)-1]$ asymptotically for large vectors $a$. *The conjectured behavior of the density $p(N)$ at a point $N < K(a)$ is*

$$p(N) = \left(\frac{N}{K(a)}\right)^{n-1}.$$

*Such a distribution would immediately imply that the semigroup $S(a)$ occupies $1/n$-th of the segment $[0;K(a)-1]$:*

$$\int_0^K (N/K)^{n-1}\,dN = K/n$$

*(the triangle fills one half of the rectangle, the parabolic triangle fills one third, and so on).*

**1999-11.** Consider the density of the semigroup $S(a)$ with multiplicities taken into account (each point is counted as many times as it has representations in the form $\langle k,a\rangle$ with $k \in \mathbb{Z}_+^n$).

Find this density $P(N)$ asymptotically for large vectors $a$. *The conjecture is that $P(N) \sim N^{n-1}$ fot all $N$ (rather than only for $N < K(a)$). Note that, for $n = 2$, both densities (taking and not taking account of multiplicities) asymptotically coincide for $N < K(a)$. It is not clear whether such a coincidence takes place for $n \geq 3$.*

**1999-12.** Complexify the group $\mathbb{Z}$ of integers employing the fact that $\mathbb{Z}$ is a braid group for two threads and, simultaneously, a dyed braid group for two threads. *The conjectured alternatives are $\mathbb{Z}$ and $\mathbb{Z}^2$.*

**1999-13.** Reflection groups and oscillatory integrals. Consider the oscillatory integral

$$I(h,\lambda) = \int_{\mathbb{R}^n} e^{iF(x,\lambda)/h}\varphi(x)\,dx, \quad F:(\mathbb{R}^n \times \mathbb{R}^m, 0) \to (\mathbb{R},0), \quad h \to 0,$$

where $\varphi: \mathbb{R}^n \to \mathbb{R}$ is a smooth function concentrated in a sufficiently small neighborhood of the origin. The *singularity index* $\beta$ of a singularity of the function $F(\cdot,0)$ at 0 is the infimum of the numbers $\gamma$ such that

$$|I(h,\lambda)| \leq C(\varphi)|h|^{\frac{1}{2}n-\gamma}$$

at all sufficiently small $|\lambda|$ under an arbitrary deformation of $F$ (the value $\beta - \frac{n}{2}$ is then called the *oscillation index*).

For simple singularities, the singularity index equals

$$\beta = \frac{1}{2} - \frac{1}{N}, \tag{1}$$

where $N$ is the *Coxeter number* of the corresponding Coxeter group (which is a finite irreducible group generated by reflections in $\mathbb{R}^{\mu}$); see ARNOLD V. I. Remarks on the stationary phase method and Coxeter numbers. *Russian Math. Surveys*, 1973, **28**(5), 19–48:

| Singularity | $A_\mu$ | $D_\mu$ | $E_6$ | $E_7$ | $E_8$ |
|---|---|---|---|---|---|
| $\beta$ | $\frac{\mu-1}{2(\mu+1)}$ | $\frac{\mu-2}{2(\mu-1)}$ | $\frac{5}{12}$ | $\frac{4}{9}$ | $\frac{7}{15}$ |
| $N$ | $\mu+1$ | $2(\mu-1)$ | 12 | 18 | 30 |

Formula (1) is also valid for boundary singularities (recall that the Coxeter number of $B_\mu$ equals $2\mu$).

**Problem:** Construct a theory of oscillatory integrals and find a similar formula for the remaining (noncrystallographic) Coxeter groups $F_4$, $G_2$, $H_3$, $H_4$, and $I_2(p)$ (whose Coxeter numbers are 12, 6, 10, 30, and $p$, respectively).

**1999-14.** Consider a family of smooth surfaces $z = f_t(x, y)$ in $\mathbb{R}^3$. Suppose that the surface corresponding to $t = 0$ is convex and, at some $t = t_\star > 0$, a hyperbolic region arises. In the computer experiment performed by A. Ortiz-Rodriguez, the line formed by the inflection points of the asymptotic curves in the hyperbolic region (*tacnodal line*) at small $t - t_\star > 0$ had the shape of the figure eight tangent to the boundary of the hyperbolic region at two singular points. Construct a rigorous theory of such figures eight.

**1999-15.** Products of matrices can be calculated by Strassen's fast matrix multiplication formula (for example, multiplying two $2 \times 2$ matrices by this formula involves 7 rather than 8 multiplications). How is this formula related to the trinity $\mathbb{R}$–$\mathbb{C}$–$\mathbb{H}$?

**1999-16.** On the plane $\mathbb{R}^2$, consider a configuration of $n$ curves diffeomorphic to straight lines. It is assumed that no two curves intersect at more than 1 point, and that all intersections of the curves are transversal.

What configurations of curves are realized by straight lines? Starting from what number $n$ of curves do deviations occur?

**1999-17. Definition.** The plane curve $\{x, y \mid x^{-2} + y^{-2} = 1\}$ is called an *anticircle.*

**Theorem 1.** The curve projectively dual to the anticircle is the astroid $\{p, q \mid p^{2/3} + q^{2/3} = 1\}$.

**Theorem 2.** The set of normals to an ellipse is the anticircle.

**Question 1.** Is there an astroid among the equidistant curves of an ellipse? *According to F. Aicardi, among the equidistant curves of an ellipse there are no curves orthogonally equivalent to the astroid. Furthermore, according to M. E. Kazarian and R. Uribe, there are no curves either projectively or affinely equivalent to the astroid*[1]. *Moreover, they proved the following: consider the affine transformation sending the four cusps of the equidistant curve to the four verticies of a fixed square. In the family obtained this way there is exactly one curve with the symmetry of a square. This only "candidate" tends to the astroid as the eccentricity of the ellipse tends to zero.*

**Question 2.** Are there multidimensional analogues of the anticircle and the astroid?

**Question 3.** Move the tangents to the ellipse along the normals at distance $s$. What are the properties of the resulting curve in the dual plane? *According to F. Aicardi, it has no cuspidal points.*

# 2000

**2000-1** (A. Ortiz-Rodriguez). How many parabolic curves (closed curves or all the curves—these are two different questions) can lie on the graph of a real polynomial of degree $D$ in two variables? This is unknown even for $D = 4$ (is it possible that there are 4 closed components?).

[1] In the Russian edition of this book, the contrary was affirmed here, which was an error.

**2000-2** (A. Ortiz-Rodriguez). How many parabolic curves can lie on a projective algebraic surface of degree $D$ in $\mathbb{R}P^3$: even the asymptotics for large $D$ is of interest (the coefficients at $D^3$ in the examples I know and in the known upper estimate differ by a factor of 20).

**2000-3.** Consider the space of hyperbolic [with the second differential of signature $(+,-)$ everywhere except at the origin] homogeneous polynomials of degree $D$ in two real variables. How many connected components does this space consist of? (*For* $D = 3$ *or* 4 *there is only one component, for* $D = 6$ *there are at least two components*; the conjectural answer grows probably with a linear rate as $D$ increases, i. e., the number of the components is of the order of $D$ for $D$ large.)

**2000-4.** Consider a generic collection of $n$ straight lines in $\mathbb{R}P^2$. How much does the number of topological classes of such collections differ from the number of topological types of collections of $n$ noncontractible circles embedded generically in $\mathbb{R}P^2$?

Similar questions are not trivial even for the affine plane, both in the case of embeddings of affine straight lines and in the case of circles—in the presence of a fixed number of intersections as well as even without intersections.

Of course, the question makes sense for straight lines in the three-dimensional space too, provided that the complexity of topological knotting of the curve configurations to be compared is bounded above.

**2000-5.** The observers assert that the number of the eruptions of the volcano of *Piton de la Fournaise* with the emission of volume less than $V$ grows like $V^{-3/2}$ as $V$ decreases [LAHAIE F., GRASSO J.-R., MARCENAC P., GIROUX S. Modélisation de la dynamique auto-organisée des éruptions volcaniques: application au comportement du Piton de la Fournaise, Réunion. *C. R. Acad. Sci. Paris, Sér.* IIa *Sci. Terre Planètes*, 1996, **323**(7), 569–574]. Are there reasonable grounds for this scaling law, similarly to the turbulence laws?

**2000-6.** The observers assert that the metabolic rate in similar organisms (such as men of different stature) is proportional to the $3/4$ power of the mass (rather than to the $2/3$ power, as the ratio of the reaction surface area to the reaction volume suggests). Are there reasonable explanations for such a fractal behavior [WEST G. B., BROWN J. H., ENQUIST B. J. A general model for the origin of allometric scaling laws in biology. *Science*, 1997, **276**(5309), 122–126]?

**2000-7.** There are observations that the number of the species (of animals, insects, birds, ...) on an island of area $S$ is proportional to $S^{1/4}$, whereas the number of the cell types in an organism with the genome of $N$ genes grows with $N$ like $N^{1/2}$. How can one explain these exponents? Compare with the Kolmogorov law, according to which the radius of the minimal but still typical brain or computer of $N$ elements grows like $N^{1/2}$ (rather than like $N^{1/3}$, as the volume argument suggests).

**2000-8.** Let a mapping of a complex projective space (or vector space) onto itself send all the complex subspaces to complex subspaces. Are there such transformations other than complex projective ones (linear ones) and their products with the complex conjugation?

*There are no other diffeomorphisms, but I do not know the answer for the case of homeomorphisms (hopefully, there are no other homeomorphisms as well). One may ask the same question even for the set-theoretic bijections (which are not forced to be homeomorphisms).*

**2000-9.** Let $\Gamma \subset \mathbb{R}^2$ be a real algebraic plane curve and $g: \mathbb{R}^2 \to \mathbb{R}$ be a polynomial. To this pair, assign the *caustic* which is a curve $C$ in another plane equipped with orthonormal coordinates $(A,B)$. The caustic consists of the points $(A,B)$ for which the restriction of the function

$$G_{A,B} = g + Ax + By$$

($x$ and $y$ being the coordinates in $\mathbb{R}^2$) to the curve $\Gamma$ possesses a degenerate critical point. For a nonsmooth curve $\Gamma$ given by the equation $f(x,y) = 0$, the critical points are defined as the zeros of the derivative $\nabla G$, while the degenerate critical points are the zeros of both $\nabla G$ and the second derivative $\nabla^2 G$; here $\nabla$ is the Hamiltonian vector field

$$f_y \frac{\partial}{\partial x} - f_x \frac{\partial}{\partial y}.$$

If $\Gamma$ is a circle ($x^2 + y^2 = 1$) then the caustic has at least 4 cusps, and its alternated length (the sum of the lengths of the segments between the cusps with alternating signs) vanishes. This follows from the Sturm–Hurwitz theorem which states that the number of zeros of the sum of a real Fourier series

$$F(t) = \sum_{n>k} \left[a_n \cos(nt) + b_n \sin(nt)\right]$$

is at least the number of zeros of the lowest harmonics entering the series with a nonzero coefficient (i. e., at least $2k+2$ zeros over the period). For instance, if the integral of $F$ vanishes ($k=0$) then there are at least two zeros (moreover, these zeros are the critical points of the primitive of $F$). This Sturm theorem proved by Hurwitz is a generalization of the Morse inequality (for the circle), because the function $F$ in the theorem can be viewed as the image of a (primitive in the extended sense) periodic function $H$ under a differential operator of degree $2k+1$:

$$F = LH$$

where

$$L = \partial(\partial^2+1)(\partial^2+4)\cdots(\partial^2+k^2), \quad \partial = (d/dt).$$

Thus, one can regard the zeros of the function $F$ as generalized critical points of the "potential" $H : \mathbb{S}^1 \to \mathbb{R}$.

The problem is to carry over the Sturm–Hurwitz theorem (and the statements on the properties of the caustic) to the case of algebraic curves $\Gamma$ other than a circle. How many singular points of the caustic are inevitable for curves $\Gamma$ of a given genus? This question arises even for singular curves $\Gamma$ of genus zero, e. g., for the degenerate elliptic curve $y^2 = x^2 + x^3$.

**2000-10.** Consider a controlled dynamical system $\dot{x} = v(x,u)$ on a compact phase space ($x \in M$) with a compact manifold of the values of the controlling parameter $u$. Let $f : M \to \mathbb{R}$ be a smooth "goal function."

Explore the phase transitions of the controls optimal on the average (i. e., those maximizing the temporal mean

$$\widehat{f} = \lim_{T\to+\infty} \frac{1}{T}\int_0^T f(x(t))\,dt$$

—either for the fixed initial point $x(0)$ or while maximizing over this parameter as well).

A *phase transition* here is defined as a nonsmooth dependence (of both the optimal strategy and the attained maximal value of the mean) on additional parameters on which the initial data of the problem (i. e., the controlled system $v$ and the goal function $f$) depend smoothly.

*Such nontrivial phase transitions are encountered even in the simplest one-dimensional case where $M = \mathbb{S}^1$ and $u \in \mathbb{S}^1$.*

**2000-11.** Study the phase transitions of the maximal mean value $\widehat{f}[\rho]$ of a smooth goal function $f : M \to \mathbb{R}$ over the choices of the mass distribution $\rho\, dx$ (with density $\rho$ with respect to the Riemannian volume $dx$) on $M$, under the condition that the density is bounded above and below by given positive smooth functions:

$$0 < r(x) \leq \rho(x) \leq R(x) < \infty$$

on $M$. Here the mean value is defined by the formula

$$\widehat{f}[\rho] = \left(\int_M f\rho\, dx\right) \Big/ \left(\int_M \rho\, dx\right).$$

**2000-12.** Given an integer matrix $A$ of order three with determinant 1 [$A \in \mathrm{SL}(3,\mathbb{Z})$], construct three eigenplanes assuming that all the eigenvalues are real, positive, and irrational. The integer points in one of the octants bounded by these three planes constitute a commutative semigroup in $\mathbb{R}^3$ while their convex hull is bounded by an infinite polyhedral surface whose vertices are integer (this surface is called the *sail* of the corresponding cubic irrational numbers).

The symmetry group of the sail in $\mathrm{SL}(3,\mathbb{Z})$ has been proved to be $\mathbb{Z}^2$, so that the quotient of the sail by the action of these symmetries turns out to be a two-torus divided into the images of the faces of the sail under the factorization (moreover, on each face that is a convex integer polygon, there were integer points which define distinguished points on the torus as well).

The problem is to calculate explicitly (e. g., using a computer and perhaps the data on cubic irrationalities published by B. N. Delone, D. K. Faddeev, and others) these torus triangulations with the images of the integer points upon them—e. g., for the first hundred of not so large matrices. The simplest example is the matrix $\begin{pmatrix} 3 & 2 & 1 \\ 2 & 2 & 1 \\ 1 & 1 & 1 \end{pmatrix}$ of the "three-dimensional golden section," the conventional golden section corresponds to the matrix $\left(\begin{smallmatrix} 2 & 1 \\ 1 & 1 \end{smallmatrix}\right)$. [1]

The interest of this "experimental" activity is due to the hope of noticing, in the result tables, some regularities which can become theorems in the sequel—for instance, on the statistics of such triangulation properties as the amount of triangular faces and other faces, the proportions of the integer lengths of the edges, those of the numbers of the edges with a common vertex, and so on. Then one

[1] The greater of the eigenvalues of this $2 \times 2$ matrix is $\phi + 2$, where $\phi = (\sqrt{5} - 1)/2$ is the golden section number.

would be able to compare such statistics with analogous statistics for other triangulations, e. g., for the sails of random octants or for the convex hulls of the sets of all the integer points in the domains bounded by random smooth surfaces, even by large spheres or ellipsoids. One may also compare the results with the partitions of the plane into the "Voronoĭ polygons" of random (arbitrary or integer) points: a Voronoĭ polygon of such a system of points is constituted by all the points on the plane for which the nearest point of the system is fixed.

By the way, while averaging in this problem, one can count the contributions of different polygons to the mean either with equal weights (which leads to an unjustifiably large contribution of small polygons since there are plenty of them) or with weights proportional to the polygon areas (which seems more reasonable to me).

Moreover, besides the distributions of the areas, the perimeter lengths, and the numbers of the vertices of the polygons (or the numbers of the sail edges with a common vertex), their joint distributions and correlations are also of interest, as well as the distributions of dimensionless parameters, e. g., the ratio of the area to the perimeter length squared (and the correlation between this ratio and the number of the vertices of the polygon).

# 2001

**2001-1** (A. Ortiz-Rodriguez)**.** Given a real polynomial $f$ in two variables $x$ and $y$, denote by $P(f)$ the set of parabolic points on the surface $\{z = f(x,y)\}$, i. e., the zero set of the Hessian $H[f] = f_{xx}f_{yy} - f_{xy}^2$. Determine the maximal number of

a) compact connected components,

b) all the connected components

of the set $P(f)$ over all the polynomials $f$ of given degree $d$. How can these connected components be mutually arranged? The first case where the answer is unknown is $d = 4$.

*The Hessian $H[f]$ of a polynomial $f$ of degree $d$ is a polynomial of degree $\leq m = 2d-4$. The Harnack inequality ensures that the parabolic set $P(f)$ has at most $N$ compact connected components, where*

$$N = \frac{(m-1)(m-2)}{2} + 1 = (d-3)(2d-5) + 1.$$

*For general polynomials of degree $2d-4$, this estimate is attained. However, it is not clear whether this estimate is attained for polynomials of degree $2d-4$ that are Hessians. The problem is the simplest case of* Hessian topology.

*There are examples of polynomials $f$ of degree $d$ for which the number of compact connected components of $P(f)$ is at least*

$$\frac{(d-1)(d-2)}{2}.$$

*So, for $d$ large, the maximal number of compact connected components of $P(f)$ lies asymptotically between $d^2/2$ and $2d^2$.* What is the true asymptotic of this number?

Similar questions on the parabolic curves are also open for such surfaces in $\mathbb{R}^3$ as the graphs of rational functions and for the graphs of the odd degree roots of real polynomials in two variables, as well as for the graphs of other single-valued real algebraic functions of a fixed degree $d$.

**2001-2** (A. Ortiz-Rodriguez)**.** Given a smooth algebraic surface $M \subset \mathbb{R}\mathrm{P}^3$, denote by $P(M)$ the set of parabolic points on $M$. Determine the maximal number of

a) connected components of the set $P(M)$ diffeomorphic to $\mathbb{S}^1$,

b) all the connected components of the set $P(M)$

over all the smooth surfaces $M$ of given degree $d$. How can these connected components be mutually arranged?

*This problem is a generalization of the previous one. It is known that the number of connected components of $P(M)$ diffeomorphic to $\mathbb{S}^1$ is at most*

$$10d^3 - 28d^2 + 4d - 3.$$

*On the other hand, there are examples of surfaces $M$ of degree $d$ for which the number of connected components of $P(M)$ diffeomorphic to $\mathbb{S}^1$ is at least*

$$\frac{d(d-1)(d-2)}{2}.$$

*So, for $d$ large, the maximal number of connected components of $P(M)$ diffeomorphic to $\mathbb{S}^1$ lies asymptotically between $d^3/2$ and $10d^3$.* What is the true asymptotic behavior of this number?

**2001-3.** Let $D$ be a real number and $(r, \varphi)$ polar coordinates in the real plane. Denote by $\mathrm{Hyp}(D)$ the set of smooth functions $F : \mathbb{S}^1 \to \mathbb{R}$ such that the homogeneous function $f(r,\varphi)=r^D F(\varphi)$ of degree $D$ is hyperbolic, i. e., its second quadratic form $d^2 f$ is of signature $(+,-)$ everywhere for $r>0$.

*For $D \geq 0$ integer, $f$ is a homogeneous polynomial of degree $D$ in $x = r\cos\varphi$, $y = r\sin\varphi$ if and only if $F$ is a trigonometric polynomial of degree $D$ and $F(\varphi+\pi) \equiv (-1)^D F(\varphi)$.*

Determine the connected components a) of the set $\mathrm{Hyp}(D)$ b) of the subset $\mathrm{Hyp}_{\mathrm{Pol}}(D)$ of $\mathrm{Hyp}(D)$ corresponding to $f$ polynomial (for $D \geq 0$ integer).

*The set* $\mathrm{Hyp}_{\mathrm{Pol}}(4)$ *is connected* (V. I. Arnold, F. Aicardi), *while the set* $\mathrm{Hyp}_{\mathrm{Pol}}(6)$ *consists of at least two connected components* (ARNOLD V.I. Astroidal geometry of hypocycloids and the Hessian topology of hyperbolic polynomials. *Russian Math. Surveys*, 2001, **56**(6), 1019–1083; Moscow: Moscow Center for Continuous Mathematical Education Press, 2001 (in Russian)). *Conjecturally, the number of connected components of* $\mathrm{Hyp}_{\mathrm{Pol}}(D)$ *grows like* $\mathrm{const}\cdot D$ *as* $D \to \infty$. *The set* $\mathrm{Hyp}(D)$ *of smooth functions has infinitely many connected components. In the polynomial case, even the number of connected components of the subset* $\mathrm{Hyp}_{\mathrm{Pol}}(D)$ *is unknown, already for* $D=6$.

**2001-4.** Let $g: \mathbb{S}^1 \to \mathbb{R}$ be a smooth function. Its *caustic* is by definition the plane curve

$$C = \bigl\{(A,B) \in \mathbb{R}^2 \bigm| \text{the function } \varphi \mapsto g(\varphi) + A\cos\varphi + B\sin\varphi \text{ is non-Morse}\bigr\}$$

(see ARNOLD V. I. Astroidal geometry of hypocycloids and the Hessian topology of hyperbolic polynomials. *Russian Math. Surveys*, 2001, **56**(6), 1019–1083; Moscow: Moscow Center for Continuous Mathematical Education Press, 2001 (in Russian)). Recall that $G: \mathbb{S}^1 \to \mathbb{R}$ is said to be *non-Morse* if there exists a point $\phi \in \mathbb{S}^1$ such that $G'(\phi) = G''(\phi) = 0$. *For instance, for $g(\varphi) = \cos(2\varphi)$ the caustic $C$ is the astroid*

$$A = -4\cos^3\phi, \quad B = 4\sin^3\phi \qquad (\phi \in \mathbb{S}^1).$$

*In this parametric equation of $C$, $\phi$ is just the point where both the first and second derivatives of* $\cos(2\varphi) + A\cos\varphi + B\sin\varphi$ *vanish.*

What curves on $\mathbb{R}^2$ are the caustics of periodic functions? *That $C$ is a caustic imposes some restrictions on the curve $C$:*

1. *A caustic has at least* 4 *cusps.*
2. *The number of cusps is even.*
3. *The alternated length of a caustic (we change sign after each cusp) is zero.*

4. *Through any point of the plane, there pass at least two tangents to the caustic.*

5. *A caustic possesses no inflection points.*

*One can also show that the caustic of a trigonometric polynomial is an algebraic curve of genus zero (see the paper cited above).*

The problem is to describe the set of restrictions complete in the following sense: each curve satisfying those restrictions is a caustic.

This problem can be generalized in several directions. First, one may consider the so-called *hypercaustic* in $\mathbb{R}^{2n}$, i. e., the curve

$$C = \Big\{ (A_1, \dots, A_n, B_1, \dots, B_n) \in \mathbb{R}^{2n} \;\Big|\; \text{the function}$$
$$G : \varphi \mapsto g(\varphi) + \sum_{k=1}^{n} \big[A_k \cos(k\varphi) + B_k \sin(k\varphi)\big] \text{ has a critical point } \phi$$
$$\text{where the derivatives } G' = G'' = \cdots = G^{(2n)} = 0 \text{ all vanish} \Big\}.$$

Second, instead of the circle $\mathbb{S}^1$ and trigonometric polynomials

$$\sum_k \big[A_k \cos(k\varphi) + B_k \sin(k\varphi)\big],$$

one can consider respectively an arbitrary curve $\Gamma \subset \mathbb{R}^2$ and polynomials on $\mathbb{R}^2$ restricted to $\Gamma$.

Apart from that, it is also possible to consider exact Lagrangian submanifolds in $T^*\mathbb{S}^1$ in place of functions (a closed curve $L \subset T^*\mathbb{S}^1$ is called an exact Lagrangian submanifold if the difference between $L$ and the zero section is the boundary of a chain of area zero).

**2001-5.** Set

$$\Sigma^{2n-1} = \Big\{ (A_1, \dots, A_n, B_1, \dots, B_n) \in \mathbb{R}^{2n} \;\Big|\; \text{the trigonometric polynomial}$$
$$\cos[(n+1)\varphi] + \sum_{k=1}^{n} \big[A_k \cos(k\varphi) + B_k \sin(k\varphi)\big] \text{ is non-Morse} \Big\}. \quad (1)$$

The discriminant $\Sigma$ divides $\mathbb{R}^{2n} = \{(A_1, \dots, A_n, B_1, \dots, B_n)\}$ into $n+1$ domains $G_2$, $G_4$, …, $G_{2n+2}$ according to the number of critical points of polynomial (1). Explore the topology and the singularities combinatorics of these domains.

*The domain $G_{2n+2}$ of trigonometric M-polynomials* (1) (*in the terminology of I. G. Petrovskiĭ*) *was examined in the paper* ARNOLD V. I. Topological classification of real trigonometric polynomials and cyclic serpents polyhedron. In: The Arnold–Gelfand Mathematical Seminars: Geometry and Singularity Theory. Editors: V. I. Arnold, I. M. Gelfand, V. S. Retakh and M. Smirnov. Boston, MA: Birkhäuser, 1997, 101–106; *the Russian translation in*: Vladimir Igorevich Arnold. Selecta–60. Moscow: PHASIS, 1997, 619–625. *This particular domain has a convex polyhedral model* (*simply a square for $n = 1$*). *It is conjectured that all these domains have polyhedral models in terms of the affine Coxeter group mirrors, similar to the descriptions of the swallowtails pyramids polyhedral models in terms of the Springer cones decompositions into the Weyl chambers for the linear Coxeter group case. But this conjecture is not confirmed yet even for small values of n.*

**2001-6.** Let $h: \mathbb{R}_+ \to \mathbb{R}_+$ be a smooth function, $h(R) > 0$ for $R > 0$ and $h(0) = 0$. Consider a curve $F$ on $\mathbb{R}^2$ with a semicubic cusp $O$. Denote by $\ell_P$ the part of the normal to $F$ at point $P$ where $F$ is smooth containing the center of curvature. Let $R_P$ be the radius of curvature of $F$ at $P$. Let $\Pi_P$ be the parabola with vertex $P$ and axis $\ell_P$ whose radius of curvature at the vertex is equal to $h(R_P)$.

Study the envelope of the family of the parabolas $\{\Pi_P\}$. *If $F$ is an astroid and $h(R) = \frac{2}{3}R$, then the family $\{\Pi_P\}$ has a smooth envelope which is tangent to $F$ at cusp $O$.* Does the family of the parabolas $\{\Pi_P\}$ possess a smooth envelope for other curves $F$ (for, possibly, other functions $h$)? *If the envelope is smooth at $O$, it is tangent to $F$ there.*

Similar problems are also interesting for the families of generic smooth curves instead of the parabolas (of curves having the same properties of the tangency to $F$ and of the curvature radius at the tangency points).

# 2002

**2002-1.** Let $f: \mathbb{R}^2 \to \mathbb{R}$ be a polynomial of degree $D$. Find the maximal number of connected components and the maximal number of closed components of the *parabolic curve* $\mathrm{Par}(f)$ of its graph (where $f_{xx}f_{yy} = f_{xy}^2$):

$$b_0(\mathrm{Par}(f)) = ?, \quad b_1(\mathrm{Par}(f)) = ?.$$

*Even for $D = 4$, it is not known whether $b_1$ attains the value 4, and the constants C in the lower and the upper bounds for large degrees D, $b_1 \sim CD^2$, differ by a factor of order of* 4:

$$(D-1)(D-2)/2 \leq b_1 \leq (2D-5)(D-3)+1.$$

**2002-2.** Let $M \subset \mathbb{RP}^3$ be a *smooth algebraic surface* of degree $D$. Find the maximal number of connected components of its parabolic line.

*The constants C in the lower and the upper bounds $CD^3$ differ by a factor of order of* 20:

$$D(D-1)(D-2)/2 \leq b_0 \leq 10D^3 - 28D^2 + 4D + 3.$$

*The lower estimates in problems 2002-1 and 2002-2 mean the existence of surfaces with many closed parabolic curves.*

**2002-3.** Let $f : \mathbb{S}^1 \to \mathbb{R}$ be a smooth function; it is called *D-hyperbolic* if the second differential $d^2 f$ of the homogeneous function $f(x,y) = r^D F(\varphi)$ (where $x = r\cos\varphi$, $y = r\sin\varphi$) is hyperbolic (has signature $(+,-)$) everywhere in $\mathbb{R}^2 \setminus \{0\}$.

Find the connected components of the space of $D$-hyperbolic functions: is the index (equal to the number of rotations of the cross $d^2 f = 0$ when the point $(x,y)$ makes one revolution around the origin) the unique invariant of the connected component? *The set of the values attained by the index is infinite and unbounded below* (*but bounded above*).

**2002-4.** For the polynomial case (where $F$ is a trigonometric polynomial and $f$ is an ordinary homogeneous polynomial of degree $D$), find the number of connected components of the set of $D$-hyperbolic polynomials. Is it growing linearly with $D$ when the latter is high?

**2002-5.** Consider a controlled dynamical system $\dot{x} = v(x,u)$, where $x$ is a point of a compact phase manifold $M$ and $u$ belongs to a compact controlling parameter manifold $U$. Let $f : M \to \mathbb{R}$ be a smooth goal function.

Study the *mean optimization problem*, maximizing the time average $\widehat{f} = \lim_{T\to+\infty} T^{-1} \int_0^T f(x(t))\,dt$ by a clever choice of the control $u(t)$ (and eventually of the initial state $x(0)$).

If the problem (i. e., the pair formed by $v$ and $f$) depends generically on some exterior parameters, then the optimization strategy and the optimal average might have singularities ("phase transitions") at the points of a *hypersurface of phase transitions* in the manifold $P$ of the values of the exterior parameter.

Find the *generic phase transitions in the mean optimization problem,* at least when the dimensions of the manifolds $M$, $U$, $P$ are not large. *The problem is open even when all these manifolds are* 1*-dimensional where there are already some nontrivial stable singularities* (*see the paper* ARNOLD V. I. Optimization in mean and phase transitions in controlled dynamical systems. *Funct. Anal. Appl.*, 2002, **36**(2), 83–92).

**2002-6.** Let $f : M \to \mathbb{R}$ be a smooth function on a compact Riemannian manifold, and $0 < r < R < \infty$ be two smooth functions on $M$. Study the *mean value optimization problem* for the space average

$$\widehat{f} = \left(\int_M f(x)\rho(x)\,dx\right) \Big/ \left(\int_M \rho(x)\,dx\right)$$

for the mass distribution defined by a density function $\rho$ with respect to the Euclidean volume element $dx$ provided that this density is restricted by the inequalities $r \leq \rho \leq R$ everywhere on $M$.

Study the generic phase transitions for the case where $f$, $r$, and $R$ depend smoothly on exterior parameters.

*It is known that the optimal strategy is* $\{$*to choose* $\rho = r$ *where* $f(x) < c$ *and* $\rho = R$ *where* $f(x) > c$ *for some constant* $c\}$, *but the study of phase transitions requires the investigation of the influence of some strange logarithmic singularities and of their regularizations in the case of even-dimensional manifolds M, as in many physical problems. See the paper* ARNOLD V. I. On a variational problem related to the phase transitions of the averages in controlled dynamical systems. In: Nonlinear Problems in Mathematical Physics I. In honour of Professor O. A. Ladyzhenskaya. Editors: M. Sh. Birman, S. Hildebrandt, V. A. Solonnikov and N. N. Ural'tseva. Dordrecht: Kluwer Acad. Publ., 2002, 23–34 (Internat. Math. Ser., 1).

**2002-7.** Let $u_0 : M^2 \to \mathbb{R}$ be a smooth "initial" function on a Riemannian manifold $M$ (the case of a 2-dimensional ball $B^2$ is already relevant). Study the *minimization problem for the Dirichlet integral* $\int_M (\nabla u)^2\,dx$, where the function $u$ is obtained from the initial function $u_0$ by an area-preserving diffeomorphism of $M$ onto itself (by an "*incompressible fluid motion*").

*The extremal function $u$ is smooth if the smooth initial mountain $u_0 : B^2 \to \mathbb{R}$, vanishing on the boundary of the ball, has just one nodegenerate (Morse) maximum inside the ball. In this case, the extremal function $u$ is the symmetrization of $u_0$ (depending only on the distance from the center of the ball).*

*But for the initial smooth mountain $u_0$ having (like the Elbrus mountain) two local maxima separated by a saddle point, the extremal function seems to have a singularity of type $|x|$ along a curve with unknown extremal function singularities at its endpoints. The problem is to study such singularities for generic $u_0$.*

**2002-8.** The $(C,B,A)$-*permutation* of the set $\{1,2,\ldots,n\}$ transports to the last place the subset $A = \{1,2,\ldots,a\}$ preceded by the transported set $B = \{a+1, \ldots, a+b\}$ while the starting position is occupied by $C = \{a+b+1, \ldots, n\}$.

Some of these $(n-1)(n-2)/2$ permutations permute *cyclically* (like the addition of a constant to the residues $\bmod n$), and some of these cyclic permutations are *transitive* (like the addition of the constant 1).

Find the proportion of both the cyclic and the transitive cyclic permutations among the $(C,B,A)$-permutations for large $n$.

More generally, starting from a permutation of $k$ elements, one defines a permutation of the set $\{1,\ldots,n\}$ from its decomposition into $k$ segments $\{a_i + 1, \ldots, a_{i+1} - 1\}$. The problem is to study the statistics of the Young diagrams formed by the cycle lengths of the resulting permutations, for the case of large $n$ and random decompositions of $n$ into $k$ parts.

**2002-9.** A mapping $\mathbb{C}^n \to \mathbb{C}^n$ (or $\mathbb{C}\mathrm{P}^n \to \mathbb{C}\mathrm{P}^n$) is called a *pseudocomplex* mapping if it sends complex subspaces to complex subspaces (one may consider separately the cases of vector, affine or projective subspaces—all the three versions are interesting).

A real diffeomorphism $\mathbb{C}\mathrm{P}^2 \to \mathbb{C}\mathrm{P}^2$ is pseudocomplex if and only if either it, or its product with the complex conjugation, is a complex projective mapping (and similarly for the other versions and other $n$'s).

Do there exist other pseudocomplex homeomorphisms? Other pseudocomplex bijections?

*These questions should have been studied by Hilbert as a part of axiomatic projective geometry, but his school seems to have missed these foundational problems.*

**2002-10.** To formulate quaternionic versions of the questions in problem *2002-9*, one should distinguish the left subspaces and the right subspaces. I would suggest studying those mappings which send left and right subspaces onto left and right subspaces (with a left one sent onto a right one also permitted).

**2002-11.** The complexification and quaternionization paradigm had been used by me many times starting from its invention in ARNOLD V. I. Distribution of ovals of real plane algebraic curves, involutions of four-dimensional smooth manifolds, and the arithmetic of integral quadratic forms. *Funct. Anal. Appl.*, 1971, **5**(3), 169–176; *the Russian original is reprinted in*: Vladimir Igorevich Arnold. Selecta–60. Moscow: PHASIS, 1997, 175–187 (see, for instance, ARNOLD V. I. Polymathematics: is mathematics a single science or a set of arts? In: Mathematics: Frontiers and Perspectives. Editors: V. I. Arnold, M. Atiyah, P. Lax and B. Mazur. Providence, RI: Amer. Math. Soc., 2000, 403–416, and ARNOLD V. I. Symplectization, complexification and mathematical trinities. In: The Arnoldfest. Proceedings of a conference in honour of V. I. Arnold for his sixtieth birthday (Toronto, 1997). Editors: E. Bierstone, B. A. Khesin, A. G. Khovanskiĭ and J. E. Marsden. Providence, RI: Amer. Math. Soc., 1999, 23–37 (Fields Inst. Commun., 24)).

For instance, it is now proved that the complex version of the tetrahedron is the octahedron: ${}^{\mathbb{C}}A_3 = B_3$, see ARNOLD V. I. Complexification of tetrahedron and pseudoprojective transformations. *Funct. Anal. Appl.*, 2001, **35**(4), 241–246.

Now the problem is to prove my old conjecture that its quaternionic version is the icosahedron:

$$ {}^{\mathbb{H}}A_3 = H_3, \quad {}^{\mathbb{C}}B_3 = H_3. $$

Perhaps one should start with the easier plane versions:

$$ {}^{\mathbb{C}}A_2 = B_2, \quad {}^{\mathbb{H}}A_2 = H_2, \quad {}^{\mathbb{C}}B_2 = H_2 $$

relating the symmetry groups of the triangle, the square, and the pentagon.

*The difficulty of all this subject lies in its nonmathematical character: the problem is to find the definition of the informal quaternionization operation rather than to prove any ready mathematical statement.*

**2002-12.** The *caustic of a periodic function* $g : \mathbb{S}^1 \to \mathbb{R}$ is the curve in the plane $\mathbb{R}^2$ of the functions

$$ G_{A,B} : \mathbb{S}^1 \to \mathbb{R}, \quad G_{A,B}(\varphi) = g(\varphi) + A\cos\varphi + B\sin\varphi, $$

consisting of those functions which are not Morse:

$$\{(A,B)\in\mathbb{R}^2 : \exists\varphi : G'_{A,B}(\varphi)=G''_{A,B}(\varphi)=0\}.$$

The caustics of generic periodic functions have many peculiar properties: for instance, each caustic has at least four cusps, and its alternative length (the alternating sum of the lengths of its segments between the cusps) vanishes. The cusps of a caustic having just 4 cusps form a parallelogram (and the barycenters of the odd and of the even cusps coincide if there are more than 4 cusps).

The problem is to replace smooth periodic functions in this theory with exact Lagrangian submanifolds of the phase cylinder $T^*\mathbb{S}^1$. Such a submanifold corresponding to a function is the graph of its differential. A general Lagrangian submanifold needs not be a section of the cotangent bundle, and the graph of the corresponding "*multivalued potential*" function needs not be an immersed curve: it may have cusps.

It is interesting to understand, which would be the four-cusp property version for the caustics of such exact Lagrangian submanifolds, and what would happen to the Sturm–Hurwitz theorem on the zeros of Fourier series (being the infinitesimal version of the caustics' cusps theorem) for such extended "multivalued periodic functions."

**2002-13.** The theory of caustics of periodic functions and Lagrangian submanifolds discussed in problem *2002-12* depends on the functions $x=\cos\varphi$ and $y=\sin\varphi$ on the circle $x^2+y^2=1$. Replacing the circle with a different curve, say, with an algebraic curve $C$: $f(x,y)=0$, and $g$ with the restriction to $C$ of a function on the plane, say, of a polynomial $P(x,y)$, we define the *C-caustic* (as the curve of non-Morse restrictions of the functions $P+Ax+By$ to $C$).

The problem is now to extend the Sturm–Hurwitz theorems on Fourier series (as well as their extensions described in problem *2002-12* and, in more detail, in the works ARNOLD V. I. Astroidal geometry of hypocycloids and the Hessian topology of hyperbolic polynomials. *Russian Math. Surveys*, 2001, **56**(6), 1019–1083; Moscow: Moscow Center for Continuous Mathematical Education Press, 2001 (in Russian)) to the $C$-caustics associated with more general curves $C$ than the circle used in problem *2002-12.*

**2002-14.** Study the triangulations of the torus $\mathbb{T}^2$ associated with cubic algebraic number fields (by the theory of higher-dimensional continued fractions). One starts with a matrix $A \in \mathrm{SL}(3,\mathbb{Z})$ having 3 positive eigenvalues. The 3 corresponding invariant planes divide $\mathbb{R}^3$ into 8 invariant octants. Each open octant contains the semigroup of its integer points. The boundary of the convex hull of this set of integer points, $\mathbb{Z}^3 \cap$ octant, is called the *sail*. The sail is invariant under $A$. It is an (infinite) polyherdal surface whose faces are bounded by convex compact polygons. It had been proved (by Dirichlet and H. Tsuchihashi—see TSUCHIHASHI H. Higher-dimensional analogues of periodic continued fractions and cusp singularities. *Tôhoku Math. J., Ser.* 2, 1983, **35**(4), 607–639) that the sail is invariant under the action of the commutative subgroup $\mathbb{Z}^2$ of $\mathrm{SL}(3,\mathbb{Z})$ formed by the matrices with the same eigenvalues.

The torus dealt with in this problem is the quotient space

$$\mathbb{T}^2 = (\text{the sail of } A)/\mathbb{Z}^2.$$

It is divided into the images of the faces of the sail under factorization. Each image contains some "integer points" (the images of the integer points of the face). Thus, we have associated a geometric object to $A$: a decomposition of $\mathbb{T}^2$ into "convex polygons" containing "integer points."

The problem is to understand which decompositions of $\mathbb{T}^2$ (and which sets of "distinguished integer points") can be obtained in this way from various matrices $A$.

**2002-15.** While comparing problem *2002-14* with the situation for *ordinary continued fractions* (where there are no restrictions on "a triangulation of the torus" and on "its distinguished points," since every finite sequence of integers is a period of some quadratic irrational number), I must repeat the old interesting problem (see problem *1993-11*) to compare the statistics of elements of the period of a random irrational number with that of the eigenvalues of a random matrix in $\mathrm{SL}(2,\mathbb{Z})$ with real eigenvalues.

I mean first to consider integer points $(p,q)$ defining quadratic equations $x^2+px+q=0$ whose roots are real, such that $p^2+q^2 \leq N$. Among the elements of the period of the continued fractions of the roots of this equation, we consider the proportion of 1's, of 2's, and so on. The limit of the proportion of $k$'s for $N$ tending to infinity is called the *k's statistic* for random quadratic irrational numbers.

The statistic for periods of the eigenvalues of random matrices is defined similarly, starting with those integer matrices $\left(\begin{smallmatrix} a & b \\ c & d \end{smallmatrix}\right)$ for which $ad - bc = 1$, the eigenvalues are real and $a^2+b^2+c^2+d^2 \leq N$.

**2002-16.** The modular group $\mathrm{SL}(2,\mathbb{Z})$ acts on the Lobachevskian plane as well as on the de Sitter world (the exterior domain of the disk of the Klein model for the Lobachevskian plane), see problem *1996-15* and book ARNOLD V. I. Arithmetics of Binary Quadratic Forms, Symmetry of Their Continued Fractions, and Geometry of Their de Sitter World. Moscow: Moscow Center for Continuous Mathematical Education Press, 2002; *Bol. Soc. Brasil. Mat.* (*N. S.*), 2003, **34**(1), 1–42. The orbits of this group are discrete in the Lobachevskian plane (but accumulate to the "absolute" circle) and are everywhere dense in the de Sitter world.

a) How precisely does an orbit inside the "absolute" circle accumulate to this circle?

b) How is an orbit outside the "absolute" circle distributed? Is there any kind of ergodicity and equipartition, as for the products of 2 rotations of the sphere or of the plane (see problems *1963-6–1963-12* for the details)?

The relation of the Lobachevsky and de Sitter geometry to the arithmetic and albegra of groups $\mathrm{SL}(2,\mathbb{Z}_p)$ is described in the paper ARNOLD V. I. Fermat dynamics, matrix arithmetics, finite circles and finite Lobachevsky planes. *Funct. Anal. Appl.*, 2004, **38**(1), 20 pp., where, for instance, the finite Lobachevky plane $\bmod p$ and "upper" half-plane containing $p(p-1)/2$ points are treated.

**2002-17.** The modular group $\mathrm{SL}(2,\mathbb{Z})$ acts on the set $\mathbb{Z}^3$ of binary quadratic forms $mx^2+ny^2+kxy$ with integer coefficients $m$, $n$, $k$. The number $h(D)$ of orbits of this action on the set of the forms with a fixed negative value of the determinant $D=4mn-k^2$ is finite, see Theorem 13 in the book ARNOLD V. I. Arithmetics of Binary Quadratic Forms, Symmetry of Their Continued Fractions, and Geometry of Their de Sitter World. Moscow: Moscow Center for Continuous Mathematical Education Press, 2002. Explore the function $h(D)$. What is the asymptotic behavior of $h(D)$ as $D \to -\infty$?

**2002-18.** A binary quadratic form $f(x,y)=mx^2+ny^2+kxy$ with integer coefficients $m$, $n$, $k$ is said to be *perfect* if the set $S=f(\mathbb{Z}^2)\subset\mathbb{Z}$ of the values of this form on $\mathbb{Z}^2$ is a multiplicative semigroup (i. e., $uv\in S$ whenever $u\in S$ and $v\in S$). Perfect quadratic forms were studied in the book ARNOLD V. I. Arithmetics of Binary Quadratic Forms, Symmetry of Their Continued Fractions, and Geometry of Their de Sitter World. Moscow: Moscow Center for Continuous Mathematical Education Press, 2002. What is the probability that a randomly chosen binary quadratic form with integer coefficients is perfect? If this form is perfect, what can one say about the structure of the semigroup of its values? *The product of three values is always a value, as it was proved in the book quoted above.*

**2002-19.** If positive integers $a$ and $n > 1$ are mutually prime then $a^{\varphi(n)} \equiv 1 \mod n$ where $\varphi(n)$ is the number of positive integers that are less than $n$ and mutually prime with $n$ (the Euler theorem). Explore the following problem: For what divisors $N$ of $\varphi(n)$ does the relation $a^{\varphi(n)/N} \equiv 1 \mod n$ take place? The relation $a^{\varphi(n)/N} \equiv -1 \mod n$? The case $a = 2$ (and $n$ odd) is already highly nontrivial.

**2002-20.** Let two positive integers $a$ and $n$ be mutually prime. To what extent is the sequence $t \mapsto a^t \mod n$ ($t \geq 1$ integer) random?

**2002-21.** Examine the sequence $\varphi(n)/n$ ($n > 1$ integer) where the Euler function $\varphi(n)$ is the number of positive integers that are less than $n$ and mutually prime with $n$.

According to Gauss, the probability of that two randomly chosen integers are mutually prime is equal to $6/\pi^2$. This implies that $\varphi(n)/n$ tends to $6/\pi^2$ as $n \to \infty$ in a certain weak sense [which is also confirmed by calculations of $\varphi(n)/n$ for not so large $n$]. Determine a rigorous meaning of this statement and prove it. What are the "oscillations" and the "variance" (and other probabilistic characteristics) of the sequence $\varphi(n)/n$?

**2002-22** (the Fermat–Euler dynamical system)**.** Let $n$ be a large odd integer, and let $\Gamma = \Gamma(n)$ be the set of the $\varphi(n)$ residues $\bmod n$ that are mutually prime with $n$, $\varphi$ being the Euler function.

The doubling mapping $x \mapsto 2x$ acts on $\Gamma$ with $N$ orbits of equal lengths, $l = \varphi(n)/N$. Is the set of $l$ residues forming one orbit asymptotically random if $n$ becomes large?

*For a truly random finite sequence of $l$ elements of a set of $m$ elements, the absence of any repetition seems rather probable for short sequences, where $l^2$ is small with respect to $m$, and seems rather improbable for long sequences, $l^2$ being large with respect to $m$. Indeed, the number of choices of $l$ distinct elements is $C = m(m-1)(m-2)\cdots(m-l+1)$, while the total number of the unrestricted choices is $T = m^l$. Hence, $C/T = \prod_{k=0}^{l-1}(1-k/m)$, $\ln(C/T) = \sum_{k=0}^{l-1}\ln(1-k/m) \sim -\sum_{k=0}^{l-1} k/m \sim -l^2/2m$.*

*Thus, the ratio of the orbit period $l = \varphi(n)/N$ and the number $\varphi(n)$ of all the possible values of elements of the orbit might indicate the randomness degree* *of the geometric progression* $\{2^i\}$ mod $n$: *large values of* $l^2$ *with respect to* $\varphi(n)$ (*i. e., the smallness of the divisor* $N(n)$ *of* $\varphi(n)$ *with respect to the square root of* $n$) *would indicate some nonrandomness.*

*The calculations of* $N(n)$ *for odd integers* $n \leq 512$ *show rather an average linear growth* (*say,* $N(511) = 48$ *while* $\varphi(511) = 432$).

*For more details on the Fermat–Euler dynamical system see*: ARNOLD V. I. Fermat–Euler dynamical systems and the statistics of arithmetics of geometric progressions. *Funct. Anal. Appl.*, 2003, **37**(1), 1–15; ARNOLD V. I. Ergodic and arithmetic properties of geometric progression's dynamics. *Moscow Math. J.*, 2004, to appear; ARNOLD V. I. Euler Groups and the Arithmetic of Geometric Progressions. Moscow: Moscow Center for Continuous Mathematical Education Press, 2003 (in Russian); ARNOLD V. I. Topology and statistics of formulae of arithmetics. *Russian Math. Surveys*, 2003, **58**(4), 637–664; ARNOLD V. I. The topology of algebra: combinatorics of squaring. *Funct. Anal. Appl.*, 2003, **37**(3), 177–190; ARNOLD V. I. Fermat dynamics, matrix arithmetics, finite circles and finite Lobachevsky planes. *Funct. Anal. Appl.*, 2004, **38**(1), 20 pp.

# 2003

I hope the problems of the list below might be useful even for beginners, being however open and allowing both experimental discoveries of new facts and creation of general theories.

**2003-1. Fermat–Euler statistics and number-theoretic turbulence.** Consider the sequence $\{a^t\}$, $t = 1, 2, \ldots$, of residues mod $n$ (where the integers $a$ and $n$ are relatively prime, $(a, n) = 1$). Fermat and Euler proved that this sequence is periodic. Denote by $T(n, a)$ its (minimal) period such that $a^T \equiv 1 \pmod{n}$. It behaves very irregularly.

**Examples:** $T(509, 2) = 508$, $T(511, 2) = 9$.

Billions of experiments (mostly by F. Aicardi) showed that *the average growth rate of* $T(n)$ (as $n \to \infty$) *is asymptotically* $Cn/\log n$. For smaller values of $n$, the slower growth rate $Cn^{7/8}$ was observed.

The problem is to prove (or disprove) this asymptotic behavior (at least for $a = 2$, $n$ being odd).

The words "average growth rate of $A(n)$ is asymptotically equal to $B(n)$" mean that

$$\lim_{n\to\infty} \frac{\sum\limits_{k=1}^{n} A(k)}{\sum\limits_{k=1}^{n} B(k)} = 1.$$

One also says in this case that *$B$ is the Cesàro asymptotic of $A$.*

**Example:** The Cesàro asymptotic of $\sin^2(\pi n/2)$ is $1/2$.

A discussion of this problem can be found in the article: ARNOLD V. I. Topology and statistics of formulae of arithmetics. *Russian Math. Surveys*, 2003, **58**(4), 637–664. This article also contains a discussion of the relation of the study of this average growth to the discovery of turbulence laws by Kolmogorov: his methods were used in Aicardi's work.

**2003-2. Randomness of arithmetic progressions.** Consider the sequence $\{at\}$, $t = 1, 2, \ldots, T$, of residues $\bmod n$ ($a$ and $n$ being relatively prime, $(a,n) = 1$). The problem is to study the statistics of the distribution of this sequence of $T$ elements in the finite circle $\mathbb{Z}_n = \mathbb{Z}/n\mathbb{Z}$ of residues $\bmod n$.

As a randomness characteristic of a set of $T$ points of the finite circle of $n$ points, we consider the sum of the squares of the lengths of the $T$ arcs into which the circle is divided by $T$ chosen points,

$$R = \sum_{i=1}^{T} a_i^2, \quad \sum a_i = n.$$

To avoid the influence of the scale $n$, we consider the reduced dimensionless number

$$r = \frac{R}{n^2}, \quad \frac{1}{T} \leq r \leq 1.$$

To eliminate the influence of the parameter $T$, we reduce the randomness characteristic once more, dividing $r$ by its minimal value:

$$s = \frac{r}{r_{\min}} = Tr, \quad 1 \leq s \leq T.$$

These characteristics were introduced and studied in the article quoted in problem *2003-1*.

**Example:** The minimal value $s_0 = 1$ of the binormalized randomness parameter $s$ is attained by the choice of the $T$ vertices of a regular $T$-gon (an army line distribution of points at equal distances). The maximal value $s_1 = T$ corresponds to the degenerate choice of a cluster of $T$ points at the same place.

The random choice (of independent uniformly distributed points in a circle) leads to the "freedom-liking" value $s_* = 2T/(T+1)$ of the birandomness parameter, which is close to 2 when there are many points.

Measuring the value of $s$ for a given set, one can evaluate some kind of degree of its randomness. Namely, the observation $s < s_*$ is a sign of some mutual repulsion of points of the set (reaching the minimal value $s_0 = 1$ of the parameter $s$ for maximal repulsion, leading to an army line formation).

Similarly, the observation of larger values of the binormalized randomness parameter, $s > s_*$, is a sign of some mutual attraction of points of the set (reaching the maximal value of the parameter $s = T$ for the strongest attraction, leading to a cluster formation).

Intermediate values of the binormalized randomness parameter $s$ (close to $s_*$) are a sign of independence of the $T$ points of the set which is, in this case, "more random" than in the cases of repulsion or attraction.

It is interesting to compare the values of the binormalized randomness parameter $s$ for different sets of residues mod $n$. The experiments with the full periods $T$ of geometric progressions of residues, discussed above in problem *2003-1*, have shown mostly some repulsion ($s \approx 1.5$), but no theorem on such a repulsion is known.

Problem *2003-2* is *to study* (at least experimentally) *the values of the binormalized randomness parameter $s$ for the $T$ points of the arithmetic progression $\{at\}$, $t = 1, 2, \ldots, T$, of residues modulo $n$ (integers $a$ and $n$ being relatively prime).*

**Remark:** One may predict a significant difference in the answers depending on the choice of the length $T$ of the progression, at least in the following two cases.

1) One may choose $T$ randomly, say between 1 and $n/2$, and study the distribution of the values $s(n, a, T)$ for large values of $n$ and random independent choices of $a$ and $T$.

2) One may choose $T(n, a)$ to be one of the denominators of the continued fraction approximation of the number

$$\frac{a}{n} = x_0 + \cfrac{1}{x_1 + \cfrac{1}{x_2 + \cdots}} = [x_0, x_1, x_2, \ldots].$$

The approximation is provided by the truncated continued fraction

$$\frac{p_k}{q_k} = [x_0, x_1, \dots, x_k],$$

and the proposal is to choose $T = q_k$ (for some randomly chosen $k$).

One might conjecture that there is more repulsion and army line structure in the second case, but there are yet no theorems confirming it, and it would be interesting to verify this conjecture by numerical experiments.

One proposal is to calculate $s(n,a,k)$ for all the continued fraction $k$-approximations of all the fractions $a/n$, $1 \le n \le N$, trying to guess the behavior of the distribution of the values of $s$ for $N \to \infty$.

One might also try to suggest the natural science conjectures on these behaviors (in both cases 1) and 2)), using the Gauss–Kuz′min distribution of the elements $x_i$ of the continued fractions of random real numbers. It would be difficult to deduce the behavior of $s$ from this distribution in a rigorous way; therefore I suggest rather a semi-empirical study in which one uses such intuitively probable things as the independence of different prime numbers with no proof (which might rely on deep number theory involving the Riemann zeta function conjecture).

Nonrigorous deductions freely using these unproved properties of number theory sometimes would become correct proofs many years later (as it happened to Legendre's discovery of the distribution of prime numbers proved only by Hadamard and Vallée Poussin).

Continuing with numerical experiments, one might also study many other sequences of residues of, say, ($T$ subsequent prime numbers) mod$(n = 100)$. The choice $n = 100$ facilitates finding these residues in the tables of primes.

For the theory of continued fractions, one might see, for instance, the book ARNOLD V. I. Continued Fractions. Moscow: Moscow Center for Continuous Mathematical Education Press, 2001, 40 pp. (in Russian).

One might try to apply the Gauss–Kuz′min statistics of continued fractions to the study of geometric progressions of residues, for example, $\{2^t \pmod{n}\}$, reducing it to the arithmetic progression of logarithms $\{t \ln 2\}$ and investigating their distribution in the "random" intervals $\big(\ln n + \ln k; \ln n + \ln(k+1)\big)$ of lengths $b = \ln((k+1)/k) \approx 1/k$.

Supposing the randomness of the real number $(\ln 2)/b$ and applying its continued fraction statistics, one might come to some (empirical) conjectures on the distributions of elements of arithmetic progressions of logarithms in the above "random" interval of length $b$ (for random $k$). The properties of these distributions might be interpreted (while exponentiating) as those of the distributions of the residues $\{2^t \pmod{n}\}$ (which should then be compared with the numerically observed properties).

The proofs of the "randomness" properties conjectured for these studies might be much more difficult than their fearless applications, which might immediately lead to the conjectures to be verified numerically (ignoring missing foundations, such as "randomness" proofs, which might appear centuries later, as happened to the Legendre statistics of the prime numbers distribution).

**2003-3. Modular groups and their Kepler cubes.** Consider the group $G = \mathrm{SL}(2, \mathbb{Z}_p)$ consisting of the matrices of order 2, whose elements are residues mod $p$ and determinant is equal to 1. This group consists of $p(p^2 - 1)$ matrices.

We associate with a finite group a directed graph (called its *monad*), whose vertices are elements of the group and arrows lead just from each element to its square (one arrow leaving each vertex).

**Example:** The monad of the cyclic group $\mathbb{Z}_4$ of order 4 is

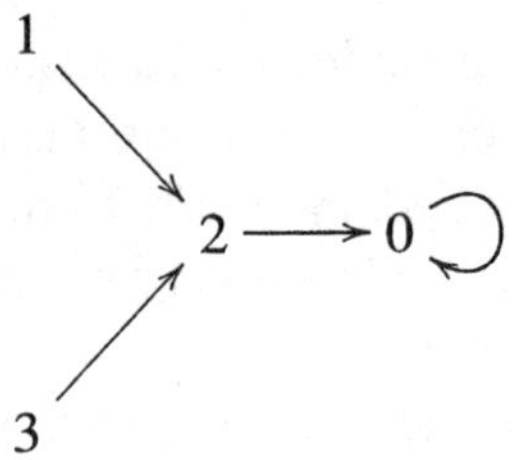


(denoting the elements additively as the residues mod 4).

**Problem:** Find the monad of the group $G = \mathrm{SL}(2, \mathbb{Z}_p)$ (for any prime number $p$).

**Example:** For $p = 3$, the monad $[G]$ consists of five components: one rooted tree $T_{1,1,6}$ and four components $A_2$.

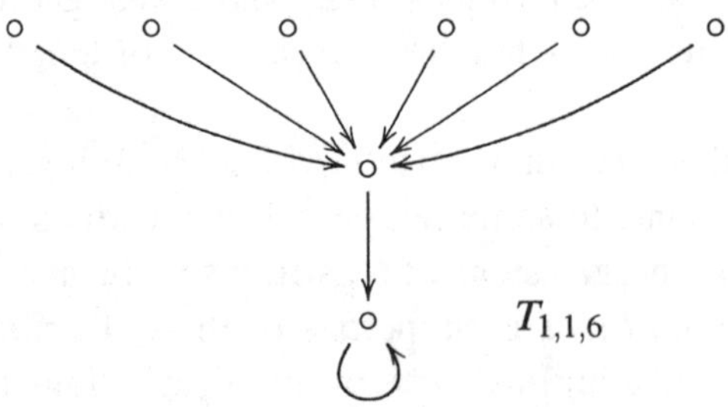

$T_{1,1,6}$

$A_2$

Each $A_2$-graph is a cycle of length 2 with trees of length 1 attached to its vertices. The $T$-component is a subgroup isomorphic to the group $\{\pm 1, \pm i, \pm j, \pm k\}$ of quaternionic units.

For $p = 5$, the monad $[G]$ consists of 17 components: one rooted tree with 32 vertices, 10 components $A_2$, and 6 components $A_4$ (the directed graph $A_m$ being a cycle of length $m$ with trees of length 1 attached to each of its vertices, thus having in all $2m$ vertices):

$$[G] = T_{1,1,30} \sqcup (10A_2) \sqcup (6A_4).$$

The 5 *Kepler cubes* are inscribed in a dodecahedron. The vertices of any Kepler cube are some 8 of the 20 vertices of the dodecahedron. The 12 edges of any Kepler cube are 12 diagonals of the 12 pentagonal faces of the dodecahedron (one diagonal of each face).

The group $G = \mathrm{SL}(2, \mathbb{Z}_5)$ contains 5 “Hamilton subgroups” isomorphic to the Hamilton group of the eight quaternionic units $\{\pm 1, \pm i, \pm j, \pm k\}$. These subgroups lie in the tree $T_{1,1,30}$ of the monad. Each Hamilton subgroup consists of the lower floor elements 1, $-1$ of the tree and of six elements of order 4 lying on the highest floor. These 5 disjoint sextuples cover all 30 elements of the highest floor of the tree.

The vertices of the dodecahedron associated with $G$ are the 20 third order elements of $G$ (forming the cycles of the 10 components $A_2$ of the monad). The pentagonal faces of the dodecahedron correspond to the 5th (and 10th) order elements of $G$ (forming the $A_4$-components of the monad).

The 5 Kepler cubes are related to the 5 Hamilton subgroups in the following way. Fix a Hamilton subgroup. Consider the subgroup of $G$ formed by the elements such that, conjugating $G$ by these elements, one transforms the chosen Hamilton subgroup into itself. This isotropy subgroup contains 24 elements: the 8 elements of the Hamilton subgroup, 8 elements of order 3, being the vertices of the Kepler cube associated with the chosen Hamilton subgroup, and 8 elements of order 6, forming an opposite cube (whose matrices are obtained from the Kepler cube matrices, multiplying them by $-1$). All vertices of the Kepler cubes belong to the ten $A_2$-components of the monad.

The vertices of any Kepler cube are obtained from one of them, conjugating it by the elements of the corresponding Hamilton subgroup. These 8 conjugations provide 4 of the 8 vertices of the cube; the other 4 are the inverses of these (or their squares, since all elements of the Kepler cube are of order 3).

One may say that the preceding construction interprets the mysterious informal expression of the Kepler cube vertices in terms of one of them, $A$, as being the 8 matrices $A^h$, $h \in \{\pm 1, \pm i, \pm j, \pm k\}$. Namely, $h = 1$ and $h = -1$ provide $A$ and $A^{-1}$, while the meaning of $A^i$ and of the others is explained by the above construction of conjugations generating the Kepler cube from the corresponding Hamilton subgroup.

Let us return to the part of the problem devoted to $p$-Kepler cubes. It requires *to extend the theory of Kepler cubes of the group* $G = \mathrm{SL}(2, \mathbb{Z}_5)$ *described above, replacing* 5 *by a larger prime number* $p$.

One of the first results in this direction is the following study of tree complexity. Consider a chain of $k$ consecutive squarings,

$$A_0, \quad A_1 = A_0^2, \quad A_2 = A_1^2, \quad \ldots, \quad A_k = A_{k-1}^2 = 1.$$

It is represented in the monad as a chain

$$A_0 \to A_1 \to A_2 \to \cdots \to A_k = 1$$

of length $k$ (containing $k+1$ elements of the group).

**Theorem:** A nondegenerate chain of length $k$ in $G = \mathrm{SL}(2, \mathbb{Z}_p)$ exists if and only if $p \equiv \pm 1 \pmod{2^k}$.

The nondegeneracy means the absence of the matrix 1 (except at the last place): $A_{k-1} \neq 1, A_k = 1$.

**Example:** Chains with $k = 2$ exist for $p = 5, 11, 13$. Chains with $k = 4$ exist for $p = 17, 47$.

Perhaps, these chains of squarings provide the generalizations of Kepler cubes to the case $p > 5$.

The combinatorics of some generalized Kepler cubes for $p = 7$ is described in the article ARNOLD V. I. Topology and statistics of formulae of arithmetics. *Russian Math. Surveys*, 2003, **58**(4), 637–664. This article also explains the relation of the combinatorics of these cubes to the generalized four-color problem on a toroidal surface and to some Riemannian surfaces associated with the monad of $G$.

The *Riemannian surface of the monad of* $G$ is constructed as a complex of dimension 2, whose vertices are the third order elements of $G$, but a relevant interpretation of complex structure in terms of combinatorial group theory of $G$ is still missing.

It would also be interesting to extend this study of monads and Riemannian surfaces to the case of *complex versions of finite modular groups*, $G_{\mathbb{C}} = \mathrm{SL}(2, \mathbb{C}\mathbb{Z}_p)$, where $\mathbb{C}\mathbb{Z}_p$ is the ring of complex integers modulo $p$:

$$\mathbb{C}\mathbb{Z}_p = \{a + bi : a \in \mathbb{Z}_p,\ b \in \mathbb{Z}_p\}.$$

**2003-4. Matrix version of Fermat's small theorem.** This recent theorem asserts the congruence of traces:

$$(\mathrm{tr}A)^p - \mathrm{tr}(A^p) = pF;$$

here $p$ is a prime number, $A$ is any unimodular matrix of integers, $A \in \mathrm{SL}(n, \mathbb{Z})$, the ratio $F$ (the result of the division of the difference by $p$) being a polynomial in the variables $\sigma_1, \ldots, \sigma_p$ with integer coefficients, where $\sigma_i$ are the coefficients of the characteristic polynomial of $A$ at the terms of degree $\leq p$:

$$\sigma_1 = \sum_i \lambda_i, \quad \sigma_2 = \sum_{i<j} \lambda_i \lambda_j, \quad \ldots, \quad \sigma_p.$$

An equivalent statement is the formula

$$(\lambda_1 + \cdots + \lambda_n)^p - (\lambda_1^p + \cdots + \lambda_n^p) = pF(\sigma_1(\lambda), \ldots, \sigma_p(\lambda)).$$

The proofs are available in the paper: ARNOLD V. I. Matrix Fermat theorem, finite circles and finite Lobachevsky plane. *Funct. Anal. Appl.*, 2004, **38**(1), 20 pp.

**Example 1:** The "cube of the sum" formula

$$(a+b)^3 - (a^3 + b^3) = 3(a+b)ab$$

can be extended to the identity

$$(a + b + \cdots + z)^3 - (a^3 + b^3 + \cdots + z^3) = 3(UV - W),$$

where $U = \sigma_1 = a + \cdots + z$, $V = \sigma_2 = ab + \cdots + yz$, $W = \sigma_3 = abc + \cdots + xyz$.

**Example 2:** For the identity matrix $A = 1$ of order $n$, our congruence takes the form

$$n^p - n = pF,$$

implying Fermat's small theorem: $n^{p-1} \equiv 1 \ (p)$ if $(n, p) = 1$.

The Newton polynomials $F$ with integer coefficients have (in the cases $p \leq 7$ for which I have explicitly calculated them) rather small coefficients, mostly equal to $\pm 1$ (neglecting the 0 coefficients implied by the weighted homogeneity of the polynomial $F$, the weight of its variable $\sigma_i$ being equal to $i$). Another rather curious property of these coefficients is the equilibrium between the numbers of positive and negative coefficients ($+UV$ and $-W$ above) whose total sum is 0 for $p > 2$.

The present problem is *to develop the theory of these strange polynomials $F$ and to generalize the traces congruence to other symmetric functions, for instance, to the congruences for the other coefficients of the characteristic polynomial of the power $A^p$* (or, in a matrix-free formulation, to other Tschirnhausen transformations).

The *Euler extension of Fermat's small theorem*, replacing the prime $p$ by any integer $m$ and replacing the Fermat congruence $a^{p-1} \equiv 1 \ (p)$ by the Euler congruence $a^m \equiv a^{m-\varphi(m)} \ (m)$, might also have *matrix versions*, which would be interesting even as conjectures guessed from calculating many examples. The simplest example in this direction replaces the prime $p$ by any integer $m$ and the difference $(a+b)^p - (a^p + b^p)$ by $\sum C_m^k a^k b^{m-k}$, the integer $k$ in the sum being relatively prime with $m$: $(k,m) = 1$. The claim is the set of the congruences: $C_m^k$ is divisible by $m$, $C_{m-1}^{k-1}$ is divisible by $k$ (provided that $(k,m) = 1$).

**Example:** For $m = 9$, the binomial coefficients $C_9^k = \mathbf{1}, 9, 36, \mathbf{84}, 126, 126, \mathbf{84}, 36, 9, 1$ are divisible by 9 if $k \neq 0, 3, 6, 9$.

For $k = 3$, the binomial coefficients $C_{m-1}^2 = \mathbf{1}, 3, 6, \mathbf{10}, 15, 21, \mathbf{28}, 36, \ldots$ are divisible by 3 if $m \neq 3, 6, 9, \ldots$

**Proof of the congruences $C_m^k \equiv 0 \ (m)$ and $C_{m-1}^{k-1} \equiv 0 \ (k)$:** Consider a subset $X$ of $k$ elements in the set $\mathbb{Z}_m$ of residues mod $m$. The translations $t_a : x \mapsto x + a \ (a = 1, \ldots, m)$ move $X$ into other subsets of $k$ elements, which are all different if $(k,m) = 1$, since otherwise we would obtain a bijection $t_a$ of $X$ to itself whose period $T$ would be a common divisor of $m$ and $k$. Thus we get a free action $t_a$ of the group $\mathbb{Z}_m$ on the set of its subsets $X$ of $k$ elements, whose number $C_m^k$ is therefore divisible by $m$.

The divisibility of $C_{m-1}^{k-1}$ by $k$ follows, since $C_m^k = \frac{m}{k} C_{m-1}^{k-1}$ and $(k,m) = 1$.

The question is therefore *to generalize these results from the binomial case to the multinomial coefficients of $(a + \cdots + z)^m$.*

Among recent peculiar discoveries in this direction, we note the strange degree $x(a,b)$ of the prime $p$, to whose power the difference of binomial coefficients is divisible,

$$C_{pa}^{pb} - C_a^b = p^{x(a,b)}z \quad \text{(where } (p,z) = 1\text{)}.$$

It seems (empirically) that $x$ is an averagely growing function, but its asymptotic behavior is known only empirically, showing also the strange (and unproved) independence of the values $x(p^m+1,b)$ (and sometimes of $x(pm+1,b)$) of the value $b$ of the second argument.

**Example:** For $p=3$, the starting values of the degree $x(a,b)$ form a Pascal-type triangle (the rows $a=2,3,\ldots,7$ being shown, $b=1,\ldots,a-1$ in each row $a$):

2
4 4
3 3 3
2 3 3 2
4 4 5 4 4
3 3 3 3 3 3

For $p=5$, the rows ($a=2,3,\ldots,11$) of the values $x(a,b)$ are:

3
3 3
3 3 3
5 5 5 5
4 4 4 4 4
3 4 4 4 4 3
3 3 4 4 4 3 3
3 3 3 4 4 3 3 3
5 5 5 5 6 5 5 5 5
4 4 4 4 4 4 4 4 4 4

The double periodicity,

$$x(a+p,b) = x(a,b) = x(a+p,b+p),$$

observed in this table, is partially preserved for other prime numbers $p$.

For $p=7$, the rows ($a=2,3,\dots,15$) of the values $x(a,b)$ are:

$$
\begin{array}{c}
3\\
3\quad 3\\
3\quad 3\quad 3\\
3\quad 3\quad 3\quad 3\\
3\quad 3\quad 3\quad 3\quad 3\\
5\quad 5\quad 5\quad 5\quad 5\quad 5\\
4\quad 4\quad 4\quad 4\quad 4\quad 4\quad 4\\
3\quad 4\quad 4\quad 4\quad 4\quad 4\quad 4\quad 3\\
3\quad 3\quad 4\quad 4\quad 4\quad 4\quad 4\quad 3\quad 3\\
3\quad 3\quad 3\quad 4\quad 4\quad 4\quad 4\quad 3\quad 3\quad 3\\
3\quad 3\quad 3\quad 3\quad 4\quad 4\quad 4\quad 3\quad 3\quad 3\quad 3\\
3\quad 3\quad 3\quad 3\quad 3\quad 4\quad 4\quad 3\quad 3\quad 3\quad 3\quad 3\\
5\quad 5\quad 5\quad 5\quad 5\quad 5\quad 6\quad 5\quad 5\quad 5\quad 5\quad 5\quad 5\\
4\quad 4\quad 4\quad 4\quad 4\quad 4\quad 4\quad 4\quad 4\quad 4\quad 4\quad 4\quad 4\quad 4
\end{array}
$$

The functions $x(mp^r+1,b)$ of $m$ and $r$ are rather peculiar, and I leave to the reader the pleasure of finding them, thus continuing, at least empirically, the Fermat experimental way of discovery of mathematical facts (see the tables below).

All these facts can be extended from the binomial coefficients to the multinomial ones, as well as to the coefficients $K$ of the Newton–Girard formula (taking into account that, for $p=2$, one should replace the difference $C_{pa}^{pb}-C_a^b$ by the twisted difference $C_{pa}^{pb}-(-1)^b C_a^b$).

The table of the first values of $x(p^r+1,b)$ (independent of $b$) is:

| $r$ \ $p$ | 2 | 3 | 5 | 7 | 9 | 11 | 13 | 17 | 19 | 23 |
|---|---|---|---|---|---|---|---|---|---|---|
| 1 | 2 | 3 | 4 | 4 | 4 | 4 | 4 | 4 | 4 | 4 |
| 2 | 3 | 4 | 5 | 5 | 5 | 5 | 5 | 5 | 5 | 5 |
| 3 | 4 | 5 | 6 | 6 | 6 | 6 | 6 | 6 | 6 | 6 |
| 4 | 5 | 6 | 7 | 7 | 7 | 7 | 7 | 7 | 7 | 7 |
| 5 | 6 | 7 | 8 | 8 | 8 | 8 | 8 | 8 | 8 | 8 |
| 6 | 7 | 8 | 9 | 9 | 9 | 9 | 9 | 9 | 9 | 9 |
| 7 | 8 | 9 | 10 | 10 | 10 | 10 | 10 | 10 | 10 | 10 |
| 8 | 9 | 10 | 11 | 11 | 11 | 11 | 11 | 11 | 11 | 11 |
| 9 | 10 | 11 | 12 | 12 | 12 | 12 | 12 | 12 | 12 | 12 |

We see that $x(p^r+1,b)=r+3$ for $p>3$ (being $r+2$ for $p=3$ and $r+1$ for $p=2$), but this is proved only by some millions of examples.

The tables of the first values of $x(mp^r + 1, b)$ (independent of $b$) are given and discussed below (for $p = 3, 5, 7, 11$).

Case $p = 3$:

| $m$ \ $r$ | 1 | 2 | 3 | 4 | 5 | 6 | 7 | 8 | 9 | 10 |
|---|---|---|---|---|---|---|---|---|---|---|
| 1 | 3 | 4 | 5 | 6 | 7 | 8 | 9 | 10 | 11 | 12 |
| 2 | 3 | 4 | 5 | 6 | 7 | 8 | 9 | 10 | 11 | 12 |
| 3 | 4 | 5 | 6 | 7 | 8 | 9 | 10 | 11 | 12 | 13 |
| 4 | 3 | 4 | 5 | 6 | 7 | 8 | 9 | 10 | 11 | 12 |
| 5 | 3 | 4 | 5 | 6 | 7 | 8 | 9 | 10 | 11 | 12 |
| 6 | 4 | 5 | 6 | 7 | 8 | 9 | 10 | 11 | 12 | 13 |
| 7 | 3 | 4 | 5 | 6 | 7 | 8 | 9 | 10 | 11 | 12 |
| 8 | 3 | 4 | 5 | 6 | 7 | 8 | 9 | 10 | 11 | 12 |
| 9 | 5 | 6 | 7 | 8 | 9 | 10 | 11 | 12 | 13 | 14 |
| 10 | 3 | 4 | 5 | 6 | 7 | 8 | 9 | 10 | 11 | 12 |
| 11 | 3 | 4 | 5 | 6 | 7 | 8 | 9 | 10 | 11 | 12 |
| 12 | 4 | 5 | 6 | 7 | 8 | 9 | 10 | 11 | 12 | 13 |
| 13 | 3 | 4 | 5 | 6 | 7 | 8 | 9 | 10 | 11 | 12 |
| 14 | 3 | 4 | 5 | 6 | 7 | 8 | 9 | 10 | 11 | 12 |
| 15 | 4 | 5 | 6 | 7 | 8 | 9 | 10 | 11 | 12 | 13 |

Case $p = 5$:

| $m$ \ $r$ | 1 | 2 | 3 | 4 | 5 | 6 | 7 | 8 | 9 | 10 |
|---|---|---|---|---|---|---|---|---|---|---|
| 1–4, 6–9, 11–14 | 4 | 5 | 6 | 7 | 8 | 9 | 10 | 11 | 12 | 13 |
| 5, 10, 15 | 5 | 6 | 7 | 8 | 9 | 10 | 11 | 12 | 13 | 14 |

Case $p = 7$:

| $m$ \ $r$ | 1 | 2 | 3 | 4 | 5 | 6 | 7 | 8 | 9 | 10 |
|---|---|---|---|---|---|---|---|---|---|---|
| 1–6, 8–13, 15 | 4 | 5 | 6 | 7 | 8 | 9 | 10 | 11 | 12 | 13 |
| 7, 14 | 5 | 6 | 7 | 8 | 9 | 10 | 11 | 12 | 13 | 14 |

Case $p = 11$:

| $m$ \ $r$ | 1 | 2 | 3 | 4 | 5 | 6 | 7 | 8 | 9 | 10 |
|---|---|---|---|---|---|---|---|---|---|---|
| 1–10, 12–15 | 4 | 5 | 6 | 7 | 8 | 9 | 10 | 11 | 12 | 13 |
| 11 | 5 | 6 | 7 | 8 | 9 | 10 | 11 | 12 | 13 | 14 |

One sees that $x(mp^r+1,b)=r+3$ for $(m,p)=1$, $p\geq 3$.

Here is the table of the function $y(a,b)$ defined by the twisted difference

$$C_{2a}^{2b}-(-1)^b C_a^b = 2^{y(a,b)}z, \quad (2,z)=1$$

(the rows correspond to $a=2,3,\dots,20$, $1\leq b<a$):

| | | | | | | | | | | | | | | | | | | |
|---|---|---|---|---|---|---|---|---|---|---|---|---|---|---|---|---|---|---|
| 3 | | | | | | | | | | | | | | | | | | |
| 1 | 2 | | | | | | | | | | | | | | | | | |
| 5 | 6 | 5 | | | | | | | | | | | | | | | | |
| 1 | 3 | 2 | 3 | | | | | | | | | | | | | | | |
| 3 | 5 | 4 | 5 | 3 | | | | | | | | | | | | | | |
| 1 | 2 | 1 | 3 | 1 | 2 | | | | | | | | | | | | | |
| 7 | 8 | 7 | 9 | 7 | 8 | 7 | | | | | | | | | | | | |
| 1 | 4 | 3 | 4 | 2 | 4 | 3 | 4 | | | | | | | | | | | |
| 3 | 6 | 5 | 6 | 4 | 6 | 5 | 6 | 3 | | | | | | | | | | |
| 1 | 2 | 1 | 4 | 2 | 3 | 2 | 4 | 1 | 2 | | | | | | | | | |
| 5 | 6 | 5 | 8 | 6 | 7 | 6 | 8 | 5 | 6 | 5 | | | | | | | | |
| 1 | 3 | 2 | 3 | 1 | 4 | 3 | 4 | 1 | 3 | 2 | 3 | | | | | | | |
| 3 | 5 | 4 | 5 | 3 | 6 | 5 | 6 | 3 | 5 | 4 | 5 | 3 | | | | | | |
| 1 | 2 | 1 | 3 | 1 | 2 | 1 | 4 | 1 | 2 | 1 | 3 | 1 | 2 | | | | | |
| 9 | 10 | 9 | 11 | 9 | 10 | 9 | 12 | 9 | 10 | 9 | 11 | 9 | 10 | 9 | | | | |
| 1 | 5 | 4 | 5 | 3 | 5 | 4 | 5 | 2 | 5 | 4 | 5 | 3 | 5 | 4 | 5 | | | |
| 3 | 7 | 6 | 7 | 5 | 7 | 6 | 7 | 4 | 7 | 6 | 7 | 5 | 7 | 6 | 7 | 3 | | |
| 1 | 2 | 1 | 5 | 3 | 4 | 3 | 5 | 2 | 3 | 2 | 5 | 3 | 4 | 3 | 5 | 1 | 2 | |
| 5 | 6 | 5 | 9 | 7 | 8 | 7 | 9 | 5 | 7 | 6 | 9 | 7 | 8 | 7 | 9 | 5 | 6 | 5 |

The apparent "double periodicity" of this function would relate to $y(a,b)$ the values $y(a+8,b)$ and $y(a+8,b+8)$, but $y(16,1)=9\neq 7=y(8,1)$.

It is, for instance, formally unknown whether the values $x(a,b)$ are bounded or unbounded (neither for a fixed prime $p$ nor totally) and whether the small values (like 2) appear finitely or infinitely many times. Such questions are also open for the multinomial coefficients (where, sometimes, the answers might be different and the results might be easier than for the particular case of binomial coefficients).

**2003-5. Weak asymptotics of Frobenius numbers.** The *Frobenius number* $K(a_1,\dots,a_s)$, defined by $s$ integers $a_i$ having no common divisor altogether, is the least integer which, together with all greater integers, belong to the additive semigroup $A$ of integers generated by the summands $a_i$:

$$A = \{k_1a_1 + \cdots + k_sa_s : k_i \geq 0,\ k_i \in \mathbb{Z}\}.$$

**Example:** For $s = 2$, $a_1 = 3$, $a_2 = 5$ the semigroup is

$$A = \{0, 3, 5, 6, 8, 9, 10, \ldots\},$$

hence the Frobenius number is $K(3,5) = 8$.

The Frobenius number of two (relatively prime) numbers was calculated by J. J. Sylvester: $K(a,b) = (a-1)(b-1)$ (*Educational Times*, 1884, **41**, p. 21).

The problem is to find the asymptotic behavior of the very irregular function $K(a)$ of $s > 2$ variables for large vectors $a \in \mathbb{Z}^s$.

The conjectural averaged behavior is

$$K \sim C \cdot \sqrt[s-1]{\prod a_i}, \quad C = \sqrt[s-1]{(s-1)!}$$

(for instance, $K(a,b,c) \sim \sqrt{2abc}$).

The averaging means the following construction (or its generalization). Let us replace the vector $a$ by a neighborhood $U$ of radius $r$ of the scaled vector $Na \in \mathbb{Z}^s$. Now replace the value $K(a)$ by the (arithmetic) mean $\widehat{K}_N$ of the values $K(b)$ of Frobenius number at the points $b$ of the neighborhood $U$ whose components $b_i$ have no common divisor greater than 1.

The conjecture is that, *for growing values of N, the mean values $\widehat{K}_N$ have a limit (probably provided by the conjectured formula above): $\lim_{N\to\infty} \widehat{K}_N$ should grow as* const $\cdot \left(\prod a_i\right)^{1/(s-1)}$ *for large a.*

I have fixed above the averaging radius $r$, but one might also choose some growth rate $r(N)$, where $r(N)/N$ tends to 0 when $N$ tends to infinity.

More on the weak asymptotic behavior is provided in the article: ARNOLD V. I. Weak asymptotics for the numbers of solutions of Diophantine problems. *Funct. Anal. Appl.*, 1999, **33**(4), 292–293. Applications of Frobenius numbers to representation theory are discussed in the article: ARNOLD V. I. Frequent representations. *Moscow Math. J.*, 2003, **3**(4), 14 pp.

**2003-6. Frequent representations.** Consider a unitary representation of a finite group in Hermitian space $\mathbb{C}^N$. The representation is called *frequent*, if the dimension of the variety of those representations in the same space, which are unitary equivalent to the given one, has the maximal possible value.

I have proved that *the multiplicities of irreducible summands in a frequent representation of a finite group are asymptotically proportional to their dimensions when N tends to infinity.*

The problem is *to find similar asymptotic proportions for orthogonal (and quaternionic) representations of finite groups.*

One might also try to extend the result to the case of *infinite groups*, taking into account the decomposition of the *regular representation* in the space of functions on the group (which contains, for a finite group, each irreducible component as many times as its dimension).

Regular representations might be decomposed into irreducible ones in the case of an infinite group too, and one might try *to replace sums by infinite series* (in the discrete case) or by *integrals* (in the case of compact Lie groups).

As an intermediate problem in this direction, one might study *the asymptotic behavior of the multiplicities of irreducible representations in the spectrum of the Laplace operator on such groups as* $\mathrm{SO}(3)$ *or* $\mathbb{S}^3 = \mathrm{Spin}(3) = \mathrm{SU}(2)$.

The rotated eigenfunction remains an eigenfunction (with the same eigenvalue), and therefore the function space is decomposed as the orthogonal sum of spaces of irreducible representations, corresponding to different eigenvalues.

Considering the first $M$ eigenvalues, one obtains a set of multiplicities of irreducible components, and the problem is *to find the behavior of the ratios of these multiplicities for M tending to infinity.*

In the case of a finite symmetry group, this proportion seems to be asymptotically the same as in a frequent representation (I observed it for millions of eigenfunctions, studying quasimodes in 1972 and working in magnetohydrodynamics in the 1980s). This *choice of frequent representations* (made by Nature, arranging the eigenfunctions symmetry) *seems to take place for any elliptic system with a (finite) symmetry, say, for the Laplace operators on compact Riemannian manifolds having a symmetry group such as the ordinary ellipsoids.*

However, some attempts of my students (to whom I had shown this phenomenon discovered by me in many examples) to prove it as a general theorem were not very successful, and therefore I repeat here *the natural representation frequency conjecture for eigenfunctions of symmetric systems* as a problem to work on (even in the case of finite groups and unitary representations where it is simpler).

The first examples of frequent representations are published in the article: ARNOLD V. I. Modes and quasimodes. *Funct. Anal. Appl.*, 1972, **6**(2), 94–101; *the Russian original is reprinted in*: Vladimir Igorevich Arnold. Selecta–60. Moscow: PHASIS, 1997, 189–202.

*Magnetohydrodynamical applications* (*to the Sakharov–Zeldovich fast dynamo problem*) are published in the paper: ARNOLD V. I. Evolution of a magnetic field under the action of drift and diffusion. In: Some Problems in Modern Analysis. To the memory of V. M. Alexeev. Editor: V. M. Tikhomirov. Moscow: Moscow University Press, 1984, 8–21 (in Russian). See also the survey article: ARNOLD V. I. Remarks on the perturbation theory for Mathieu type problems. *Russian Math. Surveys*, 1983, **38**(4), 215–233.

The dimensions of the spaces of eigenfunctions considered numerically in these studies reached many millions, the symmetry group being that of the cube in the magnetohydrodynamical paper.

A recent study of frequent unitary representations, providing the proof of the asymptotic proportion of irreducible components for the case of finite groups is published in the article: ARNOLD V. I. Frequent representations. *Moscow Math. J.*, 2003, **3**(4), 14 pp.

**2003-7. Symmetric group representations and asymptotic statistics of Young diagrams.** Irreducible representations of the finite symmetric group $S(n)$ are labeled (by Frobenius) by their Young diagrams of area $n$, representing the *partitions*

$$n = a_1 + \cdots + a_s$$

of the parameter $n$ into natural summands. One usually orders the summands, supposing that $a_1 \geq a_2 \geq \cdots \geq a_s$, representing a partition by its *Young diagram D* consisting of $n$ unit squares in the plane. The first (longest) row consists of $a_1$ squares filling the rectangle

$$0 \leq x \leq a_1, \quad -1 \leq y \leq 0.$$

The next row (below the first one) consists of $a_2$ squares filling the rectangle

$$0 \leq x \leq a_2, \quad -2 \leq y \leq -1,$$

and so on, up to the last (shortest) row filling the rectangle

$$0 \leq x \leq a_s, \quad -s \leq y \leq -s+1.$$

The corresponding unitary irreducible representation lies in the Hermitian space $\mathbb{C}^{a(D)}$, its dimension $a(D)$ being equal to the number of *monotonic fillings* of the $n$ squares of the diagram $D$ by $n$ numbers $\{1, \ldots, n\}$. A filling is monotonic if these numbers are decreasing along rows and columns of the diagram (that is, while $x$ or $-y$ are growing in our notation).

**Example:** The partition $5 = 3+2$ provides 5 monotonic fillings

$$\begin{pmatrix} 5 & 4 & 3 \\ 2 & 1 & \end{pmatrix}, \quad \begin{pmatrix} 5 & 4 & 2 \\ 3 & 1 & \end{pmatrix}, \quad \begin{pmatrix} 5 & 4 & 1 \\ 3 & 2 & \end{pmatrix}, \quad \begin{pmatrix} 5 & 3 & 2 \\ 4 & 1 & \end{pmatrix}, \quad \begin{pmatrix} 5 & 3 & 1 \\ 4 & 2 & \end{pmatrix},$$

therefore $a(3+2) = 5$. The maximal value of the dimension of the representation is attained at the partition $5 = 3+1+1$ and is equal to 6.

Vershik and Kerov proved a remarkable property of the *maximal dimension* value $\overline{a}(n)$ (of the dimensions $a(D)$ of the representations corresponding to all Young diagrams $D$ of area $n$):

**Theorem:** *The asymptotic behavior of the maximal dimension value $\overline{a}(n)$ for $n \to \infty$ coincides with that of the average dimension value, taking the averaging along all Young diagrams $D$ of area $n$ weighted with weights $w$ proportional to the squares of the dimensions of representations, $w(D) \sim a^2(D)$.*

This theorem on the coincidence of the maximum with the average implies that there exist a lot of Young diagrams $D$ for which the dimension $a(D)$ is very close to the maximal value $\overline{a}(n)$.

The problem is *to evaluate the number of Young diagrams $D$ of area $n$ for which*

$$a(D) > \overline{a}(n) - C$$

(*at least asymptotically for large $n$, at least averaging in $n$, and eventually with some dependence of $C$ on $n$, for instance, counting the diagrams $D$ satisfying the condition $a(D) > \overline{a}(n)(1-c_1)$*).

The goal of these studies is to apply them to understanding the asymptotic behavior of decompositions of frequent representations into irreducible ones, for which one also needs to study, using the Frobenius numbers, the semigroup generated by the dimensions of irreducible representations (see problem *2003-5*).

The Vershik–Kerov theorem is proved in the article: VERSHIK A. M., KEROV S. V. Asymptotics of maximal and typical dimensions of irreducible representations of a symmetric group. *Funct. Anal. Appl.*, 1985, **19**(1), 21–31.

A more or less classical *explicit formula for generalized Catalan numbers* $a(D)$ is proved, for instance, in the article: ARNOLD V. I. Frequent representations. *Moscow Math. J.*, 2003, **3**(4), 14 pp. Its strange proof is based on singularity theory (and on the Euler–Jacobi complex residue formula asserting the vanishing of the sum of the inverse values of the Jacobian of a mapping $\mathbb{C}^r \to \mathbb{C}^r$ over all preimages of a generic point).

The answer is

$$a(D) = \frac{n!}{h_1! \cdots h_n!} \prod_{i<j} (h_i - h_j),$$

where $h_i = a_i + s - i$ are the lengths of "hooks."

Most results of the study of symmetric groups admit natural extensions to *all Coxeter groups* (at least to the 4 classical series $A_k, B_k, C_k, D_k$, but in many cases to other crystallographic groups $E_6, E_7, E_8, F_4, G_2$ and even to noncrystallographic ones $I_2(p), H_3, H_4$).

Unfortunately, neither the Vershik–Kerov theorem quoted above nor other remarkable results of Vershik's theory on the asymptotic statistics of representations of symmetric groups were extended to the case of other Coxeter groups. See, for instance, the recent survey book: Asymptotic Combinatorics with Applications to Mathematical Physics (St. Petersburg, 2001). Editor: A. M. Vershik. Berlin: Springer, 2003 (Lecture Notes in Math., 1815).

I propose therefore as a problem the suggestion to find *conjectural forms of these extensions* (at least for the 4 classical series $A, B, C, D$). Such an extension might provide a better understanding of the results which are already known for the classical series $A$ of symmetric groups.

For instance, Vershik discovered that *a typical Young diagram of large area n, being observed from a distant location, looks in most cases like a standard universal curvilinear astroidal triangle*

$$\{(x,y) : f(x) + f(y) \leq 1,\ x \geq 0,\ y \geq 0\}$$

*for some special function* $f$ (explicitly calculated by him) *growing from* $f(0) = 0$ *to* $f(1) = 1$.

Namely, the image of the Young diagram under some motion and homothety (reducing its area $n$ to the constant area of universal astroidal triangle) almost coincides with this triangular domain, the difference being small for large $n$.

Here the *majority* (of "*typical*" Young diagrams) is understood in the sense of weighting, whose weights $w(D)$ are proportional to $a^2(D)$, as in the Vershik–Kerov theorem.

The problem is to find universal domains of this type for other Coxeter groups (or complex simple Lie algebras).

**2003-8. Elliptic integrals and functions, their topological nonelementarity and topological Galois theory.** The fact that elliptic functions and elliptic integrals do not belong to the set of elementary functions is well known. The problem is to prove that this event has a topological nature: *they are not topologically equivalent to any elementary function.*

I proved in 1963 the corresponding version of the Abel theorem on the unsolvability of algebraic equations of degree 5 (or higher). These equations (for instance, the equation $x^5 + ax + 1 = 0$) are topologically unsolvable in radicals: no one (complex) function, topologically equivalent to the algebraic 5-valued function $x(a)$, is a finite combination of radicals and univalent functions (for instance, rational functions).

The topological proof of this *topological unsolvability* theorem (based on the topological complexity of the monodromy group and on the topology of Riemannian surfaces) was proved in my lectures of 1964 for Moscow high school students (which started with the definitions of complex numbers, fundamental groups and Riemannian surfaces). Notes of these lectures were later published by one of the students as the book: ALEXEEV V. B. Abel's Theorem in Problems. Moscow: Nauka, 1976 (in Russian); *the French translation*: Cassini, to appear.

In my lectures, I attributed these topological results to Abel, mentioning also his parallel results on topological unsolvability of differential equations and on topological nonelementarity of integrals (proved by the topological complexity of the branchings of the corresponding multivalued complex functions).

However, my students, trying to find the exact statements and proofs in Abel's works, never discovered them, and thus *the problem of proving the topological nonelementarity of elliptic integrals and functions remains open.*

An elliptic integral is sampled by, for instance, the time function for the Newton equation with cubic potential,

$$t(X) = \int_0^X \frac{dx}{\sqrt{f(x)}}, \quad f(x) = x^3 + ax + b$$

(say, for $a = 1$, $b = 0$).

The claim is that *this multivalued function of $X$ is not topologically equivalent to any elementary function* (which is a finite combination of rational functions, radicals, exponential or logarithmic functions, trigonometric or inverse trigonometric functions), and that this remains true even if one also allows any univalent functions in combination.

An elliptic function is a meromorphic doubly periodic function, such as the Weierstrass $\wp$-function $\wp = X(t)$ (inverse to the preceding elliptic integral).

My 1963 conjecture (attributed by me to Abel) claims that *this function is also topologically nonequivalent to any elementary function* (*and that this remains true even if one extends the class of elementary functions by adding combinations with any univalent functions of a finite number of variables*).

Both statements (on elliptic integrals and elliptic functions) are different, and I am unable to reduce one of them to the other.

My 1963 reasoning provides a topological proof of the nonelementary character of the dependence of the periods of integrals on parameters (such as $a$ and $b$ above). Namely, the (modular) monodromy group of this multivalued function is not possible for any elementary function (and hence for any function topologically equivalent to an elementary one).

I would also mention a project of an (ugly) proof of the topological nonelementarity of an elliptic function. Suppose that $\wp$ is reduced to an elementary function $f$ by a homeomorphism $h$ (so that $f(x) = \wp(h(z))$). Then one should prove that $h$ must be holomorphic (since $f$ and $\wp$ are). Next, one should deduce from the Riemann classification of holomorphic simply connected domains that $h$ should be an affine transformation, and therefore that $f$ is an elliptic function (whenever it is topologically equivalent to such a function). The final step is to use the well known fact of the nonelementarity of the elliptic function $\wp$ itself (whose ugly proof ought to be replaced by a topological one, still missing).

To study the left-right topological nonequivalence of elliptic functions to elementary ones, one should first prove that an action of the group $\mathbb{Z}^2$ on a complex plane domain by holomorphic diffeomorphisms, which is topologically equivalent to the standard action by the translations, is in fact holomorphically equivalent to an (eventually different) action by some translations.

Such a reasoning forces any meromorphic function, which is topologically left-right equivalent to an elliptic function, to be genuinely elliptic (and hence nonelementary).

Similarly, for an algebraic mapping $a: X \to Y$ between closed Riemannian surfaces, one should try to prove that any perturbed holomorphic mapping belongs to the same family of algebraic mappings $\tilde{a}: \tilde{X} \to \tilde{Y}$.

Applying this algebraicity result to the "automorphic mapping" $f: G \to Y$, $f = a \circ \pi$ ($\pi: G \to X$ being a covering), one hopes to deduce the automorphic character of the perturbed meromorphic mapping $\tilde{f}: \tilde{G} \to \tilde{Y}$, $\tilde{f} = \tilde{a} \circ \tilde{\pi}$, provided that it is left-right topologically equivalent to the nonperturbed automorphic mapping $f$.

On the other hand, one might suppose that the situation is different in higher dimensions and that the automorphic character of the nonperturbed mapping $f$

might be disturbed by a perturbation $\tilde{f}$ preserving the left-right topological equivalence class of $f$, the perturbed mapping $\tilde{f}$ being no longer automorphic.

In 1964, I attributed the preceding theory to Abel, asking my students to publish the resulting "Topological Galois Theory" formally, but they have not succeeded yet.

**2003-9. Cubic irrationals and tori triangulations related to higher-dimensional continued fractions.** Consider a unimodular matrix with integer elements,

$$A \in G = \mathrm{SL}(n, \mathbb{Z}),$$

whose eigenvalues are different positive irrational numbers. The $n$ eigenplanes divide the space $\mathbb{R}^n$ into $2^n$ ortants invariant under the action of $A$.

This action also preserves the integer lattice $\mathbb{Z}^n \subset \mathbb{R}^n$, and hence preserves its intersection with each ortant.

Each of these intersections is an additive semigroup of integer vectors. Its convex hull is bounded by an infinite polyhedral surface called the *sail.*

The sail is invariant under the action of $A$, but it is also invariant under the action of other linear mappings belonging to $G$ and having the same eigenplanes (with positive eigenvalues). Such mappings form a commutative symmetry group $H$ of the sail (isomorphic to $\mathbb{Z}^{n-1}$); I use to call it "the integer version of the Cartan subgroup," since its matrices are diagonal matrices with the same eigenbasis.

Now consider the quotient variety (the sail)$/H$. For $n = 2$, it is a circle divided into segments $s_1, \ldots, s_k$, and each segment is equipped with the "integer points" (being the images of lattice points of the sail). Let $s_i$ be decomposed by these integer points into $a_i$ void arcs. Then the sail is associated with the periodic continued fraction with period $[a_1, \ldots, a_k]$, equal to the number $\lambda = a_1 + 1/(a_2 + \cdots + 1/(a_k + 1/\lambda) \cdots)$.

This number $\lambda$ describes the eigendirection of the second order matrix $A$ in suitable $\mathrm{SL}(2, \mathbb{Z})$-integer coordinates of $\mathbb{R}^2 \supset \mathbb{Z}^2$.

Similarly, for $n > 2$ we obtain a decomposition of a torus $T^{n-1}$ into faces (which are the images of convex faces of the sail) and their intersections, each containing "integer points" (being the images of those of the sail). This structure is called "the $(n-1)$-dimensional periodic continued fraction" (of the initial operator $A \in G$).

The problem is to find out, *which "triangulations"* (decompositions into faces) *and which sets of "integer points" are possible for the matrices of order* 3, $A \in \mathrm{SL}(3, \mathbb{Z})$.

In the case of ordinary continued fractions ($n = 2$), any period $[a_1, \dots, a_k]$ is possible, but the situation becomes less clear already for cubic irrationals ($n = 3$).

**Example:** The golden ratio matrix $\left(\begin{smallmatrix}2&1\\1&1\end{smallmatrix}\right)$ has a natural extension to an arbitrary dimension $n$, which is the matrix

$$A = \begin{pmatrix} 3 & 2 & 1 \\ 2 & 2 & 1 \\ 1 & 1 & 1 \end{pmatrix}$$

for $n = 3$ and similarly

$$A = \begin{pmatrix} n & n-1 & n-2 & \dots \\ n-1 & n-1 & n-2 & \dots \\ n-2 & n-2 & n-2 & \dots \\ \vdots & \vdots & \vdots & \ddots \end{pmatrix}$$

for any $n$. The corresponding triangulation (for $n = 3$) divides the 2-torus $T^2$ into 2 triangles represented by the decomposition of the square by a diagonal: . There are no integer points other than the four vertices of the square (representing a single point of the torus).

The problem involves three questions on these $T^2$-triangulations associated with cubic irrationals (being the eigenvalues of the characteristic polynomial of $A$).

1) *Which triangulations are possible* (first forgetting the integer points, but then taking them into account in the classification of triangulated structures)?

2) *Which triangulations are typical* (also either forgetting the integer points or taking them into account)?

For instance, consider all those matrices in $G$ whose elements are not too large (say, $\|A\| \leq R$) and eigenvalues are all positive.

Count the numbers of the triangles, quadruples, pentagons (and so on) of their triangulated structures. What is the asymptotics (for $R$ tending to infinity) of the ratios of these numbers? Are there more triangles than quadrangles?

Similar statistics are interesting for the numbers of integer points (on faces and edges), for the number of faces meeting at a vertex, for integral lengths of edges, integral areas of faces, integral dihedral angles between the faces along an edge, integral solid angles at vertices, and so on.

Even without a study of asymptotics for $R \to \infty$ (which might, however, be easier than a detailed study for fixed $R$ which is, say, less than 10), the tables of answers to the preceding questions on frequency might be interesting even for small values of $R$.

Later, one might compare these empirical tables with the universal statistics for the sails of arbitrary triangular pyramids described by Kontsevich and Sukhov, with no relation to integer matrices $A$: it would be interesting to see whether the proportions described above converge, as $R \to \infty$, to the proportions for random pyramids (whose existence and independence of a pyramid for almost all pyramids was proved by Kontsevich and Sukhov). Unfortunately, they present no values for these proportions (such as the ratio of the frequency of the triangular case to that of the quadrangular case), whose existence was proved in their article.

Details of these studies are described in the book: Pseudoperiodic Topology. Editors: V. Arnold, M. Kontsevich and A. Zorich. Providence, RI: Amer. Math. Soc., 1999 (AMS Transl., Ser. 2, 197; Adv. Math. Sci., 46).

The first study of periodic higher-dimensional continued fractions is published in the paper: TSUCHIHASHI H. Higher-dimensional analogues of periodic continued fractions and cusp singularities. *Tôhoku Math. J., Ser.* 2, 1983, **35**(4), 607–639.

The simplest triangulations of $T^2$ (into two triangles) were described by E. Korkina in: KORKINA E. I. Two-dimensional continued fractions. The simplest examples. *Proc. Steklov Inst. Math.*, 1995, **209**, 124–144.

More discussions of higher-dimensional continued fractions (and of their relations to many objects, such as the classification of graded commutative algebras and to the theory of Gröbner bases) are in the book ARNOLD V. I. Continued Fractions. Moscow: Moscow Center for Continuous Mathematical Education Press, 2001, 40 pp. (in Russian) ("Mathematical Education" Library, 14).

To the preceding questions on possible triangulations and frequent triangulations, I add the following question on equivalence:

3) *What are common features of those triangulations which correspond to isomorphic fields of algebraic numbers*?

Knowing a triangulation (equipped with the "integer points"), how to describe all the triangulations associated with matrices $A$ having the same characteristic polynomial (or isomorphic fields of algebraic numbers)? Is the number of such "algebraically equivalent" triangulations finite (for a given initial triangulation)?

Anyway, the explicit calculation of algebraic invariants of the field, generated by the eigenvalues of $A$, in terms of the corresponding triangulation might be useful for many purposes, even if it does not answer the preceding equivalence question.

For instance, one might study the arising combinatorial invariants of sails also for the nonperiodic case, where the algebraic invariants of the periodic case might generate interesting ergodic geometry type asymptotics of the finite part of the sail combinatorics.

In the case of usual continuous fractions, the celebrated Lagrange theorem says that the periodicity of the continued fraction of a number is equivalent to the fact that it is a quadratic irrational number.

For the case of higher-dimensional continued fractions, the deduction of periodicity from algebraic origin (based on the Dirichlet theorem on the units) is provided in Tsuchihashi's article quoted above.

The inverse theorem is available at present only partially. In the article KORKINA E. I. La périodicité des fractions continues multidimensionnelles. *C. R. Acad. Sci. Paris, Sér.* I *Math.*, 1994, **319**(8), 777–780, some theorem, implying the deduction of the algebraic origin of a pyramid from the topological periodicity of a sail's structure, is announced, but, unfortunately, its complete proof is still missing.

I am mentioning this theorem here (trying also to urge the publication of the proof), since the derivation of the matrix $A$ from the period of the sail's triangulation, leading to the proof of this theorem of Korkina, might help to understand which triangulated structures (on $T^2$) can be obtained from suitable matrices $A$ of order 3.

The first steps and first numerical experiments in the direction of questions 1)–3) posed above were made by O. Karpenkov (Moscow State University, 2003). He mostly studied Sylvester matrices

$$A = \begin{pmatrix} 0 & 1 & 0 \\ 0 & 0 & 1 \\ 1 & a & b \end{pmatrix}$$

finding a lot of triangulations. He proved, however, that some of the relevant matrices cannot be reduced to any Sylvester form (it is unknown whether such matrices are exceptional or rather typical).

It is formally unknown neither whether the number of topologically nonequivalent triangulations of $T^2$, associated with the matrices corresponding to isomorphic cubic fields, is finite or infinite, nor whether the number of fields corresponding to isomorphic triangulations is finite or infinite.

Even the infiniteness of the total set of the triangulations classes associated to all the matrices of $\mathrm{SL}(3, \mathbb{Z})$ is (formally) unproved.

The triangulations that do not correspond to any such matrix are (formally) unknown: are they typical or exceptional?

It is also formally unknown which of the classification problems discussed above are algorithmically solvable, and which ones are not (even for the reducibility of the Frobenius form).

There is no published information on the (undoubtedly existing) correlations between the sails in different ortants, into which the 3 eigenplanes divide $\mathbb{R}^3$ (between the $2^{n-1}$ different sails in the $2^n$ ortants of $\mathbb{R}^n$, for $n > 2$).

Even if some of the questions formulated above happen to be not difficult, the subjects are so fundamental that the answers should be published.

**2003-10. Extensions of Courant's theorem on the topological structure of eigenfunction nodes.** Let $u$ be an eigenfunction of the Laplace operator, $\Delta u = \lambda u$, on a connected compact manifold (eventually with a boundary and suitable boundary conditions).

The Courant theorem asserts that, ordering the eigenfunctions by decrease of eigenvalues (tending to $-\infty$), one has the following restriction on the topological structure for the $n$-th eigenfunction: *the number of connected components, into which the zeros hypersurface of the n-th component divides the manifold, is at most n*:

$$b_0(\{x : u_n(x) \neq 0\}) \leq n.$$

**Problem:** Extend this theorem to the case of systems of equations, describing oscillations of the sections of fibrations whose fiber has dimension $m > 1$.

In this case, the zeros set would be generically a submanifold of codimension $m$, and one might study the dependence of its homology or homotopy properties on the number $n$, ordering the eigenvalues, such as $b_1(\{x : u_n(x) \neq 0\}$ or $\pi_1$, for $m = 2$; this corresponds to $H^{m-1}$ in the general case. The system of two equations

$$\Delta u = \lambda u, \quad \Delta v = \lambda v, \qquad \text{for } (u(x,y), v(x,y)),$$

is the simplest example where one might even try to imitate arguments from Courant's proof, connecting points of common zeros by curves along which the vectors $(u(x,y), v(x,y))$ are parallel, and replacing, in the domains bounded by these curves, the eigenvector $(u, v)$ by its "reflected" version.

Similarly, one might consider the case of one independent variable and two dependent ones, trying to majorize, for a fixed number of the eigenvalue, the rotation number of the plane vector $(u(x), v(x))$ along the base circle $\{x \pmod{2\pi}\}$ (supposing the normal fibration of an oscillating circle to be trivial).

Similar homotopic questions occur every time when the dimension of the normal bundle (fiber) exceeds the dimension of the base (space of independent variables); in this case, the zeros set of the eigenvector is generically empty, but the homology or homotopy class of the spherical bundle section might be evaluated in terms of the number of the eigenvalue.

**2003-11. Generalizations of gravitational and Coulomb fields.** Consider the space of Hermitian matrices of order $n$ as a real vector space $\mathbb{R}^N$ where $N = n^2$. We consider them as matrices of Hermitian operators in the Hermitian space $\mathbb{C}^n$.

Consider the subvariety $\Sigma \subset \mathbb{R}^N$ formed by the matrices having a multiple eigenvalue. The codimension of this real algebraic variety is equal to 3 (look at the case $n = 2$ where $\Sigma$ is the set of real scalar matrices).

The complement of $\Sigma$ in $\mathbb{R}^N$ is the base space of the $n$ eigenvectors fibrations (we normalize the eigenvectors to have norm 1, making each fiber a circle $\mathbb{S}^1$): they are formed by the normalized eigenvectors of the base space operator.

These fibrations have natural "adiabatic connections," transporting an eigenvector to the closest vector of the neighboring fiber (corresponding to a perturbed operator).

Take, for instance, the case $n = 2$ reducing the space $\mathbb{R}^4$ of Hermitian matrices to the space $\mathbb{R}^3$ of traceless ones. In this 3-space, the multiple spectrum variety $\Sigma$ is reduced to one point (the origin), and the connection's curvature is a closed 2-form in $\mathbb{R}^3 \setminus \Sigma$.

This 2-form might be interpreted as a divergence-free vector field (using the volume element in $\mathbb{R}^3$). This vector field is smooth in $\mathbb{R}^3$ outside the origin; at the origin, it has a singularity. The latter is a Newton or Coulomb type singularity: the vector is directed along the radius-vector and its length is inverse proportional to the squared distance to the origin.

The problem is to extend this field to the case of matrices of higher order $n$. Say, if $n = 3$ the space of traceless matrices is $\mathbb{R}^8$, and the variety of multiple spectrum operators $\Sigma_2$ in it is 5-dimensional. It contains a subvariety $\Sigma_3$ of the operators having triple eigenvalues, which is just the origin.

The cone $\Sigma_2$ can be described in terms of its intersection with the sphere $\mathbb{S}^7$, bounding a neighborhood of $\Sigma_3$. This 4-dimensional intersection consists of two components (defined by the equations $\lambda_1 = \lambda_2$ and $\lambda_2 = \lambda_3$ for the usual ordering $\lambda_1 \leq \lambda_2 \leq \lambda_3$ of the eigenvalues).

The three curvature 2-forms of the three eigenvectors fibrations define the product closed 6-form corresponding to a vector field. This field has Newton or Coulomb singularities along $\Sigma_2$, but it has a worse singularity at the triple spectrum variety $\Sigma_3$.

Hence we arrive at the problem: study these $\Sigma_3$-generalizations of Newton or Coulomb forces (as well as their higher-dimensional versions for $n > 3$ and their quaternionic versions associated with multiple spectra of hyper-Hermitian forms).

The study of stratification singularities of Hermitian forms started in the article ARNOLD V. I. Modes and quasimodes. *Funct. Anal. Appl.*, 1972, **6**(2), 94–101; *the Russian original is reprinted in*: Vladimir Igorevich Arnold. Selecta–60. Moscow: PHASIS, 1997, 189–202. More details are described in the articles: ARNOLD V. I. Remarks on eigenvalues and eigenvectors of Hermitian matrices, Berry phase, adiabatic connections and quantum Hall effect. *Selecta Math.* (*N. S.*), 1995, **1**(1), 1–19; *the Russian translation in*: Vladimir Igorevich Arnold. Selecta–60. Moscow: PHASIS, 1997, 583–604; ARNOLD V. I. Relatives of the quotient of the complex projective plane by complex conjugation. *Proc. Steklov Inst. Math.*, 1999, **224**, 46–56; ARNOLD V. I. Polymathematics: is mathematics a single science or a set of arts? In: Mathematics: Frontiers and Perspectives. Editors: V. I. Arnold, M. Atiyah, P. Lax and B. Mazur. Providence, RI: Amer. Math. Soc., 2000, 403–416; CEREMADE (UMR 7534), Université Paris-Dauphine, № 9911, 10/03/1999.

The suggestion to study $\Sigma_3$-generalized Newton or Coulomb vector fields originated from M. Berry, while he was discussing the results of the above paper in *Selecta Math.* with the author who had related the Radon adiabatic connection (1918) to the theories of Berry phase described by Rytov (1938) and Ishlinsky (1952): RYTOV S. M. Sur la transition de l'optique ondulatoire a l'optique geométrique. *C. R.* (*Dokl.*) *Acad. Sci. USSR* (*N. S.*), 1938, **18**(2), 263–266; ISHLINSKY A. YU. Mechanics of special gyroscopic systems. Kiev: Academy of Sciences of Ukrainian SSR, 1952 (in Russian); *reproduced in*: Orientations, Gyroscopes and Inertial Navigation. Moscow: Nauka, 1976, p. 195.

**2003-12. Finite order projective line geometry.** The finite projective line $\mathrm{P}(\mathbb{Z}_p)$ ($p$ being a prime number) is formed by the one-dimensional subspaces of the 2-dimensional vector space $\mathbb{Z}_p^2$:

$$\mathrm{P}(\mathbb{Z}_p) = \frac{\mathbb{Z}_p^2 \setminus 0}{\mathbb{Z}_p \setminus 0}, \quad \mathbb{Z}_p = \mathbb{Z}/p\mathbb{Z}.$$

It consists of $p+1$ points which can be denoted by an affine coordinate taking the values

$$x \in \{0, 1, \ldots, p-1; \infty\}.$$

The finite $p$-projective line P consists of $p+1$ points:

$$\mathrm{P}(\mathbb{Z}_p) = \mathbb{Z}_p \sqcup \{\infty\}.$$

It is decomposed here into the affine part and the completing point at infinity. This decomposition depends on the coordinate choice and is not included in the structure of the projective line.

The group of unimodular $p$-matrices of order 2,

$$G = \mathrm{SL}(2, \mathbb{Z}_p),$$

acts on $\mathrm{P}(\mathbb{Z}_p)$ permuting one-dimensional vector subspaces of the plane $\mathbb{Z}_p^2$. This action is naturally reduced to the action of the quotient projective group

$$\mathrm{P}G = \mathrm{PSL}(2, \mathbb{Z}_p) = G/\{\pm 1\},$$

which is half as large as $G$ (having only $p(p^2-1)/2$ elements).

The permutations of the $p+1$ points of P forming the projective group P$G$ are even, but they form a small subgroup of the group of all $(p+1)!/2$ even permutations of the $p+1$ points of P.

**Example:** For $p = 5$, there are 360 even permutations of the 6 points of P, and only 60 elements in the projective group P$G$.

The problem is *to combinatorially describe the projective permutations* (*or the geometric structure of* P *preserved by these permutations of its* $p+1$ *elements*).

The projective transformations preserving one point of P form a subgroup of the $p$-affine transformations, having naturally the structure of *p-Lobachevskian plane*. In the affine coordinate, the fixed point being $\infty$, these affine transformations are

$$\{x \mapsto ax + b\}, \quad a = c^2, \quad c \neq 0.$$

Their number is $p(p-1)/2$. The condition that the coefficient $a$ is a quadratic residue is the mod $p$ version of the choice of upper half-plane ($a > 0$) in the real Lobachevskian plane model. Inequalities should be replaced by the quadratic residue properties in $p$-calculus.

The combinatorial properties of these special (affine) permutations are not too difficult to understand in terms of the Lobachevsky geometry.

However, to transfer this description to the case of other choices of fixed points (better to avoid mentioning the fixed points in the description) is not easy, and I needed long explicit calculations to do it even for $p = 5$, which case leads to nice answers. In this case, the projective group permuting the 6 points of P is isomorphic to the group of even permutations of some 5 objects (which are the *Kepler cubes* of the corresponding *dodecahedron surface*).

But I do not know whether one might find such a nice description of the group of projective permutations of a finite projective line for greater values of $p$, except $p = 7$ where one is led to the theory of non-simply-connected regular polyhedra instead of the dodecahedron.

The ratio of the order of the group of even permutations of the $p+1$ points of P to the order of the group of projective permutations on P is equal to

$$\frac{(p+1)!}{2} : \frac{p(p^2+1)}{2} = (p-2)!,$$

which might be interpreted as the number of *cyclic orders* of $p-1$ elements.

It would therefore be nice to find a set $X$ of $p-1$ elements associated with P whose cyclic order is preserved by the projective permutations (being disturbed by other even permutations of the $p+1$ points of P acting naturally on $X$).

To construct $X$, one might delete some two chosen points of P. The affine coordinate, for which the two chosen points are 0 and $\infty$, provides the natural cyclic order on $X$ (preserved by the projective permutations fixing the chosen points).

For a different choice of two points, one obtains a new cyclic order on a new set of $p-1$ points of P. The problem is *to combinatorially describe this new cyclic order in terms of the old one.*

One can relate the dodecahedron structure associated with the group $G = \mathrm{SL}(2, \mathbb{Z}_p)$, $p = 5$, to the finite projective line P by the following construction starting with any choice of a prime number $p$.

Let us find the elements of order $p$ in $G$. An example is given by the Jordan matrix

$$J = \begin{pmatrix} 1 & 1 \\ 0 & 1 \end{pmatrix}, \qquad J^p = 1 \quad \text{in } G.$$

The solutions of the equation $A^p = 1$ in $G$ are just the matrices with trace 2 in $G$. Their number is $p^2$ (including the $p^2-1$ matrices of order $p$ and the identity matrix 1). The elements of order $p = 5$ describe the rotations of the pentagonal faces of the dodecahedron constructed from $G$ with $p = 5$. One should generalize this geometry to the case $p > 5$. For $p = 7$, the dodecahedron must be replaced by a polyhedral surface of genus 3.

All elements of order $p$ in $G$ are conjugated either to $J$ (there are $(p^2-1)/2$ such matrices) or to $J^b$, $b$ being a quadratic nonresidue mod $p$ (the same number of matrices).

An element of order $p$ is sent to a point of the finite $p$-projective line P, the element being associated with the direction of its eigenvector. The preimage of a point contains $p-1$ elements of order $p$ and is equipped with a natural cyclic order.

In the case $p=5$, the six points of P are interpreted as the six pairs of parallel faces of the dodecahedron. However, to obtain the combinatorial description of their projective permutations (for $p=5$), one has to permute the five Kepler cubes which are missing in the general case $p>5$ (except the case $p=7$ studied in my paper in *Russian Math. Surveys*, 2003 quoted above).

*You are never sure*
*whether or not a problem is good*
*unless you actually solve it.*

Mikhail Gromov

# Comments

# 1956

## ▽ 1956-1

H This is a problem mentioned in the author's preface to the first (Russian) edition, see page VIII. It was thoroughly examined in the paper [1] where its origin was nevertheless affirmed as unknown.

[1] YASCHENKO I. V. Make your dollar bigger now!!! *Math. Intelligencer*, 1998, **20**(2), 38–40.

## △ 1956-1 — *N. P. Dolbilin*

R Alexei Tarasov has shown [1] that a rectangle admits a realizable folding with arbitrarily large perimeter. A realizable folding means that it could be realized in such a way as if the rectangle were made of infinitely thin but absolutely nontensile paper. Thus, a folding is a map $f: B \to \mathbb{R}^2$ which is isometric on every polygon of some subdivision of the rectangle $B$. Moreover, the folding $f$ is realizable as a piecewise isometric homotopy which, in turn, can be approximated by some isotopy of space (which corresponds to the impossibility of self-intersection of a paper sheet during the folding process).

[1] TARASOV A. On Arnold's problem on a "folded rouble", in preparation.

# 1958

## ▽ 1958-1 — *V. I. Arnold*

H This problem was published in the paper [1] (p. 178). Some asymptotics have even been calculated explicitly up to the present date (by A. V. Zorich and M. L. Kontsevich, see [2, 3]).

One might consider the purely combinatorial version of the problem, studying the $(C,B,A)$-permutations of the finite sets $\{1,2,\ldots,n\}$ (where $A = \{1,2,\ldots,a\}$, $B = \{a+1, a+2, \ldots, a+b\}$, $C = \{a+b+1, \ldots, n\}$).

The total number of such permutations is equal to $(n-1)(n-2)/2$. Some of them are rotations (isomorphic to the addition of a constant to the residues mod $n$). But it is not clear what proportion of those rotations are isomorphic to $(C,B,A)$-permutations (and of those which are transitive, i. e., have only one orbit and are isomorphic to the addition of 1).

For small values of $n$, the numbers of such $(C,B,A)$-permutations are:

| $n$ | 3 | 4 | 5 | 6 | 7 | 8 | 9 | 10 | 11 |
|---|---|---|---|---|---|---|---|---|---|
| all $(C,B,A)$-permutations | 1 | 3 | 6 | 10 | 15 | 21 | 28 | 36 | 45 |
| rotations | 0 | 2 | 2 | 8 | 6 | 16 | 16 | 26 | 22 |
| transitive rotations | 0 | 2 | 2 | 8 | 6 | 14 | 16 | 24 | 22 |

The number of all transitive rotations is only $1/n$-th of the number $n!$ of all permutations of $n$ elements and, it seems, most of these $(n-1)!$ rotations are transitive. But the statistics for the $(C,B,A)$-permutations looks differently from that for the general ones.

[1] ARNOLD V. I. Small denominators and problems of stability of motion in classical and celestial mechanics. *Russian Math. Surveys*, 1963, **18**(6), 85–191.

[2] KONTSEVICH M. L. Lyapunov exponents and Hodge theory. In: The Mathematical Beauty of Physics (Saclay, 1996). A memorial volume for Claude Itzykson. Editors: J. M. Drouffe and J. B. Zuber. River Edge, NJ: World Scientific, 1997, 318–322. (Adv. Ser. Math. Phys., 24.)

[3] ZORICH A. V. How do the leaves of a closed 1-form wind around a surface? In: Pseudoperiodic Topology. Editors: V. Arnold, M. Kontsevich and A. Zorich. Providence, RI: Amer. Math. Soc., 1999, 135–178. (AMS Transl., Ser. 2, 197; Adv. Math. Sci., 46.)

## △▽ 1958-1 — *M. L. Kontsevich*

H The subject has a long history, with several examples given by A. Katok, and a general theorem of H. Masur and W. Veech which says that for generic length of intervals the transformation is ergodic.

## △ 1958-1 — *A. V. Zorich*

R A map of an interval to itself obtained by cutting the initial interval $X$ into $n$ subintervals and putting them on $X$ in a different order without overlaps and preserving the orientation of pieces is called now an *interval exchange transformation*. The problem was suggested by V. A. Rokhlin (with a reference to V. I. Arnold) in 1959 to the members of the seminar of V. A. Rokhlin and Ya. G. Sinaĭ. The first progress was obtained by V. I. Oseledets who observed that interval exchange transformations may have quite unexpected spectral properties (see below). As a matter of fact, the term "interval exchange transformation" first appears in the paper of V. I. Oseledets [15] of 1966; however, this notion was translated into English differently.

An interval exchange transformation is a *parabolic* dynamical system and, in particular, it has rather moderate chaotic behavior. It is *never* mixing (A. Katok, about 1975, see [6]), though typically topologically weakly mixing (A. Nogueira and D. Rudolph [14], 1997).

Up to the end of the 1960s interval exchange transformations were mostly discussed in folklore, often under different names. For example, in his paper [19] of 1969, W. A. Veech studies skew products over rotation of the circle and, morally, gives a particular example of an interval exchange $T$ which is minimal (every orbit of $T$ is dense in the interval), but which is not uniquely ergodic (some orbits prefer to stay in one subset of the interval, while the other orbits prefer to stay in the complementary subset).

Interval exchange transformations attracted considerable interest in the middle of the 1970s. One of the challenges was the following conjecture of M. Keane of 1975. By definition, an interval exchange transformation $T$ preserves the Lebesgue measure on the interval. M. Keane conjectured in [7] that, under almost any choice of lengths of subintervals, the interval exchange transformation is ergodic with respect to the Lebesgue measure; moreover, he conjectured that the Lebesgue measure generically is *unique* ergodic measure. Here one assumes, of course, some natural constraints on the permutation $\pi$ which rearranges the order of subintervals under exchange. Several examples (of V. I. Oseledets, and the one of W. A. Veech mentioned above; an example of H. Keynes and D. Newton [9], 1976, and of M. Keane [7], 1977) showed that minimal and non uniquely ergodic interval exchange transformations do really exist. The number of ergodic measures is, however, always bounded (the initial result of V. I. Oseledets was sharpened by W. A. Veech [20], 1978; see also the works of A. Katok [5] (upper bound), 1973, and E. Sataev [17] (sharpness), 1975, for the analogous result concerning surface flows).

The developments in the study of surface flows and surface foliations influenced, in particular, by the works of W. Thurston, and a clear understanding that surface flows and foliations are intimately related to the interval exchange transformations, gave additional motivation for further research in this area. Considering the first return map defined by a conservative flow on a torus to an interval transversal to the flow (as in the paper of V. I. Arnold [1] of 1963) one gets an interval exchange transformation of 2 or 3 intervals. Considering the first return map for conservative flows on surfaces of higher genera one gets general interval exchange transformations. Conversely, constructing a suspension over an interval exchange transformation one can always realize an interval exchange transformation as a first return map defined by a surface flow. The genus of the surface (and even more subtle topological information) are completely determined by the permutation $\pi$.

The conjecture of M. Keane was proved independently by H. Masur [12] and by W. A. Veech [21] in 1982. Both proofs are based on applying some *renormalization procedure*. In the proof by W. A. Veech, it is the *Rauzy induction*, introduced by G. Rauzy [16], 1979. The renormalization procedure in this context is a dynamical system acting on the space of all interval exchange transformations sharing the same combinatorics. The key point is that the action is ergodic with respect to some natural measure on the space of interval exchange transformations. Actually, this dynamical system is in its turn very closely related to the *Teichmüller geodesic flow* on the *moduli spaces of Abelian differentials*.

The use of Teichmüller theory and the fundamental result of H. Masur [12] and W. A. Veech [21] on ergodicity of the Teichmüller geodesic flow proved to be an extremely powerful instrument in the study of interval exchange transformations in the subsequent two decades. As an example one can consider the following description of the error term for the ergodic averages of almost all interval exchange transformations. Let the length of the interval be normalized to one, $|X| = 1$. Consider a long piece of an orbit $x, T(x), T^2(x), \dots, T^{n-1}(x)$ of an interval exchange transformation. By ergodicity of $T$, the number of visits of this trajectory to a subinterval $X_i$ equals approximately $n \cdot |X_i|$. It is proved (A. Zorich [22], 1997) that the error term is about $n^\nu$, where the number $\nu < 1$ depends neither on the point $x$, nor on the lengths $|X_i|$ of the subintervals. It depends only on the permutation $\pi$ and, actually, the number $1 + \nu$ is the second Lyapunov exponent of the Teichmüller geodesic flow. A more precise statement of the error term involves the higher Lyapunov exponents, A. Zorich [23], 1999.

Huge classes of permutations, so-called *extended Rauzy classes*, share the same values of $\nu_i$. The extended Rauzy classes correspond naturally to connected components of the moduli spaces of Abelian differentials; they are classified by M. Kontsevich and A. Zorich [11], 2002. Surprisingly, the sums $\nu_1 + \cdots + \nu_g$ of Lyapunov exponents give rational numbers (conjecture of M. Kontsevich [10], 1997; strongly supported now by numerous implicit rigorous computations). Recently G. Forni [4], 2002, proved the conjecture of M. Kontsevich and A. Zorich that the top $g$ Lyapunov exponents are strictly positive; he also generalized the topological statement concerning the deviation of a trajectory from its asymptotic direction (A. Zorich [23], 1999) getting a similar result for the ergodic averages along surface flows.

The topological dynamics of a measured foliation on a surface and the dynamics of a corresponding interval exchange transformation are almost equivalent. Studying *topological* dynamics, one parametrizes the leaves of a foliation (or trajectories of a flow) by the length in some Riemannian metric, or by some equivalent parametrization. Note that, working with a flow defined by a multivalued Hamiltonian, one uses a completely different parametrization of trajectories, which may drastically change the dynamics of the flow. For example, K. M. Khanin and Ya. G. Sinaĭ [18], 1992, proved that in *Hamiltonian* parametrization some wide class of flows on a torus has components where the flow is mixing, while the first return map of such flow to a closed transversal corresponds to a rotation of the circle, which is *never* mixing.

There are some interesting results concerning dynamics of interval exchange transformations when the lengths of the subintervals under exchange have special arithmetic properties, see, for example, the Ph. D. thesis of P. Arnoux [2], 1981, and the paper of M. Boshernitzan and C. Carroll [3], 1997.

Interval exchange transformations with flips, when the map changes the orientation of at least one subinterval, turned out to be less interesting in the following sense: A. Nogueira [13], 1989, proved that almost all such interval exchange transformations have an open domain of periodic points.

I could not mention many interesting results in this area for it is huge: an incomplete bibliography already contains more than a hundred papers, and the theory of interval exchange transformations continues to develop.

[1] ARNOLD V. I. Small denominators and problems of stability of motion in classical and celestial mechanics. *Russian Math. Surveys*, 1963, **18**(6), 85–191.

[2] ARNOUX P. Thèse, Université de Reims, 1981.

[3] BOSHERNITZAN M. D., CARROLL C. R. An extension of Lagrange's theorem to interval exchange transformations over quadratic fields. *J. Anal. Math.*, 1997, **72**, 21–44.

[4] FORNI G. Deviation of ergodic averages for area-preserving flows on surfaces of higher genus. *Ann. Math., Ser.* 2, 2002, **155**(1), 1–103.

[5] KATOK A. Invariant measures of flows on orientable surfaces. *Sov. Math. Dokl.*, 1973, **14**, 1104–1108.

[6] KATOK A. B., SINAĬ YA. G., STEPIN A. M. Theory of dynamical systems and general transformation groups with invariant measure. In: Itogi Nauki i Tekhniki VINITI. Mathematical Analysis, Vol. 13. Moscow: VINITI, 1975, 129–262 (in Russian). [*The English translation*: *J. Sov. Math.*, 1977, **7**, 974–1065.]

[7] KEANE M. Interval exchange transformations. *Math. Z.*, 1975, **141**, 25–31.

[8] KEANE M. Non-ergodic interval exchange transformations. *Israel J. Math.*, 1977, **26**(2), 188–196.

[9] KEYNES H., NEWTON D. A "minimal," non-uniquely ergodic interval exchange transformation. *Math. Z.*, 1976, **148**(2), 101–105.

[10] KONTSEVICH M. L. Lyapunov exponents and Hodge theory. In: The Mathematical Beauty of Physics (Saclay, 1996). A memorial volume for Claude Itzykson. Editors: J. M. Drouffe and J. B. Zuber. River Edge, NJ: World Scientific, 1997, 318–322. (Adv. Ser. Math. Phys., 24.)

[11] KONTSEVICH M. L., ZORICH A. V. Connected components of the moduli spaces of Abelian differentials with prescribed singularities. [*Internet*: `http://www.arXiv.org/abs/math.GT/0201292`]

[12] MASUR H. Interval exchange transformations and measured foliations. *Ann. Math., Ser.* 2, 1982, **115**(1), 169–200.

[13] NOGUEIRA A. Almost all interval exchange transformations with flips are nonergodic. *Ergod. Theory Dynam. Systems*, 1989, **9**(3), 515–525.

[14] NOGUEIRA A., RUDOLPH D. Topological weak-mixing of interval exchange maps. *Ergod. Theory Dynam. Systems*, 1997, **17**(5), 1183–1209.

[15] OSELEDETS V. I. The spectrum of ergodic automorphisms. *Sov. Math. Dokl.*, 1966, **7**, 776–779.

[16] RAUZY G. Échanges d'intervalles et transformations induites. *Acta Arithm.*, 1979, **34**(4), 315–328.

[17] SATAEV E. A. On the number of invariant measures for flows on orientable surfaces. *Math. USSR, Izv.*, 1975, **9**, 813–830.

[18] SINAĬ YA. G., KHANIN K. M. Mixing for some classes of special flows over rotations of the circle. *Funct. Anal. Appl.*, 1992, **26**(3), 155–169.

[19] VEECH W. A. Strict ergodicity in zero dimensional dynamical systems and the Kronecker–Weyl theorem mod 2. *Trans. Amer. Math. Soc.*, 1969, **140**, 1–33.

[20] VEECH W. A. Interval exchange transformations. *J. Anal. Math.*, 1978, **33**, 222–272.

[21] VEECH W. A. Gauss measures for transformations on the space of interval exchange maps. *Ann. Math., Ser.* 2, 1982, **115**(1), 201–242.

[22] ZORICH A. V. Deviation for interval exchange transformations. *Ergod. Theory Dynam. Systems*, 1997, **17**(6), 1477–1499.

[23] ZORICH A. V. How do the leaves of a closed 1-form wind around a surface? In: Pseudoperiodic Topology. Editors: V. Arnold, M. Kontsevich and A. Zorich. Providence, RI: Amer. Math. Soc., 1999, 135–178. (AMS Transl., Ser. 2, 197; Adv. Math. Sci., 46.)

## ▽ 1958-2 — *S. A. Bogatyĭ*

*A word will not be lost and obsolete*
*if the're still many people for its repeat.*

— Hesiodes

ℛ The theory of equihedral[1] tetrahedra (i. e., tetrahedra with congruent faces) provides numerous arguments and statements in elemetary solid geometry which are themselves a wonderful beauty and moreover give a basis for various criteria for the isometry property of a mapping of a Euclidean space (of any dimension $\geq 3$) into itself.

At the moment, over 100 equivalent conditions are known that determine equihedral tetrahedra and deal with various elements of a tetrahedron. Let us recollect some of them.

**Theorem.** *For a tetrahedron, the following are equivalent*:

(1) *the lengths of skew edges are equal,*
(2) *all four faces are congruent,*
(3) *all four faces have equal perimeters,*
(4) *the products of edge lengths for each face are equal,*
(5) *all four faces have the same area,*
(6) *the circumscribed radii for all faces are equal,*
(7) *the lengths of edges meeting at each vertex have the same sum,*
(8) *the sum of planar angles at each vertex equals* 180°,
(9) *the development of the tetrahedron is an acute-angled triangle with drawn lines joining the midpoints of sides,*
(10) *the tetrahedron has three symmetry axes,*
(11) *the sum of cosines of all dihedral angles of the tetrahedron equals* 2,
(12) *the sum of cosines of the dihedral angles adjoining each face is* 1,
(13) *the sum of cosines of the dihedral angles adjoining each vertex is* 1,
(14) *alternate dihedral angles are equal,*
(15) *all four trihedral angles of the tetrahedron are congruent,*
(16) *all four trihedral angles of the tetrahedron have the same solid measure,*
(17) *the circumscribed sphere center coincides with the barycenter,*
(18) *the circumscribed sphere center coincides with the Fermat–Torricelli point,*
(19) *the centers of the circumscribed and the inscribed sphere coincide,*

[1] This term seems to be more appropriate than the term “isosceles” that is commonly used for such tetrahedra; they apparently conform to equilateral (regular) triangles rather than to isosceles ones.

(20) *the barycenter coincides with the Fermat–Torricelli point,*
(21) *the barycenter coincides with the inscribed sphere center,*
(22) *the inscribed sphere touchs each face at its circumscribed center,*
(23) *every escribed sphere touchs the corresponding face at its orthocenter,*
(24) *all four escribed spheres have equal radii,*
(25) *the planes that pass through each edge parallel to the skew edge form a rectangular parallelepiped* (*called* circumscribed),
(26) *the four new vetrices of the circumscribed parallelepiped are the four escribed sphere centers,*
(27) *for every edge, the ratio of its length and the sine of the adjoint dihedral angle is the same,*
(28) *for each face, the circumscribed center is the midpoint of the segment joining the orthocenter of this face and the projection of its opposite vertex,*
(29) *there exists a homothety that takes each vertex of the tetrahedron to the center of the corresponding escribed sphere,*
(30) *if a point is equidistant from the planes of all faces of the tetrahedron, then it is the center either of the inscribed sphere or of one of escribed spheres.*

Now it is difficult to find out the real authorship of various discoveries in the theory of equihedral tetrahedra, especially after they have been involved in the geometry course of the Kolmogorov school at Moscow University. Apparently, the book [4] contains the first substantive exposition of the theory of equihedral tetrahedra. Just after their discovery equihedral tetrahedra gained a wide knowledge [1, 2, 7, 10, 12, 14, 16, 17]. For instance, already in 1959 and 1960 problems on equihedral tetrahedra appeared at the written entrance exams to the Moscow Institute of Physics and Technology [6] (p. 108, Problem 111; p. 209, Problem 127), and in 1974 and 1994 such problems were similarly used at the Faculty of Computer Science of Moscow University [11, 15].[2] The theory of equihedral tetrahedra spread widely over the World; in 1964 it was included in the monograph [1] on solid geometry, and this book is now usually quoted as the origin. In connection with condition (6) one should note that there exists a non-equihedral tetrahedron whose faces have equal inscribed radii [5]. Condition (10) readily yields that in an equihedral tetrahedron every point depending symmetrically on the vertices (e. g.,

[2] I had certainly once proposed this problem on equihedral tetrahedra to the High School Moscow Mathematical Olympiad in the 1950s, knowing no predecessors, and it was solved there by many students. — *V. I. Arnold.*

the point for which the square root sum of the distances to the vertices is minimal) coincides with the intersection point of medians. This condition also plays an important role in the study of equihedral tetrahedra in the Lobachevskian space.

ℜ Starting from works by J. A. Lester [9] and H. Lenz [8], equihedral tetrahedra are used to obtain nontraditional characterizations of isometries of the Euclidean space, see also [3, 13]. For example: *If a mapping* $f: \mathbb{R}^n \to \mathbb{R}^n$, *where* $4 \leq n < \infty$, *takes every three points forming a regular triangle of area* 1 *to three points forming a triangle of area* 1, *then* $f$ *is an isometry. If a mapping* $f: \mathbb{R}^n \to \mathbb{R}^n$, *where* $3 \leq n < \infty$, *takes every three points forming a triangle of area* 1 *to three points forming a triangle of area* 1, *then* $f$ *is an isometry. If a mapping* $f: \mathbb{R}^n \to \mathbb{R}^n$, *where* $3 \leq n < \infty$, *takes every three points forming a regular triangle of perimeter* 1 *to three points forming a triangle of perimeter* 1, *then* $f$ *is an isometry.*

The theory of equihedral tetrahedra is an exclusively three-dimensional phenomenon. This is, to some extent, a manifestation of the fact that the full graph with four vertices has a unique nontrivial partition into nonseparable homogeneous subgraphs. The question on the "correct" many-dimensional analog of the theory of equihedral tetrahedra seems to be open; in particular, the extensions of relevant theorems to the case of higher dimension do not have such a perfect form.

[1] ALTSCHILLER-COURT N. Modern Pure Solid Geometry. New York: Chelsea, 1964.

[2] BOGATYĬ S. A. Equihedral Tetrahedra. Moscow: Moscow Center for Continuous Mathematical Education Press, to appear.

[3] BOGATAYA S. I., BOGATYĬ S. A., FROLKINA O. D. Affinity of volume-preserving mappings. *Moscow Univ. Math. Bull.*, 2001, **56**(6), 8–13.

[4] COUDERC P., BALLICIONI A. Premier livre du tétraèdre. Paris: Gauthier-Villars, 1953.

[5] DUBROVSKIĬ V. N. One more definition of an equihedral tetrahedron? *Kvant*, 1983, № 7, 51, 63 (in Russian).

[6] KALININ A. YU., TERESHIN D. A. Stereometry–11. Moscow: Moscow Institute of Physics and Technology Press, 2001 (in Russian).

[7] KUPITZ Y. S., MARTINI H. The Fermat–Torricelli point and isosceles tetrahedra. *J. Geometry*, 1994, **49**(1–2), 150–162.

[8] LENZ H. Über einen Satz von June Lester zur Charakterisierung euklidischer Bewegungen. *J. Geometry*, 1987, **28**(2), 197–201.

[9] LESTER J. A. Martin's theorem for Euclidean $n$-space and a generalization to the perimeter case. *J. Geometry*, 1986, **27**(1), 29–35.

[10] MATIZEN V. E. Equihedral and skeleton tetrahedra. *Kvant*, 1983, № 7, 34–38; Appendix to *Kvant*, 1995, № 1, 74–81 (in Russian).

[11] OVCHINNIKOV S., SHARYGIN I. F. Nonstandard problems in stereometry. *Kvant*, 1979, № 6, 33–38, 58 (in Russian).

[12] PRASOLOV V. V., SHARYGIN I. F. Problems in Stereometry. Moscow: Nauka, 1989 (in Russian).

[13] PRASOLOV V. V., TIKHOMIROV V. M. Geometry. Moscow: Moscow Center for Continuous Mathematical Education Press, 1997 (in Russian).

[14] Problem M9. *Kvant*, 1970, № 10, 42–44 (in Russian).

[15] SHARYGIN I. F. Adding on a tetrahedron. *Kvant*, 1976, № 1, 61–64 (in Russian).

[16] SHARYGIN I. F. Problem M353. *Kvant*, 1976, № 7, 31–32 (in Russian).

[17] VASIL′EV N. B. Problem M319. *Kvant*, 1975, № 12, 39–41 (in Russian).

## △▽ 1958-2 – *S. M. Gusein-Zade*

Now this problem is used in exercise books on analytic geometry for the first year students. See, e. g., [1], Problem 131.

[1] Problem Book in Analytic Geometry and Linear Algebra. Editor: Yu. M. Smirnov. Moscow: Physical and Mathematical Literature Publ., 2000 (in Russian).

## △ 1958-2 – *M. L. Kontsevich*

This problem was used as a “killer problem” given to Jewish candidates to the Mekh-mat. In [1] this problem was listed with the authorship of Yu. V. Nesterenko, and the year of the first appearance is 1974. The solution of the problem can be found in the preprint [2] by Ilan Vardi.

[1] SHEN′ A. Entrance examinations to the Mekh-mat. *Math. Intelligencer*, 1994, **16**(4), 6–10.

[2] VARDI I. Mekh-mat entrance examinations problems. Preprint, Institut des Hautes Études Scientifiques, M/00/06.
[*Internet*: `http://www.ihes.fr/PREPRINTS/M00/M00-06.ps.gz`]

## 1958-3 – *S. Yu. Yakovenko*

This question reappeared with some modification in [1], see also [2, 3], in connection with the problems on complexity of dynamical intersections.

In one of the formulations it is suggested to estimate the number of intersections between a fixed variety $Y$ and the saturation of another variety $X$ by trajectories of length $\leq N$ of a polynomial vector field in $\mathbb{R}^n$, with $\dim X + \dim Y = n-1$.

This problem was solved for $\dim X = 0$, when it reduces to the question on the number of intersections between an integral curve of a polynomial vector field, and an algebraic hypersurface. The bound, obtained by D. Novikov and S. Yakovenko [4, 5], depends polynomially on the magnitude of the coefficients and the "size" of the integral curve, while the power exponent is a computable but enormously fast growing function of the dimension $n$ and the degree of the field.

This result also holds in the complex space and can be applied to Picard–Fuchs equations for Abelian integrals. This yields some explicit bounds for the infinitesimal Hilbert problem, see the comment to problem *1978-6*.


[1] ARNOLD V. I. Dynamics of complexity of intersections. *Bol. Soc. Brasil. Mat.* (*N. S.*), 1990, **21**(1), 1–10. [*The Russian translation in*: Vladimir Igorevich Arnold. Selecta–60. Moscow: PHASIS, 1997, 489–499.]

[2] ARNOLD V. I. Dynamics of intersections. In: Analysis, et cetera. Research papers published in honor of Jürgen Moser's 60th birthday. Editors: P. H. Rabinowitz and E. Zehnder. Boston, MA: Academic Press, 1990, 77–84.

[3] ARNOLD V. I. Bounds for Milnor numbers of intersections in holomorphic dynamical systems. In: Topological Methods in Modern Mathematics. Proceedings of the symposium in honor of John Milnor's sixtieth birthday (Stony Brook, NY, 1991). Editors: L. R. Goldberg and A. V. Phillips. Houston, TX: Publish or Perish, 1993, 379–390.

[4] NOVIKOV D. I., YAKOVENKO S. YU. Meandering of trajectories of polynomial vector fields in the affine $n$-space. *Publ. Mat.*, 1997, **41**(1), 223–242.

[5] NOVIKOV D. I., YAKOVENKO S. YU. Trajectories of polynomial vector fields and ascending chains of polynomial ideals. *Ann. Inst. Fourier* (*Grenoble*), 1999, **49**(2), 563–609.


# 1959

## 1959-1

See the comment to problem *1972-20*.

# 1963

## ▽ 1963-1

ℋ This is a problem in paper [1] (p. 179: Problem I).

[1] ARNOLD V. I. Small denominators and problems of stability of motion in classical and celestial mechanics. *Uspekhi Mat. Nauk*, 1963, **18**(6), 91–192 (in Russian). [*The English translation*: *Russian Math. Surveys*, 1963, **18**(6), 85–191.]

## △▽ 1963-1 – *V. I. Arnold*

ℛ At the Kolmogorov centenary conference in Moscow (June 2003) Jonh Mather announced his new results on the diffusion genericity in systems with 2-dimensional Kolmogorov tori in the 5-dimensional space (for positive definite twisting quadratic forms), but his proofs are not yet published. However, he has recently sent me a very interesting preprint [1] saying:

"We announce a proof of the existence of Arnold diffusion for a large class of small perturbations of integrable Hamiltonian systems with positive normal torsion in the case of time periodic systems in two degrees of freedom and in the case of autonomous systems in three degrees of freedom."

[1] MATHER J. Arnold diffusion I: Announcement of the results. Preprint, Princeton University, November 25, 2002, 20 pp.

## △ 1963-1 – *M. B. Sevryuk* Also: *1966-3, 1994-33*

ℛ According to the Kolmogorov–Arnold–Moser (KAM) theory, the action variables in nearly integrable Hamiltonian systems with $1\frac{1}{2}$ or 2 degrees of freedom [1] change slightly during infinite time intervals (provided that the unperturbed integrable system satisfies appropriate nondegeneracy conditions) and just undergo oscillations with an amplitude of the order of $\sqrt{\varepsilon}$ where $0 < \varepsilon \ll 1$ is the perturbation parameter. On the other hand, in nearly integrable Hamiltonian systems with

---

[1] Recall that a Hamiltonian system with $n + \frac{1}{2}$ degrees of freedom ($n \in \mathbb{N}$) is by definition either a symplectomorphism of a $2n$-dimensional symplectic manifold or a nonautonomous Hamiltonian system of differential equations with the $2n$-dimensional phase space and right-hand side periodic in time.

$k \geq 2\frac{1}{2}$ degrees of freedom, the action variables $I$ may *a priori* exhibit considerable changes: for some solutions, the difference $|I(t)-I(0)|$ can attain large values (of the order of 1) for $|t|$ large. In 1964, V. I. Arnold [1] constructed his famous example of an analytic Hamiltonian system (close to a nondegenerate integrable system) with $2\frac{1}{2}$ degrees of freedom where such an evolution does take place. [2] B. V. Chirikov [18] coined, for this evolution, the physical term *the Arnold diffusion*. In preprint [18], the first general estimates for the diffusion rate were also evaluated, confirmed in numerical experiments reported in [34]. [3]

The diffusion rate in Arnold's example is of the order of $\exp(-1/\sqrt{\varepsilon})$, i. e., is exponentially small with respect to the perturbation parameter. According to the Nekhoroshev theorem (see the comment to problem *1966-2*), the averaged diffusion rate in analytic systems is always majorized by a quantity of the order of $\exp(-\varepsilon^{-a})$ for some $a > 0$, provided that suitable nondegeneracy conditions are met. For systems with $n \in \mathbb{N}$ degrees of freedom ($n \geq 3$), this estimate holds for $a = \frac{1}{2n}$.

For almost 40 years which have elapsed since paper [1] was published, the phenomenon of the Arnold diffusion has been studied and discussed by many authors and has been examined in various numerical experiments. Of the works of a physical nature considering the Arnold diffusion and related questions of the evolution in degenerate systems, mention is made here of the articles and books [19–21, 34, 47, 62, 73, 79–83] while "more mathematical" works are exemplified by papers [28, 41, 42, 48–50, 53, 54, 61, 71] (of all these works, papers [19,28,41,50,61] have contributed greatly to the development of our ideas on instability in nearly integrable Hamiltonian systems). In more recent several years, attention to the Arnold diffusion has increased considerably, leading to numerous interesting results and papers. Many of them have given rise to intensive discussions in mathematical periodicals, at conferences and on the Internet, see [3–11, 13–17, 22–25, 29–33, 35–40, 44–46, 51, 52, 55–60, 63–66, 70, 72, 74–78] (this list is definitely incomplete). For simplified diffusion models see, e. g., [12, 26].

Some of the recent papers devoted to the diffusion problem are in error, see, e. g., *Math. Reviews* 97g:58063, 98i:58094, 99f:58175, 2000b:37062, 2001j:37101. In particular, in review 99f:58175 of article [66] in *Math. Reviews* written by one of the authors (M. Rudnev), a "serious mistake" in the paper is pointed out: estimate (67) which is key for the proof of Theorem 2.1 is in error.

---

[2] See also [2], Ch. 4, § 23; p. 109–114.

[3] Note also that the descriptive term *the KAM theory* was first used in works [43, 80]. The author of the present comment is very grateful to Professor Chirikov for this information on the origin of the terms "Arnold diffusion" and "KAM theory."

Several counterexamples to the homoclinic splitting claimed in [66] are presented in note [38] (see also the erratum to [66] as well as papers [67,68]). Preprint [52] (a supplement to paper [51]) contains a critical survey of some recent papers devoted to the Arnold diffusion; see also Featured Reviews [69] (this is the review of paper [16]) and [27] (this is the review of paper [59]) in *Math. Reviews*.

Despite the efforts of many authors, the presence of diffusion in *generic* systems with $k \geq 2\frac{1}{2}$ degrees of freedom has not, as far as the author of this comment knows, been proved yet. In particular, the conjecture stated in problems *1966-3* and *1994-33* has been neither proved nor disproved although no expert is in serious doubt about its rightness.


[1] ARNOLD V. I. On the instability of dynamical systems with many degrees of freedom. *Sov. Math. Dokl.*, 1964, **5**(3), 581–585. [*The Russian original is reprinted in*: Vladimir Igorevich Arnold. Selecta–60. Moscow: PHASIS, 1997, 61–67.]

[2] ARNOLD V. I., AVEZ A. Ergodic Problems of Classical Mechanics, 2nd edition. Redwood City, CA: Addison-Wesley, 1989. [*The French original* 1967.] [*The first English edition* 1968.]

[3] BERNARD P. Perturbation d'un hamiltonien partiellement hyperbolique. *C. R. Acad. Sci. Paris, Sér.* I *Math.*, 1996, **323**(2), 189–194.

[4] BERTI M. Some remarks on a variational approach to Arnold's diffusion. *Discrete Contin. Dynam. Systems*, 1996, **2**(3), 307–314.

[5] BERTI M., BOLLE P. Diffusion time and splitting of separatrices for nearly integrable isochronous Hamiltonian systems. *Atti Accad. Naz. Lincei Cl. Sci. Fis. Mat. Natur. Rend. Lincei, Ser.* 9*, Mat. Appl.*, 2000, **11**(4), 235–243.

[6] BERTI M., BOLLE P. Fast Arnold diffusion in systems with three time scales. *Discrete Contin. Dynam. Systems*, 2002, **8**(3), 795–811.

[7] BERTI M., BOLLE P. A functional analysis approach to Arnold diffusion. *Ann. Institut Henri Poincaré, Analyse non linéaire*, 2002, **19**(4), 395–450.

[8] BESSI U. An approach to Arnold's diffusion through the calculus of variations. *Nonlinear Anal. Theory Methods Appl.*, 1996, **26**(6), 1115–1135.

[9] BESSI U. Arnold's diffusion with two resonances. *J. Differ. Equations*, 1997, **137**(2), 211–239.

[10] BESSI U. Arnold's example with three rotators. *Nonlinearity*, 1997, **10**(3), 763–781.

[11] BESSI U., CHIERCHIA L., VALDINOCI E. Upper bounds on Arnold diffusion times via Mather theory. *J. Math. Pures Appl., Sér.* 9, 2001, **80**(1), 105–129.

[12] BOLOTIN S. V., TRESHCHËV D. V. Unbounded growth of energy in nonautonomous Hamiltonian systems. *Nonlinearity*, 1999, **12**(2), 365–388.

[13] CHIERCHIA L. On the stability problem for nearly-integrable Hamiltonian systems. In: Seminar on Dynamical Systems (St. Petersburg, 1991). Editors: S. B. Kuksin, V. F. Lazutkin and J. Pöschel. Basel: Birkhäuser, 1994, 35–46.

[14] CHIERCHIA L. Arnold instability for nearly-integrable analytic Hamiltonian systems. In: Variational and Local Methods in the Study of Hamiltonian Systems (Trieste, 1994). Editors: A. Ambrosetti and G. F. Dell'Antonio. River Edge, NJ: World Scientific, 1995, 17–33.

[15] CHIERCHIA L. Non-degenerate 'Arnold diffusion'. Preprint, archived in `mp_arc@math.utexas.edu` #96-137.

[16] CHIERCHIA L., GALLAVOTTI G. Drift and diffusion in phase space. *Ann. Institut Henri Poincaré, Physique théorique*, 1994, **60**(1), 1–144; erratum: 1998, **68**(1), 135.

[17] CHIERCHIA L., VALDINOCI E. A note on the construction of Hamiltonian trajectories along heteroclinic chains. *Forum Math.*, 2000, **12**(2), 247–255.

[18] CHIRIKOV B. V. Research in the theory of nonlinear resonance and stochasticity. Preprint of the Novosibirsk Institute for Nuclear Physics, the USSR Academy of Sciences, 1969, № 267 (in Russian). [*The English translation*: *CERN Transl.*, 1971, № 71-40.]

[19] CHIRIKOV B. V. A universal instability of many-dimensional oscillator systems. *Phys. Rep.*, 1979, **52**(5), 263–379.

[20] CHIRIKOV B. V., FORD J., VIVALDI F. Some numerical studies of Arnold diffusion in a simple model. In: Nonlinear Dynamics and the Beam-Beam Interaction (New York, 1979). Editors: M. Month and J. C. Herrera. New York: American Institute of Physics, 1980, 323–340. (AIP Conference Proceedings, 57.)

[21] CHIRIKOV B. V., VECHESLAVOV V. V. KAM integrability. In: Analysis, et cetera. Research papers published in honor of Jürgen Moser's 60th birthday. Editors: P. H. Rabinowitz and E. Zehnder. Boston, MA: Academic Press, 1990, 219–236.

[22] CHIRIKOV B. V., VECHESLAVOV V. V. Arnold diffusion in large systems. *Zh. Eksp. Teor. Fiz.*, 1997, **112**(3), 1132–1146.

[23] CRESSON J. A $\lambda$-lemma for partially hyperbolic tori and the obstruction property. *Lett. Math. Phys.*, 1997, **42**(4), 363–377.

[24] CRESSON J. Temps d'instabilité des systèmes hamiltoniens initialement hyperboliques. *C. R. Acad. Sci. Paris, Sér.* I *Math.*, 2001, **332**(9), 831–834.

[25] CRESSON J. The transfer lemma for Graff tori and Arnold diffusion time. *Discrete Contin. Dynam. Systems*, 2001, **7**(4), 787–800.

[26] DELSHAMS A., DE LA LLAVE R., SEARA T. M. A geometric approach to the existence of orbits with unbounded energy in generic periodic perturbations by a potential of generic geodesic flows of $\mathbb{T}^2$. *Commun. Math. Phys.*, 2000, **209**(2), 353–392.

[27] DIACU F. N. Featured Review 97h:70014. *Math. Reviews*, 1997.

[28] DOUADY R. Stabilité ou instabilité des points fixes elliptiques. *Ann. Sci. École Norm. Sup., Sér.* 4, 1988, **21**(1), 1–46.

[29] EFTHYMIOPOULOS CH., VOGLIS N., CONTOPOULOS G. Diffusion and transient spectra in a 4-dimensional symplectic mapping. In: Analysis and Modelling of Discrete Dynamical Systems (Aussois, 1996). Editors: D. Benest and C. Froeschlé. Amsterdam: Gordon and Breach Sci. Publ., 1998, 91–106. (Advances Discrete Math. Appl., 1.)

[30] FONTICH E., MARTÍN P. Construction of some unstable symplectic maps. In: Proceedings of the 2nd Catalan Days on Applied Mathematics (Odeillo, 1995). Editors: M. Sofonea and J.-N. Corvellec. Perpignan: Presses Univ. Perpignan, 1995, 95–103.

[31] FONTICH E., MARTÍN P. Arnold diffusion in perturbations of analytic exact symplectic maps. *Nonlinear Anal.*, 2000, **42**(8), 1397–1412.

[32] FONTICH E., MARTÍN P. Differentiable invariant manifolds for partially hyperbolic tori and a lambda lemma. *Nonlinearity*, 2000, **13**(5), 1561–1593.

[33] FONTICH E., MARTÍN P. Arnold diffusion in perturbations of analytic integrable Hamiltonian systems. *Discrete Contin. Dynam. Systems*, 2001, **7**(1), 61–84.

[34] GADIYAK G. V., IZRAILEV F. M., CHIRIKOV B. V. Numerical experiments on the universal instability in nonlinear oscillator systems (Arnold diffusion). In: Proceedings of the 7th Intern. Conf. on Nonlinear Oscillations (Berlin, 1975); 1977, Vol. II, **1**, 315 (in Russian).

[35] GALLAVOTTI G. Arnold's diffusion in isochronous systems. *Math. Phys. Anal. Geom.*, 1999, **1**(4), 295–312.

[36] GALLAVOTTI G. Hamilton–Jacobi's equation and Arnold's diffusion near invariant tori in a priori unstable isochronous systems. *Rend. Semin. Mat. Univ. Politec. Torino*, 1999, **55**(4), 291–302.

[37] GALLAVOTTI G., GENTILE G., MASTROPIETRO V. Hamilton–Jacobi equation, heteroclinic chains and Arnold diffusion in three time scale systems. *Nonlinearity*, 2000, **13**(2), 323–340.

[38] GALLAVOTTI G., GENTILE G., MASTROPIETRO V. On homoclinic splitting problems. *Physica D*, 2000, **137**(1–2), 202–204.

[39] HALLER G. Diffusion at intersecting resonances in Hamiltonian systems. *Phys. Lett. A*, 1995, **200**(1), 34–42.

[40] HALLER G. Fast diffusion and universality near intersecting resonances. In: Hamiltonian Systems with Three or More Degrees of Freedom (S'Agaró, 1995). Editor: C. Simó. Dordrecht: Kluwer Acad. Publ., 1999, 391–397. (NATO Adv. Sci. Inst. Ser. C Math. Phys. Sci., 533.)

[41] HOLMES P. J., MARSDEN J. E. Mel′nikov's method and Arnold diffusion for perturbations of integrable Hamiltonian systems. *J. Math. Phys.*, 1982, **23**(4), 669–675.

[42] HOLMES P. J., MARSDEN J. E. Horseshoes and Arnold diffusion for Hamiltonian systems on Lie groups. *Indiana Univ. Math. J.*, 1983, **32**(2), 273–309.

[43] Izrailev F. M., Chirikov B. V. Stochasticity of the simplest dynamical model with divided phase space. Preprint of the Novosibirsk Institute for Nuclear Physics, the USSR Academy of Sciences, 1968, № 191 (in Russian).

[44] Kozlov V. V., Moshchevitin N. G. On diffusion in Hamiltonian systems. *Moscow Univ. Mech. Bull.*, 1997, **52**(5), 18–22.

[45] Kozlov V. V., Moshchevitin N. G. Diffusion in Hamiltonian systems. *Chaos*, 1998, **8**(1), 245–247.

[46] Liao X., Saari D. G., Xia Zh. Instability and diffusion in the elliptic restricted three-body problem. *Celest. Mech. Dynam. Astronom.*, 1998, **70**(1), 23–39.

[47] Lichtenberg A. J., Lieberman M. A. Regular and Chaotic Dynamics, 2nd edition. New York: Springer, 1992.

[48] Lieberman M. A. Arnold diffusion in Hamiltonian systems with three degrees of freedom. In: Nonlinear Dynamics (New York, 1979). Editor: R. H. G. Helleman. New York: New York Acad. Sci., 1980, 119–142. (Annals New York Acad. Sci., 357.)

[49] Lochak P. Effective speed of Arnold's diffusion and small denominators. *Phys. Lett. A*, 1990, **143**(1–2), 39–42.

[50] Lochak P. Canonical perturbation theory via simultaneous approximation. *Russian Math. Surveys*, 1992, **47**(6), 57–133.

[51] Lochak P. Arnold diffusion; a compendium of remarks and questions. In: Hamiltonian Systems with Three or More Degrees of Freedom (S'Agaró, 1995). Editor: C. Simó. Dordrecht: Kluwer Acad. Publ., 1999, 168–183. (NATO Adv. Sci. Inst. Ser. C Math. Phys. Sci., 533.)

[52] Lochak P. Supplement to "Arnold diffusion: a compendium of remarks and questions." Preprint, archived in mp_arc@math.utexas.edu # 98-293.

[53] MacKay R. S. Transition to chaos for area-preserving maps. In: Nonlinear Dynamics Aspects of Particle Accelerators. Editors: J. M. Jowett, M. Month and S. Turner. Berlin: Springer, 1986, 390–454. (Lecture Notes in Phys., 247.)

[54] MacKay R. S., Meiss J. D., Percival I. C. Transport in Hamiltonian systems. *Physica D*, 1984, **13**(1–2), 55–81.

[55] Marco J.-P. Transition le long des chaînes de tores invariants pour les systèmes hamiltoniens analytiques. *Ann. Institut Henri Poincaré, Physique théorique*, 1996, **64**(2), 205–252.

[56] Marco J.-P. Dynamics in the vicinity of double resonances. In: Proceedings of the IV Catalan Days of Applied Mathematics (Tarragona, 1998). Editors: C. García, C. Olivé and M. Sanromà. Tarragona: Univ. Rovira Virgili, 1998, 123–137.

[57] Marco J.-P. Transition orbits and transition times along chains of hyperbolic tori. In: Hamiltonian Systems with Three or More Degrees of Freedom (S'Agaró, 1995). Editor: C. Simó. Dordrecht: Kluwer Acad. Publ., 1999, 480–484. (NATO Adv. Sci. Inst. Ser. C Math. Phys. Sci., 533.)

[58] MASTROPIETRO V. Arnold diffusion and the d'Alembert precession problem. *Reg. Chaot. Dynamics*, 2001, **6**(4), 355–375.

[59] MOECKEL R. Transition tori in the five-body problem. *J. Differ. Equations*, 1996, **129**(2), 290–314.

[60] MORBIDELLI A. Chaotic diffusion in celestial mechanics. *Reg. Chaot. Dynamics*, 2001, **6**(4), 339–353.

[61] NEKHOROSHEV N. N. An exponential estimate of the stability time for Hamiltonian systems close to integrable ones, I. *Russian Math. Surveys*, 1977, **32**(6), 1–65.

[62] PERCIVAL I. C. Chaos in Hamiltonian systems. *Proc. Roy. Soc. London, Ser. A*, 1987, **413**(1844), 131–143.

[63] PERFETTI P. Fixed point theorems in the Arnold model about instability of the action-variables in phase-space. *Discrete Contin. Dynam. Systems*, 1998, **4**(2), 379–391.

[64] PUMARIÑO A., VALLS C. Three time scales systems exhibiting persistent Arnold's diffusion. Preprint, archived in `mp_arc@math.utexas.edu` #99-311.

[65] RUDNEV M., WIGGINS S. On the use of the Mel′nikov integral in the Arnold diffusion problem. Preprint, archived in `mp_arc@math.utexas.edu` #97-494.

[66] RUDNEV M., WIGGINS S. Existence of exponentially small separatrix splittings and homoclinic connections between whiskered tori in weakly hyperbolic near-integrable Hamiltonian systems. *Physica D*, 1998, **114**(1–2), 3–80; erratum: 2000, **145**(3–4), 349–354.

[67] RUDNEV M., WIGGINS S. On a partially hyperbolic KAM theorem. *Reg. Chaot. Dynamics*, 1999, **4**(4), 39–58.

[68] RUDNEV M., WIGGINS S. On a homoclinic splitting problem. *Reg. Chaot. Dynamics*, 2000, **5**(2), 227–242.

[69] SEVRYUK M. B. Featured Review 95b:58056. *Math. Reviews*, 1995.

[70] SIMÓ C., VALLS C. A formal approximation of the splitting of separatrices in the classical Arnold's example of diffusion with two equal parameters. *Nonlinearity*, 2001, **14**(6), 1707–1760.

[71] TENNYSON J. L., LIEBERMAN M. A., LICHTENBERG A. J. Diffusion in near-integrable Hamiltonian systems with three degrees of freedom. In: Nonlinear Dynamics and the Beam-Beam Interaction (New York, 1979). Editors: M. Month and J. C. Herrera. New York: American Institute of Physics Press, 1980, 272–301. (AIP Conference Proceedings, 57.)

[72] VALDINOCI E. Families of whiskered tori for a-priori stable/unstable Hamiltonian systems and construction of unstable orbits. *Math. Phys. Electron. J.*, 2000, **6**, Paper 2, 30 pp. (electronic).

[73] VIVALDI F. Weak instabilities in many-dimensional Hamiltonian systems. *Rev. Modern Phys.*, 1984, **56**(4), 737–754.

[74] Wood B. P., Lichtenberg A. J., Lieberman M. A. Arnold and Arnold-like diffusion in many dimensions. *Physica D*, 1994, **71**(1–2), 132–145.

[75] Xia Zh. Arnold diffusion in the elliptic restricted three-body problem. *J. Dyn. Differ. Equations*, 1993, **5**(2), 219–240.

[76] Xia Zh. Arnold diffusion and oscillatory solutions in the planar three-body problem. *J. Differ. Equations*, 1994, **110**(2), 289–321.

[77] Xia Zh. Arnold diffusion with degeneracies in Hamiltonian systems. In: Dynamical Systems and Chaos (Hachioji, 1994), Vol. 1: Mathematics, Engineering and Economics. Editors: N. Aoki, K. Shiraiwa and Y. Takahashi. River Edge, NJ: World Scientific, 1995, 278–285.

[78] Xia Zh. Arnold diffusion: a variational construction. In: Proceedings of the International Congress of Mathematicians, Vol. II (Berlin, 1998). *Doc. Math.*, 1998, Extra Vol. II, 867–877 (electronic).

[79] Zaslavskiĭ G. M. Stochasticity of Dynamical Systems. Moscow: Nauka, 1984 (in Russian).

[80] Zaslavskiĭ G. M., Chirikov B. V. Stochastic instability of nonlinear oscillations. *Uspekhi Fiz. Nauk*, 1971, **105**(1), 3–39 (in Russian, for the English translation see *Physics–Uspekhi*).

[81] Zaslavskiĭ G. M., Sagdeev R. Z., Usikov D. A., Chernikov A. A. Minimal chaos, stochastic web, and structures with symmetry of "quasicrystal" type. *Uspekhi Fiz. Nauk*, 1988, **156**(2), 193–251 (in Russian, for the English translation see *Physics–Uspekhi*).

[82] Zaslavskiĭ G. M., Sagdeev R. Z., Usikov D. A., Chernikov A. A. Weak Chaos and Quasi-regular Patterns. Cambridge: Cambridge University Press, 1991. (Cambridge Nonlinear Sci. Ser., 1.) [*The Russian original* 1991.]

[83] Zaslavskiĭ G. M., Zakharov M. Yu., Neĭshtadt A. I., Sagdeev R. Z., Usikov D. A., Chernikov A. A. Multidimensional Hamiltonian chaos. *Zh. Eksp. Teor. Fiz.*, 1989, **96**(5), 1563–1586 (in Russian).

## ▽ 1963-2

This is a problem in paper [1] (p. 179–180).

[1] Arnold V. I. Small denominators and problems of stability of motion in classical and celestial mechanics. *Uspekhi Mat. Nauk*, 1963, **18**(6), 91–192 (in Russian). [*The English translation*: *Russian Math. Surveys*, 1963, **18**(6), 85–191.]

## △ 1963-2 — *M. B. Sevryuk*

The presence of infinitely many periodic orbits in each neighborhood of the elliptic fixed point 0 of a generic analytic area-preserving mapping $(\mathbb{R}^2, 0) \hookleftarrow$ was established no later than in 1927 by G. D. Birkhoff [1] (Ch. 6, §§ 1–4; p. 150–165). In 1955, J. Moser [4] proved that by an arbitrarily small change of the coefficients of the Taylor series at the elliptic fixed point 0 of any analytic area-preserving mapping $(\mathbb{R}^2,0) \hookleftarrow$, one can obtain an analytic area-preserving diffeomorphism which has infinitely many *isolated*—hyperbolic and elliptic—periodic orbits in each neighborhood of the point 0.

The complete solution of the problem was presented in 1973 by E. Zehnder [6]. He proved that diffeomorphisms having infinitely many homoclinic points in each neighborhood of the point 0 constitute a residual (in the Baire sense) set in the space (equipped with some natural topology) of area-preserving mappings $(\mathbb{R}^2,0) \hookleftarrow$ of an arbitrary fixed smoothness class $C^r$, $1 \le r \le \omega$, with the elliptic fixed point 0. Recall that a *homoclinic* point is by definition a point of *transversal* intersection of stable $W^s(p)$ and unstable $W^u(q)$ separatrices of hyperbolic periodic points $p$ and $q$ lying in the same orbit of the mapping in question. The presence of homoclinic points is equivalent to separatrix splitting. Recall also that a subset of a topological space is said to be *residual* (in the sense of Baire) if this subset contains a countable intersection of open everywhere dense sets. Elements of such subset are sometimes said to be Baire *generic*. The inequality $1 \le r \le \omega$ includes the cases $r \in \mathbb{N}$, $r = \infty$, and $r = \omega$ ($C^\omega$ means real analyticity).

In paper [3], Zehnder's result was carried over to the spaces of analytic area-preserving mappings $(\mathbb{R}^2,0) \hookleftarrow$ with *prescribed* eigenvalues $e^{\pm 2\pi i\alpha}$ of the linearization at the elliptic fixed point 0. This paper contains also an analogous result for diffeomorphisms possessing so-called nondegenerate cantori (Aubry–Mather sets) in each neighborhood of the point 0.

Of survey works on this topic, we mention the book [5] (especially Ch. III, § 6; p. 99–107) and the paper [2].

[1] BIRKHOFF G. D. Dynamical Systems, 2nd edition. Providence, RI: Amer. Math. Soc., 1966. (Amer. Math. Soc. Colloquium Publ., 9.) [*The first edition* 1927.]

[2] CHENCINER A. La dynamique au voisinage d'un point fixe elliptique conservatif: de Poincaré et Birkhoff à Aubry et Mather. In: *Séminaire Bourbaki*, 1983–84, **622**; *Astérisque*, 1985, **121–122**, 147–170.

[3] GENECAND C. Transversal homoclinic orbits near elliptic fixed points of area-preserving diffeomorphisms of the plane. In: Dynamics Reported. Expositions in Dynamical Systems, Vol. 2. Editors: C. K. R. T. Jones, U. Kirchgraber and H.-O. Walther. Berlin: Springer, 1993, 1–30. (Dynam. Report.: Expos. Dynam. Syst., N. S., 2.)

[4] MOSER J. Nonexistence of integrals for canonical systems of differential equations. *Commun. Pure Appl. Math.*, 1955, **8**(3), 409–436.

[5] MOSER J. Stable and Random Motions in Dynamical Systems, with Special Emphasis on Celestial Mechanics. Princeton, NJ: Princeton University Press, 1973. (Ann. Math. Studies, 77.)

[6] ZEHNDER E. Homoclinic points near elliptic fixed points. *Commun. Pure Appl. Math.*, 1973, **26**(2), 131–182.

## ▽ 1963-3

H This is a problem in paper [1] (p. 180: Problem II).

[1] ARNOLD V. I. Small denominators and problems of stability of motion in classical and celestial mechanics. *Uspekhi Mat. Nauk*, 1963, **18**(6), 91–192 (in Russian). [*The English translation*: *Russian Math. Surveys*, 1963, **18**(6), 85–191.]

## △ 1963-3 — *M. B. Sevryuk*

R The perturbation sizes allowed by the rigorous (or, as one says, "analytic") proofs of various theorems in the KAM theory are, as a rule, very small—they are usually orders of magnitude smaller than the true perturbation threshold $\mu_*$ (which can be found numerically or from a combination of computer calculations and "analytic" reasoning) above which most invariant tori break up, see, e. g., [1, 2, 5–8, 11–22, 24–29, 37, 38, 40, 41]. This circumstance sometimes gives rise to the assertion that the KAM theory is not very suitable for practical purposes [38, 39], and one even speaks of its "numerical inadequacies" ([38], p. 135). In V. I. Arnold's papers [3, 4], bounded motions in planetary systems were constructed for the case where the masses and eccentricities of the planets are sufficiently small. J. Pöschel emphasized: "Concerning the Solar System Arnold [4] demonstrated the prevalence of quasi-periodic motions—provided the planets are of the size of tennis balls. Strictly speaking, the stability question is still open" ([39], p. 13).

A justification for the KAM theory from this viewpoint can be found in, e. g., [10], Section 2.7 (also see, e. g., [9], § 1.2). One of the most important conclusions of the KAM theory is that in the phase space of a generic Hamiltonian system with $n$ degrees of freedom, Cantor families of invariant tori of various dimensions $2 \leq m \leq n$ can occur, the $2m$-dimensional Hausdorff measure (the Lebesgue measure for $m = n$) of the union of these tori being positive. The existence of such families of tori requires no integrability, no special symmetries, and no conditions of the "equality type." Roughly speaking, the presence of families of invariant tori carrying quasi-periodic flows is a "codimension zero" property. In particular, the conjecture on the ergodicity of a generic Hamiltonian system on the energy level hypersurfaces (widespread in the physical literature up to the sixties) fails, cf. [34]. The statement that a certain property is of "codimension zero" cannot be verified by *any* computer calculations. Of course, one can often easily find invariant tori in the given system numerically, but in this case, one cannot guarantee that the invariant manifolds detected are genuine (e. g., that the action variables corresponding to an initial point on such a torus are indeed preserved *for ever* rather than just for a very long time). In turn, to establish the "typicality" of the families of tori, a precise estimate of the perturbation magnitude $\mu_*$ for which many invariant tori of the initial integrable system still survive is *entirely irrelevant*. The only important fact is that $\mu_* > 0$.

On the other hand, if we already know that the existence of families of invariant tori is a property of "codimension zero," then it is more suitable and expedient to look for the answers to all the quantitative questions (like the adequate estimate of $\mu_*$) numerically.

The scenario of the decay of the invariant torus with the given frequency vector as the perturbation grows is rather complicated and includes, e. g., the gradual loss of smoothness of the torus in question and its transformation into a so-called cantorus. Apart from papers [1, 2, 5–8, 11–22, 24–29, 37, 38, 40, 41], we quote here just several surveys [23, 30–33, 35, 36] treating the break-up of invariant tori and the increase in chaos in Hamiltonian systems.


[1] Abad J. J., Koch H. Renormalization and periodic orbits for Hamiltonian flows. *Commun. Math. Phys.*, 2000, **212**(2), 371–394.

[2] Abad J. J., Koch H., Wittwer P. A renormalization group for Hamiltonians: numerical results. *Nonlinearity*, 1998, **11**(5), 1185–1194.

[3] Arnold V. I. On the classical perturbation theory and the problem of stability of planetary systems. *Sov. Math. Dokl.*, 1962, **3**(4), 1008–1012. [*The Russian original is reprinted in*: Vladimir Igorevich Arnold. Selecta–60. Moscow: PHASIS, 1997, 39–45.]

[4] Arnold V. I. Small denominators and problems of stability of motion in classical and celestial mechanics. *Russian Math. Surveys*, 1963, **18**(6), 85–191.

[5] Berretti A., Gentile G. Scaling properties for the radius of convergence of a Lindstedt series: the standard map. *J. Math. Pures Appl., Sér.* 9, 1999, **78**(2), 159–176.

[6] Berretti A., Gentile G. Scaling properties for the radius of convergence of Lindstedt series: generalized standard maps. *J. Math. Pures Appl., Sér.* 9, 2000, **79**(7), 691–713.

[7] Berretti A., Gentile G. Bryuno function and the standard map. *Commun. Math. Phys.*, 2001, **220**(3), 623–656.

[8] BONETTO F., GENTILE G. On a conjecture for the critical behaviour of KAM tori. *Math. Phys. Electron. J.*, 1999, **5**, Paper 4, 8 pp. (electronic).

[9] BROER H. W., DUMORTIER F., VAN STRIEN S. J., TAKENS F. Structures in Dynamics (Finite Dimensional Deterministic Studies). Amsterdam: North-Holland, Elsevier, 1991. (Studies Math. Phys., 2.)

[10] BROER H. W., HUITEMA G. B., SEVRYUK M. B. Quasi-Periodic Motions in Families of Dynamical Systems: Order amidst Chaos. Berlin: Springer, 1996. (Lecture Notes in Math., 1645.)

[11] CELLETTI A. Rigorous and numerical determination of rotational invariant curves for the standard map. In: Analysis and Modelling of Discrete Dynamical Systems (Aussois, 1996). Editors: D. Benest and C. Froeschlé. Amsterdam: Gordon and Breach Sci. Publ., 1998, 149–180. (Advances Discrete Math. Appl., 1.)

[12] CELLETTI A., CHIERCHIA L. Rigorous estimates for a computer-assisted KAM theory. *J. Math. Phys.*, 1987, **28**(9), 2078–2086.

[13] CELLETTI A., CHIERCHIA L. Construction of analytic KAM surfaces and effective stability bounds. *Commun. Math. Phys.*, 1988, **118**(1), 119–161.

[14] CELLETTI A., CHIERCHIA L. A constructive theory of Lagrangian tori and computer-assisted applications. In: Dynamics Reported. Expositions in Dynamical Systems, Vol. 4. Editors: C. K. R. T. Jones, U. Kirchgraber and H.-O. Walther. Berlin: Springer, 1995, 60–129. (Dynam. Report.: Expos. Dynam. Syst., N. S., 4.)

[15] CELLETTI A., CHIERCHIA L. KAM stability estimates in Celestial Mechanics. *Planetary Space Sci.*, 1998, **46**(11–12), 1433–1440.

[16] CELLETTI A., FALCOLINI C., PORZIO A. Rigorous KAM stability statements for nonautonomous one-dimensional Hamiltonian systems. *Rend. Semin. Mat. Univ. Politec. Torino*, 1987, **45**(1), 43–70.

[17] CELLETTI A., FALCOLINI C., PORZIO A. Rigorous numerical stability estimates for the existence of KAM tori in a forced pendulum. *Ann. Institut Henri Poincaré, Physique théorique*, 1987, **47**(1), 85–111.

[18] CELLETTI A., FROESCHLÉ C. On the determination of the stochasticity threshold of invariant curves. *Intern. J. Bifurcation and Chaos*, 1995, **5**(6), 1713–1719.

[19] CELLETTI A., GIORGILLI A., LOCATELLI U. Improved estimates on the existence of invariant tori for Hamiltonian systems. *Nonlinearity*, 2000, **13**(2), 397–412.

[20] CHANDRE C., GOVIN M., JAUSLIN H. R. Kolmogorov–Arnold–Moser renormalization-group approach to the breakup of invariant tori in Hamiltonian systems. *Phys. Rev. E, Ser.* 3, 1998, **57**(2), part A, 1536–1543.

[21] CHANDRE C., GOVIN M., JAUSLIN H. R., KOCH H. Universality for the breakup of invariant tori in Hamiltonian flows. *Phys. Rev. E, Ser.* 3, 1998, **57**(6), 6612–6617.

[22] CHANDRE C., LASKAR J., BENFATTO G., JAUSLIN H. R. Determination of the threshold of the break-up of invariant tori in a class of three frequency Hamiltonian systems. *Physica D*, 2001, **154**(3–4), 159–170.

[23] CHENCINER A. La dynamique au voisinage d'un point fixe elliptique conservatif: de Poincaré et Birkhoff à Aubry et Mather. In: *Séminaire Bourbaki*, 1983–84, **622**; *Astérisque*, 1985, **121–122**, 147–170.

[24] GALLAVOTTI G., GENTILE G., MASTROPIETRO V. Field theory and KAM tori. *Math. Phys. Electron. J.*, 1995, **1**, Paper 5, 13 pp. (electronic).

[25] GENTILE G., MASTROPIETRO V. A possible mechanism for the KAM tori breakdown. In: Hamiltonian Systems with Three or More Degrees of Freedom (S'Agaró, 1995). Editor: C. Simó. Dordrecht: Kluwer Acad. Publ., 1999, 372–376. (NATO Adv. Sci. Inst. Ser. C Math. Phys. Sci., 533.)

[26] GREENE J. M. A method for determining a stochastic transition. *J. Math. Phys.*, 1979, **20**(6), 1183–1201.

[27] KOCH H. A renormalization group for Hamiltonians, with applications to KAM tori. *Ergod. Theory Dynam. Systems*, 1999, **19**(2), 475–521.

[28] KOCH H. On the renormalization of Hamiltonian flows, and critical invariant tori. *Discrete Contin. Dynam. Systems*, 2002, **8**(3), 633–646.

[29] LOCATELLI U., FROESCHLÉ C., LEGA E., MORBIDELLI A. On the relationship between the Bruno function and the breakdown of invariant tori. *Physica D*, 2000, **139**(1–2), 48–71.

[30] MACKAY R. S. Transition to chaos for area-preserving maps. In: Nonlinear Dynamics Aspects of Particle Accelerators. Editors: J. M. Jowett, M. Month and S. Turner. Berlin: Springer, 1986, 390–454. (Lecture Notes in Phys., 247.)

[31] MACKAY R. S. Renormalisation in Area-Preserving Maps. River Edge, NJ: World Scientific, 1993.

[32] MACKAY R. S., MEISS J. D., PERCIVAL I. C. Transport in Hamiltonian systems. *Physica D*, 1984, **13**(1–2), 55–81.

[33] MACKAY R. S., PERCIVAL I. C. Converse KAM: theory and practice. *Commun. Math. Phys.*, 1985, **98**(4), 469–512.

[34] MARKUS L., MEYER K. R. Generic Hamiltonian dynamical systems are neither integrable nor ergodic. *Mem. Amer. Math. Soc.*, 1974, **144**, 1–52.

[35] MATHER J. N., FORNI G. Action minimizing orbits in Hamiltonian systems. In: Transition to Chaos in Classical and Quantum Mechanics. Editor: S. Graffi. Berlin: Springer, 1994, 92–186. (Lecture Notes in Math., 1589.)

[36] MEISS J. D. Symplectic maps, variational principles, and transport. *Rev. Modern Phys.*, 1992, **64**(3), 795–848.

[37] OLVERA A., SIMÓ C. An obstruction method for the destruction of invariant curves. *Physica D*, 1987, **26**(1–3), 181–192.

[38] PERCIVAL I. C. Chaos in Hamiltonian systems. *Proc. Roy. Soc. London, Ser. A*, 1987, **413**(1844), 131–143.

[39] PÖSCHEL J. On small divisors with spatial structure. Habilitationsschrift, Rheinische Friedrich-Wilhelms-Universität, Bonn, 1989.

[40] WILBRINK J. Erratic behavior of invariant circles in standard-like mappings. *Physica D*, 1987, **26**(1–3), 358–368.

[41] YAMAGUCHI Y., TANIKAWA K. A remark on the smoothness of critical KAM curves in the standard mapping. *Progress Theor. Phys.*, 1999, **101**(1), 1–24.

## 1963-4

H This is a problem in paper [1] (p. 180–181: Problem III).

[1] ARNOLD V. I. Small denominators and problems of stability of motion in classical and celestial mechanics. *Uspekhi Mat. Nauk*, 1963, **18**(6), 91–192 (in Russian). [*The English translation*: *Russian Math. Surveys*, 1963, **18**(6), 85–191.]

R See the comment to problem *1972-20.*

## 1963-5

H This is a problem in paper [1] (p. 181–182: Problem IV).

[1] ARNOLD V. I. Small denominators and problems of stability of motion in classical and celestial mechanics. *Uspekhi Mat. Nauk*, 1963, **18**(6), 91–192 (in Russian). [*The English translation*: *Russian Math. Surveys*, 1963, **18**(6), 85–191.]

R See the comment to problem *1972-21.*

## ▽ 1963-6

H This is a problem in paper [1a] (§ 4, 1°; see also [1b], p. 52).

[1a] ARNOLD V. I., KRYLOV A. L. Uniform distribution of points on a sphere and some ergodic properties of solutions of linear ordinary differential equations in a complex region. *Dokl. Akad. Nauk SSSR*, 1963, **148**(1), 9–12 (in Russian). [*The English translation*: *Sov. Math. Dokl.*, 1963, **4**(1), 1–5.]

*Reprinted in*:

[1b] Vladimir Igorevich Arnold. Selecta–60. Moscow: PHASIS, 1997, 47–53.

## △▽ 1963-6 — *V. I. Arnold*

ℋ Let me briefly recollect only some of the extensions of the discussed problem that I dealt with (here open questions are of a great interest).

0. The asymptotics of interval permutations (1963, see [1]).

1. Estimates for the error of averaging method (my main accomplishments are presented in paper [2]). The research was continued by A. I. Neĭshtadt and V. I. Bakhtin, for the survey see books [4, 17].

2. Ergodic treatment of the Hopf invariant of a magnetic field and its applications to magnetic hydrodynamics [3]; continued by S. P. Novikov [21], B. A. Khesin, M. Freedman and others, for the survey see book [16] and paper [9].

Here is an interesting open question—give an ergodic (with averaging on four foliations) treatment for Novikov's generalization of the Hopf invariant to the Whitehead pseudoproduct: starting from two 2-forms $\alpha$, $\beta$ on $M^4$ satisfying the relations $\alpha^2 = \beta^2 = \alpha\beta = 0$, he constructs "products" which should be interpreted as ergodic type averages describing the 2-foliation $\alpha = 0$, the 2-foliation $\beta = 0$, the 1-foliation $\alpha = \beta = 0$, and the 3-foliation of the distribution $(\alpha = 0) \oplus (\beta = 0)$.

3. Dynamics of complexity of intersections $A^n X^k \cap Y^l$ in $M^{k+l+3}$ for $n \to \infty$ [7]. Here, for example, generalizations of the results from diffeomorphisms $A : M \to M$ to smooth mappings still are only conjectures (probably a lot of them hold for the growth of the number of periodic points, $A^n p = p$, for $n \to \infty$, but has not been proved).

The major ideas here are averaging on parameters in combination with the asymptotics on the growth of the "time" $n$. Here many things remain conjectural for the asymptotics of the number of homoclinic points. There are even local tough problems (see [8]).

4. Averaging in pseudoperiodic topology and phase transitions of the dependence of these averages on parameters. This was initiated in paper [5] (see also [6]), and continued by S. M. Gusein-Zade [18, 19]. Recently, in [14], I published Harnack type upper estimations of the averaged topological invariants of pseudoperiodic functions and varieties in terms of the periodic part Newton polyhedra (trigonometric polynomials degrees).

5. Averaging of the statistic of a sail (the convex hull of the set of lattice points in a simplicial cone) on the ball dimension.

This problem was reduced to ergodic theory of the action of $(\mathbb{R}^*)^{n-1}$ on $\mathrm{SL}(n+1,\mathbb{R})/\mathrm{SL}(n+1,\mathbb{Z})$ by M. L. Kontsevich and Yu. M. Sukhov [20]. Unfortunately, they proved only the existence of answers to my problems (and their invariance of an "almost arbitrary" cone), and the principal "physical" question on the behavior of the averages (and their comparison with those for more random polyhedra, for instance, on the mean number of lattice points on the edges of a sail) remained open. See the survey in my book [12].

6. Analogous "ergodic" questions seem to be still open for "random" quadratic irrationalities, say, on the statistic of elements of the period of a continued fraction $\lambda$ [for $\lambda^2+p\lambda+q=0$, $p^2+q^2\leq N^2$, $(p,q)\in\mathbb{Z}^2$, $N\to\infty$, or for $\det(A-\lambda\mathbf{1})=0$, $A\in\mathrm{SL}(2,\mathbb{Z})$, $\|A\|\leq N\to\infty$; in both cases $\lambda\in\mathbb{R}$].

Conjecturally, the averages are the same as those for "random" $\lambda\in\mathbb{R}$; but even the existence of a limit for $N\to\infty$ has not been proved.

It goes without saying that the same questions are interesting for algebraic numbers of degree $n$ and matrices from $\mathrm{SL}(n,\mathbb{Z})$, but for item 6 even computer experiments have not been made (for example, what triangulations of the torus $\mathbb{T}^2$ can be obtained by factorization on the symmetry group $\mathbb{Z}^2$ from a sail in $\mathbb{R}^3$ or from cubic fields).

7. There are also several works on "optimal control on the average"—phase transitions of its ergodic asymptotics etc., such as [13,15]. Only ergodically simple cases are considered there (say, the dynamics on $\mathbb{S}^1$), but crafty phase transitions already appear. And for more complicated ergodic properties, welcome are proofs even of the existence that reveal neither essential matter nor phase transitions ...

8. There is a new large-scope section of the theory of averaging, with not so many theorems yet, but, in my opinion, with new important views on problem statements (however, V. V. Kozlov affirms that similar statements had already been suggested by Poincaré for the purpose of justification of thermodynamics).

I gave this trend the name "theory of weak asymptotics" and propagated it in paper [10]. Probably numerous conjectures are the most interesting here.

9. I can also recall ergodic type conjectures and theorems on the distributions of the first digits of the areas of countries in the World (the distribution observed here is the same as that of the first digits of the deuce powers—about 30 % of units, 16 % of deuces and so on).

For example, if a country of area $S$ after each time period 1 splits into two countries of area $S/2$ with probability $p$, and unites with another country of area $S$ with probability $q$, then this very distribution establishes itself (rather rapidly) even if decays and unifications are given by more complicated matrices (splitting into three $S/3$, or into an $S/2$ and two $S/4$ etc.; and even if only neighbors can be united—precise formulations with the evolution of the neighbor graphs are available). A survey can be found in [11].

[1] ARNOLD V. I. Small denominators and problems of stability of motion in classical and celestial mechanics. *Russian Math. Surveys*, 1963, **18**(6), 85–191.

[2] ARNOLD V. I. Conditions for the applicability and estimate of the error of an averaging method for systems which pass through the states of resonance in the course of their evolution. *Sov. Math. Dokl.*, 1965, **6**, 331–334. [*The Russian original is reprinted in*: Vladimir Igorevich Arnold. Selecta–60. Moscow: PHASIS, 1997, 69–74.]

[3] ARNOLD V. I. The asymptotic Hopf invariant and its applications. In: Proceedings of the All-Union School on Differential Equations with Infinitely Many Independent Variables and on Dynamical Systems with Infinitely Many Degrees of Freedom (Dilizhan, May 21 – June 3, 1973). Yerevan: AS of Armenian SSR, 1974, 229–256 (in Russian). [*The English translation*: *Selecta Math. Sov.*, 1986, **5**(4), 327–345.] [*The Russian original is reprinted and supplemented in*: Vladimir Igorevich Arnold. Selecta–60. Moscow: PHASIS, 1997, 215–236.]

[4] ARNOLD V. I. Geometrical Methods in the Theory of Ordinary Differential Equations, 2nd edition. New York: Springer, 1988. (Grundlehren der Mathematischen Wissenschaften, 250.) [*The Russian original* 1978.]

[5] ARNOLD V. I. Remarks on quasicrystallic symmetries. *Physica D*, 1988, **33**(1–3), 21–25.

[6] ARNOLD V. I. Remarks on quasicrystallic symmetry. Appendix B in: KLEIN F. Lectures on the Icosahedron and Solution of Equations of Fifth Degree. Moscow: Nauka, 1989, 291–300 (in Russian). [*Reprinted in*: Vladimir Igorevich Arnold. Selecta–60. Moscow: PHASIS, 1997, 413–423.]

[7] ARNOLD V. I. Dynamics of complexity of intersections. *Bol. Soc. Brasil. Mat. (N. S.)*, 1990, **21**(1), 1–10. [*The Russian translation in*: Vladimir Igorevich Arnold. Selecta–60. Moscow: PHASIS, 1997, 489–499.]

[8] ARNOLD V. I. Bounds for Milnor numbers of intersections in holomorphic dynamical systems. In: Topological Methods in Modern Mathematics. Proceedings of the symposium in honor of John Milnor's sixtieth birthday (Stony Brook, NY, 1991). Editors: L. R. Goldberg and A. V. Phillips. Houston, TX: Publish or Perish, 1993, 379–390.

[9] ARNOLD V. I. Mathematical methods in classical physics. In: Trends and Perspectives in Applied Mathematics. Editors: F. John, J. E. Marsden, and L. Sirovich. New York: Springer, 1994, 1–20. (Appl. Math., 100.) [*The Russian translation in*: Vladimir Igorevich Arnold. Selecta–60. Moscow: PHASIS, 1997, 553–575.]

[10] ARNOLD V. I. Weak asymptotics for the numbers of solutions of Diophantine problems. *Funct. Anal. Appl.*, 1999, **33**(4), 292–293.

[11] ARNOLD V. I. The statistic of the first digits of the deuce powers and repartition of the World. In: Soft and Rigid Mathematical Models. Moscow: Moscow Center for Continuous Mathematical Education Press, 2000, 22–26 (in Russian). [*The German translation*: Die Mathematik und die Neue Teilung der Welt. *Hamburger Mosaik*, 2002, No. 5, 65–68.]

[12] ARNOLD V. I. Continued Fractions. Moscow: Moscow Center for Continuous Mathematical Education Press, 2001 (in Russian). ("Mathematical Education" Library, 14.)

[13] ARNOLD V. I. Optimization in mean and phase transitions in controlled dynamical systems. *Funct. Anal. Appl.*, 2002, **36**(2), 83–92.

[14] ARNOLD V. I. The longest curves of given degree and the quasicrystallic Harnack theorem in pseudoperiodic topology. *Funct. Anal. Appl.*, 2002, **36**(3), 165–171.

[15] ARNOLD V. I. On a variational problem connected with phase transitions of means in controllable dynamical systems. In: Nonlinear Problems in Mathematical Physics and Related Topics I. In honour of Professor O. A. Ladyzhenskaya. Editors: M. Sh. Birman, S. Hildebrandt, V. A. Solonnikov and N. N. Ural′tseva. Dordrecht: Kluwer Acad. Publ., 2002, 23–34. (Internat. Math. Ser., 1.)

[16] ARNOLD V. I., KHESIN B. A. Topological Methods in Hydrodynamics. New York: Springer, 1998. (Appl. Math. Sci., 125.)

[17] ARNOLD V. I., KOZLOV V. V., NEĬSHTADT A. I. Mathematical Aspects of Classical and Celestial Mechanics, 2nd edition. Berlin: Springer, 1993. (Encyclopædia Math. Sci., 3; Dynamical Systems, III.) [*The Russian original* 1985.] [*The second, revised and supplemented, Russian edition* 2002.]

[18] GUSEIN-ZADE S. M. The number of critical points of a quasiperiodic potential. *Funct. Anal. Appl.*, 1989, **23**(2), 129–130.

[19] GUSEIN-ZADE S. M. On the topology of quasiperiodic functions. In: Pseudoperiodic Topology. Editors: V. Arnold, M. Kontsevich and A. Zorich. Providence, RI: Amer. Math. Soc., 1999, 1–7. (AMS Transl., Ser. 2, 197; Adv. Math. Sci., 46.)

[20] KONTSEVICH M. L., SUKHOV YU. M. Statistics of Klein polyhedra and multidimensional continued fractions. In: Pseudoperiodic Topology. Editors: V. Arnold, M. Kontsevich and A. Zorich. Providence, RI: Amer. Math. Soc., 1999, 9–27. (AMS Transl., Ser. 2, 197; Adv. Math. Sci., 46.)

[21] NOVIKOV S. P. The analytic generalized Hopf invariant. Multivalued functionals. *Russian Math. Surveys*, 1984, **39**(5), 113–124.

## △ 1963-6 — *R. I. Grigorchuk*

R These theorems fail for the given way of averaging [1]. They become true if: the sequence of averages $f_n$ is replaced with the sequence of Cesàro averages $\frac{1}{n}\sum_{i=0}^{n-1} f_i$, and the group $\Gamma$ is free or close to free (say, hyperbolic), and some additional constraints providing regular asymptotics of the growth function $N(n)$ [2–4, 8] are imposed. Further information (and bibliography) about ergodic theorems for noncommutative transformation groups is contained in [5–7,9].

[1] BEWLEY T. Sur l'application des théorèmes ergodiques aux groupes libres de transformations: un contre-exemple. *C. R. Acad. Sci. Paris, Sér. A–B*, 1970, **270**, A1533–A1534.

[2] BUFETOV A. I. Operator ergodic theorems for actions of free semigroups and groups. *Funct. Anal. Appl.*, 2000, **34**(4), 239–251.

[3] GRIGORCHUK R. I. An individual ergodic theorem for free group actions. In: XII School on Operator Theory in Function Spaces (Tambov, 1987). Part 1. Tambov: Tambov State Pedagogical Institute, 1987, 57 (in Russian).

[4] GRIGORCHUK R. I. Ergodic theorems for the actions of a free group and a free semigroup. *Math. Notes*, 1999, **65**(5), 654–657.

[5] KRENGEL W. Ergodic Theorems. Berlin – New York: Walter de Gruyter, 1986.

[6] LINDENSTRAUSS E. Pointwise theorems for amenable groups. *Electron. Res. Announc. Amer. Math. Soc.*, 1999, **5**, 82–90 (electronic); *Invent. Math.*, 2001, **146**(2), 259–295.

[7] NEVO A. On discrete groups and pointwise ergodic theory. In: Random Walks and Discrete Potential Theory (Cortona, 1997). Cambridge: Cambridge University Press, 1999, 279–305. (Sympos. Math., 39.)

[8] NEVO A., STEIN E. M. A generalization of Birkhoff's pointwise ergodic theorem. *Acta Math.*, 1994, **173**, 135–154.

[9] TEMPELMAN A. Ergodic Theorems for Group Actions. Informational and Thermodynamical Aspects. Dordrecht: Kluwer Acad. Publ., 1992. (Math. Appl., 78.) [*Revised from the Russian original* 1986.]

## ▽ 1963-7

This is a problem in paper [1a] (§ 4, 2°; see also [1b], p. 52).

[1a] ARNOLD V. I., KRYLOV A. L. Uniform distribution of points on a sphere and some ergodic properties of solutions of linear ordinary differential equations in a complex region. *Dokl. Akad. Nauk SSSR*, 1963, **148**(1), 9–12 (in Russian). [*The English translation*: *Sov. Math. Dokl.*, 1963, **4**(1), 1–5.]

*Reprinted in*:

[1b] Vladimir Igorevich Arnold. Selecta–60. Moscow: PHASIS, 1997, 47–53.

## △ 1963-7 – *R. I. Grigorchuk*

It is unknown if the result extends to arbitrary (infinite) finitely generated groups (probably yes!). However, it is proved that it extends on hyperbolic groups satisfying certain conditions that yield the asymptotic form $Ca^n$ where $a$, $c$ are the constants from the growth function $N(n)$ of the group $\Gamma$ [1].

[1] GRIGORCHUK R. I. On the uniform distribution of orbits of actions of hyperbolic groups, in preparation.

## ▽ 1963-8

H This is a problem in paper [1a] (§ 4, 3°; see also [1b], p. 52).

[1a] ARNOLD V. I., KRYLOV A. L. Uniform distribution of points on a sphere and some ergodic properties of solutions of linear ordinary differential equations in a complex region. *Dokl. Akad. Nauk SSSR*, 1963, **148**(1), 9–12 (in Russian). [*The English translation*: *Sov. Math. Dokl.*, 1963, **4**(1), 1–5.]

*Reprinted in*:

[1b] Vladimir Igorevich Arnold. Selecta–60. Moscow: PHASIS, 1997, 47–53.

## △ 1963-8 — *R. I. Grigorchuk*

R Probably yes, although there are only preliminary results in this direction for the case of the Euclidean plane [1–3]. Unfortunately, in all these papers the group $G$ generated by two motions $A$, $B$ of the plane is considered either as the free semigroup with four generators $A$, $A^{-1}$, $B$, $B^{-1}$ (in [1,2]) or as the free group with two generators $A$, $B$, whereas $\Gamma/N$ is a solvable group ($N$ is the action's kernel). Hence, the action of $\Gamma$ is not exact, and each point of an orbit contributes according to its multiplicity. One of substantial difficulties concerning the complete solution of the problem on uniform distribution on the plane lies in the question on the asymptotic behavior for $n \to \infty$ of the growth function $N(n)$ of a 2-generated free solvable group of length 2.

[1] GUIVARCH Y. Equirepartition dans les espaces homogènes. In: Théorie ergodique (Actes Journées Ergodiques, Rennes, 1973/1974). Editors: J. P. Couze and M. S. Keane. New York: Springer, 1976, 131–142. (Lecture Notes in Math., 532.)

[2] KAZHDAN D. A. The uniform distribution on the plane. *Trudy Moskov. Mat. Obshch.*, 1965, **14**, 299–305 (in Russian).

[3] VOROBETS YA. B. On the uniform distribution of orbits of free group and semigroup actions on a plane. *Proc. Steklov Inst. Math.*, 2000, **231**, 59–89.

## ▽ 1963-9

H This is a problem in paper [1a] (§ 4, 4°; see also [1b], p. 52).

[1a] ARNOLD V. I., KRYLOV A. L. Uniform distribution of points on a sphere and some ergodic properties of solutions of linear ordinary differential equations in a complex region. *Dokl. Akad. Nauk SSSR*, 1963, **148**(1), 9–12 (in Russian). [*The English translation*: *Sov. Math. Dokl.*, 1963, **4**(1), 1–5.]

*Reprinted in*:
[1b] Vladimir Igorevich Arnold. Selecta–60. Moscow: PHASIS, 1997, 47–53.

## △ 1963-9 – *R. I. Grigorchuk*

R For the case where $G$ is the isometry group of the $n$-dimensional hyperbolic space, the individual ergodic theorem for radial averagings was proved in [2]; for Heisenberg groups—in [5]; for simple Lie groups of real rank 1—in [4]; for semisimple Lie groups with finite center and without compact factors having Kazhdan's $T$-property—in [3]; for connected semisimple groups with finite center and without nontrivial compact factors—in [1]. In all these papers, various versions of individual ergodic theorems in the spaces $L^p$ are proved; here $p$ runs over different intervals but always $p > 1$. There they mostly use averaging on a sequence of sets projecting into spheres or balls of the space $G/K$ (with respect to the invariant Riemannian metric) where $K$ is the maximal compact subgroup. The case $p = 1$ seems to remain open.

[1] MARGULIS G. A., NEVO A., STEIN E. Analogs of Wiener's ergodic theorems for semisimple Lie groups, II. *Duke Math. J.*, 2000, **103**(2), 233–259.

[2] NEVO A. Pointwise ergodic theorems for radial averages on simple Lie groups, I. *Duke Math. J.*, 1994, **76**(1), 113–140.

[3] NEVO A. Pointwise ergodic theorems for radial averages on simple Lie groups, II. *Duke Math. J.*, 1997, **86**(2), 239–259.

[4] NEVO A., STEIN E. Analogs of Wiener's ergodic theorems for semisimple groups, II. *Ann. Math., Ser.* 2, 1997, **145**(3), 565–595.

[5] NEVO A., THANGARELU S. Pointwise ergodic theorems for radial averages on the Heisenberg group. *Adv. Math.*, 1997, **127**(2), 307–334.

## 1963-10

H This is a problem in paper [1a] (§ 4, 5°; see also [1b], p. 52).

[1a] ARNOLD V. I., KRYLOV A. L. Uniform distribution of points on a sphere and some ergodic properties of solutions of linear ordinary differential equations in a complex region. *Dokl. Akad. Nauk SSSR*, 1963, **148**(1), 9–12 (in Russian). [*The English translation*: *Sov. Math. Dokl.*, 1963, **4**(1), 1–5.]

*Reprinted in*:
[1b] Vladimir Igorevich Arnold. Selecta–60. Moscow: PHASIS, 1997, 47–53.

## 1963-11

H This is a problem in paper [1a] (§ 4, 6°; see also [1b], p. 53).

[1a] ARNOLD V. I., KRYLOV A. L. Uniform distribution of points on a sphere and some ergodic properties of solutions of linear ordinary differential equations in a complex region. *Dokl. Akad. Nauk SSSR*, 1963, **148**(1), 9–12 (in Russian). [*The English translation*: *Sov. Math. Dokl.*, 1963, **4**(1), 1–5.]

*Reprinted in*:

[1b] Vladimir Igorevich Arnold. Selecta–60. Moscow: PHASIS, 1997, 47–53.

## 1963-12

H This is a problem in paper [1a] (§ 4, 7°; see also [1b], p. 53).

[1a] ARNOLD V. I., KRYLOV A. L. Uniform distribution of points on a sphere and some ergodic properties of solutions of linear ordinary differential equations in a complex region. *Dokl. Akad. Nauk SSSR*, 1963, **148**(1), 9–12 (in Russian). [*The English translation*: *Sov. Math. Dokl.*, 1963, **4**(1), 1–5.]

*Reprinted in*:

[1b] Vladimir Igorevich Arnold. Selecta–60. Moscow: PHASIS, 1997, 47–53.

# 1965

## 1965-1

H This is a problem in paper [1a] (Remarque A; see also [1b], p. 85: Remark A).

[1a] ARNOLD V. I. Sur une propriété topologique des applications globalement canoniques de la mécanique classique. *C. R. Acad. Sci. Paris*, 1965, **261**(19), 3719–3722.

*The Russian translation in*:

[1b] Vladimir Igorevich Arnold. Selecta–60. Moscow: PHASIS, 1997, 81–86.

R See the comment to problem *1972-33*.

### 1965-2

This is a problem in paper [1a] (Remarque C; see also [1b], p. 85: Remark C).

[1a] ARNOLD V. I. Sur une propriété topologique des applications globalement canoniques de la mécanique classique. *C. R. Acad. Sci. Paris*, 1965, **261**(19), 3719–3722.

*The Russian translation in*:

[1b] Vladimir Igorevich Arnold. Selecta–60. Moscow: PHASIS, 1997, 81–86.

See the comment to problem *1972-33*.

### 1965-3

This is a problem in paper [1a] (Remarque D; see also [1b], p. 86: Remark D).

[1a] ARNOLD V. I. Sur une propriété topologique des applications globalement canoniques de la mécanique classique. *C. R. Acad. Sci. Paris*, 1965, **261**(19), 3719–3722.

*The Russian translation in*:

[1b] Vladimir Igorevich Arnold. Selecta–60. Moscow: PHASIS, 1997, 81–86.

See the comment to problem *1972-33*.

## 1966

### 1966-1

This is a problem in paper [1a] (§ 2, Problem 2; see also [1b], p. 97).

[1a] ARNOLD V. I. The stability problem and ergodic properties of classical dynamical systems. In: Proceedings of the International Congress of Mathematicians (Moscow, 1966). Moscow: Mir, 1968, 387–392 (in Russian). [*The English translation*: *AMS Transl., Ser.* 2, 1968, **70**, 5–11.]

*The original is reprinted in*:

[1b] Vladimir Igorevich Arnold. Selecta–60. Moscow: PHASIS, 1997, 95–101.

See the comments to problems *1972-9* and *1972-10*.

## ▽ 1966-2

ℋ This is a problem in paper [1a] (§ 2, Problem 2; see also [1b], p. 97).

[1a] ARNOLD V. I. The stability problem and ergodic properties of classical dynamical systems. In: Proceedings of the International Congress of Mathematicians (Moscow, 1966). Moscow: Mir, 1968, 387–392 (in Russian). [*The English translation*: *AMS Transl., Ser.* 2, 1968, **70**, 5–11.]

*The original is reprinted in*:

[1b] Vladimir Igorevich Arnold. Selecta–60. Moscow: PHASIS, 1997, 95–101.

## △ 1966-2 — *M. B. Sevryuk*

ℛ In *analytic* nearly integrable Hamiltonian systems, the evolution of the action variables vanishes generically in any order of the perturbation theory and, moreover, it is *exponentially small* with respect to the perturbation parameter. This result constitutes the famous *Nekhoroshev theorem* [47, 49–51] and is sometimes called *effective stability* of the action variables (cf. [15, 17, 22, 32, 35] as well as [29]). One can formulate this theorem more precisely as follows. Consider a Hamiltonian system with $n$ degrees of freedom and the analytic Hamilton function

$$H = H_0(I) + \varepsilon H_1(I, \varphi, \varepsilon), \tag{1}$$

where $(I, \varphi) \in G \times \mathbb{T}^n$ are the action–angle variables ($G$ being a domain in $\mathbb{R}^n$) while $0 \le \varepsilon \le 1$ is the perturbation parameter. Let the unperturbed Hamilton function $H_0(I)$ satisfy certain nondegeneracy conditions called *steepness*. Then there exist positive constants $a$, $b$, $R_*$, $K_*$, and $\varepsilon_*$ such that for any solution $\varphi(t)$, $I(t)$ of the Hamiltonian system of equations

$$\begin{aligned} \frac{d\varphi}{dt} &= \omega(I) + \varepsilon \frac{\partial H_1(I, \varphi, \varepsilon)}{\partial I}, \qquad \omega(I) := \frac{\partial H_0(I)}{\partial I}, \\ \frac{dI}{dt} &= -\varepsilon \frac{\partial H_1(I, \varphi, \varepsilon)}{\partial \varphi} \end{aligned} \tag{2}$$

afforded by Hamilton function (1), the inequality

$$|I(t) - I(0)| \le R_* \varepsilon^b \quad \text{for} \quad |t| \le \exp(K_* \varepsilon^{-a}) \tag{3}$$

holds provided that $0 < \varepsilon \le \varepsilon_*$.

The constants $a$ and $b$ in inequality (3) are called the *stability exponents*, of them, exponent $a$ being the most important one. These exponents depend on the number $n$ of degrees of freedom and the so-called *steepness indices* of the function $H_0(I)$ and vanish as $n \to \infty$. One usually calls the quantity $\mathcal{T}(\varepsilon) = \exp(K_* \varepsilon^{-a})$ the *stability time*, the distance $R(\varepsilon) = R_* \varepsilon^b$ the *radius of confinement*, and the constant $\varepsilon_*$ the *threshold of validity*.

This theorem was announced by N. N. Nekhoroshev [47] in 1971. The complete proof was given in papers [50, 51]. In Nekhoroshev's works, the exponent $a$ had asymptotics $\mathrm{const}/n^2$. In the mid eighties, G. Benettin, L. Galgani, G. Gallavotti, and A. Giorgilli [10, 11, 23] examined the case of *convex* functions $H_0(I)$ (see below) in more detail. In 1993, J. Pöschel [58] obtained the value $a = 1/(2n)$ for *quasi-convex* functions $H_0(I)$ (see below). On the other hand, P. Lochak published in 1992 an essentially new proof of the Nekhoroshev theorem for quasi-convex unperturbed Hamilton functions (see [38, 40, 41] as well). In the works by A. Delshams and P. Gutiérrez [16, 17], the Kolmogorov theorem on the persistence of invariant tori and the Nekhoroshev theorem are proved in parallel.

The steepness condition is very weak: the nonsteep functions $H_0(I)$ constitute a set of infinite codimension in an appropriate functional space [48, 50]. For the precise definition and/or detailed discussion of the steepness property, see Nekhoroshev's papers [47–51]. Yu. S. Il'yashenko [31] obtained the following sufficient condition for steepness:

**Theorem** [31]. *Let the unperturbed analytic Hamilton function $H_0(I)$ be defined in some neighborhood of the closure of a bounded domain $G \subset \mathbb{R}^n$. Assume that $H_0$ does not possess critical points while the restriction of $H_0$ to any affine subspace of the space $\mathbb{R}^n$ of each dimension from 1 to $n-1$ has $\mathbb{C}$-isolated critical points only. Then the function $H_0$ is steep in G.*

Up to now, the precise values of the steepness indices have been determined for typical functions $H_0(I)$ only in the cases of two and three degrees of freedom [36] (see the comment to problem *1978-3*).

Among all steep functions $H_0(I)$, the "steepest" ones are convex and quasi-convex functions.

**Definition.** An unperturbed Hamilton function $H_0(I)$, for which the frequency vector $\omega(I) = \partial H_0(I)/\partial I$ nowhere vanishes, is said to be *convex* if there exists a constant $c > 0$ such that

$$\left\langle \eta, \frac{\partial \omega(I)}{\partial I} \eta \right\rangle \geq c|\eta|^2 \tag{4}$$

for all $I \in \overline{G}$ and $\eta \in \mathbb{R}^n$ (the angular brackets here denote the scalar product in $\mathbb{R}^n$). A function $H_0(I)$ is said to be *quasi-convex* if there exists a constant $c > 0$ such that inequality (4) holds whenever $I \in \overline{G}$, $\eta \in \mathbb{R}^n$, and $\langle \omega(I), \eta \rangle = 0$.

One easily verifies that quasi-convexity of a function $H_0$ means convexity of its level hypersurfaces $\{H_0 = \text{const}\}$.

For quasi-convex unperturbed Hamilton functions, the Nekhoroshev estimate (3) turns out to hold for $a = b = 1/(2n)$ [16,17,38,40,41,58]. Moreover, one can take

$$a = \frac{\mu}{2n}, \qquad b = \frac{1-\mu}{2} + \frac{\mu}{2n}$$

for any $0 < \mu \leq 1$ [38, 58]. These values of the stability exponents seem to be optimal (cf. [64]). On the "physical level of rigor," the value $a \approx 1/(2n)$ was obtained no later than in 1979 in paper [14]. On the other hand, the estimates of the stability exponents $a$ and $b$ can be improved considerably near *resonant* unperturbed tori [37,38] (see [40,41,49,58] as well).

For nonsteep unperturbed Hamilton functions, the action variables $I$ in system (2) can change with rate $\sim \varepsilon$ [50]. However, exponential estimates on the evolution rate for the action variables may sometimes be obtained for highly degenerate (and even linear) functions $H_0(I)$ as well, provided that the perturbation "removes" the degeneracy [39]. To be more precise, paper [39] treats Hamilton functions (1) of the form

$$H(I, \varphi, \varepsilon) = H_0(I) + \varepsilon H_{1,1}(I) + \varepsilon^2 H_{1,2}(I, \varphi)$$

with $H_0(I)$ linear and $H_{1,1}(I)$ convex. An exponential estimate on the evolution rate for variables $I$ in the case of the linear function $H_0(I) = \langle \omega_0, I \rangle$ takes place also when the perturbation is arbitrary but the constant frequency vector $\omega_0$ is Diophantine [11,15,22].

Exponential estimates on the evolution rate for the action variables in the case of degenerate unperturbed Hamilton functions are especially important in the problems of Celestial Mechanics. Of numerous recent works devoted to employing various analogues of the Nekhoroshev theorem in the studies of the stability of the Solar System, we mention [8, 12, 13, 19, 25, 26, 30, 34, 35, 46, 54, 55, 63]. In papers [54,55], the Lochak method was used.

The estimates of Nekhoroshev's original papers [47,50,51] were somewhat refined in preprints [61,62].

Another aspect of the effective absence of evolution of the action variables in nearly integrable Hamiltonian systems is the so-called *superexponential "stickiness"* of the Kolmogorov tori discovered by A. Morbidelli and A. Giorgilli [45] in

1995. It turns out that if the unperturbed Hamilton function is quasi-convex then all the trajectories in a perturbed system, starting at a distance $0 < \rho \leq \rho_*$ from a Kolmogorov torus with a Diophantine frequency vector, remain close to this torus for an exceedingly long time of the order of

$$\exp\left\{\exp\left[\left(\frac{\rho_*}{\rho}\right)^r\right]\right\},$$

provided that $\varepsilon$ is sufficiently small. Here $\rho_* > 0$ is a certain constant independent of $\varepsilon$, while $r > 0$ is another constant determined by the arithmetical properties of the frequency vector of the torus in question. Note that the small parameter here is the initial distance from the invariant torus rather than the perturbation magnitude.

In earlier works [43,57] devoted to the "stickiness" of the Kolmogorov tori, the authors obtained only an exponential estimate on the "confinement time." In papers [21,24,42,44], A. Morbidelli and A. Giorgilli established the presence of a hierarchy of domains $\{\mathcal{G}_m\}_{m\in\mathbb{N}}$, $\mathcal{G}_{m+1} \subset \mathcal{G}_m$ for each $m$, with increasing stability characteristics in the phase space $\mathcal{G} = G \times \mathbb{T}^n$ of a nearly integrable Hamiltonian system. The exponential "stickiness" of invariant tori of dimensions smaller than the number of degrees of freedom was proved in works [18,21,22,28,33,56,60].

Very recently, the so-called "analytically filtered Fourier analysis" of chaotic motions in analytic nearly integrable Hamiltonian systems was introduced ([27], see also [29]) which has led to a spectral formulation of the Nekhoroshev theorem [27,29].

The Nekhoroshev theorem can be carried over *mutatis mutandis* to *infinite dimensional* Hamiltonian systems. Here, we would confine ourselves to pointing out several important references: [1–7,9,20,52,53,59].


[1] BAMBUSI D. A Nekhoroshev-type theorem for the Pauli–Fierz model of classical electrodynamics. *Ann. Institut Henri Poincaré, Physique théorique*, 1994, **60**(3), 339–371.

[2] BAMBUSI D. Exponential stability of breathers in Hamiltonian networks of weakly coupled oscillators. *Nonlinearity*, 1996, **9**(2), 433–457.

[3] BAMBUSI D. Long time stability of some small amplitude solutions in nonlinear Schrödinger equations. *Commun. Math. Phys.*, 1997, **189**(1), 205–226.

[4] BAMBUSI D. Nekhoroshev theorem for small amplitude solutions in nonlinear Schrödinger equations. *Math. Z.*, 1999, **230**(2), 345–387.

[5] BAMBUSI D. On long time stability in Hamiltonian perturbations of non-resonant linear PDEs. *Nonlinearity*, 1999, **12**(4), 823–850.

[6] BAMBUSI D., GIORGILLI A. Exponential stability of states close to resonance in infinite-dimensional Hamiltonian systems. *J. Stat. Phys.*, 1993, **71**(3–4), 569–606.

[7] BAMBUSI D., NEKHOROSHEV N. N. A property of exponential stability in nonlinear wave equations near the fundamental linear mode. *Physica D*, 1998, **122**(1–4), 73–104.

[8] BENETTIN G., FASSÒ F., GUZZO M. Nekhoroshev-stability of $L_4$ and $L_5$ in the spatial restricted three-body problem. *Reg. Chaot. Dynamics*, 1998, **3**(3), 56–72; erratum: 1998, **3**(4), 48.

[9] BENETTIN G., FRÖHLICH J., GIORGILLI A. A Nekhoroshev-type theorem for Hamiltonian systems with infinitely many degrees of freedom. *Commun. Math. Phys.*, 1988, **119**(1), 95–108.

[10] BENETTIN G., GALGANI L., GIORGILLI A. A proof of Nekhoroshev's theorem for the stability times in nearly integrable Hamiltonian systems. *Celest. Mech.*, 1985, **37**(1), 1–25.

[11] BENETTIN G., GALLAVOTTI G. Stability of motions near resonances in quasi-integrable Hamiltonian systems. *J. Stat. Phys.*, 1986, **44**(3–4), 293–338.

[12] CELLETTI A., CHIERCHIA L. On the stability of realistic three-body problems. *Commun. Math. Phys.*, 1997, **186**(2), 413–449.

[13] CELLETTI A., FERRARA L. An application of the Nekhoroshev theorem to the restricted three-body problem. *Celest. Mech. Dynam. Astronom.*, 1996, **64**(3), 261–272.

[14] CHIRIKOV B. V. A universal instability of many-dimensional oscillator systems. *Phys. Rep.*, 1979, **52**(5), 263–379.

[15] DELSHAMS A., GUTIÉRREZ P. Effective stability for nearly integrable Hamiltonian systems. In: Proceedings of Intern. Conf. on Differential Equations ($Equa\frac{\partial i}{\partial t}ff$ *91*) (Barcelona, 1991), Vol. 1. Editors: C. Perelló, C. Simó and J. Solà-Morales. Singapore: World Scientific, 1993, 415–420.

[16] DELSHAMS A., GUTIÉRREZ P. Nekhoroshev and KAM theorems revisited via a unified approach. In: Hamiltonian Mechanics: Integrability and Chaotic Behavior (Toruń, 1993). Editor: J. Seimenis. New York: Plenum Press, 1994, 299–306. (NATO Adv. Sci. Inst. Ser. B Phys., 331.)

[17] DELSHAMS A., GUTIÉRREZ P. Effective stability and KAM theory. *J. Differ. Equations*, 1996, **128**(2), 415–490.

[18] FASSÒ F., GUZZO M., BENETTIN G. Nekhoroshev-stability of elliptic equilibria of Hamiltonian systems. *Commun. Math. Phys.*, 1998, **197**(2), 347–360.

[19] GABERN F., JORBA À. A restricted four-body model for the dynamics near the Lagrangian points of the Sun–Jupiter system. *Discrete Contin. Dynam. Systems, Ser. B*, 2001, **1**(2), 143–182.

[20] GIORGILLI A. Energy equipartition and Nekhoroshev-type estimates for large systems. In: Hamiltonian Dynamical Systems: History, Theory, and Applications (Cincinnati, 1992). Editors: H. S. Dumas, K. R. Meyer and D. S. Schmidt. New York: Springer, 1995, 147–161. (The IMA Volumes in Math. and Appl., 63.)

[21] GIORGILLI A. On the problem of stability for near to integrable Hamiltonian systems. In: Proceedings of the International Congress of Mathematicians, Vol. III (Berlin, 1998). *Doc. Math.*, 1998, Extra Vol. III, 143–152 (electronic).

[22] GIORGILLI A., DELSHAMS A., FONTICH E., GALGANI L., SIMÓ C. Effective stability for a Hamiltonian system near an elliptic equilibrium point, with an application to the restricted three body problem. *J. Differ. Equations*, 1989, **77**(1), 167–198.

[23] GIORGILLI A., GALGANI L. Rigorous estimates for the series expansions of Hamiltonian perturbation theory. *Celest. Mech.*, 1985, **37**(2), 95–112.

[24] GIORGILLI A., MORBIDELLI A. Invariant KAM tori and global stability for Hamiltonian systems. *Z. Angew. Math. Phys.*, 1997, **48**(1), 102–134.

[25] GIORGILLI A., SKOKOS CH. On the stability of the Trojan asteroids. *Astron. Astrophys.*, 1997, **317**, 254–261.

[26] GUZZO M. Nekhoroshev stability of quasi-integrable degenerate Hamiltonian systems. *Reg. Chaot. Dynamics*, 1999, **4**(2), 78–102.

[27] GUZZO M., BENETTIN G. A spectral formulation of the Nekhoroshev theorem and its relevance for numerical and experimental data analysis. *Discrete Contin. Dynam. Systems, Ser. B*, 2001, **1**(1), 1–28.

[28] GUZZO M., FASSÒ F., BENETTIN G. On the stability of elliptic equilibria. *Math. Phys. Electron. J.*, 1998, **4**, Paper 1, 16 pp. (electronic).

[29] GUZZO M., LEGA E., FROESCHLÉ C. On the numerical detection of the effective stability of chaotic motions in quasi-integrable systems. *Physica D*, 2002, **163**(1–2), 1–25.

[30] GUZZO M., MORBIDELLI A. Construction of a Nekhoroshev like result for the asteroid belt dynamical system. *Celest. Mech. Dynam. Astronom.*, 1996/97, **66**(3), 255–292.

[31] IL'YASHENKO YU. S. A steepness test for analytic functions. *Russian Math. Surveys*, 1986, **41**(1), 229–230.

[32] JORBA À., SIMÓ C. Effective stability for periodically perturbed Hamiltonian systems. In: Hamiltonian Mechanics: Integrability and Chaotic Behavior (Toruń, 1993). Editor: J. Seimenis. New York: Plenum Press, 1994, 245–252. (NATO Adv. Sci. Inst. Ser. B Phys., 331.)

[33] JORBA À., VILLANUEVA J. On the normal behaviour of partially elliptic lower-dimensional tori of Hamiltonian systems. *Nonlinearity*, 1997, **10**(4), 783–822.

[34] JORBA À., VILLANUEVA J. Numerical computation of normal forms around some periodic orbits of the restricted three-body problem. *Physica D*, 1998, **114**(3–4), 197–229.

[35] JORBA À., VILLANUEVA J. Effective stability around periodic orbits of the spatial RTBP. In: Hamiltonian Systems with Three or More Degrees of Freedom (S'Agaró, 1995). Editor: C. Simó. Dordrecht: Kluwer Acad. Publ., 1999, 628–632. (NATO Adv. Sci. Inst. Ser. C Math. Phys. Sci., 533.)

[36] LANDIS E. E. Uniform steepness indices. *Uspekhi Mat. Nauk*, 1986, **41**(4), 179 (in Russian).

[37] LOCHAK P. Canonical perturbation theory via simultaneous approximation. *Russian Math. Surveys*, 1992, **47**(6), 57–133.

[38] LOCHAK P. Hamiltonian perturbation theory: periodic orbits, resonances and intermittency. *Nonlinearity*, 1993, **6**(6), 885–904.

[39] LOCHAK P. Stability of Hamiltonian systems over exponentially long times: the near-linear case. In: Hamiltonian Dynamical Systems: History, Theory, and Applications (Cincinnati, 1992). Editors: H. S. Dumas, K. R. Meyer and D. S. Schmidt. New York: Springer, 1995, 221–229. (The IMA Volumes in Math. and Appl., 63.)

[40] LOCHAK P., NEĬSHTADT A. I. Estimates of stability time for nearly integrable systems with a quasiconvex Hamiltonian. *Chaos*, 1992, **2**(4), 495–499.

[41] LOCHAK P., NEĬSHTADT A. I., NIEDERMAN L. Stability of nearly integrable convex Hamiltonian systems over exponentially long times. In: Seminar on Dynamical Systems (St. Petersburg, 1991). Editors: S. B. Kuksin, V. F. Lazutkin and J. Pöschel. Basel: Birkhäuser, 1994, 15–34.

[42] MORBIDELLI A. Bounds on diffusion in phase space: connection between Nekhoroshev and KAM theorems and superexponential stability of invariant tori. In: Hamiltonian Systems with Three or More Degrees of Freedom (S'Agaró, 1995). Editor: C. Simó. Dordrecht: Kluwer Acad. Publ., 1999, 514–517. (NATO Adv. Sci. Inst. Ser. C Math. Phys. Sci., 533.)

[43] MORBIDELLI A., GIORGILLI A. Quantitative perturbation theory by successive elimination of harmonics. *Celest. Mech. Dynam. Astronom.*, 1993, **55**(2), 131–159.

[44] MORBIDELLI A., GIORGILLI A. On a connection between KAM and Nekhoroshev's theorems. *Physica D*, 1995, **86**(3), 514–516.

[45] MORBIDELLI A., GIORGILLI A. Superexponential stability of KAM tori. *J. Stat. Phys.*, 1995, **78**(5–6), 1607–1617.

[46] MORBIDELLI A., GUZZO M. The Nekhoroshev theorem and the asteroid belt dynamical system. *Celest. Mech. Dynam. Astronom.*, 1996/97, **65**(1–2), 107–136.

[47] NEKHOROSHEV N. N. On the behavior of Hamiltonian systems close to integrable ones. *Funct. Anal. Appl.*, 1971, **5**(4), 338–339.

[48] NEKHOROSHEV N. N. Stable lower estimates for smooth mappings and the gradients of smooth functions. *Math. USSR, Sb.*, 1973, **19**(3), 425–467.

[49] NEKHOROSHEV N. N. The method of successive canonical changes of variables. An addendum to: MOSER J. Lectures on Hamiltonian Systems. Moscow: Mir, 1973, 150–164 (in Russian).

[50] NEKHOROSHEV N. N. An exponential estimate of the stability time for Hamiltonian systems close to integrable ones, I. *Russian Math. Surveys*, 1977, **32**(6), 1–65.

[51] NEKHOROSHEV N. N. An exponential estimate of the stability time for Hamiltonian systems close to integrable ones, II. *Trudy Semin. Petrovskogo*, 1979, **5**, 5–50 (in Russian). [*The English translation in*: Topics in Modern Mathematics. Editor: O. A. Oleĭnik. New York: Consultant Bureau, 1985, 1–58. (Petrovskiĭ Semin., 5.)]

[52] NEKHOROSHEV N. N. Exponential stability of the approximate fundamental mode of a nonlinear wave equation. *Funct. Anal. Appl.*, 1999, **33**(1), 69–71.

[53] NEKHOROSHEV N. N. Strong stability of the approximate fundamental mode of a nonlinear string equation. *Trudy Moskov. Mat. Obshch.*, 2002, **63**, 166–236 (in Russian, for the English translation see *Trans. Moscow Math. Soc.*).

[54] NIEDERMAN L. Stability over exponentially long times in the planetary problem. In: From Newton to Chaos: Modern Techniques for Understanding and Coping with Chaos in $N$-body Dynamical Systems (Cortina d'Ampezzo, 1993). Editors: A. E. Roy and B. A. Steves. New York: Plenum Press, 1995, 109–118. (NATO Adv. Sci. Inst. Ser. B Phys., 336.)

[55] NIEDERMAN L. Stability over exponentially long times in the planetary problem. *Nonlinearity*, 1996, **9**(6), 1703–1751.

[56] NIEDERMAN L. Nonlinear stability around an elliptic equilibrium point in a Hamiltonian system. *Nonlinearity*, 1998, **11**(6), 1465–1479.

[57] PERRY A. D., WIGGINS S. KAM tori are very sticky: rigorous lower bounds on the time to move away from an invariant Lagrangian torus with linear flow. *Physica D*, 1994, **71**(1–2), 102–121.

[58] PÖSCHEL J. Nekhoroshev estimates for quasi-convex Hamiltonian systems. *Math. Z.*, 1993, **213**(2), 187–216.

[59] PÖSCHEL J. On Nekhoroshev estimates for a nonlinear Schrödinger equation and a theorem by Bambusi. *Nonlinearity*, 1999, **12**(6), 1587–1600.

[60] PÖSCHEL J. On Nekhoroshev's estimate at an elliptic equilibrium. *Internat. Math. Res. Notices*, 1999, **4**, 203–215.

[61] ROSIKOV YU. V. The normalization of a perturbed system of Hamiltonian equations in the resonant case. Preprint of the Moscow Space Research Institute, Russia Academy of Sciences, 1994, № 1892, 37 pp. (in Russian).

[62] ROSIKOV YU. V. A refinement of the exponential estimate on the stability time in nearly integrable Hamiltonian systems. Preprint of the Moscow Space Research Institute, Russia Academy of Sciences, 1994, № 1893, 34 pp. (in Russian).

[63] STEICHEN D., GIORGILLI A. Long time stability for the main problem of artificial satellites. *Celest. Mech. Dynam. Astronom.*, 1997/98, **69**(3), 317–330.

[64] VECHESLAVOV V. V., CHIRIKOV B. V. What is the rate of the Arnold diffusion equal to? Preprint of the Novosibirsk Institute for Nuclear Physics, the USSR Academy of Sciences, 1989, № 72 (in Russian).

## 1966-3

ℋ This is a problem in paper [1a] (§ 3, Conjecture; see also [1b], p. 98).

[1a] ARNOLD V. I. The stability problem and ergodic properties of classical dynamical systems. In: Proceedings of the International Congress of Mathematicians (Moscow, 1966). Moscow: Mir, 1968, 387–392 (in Russian). [*The English translation*: *AMS Transl., Ser.* 2, 1968, **70**, 5–11.]

*The original is reprinted in*:

[1b] Vladimir Igorevich Arnold. Selecta–60. Moscow: PHASIS, 1997, 95–101.

ℛ See the comment to problem *1963-1*.

## 1966-4

ℋ This is a problem in paper [1a] (§ 4; see also [1b], p. 99). The conjectures on the number of fixed points of symplectomorphisms were first formulated by V. I. Arnold in paper [2a] (see also [2b]), see problems *1965-1 – 1965-3*.

[1a] ARNOLD V. I. The stability problem and ergodic properties of classical dynamical systems. In: Proceedings of the International Congress of Mathematicians (Moscow, 1966). Moscow: Mir, 1968, 387–392 (in Russian). [*The English translation*: *AMS Transl., Ser.* 2, 1968, **70**, 5–11.]

*The original is reprinted in*:

[1b] Vladimir Igorevich Arnold. Selecta–60. Moscow: PHASIS, 1997, 95–101.

[2a] ARNOLD V. I. Sur une propriété topologique des applications globalement canoniques de la mécanique classique. *C. R. Acad. Sci. Paris*, 1965, **261**(19), 3719–3722.

*The Russian translation in*:

[2b] Vladimir Igorevich Arnold. Selecta–60. Moscow: PHASIS, 1997, 81–86.

ℛ See the comment to problem *1972-33*.

## 1966-5

ℋ This is a problem in paper [1a] (§ 4; see also [1b], p. 99). The conjectures on the number of fixed points of symplectomorphisms were first formulated by V. I. Arnold in paper [2a] (see also [2b]), see problems *1965-1 – 1965-3*.

[1a] ARNOLD V. I. The stability problem and ergodic properties of classical dynamical systems. In: Proceedings of the International Congress of Mathematicians (Moscow, 1966). Moscow: Mir, 1968, 387–392 (in Russian). [*The English translation*: *AMS Transl., Ser.* 2, 1968, **70**, 5–11.]

*The original is reprinted in*:

[1b] Vladimir Igorevich Arnold. Selecta–60. Moscow: PHASIS, 1997, 95–101.

[2a] ARNOLD V. I. Sur une propriété topologique des applications globalement canoniques de la mécanique classique. *C. R. Acad. Sci. Paris*, 1965, **261**(19), 3719–3722.

*The Russian translation in*:

[2b] Vladimir Igorevich Arnold. Selecta–60. Moscow: PHASIS, 1997, 81–86.

R See the comment to problem *1972-33*.

## 1966-6

H This is a problem in paper [1a] (§ 4, Problem; see also [1b], p. 100).

[1a] ARNOLD V. I. The stability problem and ergodic properties of classical dynamical systems. In: Proceedings of the International Congress of Mathematicians (Moscow, 1966). Moscow: Mir, 1968, 387–392 (in Russian). [*The English translation*: *AMS Transl., Ser.* 2, 1968, **70**, 5–11.]

*The original is reprinted in*:

[1b] Vladimir Igorevich Arnold. Selecta–60. Moscow: PHASIS, 1997, 95–101.

# 1968

## 1968-2

R See the comment to problem *1976-12* by S. L. Tabachnikov.

# 1969

### 1969-1 – *V. D. Sedykh*

The invariance of the volume of any polyhedron under its flexure was proved in [2] (the idea belongs to I. Sabitov who proved previously the indicated conjecture for polyhedra that are homeomorphic to a sphere [3]).

The same issue of the journal *Beiträge ...* contains paper [1] by V. A. Aleksandrov where a counterexample to the analogous statement in the spherical space is given.

[1] ALEKSANDROV V. A. An example of a flexible polyhedron with nonconstant volume in the spherical space. *Beiträge zur Algebra und Geometrie*, 1997, **38**(1), 11–18.

[2] CONNELLY R., SABITOV I., WALZ A. The bellows conjecture. *Beiträge zur Algebra und Geometrie*, 1997, **38**(1), 1–10.

[3] SABITOV I. KH. On the problem of the invariance of the volume of a deformable polyhedron. *Russian Math. Surveys*, 1995, **50**(2), 451–452.

### 1969-2

See the comment to problem *1998-5*.

# 1970

### 1970-1 – *M. B. Sevryuk*

For every pair $(M,G)$ with $M$ a manifold and $G$ a Lie group acting on this manifold, one can define the concept of a *versal unfolding* (also called *versal deformation*) of an arbitrary element $m \in M$ with respect to the action of the group $G$. The problem concerns the simplest (but very important) case where $M$ is a subset of the space $\mathrm{gl}(N,\mathbb{D}) = \mathbb{D}^{N^2}$ of matrices of order $N$ over a (not necessarily commutative) field $\mathbb{D}$ while $G$ is a subgroup of the group $\mathrm{GL}(N,\mathbb{D})$ of nonsingular matrices of order $N$ (it is supposed that $G$ acts by conjugation and leaves $M$ invariant). Below we list the most important works devoted to versal unfoldings of matrices.

Versal unfoldings of arbitrary complex matrices, i. e., for

$$M = \mathrm{gl}(N,\mathbb{C}), \quad G = \mathrm{GL}(N,\mathbb{C})$$

were constructed by V. I. Arnold in paper [1] and are discussed in detail in his works [2, 3] as well.

Versal unfoldings of arbitrary real matrices:

$$M = \mathrm{gl}(N,\mathbb{R}), \quad G = \mathrm{GL}(N,\mathbb{R})$$

were constructed by D. M. Galin in note [10].

The author of the present comment has failed to find a description of versal unfoldings of arbitrary quaternionic matrices:

$$M = \mathrm{gl}(N,\mathbb{H}) = \mathrm{u}^*(2N), \quad G = \mathrm{GL}(N,\mathbb{H}) = \mathrm{U}^*(2N)$$

in the literature. Note, however, that one should apply the standard concepts of matrix algebra to quaternionic matrices with care since the field of quaternions is not commutative. The space $\mathrm{gl}(N,\mathbb{H})$ is a Lie algebra over $\mathbb{R}$.

Before proceeding to versal unfoldings of the elements of the classical Lie and Jordan algebras, recall some definitions. Let $\mathbb{D}$ be one of the fields $\mathbb{R}$, $\mathbb{C}$, or $\mathbb{H}$. Let also $\sigma\colon \mathbb{D} \to \mathbb{D}$ be either the identity transformation, or the complex conjugation involution

$$\sigma\colon\ z = a + bi \mapsto \bar{z} = a - bi$$

(in the case $\mathbb{D} = \mathbb{C}$), or the quaternionic conjugation involution

$$\sigma\colon\ r = a + bi + cj + dk \mapsto \bar{r} = a - bi - cj - dk$$

(in the case $\mathbb{D} = \mathbb{H}$). Consider the following involutions acting on $\mathrm{gl}(N,\mathbb{D})$:

$$L \mapsto L^{\mathrm{c}} = \sigma(L^{\mathrm{t}})$$

(the superscript t means taking the transposed matrix) and

$$L \mapsto \alpha(L) = KL^{\mathrm{c}}K^{-1}, \qquad K \text{ is nonsingular}, \quad K^{\mathrm{c}} = \varepsilon K, \quad \varepsilon = \pm 1$$

($K \in \mathrm{GL}(N,\mathbb{D})$ being a fixed matrix subject to the conditions indicated). The space of $\alpha$-skew-symmetric matrices

$$M_- = \{X \in \mathrm{gl}(N,\mathbb{D}) \mid \alpha(X) = -X\}$$

is closed with respect to commutation $[X,Y] = XY - YX$ and is isomorphic to one of the classical Lie algebras listed in the table below. The space of $\alpha$-symmetric matrices

$$M_+ = \{X \in \mathrm{gl}(N,\mathbb{D}) \mid \alpha(X) = X\}$$

is closed with respect to symmetrized multiplication $X \circ Y = XY + YX$ and is isomorphic to one of the classical Jordan algebras. Recall that a *Jordan algebra* [1] is a commutative (but not necessarily associative) algebra with the identity $(x^2y)x \equiv x^2(yx)$. Any associative algebra becomes a Jordan algebra if one proceeds to the new multiplication $x \circ y = xy + yx$, just as any associative algebra becomes a Lie algebra if one proceeds to the new multiplication $[x,y] = xy - yx$. The space

$$G = \{A \in \mathrm{GL}(N,\mathbb{D}) \mid \alpha(A) = A^{-1}\}$$

is closed with respect to matrix multiplication and is a Lie group. Both the spaces $M_-$ and $M_+$ are invariant with respect to the action of $G$ on $\mathrm{gl}(N,\mathbb{D})$ by conjugation.

All the possible involutions $\alpha$ up to the natural equivalence relation and the corresponding nomenclature of the classical Lie algebras $M_-$ and Lie groups $G$ are presented in the table (the last column of this table shows the field $\{\eta \in \mathbb{D} \mid \sigma(\eta) = \eta\}$ over which one has to consider the given Lie algebra and the corresponding Jordan algebra).

| field $\mathbb{D}$ | $\sigma(\eta)$ | $\varepsilon$ | signature $K$ | algebra $M_-$ | group $G$ | $N$ | field Fix $\sigma$ |
|---|---|---|---|---|---|---|---|
| $\mathbb{R}$ | $\eta$ | 1 | $(p,q)$ | $\mathrm{o}(p,q)$ | $\mathrm{O}(p,q)$ | $p+q$ | $\mathbb{R}$ |
| $\mathbb{R}$ | $\eta$ | $-1$ | | $\mathrm{sp}(2n,\mathbb{R})$ | $\mathrm{SP}(2n,\mathbb{R})$ | $2n$ | $\mathbb{R}$ |
| $\mathbb{C}$ | $\eta$ | 1 | | $\mathrm{o}(N,\mathbb{C})$ | $\mathrm{O}(N,\mathbb{C})$ | $N$ | $\mathbb{C}$ |
| $\mathbb{C}$ | $\eta$ | $-1$ | | $\mathrm{sp}(2n,\mathbb{C})$ | $\mathrm{SP}(2n,\mathbb{C})$ | $2n$ | $\mathbb{C}$ |
| $\mathbb{C}$ | $\bar{\eta}$ | 1 | $(p,q)$ | $\mathrm{u}(p,q)$ | $\mathrm{U}(p,q)$ | $p+q$ | $\mathbb{R}$ |
| $\mathbb{H}$ | $\bar{\eta}$ | 1 | $(p,q)$ | $\mathrm{sp}(p,q)$ | $\mathrm{SP}(p,q)$ | $p+q$ | $\mathbb{R}$ |
| $\mathbb{H}$ | $\bar{\eta}$ | $-1$ | | $\mathrm{o}^*(2n)$ | $\mathrm{O}^*(2n)$ | $n$ | $\mathbb{R}$ |

All the necessary details and proofs (or references) are given in papers [5,6,19,23] (the table above is reproduced from [6]).

[1] The theory of Jordan algebras is expounded in detail in books [4,14].

Note that one often writes just $\mathrm{sp}(2n)$ instead of $\mathrm{sp}(2n,\mathbb{R})$, whereas the notations $\mathrm{o}(N,0)$ and $\mathrm{u}(N,0)$ are often replaced with $\mathrm{o}(N)$ and $\mathrm{u}(N)$, respectively (similar conventions hold for the corresponding Lie groups as well).

Versal unfoldings of real Hamiltonian matrices:

$$M = \mathrm{sp}(2n,\mathbb{R}), \quad G = \mathrm{SP}(2n,\mathbb{R})$$

($M$ is the space of real Hamiltonian matrices of the given even order $2n$, while $G$ is the group of real symplectic linear operators) were found independently by D. M. Galin, by three Canadian mathematicians and physicists J. Patera, C. Rousseau, D. Schlomiuk, and by H. Koçak. Papers [11, 15] were devoted exclusively to the algebra $\mathrm{sp}(2n,\mathbb{R})$. It is worthwhile to note that the formulae for the number of parameters in miniversal unfoldings pointed out in these two works seem to be different but are in fact equivalent. In paper [20], versal unfoldings for the elements of all the classical real Lie algebras $\mathrm{o}(p,q)$, $\mathrm{sp}(2n,\mathbb{R})$, $\mathrm{u}(p,q)$, $\mathrm{sp}(p,q)$, and $\mathrm{o}^*(2n)$ were constructed.

Versal unfoldings for the elements of all the classical complex Lie algebras $\mathrm{o}(N,\mathbb{C})$ and $\mathrm{sp}(2n,\mathbb{C})$ were found by J. Patera and C. Rousseau [17].

Versal unfoldings for the elements of all the classical Jordan algebras corresponding to the classical Lie algebras $\mathrm{o}(p,q)$, $\mathrm{sp}(2n,\mathbb{R})$, $\mathrm{o}(N,\mathbb{C})$, $\mathrm{sp}(2n,\mathbb{C})$, $\mathrm{u}(p,q)$, $\mathrm{sp}(p,q)$, and $\mathrm{o}^*(2n)$ were constructed in paper [18].

Versal unfoldings of real equivariant Hamiltonian matrices [in this case, $M \subset \mathrm{sp}(2n,\mathbb{R})$ is the space of the real Hamiltonian matrices of order $2n$ that commute with the action on $\mathbb{R}^{2n}$ of the given compact Lie group $\Gamma$ preserving the symplectic structure, while $G \subset \mathrm{SP}(2n,\mathbb{R})$ is the group of the real symplectic linear operators that commute with the action of $\Gamma$] were described in work [16].

Versal unfoldings of real infinitesimally reversible matrices (here $M$ is the space of the real matrices that anti-commute with the given involutive matrix $R$, while $G$ is the group of the linear operators that commute with $R$) were constructed independently by M. B. Sevryuk [21] and C.-W. Shih [22].

Versal unfoldings of matrices that are simultaneously Hamiltonian and infinitesimally reversible (one assumes that the reversing linear involution $R$ is antisymplectic, i. e., satisfies the identity $\langle Rx, Ry\rangle \equiv -\langle x,y\rangle$, where $\langle\cdot,\cdot\rangle$ is the linear symplectic structure) were found in work [24].

Versal unfoldings of Hamiltonian matrices, infinitesimally reversible matrices, and matrices that are simultaneously Hamiltonian and infinitesimally reversible were treated from a unified viewpoint by I. Hoveijn [13].

In works [9, 12], M. I. García and co-authors constructed versal unfoldings of *pairs* of complex matrices of sizes $n \times n$ and $n \times m$ with respect to a certain action of the so-called state feedback group.

In work [7] (see also [8]), the authors found versal unfoldings of *pairs* of complex matrices $(A, B)$ of the same size $m \times n$, i. e., those of *matrix pencils* $A - \lambda B$ with respect to the natural action of the group $\mathrm{GL}(m, \mathbb{C}) \times \mathrm{GL}(n, \mathbb{C})$: $A - \lambda B \xmapsto{(P,Q)} P^{-1}(A - \lambda B)Q$, where $P \in \mathrm{GL}(m, \mathbb{C})$, $Q \in \mathrm{GL}(n, \mathbb{C})$.


[1] ARNOLD V. I. On matrices depending on parameters. *Russian Math. Surveys*, 1971, **26**(2), 29–43. [*The Russian original is reprinted in*: Vladimir Igorevich Arnold. Selecta–60. Moscow: PHASIS, 1997, 155–173.]

[2] ARNOLD V. I. Lectures on bifurcations and versal families. *Russian Math. Surveys*, 1972, **27**(5), 54–123.

[3] ARNOLD V. I. Geometrical Methods in the Theory of Ordinary Differential Equations, 2nd edition. New York: Springer, 1988, § 30. (Grundlehren der Mathematischen Wissenschaften, 250.) [*The Russian original* 1978.]

[4] BRAUN H., KOECHER M. Jordan-Algebren. Berlin: Springer, 1966.

[5] BURGOYNE N., CUSHMAN R. Conjugacy classes in linear groups. *J. Algebra*, 1977, **44**(2), 339–362.

[6] DJOKOVIĆ D. Ž., PATERA J., WINTERNITZ P., ZASSENHAUS H. Normal forms of elements of classical real and complex Lie and Jordan algebras. *J. Math. Phys.*, 1983, **24**(6), 1363–1374.

[7] EDELMAN A., ELMROTH E., KÅGSTRÖM B. A geometric approach to perturbation theory of matrices and matrix pencils. Part I: versal deformations. *SIAM J. Matrix Anal. Appl.*, 1997, **18**(3), 653–692.

[8] EDELMAN A., ELMROTH E., KÅGSTRÖM B. A geometric approach to perturbation theory of matrices and matrix pencils. Part II: a stratification-enhanced staircase algorithm. *SIAM J. Matrix Anal. Appl.*, 1999, **20**(3), 667–699.

[9] FERRER J., GARCÍA M. I., PUERTA F. Brunowsky local form of a holomorphic family of pairs of matrices. *Linear Algebra Appl.*, 1997, **253**, 175–198.

[10] GALIN D. M. On real matrices depending on parameters. *Uspekhi Mat. Nauk*, 1972, **27**(1), 241–242 (in Russian).

[11] GALIN D. M. Versal deformations of linear Hamiltonian systems. *Trudy Semin. Petrovskogo*, 1975, **1**, 63–74 (in Russian). [*The English translation*: *AMS Transl., Ser.* 2, 1982, **118**, 1–12.]

[12] GARCÍA-PLANAS M. I. Versal deformations of pairs of matrices. *Linear Algebra Appl.*, 1992, **170**, 194–200.

[13] HOVEIJN I. Versal deformations and normal forms for reversible and Hamiltonian linear systems. *J. Differ. Equations*, 1996, **126**(2), 408–442.

[14] JACOBSON N. Structure and representations of Jordan algebras. Providence, RI: Amer. Math. Soc., 1968. (Amer. Math. Soc. Colloquium Publ., 39.)

[15] KOÇAK H. Normal forms and versal deformations of linear Hamiltonian systems. *J. Differ. Equations*, 1984, **51**(3), 359–407.

[16] MELBOURNE I. Versal unfoldings of equivariant linear Hamiltonian vector fields. *Math. Proc. Cambridge Phil. Soc.*, 1993, **114**(3), 559–573.

[17] PATERA J., ROUSSEAU C. Complex orthogonal and symplectic matrices depending on parameters. *J. Math. Phys.*, 1982, **23**(5), 705–714.

[18] PATERA J., ROUSSEAU C. Versal deformations of elements of classical Jordan algebras. *J. Math. Phys.*, 1983, **24**(6), 1375–1380.

[19] PATERA J., ROUSSEAU C., SCHLOMIUK D. Dimensions of orbits and strata in complex and real classical Lie algebras. *J. Math. Phys.*, 1982, **23**(4), 490–494.

[20] PATERA J., ROUSSEAU C., SCHLOMIUK D. Versal deformations of elements of real classical Lie algebras. *J. Phys. A: Math. Gen.*, 1982, **15**(4), 1063–1086.

[21] SEVRYUK M. B. Linear reversible systems and their versal deformations. *J. Sov. Math.*, 1992, **60**(5), 1663–1680. [*The Russian original*: *Trudy Seminara im. I. G. Petrovskogo*, 1991, **15**, 33–54.]

[22] SHIH C.-W. Normal forms and versal deformations of linear involutive dynamical systems. *Chinese J. Math.*, 1993, **21**(4), 333–347.

[23] WALL G. E. On the conjugacy classes in the unitary, symplectic and orthogonal groups. *J. Austral. Math. Soc.*, 1963, **3**(1), 1–62.

[24] WAN Y.-H. Versal deformations of infinitesimally symplectic transformations with antisymplectic involutions. In: Singularity Theory and its Applications, Part II. Editors: M. Roberts and I. Stewart. Berlin: Springer, 1991, 301–320. (Lecture Notes in Math., 1463.)

## 1970-2

ℜ According to [3], the "center–focus" problem is trivial (and, moreover, from I. Bendixson's results it follows that the general problem of topological classification of equilibria of systems $\dot{x} = v(x)$ in $\mathbb{R}^n$ is trivial for $n = 2$); see also papers [1, 2].

[1] ARNOLD V. I. Algebraic unsolvability of the problem of Lyapunov stability and the problem of topological classification of singular points of an analytic system of differential equations. *Funct. Anal. Appl.*, 1970, **4**(3), 173–180.

[2] ARNOLD V. I. Algebraic unsolvability of the problem of stability and the problem of topological classification of singular points of analytic systems of differential equations. *Uspekhi Mat. Nauk*, 1970, **25**(2), 265–266 (in Russian).

[3] ARNOLD V. I. Local problems of analysis. *Moscow Univ. Math. Bull.*, 1970, **25**(2), 77–80.

## 1970-3 — *V. I. Arnold*

ℋ Let $(M^{2n}, \omega^2)$ be a symplectic manifold. In paper [1], the following question is posed: *does every Hamiltonian field on m with the rotation class* 0 *have a single-valued Hamiltonian*? Equivalent statements are: *does every homology class* $H_1(M, \mathbb{R})$ *contain a Hamiltonian field*? and *is the operator of multiplication by* $(\omega^2)^{n-1}$ *an isomorphism* $H^1(M, \mathbb{R}) \to H^{2n-1}(M, \mathbb{R})$? The answer is positive if $(M, \omega^2)$ admits a Kähler structure.

ℛ The problem has been solved by W. Thurston: the mapping $H^1 \to H^{2n-1} \sim H^1$ need not be an isomorphism (for example, if $\dim H_1 = 2k$).

[1] ARNOLD V. I. One-dimensional cohomologies of Lie algebras of nondivergent vector fields and rotation numbers of dynamic systems. *Funct. Anal. Appl.*, 1969, **3**(4), 319–321. [*The Russian original is reprinted in*: Vladimir Igorevich Arnold. Selecta–60. Moscow: PHASIS, 1997, 147–150.]

## 1970-5 — *M. B. Sevryuk*

ℛ Given an arbitrary point $x \in \mathbb{R}^n$, denote by $W(x)$ the set of numbers $w > 0$ such that the inequality

$$|q \cdot x + q_0| < |q|^{-w} \tag{1}$$

has infinitely many integer solutions ($q \in \mathbb{Z}^n \setminus \{0\}$, $q_0 \in \mathbb{Z}$), where

$$q \cdot x = q_1 x_1 + \cdots + q_n x_n$$

means the scalar product of vectors and

$$|q| = \max(|q_1|, \ldots, |q_n|) \geq 1$$

is the $l_\infty$-norm of vector $q$. On the other hand, denote by $T(x)$ the set of numbers $\tau > 0$ for which there exists $\gamma > 0$ (dependent on $\tau$) such that the inequality

$$|q \cdot x + q_0| \geq \gamma |q|^{-\tau} \tag{2}$$

holds for all ($q \in \mathbb{Z}^n \setminus \{0\}$, $q_0 \in \mathbb{Z}$). One easily verifies that

$$\sup W(x) = \inf T(x)$$

for each $x \in \mathbb{R}^n$. Denote the number $\sup W(x) = \inf T(x)$ by $\nu(x)$. The classical Dirichlet theorem (of 1842) on simultaneous approximations implies that $n \in W(x)$

for all $x \in \mathbb{R}^n$, so that the inequality $\nu(x) \geq n$ is always valid. On the other hand, for almost all $x \in \mathbb{R}^n$ (in the sense of the Lebesgue measure), the equality $\nu(x) = n$ holds (this easily follows from the Borel–Cantelli lemma). For the corresponding proofs and bibliography see, e. g., works [12, 14, 15].

The points $x \in \mathbb{R}^n$ with $\nu(x) < +\infty$ are said to be *Diophantine* (or *Diophantine normal*). As was explained above, almost all the points $x \in \mathbb{R}^n$ are Diophantine (with the Diophantine "exponent" $\nu(x) = n$). The linear dependence of numbers $x_1, \ldots, x_n, 1$ over $\mathbb{Q}$ is a sufficient (but not necessary) condition for the equality $\nu(x) = +\infty$ [which means that $W(x)$ coincides with the set of all positive numbers whereas $T(x)$ is empty].

Many problems in mathematics and mathematical physics involving "small divisors" require examining the Diophantine approximations on the submanifolds of space $\mathbb{R}^n$. This thesis was first formulated by V. I. Arnold in 1968 in his lecture "Diophantine approximations in analysis" at a symposium on number theory in the city of Vladimir, see [10]. The following statement holds:

**Theorem 1.** *Almost all the points of a generic smooth submanifold $M$ in $\mathbb{R}^n$ (of any positive dimension) are Diophantine. To be more precise, there exists a number $\nu_M$, $n \leq \nu_M < +\infty$, such that $\nu(x) \leq \nu_M$ for almost all the points $x \in M$.*

This theorem was first proved by A. S. Pyartli in paper [11] (one of the main lemmas in this article was due to G. A. Margulis). In numerous works following Pyartli's landmark paper, the genericity conditions imposed on the submanifold $M$ were refined, the estimates of $\nu_M$ were improved, and analogues of inequalities (1) and (2) for the quantities $|q \cdot x|$ as well as $|q \cdot x + q_0 + f(x)|$ and $|q \cdot x + f(x)|$ ($f$ being a smooth function on the submanifold) were considered, see, e. g., [2–4, 6, 13, 16].

The Diophantine approximations on submanifolds were first exploited in the KAM theory by Ī. O. Parasyuk in paper [9] (Parasyuk used Pyartli's results) and in the averaging theory, by V. I. Bakhtin in paper [2], see also a discussion in book [1].

Theorem 1 (and all its versions where the question of the optimal value of quantity $\nu_M$ is not raised) is rather easy. The precise value of $\nu_M$ is usually inessential for dynamical applications (it affects only the necessary smoothness of the right-hand sides of the equations). In the works by V. I. Bakhtin, V. I. Bernik, H. W. Broer, G. B. Huitema, A. S. Pyartli, M. B. Sevryuk, and Zh. Xia [2–4, 6, 11, 13, 16] cited above, the problem of calculating the optimal value of $\nu_M$ was not considered. In fact, for generic analytic submanifolds $M \subset \mathbb{R}^n$, the equality $\nu_M = n$ holds (i. e., the Diophantine "exponent" is the same as in the ambient space). Moreover, the conjecture is very likely that this equality is valid for sufficiently smooth non-analytic generic submanifolds as well. However, even for special classes of submanifolds $M$, the proof of the equality $\nu_M = n$ is immeasurably more complicated than just an existence proof for some $\nu_M < +\infty$ is. Submanifolds $M \subset \mathbb{R}^n$ with $\nu_M = n$ (they are said to be *extremal*) are treated in problem *1972-22*.

The author of the present comment is unaware of any works studying the connections of Diophantine approximations on submanifolds with bifurcations of those submanifolds in $k$-parameter families.

Note finally that the sets of points $x$ subject to inequalities of form (1) or (2) can be investigated not only from the viewpoint of the Lebesgue measure but also from the viewpoint of the Hausdorff dimension, see book [5]. For open domains of the Euclidean space, such a problem was considered in detail in paper [8] and on submanifolds, in paper [7].

[1] ARNOLD V. I. Geometrical Methods in the Theory of Ordinary Differential Equations, 2nd edition. New York: Springer, 1988, § 18. (Grundlehren der Mathematischen Wissenschaften, 250.) [*The Russian original* 1978.]

[2] BAKHTIN V. I. Averaging in multifrequency systems. *Funct. Anal. Appl.*, 1986, **20**(2), 83–88.

[3] BAKHTIN V. I. Diophantine approximations on images of mappings. *Dokl. Akad. Nauk Beloruss. SSR*, 1991, **35**(5), 398–400 (in Russian).

[4] BERNIK V. I. Diophantine approximations on differentiable manifolds. *Dokl. Akad. Nauk Beloruss. SSR*, 1989, **33**(8), 681–683 (in Russian).

[5] BERNIK V. I., MEL′NICHUK YU. V. Diophantine Approximations and Hausdorff Dimension. Minsk: Nauka i Tekhnika, 1988 (in Russian).

[6] BROER H. W., HUITEMA G. B., SEVRYUK M. B. Quasi-Periodic Motions in Families of Dynamical Systems: Order amidst Chaos. Berlin: Springer, 1996. (Lecture Notes in Math., 1645.)

[7] DODSON M. M., PÖSCHEL J., RYNNE B. P., VICKERS J. A. G. The Hausdorff dimension of small divisors for lower-dimensional KAM-tori. *Proc. Roy. Soc. London, Ser. A*, 1992, **439**(1906), 359–371.

[8] DODSON M. M., VICKERS J. A. G. Exceptional sets in Kolmogorov–Arnold–Moser theory. *J. Phys. A: Math. Gen.*, 1986, **19**(3), 349–374.

[9] PARASYUK Ī. O. Persistence of quasi-periodic motions in reversible multifrequency systems. *Dokl. Akad. Nauk Ukrain. SSR, Ser. A*, 1982, **9**, 19–22 (in Russian).

[10] POSTNIKOV A. G., FREĬMAN G. A. Inter-college symposium on the number theory (Vladimir, June of 1968). *Uspekhi Mat. Nauk*, 1969, **24**(1), 235–237 (in Russian).

[11] Pyartli A. S. Diophantine approximations on submanifolds of the Euclidean space. *Funct. Anal. Appl.*, 1969, **3**(4), 303–306.

[12] Schmidt W. M. Diophantine Approximation. Berlin: Springer, 1980. (Lecture Notes in Math., 785.)

[13] Sevryuk M. B. The iteration-approximation decoupling in the reversible KAM theory. *Chaos*, 1995, **5**(3), 552–565.

[14] Sprindžuk V. G. Metric Theory of Diophantine Approximations. New York: John Wiley, 1979. [*The Russian original* 1977.]

[15] Sprindžuk V. G. Achievements and problems in the theory of Diophantine approximations. *Russian Math. Surveys*, 1980, **35**(4), 1–80.

[16] Xia Zh. Existence of invariant tori in volume-preserving diffeomorphisms. *Ergod. Theory Dynam. Systems*, 1992, **12**(3), 621–631.

## 1970-6 – *A. M. Lukatskiĭ*

ℛ A study of the equations in variations for the hydrodynamical Euler equation was performed by G. Misiołek. He distinguished the cases where the geodesics have no conjugate points. According to his theorem of 1993 [3], flows of an ideal incompressible fluid with constant pressure over a Riemannian manifold of nonpositive curvature are characterized by geodesics without conjugate points on the group of volume-preserving diffeomorphisms. Typical examples of that kind are flows with velocity vector fields—simple harmonic on tori. The existence of conjugate points for geodesics on the group of volume-preserving diffeomorphisms of a compact manifold is usually due to two-dimensional directions passing through the initial velocity field of the geodesic and having positive sectional curvatures.

Such an example of geodesic with conjugate points was also firstly constructed by Misiołek [4] for the flow on a 2-torus having the stream function $\cos 6x \cos 2x$ (the existence of two-dimensional directions with negative sectional curvature had been already pointed out by Arnold in [1]). Misiołek also noticed the existence two-dimensional directions with positive sectional curvatures for the *ABC*-flows on a 3-torus.

For manifolds with nonvanishing curvature, an example of a flow with conjugate points is given by a rotation of a 2-sphere, which has been also proved by Misiołek (the existence of two-dimensional directions with positive sectional curvatures for that flow was established by Lukatskiĭ [2]).

[1] ARNOLD V. I. Sur la géométrie différentielle de groupes de Lie de dimension infinie and ses applications à l'hydrodynamique des fluides parfaits. *Ann. Inst. Fourier (Grenoble)*, 1966, **16**(1), 319–361.

[2] LUKATSKIĬ A. M. Curvature of groups of diffeomorphisms preserving the measure of the 2-sphere. *Funct. Anal. Appl.*, 1979, **13**(3), 174–177.

[3] MISIOŁEK G. Stability of flows of ideal fluids and the geometry of the group of diffeomorphisms. *Indiana Univ. Math. J.*, 1993, **42**(1), 215–235.

[4] MISIOŁEK G. Conjugate points in $\mathcal{D}_\mu(\mathbb{T}^2)$. *Proc. Amer. Math. Soc.*, 1996, **124**(3), 977–982.

## 1970-7 – *A. M. Lukatskiĭ*

The sectional curvatures of the group SDiff$(S^2)$ were firstly calculated by Lukatskiĭ [3] for the vector field $h$ of a rotation around the axis $Z$ and for the generalized tradewind flow $zh$ on $S^2$. He proved the non-negativeness (and, in a regular, the positiveness) of the sectional curvatures of two-dimensional directions passing across $h$. In the case of the generalized tradewind flow, on the contrary, the analogous curvatures generally are negative. Then, during the calculation of the curvature tensor of diffeomorphism groups, Lukatskiĭ established the negativeness of sectional curvatures for a wider class of vector fields on $S^2$, e. g. for vector fields of the form $f(z)h$. Then Arakelyan and Savvidy calculated the sectional curvatures of the group SDiff$(S^2)$ using the Klebsch–Gordan coefficients. Rather recently, K. Yoshida performed the most complete analysis of sectional curvatures of this group [6].

For the $n$-dimensional torus the sectional curvatures of the group SDiff$(\mathbb{T}^n)$ have been calculated firstly by Lukatskiĭ, as well as the Ricci curvature of this group [4]. For the 3-dimensional torus T. Kambe, F. Nakamura and Y. Hattori investigated the curvature of the $ABC$-field on $\mathbb{T}^3$. They established that such sectional curvatures do not depend on the values of parameters $A$, $B$, $C$.

[1] ARAKELYAN T. A., SAVVIDY G. K. Geometry of a group of area-preserving diffeomorphisms. *Phys. Lett. B*, 1989, **223**(1), 41–46.

[2] KAMBE T., NAKAMURA F., HATTORI Y. Kinematical instability and line-stretching in relation to the geodesics of fluid motion. In: Topological Aspects of the Dynamics of Fluids and Plasmas. Editors: H. K. Moffatt, G. M. Zaslavsky, P. Comte and M. Tabor. Dordrecht: Kluwer Acad. Publ., 1992, 493–504. (NATO Adv. Sci. Inst. Ser. E Appl. Sci., 218.)

[3] LUKATSKIĬ A. M. Curvature of groups of diffeomorphisms preserving the measure of the 2-sphere. *Funct. Anal. Appl.*, 1979, **13**(3), 174–177.

[4] LUKATSKIĬ A. M. Curvature of the group of measure-preserving diffeomorphisms of the $n$-dimensional torus. *Sib. Math. J.*, 1984, **25**(6), 893–903.

[5] LUKATSKIĬ A. M. Structure of the curvature tensor of the group of measure-preserving diffeomorphisms of a compact two-dimensional manifold. *Sib. Math. J.*, 1988, **29**(6), 947–951.

[6] YOSHIDA K. Riemannian curvature on the group of area-preserving diffeomorphisms (motions of fluid) of 2-sphere. *Physica D*, 1997, **100**(3–4), 377–389.

## 1970-8

See paper [1].

[1] FADDEEV L. D. On the theory of stability for stationary plane-parallel currents of an ideal fluid. *Zap. Nauch. Semin. Leningrad. Otd. Mat. Inst. Steklova*, 1971, **21**, 164–172 (in Russian). (Boundary Problems of Mathematical Physics and Related Questions of the Function Theory, 5.)

## 1970-9 – *A. M. Lukatskiĭ*

In two-dimensional hydrodynamics, sufficient conditions for the fixed sign of the second variation of the energy form $\delta^2(E)$ were established in Arnold's papers (for example, for the currents with plane-parallel section). See [1], Ch. II, Sect. 4.

The question about the indices in the two-dimensional case remains open. In the three-dimensional, case the second variation of the energy form $\delta^2(E)$ is not definite (the indices of inertia are equal to $(\infty,\infty)$), see [1], Ch. II, Sect. 5.

[1] ARNOLD V. I., KHESIN B. A. Topological Methods in Hydrodynamics. New York: Springer, 1998. (Appl. Math. Sci., 125.)

## 1970-10 – *V. I. Arnold, B. A. Khesin*

N. A. Nikishin [1] and C. P. Simon [2] independently proved that an arbitrary symplectomorphism of the sphere $S^2$ has at least two geometrically different fixed points (this had been conjectured by A. I. Schnirelmann). In both papers [1,2] the main lemma is the statement that the index of an isolated fixed point 0 of a symplectomorphism $(\mathbb{R}^2,0) \to (\mathbb{R}^2,0)$ does not exceed 1.

In paper [2] it is also mentioned that any gradient or divergence-free vector field on a two-dimensional sphere $S^2$ has at least two (geometrically different) singular points. The latter fact is almost evident. Indeed, a smooth divergence-free vector field on a two-dimensional sphere is Hamiltonian. Any potential (for the case of a gradient field) or Hamiltonian function (for the case of a Hamiltonian field) on a compact manifold has at least two critical points, corresponding to its maximum and minimum. Therefore the corresponding gradient or Hamiltonian (i. e., skew-gradient) field has at least two singular points.

[1] NIKISHIN N. A. Fixed points of diffeomorphisms of two-dimensional spheres preserving oriented area. *Funct. Anal. Appl.*, 1974, **8**(1), 77–79.

[2] SIMON C. P. A bound for the fixed-point index of an area-preserving map with applications to mechanics. *Invent. Math.*, 1974, **26**(3), 187–200.

### 1970-11 — *V. A. Vassiliev*

ℛ This group is trivial if $n \neq 2$, and is isomorphic to $\mathbb{Z}^N$, $N = (m-1)^3$, for $n = 2$. More generally, for any $k \geq 2$ the group $\pi_k(\mathbb{C}\mathrm{P}^n \setminus V)$ is isomorphic to the $k$-th homotopy group of the wedge of $(m-1)^{n+1}$ $n$-dimensional spheres.

**Proof.** Let $W \subset \mathbb{C}^{n+1}$ be the conical hypersurface, such that $V$ is the projectivization of $W$. Then the space $\mathbb{C}^{n+1} \setminus W$ is homeomorphic to $(\mathbb{C}\mathrm{P}^n \setminus V) \times \mathbb{C}^*$. Indeed, we have the fiber bundle $\mathbb{C}^{n+1} \setminus W \to \mathbb{C}\mathrm{P}^n \setminus V$ with fiber $\mathbb{C}^*$. The unique obstruction to the triviality of this bundle is the first Chern class of the tautological bundle over $\mathbb{C}\mathrm{P}^n$, whose restriction to $\mathbb{C}\mathrm{P}^n \setminus V$ is, of course, trivial.

In particular, $\pi_k(\mathbb{C}\mathrm{P}^n \setminus V) = \pi_k(\mathbb{C}^{n+1} \setminus W)$ if $k > 1$.

Also, we have the Milnor fibration $\mathbb{C}^{n+1} \setminus W \to \mathbb{C}^*$ given by the homogemeous function $f$ distinguishing the cone $W$. The exact sequence of this fibration gives us $\pi_k(\mathbb{C}^{n+1} \setminus W) = \pi_k(f^{-1}(1))$ for $k > 1$. But $f^{-1}(1)$ is the Milnor fiber homotopy equivalent to the wedge of $(m-1)^{n+1}$ copies of the $n$-dimensional sphere.

### 1970-13 — *V. A. Vassiliev* Also: *1981-13*

ℛ The rational cohomology ring of this space is the same as of the group $\mathrm{PGL}(\mathbb{C}\mathrm{P}^2)$, in particular its Poincaré polynomial is equal to $(1+t^3)(1+t^5)$. This group acts on this space with at most finite stationary groups; the embedding of any orbit of this action into entire space is a (rational) homology equivalence.

More generally, let $P(d,n)$ be the Poincaré polynomial of the homology group of the space of nonsingular hypersurfaces of degree $d$ in $\mathbb{C}\mathrm{P}^n$. Then we have:

$$P(3,2) = (1+t^3)(1+t^5) \qquad \text{(see above)},$$
$$P(3,3) = (1+t^3)(1+t^5)(1+t^7)$$

(i. e., we again have the homology equivalence with the corresponding projective linear group),

$$P(4,2) = (1+t^3)(1+t^5)(1+t^6),$$
$$P(5,2) = (1+t^3)(1+t^5).$$

The first three formulae are proved in [4], the formula for $P(5,2)$ is an unpublished result of A. Gorinov, 2001.

This calculation uses the topological study of discriminant sets (initiated in [1]), i. e., of the complementary sets of singular objects.

The additional generator of degree 6 in the formula for $P(4,2)$ is induced from a homology class of the moduli space of curves of genus 3 (calculated in [3]).

A "real" counterpart of this problem is the rigid isotopy classification of algebraic hypersurfaces. For some results in this theory, also based on the theory of discriminants, see [2].

[1] ARNOLD V. I. On some topological invariants of algebraic functions. *Trans. Moscow Math. Soc.*, 1970, **21**, 30–52.

[2] KHARLAMOV V. M. Rigid isotopy classification of real plane curves of degree 5. *Funct. Anal. Appl.*, 1981, **15**(1), 73–74.

[3] LOOIJENGA E. J. N. Cohomology of $\mathcal{M}_3$ and $\mathcal{M}_3^1$. In: Mapping Class Groups and Moduli Spaces of Riemann Surfaces (Göttingen/Seattle, WA, 1991). Editors: C.-F. Bödigheimer and R. M. Hain. Providence, RI: Amer. Math. Soc., 1993, 205–228. (Contemp. Math., 150.)

[4] VASSILIEV V. A. How to calculate homology groups of spaces of nonsingular algebraic projective hypersurfaces. *Proc. Steklov Inst. Math.*, 1999, **225**, 121–140.

## ▽ 1970-14 – *M. L. Kontsevich*

Spaces of knots seem to have homotopy types of a finite $CW$-complex. A. Hatcher conjectured in [2] that the space of knots in $\mathbb{S}^3$ of a given knot type $K$ is homotopoically equivalent to a finite-dimensional manifold of the form $(X_K \times$

SO(4))/$\Gamma_K$ where $\Gamma_K$ is a finite group and $X_K$ is the product of a torus of some dimension and a number of configuration spaces $C_n$ of ordered $n$-tuples of distinct points in $\mathbb{R}^2$. In general, for knots with hyperbolic complements the space will be SO(4)/$\Gamma_K$ where $\Gamma_K$ is the finite group of isometries of $\mathbb{S}^3 - K$, and configuration spaces appear for composite and satellite knots.

In general, I conjectured in the 1990s (as a comment to the Thurston classification program in 3-dimensional topology) that for any connected compact 3-dimensional manifold $M$ whose boundary contains $S^2$ as a component, the classifying space $\mathrm{BDiff}(M, \mathrm{rel}\, S^2)$ of the diffeomorphisms identical on the given $S^2$-component of the boundary, has the homotopy type of a finite $CW$-complex. A version of this conjecture for the case of irreducible 3-manifolds was proved by A. Hatcher in [1].

[1] Hatcher A., McCullough D. Finiteness of classifying spaces of relative diffeomorphism groups of 3-manifolds. *Geometry & Topology*, 1997, **1**, 91–109 (electronic).
[*Internet*: `http:/www.arXiv.org/abs/math.GT/9712260`]

[2] Hatcher A. Spaces of knots.
[*Internet*: `http:/www.arXiv.org/abs/math.GT/9909095`]

### △ 1970-14 — *V. A. Vassiliev*

The fundamental groups of spaces of toric knots in $S^3$ were calculated by D. Goldsmith [1].

A. Hatcher [2] has calculated the homotopy types (in particular fundamental groups) of all components of spaces of toric knots and hyperbolic knots in $\mathbb{S}^3$. His work contains also nice conjectures concerning other components of spaces of knots.

A general approach to the calculation of cohomology groups of spaces of knots in $\mathbb{R}^n$, $n \geq 3$, was proposed in [4]; if $n > 3$ it calculates all these groups. However the real calculations in higher dimensions are of the same complexity as in the initial classical case concerning invariants of knots in $\mathbb{R}^3$.

The first such non-trivial one-dimensional cohomology class of finite degree (equal to 3) was found in a computer experiment by D. Teiblum and V. Turchin, see [5].

For further results on such cohomology classes see [3, 6].

The results of [3] prove that the simultaneous study of all cohomology classes of spaces of knots is a more natural problem than just the study of knot invariants. Indeed, the rings of finite degree cohomology classes of these spaces admit an elegant algebraic description, from which the 0-dimensional part (responsible for the invariants) is obtained by easy factorization.

[1] GOLDSMITH D. Motions of links in the 3-sphere. *Bull. Amer. Math. Soc.*, 1974, **80**, 62–66; *Math. Scand.*, 1982, **50**, 167–205.

[2] HATCHER A. Topological moduli spaces of knots. [*Internet*: `http://www.math.cornell.edu/~hatcher/Papers/`]

[3] TOURTCHINE V. Sur l'homologie des espaces des nœuds non-compacts. [*Internet*: `http:/www.arXiv.org/abs/math.QA/0010017`]

[4] VASSILIEV V. A. Cohomology of knot spaces. In: Theory of Singularities and its Applications. Editor: V. I. Arnold. Providence, RI: Amer. Math. Soc., 1990, 23–69. (Adv. Sov. Math., 1.)

[5] VASSILIEV V. A. Topology of two-connected graphs and homology of spaces of knots. In: Differential and Symplectic Topology of Knots and Curves. Editor: S. Tabachnikov. Providence, RI: Amer. Math. Soc., 1999, 253–286. (AMS Transl., Ser. 2, 190; Adv. Math. Sci., 42.)

[6] VASSILIEV V. A. Combinartorial formulae for cohomology of knot spaces. *Moscow Math. J.*, 2001, **1**(1), 91–123.

## ▽ 1970-15 – *V. V. Goryunov*

Also: *1995-1, 1995-2, 1996-8, 1996-13*

R Singularity theory provides rather convenient realizations of certain Hurwitz spaces. The very first example of this kind is the miniversal deformation of the function singularity $A_k$ which coincides with the space of rational functions with just one pole, of order $k+1$. See [1–3] for some other models of similar type.

The dimension of the moduli space of meromorphic functions on curves of fixed genus and with fixed orders of poles is equal to the number of finite critical values of a generic function of this kind. One of reflections of this in singularity theory is the coincidence of the Milnor and Tyurina numbers for functions on curves in $\mathbb{C}^3$ proved by D. Mond and D. van Straten [4]. These numbers also coincide in the case of functions on Gorenstein curves in $\mathbb{C}^4$ [5].

[1] GORYUNOV V. V. Functions on space curves. *J. London Math. Soc., Ser.* 2, 2000, **61**(3), 807–822.

[2] GORYUNOV V. V. Simple functions on space curves. *Funct. Anal. Appl.*, 2000, **34**(2), 129–132.

[3] Goryunov V. V., Lando S. K. On enumeration of meromorphic functions on the line. In: The Arnoldfest. Proceedings of a conference in honour of V. I. Arnold for his sixtieth birthday (Toronto, 1997). Editors: E. Bierstone, B. A. Khesin, A. G. Khovanskiĭ and J. E. Marsden. Providence, RI: Amer. Math. Soc., 1999, 209–223. (Fields Inst. Commun., 24.)

[4] Mond D., van Straten D. Milnor number equals Tjurina number for functions on space curves. *J. London Math. Soc., Ser.* 2, 2001, **63**(1), 177–187.

[5] van Straten D. Private communication.

## △ ▽ 1970-15 – *S. K. Lando*

Also: *1995-1, 1995-2, 1996-8, 1996-13*

This group of problems is related to the geometry of spaces of meromorphic functions on complex curves, and to the topological classification of meromorphic functions. The problem of topological classification of meromorphic functions with fixed branching data was first posed by A. Hurwitz [13]. Two holomorphic functions $f_1 : C_1 \to \mathbb{C}P^1$, $f_2 : C_2 \to \mathbb{C}P^2$, where $C_1$, $C_2$ are connected smooth closed complex curves, are considered to be *isomorphic* if there exists a homeomorphism $h : C_1 \to C_2$ such that $f_1 = f_2 \circ h$. Obviously, isomorphic meromorphic functions are defined on complex curves of the same genus, and they have the same degree, coinciding branching points in the image (critical values), and coinciding branching data at each of these points. If we fix branching data, then the set of isomorphism classes becomes finite. Hence the problem of topological classification can be understood as the enumeration problem for these classes.

There are different languages in which the problem can be expressed: it can be reformulated as an enumeration problem for sets of permutations belonging to given conjugacy classes in the symmetric group with the identity product, an enumeration problem for graphs on surfaces [7,14,22] (for example, the classification of generic polynomials gives a proof of the Cayley theorem on enumeration of marked trees, see [15]), or a problem about the degree of the Lyashko–Looijenga mapping. The *Lyashko–Looijenga mapping* [2, 15, 16] associates to a meromorphic function the set of its critical values. It is closely related to the geometry of the spaces of meromorphic functions. The idea of using this mapping in the classification problem for meromorphic functions belongs to Arnold [1].

In [13] A. Hurwitz enumerated (without proof) meromorphic functions on rational curves such that all their critical values but one are simple. This result was rediscovered in [10] (in terms of permutations), and another proof, based on the geometry, is given in [8] (the reader must be careful, however: the proof contains a serious gap). Meromorphic functions with one degenerate critical value on curves of arbitrary genera are enumerated by means of the intersection theory on the moduli space of curves [5, 6, 12]. The corresponding formula leads to the explicit generalization of the Hurwitz result for the case of genus 1 (see also [11]). An enumeration formula for the general case expressing the required number as a certain sum over all characters of the symmetric group is given in [17, 18], but it is too cumbersome to make use of.

An explicit answer in the case of more than one degenerate critical value is known only in few cases. In particular, all polynomials with given branching data at finite points are classified [9] (see also a geometric proof in [14]). Formulas for the *transversal* multiplicity of the Lyashko–Looijenga mapping are given in [23].

[1] ARNOLD V. I. Topological classification of trigonometric polynomials and combinatorics of graphs with an equal number of vertices and edges. *Funct. Anal. Appl.*, 1996, **30**(1), 1–14.

[2] ARNOLD V. I. Critical points of functions and the classification of caustics. *Uspekhi Mat. Nauk*, 1974, **29**(3), 243–244 (in Russian). [*Reprinted in*: Vladimir Igorevich Arnold. Selecta–60. Moscow: PHASIS, 1997, 213–214.]

[3] CRESCIMANNO M., TAYLOR W. Large *N* phases of chiral $QCD_2$. *Nuclear Phys. B*, 1995, **437**(1), 3–24.

[4] DUBROVIN B. A. Geometry of 2D topological field theories. In: Integrable Systems and Quantum Groups (Montecatini Terme, 1993). Editors: M. Francaviglia and S. Greco. Berlin: Springer *and* Florence: Centro Internazionale Matematico Estivo, 1996, 120–348. (Lecture Notes in Math., 1620; Fondazione C. I. M. E.)

[5] EKEDAHL T., LANDO S. K., SHAPIRO M. Z., VAINSHTEIN A. D. On Hurwitz numbers and Hodge integrals. *C. R. Acad. Sci. Paris, Sér.* I *Math.*, 1999, **328**(12), 1175–1180.

[6] EKEDAHL T., LANDO S. K., SHAPIRO M. Z., VAINSHTEIN A. D. Hurwitz numbers and intersections on moduli spaces of curves. *Invent. Math.*, 2001, **146**(2), 297–327.

[7] EL MARRAKI M., HANUSSE N., ZIPPERER J., ZVONKINE A. D. Cacti, braids and complex polynomials. *Sémin. Lotharing. de Combinatoire*, 1997, **37**, Art. B37b, 36 pp. (electronic).
[*Internet*: `http://www-irma.u-strasbg.fr/EMIS/journals/SLC`]

[8] GORYUNOV V. V., LANDO S. K. On enumeration of meromorphic functions on the line. In: The Arnoldfest. Proceedings of a conference in honour of V. I. Arnold for his sixtieth birthday (Toronto, 1997). Editors: E. Bierstone, B. A. Khesin, A. G. Khovanskiĭ and J. E. Marsden. Providence, RI: Amer. Math. Soc., 1999, 209–223. (Fields Inst. Commun., 24.)

[9] GOULDEN I. P., JACKSON D. M. The combinatorial relationship between trees, cacti and certain connection coefficients for the symmetric group. *European J. Combinatorics*, 1992, **13**(5), 357–365.

[10] GOULDEN I. P., JACKSON D. M. Transitive factorization into transpositions, and holomorphic mappings on the sphere. *Proc. Amer. Math. Soc.*, 1997, **125**(1), 51–60.

[11] GOULDEN I. P., JACKSON D. M., VAINSHTEIN A. D. The number of ramified coverings of the sphere by the torus and surfaces of higher genera. [*Internet*: `http://www.arXiv.org/abs/math.AG/9902125`]

[12] GRABER T., PANDHARIPANDE R. Localization of virtual classes. [*Internet*: `http://www.arXiv.org/abs/math.AG/9708001`]

[13] HURWITZ A. Über Riemann'sche Flächen mit gegebenen Verzweigungspunkten. *Math. Ann.*, 1891, **39**(1), 1–61.

[14] LANDO S. K., ZVONKINE D. A. On multiplicities of the Lyashko–Looijenga mapping on the discriminant strata. *Funct. Anal. Appl.*, 1999, **33**(3), 178–188.

[15] LOOIJENGA E. J. N. The complement of the bifurcation variety of a simple singularity. *Invent. Math.*, 1974, **23**(2), 105–116.

[16] LYASHKO O. V. Geometry of bifurcation diagrams. In: Itogi Nauki i Tekhniki VINITI. Current Problems in Mathematics, Vol. 22. Moscow: VINITI, 1983, 99–129 (in Russian). [The English translation: *J. Sov. Math.*, 1984, **27**, 2736–2759.]

[17] MEDNYKH A. D. Determination of the number of nonsingular coverings of a compact Riemann surface. *Sov. Math. Dokl.*, 1978, **239**(2), 269–271.

[18] MEDNYKH A. D. Branched coverings of Riemann surfaces whose branch orders coincide with the multiplicity. *Commun. Algebra*, 1990, **18**(5), 1517–1533.

[19] NATANZON S. M. Topology of 2-dimensional coverings and meromorphic functions on real and complex algebraic curves. *Trudy Semin. Vekt. Tenz. Anal.*, 1988, **23**, 79–103 (in Russian) [*The English translation*: *Selecta Math. Sov.*, 1993, **12**(3), 251–291.]

[20] STREHL V. Minimal transitive products of transpositions: the reconstruction of a proof by A. Hurwitz. *Sémin. Lotharing. de Combinatorie*, 1996, **37**, Art. B37c, 12 pp. (electronic). [*Internet*: `http://www-irma.u-strasbg.fr/EMIS/journals/SLC`]

[21] SHAPIRO B. Z., SHAPIRO M. Z., VAINSHTEIN A. D. Ramified coverings of $S^2$ with one degenerate branching point and enumeration of edge-oriented graphs. In: Topics in Singularity Theory. V. I. Arnold's 60th Anniversary Collection. Editors: A. Khovanskiĭ, A. Varchenko and V. Vassiliev. Providence, RI: Amer. Math. Soc., 1997, 219–227. (AMS Transl., Ser. 2, 180; Adv. Math. Sci., 34.)

[22] ZDRAVKOVSKA S. The topological classification of polynomial mappings. *Uspekhi Mat. Nauk*, 1970, **25**(4), 179–180 (in Russian).

[23] ZVONKINE D. A. Multiplicities of the Lyashko–Looijenga map on its strata. *C. R. Acad. Sci. Paris, Sér.* I *Math.*, 1997, **324**(12), 1349–1353.

## △▽ 1970-15 – *S. M. Natanzon*

Also: *1995-1, 1995-2, 1996-8, 1996-13*

The space $H_{g,n}$ of all meromorphic functions $(f: P \to \mathbb{CP}^1)$ of degree $n$ on Riemann surfaces $P$ of genus $g$ ($\dim_{\mathbb{C}} H_{g,n} = 2g + 2n - 2$) is called a Hurwitz space. We assume here that functions $(f: P \to \mathbb{CP}^1)$ and $(\widetilde{f}: \widetilde{P} \to \mathbb{CP}^1)$ are the same if and only if there exist holomorphic maps $\psi_1: \widetilde{P} \to P$ and $\psi_2: \mathbb{CP}^1 \to \mathbb{CP}^1$ such that $f\psi_1 = \psi_2\widetilde{f}$. The connection of the Hurwitz space was proved by Hurwitz in [6]. It has different compactifications: algebra-geometric [5], function-theoretic [2], geometric [16], stable [8]. In [16] the Euler characteristic of the geometric compactification $N_{g,n}$ of $H_{g,x}$ was found.

The correspondence "a meromorphic function" $\mapsto$ "the set of its critical values" generates the Lyashko–Looijenga map $\varphi: N_{g,n} \to \mathbb{CP}^m$. The natural stratification of $\mathbb{CP}^m$ by Schubert cells generates (by means of $\varphi$) the stratifications of $N_{g,n}$ and $H_{g,n}$. Connected components of these stratifications are classes of topological equivalence [10]. We assume here that functions $f: P \to \mathbb{CP}^1$ and $\widetilde{f}: \widetilde{P} \to \mathbb{CP}^1)$ are topological equivalences if and only if there exist homeomorphisms $\psi_1: \widetilde{P} \to P$ and $\psi_2: \mathbb{CP}^1 \to \mathbb{CP}^1$ such that $f\psi_1 = \psi_2\widetilde{f}$.

An automorphism $\alpha \in \mathrm{Aut}(\mathbb{CP}^1)$ acts on any connected component $M$ of the stratification by the rule: $\alpha(f: P \to \mathbb{CP}^1) = (\alpha f: P \to \mathbb{CP}^1)$. According to [12], $M/\mathrm{Aut}(\mathbb{CP}^1) \cong \mathbb{R}^m/\mathrm{Mod}_M$, where $\mathrm{Mod}_M$ is a discrete group of homeomorphisms. For spaces of trigonometric polynomials this was proved in [1]. A description of the group $\mathrm{Mod}_M$ for the spaces of meromorphic functions in general position is contained in [10]. Generalisations of these theorems for arbitrary morphisms of Riemann surfaces $f: L_1 \to P_2$ are contained in [14].

A special role belongs to the subset $H_{g,n}(n_1, \ldots, n_k) \subset H_{g,n}$, that is, the set of all meromorphic functions with simple finite critical values and divisors of poles in form of $n_1 p_1 + \cdots + n_k p_k$ $(p_i \neq p_j)$. According to [3] $H_{g,n}(n_1, \ldots, n_k)$ has a natural structure of Frobenius manifold. The connections of $H_{g,n}(n_1, \ldots, n_k)$ was proved in [9, 11]. In [17] there are described all connected components of the spaces $H_{g,n}(n_1, \ldots, n_k \mid m_1, \ldots, m_r) \subset H_{g,n}$, that is, the set of all meromorphic functions with simple finite nonzero critical values and divisors in the form of $(m_1 q_1 + \cdots + m_r q_r) - (n_1 p_1 + \cdots + n_k p_k)$. In [4] a homotopical type of the space of rational functions $H_{0,n}$ is found. Connected components of the space of polynomials were investigated in [7].

A real analog of $H_{g,n}$ is a space $\mathbb{R}H_{g,n}$ of all real meromorphic functions of genus $g$ and degree $n$. By definition a real meromorphic function is $(P, \tau, f)$, where $(f: P \to \mathbb{CP}^8) \subset H_{g,n}$ and $\tau: P \to P$ is a antiholomorphic involution such that $f\tau =$

$\bar{f}$. All connected components of $\mathbb{R}H_{g,n}$ are found in [11]. A theorem of Sturm–Hurwitz type (in the sense of Arnold) about real zeros for real functions from $H_{g,n}(n_1, \ldots, n_k)$ is contained in [13]. A real analog of the space $H_{g,n}(n_8, \ldots, n_k \mid m_1, \ldots, m_r)$ and its connected components was found in [15].

[1] ARNOLD V. I. Topological classification of trigonometric polynomials and combinatorics of graphs with an equal number of vertices and edges. *Funct. Anal. Appl.*, 1996, **30**(1), 1–14.

[2] DIAZ S., EDIDIN D. Towards the homology of Hurwitz spaces. *J. Differ. Geom.*, 1996, **43**(1), 66–98.

[3] DUBROVIN B. A. Geometry of 2D topological field theories. In: Integrable Systems and Quantum Groups (Montecatini Terme, 1993). Editors: M. Francaviglia and S. Greco. Berlin: Springer, 1996, 120–348. (Lecture Notes in Math., 1620.)

[4] GUEST M. A., KOZLOWSKI A., MURAYAMA M., YAMAGUCHI K. The homotory type of the space of rational functions. *J. Math. Kyoto Univ.*, 1995, **35**(4), 631–638.

[5] HARRIS J., MUMFORD D. On the Kodaira dimension of the moduli space of curves. *Invent. Math.*, 1982, **67**(1), 23–86.

[6] HURWITZ A. Über Riemann'sche Flächen mit gegebenen Verzweigungspunkten. *Math. Ann.*, 1891, **39**(1), 1–61.

[7] KHOVANSKIĬ A. G., ZDRAVKOVSKA S. Branched covers of $S^2$ and braid groups. *J. Knot Theory Ramifications*, 1996, **5**(1), 55–75.

[8] KONTSEVICH M., MANIN YU. Gromov–Witten classes, quantum cohomology and enumerative geometry. *Commun. Math. Phys.*, 1994, **164**(3), 525–562.

[9] NATANZON S. M. Spaces of real meromorphic functions on real algebraic curves. *Sov. Math. Dokl.*, 1984, **30**(3), 724–726.

[10] NATANZON S. M. Uniformization of spaces of meromorphic functions. *Sov. Math. Dokl.*, 1986, **33**(2), 487–490.

[11] NATANZON S. M. Topology of 2-dimensional coverings and meromorphic functions on real and complex algebraic curves. *Trudy Semin. Vekt. Tenz. Anal.*, 1988, **23**, 79–103 (in Russian) [*The English translation*: *Selecta Math. Sov.*, 1993, **12**(3), 251–291.]

[12] NATANZON S. M. Spaces of meromorphic functions on Riemann surfaces. In: Topics in Singularity Theory. V. I. Arnold's 60th Anniversary Collection. Editors: A. Khovanskiĭ, A. Varchenko and V. Vassiliev. Providence, RI: Amer. Math. Soc., 1997, 175–180. (AMS Transl., Ser. 2, 180; Adv. Math. Sci., 34.)

[13] NATANZON S. M. Moduli of real algebraic surfaces and their superanalogues. Differentials, spinors and Jacobians of real curves. *Russian Math. Surveys*, 1999, **54**(6), 1091–1147.

[14] NATANZON S. M. Moduli of Riemann surfaces, Hurwitz-type spaces, and their superanalogues. *Russian Math. Surveys*, 1999, **54**(1), 61–117.

[15] NATANZON S. M., SHADRIN S. V. Topological classification of unitary functions of arbitrary genus. *Russian Math. Surveys*, 2000, **55**(6), 1163–1164.

[16] NATANZON S. M., TURAEV V. G. A compactification of the Hurwitz space. *Topology*, 1999, **38**(4), 889–914.

[17] WAJNRYB B. Orbits of Hurwitz action for coverings of a sphere with two special fibers. *Indag. Math. (N. S.)*, 1996, **7**(4), 549–558.

## △ 1970-15 – *D. A. Zvonkine*

Also: *1995-1, 1995-2, 1996-8, 1996-13*

ℛ The space of meromorphic functions $f$ of degree $n$ on a Riemann surface of genus $g$ is stratified according to the multiplicities of critical points and values. The first question on the geometry of the discriminant is to find the multiplicity of the generalized Lyashko–Looijenga map (the LL map) on a given stratum of the discriminant, or the transversal multiplicity of a less degenerate stratum with respect to a more degenerate one.

The transversal multiplicity is easier to calculate, because it does not depend on the global structure of the strata, and it is known in all cases (see [8]).

The multiplicity of the LL map is explicitly known in two cases: first, for polynomials ($g = 0$, the function $f$ has a unique pole—see [5]), and second, for rational functions with any poles, but only simple critical values ($g = 0$, $f$ has poles of arbitrary multiplicities, but only simple critical values—see [3]).

Finding the multiplicity of the LL map is equivalent to counting ramified covers of the sphere with given ramification type, or to counting some labeled graphs embedded into the Riemann surface, or to counting lists of permutations with given lengths of cycles and a given product.

In the two above cases where the multiplicity is known, the labeled graphs can be counted by combinatorial methods. For the case of polynomials, see [2]. For rational functions with simple critical values, the answer was first given by Hurwitz without a proof. For a combinatorial proof, see [3].

The multiplicity of the LL map can also be found by algebro-geometric methods. For meromorphic functions with simple critical values (but any poles) on Riemann surfaces of *arbitrary* genus, the multiplicity can be expressed as an intergal of Chern classes of some vector bundles over the moduli space of Riemann surfaces [1]. This is due to the fact that the space of meromorphic functions on a Riemann surface is itself a vector bundle over the moduli space.

Combining algebra-geometric and combinatorial approaches, A. Okounkov and R. Pandharipande obtained a new proof of Witten's conjecture (see [7]), and some new results on Gromov–Witten invariants of the Riemann sphere are to appear shortly.

Finally, computing the multiplicity of the LL map using lists of permutations, leads to a general formula, applicable in all cases, that expresses the multiplicity as a sum over all the irreducible representations of the symmetric group $S_n$ (see [6]).

Using this formula, a more explicit result was obtained for meromorphic functions with a unique pole and arbitrary multiplicities of critical values on surfaces of any genus [4].


[1] Ekedahl T., Lando S., Shapiro M., Vainshtein A. Hurwitz numbers and intersections on moduli spaces of curves. *Invent. Math.*, 2001, **146**(2), 297–327.

[2] Goulden I. P., Jackson D. M The combinatorial relationship between trees, cacti and certain connection coefficients for the symmetric group. *European J. Combinatorics*, 1992, **13**(5), 357–365.

[3] Goulden I. P., Jackson D. M. Transitive factorization into transpositions, and holomorphic mappings on the sphere. *Proc. Amer. Math. Soc.*, 1997, **125**(1), 51–60.

[4] Goupil A., Schaeffer G. Factoring *n*-cycles and counting maps of given genus. *European J. Combinatorics*, 1998, **19**(7), 819–834.

[5] Lando S. K., Zvonkine D. A. On multiplicities of the Lyashko–Looijenga mapping on the discriminant strata. *Funct. Anal. Appl.*, 1999, **33**(3), 178–188.

[6] Mednykh A. D. Nonequivalent coverings of Riemann surfaces with a prescribed ramification type. *Sib. Math. J.*, 1984, **25**(4), 606–625.

[7] Okounkov A., Pandharipande R. Gromov–Witten theory, Hurwitz numbers, and matrix models, I.
[*Internet*: `http://www.arXiv.org/abs/math.AG/0101147`]

[8] Zvonkine D. A. Transversal multiplicities of the Lyashko–Looijenga map. *C. R. Acad. Sci. Paris, Sér.* I *Math.*, 1997, **325**(6), 589–594.


## 1970-16

H This is a problem in paper [3], see also [1, 2].


[1] Arnold V. I. Algebraic unsolvability of the problem of Lyapunov stability and the problem of topological classification of singular points of an analytic system of differential equations. *Funct. Anal. Appl.*, 1970, **4**(3), 173–180.

[2] Arnold V. I. Algebraic unsolvability of the problem of stability and the problem of topological classification of singular points of analytic systems of differential equations. *Uspekhi Mat. Nauk*, 1970, **25**(2), 265–266 (in Russian).

[3] Arnold V. I. Local problems of analysis. *Moscow Univ. Math. Bull.*, 1970, **25**(2), 77–80.

# 1971

## 1971-1 – *R. I. Bogdanov*

R In note [1] this problem was considered for generic diffeomorphisms $(\mathbb{R}^2, 0) \hookleftarrow$ (to be more precise, diffeomorphisms with finite modality) of class $C^\infty$. The answer to the question posed in the problem is positive for the case $k = 2$.

[1] BOGDANOV R. I. Factorization of diffeomorphisms over phase portraits of vector fields on the plane. *Funct. Anal. Appl.*, 1997, **31**(2), 126–128.

## 1971-2 – *M. B. Mishustin*

H The conjecture was proved by A. S. Pyartli in [9] for simple resonances with heavy restrictions on eigenvalues. Bruno in [4] suggested necessary conditions for the conjeccture's validity, as well as counterexamples to it. Many cases are covered neither by Pyartli's sufficiency nor by Bruno's necessity. Yu. S. Il′yashenko and A. S. Pyartli in [6] construct invariant manifolds in some of these cases.

In [2] Arnold establishes a relation between bifurcation of invariant manifolds and geometry of neighborhoods of elliptic curves. Book [3] contains (in its Sections 27 and 36) surveys of this theory and of its applications.

Since that time bifurcations of invariant manifolds have been studied in many works for particular and neighbor cases, see, say, [5,8] to mention just two... Unfortunately, the author of the present comment cannot point to works either somehow classifying this activity or generalizing results of [9] and [4].

There were also numerical methods developed, studying invariant manifolds, for example, see [7] or a scientific report of Siberian Branch of Russian Academy of Sciences in 2000 [10]. Some of them appear in mathematical packages, for example, DDE-BIFTOOL for MathLab. So, the conjecture of invariant manifolds can now be regarded much more like a method of research.

[1] ARNOLD V. I. Remarks on singularities of finite codimension in complex dynamical systems. *Funct. Anal. Appl.*, 1969, **3**(1), 1–5. [*The Russian original is reprinted in*: Vladimir Igorevich Arnold. Selecta–60. Moscow: PHASIS, 1997, 129–137.]

[2] ARNOLD V. I. Bifurcations of invariant manifolds of differential equations and normal forms in neighborhoods of elliptic curves. *Funct. Anal. Appl.*, 1976, **10**(4), 249–259.

[3] ARNOLD V. I. Geometrical Methods in the Theory of Ordinary Differential Equations, 2nd edition. New York: Springer, 1988. (Grundlehren der Mathematischen Wissenschaften, 250.) [*The Russian original* 1978.]

[4] BRUNO A. D. Normal form of differential equations with a small parameter. *Math. Notes*, 1974, **16**, 832–836.

[5] BURLAKOVA L. A., IRTEGOV V. D., NOVIKOV M. A. Stability and bifurcations of invariant manifolds of systems of bodies in Newton force field. In: International Aerospace Congress (Moscow, 1994). Moscow: Petrovka, 1994, 232–236.

[6] IL'YASHENKO YU. S., PYARTLI A. S. Materialization of Poincaré resonances and divergence of normalizing series. *Trudy Semin. Petrovskogo*, 1981, **7**, 3–49 (in Russian). [*The English translation*: *J. Sov. Math.*, 1985, **31**, 3053–3092.]

[7] JOHNSON M. E., JOLLY M. S., KEVREKIDIS I. G. Two-dimensional invariant manifolds and global bifurcations: some approximation and visualization studies. *Numerical Algorithms*, 1997, **14**(1–3), 125–140.

[8] LLIBRE J., NUÑES A. Separatrix surfaces and invariant manifolds of a class of integrable Hamiltonian systems and their perturbations. *Mem. Amer. Math. Soc.*, 1994, **107**, viii + 191 pp.

[9] PYARTLI A. S. Birth of complex invariant manifolds close to a singular point of a parametrically dependent vector field. *Funct. Anal. Appl.*, 1972, **6**(4), 339–340.

[10] Siberian Branch of Russian Academy of Sciences. Physical and Mathematical Sciences. Scientific report, 2000 (in Russian).
[*Internet*: `http://www.sbras.ru/win/sbras/rep/2000/fiz-mat/fmn1.html`]

## 1971-3

In addition to the papers listed in the problem statement, see also paper [1].

[1] ARNOLD V. I. Problèmes résolubles et problèmes irrésolubles analytiques et géométriques. In: Passion des Formes. Dynamique Qualitative Sémiophysique et Intelligibilité. Dédié à R. Thom. Fontenay-St Cloud: ENS Éditions, 1994, 411–417; In: Formes et Dynamique, Renaissance d'un Paradigme. Hommage à René Thom. Paris: Eshel, 1995. [*The Russian translation in*: Vladimir Igorevich Arnold. Selecta–60. Moscow: PHASIS, 1997, 577–582.]

## 1971-4 — *M. B. Sevryuk* Also: *1976-29*

This is the problem on the converse of the classical Lagrange–Dirichlet theorem: *The equilibrium* 0 *of the system* $\ddot{x} = -\partial U/\partial x$, $x \in \mathbb{R}^n$, *is stable if the potential* $U$ *attains a strict local minimum at the critical point* 0 (proof: one can take the total mechanical energy as a Lyapunov function). The question whether the converse theorem holds was raised by A. M. Lyapunov, see [11] (Ch. I, n° 16 and Ch. II, n° 25) and [12]. Generally speaking, the hypotheses of the Lagrange–Dirichlet theorem are not necessary for stability even for one-degree-of-freedom systems, and the Painlevé–Wintner $C^\infty$-counterexample $\{U(x) = \exp(-x^{-2}) \cos x^{-1}$ for $x \neq 0$ and $U(0) = 0\}$ is well known. The problem on the converse of the Lagrange–Dirichlet theorem makes therefore sense only under one or another additional assumptions (e. g., that of analyticity of the potential).

The problem on the converse of the Lagrange–Dirichlet theorem is considered in a rich body of literature, see, e. g., monographs and surveys [1, 2, 9, 15, 16]. Here we mention only important works by V. V. Kozlov and V. P. Palamodov [3–8, 10, 13, 14]. In particular, in paper [13], V. P. Palamodov proved the converse Lagrange–Dirichlet theorem for systems with two degrees of freedom in the case of analytic potentials $U$ or infinitely differentiable potentials for which the critical point 0 is of finite multiplicity. In work [14], Palamodov announced (along with a sketch of the proof) a complete solution of the problem on the converse of the Lagrange–Dirichlet theorem in the case of analytic potentials for an arbitrary number of degrees of freedom.

The statement on the instability of equilibria of systems with *harmonic* potentials $U$ (i. e., potentials satisfying the Laplace equation $\Delta U = 0$) is a particular case of the converse Lagrange–Dirichlet theorem. In particular, the following Irnshaw theorem holds (see, e. g., [17]): *An equilibrium of a system of electric charges in a stationary electric field is always unstable.*


[1] ARNOLD V. I., KOZLOV V. V., NEĬSHTADT A. I. Mathematical Aspects of Classical and Celestial Mechanics, 2nd edition. Berlin: Springer, 1993, Ch. 7, § 5. (Encyclopædia Math. Sci., 3; Dynamical systems, III.) [*The Russian original* 1985.] [*The second, revised and supplemented, Russian edition* 2002.]

[2] KARAPETYAN A. V., RUMYANTSEV V. V. Stability of Conservative and Dissipative Systems. Itogi Nauki i Tekhniki VINITI. General Mechanics, Vol. 6. Moscow: VINITI, 1983 (in Russian).

[3] KOZLOV V. V. Instability of an equilibrium in a potential field. *Russian Math. Surveys*, 1981, **36**(1), 238–239.

[4] KOZLOV V. V. On the instability of an equilibrium in a potential field. *Russian Math. Surveys*, 1981, **36**(3), 256–257.

[5] KOZLOV V. V. A conjecture on the existence of asymptotic motions in classical mechanics. *Funct. Anal. Appl.*, 1982, **16**(4), 303–304.

[6] KOZLOV V. V. Asymptotic solutions of the equations of classical mechanics. *J. Appl. Math. Mech.*, 1982, **46**(4), 454–457.

[7] Kozlov V. V. Asymptotic motions and the problem on the converse of the Lagrange–Dirichlet theorem. *J. Appl. Math. Mech.*, 1986, **50**(6), 719–725.

[8] Kozlov V. V. On a problem by Kelvin. *J. Appl. Math. Mech.*, 1989, **53**(1), 133–135.

[9] Kozlov V. V., Furta S. D. Asymptotics of the Solutions of Strongly Nonlinear Systems of Differential Equations. Moscow: Moscow University Press, 1996 (in Russian).

[10] Kozlov V. V., Palamodov V. P. On the asymptotic solutions of the equations of classical mechanics. *Sov. Math. Dokl.*, 1982, **25**(2), 335–339.

[11] Lyapunov A. M. The General Problem of the Stability of Motion. London: Taylor & Francis, 1992. [*The Russian original* 1892.] [*The first French translation* 1907.]

[12] Lyapunov A. M. On the instability of an equilibrium in some cases where the force function is not a maximum. In: Collected Papers, Vol. II. Moscow–Leningrad: the USSR Academy of Sciences Press, 1956, 391–400 (in Russian). [*The original publication* 1897.]

[13] Palamodov V. P. On the stability of an equilibrium in a potential field. *Funct. Anal. Appl.*, 1977, **11**(4), 277–289.

[14] Palamodov V. P. Stability of motion and algebraic geometry. In: Dynamical Systems in Classical Mechanics. Editor: V. V. Kozlov. Providence, RI: Amer. Math. Soc., 1995, 5–20. (AMS Transl., Ser. 2, 168; Adv. Math. Sci., 25.)

[15] Rouche N., Habets P., Laloy M. Stability Theory by Liapunov's Direct Method. New York: Springer, 1977. (Appl. Math. Sci., 100.)

[16] Rumyantsev V. V., Sosnitskiĭ S. P. On the instability of an equilibrium of holonomic conservative systems. *J. Appl. Math. Mech.*, 1993, **57**(6), 1101–1122.

[17] Tamm I. E. Basic Electricity Theory, 10th edition. Moscow: Nauka, 1989, Ch. I, § 19 (in Russian).

## 1971-9 — *S. Yu. Yakovenko*

One of the possible variants is the dynamics of intersections discussed in problem *1988-6* (commentary), see also problems *1988-7*, *1989-2*, *1990-1*, *1990-20*, *1990-21*, *1992-12–1992-14*, *1994-45–1994-50*, where mostly the case of generic smooth maps is considered [1–3].

Yet it is the algebraicity of the discrete time dynamical system that should also play an important role. The straightforward generalization, "estimate the number of periodic points of period $n$ in terms of the degree and $n$," is trivial: the union of all $n$-periodic orbits is an algebraic subvariety for any finite $n$, and its complexity can be easily estimated. For instance, if this set is discrete, than the number of its points grows exponentially in $n$ by virtue of the Bézout theorem.

It is the nonalgebraicity of solutions (limit cycles) of planar polynomial vector fields, that makes them so difficult to track. Thus a "proper" Hilbert-type question for discrete time systems should involve infinite aperiodic orbits of polynomial maps. In particular, one might try to begin by estimating "nonalgebraicity" of infinite orbits. To do this, a numeric measure for this is to be introduced and bounded from above in terms of the degree of the polynomial map.

One such characteristic can be easily described. What can be the maximal time during which an orbit may stay on a given algebraic subvariety, without being forced to stay on it forever? This question is a discrete time analog of the question on the maximal order of tangency between trajectories of a polynomial vector field and an algebraic hypersurface, the problem posed by J.-J. Risler in connection with control problems [6].

The discrete time problem was solved by D. Novikov and S. Yakovenko in [5] for dimension-preserving polynomial maps. The continuous time problem was solved by A. Gabrielov and A. Khovanskiĭ [4] who gave an exponential bound for the maximal order of tangency.

[1] Arnold V. I. Dynamics of complexity of intersections. *Bol. Soc. Brasil. Mat.* (*N. S.*), 1990, **21**(1), 1–10. [*The Russian translation in*: Vladimir Igorevich Arnold. Selecta–60. Moscow: PHASIS, 1997, 489–499.]

[2] Arnold V. I. Dynamics of intersections. In: Analysis, et cetera. Research papers published in honor of Jürgen Moser's 60th birthday. Editors: P. H. Rabinowitz and E. Zehnder. Boston, MA: Academic Press, 1990, 77–84.

[3] Arnold V. I. Bounds for Milnor numbers of intersections in holomorphic dynamical systems. In: Topological Methods in Modern Mathematics. Proceedings of the symposium in honor of John Milnor's sixtieth birthday (Stony Brook, NY, 1991). Editors: L. R. Goldberg and A. V. Phillips. Houston, TX: Publish or Perish, 1993, 379–390.

[4] Gabrielov A. M., Khovanskiĭ A. G. Multiplicity of a Noetherian intersection. In: Geometry of Differential Equations. Editors: A. G. Khovanskiĭ, A. N. Varchenko and V. A. Vassiliev. Providence, RI: Amer. Math. Soc., 1998, 119–130. (AMS Transl., Ser. 2, 186; Adv. Math. Sci., 39.)

[5] Novikov D. I., Yakovenko S. Yu. Trajectories of polynomial vector fields and ascending chains of polynomial ideals. *Ann. Inst. Fourier* (*Grenoble*), 1999, **49**(2), 563–609.

[6] Risler J.-J. A bound for the degree of nonholonomy in the plane. *Theoret. Comput. Sci.*, 1996, **157**(1), 129–136.

## 1971-11 — *A. M. Lukatskiĭ*

Also: *1989-19, 1992-11, 1994-26, 1994-27*

ℜ There are two Kolmogorov's 1958 conjectures on the behavior of the dimension of minimal attractors (i. e., the attractor which does not contain a smaller attractor) when its Reynolds number $R$ tends to the infinity:

a) weak

$$\max_{\min \mathrm{Attr}} \dim \min \mathrm{Attr} \to \infty \quad \text{for } \nu \to 0;$$

b) strong

$$\min_{\min \mathrm{Attr}} \dim \min \mathrm{Attr} \to \infty \quad \text{for } \nu \to 0;$$

(see [1], Ch. I) where $\nu$ is the (kinematic) viscosity of a current.

The primary upper bound (by Ladyzhenskaya, Il′yashenko, and Chetaev [3–5, 8]) of the dimension of a maximal attractor for the Navier–Stokes equation on two-dimensional torus by means of the viscosity $\nu$ had the form:

$$\dim \mathrm{Attr} \le \frac{\mathrm{const}}{\nu^4}.$$

Also, Témam's estimate [9] in an arbitrary two-dimensional domain $M$ is known:

$$\dim \mathrm{Attr} \le c(M)R, \qquad R = \frac{\|f\|_{L^2}}{\lambda_1 \nu^2}$$

(here $f$ is the exterior force and $\lambda_1$ is the first eigenvalue of the Stokes operator), which was then rewritten by Il′yin in another form [6] (see also [2]).

The best known upper bound of the dimesion of a maximal attractor was obtained by Il′yin [7] (for the Navier–Stokes equation in a two-dimensional domain $M$ in the presence of an exterior force $f$) and has the following form:

$$\dim \mathrm{Attr} \le \frac{1}{\pi} \|f\|_{L^2} \frac{\mathrm{vol}(M)}{\nu^2}.$$

Additional literature is given in book [1].

[1] Arnold V. I., Khesin B. A. Topological Methods in Hydrodynamics. New York: Springer, 1998. (Appl. Math. Sci., 125.)

[2] Chepyzhov V. V., Vishik M. I. A Hausdorff dimension estimate for kernel sections of nonautonomous evolution equations. *Indiana Univ. Math. J.*, 1993, **42**(3), 1057–1076.

[3] IL'YASHENKO YU. S. Weakly contracting systems and attractors of the Galerkin approximations of the Navier–Stokes equations on a two-dimensional torus. *Uspekhi Mekhaniki*, 1982, **5**(1–2), 31–63 (in Russian). [*The English translation*: *Selecta Math. Sov.*, 1992, **11**(3), 203–239.]

[4] IL'YASHENKO YU. S. On the dimension of attractors of *k*-contracting systems in an infinite-dimensional space. *Moscow Univ. Math. Bull.*, 1983, **38**(3), 61–69.

[5] IL'YASHENKO YU. S., CHETAEV A. N. On the dimension of attractors for a class of dissipative systems. *J. Appl. Math. Mech.*, 1983, **46**(3), 290–295.

[6] IL'YIN A. A. Partly dissipative semigroups generated by the Navier–Stokes system on two-dimensional manifolds, and their attractors. *Sb. Math.*, 1994, **78**(1), 47–76.

[7] IL'YIN A. A. Attractors for Navier–Stokes equations in domains with finite measure. *Nonlinear Anal.*, 1996, **27**(5), 605–616.

[8] LADYZHENSKAYA O. A. The finite-dimensionality of bounded invariant sets for the Navier–Stokes system and other dissipative systems. *Zap. Nauch. Semin. Leningrad. Otd. Mat. Inst. Steklova*, 1982, **115**, 137–155 (in Russian). (Boundary Problems in Mathematical Physics and Related Questions of the Function Theory, 14.)

[9] TÉMAM R. Infinite-Dimensional Dynamical Systems in Mechanics and Physics. New York: Springer, 1988. (Appl. Math. Sci., 68.)

# 1972

## 1972-2 — *S. V. Chmutov*

ℛ The problem was solved by A. M. Gabrielov [1] and E. Looijenga [2].

[1] GABRIELOV A. M. Intersection matrices for certain singularities. *Funct. Anal. Appl.*, 1973, **7**(3), 182–193.

[2] LOOIJENGA E. J. N. On the semi-universal deformation of a simple-elliptic hypersurface singularity. II. The discriminant. *Topology*, 1978, **17**(1), 23–40.

## 1972-3 — *V. D. Sedykh* Also: *1978-1, 1979-2, 1980-17*

ℛ Problem a) was solved by L. N. Bryzgalova for $n \le 6$ and by V. I. Matov for any $n$ where $n$ is the dimension of the parameter $x$ (see [6], and also [1–5]).

[1] ARNOLD V. I. Lectures on bifurcations in versal families. *Russian Math. Surveys*, 1972, **27**(5), 54–123.

[2] ARNOLD V. I. Catastrophe Theory. Berlin: Springer, 1992, Sect. 10. [*The Russian original* 1990.]

[3] ARNOLD V. I., VASSILIEV V. A., GORYUNOV V. V., LYASHKO O. V. Singularities. I. Local and Global Theory. Berlin: Springer, 1993, Sect. 3.3. (Encyclopædia Math. Sci., 6; Dynamical Systems, VI.) [*The Russian original* 1988.]

[4] ARNOLD V. I., GUSEIN-ZADE S. M., VARCHENKO A. N. Singularities of Differentiable Maps, Vol. I: The classification of critical points, caustics and wave fronts. Boston, MA: Birkhäuser, 1985, Sect. 10.3.5. (Monographs in Math., 82.) [*The Russian original* 1982.]

[5] LEVANTOVSKIĬ L. V. Singularities of the boundary of the stability domain. *Funct. Anal. Appl.*, 1982, **16**(1), 34–37.

[6] MATOV V. I. The topological classification of germs of the maximum and minimax functions of a family of functions in general position. *Russian Math. Surveys*, 1982, **37**(4), 127–128.

## 1972-5 – *V. N. Karpushkin*

ℛ The uniform singularity index in terms of a phase in a degenerate point is calculated through the individual index of the singularity of an oscillatory integral. Concerning individual singularity indices, see [1, 9, 10, 15]. Conjectures about uniform estimates of oscillatory integrals with individual singularity index and about the semicontinuity of singularity index were first formulated in [1].

Uniform estimates of oscillatory integrals with individual singularity index for analytic phases depending on two variables were obtained in [5, 6]. These results generalize those achieved by I. M. Vinogradov for phases depending on one variable [16], and by J. Duistermaat for simple singularities [3].

The results of Colin de Verdière from [2] were partially extended by V. N. Karpushkin for a phase in two variables [8].

D. A. Popov disproved Colin de Verdière's result for $A_3$ [14]. Estimates regarding partial perturbations of a phase were obtained in [7].

The conjecture about uniform estimates of oscillatory integrals with individual singularity index is equivalent to the conjecture about the semicontinuity of singularity index, for a linear perturbation of an $\mathbb{R}$-nondegenerate semiquasihomogeneous phase in three variables, see [11]. A. N. Varchenko showed that the conjecture about the semicontinuity of singularity index fails for $\mathbb{R}$-nondegenerate semiquasihomogeneous polynomials in $\mathbb{R}^n$ ($n \geq 3$), see [15]. Everything stated above applies to volumes (areas, lengths) when a coordinate of the intersection point of the line $x_1 = \cdots = x_n$ with the bound of the Newton polyhedron of a phase is greater than one [13].

Uniform estimates of volumes (areas) with individual singularity indices are true for all 0-modal and unimodal phases [4]. Uniform estimates with individual singularity indices probably hold for all unimodal phases except some phases of series $P_m$, $m \geq 8$, and except phases of series $\tilde{R}_m$, see [1,4–6,9,10,15].

Uniform estimates of oscillatory integrals with the index being the maximum of individual singularity indices from all adjoining singularities are true for the unimodal phases of series $\tilde{R}_m$, see [12].

[1] ARNOLD V. I. Remarks on the stationary phase method and Coxeter numbers. *Russian Math. Surveys*, 1973, **28**(5), 19–48.

[2] COLIN DE VERDIÈRE Y. Nombre de points entiers dans une famille homothétique de domaines de $\mathbb{R}^n$. *Ann. Sci. École Norm. Sup., Sér.* 4, 1977, **10**(4), 559–575.

[3] DUISTERMAAT J. Oscillatory integrals, Lagrangian immersions and unfolding of singularities. *Commun. Pure Appl. Math.*, 1974, **27**(2), 209–281.

[4] KARPUSHKIN V. N. Uniform estimates of integrals with unimodal phase. *Uspekhi Mat. Nauk*, 1983, **38**(3), 128 (in Russian).

[5] KARPUSHKIN V. N. Uniform estimates of oscillatory integrals with a parabolic or hyperbolic phase. *Trudy Semin. Petrovskogo*, 1983, **9**, 1–39 (in Russian). [*The English translation*: *J. Sov. Math.*, 1986, **33**, 1159–1188.]

[6] KARPUSHKIN V. N. A theorem concerning uniform estimates of oscillatory integrals when the phase is a function of two variables. *Trudy Semin. Petrovskogo*, 1984, **10**, 150–169 (in Russian). [*The English translation*: *J. Sov. Math.*, 1986, **35**, 2809–2826.]

[7] KARPUSHKIN V. N. Uniform estimates for oscillatory integrals and volumes under a partial deformation of a phase. In: Geometry and the Theory of Singularities in Nonlinear Equations. Voronezh: Voronezh University Press, 1987, 151–159 (in Russian). (Novoe v Global′nom Analize, 7.)

[8] KARPUSHKIN V. N. Uniform estimates for some oscillating integrals. *Sib. Math. J.*, 1989, **30**(2), 240–249.

[9] KARPUSHKIN V. N. Oscillatory integrals and volumes with semiquasihomogeneous phase. *Funct. Anal. Appl.*, 1992, **26**(1), 46–48.

[10] KARPUSHKIN V. N. Dominant term in the asymptotics of oscillatory integrals with a phase of the series $T$. *Math. Notes*, 1994, **56**(6), 1304–1305.

[11] KARPUSHKIN V. N. A remark about uniform estimates and counterexample of A. N. Varchenko. In: Some Problems of Fundamental and Applied Mathematics. Moscow: Moscow Instintute of Physics and Technology Press, 1998, 74–79 (in Russian).

[12] KARPUSHKIN V. N. Uniform estimates of oscillatory integrals with phase from the series $\tilde{R}_m$. *Math. Notes*, 1998, **64**(3), 404–406.

[13] KARPUSHKIN V. N. Uniform estimates of volumes. *Proc. Steklov Inst. Math.*, 1998, **221**, 214–220.

[14] Popov D. A. Estimates with constants for some classes of oscillatory integrals. *Russian Math. Surveys*, 1997, **52**(1), 73–145.

[15] Varchenko A. N. Newton polyhedra and estimation of oscillating integrals. *Funct. Anal. Appl.*, 1976, **10**(3), 175–196.

[16] Vinogradov I. M. The Method of Trigonometric Sums in the Number Theory. Moscow: Nauka, 1971 (in Russian).

## 1972-6 — *S. M. Gusein-Zade*

ℛ In [1] it was indicated that this result had been proved in [2].

[1] Arnold V. I. Remarks on the stationary phase method and Coxeter numbers. *Russian Math. Surveys*, 1973, **28**(5), 19–48.

[2] Tyurina G. N. The topological properties of isolated singularities of complex spaces of codimension one. *Math. USSR, Izv.*, 1968, **2**, 557–571.

## ▽ 1972-7

ℋ The transversality conjecture was formulated by V. I. Arnold in paper [1].

[1] Arnold V. I. Modes and quasimodes. *Funct. Anal. Appl.*, 1972, **6**(2), 94–101. [*The Russian original is reprinted in*: Vladimir Igorevich Arnold. Selecta–60. Moscow: PHASIS, 1997, 189–202.]

## △ 1972-7 — *Ya. M. Dymarskiĭ*

ℛ Sufficient conditions for the validity of this conjecture were obtained in [3,5] (for a family of membranes) and in [1,2] (for a family of oscillating systems parametrized by the potential). In [4], there is an example of a membrane family for which the above-mentioned conditions from [3,5] are not satisfied.

The first theorems substantiating the conjecture were proved in [6] for various one-dimensional families of oscillating systems.

Diverse aspects of the conjecture were considered by the author of the present comment in his talk at V. I. Arnold's seminar (Moscow, September 26, 2000).

[1] Dymarskiĭ Ya. M. On manifolds of self-adjoint elliptic operators with multiple eigenvalues. *Methods Funct. Anal. Topology*, 2001, **7**(2), 68–74.

[2] DYMARSKIĬ YA. M. Manifolds of eigenfunctions and potentials of a family of periodic Sturm–Liouville problems. *Ukrain. Math. J.*, 2002, **54**(8), 1251–1264.

[3] LUPO D., MICHELETTI A. M. On multiple eigenvalues of selfadjoint compact operators. *J. Math. Anal. Appl.*, 1993, **172**(1), 106–116.

[4] LUPO D., MICHELETTI A. M. A remark on the structure of the set of perturbations which keep fixed the multiplicity of two eigenvalues. *Revista Mat. Apl.*, 1995, **16**(2), 47–56.

[5] LUPO D., MICHELETTI A. M. On the persistence of the multiplicity of eigenvalues for some variational elliptic operator depending on the domain. *J. Math. Anal. Appl.*, 1995, **193**(3), 990–1002.

[6] UHLENBECK K. Generic properties of eigenfunctions. *Amer. J. Math.*, 1976, **98**(4), 1059–1078.

## 1972-8 — *V. N. Karpushkin*

Let $G$ be a finite group, and let $F$ be the set of all orthogonal representations of the group $G$ in $\mathbb{R}^N$. The set $F$ is a union of some connected components of different dimension. Each of these connected components is a smooth manifold and represents a class of equivalent representations. The problem is to investigate components of the set $F$ having maximal dimension. Representations corresponding to these components are called "the most probable," see [1].

It is, however, worthwhile to mention that Theorems 1 and 3 from [3] about the "most probable" representations of a finite group are actually *not* substantiated in that the author has not succeeded to prove Lemma 5 of the latter paper.

P.S. (*V. I. Arnold*): See also [2] for newer results on unitary representations.

[1] ARNOLD V. I. Modes and quasimodes. *Funct. Anal. Appl.*, 1972, **6**(2), 94–101. [*The Russian original is reprinted in*: Vladimir Igorevich Arnold. Selecta–60. Moscow: PHASIS, 1997, 189–202.]

[2] ARNOLD V. I. Frequent representations. *Moscow Math. J.*, 2003, **3**(4), 14 pp.

[3] KARPUSHKIN V. N. On the asymptotic behavior of eigenvalues of symmetric manifolds and on most probable representations of finite groups. *Moscow Univ. Math. Bull.*, 1974, **29**(2), 136–139.

## 1972-9 — *A. I. Neĭshtadt* Also: *1966-1*

ℛ This is the question about the application of the averaging method to the systems of the form

$$\dot{I} = \varepsilon f(I,\varphi,\varepsilon), \qquad I \in \mathbb{R}^n,$$
$$\dot{\varphi} = \omega(I) + \varepsilon g(I,\varphi,\varepsilon), \qquad \varphi \in \mathbb{T}^m.$$

Here $\varepsilon > 0$ is a small parameter, and the righthand side of the system is $2\pi$-periodic with respect to all components $\varphi_i$ of the vector $\varphi$. The variables $I$ are called *slow variables*, and the variables $\varphi$ are called *fast variables* or *phases*. The components of the vector $\omega$ are called *frequencies*. The system under consideration is called a *perturbed system in the standard form of the averaging method* or a *system with rotating phases*.

For an approximate description of the evolution of slow variables $I$ on time intervals of lengths of order $1/\varepsilon$, the averaging method prescribes to replace the slow component $I(t)$ of the solution $I(t)$, $\varphi(t)$ of the system with rotating phases by the solution $J(t)$, $J(0) = I(0)$ of the *averaged system*

$$\dot{J} = \varepsilon F(J), \quad F(J) = \frac{1}{(2\pi)^m} \int_0^{2\pi} \cdots \int_0^{2\pi} f(J,\varphi,0)\, d\varphi_1 \ldots d\varphi_m.$$

The value $\sup\limits_{0 \le t \le 1/\varepsilon} |I(t) - J(t)|$ is the error of the averaging method for the initial data $I(0)$, $\varphi(0)$ on the time interval $[0; 1/\varepsilon]$.

In the two-frequency case ($m = 2$) under the assumption that the ratio of frequencies is changing with nonzero velocity along the solutions of the averaged system, the following estimates are valid.

For any $\kappa > c\sqrt{\varepsilon}$ one can choose a set of "bad" initial conditions of measure at most $\kappa$ in such a way that, outside this set, the error of the averaging method is $O(\sqrt{\varepsilon}\ln\kappa)$ on the time interval $0 \le t \lesssim 1/\varepsilon$ provided that some additional nondegeneracy condition (the so-called condition B) is satisfied [5, 6]. Here $c$ is some positive constant. For some set of initial data of measure $\sim \sqrt{\varepsilon}$ the averaging method may not work at all (i. e., it gives an error $\sim 1$) because of captures into resonances. The further fate of phase points captured into resonance can also be described [7]. There are estimates for the cases when condition B is replaced by weaker nondegeneracy conditions [8]. If none of these nondegeneracy conditions is satisfied, then for any $\kappa > c\sqrt{\varepsilon}$ one can choose a set of initial conditions of measure at most $\kappa$ in such a way that, outside this exceptional set, the error of the averaging method is $O(\sqrt{\varepsilon}/\sqrt{\kappa})$; for analytic systems this estimate was obtained in [9], and for the case of finite smoothness it follows from a union of results of [5,6] and [3,4] (the case when there is only one slow variable was considered in [5,6]). All estimates pointed out above cannot be improved. Detailed discussion of the results expounded above is contained in books [1,2].

[1] Arnold V. I. Geometrical Methods in the Theory of Ordinary Differential Equations, 2nd edition. New York: Springer, 1988. (Grundlehren der Mathematischen Wissenschaften, 250.) [*The Russian original* 1978.]

[2] Arnold V. I., Kozlov V. V., Neĭshtadt A. I. Mathematical Aspects of Classical and Celestial Mechanics, 2nd edition. Berlin: Springer, 1993. (Encyclopædia Math. Sci., 3; Dynamical Systems, III.) [*The Russian original* 1985.] [*The second, revised and supplemented, Russian edition* 2002.]

[3] Bakhtin V. I. Averaging method in multi-frequency systems. Ph. D. Thesis, Moscow State University, 1986 (in Russian).

[4] Bakhtin V. I. Averaging in a general-position single-frequency system. *Differ. Equations*, 1991, **27**(9), 1051–1061.

[5] Neĭshtadt A. I. On some resonant problems in nonlinear systems. Ph. D. Thesis, Moscow State University, 1975 (in Russian).

[6] Neĭshtadt A. I. Passage through a resonances in the two-frequency problem. *Sov. Phys. Dokl.*, 1975, **20**(3), 189–191.

[7] Neĭshtadt A. I. Scattering by resonances. *Celest. Mech. Dynam. Astron.*, 1996/97, **65**(1–2), 1–20.

[8] Pronchatov V. E. An error estimate for the averaging method in the two-frequency problem. *Math. USSR, Sb.*, 1985, **50**(1), 241–258.

[9] Pronchatov V. E. On an error estimate for the averaging method in the two-frequency problem. *Math. USSR, Sb.*, 1989, **62**(1), 29–40.

## **1972-10** — ***A. I. Neĭshtadt*** Also: *1966-1*

ℜ The problem has been solved for two-frequency systems, see [1] and problem *1972-9*. So, now "multi-frequency" means that number of frequencies is greater than 2.

The possibility of using the averaging method in the multi-frequency case follows from the general result about averaging in systems with slow and fast motions [2] (see also [5]). Estimates for the error of the averaging method for different relations between the number of slow variables $n$ and the number of fast variables $m$ (which is equal to the number of frequencies) are contained in [3,4,6]. In particular, if $n \geq m$ and the frequency map $I \mapsto \omega(I)$ is nondegenerate, then for any $\kappa > c\sqrt{\varepsilon}$ one can choose a set of initial data of measure at most $\kappa$ in such a way that, outside this set, the error of the averaging method is $O(\sqrt{\varepsilon}/\kappa)$ on the time interval $0 \le t \lesssim 1/\varepsilon$ [6] (the notations are the same as in the comment to problem *1972-9*). The same estimate is valid if $n \ge m-1$ and the map $I \mapsto (\omega_1(I) : \omega_2(I) : \dots : \omega_m(I))$ from the space of slow variables to the projective space of ratios of frequencies is nondegenerate.

For $n < m$ the image of the domain of the slow variables space for the frequency map $I \mapsto \omega(I)$ is a surface $M$ of positive codimension in the frequency space. The deduction of the estimates for the averaging method error in this case is based on results about Diophantine approximations on $M$, see problem *1970-5*. For $n < m-1$ it is reasonable to consider also an analogous surface $M'$ in the space of ratios of frequencies.

For any $m$ and $n$ the above-mentioned estimate of the averaging method error is valid for almost all members of a generic family of frequency maps depending on a sufficiently large number of parameters [3]. This estimate is also valid, if $M$ satisfies some curvature condition given in [4]. It is shown in [3] that for generic maps $\omega$ outside some exceptional set of initial data of measure $\le \kappa$ (with prescribed $\kappa > c\sqrt{\varepsilon}$) the error of the averaging method is $O(\varepsilon^{1/(p+1)}/\kappa)$ on the time interval $0 \le t \lesssim 1/\varepsilon$ provided that $C^p_{n+p} \ge n+m$. The required genericity condition for $\omega$ is presented in [3] in an explicit form. Maps $\omega$ that do not comply with this condition belong to a codimension 1 surface in the functional space of all maps $\omega$. The above-mentioned estimates cannot be improved in the class of power estimates and under conditions that were used to derive these estimates. These conditions are imposed only on the map $\omega$. It seems plausible that it may be possible to improve these estimates, if we impose some condition of generic mutual disposition of resonant surfaces $k \cdot \omega = 0$, $k \in \mathbb{Z}^m$, and vector field of the averaged system (as it was made for two-frequency case, cf. problem *1972-9*). Therefore, the problem under discussion cannot currently be considered as a solved one.

[1] ARNOLD V. I. Conditions for the applicability and estimate of the error of an averaging method for systems which pass through the states of resonance in the course of their evolution. *Sov. Math. Dokl.*, 1965, **6**, 331–334. [*The Russian original is reprinted in*: Vladimir Igorevich Arnold. Selecta–60. Moscow: PHASIS, 1997, 69–74.]

[2] ANOSOV D. V. Averaging in systems of ordinary differential equations with rapidly oscillating solutions. *Izv. Akad. Nauk SSSR, Ser. Mat.*, 1960, **24**(5), 721–742 (in Russian).

[3] BAKHTIN V. I. Averaging in multi-frequency systems. *Funct. Anal. Appl.*, 1986, **20**(2), 83–88.

[4] DODSON M. M., RYNNE B. P., VICKERS J. A. G. Averaging in multi-frequency systems. *Nonlinearity*, 1989, **2**(1), 137–148.

[5] KASUGA T. On the adiabatic theorem for the Hamiltonian system of differential equations in the classical mechanics, I; II; III. *Proc. Japan. Acad.*, 1961, **37**(7), 366–371; 372–376; 377–382.

[6] NEĬSHTADT A. I. Averaging in multi-frequency systems, II. *Sov. Phys. Dokl.*, 1976, **21**(2), 80–82.

## 1972-11 – *V. A. Vassiliev*

ℛ The cohomology rings of braid groups of series $D$ and $C$ were calculated by V. V. Goryunov [1], and of all other braid groups (including $E$) by M. Salvetti [2].

[1] GORYUNOV V. V. Cohomology of braid groups of series $C$ and $D$. *Trans. Moscow Math. Soc.*, 1982, **42**, 233–241.

[2] SALVETTI M. The homotopy tupe of Artin groups. *Math. Res. Lett.*, 1994, **1**(5), 565–577.

## 1972-12 – *V. D. Sedykh* Also: *1973-2, 1976-32, 1981-28*

ℛ Let $M$ be a smooth closed $k$-dimensional submanifold in $\mathbb{R}^n$. The *convex hull* of a manifold $M$ is the intersection of all half-spaces containing $M$. A germ of the convex hull at a point of its boundary is called a *singularity* of the convex hull. Two singularities are said to be *equivalent* if one singularity can be transferred to the other by a suitable diffeomorphism of $\mathbb{R}^n$.

We consider the classification of singularities of the convex hulls of generic submanifolds with respect to this equivalence. *Generic* submanifolds are defined by embeddings $M \hookrightarrow \mathbb{R}^n$ which belong to a certain open dense subset in the space of all embeddings of $M$ into $\mathbb{R}^n$ equipped with the $C^\infty$-topology.

Except the trivial cases $k = 0$, $n = 1$, $n = 2$, the indicated classification is obtained only for $n = 3$.

**Theorem** [3, 5]. *Singularities of the convex hull of a smooth closed generic curve in $\mathbb{R}^3$ are equivalent to germs at zero of the sets*

$$z \geq 0, \qquad z \geq x|x|, \qquad z \geq |x|,$$

$$z \geq \min_t(t^4 + xt^2 + yt), \qquad z \geq \min^2(x, y, 0), \qquad \{z \geq \min^2(x, y, 0),\ x + y \geq 0\}.$$

**Theorem** [11]. *Singularities of the convex hull of a smooth closed generic surface in* $\mathbb{R}^3$ *are equivalent to germs at zero of the sets*

$$z \geq 0, \qquad z \geq x|x|, \qquad z \geq \rho^2(x,y),$$

*where* $\rho(x,y)$ *is the distance from a point* $(x,y)$ *to the angle* $y \geq \alpha|x|$, $\alpha > 0$, $\alpha \neq 1$.

The number $\alpha$ is a *module* (continuous invariant): singularities with different $\alpha$ are not equivalent. For $n > 3$, singularities of convex hulls have moduli as well.

**Theorem** [8]. *For* $n \geq 5$, *there are smooth closed submanifolds of codimension* 1 *and* 2 *in* $\mathbb{R}^n$ *such that some singularities of their convex hulls have functional moduli which cannot be removed by small deformations of a submanifold.*

**Theorem** [4, 6, 7]. *For any* $n \geq 4$ *and* $k = 1, \dots, n-3$, *and for any natural number N, there are smooth closed k-dimensional submanifolds in* $\mathbb{R}^n$ *such that some singularities of their convex hulls have at least N functional moduli (of k variables) which can not be removed by small deformations of a submanifold.*

Singularities of convex hulls of smooth closed two-dimensional and three-dimensional generic surfaces in $\mathbb{R}^4$ have moduli (see [4, 6]). Normal forms of some singularities of convex hulls in these dimensions were obtained in [1, 9, 10]. Moreover, it is proved in [1] that singularities of convex hulls of smooth closed generic hypersurfaces in $\mathbb{R}^4$ have no functional moduli. I think that singularities of convex hulls of smooth closed generic two-dimensional surfaces in $\mathbb{R}^4$ have no functional moduli as well (see [6]).

Problem *1972-12* is connected with the problem on the smoothness of the boundary of the vector sum of two convex bodies which are bounded by smooth hypersurfaces. For some results on this problem see [2].

[1] BOGAEVSKY I. A. Singularities of convex hulls of three-dimensional hypersurfaces. *Proc. Steklov Inst. Math.*, 1998, **221**, 71–90.

[2] KISELMAN C. O. How smooth is the shadow of a smooth convex body. *J. London Math. Soc., Ser.* 2, 1986, **33**(1), 101–109; *Serdica Math. J.*, 1986, **12**(2), 189–195.

[3] SEDYKH V. D. Singularities of the convex hull of a curve in $\mathbb{R}^3$. *Funct. Anal. Appl.*, 1977, **11**(1), 72–73.

[4] SEDYKH V. D. Moduli of singularities of convex hulls. *Russian Math. Surveys*, 1981, **36**(5), 175–176.

[5] SEDYKH V. D. Structure of the convex hull of a space curve. *Trudy Semin. Petrovskogo*, 1981, **6**, 239–256 (in Russian). [*The English translation*: *J. Sov. Math.*, 1986, **33**, 1140–1153.]

[6] SEDYKH V. D. Singularities of convex hulls. *Sib. Math. J.*, 1983, **24**(3), 447–461.

[7] SEDYKH V. D. Convex hulls and the Legendre transform. *Sib. Math. J.*, 1983, **24**(6), 923–933.

[8] SEDYKH V. D. Functional moduli of singularities of convex hulls of manifolds of codimension 1 and 2. *Math. USSR, Sb.*, 1984, **47**(1), 223–236.

[9] SEDYKH V. D. Stabilization of singularities of convex hulls. *Math. USSR, Sb.*, 1989, **63**(2), 499–505.

[10] SEDYKH V. D. The sewing of a swallowtail and a Whitney umbrella in a four-dimensional controlled system. In: Proceedings of Gubkin State Oil and Gas Academy. Moscow: Neft′ i Gaz, 1997, 58–68 (in Russian).

[11] ZAKALYUKIN V. M. Singularities of convex hulls of smooth manifolds. *Funct. Anal. Appl.*, 1978, **11**(3), 225–227.


## 1972-13 — *V. A. Vassiliev*

ℛ For any isolated semiquasihomogeneous function singularity, the modality equals the number of linearly independent elements of the local ring whose quasi-homogeneous degrees are greater than or equal to the degree of the principal part of the function.

This equality consists of two inequalities. One of them (the modality is not less than this number) was proved in [1], where it also was conjectured that this estimate is sharp. This conjecture was later proved in [3].

In particular, for the Brieskorn singularity $\sum_{i=1}^{n} x_i^{a_i}$, this number is equal to the number of points $(k_1, \ldots, k_n)$ of the integral lattice $\mathbb{Z}_+^n$ satisfying the conditions $k_i \leq a_i - 2$ for any $i$ and $\sum_{i=1}^{n} k_i/a_i \geq 1$.

A related result for $\Gamma$-nondegenerate isolated singularities of two variables: the modality of such a singularity equals the number of integral lattice points in the domain bounded by the Newton diagram and two rays issuing from the point $(2,2)$ and parallel to positive coordinate rays. This formula was also conjectured in [1] and then proved in [2].


[1] ARNOLD V. I. Normal forms of functions in neighborhoods of degenerate critical points. *Russian Math. Surveys*, 1974, **29**(2), 10–50.

[2] KUSHNIRENKO A. G. Polyèdres de Newton et nombres de Milnor. *Invent. Math.*, 1976, **32**(1), 1–31.

[3] VARCHENKO A. N. A lower bound for the codimension of the stratum $\mu = \text{const}$ in terms of the mixed Hodge structure. *Moscow Univ. Math. Bull.*, 1982, **37**(6), 30–33.

## ▽ 1972-14 — *V. V. Goryunov*

ℛ H. Knörrer showed in [5] that the complement of the discriminant of the simple zero-dimensional complete intersection $x^2 = y^2 = 0$ in $\mathbb{C}^2$ has a non-trivial second homotopy group. Perhaps this is the only negative example known up to now.

On the other hand, there exist many positive examples (see [1,2] for some of them), provided mainly by simple singularities of various classifications. The only positive examples coming from non-simple cases are those by P. Jaworski of the bifurcation diagrams of functions of the parabolic function singularities [4].

One of the interesting related problems is that of whether the discriminants of Shephard–Todd groups [7] possess the $K(\pi,1)$ property. At the moment, the question remains open just in 6 cases (groups nos. 24, 27, 29, 31, 33 and 34 of [7]) and has been answered positively in all the others [6]. The $G_{31}$ discriminant is depicted in [3].

[1] ARNOLD V. I., VASSILIEV V. A., GORYUNOV V. V., LYASHKO O. V. Singularities. I. Local and Global Theory. Berlin: Springer, 1993, Ch. 2, Sect. 5. (Encyclopædia Math. Sci., 6; Dynamical Systems, VI.) [*The Russian original* 1989.]

[2] ARNOLD V. I., VASSILIEV V. A., GORYUNOV V. V., LYASHKO O. V. Singularities. II. Classification and Applications. Berlin: Springer, 1993, Ch. 1. (Encyclopædia Math. Sci., 39; Dynamical Systems, VIII.) [*The Russian original* 1989.]

[3] GORYUNOV V. V., BAINES C. E. Cyclically equivariant function singularities and unitary reflection groups $G(2m,2,n)$, $G_9$ and $G_{31}$. *St. Petersburg Math. J.*, 2000, **11**(5), 761–774.

[4] JAWORSKI P. Distribution of critical values of miniversal deformations of parabolic singularities. *Invent. Math.*, 1986, **86**(1), 19–33.

[5] KNÖRRER H. Zum $K(\pi,1)$-Problem für isolierte Singularitäten von vollständigen Durchschnitten. *Compos. Math.*, 1982, **45**(3), 333–340.

[6] NAKAMURA T. A note on the $K(\pi,1)$ property of the orbit space of the unitary reflection group $G(m,l,n)$. *Sci. Papers College Arts Sci. Univ. Tokyo*, 1983, **33**(1), 1–6.

[7] SHEPHARD G. C., TODD J. A. Finite unitary reflection groups. *Canad. J. Math.*, 1954, **6**, 274–304.

## △ 1972-14 — *V. A. Vassiliev*

ℛ For many classification problems the complements of suitably defined bifurcation diagrams of simple (i. e., 0-modal) objects turn out to be $K(\pi,1)$-spaces; see, in particular, [1–6, 8, 9].

However, for singularities of complete intersections $\mathbb{C}^2 \to \mathbb{C}^2$ this is not the case: H. Knörrer showed in [7] that the complement of the discriminant of the complete intersection $(x^2, y^2)$ in $\mathbb{C}^2$ has non-trivial group $\pi_2$.

For non-simple singularities nothing is known to me.


[1] BRIESKORN E. Sur les groupes de tresses [d'après V. I. Arnold]. In: Séminaire Bourbaki, 24ème année (1971/1972), Exp. No. 401. Berlin: Springer, 1973, 21–44. (Lecture Notes in Math., 317.)

[2] DELIGNE P. Les immeubles de groupes de tresses généralisés. *Invent Math.*, 1972, **17**, 273–302.

[3] GORYUNOV V. V. Geometry of bifurcation diagrams of simple projections onto the line. *Funct. Anal. Appl.*, 1981, **15**(2), 77–82.

[4] GORYUNOV V. V. Projection of 0-dimensional complete intersection onto a line and the $K(\pi, 1)$-conjecture. *Russian Math. Surveys*, 1982, **37**(3), 206–208.

[5] GORYUNOV V. V. Singularities of projections of complete intersectrions. In: Itogi Nauki i Tekhniki VINITI. Current Problems in Mathematics, Vol. 22. Moscow: VINITI, 1983, 167–206 (in Russian). [*The English translation*: *J. Sov. Math.*, 1984, **27**, 2785–2811.]

[6] GORYUNOV V. V. Vector fields and functions on the discriminants of complete intersections, and bifurcation diagrams of projections. In: Itogi Nauki i Tekhniki VINITI. Current Problems in Mathematics. Newest Results, Vol. 33. Moscow: VINITI, 1988, 31–54 (in Russian). [*The English translation*: *J. Sov. Math.*, 1990, **52**(4), 3231–3245.]

[7] KNÖRRER H. Zum $K(\pi, 1)$-Problem für isolierte Singularitäten von vollständigen Durchschnitten. *Compos. Math.*, 1982, **45**(3), 333–340.

[8] LOOIJENGA E. J. N. The complement of the bifurcation variety of a simple singularity. *Invent. Math.*, 1974, **23**(2), 105–116.

[9] LYASHKO O. V. The geometry of bifurcation diagrams. *Russian Math. Surveys*, 1979, **34**(3), 209–210.


## 1972-16 – *V. I. Arnold*

The corresponding phase curve is given by the equation $q^2 + 3p^2 + |p|^3 = 4$ in the phase plane (in suitable coordinates); see papers [1–3].


[1] ROYTVARF A. A. The motion of a continuous medium in the force field with a rooted singularity. *Moscow Univ. Mech. Bull.*, 1987, **42**(1), 24–27.

[2] ROYTVARF A. A. Two-valued velocity field with a square root singularity. *Moscow Univ. Mech. Bull.*, 1988, **43**(3), 16–19.

[3] ROYTVARF A. A. On the dynamics of a one-dimensional self-gravitating medium. *Physica D*, 1994, **73**(3), 189–204.

## 1972-17

H This is a problem in V. I. Arnold's comment [1] to H. Poincaré's paper "On a geometric theorem" ("Sur un théorème de géométrie"). The conjectures on the number of fixed points of symplectomorphisms were first formulated by V. I. Arnold in paper [2a] (see also [2b]); see problems *1965-1–1965-3*.

[1] ARNOLD V. I. A comment to H. Poincaré's paper "Sur un théorème de géométrie." In: POINCARÉ H. Selected Works in Three Volumes (in Russian). Editors: N. N. Bogolyubov, V. I. Arnold and I. B. Pogrebysskiĭ. Vol. II. New methods of celestial mechanics. Topology. Number theory. Moscow: Nauka, 1972, 987–989 (in Russian).

[2a] ARNOLD V. I. Sur une propriété topologique des applications globalement canoniques de la mécanique classique. *C. R. Acad. Sci. Paris*, 1965, **261**(19), 3719–3722.

*The Russian translation in*:

[2b] Vladimir Igorevich Arnold. Selecta–60. Moscow: PHASIS, 1997, 81–86.

R See the comment to problem *1972-33*.

## 1972-18

H This is a problem in V. I. Arnold's comment [1] to H. Poincaré's paper "On a geometric theorem" ("Sur un théorème de géométrie").

[1] ARNOLD V. I. A comment to H. Poincaré's paper "Sur un théorème de géométrie." In: POINCARÉ H. Selected Works in Three Volumes (in Russian). Editors: N. N. Bogolyubov, V. I. Arnold and I. B. Pogrebysskiĭ. Vol. II. New methods of celestial mechanics. Topology. Number theory. Moscow: Nauka, 1972, 987–989 (in Russian).

R See the comment to problem *1970-10*.

## 1972-20 — *A. A. Glutsyuk, M. B. Sevryuk*

Also: *1959-1*, *1963-4*, *1994-53*

R This problem has a rich history starting from the works by H. Poincaré and A. Denjoy. The great progress achieved in examining the questions pointed out in the problem for the last 40 years has been primarily due to V. I. Arnold, M. R. Herman, J.-C. Yoccoz, and R. Pérez-Marco. In this brief comment, we shall confine ourselves to a list of their main works dealing with the topics under consideration. [1]

[1] In these works, the history of the problem is expounded and references to papers and books by other authors are given as well.

V. I. Arnold proved in 1958 (see [1,3,4] and the preliminary publication [2]) that any orientation preserving analytic diffeomorphism of a circle with Diophantine [2] rotation number $\mu$, sufficiently close to the rotation through the angle $2\pi\mu$, is analytically reducible to this rotation. He also formulated the conjecture on the existence of a subset $M \subset [0;1]$ of measure 1 such that any analytic diffeomorphism of a circle with rotation number $\mu \in M$ (not necessarily close to a rotation) is analytically reducible to the rotation through the angle $2\pi\mu$. The history of this discovery is described by V. I. Arnold in his recollections [6].

Arnold's conjecture (as well as its modified version stating the smooth reducibility of smooth—of class $C^r$ with $r_0 \leq r \leq +\infty$—diffeomorphisms of the circle) was proved by M. R. Herman in 1976 [8, 10, 12]. The set $M$ dealt with in Arnold's conjecture is the set of Diophantine numbers. Some preliminary results were obtained by Herman in [7, 9, 11]. In 1985, Herman found a much simpler proof of the theorem on the reducibility to a rotation of circle diffeomorphisms (close to a rotation) with almost every rotation number [15]. Of Herman's other papers concerning the reducibility of circle diffeomorphisms and related questions, we mention [13, 14].

M. R. Herman's results on circle diffeomorphisms were refined and improved essentially by his student J.-C. Yoccoz in works [30–34, 37]. [3]

The remaining items of the problem (those on topological obstacles to analytically straightening a circle diffeomorphism, to prolonging the annulus of reducibility of a circle diffeomorphism to a rotation, and to prolonging the reducibility disk in Siegel's problem on linearizing germs of holomorphic mappings of the complex plane; [4] here one can add the question on topological obstacles to the linearization itself in Siegel's problem) pertain to the so-called *materialization of resonances* in holomorphic dynamics. Conjectures on the existence of topological obstacles (in the form of periodic orbits) to reducibility of analytic circle diffeomorphisms and linearization of germs of holomorphic mappings $(\mathbb{C},0) \to (\mathbb{C},0)$ were formulated by V. I. Arnold in 1958 (see a detailed discussion in [5] as well

[2] Recall that a number $\mu \in \mathbb{R}$ is said to be *Diophantine* if there exist positive constants $\tau$ and $\gamma$ such that $|q\mu - p| \geq \gamma/q^\tau$ for all rational $p/q$ ($p \in \mathbb{Z}$, $q \in \mathbb{N}$). In particular, all the Diophantine numbers are irrational. The converse is not true, but non-Diophantine numbers constitute a set of measure zero.

[3] See also papers [19, 28] by K. M. Khanin and Ya. G. Sinaĭ, paper [29] by J. Stark, and papers [17, 18] by Y. Katznelson and D. S. Ornstein.

[4] The classical results in the latter problem were obtained by C. L. Siegel and A. D. Bruno, see Herman's detailed survey [16] as well as the references in J.-C. Yoccoz's and R. Pérez-Marco's works [20–27, 35, 36] cited below.

as recollections [6]). These conjectures have been proved in part and disproved in part by R. Pérez-Marco and J.-C. Yoccoz.

In note [20], R. Pérez-Marco constructed examples of germs of holomorphic mappings $(\mathbb{C},0) \to (\mathbb{C},0)$, $z \mapsto e^{2\pi i\alpha}z + O(z^2)$, with $\alpha$ irrational, that are reduced to the rotation $w \mapsto e^{2\pi i\alpha}w$ by no local holomorphic transformation $w = z + O(z^2)$ and, at the same time, have no periodic orbits other than the origin in a neighborhood of the origin. Moreover, paper [20] described all the numbers $\alpha \in [0;1] \setminus \mathbb{Q}$ for which there exist germs with the properties indicated. The set of all such $\alpha$ is of measure zero. These examples were expounded in more detail in work [23] where Pérez-Marco constructed also examples of analytic circle diffeomorphisms with irrational rotation number $\mu$ that are not reduced to the rotation through the angle $2\pi\mu$ by an analytic change of variables and, at the same time, have no periodic orbits in a complex neighborhood of the circle. In paper [26], Pérez-Marco constructed examples of conformal mappings $(\mathbb{C},0) \to (\mathbb{C},0)$, $z \mapsto e^{2\pi i\alpha}z + O(z^2)$, that are holomorphically reducible to a rotation in a neighborhood of the origin, are one-sheeted in a neighborhood of the disk of reducibility to a rotation, and have no periodic orbits other than the origin. Moreover, the boundary of the reducibility disk in these examples is a $C^\infty$-smooth Jordan curve. As far as the authors of the present comment know, the question whether the numbers $\alpha$ for which such mappings exist constitute a set of measure zero is still open.

Of other papers by Pérez-Marco, as well as of Yoccoz's papers devoted to the problems of the linearization and structure of germs of holomorphic mappings of the complex plane, we mention [21, 22, 24, 25, 27, 35, 36].

All the holomorphic nonreducibility examples with no neighboring periodic orbits, constructed by Pérez-Marco, seem to be exceptional, i. e., to belong to a very small set of those holomorphic mappings which have a given rotation number. But the corresponding exceptionality theorem has not yet been formally proved.


[1] ARNOLD V. I. On mappings of the circle onto itself. Diploma Thesis, Faculty of Mechanics and Mathematics of Moscow State University, 1959 (in Russian).

[2] ARNOLD V. I. On analytic mappings of the circle onto itself. *Uspekhi Mat. Nauk*, 1960, **15**(2), 212–214 (in Russian).

[3] ARNOLD V. I. Small denominators I. Mappings of the circumference onto itself. *Izv. Akad. Nauk SSSR, Ser. Mat.*, 1961, **25**(1), 21–86; corrigenda: 1964, **28**(2), 479–480 (in Russian). [*The English translation*: *AMS Transl., Ser.* 2, 1965, **46**, 213–284.]

[4] ARNOLD V. I. Geometrical Methods in the Theory of Ordinary Differential Equations, 2nd edition. New York: Springer, 1988, §§ 11–12. (Grundlehren der Mathematischen Wissenschaften, 250.) [*The Russian original* 1978.]

[5] Arnold V. I. Sur quelques problèmes de la théorie des systèmes dynamiques. *Topol. Methods Nonlinear Anal.*, 1994, **4**(2), 209–225. [*The Russian translation in*: Vladimir Igorevich Arnold. Selecta–60. Moscow: PHASIS, 1997, 533–551.]

[6] Arnold V. I. From Hilbert's superposition problem to dynamical systems. In: The Arnoldfest. Proceedings of a conference in honour of V. I. Arnold for his sixtieth birthday (Toronto, 1997). Editors: E. Bierstone, B. A. Khesin, A. G. Khovanskiĭ and J. E. Marsden. Providence, RI: Amer. Math. Soc., 1999, 1–18. (Fields Institute Commun., 24.) [*The Russian version in*: Vladimir Igorevich Arnold. Selecta–60. Moscow: PHASIS, 1997, 727–740.]

[7] Herman M. R. Conjugaison $C^{\infty}$ des difféomorphismes du cercle dont le nombre de rotations satisfait à une condition arithmétique. *C. R. Acad. Sci. Paris, Sér. A–B*, 1976, **282**(10), Ai, A503–A506.

[8] Herman M. R. Conjugaison $C^{\infty}$ des difféomorphismes du cercle pour presque tout nombre de rotation. *C. R. Acad. Sci. Paris, Sér. A–B*, 1976, **283**(8), Aii, A579–A582.

[9] Herman M. R. La conjugaison des difféomorphismes du cercle à des rotations. *Bull. Soc. Math. France Suppl. Mém.*, 1976, **46**, 181–188 [Supplément au *Bull. Soc. Math. France*, 1976, **104**(2)].

[10] Herman M. R. Sur la conjugaison différentiable des difféomorphismes du cercle à des rotations. Thèse d'État, Univ. Paris-Sud, Orsay, 1976.

[11] Herman M. R. Mesure de Lebesgue et nombre de rotation. In: Geometry and Topology. Proc. III Latin Amer. School of Math. (Rio de Janeiro, 1976). Editors: J. Palis and M. do Carmo. Berlin: Springer, 1977, 271–293. (Lecture Notes in Math., 597.)

[12] Herman M. R. Sur la conjugaison différentiable des difféomorphismes du cercle à des rotations. *Inst. Hautes Études Sci. Publ. Math.*, 1979, **49**, 5–233.

[13] Herman M. R. Résultats récents sur la conjugaison différentiable. In: Proceedings of the International Congress of Mathematicians (Helsinki, 1978), Vol. 2. Editor: O. Lehto. Helsinki: Acad. Sci. Fennica, 1980, 811–820.

[14] Herman M. R. Sur les difféomorphismes du cercle de nombre de rotation de type constant. In: Conference on Harmonic Analysis in Honor of Antoni Zygmund (Chicago, 1981), Vol. II. Editors: W. Beckman, A. P. Calderón, R. Fefferman and P. W. Jones. Belmont: Wadsworth, 1983, 708–725.

[15] Herman M. R. Simple proofs of local conjugacy theorems for diffeomorphisms of the circle with almost every rotation number. *Bol. Soc. Brasil. Mat. (N. S.)*, 1985, **16**(1), 45–83.

[16] Herman M. R. Recent results and some open questions on Siegel's linearization theorem of germs of complex analytic diffeomorphisms of $\mathbb{C}^n$ near a fixed point. In: VIIIth Intern. Congress on Mathematical Physics (Marseille, 1986). Editors: M. Mebkhout and R. Sénéor. Singapore: World Scientific, 1987, 138–184.

[17] Katznelson Y., Ornstein D. S. The differentiability of the conjugation of certain diffeomorphisms of the circle. *Ergod. Theory Dynam. Systems*, 1989, **9**(4), 643–680.

[18] KATZNELSON Y., ORNSTEIN D. S. The absolute continuity of the conjugation of certain diffeomorphisms of the circle. *Ergod. Theory Dynam. Systems*, 1989, **9**(4), 681–690.

[19] KHANIN K. M., SINAÏ YA. G. A new proof of M. Herman's theorem. *Commun. Math. Phys.*, 1987, **112**(1), 89–101.

[20] PÉREZ-MARCO R. Sur la structure des germes holomorphes non linéarisables. *C. R. Acad. Sci. Paris, Sér.* I *Math.*, 1991, **312**(7), 533–536.

[21] PÉREZ-MARCO R. Centralisateurs non dénombrables de germes de difféomorphismes holomorphes non linéarisables de $(\mathbb{C},0)$. *C. R. Acad. Sci. Paris, Sér.* I *Math.*, 1991, **313**(7), 461–464.

[22] PÉREZ-MARCO R. Solution complète au problème de Siegel de linéarisation d'une application holomorphe au voisinage d'un point fixe (d'après J.-C. Yoccoz). In: *Séminaire Bourbaki*, 1991–92; *Astérisque*, 1992, **206**, Exp. No. 753, 4, 273–310.

[23] PÉREZ-MARCO R. Sur les dynamiques holomorphes non linéarisables et une conjecture de V. I. Arnold. *Ann. Sci. École Norm. Sup., Sér.* 4, 1993, **26**(5), 565–644.

[24] PÉREZ-MARCO R. Nonlinearizable holomorphic dynamics having an uncountable number of symmetries. *Invent. Math.*, 1995, **119**(1), 67–127.

[25] PÉREZ-MARCO R. Fixed points and circle maps. *Acta Math.*, 1997, **179**(2), 243–294.

[26] PÉREZ-MARCO R. Siegel disks with quasi-analytic boundary. Preprint, Université Paris-Sud.

[27] PÉREZ-MARCO R. Total convergence or general divergence in small divisors. *Commun. Math. Phys.*, 2001, **223**(3), 451–464.

[28] SINAÏ YA. G., KHANIN K. M. Smoothness of conjugacies of diffeomorphisms of the circle with rotations. *Russian Math. Surveys*, 1989, **44**(1), 69–99.

[29] STARK J. Smooth conjugacy and renormalisation for diffeomorphisms of the circle. *Nonlinearity*, 1988, **1**(4), 541–575.

[30] YOCCOZ J.-C. Centralisateur d'un difféomorphisme du cercle dont le nombre de rotation est irrationnel. *C. R. Acad. Sci. Paris, Sér. A–B*, 1980, **291**(9), A523–A526.

[31] YOCCOZ J.-C. $C^1$-conjugaison des difféomorphismes du cercle. In: Geometric Dynamics. Proc. Intern. Symp. at the IMPA (Rio de Janeiro, 1981). Editor: J. Palis, Jr. Berlin: Springer, 1983, 814–827. (Lecture Notes in Math., 1007.)

[32] YOCCOZ J.-C. Conjugaison différentiable des difféomorphismes du cercle dont le nombre de rotation vérifie une condition diophantienne. *Ann. Sci. École Norm. Sup., Sér.* 4, 1984, **17**(3), 333–359.

[33] YOCCOZ J.-C. Il n'y a pas de contre-exemple de Denjoy analytique. *C. R. Acad. Sci. Paris, Sér.* I *Math.*, 1984, **298**(7), 141–144.

[34] YOCCOZ J.-C. Centralisateurs et conjugaison différentiable des difféomorphismes du cercle. Thèse d'État, Univ. Paris-Sud, Orsay, 1985.

[35] YOCCOZ J.-C. Linéarisation des germes de difféomorphismes holomorphes de $(\mathbb{C},0)$. *C. R. Acad. Sci. Paris, Sér.* I *Math.*, 1988, **306**(1), 55–58.

[36] YOCCOZ J.-C. Théorème de Siegel, nombres de Bruno et polynômes quadratiques. In: Petits Diviseurs en Dimension 1. *Astérisque*, 1995, **231**, 3–88.

[37] YOCCOZ J.-C. Centralisateurs et conjugaison différentiable des difféomorphismes du cercle. In: Petits Diviseurs en Dimension 1. *Astérisque*, 1995, **231**, 89–242.

## 1972-21 – *M. B. Sevryuk* Also: *1963-5*

This problem deals with systems of ordinary differential equations of the form

$$\dot{\varphi} = \omega, \quad \dot{x} = A(\varphi)x; \qquad \varphi \in \mathbb{T}^n = \mathbb{R}^n/2\pi\mathbb{Z}^n, \quad x \in \mathbb{R}^N, \tag{1}$$

where $\omega$ is a constant vector with rationally independent components while $A$ is a smooth matrix-valued function on $\mathbb{T}^n$. Such systems are called *linear equations with quasi-periodic coefficients.* The manifold $\{x = 0\}$ is an invariant $n$-dimensional torus of system (1) which carries quasi-periodic motions with frequency vector $\omega$. The question is under what conditions system (1) is *reducible*, i.e., there exists a smooth change of variables $x = B(\varphi)y$ turning (1) into a linear equation with constant coefficients

$$\dot{\varphi} = \omega, \quad \dot{y} = Cy.$$

In the case of periodic coefficients $n = 1$ (i.e., where $\{x = 0\}$ is a closed trajectory of period $2\pi/|\omega|$), the classical Floquet theorem guarantees the existence of the desired change of variables $x = B(\varphi)y$ provided that the matrix-valued function $B(\varphi)$ is allowed either to range in $\mathrm{GL}(N, \mathbb{C})$ or to be $4\pi$-periodic. The conditions on $A(\varphi)$ under which the function $B(\varphi)$ can be chosen to be real-valued and $2\pi$-periodic are also known. On the other hand, our knowledge of the much more difficult case $n > 1$ is still far from being exhaustive in spite of the rich body of literature devoted to this case. The reducibility problem for linear equations with quasi-periodic coefficients is discussed in the general context of the theory of quasi-periodic motions in dynamical systems in V. I. Arnold's works [2, 3] (in book [3], a proof of the Floquet theorem is given).

In the case $n > 1$, the reducibility problem is usually considered under the following additional condition: There exist positive constants $\tau$ and $\gamma$ such that $|q \cdot \omega| \geq \gamma |q|^{-\tau}$ for all $q \in \mathbb{Z}^n \setminus \{0\}$ where $q \cdot \omega = q_1\omega_1 + \cdots + q_n\omega_n$ and $|q| = \max(|q_1|, \ldots, |q_n|)$. If this condition is met the frequency vector $\omega$ is said to be *Diophantine*. Apart from that, one often assumes that the function $A(\varphi)$ is analytic. For $N = 1$ and $\omega$ Diophantine, system (1) is always reducible; a detailed proof of this statement for $A(\varphi)$ analytic is presented in, e. g., book [5].

**Reducible systems are typical.** For each $N \geq 2$ and for a fixed Diophantine vector $\omega$, the reducible functions $A(\varphi)$ fill up a certain domain in the functional space of all the analytic matrix-valued functions on the $n$-torus. This was first shown in papers [1, 15].

**Irreducible systems are typical.** For $N \geq 2$, irreducible functions $A(\varphi)$ fill up a domain as well—at least for some vectors $\omega$ (it is unknown whether there are Diophantine vectors among such vectors $\omega$)—in the functional space of all the matrix-valued functions (even in the $C^0$-topology). This was proved in works [6, 8, 9]. Moreover, the irreducible systems (1) considered in [6, 8, 9] are reduced to a linear equation with constant coefficients not only by no linear quasi-periodic change of variables (with the same frequency vector $\omega$ or its multiple) but also by no linear almost periodic change of variables. [1] Putting it another way, there is no smooth almost periodic matrix-valued function $\widehat{B} = \widehat{B}(t)$, $\widehat{B}: \mathbb{R} \to \mathrm{GL}(N, \mathbb{R})$ such that the matrix

$$C = \widehat{B}^{-1} A(\omega t) \widehat{B} - \widehat{B}^{-1}\, d\widehat{B}/dt$$

is independent of $t$. On the other hand, the author of the present comment has failed to find in the literature examples of systems (1) irreducible in the usual sense (by quasi-periodic changes of coordinates with the same frequency vector $\omega$) but reducible by some almost periodic change of coordinates.

One of the generalizations of the concept of reducibility for systems (1) is almost reducibility. A system $\dot{x} = \widehat{A}(t)x$, $t \in \mathbb{R}$, $x \in \mathbb{R}^N$, with an arbitrary continuous coefficient matrix $\widehat{A}(t)$ is said to be *almost reducible* (in the sense of B. F. Bylov) if there exists a constant matrix $C' \in \mathrm{gl}(N, \mathbb{R})$ such that, for any $\delta > 0$, there is a differentiable change of coordinates $x = \widehat{B}_\delta(t)y$ satisfying the conditions

$$|\widehat{B}_\delta| < +\infty, \qquad |\widehat{B}_\delta^{-1}| < +\infty, \qquad |d\widehat{B}_\delta/dt| < +\infty$$

(such linear changes are called *Lyapunov changes*) and casting the system $\dot{x} = \widehat{A}(t)x$ to the system $\dot{y} = [C' + \Psi_\delta(t)]y$ with $|\Psi_\delta| < \delta$. Here

$$|M| = \sup_{t\in\mathbb{R}} \max_{i,j=1}^{N} |M_{ij}(t)| \quad \text{for} \quad M: \mathbb{R} \to \mathrm{gl}(N, \mathbb{R}).$$

For the general theory of almost reducible systems of differential equations see, e. g., monograph [7] (cf. also [14]).

[1] A function on $\mathbb{R}$ is said to be *almost periodic* (in the sense of H. Bohr) if it belongs to the closure of the space of trigonometric polynomials in the metric of uniform convergence.

**Not all the systems are almost reducible.** In paper [28], examples of systems (1) were constructed for all $n > 1, N > 1$ for which the system $\dot{x} = A(\omega t)x$ is not almost reducible.

**Almost reducible systems are typical.** In the space of linear differential equations with almost periodic coefficients, almost reducible systems are typical. For the precise formulation and a proof of this statement, see paper [29].

Irreducible systems of type (1) with discrete time (diffeomorphisms) were constructed in works [16, 17, 37]. In a recent paper [19], an interesting numerical method for exploring the linearized normal behavior of invariant curves of diffeomorphisms is invented, the rotation numbers of the curves being assumed to be irrational. In some cases, this method can detect the irreducibility of the linearized system.

The reducibility problem for systems (1) is closely connected with the asymptotic behavior of integrals of quasi-periodic functions. For simplicity, consider, e. g., the case $N = 1$. If a coordinate change $x = \widehat{B}(t)y$ reduces an equation $\dot{x} = A(\omega t)x$ to the form $\dot{y} = Cy$ with $C = \text{const}$ (here $x \in \mathbb{R}$, $y \in \mathbb{R}$), then the "general solution" of the equation $\dot{x} = A(\omega t)x$ has the form $x(t) = c_1 e^{I(t)} = c_2 \widehat{B}(t) e^{Ct}$ where

$$I(t) = \int_0^t A(\omega\tau)\, d\tau.$$

If the mean value of the function $A(\varphi)$ over the torus $\mathbb{T}^n$ vanishes then $I(t) = o(t)$ as $t \longrightarrow \infty$ according to H. Weyl's theorem on the coincidence of the temporal mean and spatial mean. Then $C = 0$ (provided that the change $x = \widehat{B}y$ is almost periodic). In turn, this implies that $I(t) = O(1)$ as $t \to \infty$. On the other hand, for many classes of non-Diophantine vectors $\omega \in \mathbb{R}^n$, there are known examples of smooth functions $A : \mathbb{T}^n \to \mathbb{R}$ with zero mean and unbounded (as $t \to \infty$) integral $I(t)$, see, e. g., [34]. If the mean value of the function $A$ vanishes but the integral $I(t)$ is not bounded as $t \to \infty$, then system (1) is reduced to an equation with constant coefficients by no almost periodic change of coordinates.

Among the latest works dealing with asymptotic properties of integrals of quasi-periodic functions, we mention papers [31–33].

Many works have been devoted to exploring the reducibility of system (1) for a nearly constant matrix $A(\varphi)$. Among those works, we mention monograph [4], and among the recent works, papers [14, 21, 22]. In work [18], sufficient conditions for the reducibility of system (1) are given without the assumption that the matrix $A(\varphi)$ is close to a constant one.

In the case of Hamiltonian systems, the reducibility problem is closely connected with the presence of sufficiently many first integrals, see [27].

In paper [20], the so-called *effective* reducibility of systems (1) is considered. In this work, it is supposed that $A = A_0 + \varepsilon Q(\varphi, \varepsilon)$ where the matrix $A_0$ is constant and $\varepsilon$ is a small parameter, and the question is examined of reducing system (1) to the form

$$\dot{\varphi} = \omega, \quad \dot{y} = [A_0^*(\varepsilon) + R^*(\varphi, \varepsilon)]y$$

with the remainder $R^*(\varphi, \varepsilon)$ exponentially small with respect to $\varepsilon$ as $\varepsilon \to 0$.

For a "singular" dependence of the matrix $A$ on the small parameter $\varepsilon$, the effective reducibility of systems (1) was explored earlier. The cases $A = \varepsilon Q(\varphi, \varepsilon)$ and $A = \varepsilon^{-1} Q(\varphi, \varepsilon)$ were considered, e. g., in paper [35] and in paper [36], respectively.

One of the important particular cases of the reducibility problem for systems (1) is the reducibility of the one-dimensional Schrödinger equation with a quasi-periodic potential. This problem was treated in, e. g., papers [10, 11, 14, 30, 36], whereas its discrete analogue was treated in works [12, 13].

Note finally that the reducibility problem can be formulated for systems of the form

$$\dot{\varphi} = \omega, \quad \dot{x} = A(\varphi)x; \qquad \varphi \in \mathbb{T}^n, \quad x \in G; \qquad A : \mathbb{T}^n \to \mathfrak{g} = T_e G, \tag{2}$$

where $G$ is an arbitrary compact Lie group and $\mathfrak{g}$ is its Lie algebra ($e$ being the unit of $G$). Here $A(\varphi)x$ should be understood as $(DR_x)A(\varphi)$ where $R_x : G \to G$ denotes the right shift of the group $G$ generated by $x$, while $DR_x : \mathfrak{g} \to T_x G$ is the induced mapping of the tangent spaces. For the results on the reducibility of systems (2) and their discrete time analogues, see, e. g., R. Krikorian's works [23–26] and L. H. Eliasson's recent paper [14]. The case $G = \mathrm{SL}(2, \mathbb{R})$ corresponds in fact to one-dimensional Schrödinger operators with quasi-periodic coefficients.

[1] ADRIANOVA L. YA. On reducibility of systems of $n$ linear differential equations with quasi-periodic coefficients. *Vestnik Leningrad. Univ., Ser. Mat., Mekh., Astron.*, 1962, **7**(2), 14–24 (in Russian).

[2] ARNOLD V. I. Small denominators and problems of stability of motion in classical and celestial mechanics. *Russian Math. Surveys*, 1963, **18**(6), 85–191 (see Ch. VI, § 2, n° 5).

[3] ARNOLD V. I. Geometrical Methods in the Theory of Ordinary Differential Equations, 2nd edition. New York: Springer, 1988, § 26. (Grundlehren der Mathematischen Wissenschaften, 250.) [*The Russian original* 1978.]

[4] BOGOLYUBOV N. N., MITROPOL′SKIĬ YU. A., SAMOĬLENKO A. M. Methods of Accelerated Convergence in Nonlinear Mechanics. New York: Springer, 1976, Ch. V. [*The Russian original* 1969.]

[5] BROER H. W., HUITEMA G. B., SEVRYUK M. B. Quasi-Periodic Motions in Families of Dynamical Systems: Order amidst Chaos. Berlin: Springer, 1996, Sect. 1.5. (Lecture Notes in Math., 1645.)

[6] BRONSHTEĬN I. U., CHERNIĬ V. F. Linear extensions satisfying the Perron condition. I; II. *Differ. Equations*, 1978, **14**(10), 1234–1243; 1980, **16**(2), 123–128.

[7] BYLOV B. F., VINOGRAD R. E., GROBMAN D. M., NEMYTSKIĬ V. V. The Theory of Lyapunov Exponents and Its Applications to Stability Problems. Moscow: Nauka, 1966, Ch. VII (in Russian).

[8] BYLOV B. F., VINOGRAD R. E., LIN V. YA., LOKUTSIEVSKIĬ O. V. On topological obstacles to block diagonalization for some exponentially split almost periodic systems. Preprint of the Moscow Institute for Applied Mathematics, the USSR Academy of Sciences, 1977, № 58 (in Russian).

[9] BYLOV B. F., VINOGRAD R. E., LIN V. YA., LOKUTSIEVSKIĬ O. V. On topological reasons for anomalous behavior of some almost periodic systems. In: Problems of Asymptotic Theory of Nonlinear Oscillations. Editors: N. N. Bogolyubov, A. Yu. Ishlinskiĭ, V. S. Korolyuk, O. B. Lykova, A. M. Samoĭlenko and A. N. Sharkovskiĭ. Kiev: Naukova Dumka, 1977, 54–61 (in Russian).

[10] DINABURG E. I., SINAĬ YA. G. On the one-dimensional Schrödinger equation with a quasi-periodic potential. *Funct. Anal. Appl.*, 1975, **9**(4), 279–289.

[11] ELIASSON L. H. Floquet solutions for the 1-dimensional quasi-periodic Schrödinger equation. *Commun. Math. Phys.*, 1992, **146**(3), 447–482.

[12] ELIASSON L. H. Reducibility and point spectrum for linear quasi-periodic skew-products. In: Proceedings of the International Congress of Mathematicians, Vol. II (Berlin, 1998). *Doc. Math.*, 1998, Extra Vol. II, 779–787 (electronic).

[13] ELIASSON L. H. On the discrete one-dimensional quasi-periodic Schrödinger equation and other smooth quasi-periodic skew products. In: Hamiltonian Systems with Three or More Degrees of Freedom (S'Agaró, 1995). Editor: C. Simó. Dordrecht: Kluwer Acad. Publ., 1999, 55–61. (NATO Adv. Sci. Inst. Ser. C Math. Phys. Sci., 533.)

[14] ELIASSON L. H. Almost reducibility of linear quasi-periodic systems. In: Smooth Ergodic Theory and Its Applications. Proc. AMS Summer Research Inst. (Seattle, WA, 1999). Editors: A. B. Katok, R. de la Llave, Ya. B. Pesin and H. Weiss. Providence, RI: Amer. Math. Soc., 2001, 679–705. (Proc. Symposia Pure Math., 69.)

[15] GEL′MAN A. E. On reducibility of a certain class of systems of differential equations with quasi-periodic coefficients. *Dokl. Akad. Nauk SSSR*, 1957, **116**(4), 535–537 (in Russian).

[16] HERMAN M. R. Construction d'un difféomorphisme minimal d'entropie topologique non nulle. *Ergod. Theory Dynam. Systems*, 1981, **1**(1), 65–76.

[17] HERMAN M. R. Une méthode pour minorer les exposants de Lyapounov et quelques exemples montrant le caractère local d'un théorème d'Arnold et de Moser sur le tore de dimension 2. *Comment. Math. Helvetici*, 1983, **58**(3), 453–502.

[18] JOHNSON R. A., SELL G. R. Smoothness of spectral subbundles and reducibility of quasi-periodic linear differential systems. *J. Differ. Equations*, 1981, **41**(2), 262–288.

[19] JORBA À. Numerical computation of the normal behaviour of invariant curves of $n$-dimensional maps. *Nonlinearity*, 2001, **14**(5), 943–976.

[20] JORBA À., RAMÍREZ-ROS R., VILLANUEVA J. Effective reducibility of quasiperiodic linear equations close to constant coefficients. *SIAM J. Math. Anal.*, 1997, **28**(1), 178–188.

[21] JORBA À., SIMÓ C. On the reducibility of linear differential equations with quasiperiodic coefficients. *J. Differ. Equations*, 1992, **98**(1), 111–124.

[22] JORBA À., SIMÓ C. On quasiperiodic perturbations of elliptic equilibrium points. *SIAM J. Math. Anal.*, 1996, **27**(6), 1704–1737.

[23] KRIKORIAN R. Réductibilité presque partout des systèmes quasi périodiques analytiques dans le cas SO(3). *C. R. Acad. Sci. Paris, Sér.* I *Math.*, 1995, **321**(8), 1039–1044.

[24] KRIKORIAN R. $C^0$-densité globale des systèmes produits-croisés sur le cercle réductibles. *Ergod. Theory Dynam. Systems*, 1999, **19**(1), 61–100.

[25] KRIKORIAN R. Réductibilité des Systèmes Produits-Croisés à Valeurs dans des Groupes Compacts. Paris: Soc. Math. France, 1999. (Astérisque, 259.)

[26] KRIKORIAN R. Réductibilité presque partout des flots fibrés quasi-périodiques à valeurs dans des groupes compacts. *Ann. Sci. École Norm. Sup., Sér.* 4, 1999, **32**(2), 187–240.

[27] KUKSIN S. B. An infinitesimal Liouville–Arnold theorem as a criterion of reducibility for variational Hamiltonian equations. *Chaos, Solitons & Fractals*, 1992, **2**(3), 259–269.

[28] MILLIONSHCHIKOV V. M. A proof of the existence of non-regular systems of linear differential equations with quasi-periodic coefficients. *Differ. Uravneniya*, 1969, **5**(11), 1979–1983 (in Russian, for the English translation see *Differ. Equations*).

[29] MILLIONSHCHIKOV V. M. On typicality of almost reducible systems with almost periodic coefficients. *Differ. Equations*, 1978, **14**(4), 448–450.

[30] MOSER J., PÖSCHEL J. An extension of a result by Dinaburg and Sinaĭ on quasiperiodic potentials. *Comment. Math. Helvetici*, 1984, **59**(1), 39–85.

[31] MOSHCHEVITIN N. G. Recent results on asymptotic behavior of integrals of quasiperiodic functions. In: Dynamical Systems in Classical Mechanics. Editor: V. V. Kozlov. Providence, RI: Amer. Math. Soc., 1995, 201–209. (AMS Transl., Ser. 2, 168; Adv. Math. Sci., 25.)

[32] MOSHCHEVITIN N. G. Multidimensional Diophantine approximations and dynamical systems. *Reg. Khaot. Dinamika*, 1997, **2**(1), 81–95 (in Russian).

[33] MOSHCHEVITIN N. G. Differential equations with almost periodic and conditionally periodic coefficients: recurrence and reducibility. *Math. Notes*, 1998, **64**(1–2), 194–201.

[34] SPRINDŽUK V. G. Quasi-periodic functions with the unbounded indefinite integral. *Dokl. Akad. Nauk Beloruss. SSR*, 1968, **12**(1), 5–9 (in Russian).

[35] TRESHCHЁV D. V. An estimate of irremovable nonconstant terms in the reducibility problem. In: Dynamical Systems in Classical Mechanics. Editor: V. V. Kozlov. Providence, RI: Amer. Math. Soc., 1995, 91–128. (AMS Transl., Ser. 2, 168; Adv. Math. Sci., 25.)

[36] TRESHCHЁV D. V. On the reducibility of the one-dimensional Schrödinger equation with quasi-periodic potential. In: Dynamical Systems in Classical Mechanics. Editor: V. V. Kozlov. Providence, RI: Amer. Math. Soc., 1995, 129–140. (AMS Transl., Ser. 2, 168; Adv. Math. Sci., 25.)

[37] WAGENER F. O. O. On the skew Hopf bifurcation. Ph. D. Thesis, University of Groningen, 1998.

## 1972-22 — *M. B. Sevryuk*

A smooth submanifold $M \subset \mathbb{R}^n$ is said to be *extremal* if, for almost all the points $x \in M$ (with respect to the Lebesgue measure on $M$), the supremum of the set of numbers $w > 0$ such that the inequality

$$|q \cdot x + q_0| < |q|^{-w} \tag{1}$$

possesses infinitely many integer solutions ($q \in \mathbb{Z}^n \setminus \{0\}$, $q_0 \in \mathbb{Z}$) is equal to $n$ [here $q \cdot x = q_1 x_1 + \cdots + q_n x_n$ and $|q| = \max(|q_1|, \ldots, |q_n|)$]. Diophantine inequalities of type (1) are considered in more detail in the comment to problem *1970-5*. [1] It is easy to show that every open domain of the space $\mathbb{R}^n$ is extremal. K. Mahler's famous conjecture of 1932 (to be more precise, the "real" part of that conjecture) consisted in that the curve $\{(t, t^2, \ldots, t^n) \mid t \in \mathbb{R}\}$ is extremal. Mahler's conjecture was proved by V. G. Sprindžuk, see [11, 12].

[1] That almost all the points of a generic smooth submanifold $M \subset \mathbb{R}^n$ of any positive dimension are Diophantine was first proved (without extremality) by A. S. Pyartli [9].

The first theorem on the extremality of general submanifolds was due to W. M. Schmidt who proved in work [10] that planar $C^3$-curves whose curvature is positive almost everywhere are extremal. The analogous theorem for curves in three-dimensional space was obtained 30 years later in papers [1,2].

The extremality of generic submanifolds $M^m \subset \mathbb{R}^n$ for each $n$ and $m \geq n/2$ was proved in works [4, 13, 14], and for $m(m+3)/2 > n$ in papers [5,6].

A detailed survey of the theory of extremal manifolds up to the end of the seventies is presented in V. G. Sprindžuk's book [14] and in his paper [15]; for an up-to-date survey see book [3] by V. I. Bernik and M. M. Dodson.

The extremality of generic submanifolds $M^m \subset \mathbb{R}^n$ without any restrictions on $m$ and $n$ was established in work [8] using ideas from the ergodic theory of flows on lattices. The results by D. Ya. Kleinbock and G. A. Margulis [8] and other recent achievements in the theory of Diophantine approximations are discussed in detail in book [3], in book [16], Ch. IV, and in survey [7].


[1] BERESNEVICH V. V., BERNIK V. I. Extremal smooth curves in the three-dimensional Euclidean space. *Dokl. Akad. Nauk Belarusi*, 1994, **38**(3), 9–12 (in Russian).

[2] BERESNEVICH V. V., BERNIK V. I. On a metrical theorem of W. Schmidt. *Acta Arithm.*, 1996, **75**(3), 219–233.

[3] BERNIK V. I., DODSON M. M. Metric Diophantine Approximation on Manifolds. Cambridge: Cambridge University Press, 1999. (Cambridge Tracts in Math., 137.)

[4] BERNIK V. I., KOVALEVSKAYA E. I. Diophantine approximations on manifolds of dimension $n$ in $\mathbb{R}^{2n}$. *Dokl. Akad. Nauk Beloruss. SSR*, 1990, **34**(12), 1061–1064 (in Russian).

[5] DODSON M. M., RYNNE B. P., VICKERS J. A. G. Metric Diophantine approximation and Hausdorff dimension on manifolds. *Math. Proc. Cambridge Phil. Soc.*, 1989, **105**(3), 547–558.

[6] DODSON M. M., RYNNE B. P., VICKERS J. A. G. Khintchine-type theorems on manifolds. *Acta Arithm.*, 1991, **57**(2), 115–130.

[7] KLEINBOCK D. YA. Some applications of homogeneous dynamics to number theory. In: Smooth Ergodic Theory and Its Applications. Proc. AMS Summer Research Inst. (Seattle, WA, 1999). Editors: A. B. Katok, R. de la Llave, Ya. B. Pesin and H. Weiss. Providence, RI: Amer. Math. Soc., 2001, 639–660. (Proc. Symposia Pure Math., 69.)

[8] KLEINBOCK D. YA., MARGULIS G. A. Flows on homogeneous spaces and Diophantine approximation on manifolds. *Ann. Math., Ser.* 2, 1998, **148**(1), 339–360.

[9] PYARTLI A. S. Diophantine approximations on submanifolds of the Euclidean space. *Funct. Anal. Appl.*, 1969, **3**(4), 303–306.

[10] SCHMIDT W. M. Metrische Sätze über simultane Approximation abhängiger Großen. *Monatsh. Math.*, 1964, **68**(2), 154–166.

[11] SPRINDŽUK V. G. A proof of Mahler's conjecture on the measure of the set of $S$-numbers. *Izvestiya Akad. Nauk SSSR, Ser. Matem.*, 1965, **29**(2), 379–436 (in Russian).

[12] SPRINDŽUK V. G. Mahler's Problem in Metric Number Theory. Providence, RI: Amer. Math. Soc., 1969. (Transl. Math. Monographs, 25.) [*The Russian original* 1967.]

[13] SPRINDŽUK V. G. The method of trigonometric sums in the metric theory of Diophantine approximations of dependent quantities. *Trudy Matem. Inst. im. V. A. Steklova*, 1972, **128**(2), 212–228 (in Russian, for the English translation see *Proc. Steklov Inst. Math.*).

[14] SPRINDŽUK V. G. Metric Theory of Diophantine Approximations. New York: John Wiley, 1979. [*The Russian original* 1977.]

[15] SPRINDŽUK V. G. Achievements and problems in the theory of Diophantine approximations. *Russian Math. Surveys*, 1980, **35**(4), 1–80.

[16] STARKOV A. N. Dynamical Systems on Homogeneous Spaces. Providence, RI: Amer. Math. Soc., 2000. (Transl. Math. Monographs, 190.) [*The Russian original* 1999.]

## ▽ **1972-23 — *A. A. Glutsyuk*** Also: *1976-30*

R The statement of problem *1972-23* is the gradient conjecture formulated by R. Thom [4]. It was proved in the joint work by K. Kurdyka and T. Mostowski (announced in [1]) and published in the joint paper by the same authors and A. Parusiński [2].

H For the historical review of the problem and previous results see papers [2, 3] and the bibliography to them.

[1] KURDYKA K. On the gradient conjecture of R. Thom. In: Seminari di Geometria. 1998–1999 (Bologna, 1997). Editor: S. Coen. Bologna: Univ. Studi Bologna, 2000, 143–151.

[2] KURDYKA K., MOSTOWSKI T., PARUSIŃSKI A. Proof of the gradient conjecture of R. Thom. *Ann. Math., Ser.* 2, 2000, **152**(3), 763–792.

[3] MOUSSU R. Sur la dynamique des gradients. Existence de variétés invariantes. *Math. Ann.*, 1997, **307**(3), 445–460.

[4] THOM R. Problèmes rencontrés dans mon parcours mathématique: un bilan. *Inst. Hautes Études Sci. Publ. Math.*, 1989, **70**, 199–214.

## △ 1972-23 — *D. I. Novikov* Also: *1976-30*

ℛ Let $x(t)$ be a trajectory of the gradient vector field $\nabla f$ of a real analytic function $f : \mathbb{R}^n \to \mathbb{R}$, and suppose that $\lim_{t\to+\infty} x(t) = x_0$ is a critical point of $f$. The "gradient conjecture of R. Thom" claims that there exists a limit of secants to this trajectory:

$$\lim_{t\to\infty} \frac{x(t) - x_0}{|x(t) - x_0|}.$$

The conjecture was proved recently in [1]. In fact, the authors prove that the trajectory $x(t)$ has finite length after a standard blow-up with center $x_0$, by using essentially only a Łojasiewicz's inequality. The proof is motivated by the classic proof of Łojasiewicz of the finitenes of the length of $x(t)$.

The conjecture on the limit existence of the tangent to a trajectory

$$\lim_{t\to\infty} \frac{\nabla f(x(t))}{|\nabla f(x(t))|}$$

is stronger: even if the derivative exists at a point, the limit of the derivatives might not exist at that point. It is still open, as well as the following (stronger) finiteness: the trajectory $x(t)$ intersects any analytic set $A$ in a finite number of points (or stays in $A$).

[1] Kurdyka K., Mostowski T., Parusiński A. Proof of the gradient conjecture of R. Thom. *Ann. Math., Ser.* 2, 2000, **152**(3), 763–792.

## ▽ 1972-26 — *V. M. Kharlamov*

ℛ A similar question is found in the sixteenth Hilbert problem. The beginning of the 70s, when Arnold includes it into his list, marks a turning point in topology of real algebraic varieties influenced, to a great extent, first, by Gudkov's classification of real plane sectics and Gudkov's general conjectures based on it, and, second, by Arnold's illuminating paper on arrangements of ovals of real algebraic plane curves (see the comment to problem *1976-36*).

To get an idea of the achievements in this domain, one can see surveys [1–5].

As an example, let us state two basic, rather sharp, bounds on the Betti numbers, noticing that, however, even for Betti numbers the whole range of their values is far from being well understood (for more details and references one can look, for example, at the survey by Degtyarev and Kharlamov [2]). For simplicity, consider a compact nonsingular hypersurface $A$ in $\mathbb{R}^m$. Denote by $\mathbb{R}A$ the set of its real points and by $\mathbb{C}A$ the set of its complex points. The so-called Petrovskiĭ–Oleĭnik inequality bounds the Euler characteristic of $\mathbb{R}A$. According to it, if $m$ is odd, $|1-\chi(\mathbb{R}A)|$ is bounded by some explicit polynomials in $n$ of degree $m$. These polynomials count the number of integral points which belong to the layer $\frac{1}{2}(m-1)n < x_1 + \cdots + x_m < \frac{1}{2}(m+1)n$ of the open cube $]0;n[^m$. The other bound, the so-called Smith–Thom inequality, states that the total Betti number $\sum b_i(\mathbb{R}A;\mathbb{Z}/2)$ is bounded by some other polynomials, which count the number of integral point in $]0;n[^m$ with $\sum x_i$ not congruent to 0 modulo $n$ and are equal, in fact, to $\sum b_i(\mathbb{C}A;\mathbb{Z}/2)$. (It is rather interesting that for compact hypersurfaces these bounds remain true even if the hypersurfaces become singular.)

[1] ARNOLD V. I., OLEĬNIK O. A. Topology of real algebraic varieties. *Moscow Univ. Math. Bull.*, 1979, **34**(6), 5–17.

[2] DEGTYAREV A., KHARLAMOV V. Topological properties of real algebraic varieties: Rokhlin's way. *Russian Math. Surveys*, 2000, **55**(4), 735–814.

[3] GUDKOV D. A. The topology of real projective algebraic varieties. *Russian Math. Surveys*, 1974, **29**(4), 1–79.

[4] VIRO O. YA. Progress in the topology of real algebraic varieties over the last six years. *Russian Math. Surveys*, 1986, **41**(3), 55–82.

[5] WILSON G. Hilbert's sixteenth problem. *Topology*, 1978, **17**(1), 53–73.

### △ 1972-26 — *V. I. Arnold* Also: *1976-35, 1976-36, 1979-17*

For more information on the subject, see problems *1979-17–1979-23*, *1980-9, 1981-23, 1983-4, 1983-5, 1985-6, 1988-2, 1989-7, 1990-5, 1991-7, 1993-8, 1993-19, 2001-1–2001-3, 2002-1, 2002-2, 2002-4*, and especially problems *1979-19* and *1985-6* on the Ragsdale conjecture.

### 1972-27 — *F. Napolitano* Also: *1976-34*

The cohomology classes of the complement of the discriminant of the universal algebraic functions $z^k + a_1 z^{k-1} + \cdots + a_{k-1} z + a_k = 0$ of $k$ variables were described by Arnold [1,2]. These classes give obstructions to the representation of algebraic functions by complete superposition. The results of Arnold imply that the universal algebraic function of $k$ variables is not representable by complete superposition of algebraic functions of less than $k - D_2(k)$ variables, where $D_2(k)$

is the number of units in the binary representation of $k$ [3]. This result has been latter improved by Lin to $k-1$ using different methods [5, 6]. Comments on this problem and its relations with other problems can be found in [4].

[1] ARNOLD V. I. On some topological invariants of algebraic functions. *Trans. Moscow Math. Soc.*, 1970, **21**, 30–52.

[2] ARNOLD V. I. Cohomology classes of algebraic functions invariant under Tschirnhausen transformations. *Funct. Anal. Appl.*, 1970, **4**(1), 74–75. [*The Russian original is reprinted in*: Vladimir Igorevich Arnold. Selecta–60. Moscow: PHASIS, 1997, 151–154.]

[3] ARNOLD V. I. Topological invariants of algebraic functions, II. *Funct. Anal. Appl.*, 1970, **4**(2), 91–98.

[4] ARNOLD V. I. From Hilbert's superposition problem to dynamical systems. In: The Arnoldfest. Proceedings of a conference in honour of V. I. Arnold for his sixtieth birthday (Toronto, 1997). Editors: E. Bierstone, B. A. Khesin, A. G. Khovanskiĭ and J. E. Marsden. Providence, RI: Amer. Math. Soc., 1999, 1–18. (Fields Inst. Commun., 24.)

[5] LIN V. YA. On superpositions of algebraic functions. *Funct. Anal. Appl.*, 1972, **6**(3), 240–241.

[6] LIN V. YA. Superpositions of algebraic functions. *Funct. Anal. Appl.*, 1976, **10**(1), 32–38.

### **1972-32** — ***V. A. Vassiliev*** Also: *1975-14*

Nothing is known to me.

### ▽ **1972-33**

This is a problem in V. I. Arnold's comment [1] to H. Poincaré's paper "On a geometric theorem" ("Sur un théorème de géométrie").

[1] ARNOLD V. I. A comment to H. Poincaré's paper "Sur un théorème de géométrie." In: POINCARÉ H. Selected Works in Three Volumes (in Russian). Editors: N. N. Bogolyubov, V. I. Arnold and I. B. Pogrebysskiĭ. Vol. II. New methods of celestial mechanics. Topology. Number theory. Moscow: Nauka, 1972, 987–989 (in Russian).

### △ **1972-33** — ***M. B. Sevryuk***

Also: *1965-1–1965-3*, *1966-4*, *1966-5*, *1972-17*, *1976-39*

This problem is the famous *Arnold conjecture* about the number of fixed points of symplectomorphisms (i. e., symplectic diffeomorphisms) homologous to the identity [1]. Various statements on fixed points of symplectomorphisms homologous to the identity were formulated by V. I. Arnold as conjectures in the mid 1960s and later in the 1970s in a series of works [1] (see problems *1965-1–1965-3*), [2] (see problems *1966-4* and *1966-5*), [3] (the present problem as well as problems *1972-17* and *1972-18*; see also problem *1970-10*), [4] (see problem *1976-39*), [5] (see problem *1976-39*), and [9]. All these conjectures are discussed in detail in Arnold's works [6–9,11]. The best known conjecture is given in the formulation of the present problem: *a symplectomorphism $F$ of a closed*[2] *symplectic manifold $M$ onto itself possesses at least as many fixed points as a smooth function on $M$ must have critical points, whenever $F$ is homologous to the identity.* This conjecture can be understood both "*algebraically*" (the numbers of the fixed/critical points are calculated counting multiplicities) and "*geometrically*" (one counts the numbers of geometrically distinct fixed/critical points). For instance, any center-of-mass-preserving symplectomorphism of the standard symplectic torus $\mathbb{T}^{2n}$ is expected to possess at least $2n+1$ geometrically distinct fixed points, and at least $4^n$ fixed points counting multiplicities. Arnold [1] proved his conjectures for symplectomorphisms which are not too far from the identity mapping. These proofs showed that these conjectures are some extensions of the Morse inequality.

Both the "algebraic" and "geometric" versions of the Arnold conjecture admit weakened forms. The weakened "algebraic" form asserts that the number of fixed points of $F$ counting multiplicities is at least the sum of the Betti numbers (over $\mathbb{Z}$) of manifold $M$. The weakened "geometric" form states that the number of geometrically distinct fixed points of $F$ is at least the Ljusternik–Schnirelmann category of manifold $M$.

The Arnold conjecture has affected greatly the development of the symplectic geometry and topology in the subsequent years. The first noticeable step towards proving the conjecture was Ya. M. Eliashberg's announcement [23] of a proof for all the closed two-dimensional surfaces. Eliashberg announced also similar statements for surfaces with boundary and for nonorientable surfaces. In their milestone paper [21] (see also [22]), C. C. Conley and E. Zehnder proved the Arnold conjecture (in both the "algebraic" and "geometric" versions) for tori $\mathbb{T}^{2n}$ of all the even dimensions with the standard symplectic structure. Conley and Zehnder introduced in [21] a new technique of constructing a certain action functional on the space of contractible loops on the manifold. This technique can be

---

[1] Recall that a symplectomorphism $F$ of a symplectic manifold $M$ is said to be *homologous to the identity* (or to *preserve the center-of-mass*, or to be *exactly homotopic to the identity*) if $F = g_0^1$ where $g_0^t$ are the phase flow maps for the time interval from 0 to $t$ for a certain (generally speaking, nonautonomous) Hamiltonian vector field on $M$.

[2] i. e., compact and without boundary

regarded as a hyperbolic analogue of the Morse theory for positive functionals. During several subsequent years, the Arnold conjecture was proved for some other symplectic manifolds and classes of manifolds by many authors [6, 7, 82] (see, e. g., [12–17, 26, 33–35, 40, 41, 45, 73–76, 78–81] as well as [18]; some of these papers contain also proofs of more general versions of the Arnold conjecture—see below).

In the late 1980s, A. Floer published a series of very important papers [29–32] (see also his earlier works [27, 28]) where he, apart from other achievements, combined the variational approach by Conley and Zehnder [21, 22] with M. L. Gromov's elliptic methods [39] and defined what has become known as the Floer (co)homology theory. This enabled him to prove the Arnold conjecture for the so-called *positive*, or *monotone*, symplectic manifolds [31]. Afterwards, Floer's landmark result was generalized by H. Hofer and D. A. Salamon [42] and by K. Ono [62] to *semi-positive*, or *weakly monotone*, manifolds (in particular, to all the symplectic manifolds of dimensions $\leq 6$), and by G. C. Lu [51–53], to products of weakly monotone manifolds (and the so-called Calabi–Yau manifolds).

A quite different approach to the Arnold conjecture was proposed by B. Fortune [34] (see also [35, 79]) who proved it for projective spaces $\mathbb{C}\mathrm{P}^n$ with the standard symplectic structure. This proof was based on the fact that $\mathbb{C}\mathrm{P}^n$ is the reduced symplectic manifold of $\mathbb{C}^{n+1}$ under the Hopf $\mathbb{S}^1$-action (the symplectic quotient $\mathbb{C}^{n+1}//\mathbb{S}^1$) and any Hamiltonian system on $\mathbb{C}\mathrm{P}^n$ is the Marsden–Weinstein reduction of an appropriate Hamiltonian system on $\mathbb{C}^{n+1}$. A. B. Givental [38] carried over Fortune's techniques to general toric manifolds, i. e., symplectic quotients $\mathbb{C}^m//\mathbb{T}^k$. L. A. Ibort and C. Martínez Ontalba [44] showed that Fortune's method is in fact universal: the fixed point problem for a symplectomorphism (homologous to the identity) of every closed symplectic manifold can be translated into a critical point problem with symmetry on loops in the space $\mathbb{R}^{2N}$ (for suitable $N$) endowed with the standard symplectic structure.

Of numerous papers by other authors devoted to the Arnold conjecture, we would quote here [47, 54, 56, 64, 70]. Many results related to the Arnold conjecture are described in, e. g., books [43, 55].

A further extension of Floer's ideas and the theory of the so-called Gromov–Witten invariants have recently led K. Fukaya – K. Ono, H. Hofer – D. A. Salamon, J. Li – G. Liu – G. Tian, Y. B. Ruan, and B. Siebert to a complete proof of the *weakened algebraic* Arnold conjecture *in the nondegenerate case*. So, the number of fixed points of any center-of-mass-preserving symplectomorphism $F: M \to M$ of an arbitrary closed symplectic manifold $M$ is no less than the sum $B(M)$ of the Betti numbers of $M$ provided that all the fixed points of $F$ are nondegenerate. Some papers with the proofs have not appeared yet. We would confine ourselves here to eight references [36, 37, 48–50, 66, 71, 72] (see also brief survey [63]).

On the other hand, Yu. B. Rudyak and J. F. Oprea [67–69] proved the *geometric* Arnold conjecture for an arbitrary closed symplectic manifold $M$ *subject to the condition* $\omega^2|_{\pi_2(M)} = 0$ where $\omega^2$ is the symplectic structure on $M$ [this condition means that $\omega^2$ vanishes on the image of the Hurewicz homomorphism $\varphi : \pi_2(M) \to H_2(M, \mathbb{Z})$ or, in simpler terms, that the integral of $\omega^2$ vanishes over the image of every smooth mapping $S^2 \to M$]. So, the number of geometrically distinct fixed points of any center-of-mass-preserving symplectomorphism $F : M \to M$ is no less than the minimal number of geometrically distinct critical points of a smooth function on $M$, provided that the closed symplectic manifold $(M, \omega^2)$ satisfies the condition $\omega^2|_{\pi_2(M)} = 0$.

More general variants of the Arnold conjecture refer to the number of intersection points for two Lagrangian submanifolds of a symplectic manifold. [3] Namely, one conjectures that, *under suitable additional hypotheses*, a closed Lagrangian submanifold $L$ of a symplectic manifold $K$ and the image $A(L)$ of $L$ under a symplectomorphism $A : K \to K$ possess at least as many intersection points as a smooth function on $L$ must have critical points (both "algebraically" and "geometrically"), provided that $A$ is homologous to the identity. [4] The following 'additional hypothesis' seems to be sufficient: the integral of the symplectic structure vanishes on every disk whose boundary lies in $L$ (see [6, 7]).

Arnold [1] proved his conjectures on the Lagrange intersections of exact Lagrangian manifolds which are not too far one from the other. He showed that this version of the Lagrange intersection theory is some extension of the Morse inequalities.

---

[3] Let $(M, \omega^2)$ be an arbitrary symplectic manifold and $p_1$, $p_2$ be the projections of $M \times M$ onto the first and second factors, respectively. Then $\widetilde{\omega}^2 = p_1^*\omega^2 - p_2^*\omega^2$ is a symplectic structure on $M \times M$ (the so-called *twisted product form*), and it is very easy to verify that a mapping $F : M \to M$ is a symplectomorphism if and only if its graph $\Gamma_F = \{(m, F(m)) \mid m \in M\}$ is a Lagrangian submanifold of $M \times M$ with respect to this symplectic structure (see [6, 7]). In particular, the diagonal $\Gamma_{\text{id}} = \{(m, m) \mid m \in M\}$ in $M \times M$ is a Lagrangian submanifold. On the other hand, the intersection points of $\Gamma_F$ and $\Gamma_{\text{id}}$ are just the points $(m, m)$ for which $F(m) = m$. Thus, the question of the number of fixed points of symplectomorphisms is a particular case of the question of the number of intersection points for Lagrangian submanifolds (cf. the comments to problem *1988-6*).

[4] This conjecture does imply the conjecture on fixed points of symplectomorphisms, if one forgets about the 'additional hypotheses' required. Indeed, in the notation of the previous footnote, a mapping $(m_1, m_2) \mapsto (m_1, F(m_2))$ of $M \times M$ onto itself is a center-of-mass-preserving symplectomorphism (with respect to $\widetilde{\omega}^2$) whenever the mapping $F : M \to M$ is a center-of-mass-preserving symplectomorphism (with respect to $\omega^2$). The reason is that if $V$ is a Hamiltonian vector field on $(M, \omega^2)$ with a Hamilton function $H : M \to \mathbb{R}$, then $(0, V)$ is a Hamiltonian vector field on $(M \times M, \widetilde{\omega}^2)$ with the Hamilton function $-H \circ p_2$.

Here we would confine ourselves to several key references. Some conjectures on Lagrangian intersections were stated already in Arnold's landmark paper [1] (see problems *1965-1* and *1965-2*), cf. a detailed discussion in [6, 7, 9]. The special case of the general conjecture formulated above in which $K$ is the cotangent bundle (with the canonical symplectic structure) of a torus and $L$ is the zero section was proved by C. C. Conley and E. Zehnder [21] and by M. Chaperon [12] (see also survey [16]). For the situation where $K$ is the cotangent bundle of an arbitrary closed manifold (and $L$ is still the zero section) the conjecture was then proved by H. Hofer [40], F. Laudenbach and J.-C. Sikorav [45], and J.-C. Sikorav [76]. The particular case of the previous result where $A$ is $C^0$-close to the identity was independently considered by A. Weinstein [78, 80, 81]. Further achievements are due to H. Hofer [41], A. Floer [30, 32], and J.-C. Sikorav [77]. Floer's (co)homology theory for Lagrangian intersections [30, 32] was essentially extended by Y.-G. Oh [57–61]. Important estimates were recently obtained by Yu. V. Chekanov [19, 20] and P. E. Pushkar′ [65]. Surveys of the theory are given in [24, 25].

Proofs of other generalizations of the Arnold conjecture—given by Yu. V. Chekanov [17, 18]—concern the number of critical points of the so-called quasifunctions [6, 7, 18]. A quasifunction on a closed manifold $W$ is defined as a Legendrian submanifold of $J^1(W, \mathbb{R})$ homotopic to the zero section $j^1 0$, here $J^1(W, \mathbb{R})$ denotes the space (equipped with the canonical contact structure) of 1-jets of functions $W \to \mathbb{R}$. One more example is the recent proof by Chekanov and Pushkar′ of the theorem on the minimal number of singularities necessary for the reversion of a wave front on the plane (see [10]).

An analogue of the Arnold conjecture can be proved for symplectomorphisms non homologous to the identity [46]. Of course, in this case fixed points may be absent, but, however, one can estimate their number in terms of the so-called Novikov homology. Similar results for the more general problem of Lagrangian intersections were obtained even earlier by Sikorav [75].

[1] ARNOLD V. I. Sur une propriété topologique des applications globalement canoniques de la mécanique classique. *C. R. Acad. Sci. Paris*, 1965, **261**(19), 3719–3722. [*The Russian translation in*: Vladimir Igorevich Arnold. Selecta–60. Moscow: PHASIS, 1997, 81–86.]

[2] ARNOLD V. I. The stability problem and ergodic properties of classical dynamical systems. In: Proceedings of the International Congress of Mathematicians (Moscow, 1966). Moscow: Mir, 1968, 387–392 (in Russian). [*The English translation*: *AMS Transl., Ser.* 2, 1968, **70**, 5–11.] [*The original is reprinted in*: Vladimir Igorevich Arnold. Selecta–60. Moscow: PHASIS, 1997, 95–101.]

[3] Arnold V. I. A comment to H. Poincaré's paper "Sur un théorème de géométrie." In: Poincaré H. Selected Works in Three Volumes (in Russian). Editors: N. N. Bogolyubov, V. I. Arnold and I. B. Pogrebysskiĭ. Vol. II. New methods of celestial mechanics. Topology. Number theory. Moscow: Nauka, 1972, 987–989 (in Russian).

[4] Arnold V. I. Fixed points of symplectic diffeomorphisms. In: Problems of present day mathematics. Editor: F. E. Browder. In: Mathematical Developments Arising from Hilbert Problems (Northern Illinois University, 1974). Part 1. Editor: F. E. Browder. Providence, RI: Amer. Math. Soc., 1976, 35–79; see XX, p. 66. (Proc. Symposia Pure Math., 28.)

[5] Arnold V. I. Some problems in the theory of differential equations. In: Unsolved Problems of Mechanics and Applied Mathematics. Moscow: Moscow University Press, 1977, 3–9 (in Russian).

[6] Arnold V. I. The first steps of symplectic topology. *Russian Math. Surveys*, 1986, **41**(6), 1–21. [*The Russian original is reprinted in*: Vladimir Igorevich Arnold. Selecta–60. Moscow: PHASIS, 1997, 365–389.]

[7] Arnold V. I. First steps of symplectic topology. In: VIIIth Intern. Congress on Mathematical Physics (Marseille, 1986). Editors: M. Mebkhout and R. Sénéor. Singapore: World Scientific, 1987, 1–16.

[8] Arnold V. I. On some problems in symplectic topology. In: Topology and Geometry. Rohlin Seminar. Editor: O. Ya. Viro. Berlin: Springer, 1988, 1–5. (Lecture Notes in Math., 1346.) [*The Russian translation in*: Vladimir Igorevich Arnold. Selecta–60. Moscow: PHASIS, 1997, 425–429.]

[9] Arnold V. I. Mathematical Methods of Classical Mechanics, 2nd edition. New York: Springer, 1989, Appendix 9. (Graduate Texts in Math., 60.) [*The Russian original* 1974.]

[10] Arnold V. I. Topological Invariants of Plane Curves and Caustics. Dean Jacqueline B. Lewis Memorial Lectures, Rutgers University. Providence, RI: Amer. Math. Soc., 1994. (University Lecture Series, 5.)

[11] Arnold V. I. Symplectic geometry and topology. *J. Math. Phys.*, 2000, **41**(6), 3307–3343.

[12] Chaperon M. Quelques questions de géométrie symplectique (d'après, entre autres, Poincaré, Arnold, Conley et Zehnder). In: *Séminaire Bourbaki*, 1982–83; *Astérisque*, 1983, **105–106**, 231–249.

[13] Chaperon M. Une idée du type "géodésiques brisées" pour les systèmes hamiltoniens. *C. R. Acad. Sci. Paris, Sér.* I *Math.*, 1984, **298**(13), 293–296.

[14] Chaperon M. An elementary proof of the Conley–Zehnder theorem in symplectic geometry. In: Dynamical Systems and Bifurcations. Proc. Intern. Workshop at the Groningen University (Groningen, 1984). Editors: B. L. J. Braaksma, H. W. Broer and F. Takens. Berlin: Springer, 1985, 1–8. (Lecture Notes in Math., 1125.)

[15] CHAPERON M. Questions de géométrie symplectique. In: Géométrie Symplectique et Mécanique (Balaruc, 1983). Editor: J.-P. Dufour. Travaux en Cours. Paris: Hermann, 1985, 30–45.

[16] CHAPERON M., ZEHNDER E. Quelques résultats globaux en géométrie symplectique. In: Séminaire Sud-Rhodanien de Géométrie. III (Lyon, 1983). Géométrie Symplectique et de Contact: autour du Théorème de Poincaré–Birkhoff. Editors: P. Dazord and N. Desolneux-Moulis. Travaux en Cours. Paris: Hermann, 1984, 51–121.

[17] CHEKANOV YU. V. Legendrian Morse theory. *Uspekhi Mat. Nauk*, 1987, **42**(4), 139 (in Russian).

[18] CHEKANOV YU. V. Critical points of quasifunctions, and generating families of Legendrian manifolds. *Funct. Anal. Appl.*, 1996, **30**(2), 118–128.

[19] CHEKANOV YU. V. Hofer's symplectic energy and Lagrangian intersections. In: Contact and Symplectic Geometry (Cambridge, 1994). Editor: C. B. Thomas. Cambridge: Cambridge University Press, 1996, 296–306. (Publ. Newton Inst., 8.)

[20] CHEKANOV YU. V. Lagrangian intersections, symplectic energy, and areas of holomorphic curves. *Duke Math. J.*, 1998, **95**(1), 213–226.

[21] CONLEY C. C., ZEHNDER E. The Birkhoff–Lewis fixed point theorem and a conjecture of V. I. Arnold. *Invent. Math.*, 1983, **73**(1), 33–49.

[22] CONLEY C. C., ZEHNDER E. A global fixed point theorem for symplectic maps and subharmonic solutions of Hamiltonian equations on tori. In: Nonlinear Functional Analysis and its Applications (University of California, 1983). Part 1. Editor: F. E. Browder. Providence, RI: Amer. Math. Soc., 1986, 283–299. (Proc. Symposia Pure Math., 45.)

[23] ELIASHBERG YA. M. Estimates of the number of fixed points of area-preserving transformations. Deposited in VINITI on January 30, 1979, №374-79, 104 p. (in Russian), see also *Ref. Zh. Mat.*, 1979, 5A491 (in Russian).

[24] ELIASHBERG YA. M., GROMOV M. L. Lagrangian intersections and the stable Morse theory. *Boll. Un. Mat. Ital. B, Ser.* 7, 1997, **11**(2), suppl., 289–326.

[25] ELIASHBERG YA. M., GROMOV M. L. Lagrangian intersection theory: finite-dimensional approach. In: Geometry of Differential Equations. Editors: A. G. Khovanskiĭ, A. N. Varchenko and V. A. Vassiliev. Providence, RI: Amer. Math. Soc., 1998, 27–118. (AMS Transl., Ser. 2, 186; Adv. Math. Sci., 39.)

[26] FLOER A. Proof of the Arnold conjecture for surfaces and generalizations to certain Kähler manifolds. *Duke Math. J.*, 1986, **53**(1), 1–32.

[27] FLOER A. Holomorphic curves and a Morse theory for fixed points of exact symplectomorphisms. In: Séminaire Sud-Rhodanien de Géométrie. VI (Lyon, 1986). Aspects Dynamiques et Topologiques des Groupes Infinis de Transformation de la Mécanique. Editors: P. Dazord, N. Desolneux-Moulis and J.-M. Morvan. Travaux en Cours, 25. Paris: Hermann, 1987, 49–60.

[28] FLOER A. Morse theory for fixed points of symplectic diffeomorphisms. *Bull. Amer. Math. Soc. (N. S.)*, 1987, **16**(2), 279–281.

[29] FLOER A. An instanton-invariant for 3-manifolds. *Commun. Math. Phys.*, 1988, **118**(2), 215–240.

[30] FLOER A. Morse theory for Lagrangian intersections. *J. Differ. Geom.*, 1988, **28**(3), 513–547.

[31] FLOER A. Symplectic fixed points and holomorphic spheres. *Commun. Math. Phys.*, 1989, **120**(4), 575–611.

[32] FLOER A. Cuplength estimates on Lagrangian intersections. *Commun. Pure Appl. Math.*, 1989, **42**(4), 335–356.

[33] FLOER A., ZEHNDER E. Fixed point results for symplectic maps related to the Arnold conjecture. In: Dynamical Systems and Bifurcations. Proc. Intern. Workshop at the Groningen University (Groningen, 1984). Editors: B. L. J. Braaksma, H. W. Broer and F. Takens. Berlin: Springer, 1985, 47–63. (Lecture Notes in Math., 1125.)

[34] FORTUNE B. A symplectic fixed point theorem for $\mathbb{C}P^n$. *Invent. Math.*, 1985, **81**(1), 29–46.

[35] FORTUNE B., WEINSTEIN A. A symplectic fixed point theorem for complex projective spaces. *Bull. Amer. Math. Soc.* (*N. S.*), 1985, **12**(1), 128–130.

[36] FUKAYA K., ONO K. Arnold conjecture and Gromov–Witten invariant. *Topology*, 1999, **38**(5), 933–1048.

[37] FUKAYA K., ONO K. Arnold conjecture and Gromov–Witten invariant for general symplectic manifolds. In: The Arnoldfest. Proceedings of a conference in honour of V. I. Arnold for his sixtieth birthday (Toronto, 1997). Editors: E. Bierstone, B. A. Khesin, A. G. Khovanskiĭ and J. E. Marsden. Providence, RI: Amer. Math. Soc., 1999, 173–190. (Fields Institute Commun., 24.)

[38] GIVENTAL A. B. A symplectic fixed point theorem for toric manifolds. In: The Floer Memorial Volume. Editors: H. Hofer, C. H. Taubes, A. Weinstein and E. Zehnder. Basel: Birkhäuser, 1995, 445–481. (Progr. Math., 133.)

[39] GROMOV M. L. Pseudo holomorphic curves in symplectic manifolds. *Invent. Math.*, 1985, **82**(2), 307–347.

[40] HOFER H. Lagrangian embeddings and critical point theory. *Ann. Institut Henri Poincaré, Analyse non linéaire*, 1985, **2**(6), 407–462.

[41] HOFER H. Ljusternik–Schnirelman theory for Lagrangian intersections. *Ann. Institut Henri Poincaré, Analyse non linéaire*, 1988, **5**(5), 465–499.

[42] HOFER H., SALAMON D. A. Floer homology and Novikov rings. In: The Floer Memorial Volume. Editors: H. Hofer, C. H. Taubes, A. Weinstein and E. Zehnder. Basel: Birkhäuser, 1995, 483–524. (Progr. Math., 133.)

[43] HOFER H., ZEHNDER E. Symplectic Invariants and Hamiltonian Dynamics. Basel: Birkhäuser, 1994, Ch. 6.

[44] IBORT L. A., MARTÍNEZ ONTALBA C. Arnold's conjecture and symplectic reduction. *J. Geom. Phys.*, 1996, **18**(1), 25–37.

[45] LAUDENBACH F., SIKORAV J.-C. Persistance d'intersection avec la section nulle au cours d'une isotopie hamiltonienne dans un fibré cotangent. *Invent. Math.*, 1985, **82**(2), 349–357.

[46] LÊ H. V., ONO K. Symplectic fixed points, the Calabi invariant and Novikov homology. *Topology*, 1995, **34**(1), 155–176.

[47] LÊ H. V., ONO K. Cup-length estimates for symplectic fixed points. In: Contact and Symplectic Geometry (Cambridge, 1994). Editor: C. B. Thomas. Cambridge: Cambridge University Press, 1996, 268–295. (Publ. Newton Inst., 8.)

[48] LI J., TIAN G. Virtual moduli cycles and Gromov–Witten invariants of general symplectic manifolds. In: Topics in Symplectic 4-Manifolds (Irvine, CA, 1996). Editor: R. J. Stern. Cambridge, MA: Intern. Press, 1998, 47–83. (First Intern. Press Lecture Series, 1.)

[49] LI J., TIAN G. Comparison of algebraic and symplectic Gromov–Witten invariants. *Asian J. Math.*, 1999, **3**(3), 689–728.

[50] LIU G., TIAN G. Floer homology and Arnold conjecture. *J. Differ. Geom.*, 1998, **49**(1), 1–74.

[51] LU G. C. Arnold conjecture for product of monotone manifolds. *Adv. Math.* (*China*), 1993, **22**(5), 463–465.

[52] LU G. C. The Arnold conjecture for a product of weakly monotone manifolds. *Chinese J. Math.*, 1996, **24**(2), 145–157.

[53] LU G. C. The Arnold conjecture for a product of monotone manifolds and Calabi–Yau manifolds. *Acta Math. Sinica* (*N. S.*), 1997, **13**(3), 381–388.

[54] MCCORD C., OPREA J. Rational Ljusternik–Schnirelmann category and the Arnold conjecture for nilmanifolds. *Topology*, 1993, **32**(4), 701–717.

[55] MCDUFF D., SALAMON D. A. Introduction to Symplectic Topology, 2nd edition. New York: Clarendon Press, Oxford University Press, 1998, Ch. 11.

[56] OH Y.-G. A symplectic fixed point theorem on $\mathbb{T}^{2n} \times \mathbb{C}\mathrm{P}^k$. *Math. Z.*, 1990, **203**(4), 535–552.

[57] OH Y.-G. Floer cohomology and Arnold–Givental's conjecture on Lagrangian intersections. *C. R. Acad. Sci. Paris, Sér.* I *Math.*, 1992, **315**(3), 309–314.

[58] OH Y.-G. Floer cohomology of Lagrangian intersections and pseudo-holomorphic disks. I. *Commun. Pure Appl. Math.*, 1993, **46**(7), 949–993; addendum: 1995, **48**(11), 1299–1302.

[59] OH Y.-G. Floer cohomology of Lagrangian intersections and pseudo-holomorphic disks. II. $(\mathbb{C}\mathrm{P}^n, \mathbb{R}\mathrm{P}^n)$. *Commun. Pure Appl. Math.*, 1993, **46**(7), 995–1012.

[60] OH Y.-G. Floer cohomology of Lagrangian intersections and pseudo-holomorphic disks. III. Arnold–Givental conjecture. In: The Floer Memorial Volume. Editors: H. Hofer, C. H. Taubes, A. Weinstein and E. Zehnder. Basel: Birkhäuser, 1995, 555–573. (Progr. Math., 133.)

[61] OH Y.-G. Relative Floer and quantum cohomology and the symplectic topology of Lagrangian submanifolds. In: Contact and Symplectic Geometry (Cambridge, 1994). Editor: C. B. Thomas. Cambridge: Cambridge University Press, 1996, 201–267. (Publ. Newton Inst., 8.)

[62] ONO K. On the Arnold conjecture for weakly monotone symplectic manifolds. *Invent. Math.*, 1995, **119**(3), 519–537.

[63] ONO K. On Arnold's conjecture for symplectic fixed points. In: Homotopy and Geometry (Warsaw, 1997). Editors: J. Oprea and A. Tralle. Warsaw: Polish Academy of Sciences, Institute of Mathematics, 1998, 13–24. (Banach Center Publ., 45.)

[64] PUSHKAR′ P. E. Generating functions of symplectomorphisms. *Funct. Anal. Appl.*, 1994, **28**(3), 198–201.

[65] PUSHKAR′ P. E. Lagrangian intersections in a symplectic space. *Funct. Anal. Appl.*, 2000, **34**(4), 288–292.

[66] RUAN Y. B. Virtual neighborhoods and pseudo-holomorphic curves. *Turkish J. Math.*, 1999, **23**(1), 161–231.

[67] RUDYAK YU. B. On analytical applications of stable homotopy (the Arnold conjecture, critical points). *Math. Z.*, 1999, **230**(4), 659–672.

[68] RUDYAK YU. B. On strict category weight, gradient-like flows, and the Arnold conjecture. *Internat. Math. Res. Notices*, 2000, **5**, 271–279.

[69] RUDYAK YU. B., OPREA J. On the Ljusternik–Schnirelmann category of symplectic manifolds and the Arnold conjecture. *Math. Z.*, 1999, **230**(4), 673–678.

[70] SCHWARZ M. A quantum cup-length estimate for symplectic fixed points. *Invent. Math.*, 1998, **133**(2), 353–397.

[71] SIEBERT B. Algebraic and symplectic Gromov–Witten invariants coincide. *Ann. Inst. Fourier (Grenoble)*, 1999, **49**(6), 1743–1795.

[72] SIEBERT B. Symplectic Gromov–Witten invariants. In: New Trends in Algebraic Geometry (Warwick, 1996). Editors: K. Hulek, F. Catanese, C. Peters and M. Reid. Cambridge: Cambridge University Press, 1999, 375–424. (London Math. Soc. Lecture Note Ser., 264.)

[73] SIKORAV J.-C. Points fixes d'un symplectomorphisme homologue à l'identité. *C. R. Acad. Sci. Paris, Sér.* I *Math.*, 1984, **299**(8), 343–346.

[74] SIKORAV J.-C. Points fixes d'une application symplectique homologue à l'identité. *J. Differ. Geom.*, 1985, **22**(1), 49–79.

[75] SIKORAV J.-C. Un problème de disjonction par isotopie symplectique dans un fibré cotangent. *Ann. Sci. École Norm. Sup., Sér.* 4, 1986, **19**(4), 543–552.

[76] SIKORAV J.-C. Problèmes d'intersections et de points fixes en géométrie hamiltonienne. *Comment. Math. Helvetici*, 1987, **62**(1), 62–73.

[77] SIKORAV J.-C. Quelques propriétés des plongements lagrangiens. *Mém. Soc. Math. France (N. S.)*, 1991, **46**, 151–167.

[78] WEINSTEIN A. $C^0$ perturbation theorems for symplectic fixed points and Lagrangian intersections. In: Séminaire Sud-Rhodanien de Géométrie. III (Lyon, 1983). Géométrie Symplectique et de Contact: autour du Théorème de Poincaré–Birkhoff. Editors: P. Dazord and N. Desolneux-Moulis. Travaux en Cours. Paris: Hermann, 1984, 140–144.

[79] WEINSTEIN A. Symplectic reduction and fixed points. In: Géométrie Symplectique et Mécanique (Balaruc, 1983). Editor: J.-P. Dufour. Travaux en Cours. Paris: Hermann, 1985, 140–148.

[80] WEINSTEIN A. Critical point theory, symplectic geometry and Hamiltonian systems. In: Proceedings of the 1983 Beijing Symposium on Differential Geometry and Differential Equations. Editors: Shan Tao Liao and S. S. Chern. Beijing: Science Press, 1986, 261–289.

[81] WEINSTEIN A. On extending the Conley–Zehnder fixed point theorem to other manifolds. In: Nonlinear Functional Analysis and its Applications (University of California, 1983). Part 2. Editor: F. E. Browder. Providence, RI: Amer. Math. Soc., 1986, 541–544. (Proc. Symposia Pure Math., 45.)

[82] ZEHNDER E. The Arnold conjecture for fixed points of symplectic mappings and periodic solutions of Hamiltonian systems. In: Proceedings of the International Congress of Mathematicians (Berkeley, 1986), Vol. 2. Editor: A. M. Gleason. Providence, RI: Amer. Math. Soc., 1987, 1237–1246.

# 1973

## 1973-2

See the comments to problems *1972-12* and *1981-28*.

## 1973-3 — *V. M. Zakalyukin*

Several more specific problems involving singularity theory and mathematical economy were formulated recently by P. Chiappori and I. Ekeland. In a series of papers (see, e. g., [1, 2]) they applied the techniques of differential forms and theory of Pfaffian systems to various models of market economy.

In particular, one of I. Ekeland's questions (related to the desaggregation problem) was to find a minimal possible number of terms in the Darboux formula $\sum p_i\, dq_i$ for a given germ of differential 1-form provided that $q_i$ are concave and $p_i$ are positive.

The question was answered (except very degenerate cases) in [3] basing on an unexpected application of the theory of Lagrangian and Legendrian mappings.

[1] CHIAPPORI P. A., EKELAND I. Problèmes d'agrégation en théorie du consommateur et calcul différentiel extérieur. *C. R. Acad. Sci. Paris, Sér.* I *Math.*, 1996, **323**(5), 565–570.

[2] EKELAND I. La modélisation mathématiques en économie. *Gaz. Math., Soc. Math. France*, 1998, **78**, 51–62.

[3] ZAKALYUKIN V. M. Concave Darboux theorem. *C. R. Acad. Sci. Paris, Sér.* I *Math.*, 1998, **327**(7), 633–638.

## 1973-4 Also: *1976-31, 1994-37*

See papers [1–4].

[1] ARNOLD V. I. Algebraic unsolvability of the problem of Lyapunov stability and the problem of topological classification of singular points of an analytic system of differential equations. *Funct. Anal. Appl.*, 1970, **4**(3), 173–180.

[2] ARNOLD V. I. Algebraic unsolvability of the problem of stability and the problem of topological classification of singular points of analytic systems of differential equations. *Uspekhi Mat. Nauk*, 1970, **25**(2), 265–266 (in Russian).

[3] ARNOLD V. I. Local problems of analysis. *Moscow Univ. Math. Bull.*, 1970, **25**(2), 77–80.

[4] ARNOLD V. I. Problèmes résolubles et problèmes irrésolubles analytiques et géométriques. In: Passion des Formes. Dynamique Qualitative Sémiophysique et Intelligibilité. Dédié à R. Thom. Fontenay-St Cloud: ENS Éditions, 1994, 411–417; In: Formes et Dynamique, Renaissance d'un Paradigme. Hommage à René Thom. Paris: Eshel, 1995. [*The Russian translation in*: Vladimir Igorevich Arnold. Selecta–60. Moscow: PHASIS, 1997, 577–582.]

## 1973-5 — *A. A. Davydov*

Implicit ordinary differential equations are important in the description of fields of characteristic directions of linear second order partial equations with two independent variables, fields of asymptotic directions and of principal curvature directions on a surface smoothly embedded in $\mathbb{R}^3$, a slow motion of relaxation type equation and others [1–4, 8, 14, 19].

For a generic first order implicit differential equation there are found its smooth (analytic, topological) normal forms near its folded regular points [6] and near its folded elementary singular points [7, 9, 10, 14] and the presence of invariants (moduli) in normal forms even under $C^0$-diffeomorphisms near a pleated singular point [7].

Recently there were found normal forms of corank one singularities of implicit equations of the form $A(x)\dot{x} = v(x)$, where $A$ is a smooth square matrix function and $(x, v(x)) \in \mathbb{R}^n$ (see [11, 12, 16–18, 20]).

The implicit first order differential equations appearing in the study of the net of asymptotic lines and the field of principal curvature directions on a smooth surface in $\mathbb{R}^3$ have singular points of special types. The respective normal forms and the bifurcations observed in generic one parametric families of such equations are described in [4, 5].

Implicit high order differential equations are not studied so well. Only some types of generic singularities of second order equations have been described [13, 15].

[1] ARNOLD V. I. Surfaces defined by hyperbolic equations. *Math. Notes*, 1988, **44**(1), 489–497. [*The Russian original is reprinted in*: Vladimir Igorevich Arnold. Selecta–60. Moscow: PHASIS, 1997, 397–412.]

[2] ARNOLD V. I. Catastrophe Theory. Berlin: Springer, 1992. [*The Russian original* 1990.]

[3] ARNOLD V. I. Contact structure, relaxational oscillations and singular points of implicit differential equations. In: Global Analysis—Studies and Applications, III. Editors: Yu. G. Borisovich and Yu. E. Gliklikh. Berlin: Springer, 1988, 173–179. (Lecture Notes in Math., 1334.)

[4] BRUCE J. W., TARI F. On the multiplicity of implicit differential equations. *J. Differ. Equations*, 1998, **148**(1), 122–147.

[5] BRUCE J. W., FLETCHER G. J., TARI F. Bifurcations of implicit differential equations. *Proc. Roy. Soc. Edinburgh, Sect. A*, 2000, **130**(3), 485–506.

[6] CIBRARIO M. Sulla riduzione a forma canonica delle equazioni lineari alle derivate parziali di secondo ordine di tipo misto. *Ist. Lombardo, Rend., Ser.* II, 1932, **65**, 889–906.

[7] DAVYDOV A. A. Normal form of a differential equation, not solvable for the derivative, in a neighborhood of a singular point. *Funct. Anal. Appl.*, 1985, **19**(2), 81–89.

[8] DAVYDOV A. A. Structural stability of control systems on orientable surfaces. *Math. USSR, Sb.*, 1992, **72**(1), 1–28.

[9] DAVYDOV A. A., ROSALES-GONZÁLEZ E. The complete classification of typical second-order linear partial-differential equations on the plane. *Dokl. Math.*, 1996, **54**(2), 669–672.

[10] DAVYDOV A. A., ORTIZ-BOBADILLA L. Smooth normal forms of folded elementary singular points. *J. Dynam. Control Systems*, 1995, **1**(4), 463–482.

[11] HAYAKAWA A., ISHIKAWA G., IZUMIYA S., YAMAGUCHI K. Classification of generic integral diagrams and first order ordinary differential equations. *Intern. J. Math.*, 1994, **5**(4), 447–489.

[12] GUZMAN A. Ph. D. Thesis, UNAM, Mexico, 1994.

[13] KOSSOWSKI M. Fiber completions, contact singularities and single valued solutions for $C^\infty$-second order ODE. *Canad. J. Math.*, 1996, **48**(4), 849–870.

[14] KUZ′MIN A. G. Nonclassical Equations of Mixed Type and Their Applications to Gas Dynamics. Basel: Birkhäuser, 1992. (Internat. Ser. Numer. Math., 109.)

[15] LEMASURIER M. Singularities of implicit second order differential equations. Ph. D. Thesis, University of Georgia, 1998.

[16] MEDVEĎ M. Qualitative properties of generalized vector. *Rivista Mat. Pura Appl.*, 1994, **15**, 7–31.

[17] PAZIJ N. D. Normal forms of transversally singular quasilinear Sobolev systems. *Differ. Equations*, 1996, **32**(6), 850–852.

[18] PAZIJ N. D. Ph. D. Thesis, Chelyabinsk State University, 1999 (in Russian).

[19] TAKENS F. Geometric aspects of non-linear R. L. C. networks. In: Dynamical Systems—Warwick 1974. Editor: A. Manning. Berlin–Heidelberg: Springer, 1975, 305–331. (Lecture Notes in Math., 468.)

[20] ZHITOMIRSKIĬ M. YA. Local normal forms for constrained systems on two manifolds. *Bol. Soc. Brasil. Mat. (N. S.)*, 1993, **24**(2), 211–232.

## 1973-7 – *S. M. Gusein-Zade*

Also: *1975-6, 1975-24, 1976-13*

The smoothness of the stratum $\mu = \text{const}$ has been proved for functions of two variables (J. Wahl, 1971; see [1]). On the other hand, for functions of more than two variables the smoothness of the stratum $\mu = \text{const}$, generally speaking, does not take place [2]. This result and related questions are discussed in book [3].

[1] BRIESKORN E. Special singularities-resolution, deformation and monodromy. Lecture notes prepared in connection with the Summer Institute on Algebraic Geometry held at Humboldt State University in Arcata, California. Providence, RI: Amer. Math. Soc., 1974.

[2] LUENGO I. The $\mu$-constant stratum is not smooth. *Invent. Math.*, 1987, **90**(1), 139–152.

[3] ARNOLD V. I., VASSILIEV V. A., GORYUNOV V. V., LYASHKO O. V. Singularities. I. Local and Global Theory. Berlin: Springer, 1993, Ch. 2, Sect. 1.11. (Encyclopædia Math. Sci., 6; Dynamical Systems, VI.) [*The Russian original* 1988.]

## 1973-8 – *S. M. Gusein-Zade*

ℛ The semicontinuity of the (proper) modality was proved by A. M. Gabrielov [1].

[1] GABRIELOV A. M. Bifurcations, Dynkin diagrams, and modality of isolated singularities. *Funct. Anal. Appl.*, 1974, **8**(2), 94–98.

## 1973-10 – *V. I. Arnold*

ℛ The problem was solved by Kushnirenko and Gabrielov, see references in [1,2].

[1] ARNOLD V. I. Remarks on the stationary phase method and Coxeter numbers. *Russian Math. Surveys*, 1973, **28**(5), 19–48.

[2] ARNOLD V. I. Normal forms of functions in neighborhoods of degenerate critical points. *Russian Math. Surveys*, 1974, **29**(2), 10–50.

## 1973-11 – *V. I. Arnold* Also: *1976-15*

ℛ The conjecture that there exists a nondegenerate quasihomogeneous function with the given weights whenever the coefficients of the Poincaré polynomial are nonnegative was disproved by B. M. Ivlev's example with the weights $(1,24,33,58,265)$, cf. [1].

[1] ARNOLD V. I., VASSILIEV V. A., GORYUNOV V. V., LYASHKO O. V. Singularities. I. Local and Global Theory. Berlin: Springer, 1993, Ch. 1, Sect. 3.4. (Encyclopædia Math. Sci., 6; Dynamical Systems, VI.) [*The Russian original* 1988.]

## 1973-15 Also: *1984-11*

ℛ See papers [1–3].

[1] ARNOLD V. I. Critical points of functions on a manifold with a boundary, the simple Lie groups $B_k$, $C_k$, $F_4$ and singularities of evolutes. *Russian Math. Surveys*, 1978, **33**(5), 99–116.

[2] ARNOLD V. I. Lagrange and Legendre cobordisms, I; II. *Funct. Anal. Appl.*, 1980, **14**(3), 167–177; **14**(4), 252–260.

[3] BARANNIKOV S. A. The framed Morse complex and its invariants. In: Singularities and Bifurcations. Editor: V. I. Arnold. Providence, RI: Amer. Math. Soc., 1994, 93–115. (Adv. Sov. Math., 21.)

## 1973-16

See the comments to problem *1975-20.*

## ▽ 1973-17 – *S. M. Gusein-Zade*

The description of this stratification does not exist. The decomposition into strata $\mu = \text{const}$ discussed in V. A. Vassiliev's comment (below) is not a stratification. The example showing this can be found in [1].

[1] GUSEIN-ZADE S. M., NEKHOROSHEV N. N. On adjacencies of singularities $A_k$ to points of the $\mu = \text{const}$ stratum of a singularity. *Funct. Anal. Appl.*, 1983, **17**(4), 312–313.

## △ 1973-17 – *V. A. Vassiliev*

The space of all isolated singularities of two variables naturally splits into the $\mu = \text{const}$ strata.

In the case of complex variables these strata are exactly the collections of irreducible germs of curves with fixed Puiseux exponents of corresponding components and equal tangency numbers between corresponding components. (This result was obtained independently by many authors, uncliding J. Wahl and A. N. Varchenko, athough probably in a different statement it was known much earlier.) All these strata are smooth. However the splitting of the space of functions into $\mu = \text{const}$ strata is not a Whitney stratification and even not a primary stratification because of effects found in [3, 4]: one of such strata can approach some particular points of another one but not approach its other points.

In the case of real functions $\mathbb{R}^2 \to \mathbb{R}$, all the strata $\mu = \text{const}$ are the real forms of complex ones. They can be characterized by the additional information on which components of the corresponding complex curve are real (i. e., have one-dimensional intersection with $\mathbb{R}^2$) and which are not (and hence are mapped by the complex conjugation to some other component of the curve).

Any complex $\mu = \text{const}$ stratum of functions $\mathbb{C}^2 \to \mathbb{C}$ contains a completely real representative (i. e., a germ of a complex curve whose components are all real in this sense), see [1, 2].

[1] A'CAMPO N. Le groupe de monodromie du déploiement des singularités isolées de courbes planes, I. *Math. Ann.*, 1975, **213**(1), 1–32.

[2] GUSEIN-ZADE S. M. Dynkin diagrams for singularities of functions of two variables. *Funct. Anal. Appl.*, 1974, **8**(4), 295–300.

[3] GUSEIN-ZADE S. M., NEKHOROSHEV N. N. On adjacencies of singularities $A_k$ to points of the $\mu = \text{const}$ stratum of a singularity. *Funct. Anal. Appl.*, 1983, **17**(4), 312–313.

[4] GUSEIN-ZADE S. M., NEKHOROSHEV N. N. Singularities of type $A_k$ on plane curves of a chosen degree. *Funct. Anal. Appl.*, 2000, **34**(3), 214–215.

## 1973-19 – *V. A. Vassiliev*

By a theorem of A'Campo [1], a complex function singularity has Morsifications with exactly two critical values if and only if it is simple (of one of classes $A_k$, $D_k$, $E_6$, $E_7$, $E_8$). On the other hand, the real simple singularity $D^+_{2k}$ has no real Morsifications (i. e., Morsifications with only real critical points) with only 2 critical values.

The (non)existence of Morsifications with $\leq k$ critical values should say much on the intersection form of the singularity and thus have nice algebraic characterizations.

A related question: does any singularity has arbitrary (not neseccarily Morse) perturbations with all numbers of critical values?

[1] A'CAMPO N. Le groupe de monodromie du déploiement des singularités isolées de courbes planes. II. In: Proceedings of the International Congress of Mathematicians (Vancouver, 1974), Vol. 1. Montreal: Canadian Mathematical Congress, 1975, 395–404.

## 1973-20

See the comment to problem *1986-12*.

## 1973-23 – *V. I. Arnold, B. A. Khesin*

The topological invariance of the asymptotic Hopf invariant [1] (or helicity) for a field on $\mathbb{S}^3$ is still an open problem. J. Gambaudo and E. Ghys in [3] (see also the survey in [2]) have established such an invariance for a certain class of vector fields in a solid torus.

One should also mention paper [4] by T. Vogel where a modified notion of the system of short paths (used in the definition of asymptotic Hopf invariant) was suggested. Namely, all limits in the definition are to be understood in the $L^1$-sense, rather than pointwise almost everywhere. With this modification one readily shows that, e. g., the system of shortest geodesics on a manifold is a system of short paths for any vector field. Thus paper [4] resolves the question emphasized in [2]: the existence of a short paths system for any (in particular, non-generic) vector field.


[1] ARNOLD V. I. The asymptotic Hopf invariant and its applications. In: Proceedings of the All-Union School on Differential Equations with Infinitely Many Independent Variables and on Dynamical Systems with Infinitely Many Degrees of Freedom (Dilizhan, May 21 – June 3, 1973). Yerevan: AS of Armenian SSR, 1974, 229–256 (in Russian). [*The English translation*: *Selecta Math. Sov.*, 1986, **5**(4), 327–345.] [*The Russian original is reprinted and supplemented in*: Vladimir Igorevich Arnold. Selecta–60. Moscow: PHASIS, 1997, 215–236.]

[2] ARNOLD V. I., KHESIN B. A. Topological Methods in Hydrodynamics. New York: Springer, 1998. (Appl. Math. Sci., 125.)

[3] GAMBAUDO J. -M., GHYS É. Enlacements asymptotiques. *Topology*, 1997, **36**(6), 1355–1379.

[4] VOGEL T. I. On the asymptotic linking number. *Commun. Math. Phys.*, to appear. [*Internet*: `http://www.arXiv.org/abs/math.DS/0011159`]


## 1973-24 — *B. A. Khesin* Also: *1977-8*

ℛ The relation of the Hopf invariant to the Ray–Singer torsion was discussed by A. Schwarz in [2], see also book [1].


[1] ARNOLD V. I., KHESIN B. A. Topological Methods in Hydrodynamics. New York: Springer, 1998. (Appl. Math. Sci., 125.)

[2] SCHWARZ A. S. The partition function of degenerate quadratic functional and Ray–Singer invariants. *Lett. Math. Phys.*, 1977/78, **2**(3), 247–252.


## 1973-25 — *B. A. Khesin, A. M. Lukatskiĭ* Also: *1991-1*

ℛ In [1, 5] it was shown that the energy of a divergence-free field is bounded from below by its asymptotic Hopf invariant (or helicity), where the latter measures the average linking of the field trajectories. Thus the presence of linked trajectories prevents a relaxation of the given field to a field with arbitrarily small energy.

On the other hand, the rotation field in the ball is an example of a field with all trajectories pairwise unlinked. A. D. Sakharov and Ya. B. Zeldovich suggested that this rotation field can be transformed by a volume-preserving diffeomorphism to a field whose energy is less than any given $\varepsilon$ and gave a heuristic construction of the transformations. This conjecture was proved by M. Freedman in 1991, see [3].

The question remains whether the presence of a linked closed trajectory for a field could provide an energy lower bound (even if the averaged linking of all trajectories totals zero) and therefore could prevent a relaxation of the field to arbitrarily small energies. Apparently, one of the the best results in this direction is as follows [4]: if a field $\xi$ in $\mathbb{R}^3$ has an invariant torus $T$ forming an essential knot of type $K$, then

$$E(\xi) \geq \left(\frac{16}{\pi \operatorname{volume}(T)}\right)^{1/3} |\operatorname{Flux}\xi|^2 (2\operatorname{genus}(K) - 1),$$

where $\operatorname{Flux}\xi$ is the flux of $\xi$ through a crossection of $T$, $\operatorname{volume}(T)$ is the volume of the solid torus, and $\operatorname{genus}(K)$ is the genus of the knot $K$. (In particular, for a nontrivial knot $\operatorname{genus}(K) \geq 1$ and hence $E(\xi) > 0$.)

In particular, it is sufficient for the field to have closed linked trajectories of the elliptic (generic) type. Such a field has a nearby invariant torus (as a matter of fact, many tori, by the KAM theory) and its energy has a nonzero lower bound. For the fields with closed linked hyperbolic trajectories or with ergodic behavior the question about the lower energy bound is still open. See details and references in [2].

[1] ARNOLD V. I. The asymptotic Hopf invariant and its applications. In: Proceedings of the All-Union School on Differential Equations with Infinitely Many Independent Variables and on Dynamical Systems with Infinitely Many Degrees of Freedom (Dilizhan, May 21 – June 3, 1973). Yerevan: AS of Armenian SSR, 1974, 229–256 (in Russian). [*The English translation*: *Selecta Math. Sov.*, 1986, **5**(4), 327–345.] [*The Russian original is reprinted and supplemented in*: Vladimir Igorevich Arnold. Selecta–60. Moscow: PHASIS, 1997, 215–236.]

[2] ARNOLD V. I., KHESIN B. A. Topological Methods in Hydrodynamics. New York: Springer, 1998. (Appl. Math. Sci., 125.)

[3] FREEDMAN M. Zeldovich's neutron star and the prediction of magnetic froth. In: The Arnoldfest. Proceedings of a conference in honour of V. I. Arnold for his sixtieth birthday (Toronto, 1997). Editors: E. Bierstone, B. A. Khesin, A. G. Khovanskiĭ and J. E. Marsden. Providence, RI: Amer. Math. Soc., 1999, 165–172. (Fields Inst. Commun., 24.)

[4] FREEDMAN M. H., HE Z. -X. Divergence-free fields: energy and asymptotic crossing number. *Ann. Math., Ser.* 2, 1991, **134**(1), 189–229.

[5] MOFFATT H. K. The degree of knottedness of tangled vortex lines. *J. Fluid Mech.*, 1969, **35**, 117–129.

## 1973-26 — *B. A. Khesin, A. M. Lukatskiĭ*

ℛ The relaxation paradox is discussed in [1]. It turns out that the limiting fields for the energy relaxation either have invariant tori or they are eigenfields for the curl operator. Usually, the latter fields are non-integrable. (Example: the $ABC$-fields on $\mathbb{T}^3$, where $A, B, C \neq 0$.) Since the fields cannot change their topology during the relaxation process, it is not clear what could be the limit of a generic field without invariant surfaces.

One possible way to resolve the relaxation paradox is to allow weak compressibility of the fluid, and then to consider the "incompressible limit," see [3]. Singularities of the fields that are the extremals of the incompressible variational problem have been studied by H. K. Moffatt and his school. See also the comment to problem *1986-12*, as well as the survey and references in [2].

[1] ARNOLD V. I. The asymptotic Hopf invariant and its applications. In: Proceedings of the All-Union School on Differential Equations with Infinitely Many Independent Variables and on Dynamical Systems with Infinitely Many Degrees of Freedom (Dilizhan, May 21 – June 3, 1973). Yerevan: AS of Armenian SSR, 1974, 229–256 (in Russian). [*The English translation*: *Selecta Math. Sov.*, 1986, **5**(4), 327–345.] [*The Russian original is reprinted and supplemented in*: Vladimir Igorevich Arnold. Selecta–60. Moscow: PHASIS, 1997, 215–236.]

[2] ARNOLD V. I., KHESIN B. A. Topological Methods in Hydrodynamics. New York: Springer, 1998. (Appl. Math. Sci., 125.)

[3] MORGULIS A., YUDOVICH V. I., ZASLAVSKY G. M. Compressible helical flows. *Commun. Pure Appl. Math.*, 1995, **48**(5), 571–582.

## 1973-27 — *A. A. Glutsyuk, S. K. Lando*

ℛ The authors of the present comment do not know whether any solution of the problem has been published.

For the singularity $A_k$ the statement was proved by E. Looijenga [4] and independently by O. V. Lyashko (see [1, 5, 6]). The analogous problem for the space of polynomials symmetric with respect to the action of a cyclic group was solved in a joint unpublished work by I. M. Pak and A. E. Postnikov, and a little later separately by A. A. Glutsyuk [3] (see also the bibliography there). The principal result of each paper is a combinatorial interpretation of the covering taking each (normalized) cyclically-symmetric polynomial to the set of its critical values and calculation of its degree. The combinatiorial interpretations obtained in each paper are completely different.

The series $D$ corresponds to the case of trigonometric polynomials with one pole of degree 1 (see [2]), and it causes no problems. The singularities of the series $E$ correspond to some strata in an appropriate space of trigonometric polynomials. The corresponding graphs were not described explicitly, a general description can be found, e. g., in [7]. Additional bibliography is given in the comments to problem *1970-15*.


[1] ARNOLD V. I. Critical points of functions and the classification of caustics. *Uspekhi Mat. Nauk*, 1974, **29**(3), 243–244 (in Russian). [*Reprinted in*: Vladimir Igorevich Arnold. Selecta–60. Moscow: PHASIS, 1997, 213–214.]

[2] ARNOLD V. I. Topological classification of trigonometric polynomials and combinatorics of graphs with an equal number of vertices and edges. *Funct. Anal. Appl.*, 1996, **30**(1), 1–14.

[3] GLUTSYUK A. A. An analogue of Cayley's theorem for cyclic symmetric connected graphs with one cycle that are associated with generalized Lyashko–Looijenga coverings. *Russian Math. Surveys*, 1993, **48**(2), 182–183.

[4] LOOIJENGA E. J. N. The complement of the bifurcation variety of a simple singularity. *Invent. Math.*, 1974, **23**(2), 105–116.

[5] LYASHKO O. V. The geometry of bifurcation diagrams. *Russian Math. Surv.*, 1979, **34**(3), 209–210.

[6] LYASHKO O. V. Geometry of bifurcation diagrams. In: Itogi Nauki i Tekhniki VINITI. Current Problems in Mathematics, Vol. 22. Moscow: VINITI, 1983, 94–129 (in Russian). [*The English translation*: *J. Sov. Math.*, 1984, **27**, 2736–2759.]

[7] ZVONKINE D. A., LANDO S. K. On multiplicities of the Lyashko–Looijenga mapping on discriminant strata. *Funct. Anal. Appl.*, 2000, **33**(3), 178–188.


# 1974

## 1974-2

ℛ See paper [1].


[1] GIVENTAL A. B. Lagrangian imbeddings of surfaces and unfolded Whitney umbrella. *Funct. Anal. Appl.*, 1986, **20**(3), 197–203.

## 1974-4 – *V. I. Arnold*

H Such a problem—on the classification of simple singularities of Lagrangian projections of singular Lagrangian manifolds—was discovered by A. B. Givental [1]. O. P. Shcherbak was the first who introduced noncrystallographic Coxeter groups in the singularity theory [2].

[1] GIVENTAL A. B. Singular Lagrangian manifolds and their Lagrangian maps. In: Itogi Nauki i Tekhniki VINITI. Current Problems in Mathematics. Newest Results, Vol. 33. Moscow: VINITI, 55–112 (in Russian). [*The English translation*: *J. Sov. Math.*, 1990, **52**(4), 3246–3278.]

[2] SHCHERBAK O. P. Wavefront and reflection groups. *Russian Math. Surveys*, 1988, **43**(3), 149–194.

## 1974-5 – *V. V. Goryunov*

R Many of Shephard–Todd's finite unitary reflection groups [6] were recently realized as monodromy groups of equivariant versions of the ordinary simple function singularities [1–4]. See [7] for an elegant explanation of this observation, based on the classification of Springer regular elements of the Weyl groups, as well as on the fact that the discriminants of the equivariant singularities coincide with the discriminants of the related Shephard–Todd groups.

As a by-product, the results of [1–4] have demonstrated that a shorter classification of elliptic singularities is in certain cases more natural than a longer one of simple singularities.

At the moment, work is in progress on realizations of some of Popov's affine complex reflection groups [5] as monodromy groups of equivariant $P_8$, $X_9$ and $J_{10}$ (V. V. Goryunov and S. H. Man).

[1] BAINES C. E. Topics in functions with symmetry. Ph. D. Thesis, University of Liverpool, 2000.

[2] GORYUNOV V. V. Unitary reflection groups associated with singularities of functions with cyclic symmetry. *Russian Math. Surveys*, 1999, **54**(5), 873–893.

[3] GORYUNOV V. V. Unitary reflection groups and automorphisms of simple hypersurface singularities. In: New developments in singularity theory (Cambridge, 2000). Editors: D. Siersma, C. T. C. Wall and V. Zakalyukin. Dordrecht: Kluwer Acad. Publ., 2001, 305–328. (NATO Sci. Ser. II Math. Phys. Chem., 21.)

[4] GORYUNOV V. V., BAINES C. E. Cyclically equivariant function singularities and unitary reflection groups $G(2m,2,n)$, $G_9$, $G_{31}$. *St. Petersburg Math. J.*, 2000, **11**(5), 761–774.

[5] Popov V. L. Discrete complex reflection groups. *Commun. Math. Inst., Rijksuniv. Utrecht*, 1982, **15**, 89 pp.

[6] Shephard G. C., Todd J. A. Finite unitary reflection groups. *Canad. J. Math.*, 1954, **6**, 274–304.

[7] Słodowy P. Simple singularities and complex reflections. In: New Developments in Singularity Theory (Cambridge, 2000). Dordrecht: Kluwer Acad. Publ., 2001, 329–348. (NATO Sci. Ser. II Math. Phys. Chem., 21.)

## 1974-6 – *V. I. Arnold, B. A. Khesin*

The progress in this area achieved up to the present date is described in works [1–8].

[1] Arnold V. I. The first steps of symplectic topology. *Russian Math. Surveys*, 1986, **41**(6), 1–21. [*The Russian original is reprinted in*: Vladimir Igorevich Arnold. Selecta–60. Moscow: PHASIS, 1997, 365–389.]

[2] Arnold V. I. First steps of symplectic topology. In: VIIIth International Congress on Mathematical Physics (Marseille, 1986). Editors: M. Mebkhout and R. Sénéor. Singapore: World Scientific, 1987, 1–16.

[3] Arnold V. I. Mysterious Mathematical Trinites. The Topological Economy Principle in Algebraic Geometry. Moscow: Moscow Center for Continuous Mathematical Education Press, 1997 (in Russian).

[4] Arnold V. I. Symplectization, complexification and mathematical trinities. In: The Arnoldfest. Proceedings of a conference in honour of V. I. Arnold for his sixtieth birthday (Toronto, 1997). Editors: E. Bierstone, B. A. Khesin, A. G. Khovanskiĭ and J. E. Marsden. Providence, RI: Amer. Math. Soc., 1999, 23–37. (Fields Institute Commun., 24.); CEREMADE (UMR 7534), Université Paris-Dauphine, № 9815, 04/03/1998.
[*Internet*: `http://www.pdmi.ras.ru/~arnsem/Arnold/arn-papers.html`]

[5] Arnold V. I. Polymathematics: is mathematics a single science or a set of arts? In: Mathematics: Frontiers and Perspectives. Editors: V. I. Arnold, M. Atiyah, P. Lax and B. Mazur. Providence, RI: Amer. Math. Soc., 2000, 403–416; CEREMADE (UMR 7534), Université Paris-Dauphine, № 9911, 10/03/1999.
[*Internet*: `http://www.pdmi.ras.ru/~arnsem/Arnold/arn-papers.html`]

[6] Khesin B. A. Informal complexification and Poisson structures on moduli spaces. In: Topics in Singularity Theory. V. I. Arnold's 60th Anniversary Collection. Editors: A. Khovanskiĭ, A. Varchenko and V. Vassiliev. Providence, RI: Amer. Math. Soc., 1997, 147–155. (AMS Transl., Ser. 2, 180; Adv. Math. Sci., 34.)

[7] Khesin B. A., Rosly A. A. Polar homology and holomorphic bundles. *Phil. Trans. Roy. Soc. London, Ser. A*, 2001, **359**, 1413–1427.

[8] KHESIN B. A., ROSLY A. A. Symplectic geometry on moduli spaces of holomorphic bundles over complex surfaces. In: The Arnoldfest. Proceedings of a conference in honour of V. I. Arnold for his sixtieth birthday (Toronto, 1997). Editors: E. Bierstone, B. A. Khesin, A. G. Khovanskiĭ and J. E. Marsden. Providence, RI: Amer. Math. Soc., 1999, 311–323. (Fields Inst. Commun., 24.)

## 1974-7 – *S. V. Chmutov*

For the group $\mathbb{Z}_2$ the problem was solved in [1]. In this case the problem is equivalent to the classification of singularities on the manifold with boundary. I. G. Shcherbak found [6] an interesting duality on the set of boundary singularities. These results were generalized to the action of the group $\mathbb{Z}_2^k$ in [7]. For dihedral groups the problem was solved by O. V. Lyashko in [4]. In some cases the classification is related to the finite groups generated by reflections (see [3–5]). The list of unimodal singularities from [4] was completed by M. B. Sevryuk [2] (the completion refers to the case of the group $A_2$ and critical point $\widetilde{A}_3$). Concerning equivariant singularities, see also [8–10].

[1] ARNOLD V. I. Critical points of functions on a manifold with boundary, the simple Lie groups $B_k$, $C_k$, $F_4$, and singularities of evolutes. *Russian Math. Surveys*, 1978, **33**(5), 99–116.

[2] ARNOLD V. I., VASSILIEV V. A., GORYUNOV V. V., LYASHKO O. V. Singularity Theory. II. Classification and Applications. Berlin: Springer, 1993. (Encyclopædia Math. Sci., 39; Dynamical Systems, VIII.) [*The Russian original* 1989.]

[3] CHMUTOV S. V., VARCHENKO A. N. Finite irreducible groups, generated by reflections, are monodromy groups of appropriate singularities. *Funct. Anal. Appl.*, 1984, **18**(3), 171–183.

[4] LYASHKO O. V. Classification of critical points of functions on a manifold with singular boundary. *Funct. Anal. Appl.*, 1983, **17**(3), 187–193.

[5] ROBERTS R. M., ZAKALYUKIN V. M. Symmetric wavefronts, caustic and Coxeter groups. In: Singularity Theory (Trieste, 1991). Editors: D. T. Lê, K. Saito and B. Teissier. River Edge, NJ: World Scientific, 1995, 594–626.

[6] SHCHERBAK I. G. Duality of boundary singularities. *Russian Math. Surveys*, 1984, **39**(2), 195–196.

[7] SHCHERBAK I. G. Singularities in the presence of symmetries. In: Topics in Singularity Theory. V. I. Arnold's 60th Anniversary Collection. Editors: A. Khovanskiĭ, A. Varchenko and V. Vassiliev. Providence, RI: Amer. Math. Soc., 1997, 189–196. (AMS Transl., Ser. 2, 180; Adv. Math. Sci., 34.)

[8] WALL C. T. C. A note on symmetry of singularities. *Bull. London Math. Soc.*, 1980, **12**(3), 169–175.

[9] WALL C. T. C. A second note on symmetry of singularities. *Bull. London Math. Soc.*, 1980, **12**(5), 347–354.

[10] WASSERMAN G. Classification of singularities with compact abelian symmetry. *Regensburg Math. Schrift*, 1977, **1**, 284 pp.

## 1974-8 – *I. A. Bogaevsky*

ℜ A moving wave front can reconstruct (undergo a perestroika) in the course of time. For instance, let us consider a perturbation spreading with a unit velocity into an ellipse. In this case, fronts are interior ellipse equidistant curves. Such a front is reconstructing as follows: it is a smooth curve at the beginning, then it acquires four cusps and two self-intersection points, these two points disappear soon, and then they appear again. After this the front becomes smooth again.

It is necessary to examine local perestroikas of a wave front moving in the $n$-dimensional space and depending on time generically. This problem was solved for $n \leq 5$ in [1]. It turned out that, with this restriction, the number of perestroikas is finite modulo diffeomorphisms of the ambient space smoothly depending on time and time shifts. Pictures of perestroikas for $n = 2$ and $n = 3$ can be found in [2–4].

A perestroika of the type $P_8^{\cdot}$ (notations of Section 21.8 in [4] are used) appears for $n = 6$ in a stable way. This perestroika is certain to have one number modulus; however it is not known whether it has functional moduli or not. Some of the perestroikas that appear stably in the case $n \geq 6$ were studied in [6, 7], but the perestroika $P_8^{\cdot}$ is not among them.

In applications, there often is a case where a moving front does not have a generic time dependence, but it is determined uniquely by the current time moment and an *initial condition*, i. e., a wave front at a particular moment. These front sets are called *evolutionary*. For example, equidistants of a smooth hypersurface form evolutionary front sets depending on the distance and this hypersurface (the initial condition) itself. A theorem of transversality often used in applications was proved in [5]; it states that a wave front evolving in time for $n \leq 5$ and a typical initial condition undergoes only perestroikas described in [1]. However, a list of perestroikas realized in this evolutionary set for given $n$ could be less than a general one. For example, if a front does not depend on time, there are no perestroikas at all.

In conclusion, let us give some rigorous definitions. A smooth manifold with a tangent hyperplane field satisfying a condition of maximal non-integrability at each point is called *contact*. A *Legendrian submanifold* of a contact manifold is a smooth integral submanifold of the maximal possible dimension (equal to the half of the dimension of a contact hyperplane). A *Legendrian fibration* is a smooth fibration whose space has a contact structure, and the fibers are Legendrian submanifolds. Let $L_t^{n-1} \subset E^{2n-1}$ be a Legendrian submanifold in the space of the Legendrian fibration $\pi: E^{2n-1} \to B^n$, depending smoothly on time $t$. A hypersurface set $\pi(L_t^{n-1}) \subset B^n$ (generally speaking, singular) is called the front set depending on time. A theorem proved in [1] is valid, if the set of Legendrian manifolds $L_t^{n-1}$ generically depends on time; and the theorem proved in [5] is applicable if $L_t^{n-1} = g_t(L_0^{n-1})$, where $L_0^{n-1} \subset E^{2n-1}$ is a typical Legendrian submanifold (an initial condition), and $g_t : E^{2n-1} \to E^{2n-1}$ is a given set of smooth diffeomorphisms preserving a contact structure and depending on time smoothly.

[1] Arnold V. I. Wave front evolution and equivariant Morse lemma. *Commun. Pure Appl. Math.*, 1976, **29**(6), 557–582; correction: 1977, **30**(6), 823. [*The Russian translation in*: Vladimir Igorevich Arnold. Selecta–60. Moscow: PHASIS, 1997, 289–318.]

[2] Arnold V. I. Singularities of Caustics and Wave Fronts. Dordrecht: Kluwer Acad. Publ., 1990. (Math. Appl., Sov. Ser., 62.)

[3] Arnold V. I. Catastrophe Theory. Berlin: Springer, 1992. [*The Russian original* 1990.]

[4] Arnold V. I., Gusein-Zade S. M., Varchenko A. N. Singularities of Differentiable Maps, Vol. I: The classification of critical points, caustics and wave fronts. Boston, MA: Birkhäuser, 1985. (Monographs in Math., 82.) [*The Russian original* 1982.]

[5] Bogaevsky I. A. Perestroikas of fronts in evolutionary families. *Proc. Steklov Inst. Math.*, 1995, **209**, 57–72.

[6] Zakalyukin V. M. Reconstructions of wave fronts depending on one parameter. *Funct. Anal. Appl.*, 1976, **10**(2), 139–140.

[7] Zakalyukin V. M. Reconstructions of fronts and caustics depending on a parameter and versality of mappings. In: Itogi Nauki i Tekhniki VINITI. Current Problems in Mathematics, Vol. 22. Moscow: VINITI, 1983, 56–93 (in Russian). [*The English translation*: *J. Sov. Math.*, 1984, **27**, 2713–2735.]

# 1975

## 1975-6 — *A. M. Gabrielov*

ℜ For homogeneous functions, see [1]. See also the comments to problems *1973-7* and *1976-16*.

[1] GABRIELOV A. M., KUSHNIRENKO A. G. Description of deformations with constant Milnor number for homogeneous functions. *Funct. Anal. Appl.*, 1975, **9**(4), 329–331.

## 1975-7 — *V. A. Vassiliev*

ℛ Yes, they can. In paper [1] two $\mu = \text{const}$ strata of function singularities $\mathbb{C}^3 \to \mathbb{C}$ were constructed, that do not intersect the set of real functions (i. e., complexifications of functions $\mathbb{R}^3 \to \mathbb{R}$). The complex conjugation sends these two strata one into the other, in particular they are topologically equivalent.

[1] VASSILIEV V. A., SERGANOVA V. V. On the number of real and complex moduli of singularities of smooth functions and matroid realizations. *Math. Notes*, 1991, **49**(1), 15–20.

## 1975-8 — *S. V. Chmutov*

ℛ This is a particular case of problems *1979-3* and *1980-11*. It was solved by A. N. Varchenko (see reference [4] in the comment to problem *1980-11*).

## 1975-9 — *V. I. Arnold*

ℛ The worst singularity indices $\beta = \frac{1}{2} - \frac{1}{N}$ also feature a curious numerology:

| | | | | | | | | | | | $k=3$ | $k=3$ | $k>3$ |
|---|---|---|---|---|---|---|---|---|---|---|---|---|---|
| $l$ | 0 | 1 | 2 | 3 | 4 | 5 | 6 | 7 | 8 | 9 | 10 | 11 | 10 |
| $N$ | 2 | 3 | 4 | 6 | 8 | 12 | $\infty$ | $\infty$ | $-24$ | $-16$ | $-12$ | $-8$ | $-6$ |

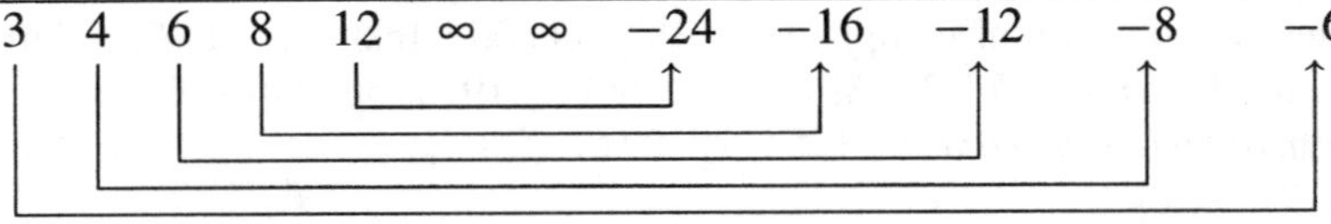

(here $l$ is the number of parameters, and $k$ is the number of variables), see Theorem XX in [1] (and [2], p. 256).

Neither results on the value $s(32)$ nor investigations of the numerology of the sequence $s(\mu)$ are known to me.

[1] ARNOLD V. I. Critical points of smooth functions and their normal forms. *Russian Math. Surveys*, 1975, **30**(5), 1–75.

[2] Vladimir Igorevich Arnold. Selecta–60. Moscow: PHASIS, 1997.

## 1975-12 — *S. M. Gusein-Zade*

It is proved that, if the zero level curve $\{f = 0\}$ of a (real) function germ $f$ is real (i. e., does not have pairs of complex conjugate components), then $f$ has a real morsification (S. M. Gusein-Zade [1], N. A'Campo [2]).

[1] GUSEIN-ZADE S. M. Dynkin diagrams for singularities of functions of two variables. *Funct. Anal. Appl.*, 1974, **8**(4), 295–300.

[2] A'CAMPO N. Le groupe de monodromie du deploiement des singularités isolées de courbes planes, I. *Math. Ann.*, 1975, **213**(1), 1–32.

## 1975-13 — *S. V. Chmutov*

The problem is closely related to *1973-19*. If the number of critical values of the perturbation is 2, then the singularity is simple (N. A'Campo, see the reference in the comment to problem *1973-19*). The similar question can be formulated for polynomials. Suppose that all $(d-1)^2$ critical points of a polynomial of degree $d$ in $n$ variables are Morse critical points. What is the minimal number of its critical values? The problem is solved only for $n = 2$. The answer is 3 (see [1, 2]).

[1] CHMUTOV S. V. Spectrum and equivariant deformations of critical points. *Uspekhi Mat. Nauk*, 1984, **39**(4), 113–114 (in Russian).

[2] CHMUTOV S. V. Extremal distributions of critical points and critical values. In: Singularity Theory (Trieste, 1991). Editors: D. T. Lê, K. Saito and B. Teissier. River Edge, NJ: World Scientific, 1995, 192–205.

## 1975-14

See the comment to problem *1972-32*.

## 1975-15 — *S. M. Gusein-Zade* Also: *1982-12*

For a singularity of two variables, sufficient conditions which guarantee the possibility to split off a singularity $A_1$ were described in [1].

[1] GUSEIN-ZADE S. M. On singularities from which an $A_1$ can be split off. *Funct. Anal. Appl.*, 1993, **27**(1), 57–60.

## ▽ 1975-17 — *V. V. Goryunov*

ℛ A limit of a series of function germs of finite multiplicity is a function of infinite multiplicity. This approach led D. Siersma to the consideration of functions with non-isolated singularities [4]. Later on the direction has been developed mainly by himself [5,6] and R. Pellikaan [2,3], as well as by M. Tibar, A. Zaharia, T. de Jong, G. Jiang (see [7] for a rather complete reference list). Some of their classificational, algebraic and topological results were surveyed in [1]. In addition, in [1] there was introduced a different viewpoint on the classification, which allows deformation of the singular set. This provided some new discriminant sets possessing the $K(\pi, 1)$ property.

[1] Arnold V. I., Vassiliev V. A., Goryunov V. V., Lyashko O. V. Singularities. II. Classification and Applications. Berlin: Springer, 1993, Ch. 1, Sect. 4. (Encyclopædia Math. Sci., 39; Dynamical Systems, VIII.) [*The Russian original* 1989.]

[2] Pellikaan R. On hypersurface singularities which are stems. *Compos. Math.*, 1989, **71**(2), 229–240.

[3] Pellikaan R. Series of isolated singularities. In: Singularities (Iowa City, IA, 1986). Editor: R. Randell. Providence, RI: Amer. Math. Soc., 1989, 241–259. (Contemp. Math., 90.)

[4] Siersma D. Isolated line singularities. In: Singularities. Part 2 (Arcata, CA, 1981). Editor: P. Orlik. Providence, RI: Amer. Math. Soc., 1983, 485–496. (Proc. Symposia Pure Math., 40.)

[5] Siersma D. Singularities with critical locus a 1-dimensional complete intersection and transversal type $A_1$, *Topology Appl.*, 1987, **27**(1), 51–73.

[6] Siersma D. Variation mappings on singularities with a 1-dimensional critical locus. *Topology*, 1991, **30**(3), 445–469.

[7] Siersma D. The vanishing topology of non isolated singularities. In: New Developments in Singularity Theory (Cambridge, 2000). Editors: D. Siersma, C. T. C. Wall and V. Zakalyukin. Dordrecht: Kluwer Acad. Publ., 2001, 447–472. (NATO Sci. Ser. II Math. Phys. Chem., 21.)

## △ 1975-17 — *S. M. Gusein-Zade*

ℛ An explanation of the notion of a series of singularities can be found in [1].

[1] Siersma D. Periodicities in Arnold's lists of singularities. In: Real and Complex Singularities (Oslo, 1976). Editor: P. Holm. Alphen aan den Rijn: Sijthoff & Noordhoff, 1977, 497–524.

## 1975-18 — *V. A. Vassiliev*

ℛ This problem was solved by O. V. Lyashko [1].

[1] LYASHKO O. V. Decompositions of simple singularities of functions. *Funct. Anal. Appl.*, 1976, **10**(2), 122–128.

## 1975-19 — *V. A. Vassiliev*

ℛ These cohomology rings have not been calculated; I expect that the answers are quite complicated.

In the case $n = 1$ N. A. Nekrasov [1] proved that these stable groups are well defined and finitely generated in any particular dimension. The similar result for arbitrary $n$ (and also stable over $n \to \infty$) is also true. This follows from almost the same considerations plus the theorems on stabilization of discriminant strata from [2, 3], and some facts on cohomology of configuration spaces.

Similar facts hold also for the stabilization of cohomology rings of caustics of isolated function singularities.

See also the comments to problems *1975-24*, *1976-28*, *1980-15*, *1985-7*, *1985-22*, and *1998-8*.

[1] NEKRASOV N. A. On the cohomology of the complement of the bifurcation diagram of the singularity $A_\mu$. *Funct. Anal. Appl.*, 1993, **27**(4), 245–250.

[2] VASSILIEV V. A. Stable cohomologies of the complements of the discriminants of deformations of singularities of smooth functions. In: Itogi Nauki i Tekhniki VINITI. Current Problems in Mathematics. Newest Results, Vol. 33. Moscow: VINITI, 1988, 3–29 (in Russian). [*The English translation*: *J. Sov. Math.*, 1990, **52**(4), 3217–3230.]

[3] VASSILIEV V. A. Complements of Discriminants of Smooth Maps: Topology and Applications, revised edition. Providence, RI: Amer. Math. Soc., 1994. (Transl. Math. Monographs, 98.)

## ▽ 1975-20 — *V. I. Arnold* Also: *1973-16*

ℛ For the curves ($m = 1$), the list consists of 1) all mappings with nonzero square term of the Taylor series at the singular point (an infinite series numbered with one index), 2) all mappings with zero square term, but nonzero cubic term, of the Taylor series (two series with three indices), 3) mappings with 6-jet $(t^4, t^6, 0, \ldots, 0)$ (seven series with one index), 4) thirty-two "sporadic" curves, see [3].

The case of plane curves ($m = 1$, $n = 2$) was investigated by Bruce and Gaffney [4].

Now something is known about both the symplectic and the contact version of the problem [1,2].

[1] ARNOLD V. I. First steps of local symplectic algebra. CEREMADE (UMR 7534), Université Paris-Dauphine, № 9902, 20/01/1999; In: Differential Topology, Infinite-Dimensional Lie Algebras, and Applications. D. B. Fuchs' 60th Anniversary Collection. Editors: A. Astashkevich and S. Tabachnikov. Providence, RI: Amer. Math. Soc., 1999, 1–8. (AMS Transl., Ser. 2, 194; Adv. Math. Sci., 44.)
[*Internet*: `http://www.pdmi.ras.ru/~arnsem/Arnold/arn-papers.html`]

[2] ARNOLD V. I. First steps of local contact algebra. CEREMADE (UMR 7534), Université Paris-Dauphine, № 9909, 10/02/99; *Canad. J. Math.*, 1999, **51**(6), 1123–1134.

[3] ARNOLD V. I. Simple singularities of curves. CEREMADE (UMR 7534), Université Paris-Dauphine, № 9906, 09/02/1999; *Proc. Steklov Inst. Math.*, 1999, **226**, 20–28.
[*Internet*: `http://www.pdmi.ras.ru/~arnsem/Arnold/arn-papers.html`]

[4] BRUCE J. W., GAFFNEY T. J. Simple singularities of mappings $(\mathbb{C},0) \to (\mathbb{C}^2,0)$. *J. London Math. Soc., Ser.* 2, 1982, **26**(3), 465–474.

### △ **1975-20** — ***V. V. Goryunov*** Also: *1973-16*

Both real and complex classifications of simple map-germs between $m$- and $n$-dimensional manifolds, $m \geq n$, were obtained in [2] (see also Appendix to [3] for a reduction of arbitrary mappings to projections). Moreover, paper [2] allows the source to be a complete intersection with an isolated singularity. A considerable part of the classification is indeed based on the $A$–$D$–$E$ function singularities and their versal deformations.

As for the case of $m < n$, in addition to V. I. Arnold's comment, I have to mention that the complete classifications of simple singularities have been obtained by now just in three more cases: for curves in the 3-space [1], for $\mathbb{R}^2 \to \mathbb{R}^3$ (by D. Mond in [7, 8]) and for $\mathbb{R}^3 \to \mathbb{R}^4$ (by K. Houston and N. Kirk in [6]). All simple germs in these cases have corank at most 1. Of course, the complex classifications are also contained in the last three papers. Also some work has been done on classifying multi-germs and corank 1 maps [4, 5, 9].

[1] GIBSON C. G., HOBBS C. A. Simple singularities of space curves. *Math. Proc. Cambridge Phil. Soc.*, 1993, **113**(2), 297–310.

[2] GORYUNOV V. V. Singularities of projections of complete intersections. In: Itogi Nauki i Tekhniki VINITI. Current Problems in Mathematics, Vol. 22. Moscow: VINITI, 1983, 167–206 (in Russian). [*The English translation*: *J. Sov. Math.*, 1984, **27**, 2785–2811.]

[3] GORYUNOV V. V. Projections of generic surfaces with boundaries. In: Theory of Singularities and its Applications. Editor: V. I. Arnold. Providence, RI: Amer. Math. Soc., 1990, 157–200. (Adv. Sov. Math., 1.)

[4] HOUSTON K. On the classification and topology of complex multi-germs of corank one and codimension one. Preprint, University of Leeds, December 2001.

[5] HOUSTON K. On the classification of real mono-germs of corank one and codimension one. Preprint, University of Leeds, February 2002.

[6] HOUSTON K., KIRK N. On the classification and geometry of corank 1 map-germs from three-space to four-space. In: Singularity Theory (Liverpool, 1996). Editors: B. Bruce and D. Mond. Cambridge: Cambridge University Press, 1999, 325–351. (London Math. Soc. Lecture Note Ser., 263.)

[7] MOND D. On the classification of germs of maps from $\mathbb{R}^2$ to $\mathbb{R}^3$. *Proc. London Math. Soc., Ser.* 3, 1985, **50**(2), 333–369.

[8] MOND D. Singularities of mappings from surfaces to 3-space. In: Singularity theory (Trieste, 1991). Editors: D. T. Lê, K. Saito and B. Teissier. River Edge, NJ: World Scientific, 1995, 509–526.

[9] WIK ATIQUE R. On the classification of multi-germs of maps from $\mathbb{C}^2$ to $\mathbb{C}^3$ under $\mathcal{A}$-equivalence. In: Real and Complex Singularities (São Carlos, 1998). Editors: J. W. Bruce and F. Tari. Boca Raton, FL: Chapman and Hall, 2000, 119–133. (Chapman and Hall/CRC Res. Notes in Math., 412.)

## 1975-21

ℛ One can treat this problem as solved in principle, see survey [1] (the section on Khovanskiĭ's "geometry of formulae").

[1] ARNOLD V. I., VARCHENKO A. N., GIVENTAL A. B., KHOVANSKIĬ A. G. Singularities of functions, wave fronts, caustics and multidimensional integrals. In: Mathematical Physics Reviews, Vol. 4. Editor: S. P. Novikov. Chur: Harwood Acad. Publ., 1984, 1–92. (Sov. Sci. Rev., Sect. C: Math. Phys. Rev., 4.)

## 1975-22

ℛ This question on the structure of twofold branched coverings is mainly solved (see, e. g., [1]).

[1] ARNOLD V. I., GUSEIN-ZADE S. M., VARCHENKO A. N. Singularities of Differentiable Maps, Vol. II: Monodromy and Asymptotics of Integrals. Boston, MA: Birkhäuser, 1988. (Monographs in Math., 83.) [*The Russian original* 1984.]

## 1975-23 — *V. I. Arnold*

ℛ For $M$-singularities, see survey [2]. I am unaware of whether the question on the existence of $M$-singularities in any $\mathbb{C}$-equivalence class is solved (maybe even it is answered negatively).

The real modality and the complex one can be different for algebraic group actions (the example was shown by E. B. Vinberg, see paper [1]).

The answer in the case of complex stratification seems to be negative, because some strata in the complex problem may not appear in the real domain at all (like the self-intersection lines of the "pyramid" $D_4$, though in this very case there exists the "purse" real form where they do appear).

See also the comment to problem *1979-6*.

[1] ARNOLD V. I. On some problems in singularity theory. In: Geometry and Analysis. Papers dedicated to the memory of V. K. Patodi. Bangalore: Indian Acad. Sci., 1980, 1–9. [*Reprinted in*: *Proc. Indian Acad. Sci. Math. Sci.*, 1981, **90**(1), 1–9.]

[2] ARNOLD V. I., OLEĬNIK O. A. Topology of real algebraic varieties. *Moscow Univ. Math. Bull.*, 1979, **34**(6), 5–17.

## 1975-24 — *V. A. Vassiliev*

ℛ The $\mu = \text{const}$ stratum of an isolated function singularity can be nonsmooth, see the comment to problem *1973-7*.

Such a stratum with a fixed value of $\mu$ is irreducible in the bifurcation diagram of any isolated singularity with trivial $(\mu+1)$-jet, i. e., whose Taylor expansion begins with the terms of order $\geq \mu+2$, see [1]. Moreover, for any finite collection of $\mu = \text{const}$ strata $K_1, \ldots, K_m$ with Milnor numbers $\mu_1, \ldots, \mu_m$ the stratum of the bifurcation diagram of the function $f = \sum_{i=1}^{n} x_i^N$, $N > n(m(2+\max \mu_i)-1)$, consisting of all functions having $m$ critical points of these types with zero critical value, also is irreducible, see [1,3].

Let $A$ be any closed semialgebraic $\mathrm{Diff}(\mathbb{C}^n)$-invariant subvariety of the jet space $J^k(\mathbb{C}^n, \mathbb{C})$ with some $k$ (e. g., the closure of a $\mu = \text{const}$ stratum considered as a subset of such a jet space with $k \geq \mu+1$). Then the cohomology rings of complements of the corresponding strata of bifurcation diagrams of function singularities stabilize to the cohomology ring of the iterated loop space $\Omega^{2n}(J^k(\mathbb{C}^n,\mathbb{C}) \setminus A)$, see [2,3].

See also problems *1975-19*, *1976-28*, *1980-15*, *1985-7*, and *1985-22*.

[1] VASSILIEV V. A. Stable cohomologies of the complements of the discriminants of deformations of singularities of smooth functions. In: Itogi Nauki i Tekhniki VINITI. Current Problems in Mathematics. Newest Results, Vol. 33. Moscow: VINITI, 1988, 3–29 (in Russian). [*The English translation*: *J. Sov. Math.*, 1990, **52**(4), 3217–3230.]

[2] VASSILIEV V. A. Topology of complements to discriminants and loop spaces. In: Theory of Singularities and its Applications. Editor: V. I. Arnold. Providence, RI: Amer. Math. Soc., 1990, 9–21. (Adv. Sov. Math., 1.)

[3] VASSILIEV V. A. Complements of Discriminants of Smooth Functions: Topology and Applications, revised edition. Providence, RI: Amer. Math. Soc., 1994. (Transl. Math. Monographs, 98.)

## 1975-25 – *V. I. Arnold*

ℜ The first part of the problem is principally solved, see papers [1,6,8,10,13]. Only a little is known on the second part (cf. papers [2–5,7,9,11,12,14–16]).

See also the comment to problem *1981-14*.

[1] ARNOLD V. I. Wave front evolution and equivariant Morse lemma. *Commun. Pure Appl. Math.*, 1976, **29**(6), 557–582; correction: 1977, **30**(6), 823. [*The Russian translation in*: Vladimir Igorevich Arnold. Selecta–60. Moscow, PHASIS, 1997, 289–318.]

[2] ARNOLD V. I. Lagrange and Legendre cobordisms, I; II. *Funct. Anal. Appl.*, 1980, **14**(3), 167–177; **14**(4), 252–260.

[3] ARNOLD V. I. On some nonlinear problems. In: Crafoord Prize in Mathematics, 1982. Crafoord Lectures. Stockholm: The Royal Swedish Academy of Sciences, 1982, 1–7. [*The Russian translation in*: Vladimir Igorevich Arnold. Selecta–60. Moscow: PHASIS, 1997, 335–344.]

[4] ARNOLD V. I. On the Newton attraction of gatherings of dust-like particles. *Uspekhi Mat. Nauk*, 1982, **37**(4), 125 (in Russian).

[5] ARNOLD V. I. On the Newton potential of hyperbolic layers. *Trudy Tbilis. Univ., Ser. Mat., Mekh., Astron.*, 1982, **232/233**(13–14), 23–29 (in Russian). [*The English translation*: *Selecta Math. Sov.*, 1985, **4**(2), 103–106.]

[6] ARNOLD V. I. Reconstructions of singularities of potential flows in a collision-free medium and caustic metamorphoses in three-dimensional space. *Trudy Semin. Petrovskogo*, 1982, **8**, 21–57 (in Russian). [*The English translation*: *J. Sov. Math.*, 1986, **32**(3), 229–257.]

[7] Arnold V. I. Some algebro-geometrical aspects of the Newton attraction theory. In: Arithmetic and Geometry. Papers dedicated to I. R. Shafarevich, Vol. II: Geometry. Basel: Birkhäuser, 1983, 1–3. (Progr. Math., 36.)

[8] Arnold V. I. Singularities of Caustics and Wave Fronts. Dordrecht: Kluwer Acad. Publ., 1990. (Math. Appl., Sov. Ser., 62.)

[9] Arnold V. I. Mathematical problems in classical physics. In: Trends and Perspectives in Applied Mathematics. Editors: F. John, J. E. Marsden and L. Sirovich. New York: Springer, 1994, 1–20. (Appl. Math. Sci., 100.) [*The Russian translation in*: Vladimir Igorevich Arnold. Selecta–60. Moscow: PHASIS, 1997, 553–575.]

[10] Arnold V. I., Baryshnikov Yu. M., Bogaevskiĭ I. A. Supplement 2 in: Gurbatov S. N., Malakhov A. N., Saichev A. I. Nonlinear Random Waves and Turbulence in Nondispersive Media: Waves, Rays, Particles. Translated from Russian. Manchester: Manchester University Press, 1991. (Nonlinear Science: Theory and Applications.)

[11] Arnold V. I., Zeldovich Ya. B., Shandarin S. F. Elements of the large scale structure of the Universe. *Uspekhi Mat. Nauk*, 1981, **36**(3), 244–245 (in Russian).

[12] Arnold V. I., Zeldovich Ya. B., Shandarin S. F. The large scale structure of the Universe. I. General properties. One- and two-dimensional models. *Geophys. Astrophys. Fluid Dynamics*, 1982, **20**(1–2), 111–130.]

[13] Bakhtin V. I. Topologically normal forms of caustic transformations of $D_\mu$-series. *Moscow Univ. Math. Bull.*, 1987, **42**(4), 63–66.

[14] Roytvarf A. A. The motion of a continuous medium in the force field with a rooted singularity. *Moscow Univ. Mech. Bull.*, 1987, **42**(1), 24–27.

[15] Roytvarf A. A. On the dynamics of a one-dimensional self-gravitating medium. *Physica D*, 1994, **73**(3), 189–204.

[16] Stanchenko S. V. Arbitrary deformations of Lagrangian and Legendrian mappings *Proc. Steklov Inst. Math.*, 1995, **209**, 191–202.

## 1975-26 — *V. I. Arnold*

In addition to the papers cited in the comment to problem *1970-1*, one should mention works [1–11,13], as well as papers [1,12,14,15] on the degeneracy of the spectral sequence for the natural stratification of the Hermitian matrix space by the multiplicities of eigenvalues and on the ring of curvature forms.

[1] Arnold V. I. Remarks on eigenvalues and eigenvectors of Hermitian matrices, Berry phase, adiabatic connections and quantum Hall effect. *Selecta Math.* (*N. S.*), 1995, **1**(1), 1–19. [*The Russian translation in*: Vladimir Igorevich Arnold. Selecta–60. Moscow: PHASIS, 1997, 583–604.]

[2] ARNOLD V. I. Relatives of the quotient of the complex projective plane by complex conjugation. *Proc. Steklov Inst. Math.*, 1999, **224**, 46–56.
[*Internet*: `http://www.pdmi.ras.ru/~arnsem/Arnold/arn-papers.html`]

[3] ARNOLD V. I. Symplectization, complexification and mathematical trinities. In: The Arnoldfest. Proceedings of a conference in honour of V. I. Arnold for his sixtieth birthday (Toronto, 1997). Editors: E. Bierstone, B. A. Khesin, A. G. Khovanskiĭ and J. E. Marsden. Providence, RI: Amer. Math. Soc., 1999, 23–37. (Fields Institute Commun., 24.); CEREMADE (UMR 7534), Université Paris-Dauphine, № 9815, 04/03/1998.
[*Internet*: `http://www.pdmi.ras.ru/~arnsem/Arnold/arn-papers.html`]

[4] ARNOLD V. I., KOZLOV V. V., NEĬSHTADT A. I. Mathematical Aspects of Classical and Celestial Mechanics, 2nd edition. Berlin: Springer, 1993. (Encyclopædia Math. Sci., 3; Dynamical Systems, III.) [*The Russian original* 1985.] [*The second, revised and supplemented, Russian edition* 2002.]

[5] MAĬLYBAEV A. A. On cones tangent to the stability domain of a family of real matrices. *Vestnik Moskov. Univ. Ser. Mat. Mekh.*, 1998, № 6, 51–54 (in Russian, for the English translation see *Moscow Univ. Math. Bull.*).

[6] MAĬLYBAEV A. A. A method for reducing families of matrices to normal forms. *Dokl. Math.*, 1999, **60**(1), 39–43.

[7] MAĬLYBAEV A. A. Reducing families of matrices to their normal forms and an application to the stability theory. *Fundam. Prikl. Mat.*, 1999, **5**(4), 1111–1133 (in Russian).

[8] MAĬLYBAEV A. A. Singularities of the boundaries of stability domains: analysis and applications. Ph. D. Thesis, Moscow State University, 1999 (in Russian).

[9] MAĬLYBAEV A. A., SEĬRANYAN A. P. On singularities of the boundary of the domain of stability. *Dokl. Phys.*, 1998, **43**(4), 248–252.

[10] MAĬLYBAEV A. A., SEĬRANYAN A. P. Singularities of the boundaries of stability domains. *J. Appl. Math. Mech.*, 1998, **62**(6), 909–920.

[11] MAĬLYBAEV A. A., SEĬRANYAN A. P. The stability domains of Hamiltonian systems. *J. Appl. Math. Mech.*, 1999, **63**(4), 545–555.

[12] POSTNIKOV A. E., SHAPIRO B. Z., SHAPIRO M. Z. Algebras of curvature forms on homogeneous manifolds. In: Differential Topology, Infinite-Dimensional Lie Algebras, and Applications. D. B. Fuchs' 60th Anniversary Collection. Editors: A. Astashkevich and S. Tabachnikov. Providence, RI: Amer. Math. Soc., 1999, 227–235. (AMS Transl., Ser. 2, 194; Adv. Math. Sci., 44.)

[13] SEĬRANYAN A. P., MAĬLYBAEV A. A. On singularities of the stability domain boundaries of Hamiltonian and gyroscopic systems. *Dokl. Phys.*, 1999, **44**(4), 251–255.

[14] SHAPIRO M. Z., VAINSHTEIN A. D. Stratification of Hermitian matrices and the Alexander mapping. *C. R. Acad. Sci. Paris, Sér.* I *Math.*, 1995, **321**(12), 1599–1604.

[15] SHAPIRO B. Z., SHAPIRO M. Z. On ring generated by Chern 2-forms on $SL_n/\mathcal{B}$. *C. R. Acad. Sci. Paris, Sér.* I *Math.*, 1998, **326**(1), 75–80.

### ▽ 1975-27 – *V. I. Arnold*

A survey of progress in this area can be found, for example, in paper [1] and in book [2], and a list of open questions—in paper [3]. A significant contribution to the investigation of asymptotic of oscillatory integrals was made by A. N. Varchenko [5] and V. N. Karpushkin [4]. Integrals of the saddle-point method were studied in works of V. A. Vassiliev [6] and Yu. M. Baryshnikov.

[1] ARNOLD V. I., VARCHENKO A. N., GIVENTAL A. B., KHOVANSKIĬ A. G. Singularities of functions, wave fronts, caustics and multidimensional integrals. In: Mathematical Physics Reviews, V. 4. Editor: S. P. Novikov. Chur: Harwood Acad. Publ., 1984, 1–92. (Sov. Sci. Rev., Sect. C: Math. Phys. Rev., 4.)

[2] ARNOLD V. I. Singularities of Caustics and Wave Fronts. Dordrecht: Kluwer Acad. Publ., 1990. (Math. Appl., Sov. Ser., 62.)

[3] ARNOLD V. I. Mathematical problems in classical physics. In: Trends and Perspectives in Applied Mathematics. Editors: F. John, J. E. Marsden and L. Sirovich. New York: Springer, 1994, 1–20. (Appl. Math. Sci., 100.) [*The Russian translation in*: Vladimir Igorevich Arnold. Selecta–60. Moscow: PHASIS, 1997, 553–575.]

[4] KARPUSHKIN V. N. A theorem concerning uniform estimates of oscillatory integrals when the phase is a function of two variables. *Trudy Semin. Petrovskogo*, 1984, **10**, 150–169 (in Russian). [*The English translation*: *J. Sov. Math.*, 1986, **35**, 2809–2826.]

[5] VARCHENKO A. N. Newton polyhedra and estimation of oscillating integrals. *Funct. Anal. Appl.*, 1976, **10**(3), 175–196.

[6] VASSILIEV V. A. Asymptotic behavior of exponential integrals in the complex domain. *Funct. Anal. Appl.*, 1979, **13**(4), 239–247.

### △ 1975-27 – *V. N. Karpushkin*

Concerning uniform estimates of oscillatory integrals and volumes, see the comments to problems *1972-5* and *1976-22.*

### 1975-28 – *V. I. Arnold*

The singularities of envelopes of families of submanifolds were studied from the given viewpoint in works [1–12].

[1] ARNOLD V. I. On the theory of envelopes. *Uspekhi Mat. Nauk*, 1976, **31**(3), 248–249 (in Russian).

[2] ARNOLD V. I. Wave front evolution and equivariant Morse lemma. *Commun. Pure Appl. Math.*, 1976, **29**(6), 557–582; correction: 1977, **30**(6), 823. [*The Russian translation in*: Vladimir Igorevich Arnold. Selecta–60. Moscow: PHASIS, 1997, 289–318.]

[3] ARNOLD V. I. Lagrangian manifolds with singularities, asymptotic rays and the unfurled swallowtail. *Funct. Anal. Appl.*, 1981, **15**(4), 235–246.

[4] ARNOLD V. I. Sweeping a caustic by the return edge of a moving front. *Uspekhi Mat. Nauk*, 1981, **36**(4), 233 (in Russian).

[5] ARNOLD V. I. Asymptotic rays in the symplectic geometry. *Uspekhi Mat. Nauk*, 1982, **37**(2), 182–183.

[6] ARNOLD V. I. Singularities in variational calculus. In: Itogi Nauki i Tekhniki VINITI. Current Problems in Mathematics, Vol. 22. Moscow: VINITI, 1983, 3–55 (in Russian). [*The English translation*: *J. Sov. Math.*, 1984, **27**, 2679–2713.]

[7] ARNOLD V. I. Singularities of systems of rays. *Russian Math. Surveys*, 1983, **38**(2), 87–176.

[8] ARNOLD V. I. Singularities of ray systems. In: Proceedings of the International Congress of Mathematicians (Warsaw, August 16–24, 1983), V. 1. Editors: Z. Ciesielski and C. Olech. Warsaw: PWN *and* Amsterdam: North-Holland, 1984, 27–49.

[9] ARNOLD V. I. Implicit differential equations, contact structures, and relaxation oscillations. *Uspekhi Mat. Nauk*, 1985, **40**(5), 188 (in Russian).

[10] ARNOLD V. I. Contact structure, relaxation oscillations and singular points of implicit differential equations. In: Geometry and the Theory of Singularities in Nonlinear Equations. Voronezh: Voronezh University Press, 1987, 3–8 (in Russian). (Novoe v Global′nom Analize, 7.) [*Reprinted in*: Vladimir Igorevich Arnold. Selecta–60. Moscow: PHASIS, 1997, 391–396.]

[11] ARNOLD V. I., VASSILIEV V. A., GORYUNOV V. V., LYASHKO O. V. Singularities. II. Classification and Applications. Berlin: Springer, 1993. (Encyclopædia Math. Sci., 39; Dynamical Systems, VIII.) [*The Russian original* 1989.]

[12] PETRIS J. E. H. Normalformen für Singularitäten von einparametrigen Flächenscharen. Dissertation, ETH № 9016, Eidgenössische Technische Hochschule Zürich, 1990, 170 S.

## 1975-29 – *A. A. Agrachev, V. I. Arnold, A. A. Davydov*

The problem is far from its complete solution. The questions related to it were considered in papers and books [1–49].

The information on typical singularities in variational problems with nonholonomic constraints, in particular, in sub-Riemannian problems, can be found in papers [1–8, 30, 32, 37–40, 46, 47]. Sub-Riemannian problems are defined on Riemannian manifolds endowed with non-involutive distributions (treated as nonholonomic constraints). The problem is to find a shortest curve connecting two points in the manifold among all integral curves of the distribution. An important feature of these problems is that the family of arbitrary short extremals started from a fixed point have envelops and self-intersections so that the initial point belongs to the closure of the caustic and the cut locus.

Typical singularities are classified in the case of the contact distribution on the 3-dimensional manifold (see [2, 6, 37]) and, partially, in the case of the quasi-contact distribution on the 4-dimensional manifold (see [32]). One of the tools is a careful study of versal deformations of the families of functions on the circle, which links the subject with many actual problems of the singularities theory.

In general, sub-Riemannian singularities are rather racy. In particular, sub-Riemannian distance for generic analytic rank $k$ distribution on the $n$-dimensional real-analytic Riemannian manifold is not a subanalytic function if

$$n \geq (k-1)\left(\frac{k^2}{3}+\frac{5k}{6}+1\right);$$

the level sets of such a function (sub-Riemannian balls) are subanalytic for $k \geq 3$ and arbitrary $n$ (see [8]).


[1] AGRACHEV A. A. Methods of control theory in nonholonomic geometry. In: Proceedings of the International Congress of Mathematicians (Zürich, 1994). Editor: S. D. Chatterji. Basel: Birkhäuser, 1995, 1473–1483.

[2] AGRACHEV A. A. Exponential mappings for contact sub-Riemannian structures. *J. Dynam. Control Systems*, 1996, **2**(3), 321–358.

[3] AGRACHEV A. A. Compactness for sub-Riemannian length-minimizers and subanalyticity. *Rend. Semin. Mat. Torino*, 1998, **56**(4), 1–12.

[4] AGRACHEV A. A., BONNARD B., CHYBA M., KUPKA I. Sub-Riemannian sphere in Martinet flat case. *J. ESAIM: Control, Optimisation and Calculus of Variations*, 1997, **2**, 377–448.

[5] AGRACHEV A. A., CHARLOT G., GAUTHIER J.-P. A., ZAKALYUKIN V. M. On sub-Riemannian caustics and wave fronts for contact distributions in the three-space. *J. Dynam. Control Systems*, 2000, **6**(3), 365–395.

[6] AGRACHEV A. A., EL-ALAOUI C., GAUTHIER J.-P. A. Sub-Riemannian metrics on $\mathbb{R}^3$. In: Geometric Control and Non-Holonomic Mechanics (Mexico City, 1996). Editors: V. Jurdjevic and R. W. Sharpe. Providence, RI: Amer. Math. Soc., 1998, 29–78. (Canad. Math. Soc. Conf. Proc., 25.)

[7] AGRACHEV A. A., GAUTHIER J.-P. A. On the Dido problem and plane isoperimetric problems. *Acta Appl. Math.*, 1999, **57**(3), 287–338.

[8] AGRACHEV A. A., GAUTHIER J.-P. A. On subanalyticity of Carnot–Carathéodory distances. *Ann. Institut Henri Poincaré, Analyse non linéaire*, 2001, **18**, 359–382.

[9] ARNOLD V. I. Some open problems of the singularity theory. In: The Theory of Cubature Formulae and Applications of the Functional Analysis to Problems in Mathematical Physics. Editor: S. V. Uspenskiĭ. Novosibirsk: Press of the Institute of Mathematics of the Siberian Branch of the USSR Academy of Sciences, 1976, 5–15 (in Russian). (Trudy Semin. S. L. Soboleva, 1.)

[10] ARNOLD V. I. Lagrange and Legendre cobordisms, I; II. *Funct. Anal. Appl.*, 1980, **14**(3), 167–177; **14**(4), 252–260.

[11] ARNOLD V. I. On some problems in singularity theory. In: Geometry and Analysis. Papers dedicated to the memory of V. K. Patodi. Bangalore: Indian Acad. Sci., 1980, 1–9. [*Reprinted in*: *Proc. Indian Acad. Sci. Math. Sci.*, 1981, **90**(1), 1–9.]

[12] ARNOLD V. I. Lagrangian manifolds with singularities, asymptotic rays and the unfurled swallowtail. *Funct. Anal. Appl.*, 1981, **15**(4), 235–246.

[13] ARNOLD V. I. Sweeping a caustic by the return edge of a moving front. *Uspekhi Mat. Nauk*, 1981, **36**(4), 233 (in Russian).

[14] ARNOLD V. I. Reconstructions of singularities of potential flows in a collision-free medium and caustic metamorphoses in three-dimensional space. *Trudy Semin. Petrovskogo*, 1982, **8**, 21–57 (in Russian). [*The English translation*: *J. Sov. Math.*, 1986, **32**(3), 229–257.]

[15] ARNOLD V. I. Singularities of Legendre varieties, of evolvents and of fronts at an obstacle. *Ergod. Theory Dynam. Systems*, 1982, **2**(3–4), 301–309.

[16] ARNOLD V. I. Singularities in variational calculus. In: Itogi Nauki i Tekhniki VINITI. Current Problems in Mathematics, Vol. 22. Moscow: VINITI, 1983, 3–55 (in Russian). [*The English translation*: *J. Sov. Math.*, 1984, **27**, 2679–2713.]

[17] ARNOLD V. I. Singularities of systems of rays. *Russian Math. Surveys*, 1983, **38**(2), 87–176.

[18] ARNOLD V. I. Singularities in the calculus of variations. *Uspekhi Mat. Nauk*, 1984, **39**(5), 256 (in Russian).

[19] ARNOLD V. I. Singularities of ray systems. In: Proceedings of the International Congress of Mathematicians (Warsaw, 1983), Vol. 1. Editors: Z. Ciesielski and C. Olech. Warsaw: PWN *and* Amsterdam: North-Holland, 1984, 27–49.

[20] ARNOLD V. I. Preface in: GRIFFITHS P. Exterior Differential Systems and the Calculus of Variations. Translated from the English original 1983. Moscow: Mir, 1986, 5–6 (in Russian).

[21] ARNOLD V. I. On the interior scattering of waves, defined by hyperbolic variational principles. *J. Geom. Phys.*, 1988, **5**(3), 305–315.

[22] ARNOLD V. I. Surfaces defined by hyperbolic equations. *Math. Notes*, 1988, **44**(1), 489–497. [*The Russian original is reprinted in*: Vladimir Igorevich Arnold. Selecta–60. Moscow: PHASIS, 1997, 397–412.]

[23] ARNOLD V. I. Singularities of Caustics and Wave Fronts. Dordrecht: Kluwer Acad. Publ., 1990. (Math. Appl., Sov. Ser., 62.)

[24] ARNOLD V. I., BARYSHNIKOV YU. M., BOGAEVSKIĬ I. A. Supplement 2 in: GURBATOV S. N., MALAKHOV A. N., SAICHEV A. I. Nonlinear Random Waves and Turbulence in Nondispersive Media: Waves, Rays, Particles. Translated from Russian. Manchester: Manchester University Press, 1991. (Nonlinear Science: Theory and Applications.)

[25] ARNOLD V. I. Sur les propriétés topologiques des projections lagrangiennes en géométrie symplectique des caustiques. CEREMADE (UMR 7534), Université Paris-Dauphine, № 9320, 14/06/1993; *Rev. Mat. Univ. Complut. Madrid*, 1995, **8**(1), 109–119. [*The Russian translation in*: Vladimir Igorevich Arnold. Selecta–60. Moscow: PHASIS, 1997, 525–532.]

[26] ARNOLD V. I. Topological Invariants of Plane Curves and Caustics. Dean Jacqueline B. Lewis Memorial Lectures, Rutgers University. Providence, RI: Amer. Math. Soc., 1994. (University Lecture Series, 5.)

[27] ARNOLD V. I. Invariants and perestroikas of plane fronts. *Proc. Steklov Inst. Math.*, 1995, **209**, 11–56.

[28] ARNOLD V. I. Topological properties of Legendre projections in contact geometry of wave fronts. *St. Petersburg Math. J.*, 1995, **6**(3), 439–452.

[29] ARNOLD V. I. Topological problems of the theory of wave propagation. *Russian Math. Surveys*, 1996, **51**(1), 1–47.

[30] BONNARD B., LAUNAY G., TRÉLAT E. The transcendence required for computing the sphere and wave front in the Martinet sub-Riemannian geometry. In: Proceedings of the International Conference Dedicated to the 90th Birthday of L. S. Pontryagin (Moscow, 1998), Vol. 3: Geometic Control Theory. Itogi Nauki i Tekhniki VINITI. Contemporary Mathematics and its Applications. Thematic Surveys, Vol. 64. Moscow: VINITI, 1999, 82–117 (in Russian).

[31] BRESSAN A., PICCOLI B. A generic classification of time-optimal planar stabilizing feedbacks. *SIAM J. Control Optimization*, 1998, **36**(1), 12–32.

[32] CHARLOT G. Quasi-contact *S-R* metrics: normal form in $\mathbb{R}^{2n}$, wave front and caustic in $\mathbb{R}^4$. *Acta Appl. Math.* (to appear).

[33] DAVYDOV A. A., JONGEN H. T. Normal forms in one-parametric optimization. *Ann. Operations Research*, 2001, **101**, 255–265.

[34] DAVYDOV A. A., ZAKALYUKIN V. M. Point singularities of the conditional minimum on a three-dimensional manifold. *Proc. Steklov Inst. Math.*, 1998, **220**, 109–125.

[35] DAVYDOV A. A., ZAKALYUKIN V. M. Classification of relative minima singularities. In: Geometry and Topology of Caustics—CAUSTICS'98 (Warsaw). Editors: S. Janeczko and V. M. Zakalyukin. Warsaw: Polish Academy of Sciences, Institute of Mathematics, 1999, 75–90. (Banach Center Publ., 50.)

[36] DAVYDOV A. A., ZAKALYUKIN V. M. The coincidence of generic singularities of solutions of extremal problems with constraints. In: Proceedings of the International Conference Dedicated to the 90th Birthday of L. S. Pontryagin (Moscow, 1998), Vol. 3: Geometic Control Theory. Itogi Nauki i Tekhniki VINITI. Contemporary Mathematics and its Applications. Thematic Surveys, Vol. 64. Moscow: VINITI, 1999, 118–143 (in Russian).

[37] EL-ALAOUI C., GAUTHIER J.-P., KUPKA I. Small sub-Riemannian balls on $R^3$. *J. Dynam. Control Systems*, 1996, **2**(3), 359–421.

[38] GE ZH. Horizontal path space and Carnot–Carathéodory metric. *Pacific J. Math.*, 1993, **161**(2), 255–286.

[39] JACQUET S. Subanalyticity of the sub-Riemannian distance. *J. Dynam. Control Systems*, 1999, **5**(3), 303–328.

[40] JACQUET S. Regularity of the sub-Riemannian distance and cut locus. In: Nonlinear Control in the Year 2000 (Paris, 2000). Editors: A. Isidori, F. Lamnabhi-Lagarrigue and W. Respondek. London: Springer, 2000, 321–332. (Lecture Notes in Control and Inform. Sci., 258.)

[41] PICCOLI B. Classification of generic singularities for the planar time-optimal synthesis. *SIAM J. Control Optimization*, 1996, **34**(6), 1914–1946.

[42] PICCOLI B. Regular time-optimal syntheses for smooth planar systems. *Rend. Semin. Mat. Univ. Padova*, 1996, **95**, 59–79.

[43] SIERSMA D. Properties of conflict sets in the plane. In: Geometry and Topology of Caustics—CAUSTICS'98 (Warsaw). Editors: S. Janeczko and V. M. Zakalyukin. Warsaw: Polish Academy of Sciences, Institute of Mathematics, 1999, 267–276. (Banach Center Publ., 50.)

[44] SIERSMA D. Voronoï diagrams and Morse theory of the distance function. In: Geometry in Present Day Science (Aarhus, 1997). Editors: O. E. Barndorff-Nielsen and E. B. Vedel Jensen. Singapore: World Scientific, 1999, 187–208.

[45] SOTOMAYOR J., SIERSMA D., GARCÍA R. Curvatures of conflict surfaces in Euclidean 3-space. In: Geometry and Topology of Caustics—CAUSTICS'98 (Warsaw). Editors: S. Janeczko and V. M. Zakalyukin. Warsaw: Polish Academy of Sciences, Institute of Mathematics, 1999, 277–285. (Banach Center Publ., 50.)

[46] SUSSMANN H. J. Optimal control and piecewise analyticity of the distance function. In: Optimization and Nonlinear Analysis (Haifa, 1990). Editors: A. Ioffe, M. Marcus and S. Reich. Harlow: Longman *and* New York: John Wiley, 1992, 298–310. (Pitman Res. Notes Math. Ser., 244.)

[47] Vershik A. M., Gershkovich V. Ya. Nonholonomic dynamical systems. Geometry of disributions and variational problems. In: Itogi nauki i Tekhniki VINITI. Current Problems in Mathematics, Vol. 16. Moscow: VINITI, 1987, 5–85 (in Russian). [*The English translation in*: Integrable Systems. Nonholonomic Dynamical Systems. Editors: V. I. Arnold and S. P. Novikov. Berlin: Springer, 1994, 1–81. (Encyclopædia Math. Sci., 16; Dynamical Systems, VII.)]

[48] Zakalyukin V. M. Reconstructions of fronts and caustics depending on a parameter and versality of mappings. In: Itogi Nauki i Tekhniki VINITI. Current Problems in Mathematics, Vol. 22. Moscow: VINITI, 1983, 56–93 (in Russian). [*The English translation*: *J. Sov. Math.*, 1984, **27**, 2713–2735.]

[49] Zakalyukin V. M. Envelopes of families of wave fronts and control theory. *Proc. Steklov Inst. Math.*, 1995, **209**, 114–123.

## 1975-30 – *V. I. Arnold*

H The problem was first posed already by Oskar II, the King of Sweden, in 1884 as one of the four challenge problems for the Prix (see [6]).

R For ordinary differential equations it was mainly solved 100 years later by A. A. Davydov [11] after preliminary works of R. Thom [12] and L. Dara [10], see also papers [2–6, 8] and books [1, 7, 9]. In the case of partial differential equations, not much is known yet. Particular results have been obtained by V. V. Lychagin and M. Ya. Zhitomirskiĭ.

[1] Arnold V. I. Geometrical Methods in the Theory of Ordinary Differential Equations, 2nd edition. New York: Springer, 1988. (Grundlehren der Mathematischen Wissenschaften, 250.) [*The Russian original* 1978.]

[2] Arnold V. I. Implicit differential equations, contact structures, and relaxation oscillations. *Uspekhi Mat. Nauk*, 1985, **40**(5), 188 (in Russian).

[3] Arnold V. I. Catastrophe theory. In: Itogi Nauki i Tekhniki VINITI. Current Problems in Mathematics, Vol. 5. Moscow: VINITI, 1986, 219–277 (in Russian). [*The English translation in*: Bifurcation Theory and Catastrophe Theory. Editor: V. I. Arnold. Berlin: Springer, 1994, 207–264. (Encyclopædia Math. Sci., 5; Dynamical Systems, V.)]

[4] Arnold V. I. Contact structure, relaxation oscillations and singular points of implicit differential equations. In: Geometry and the Theory of Singularities in Nonlinear Equations. Voronezh: Voronezh University Press, 1987, 3–8 (in Russian). (Novoe v Global′nom Analize, 7.) [*Reprinted in*: Vladimir Igorevich Arnold. Selecta–60. Moscow: PHASIS, 1997, 391–396.]

[5] Arnold V. I. Bifurcations and singularities in mathematics and mechanics. In: Theoretical and Applied Mechanics (XVII IUTAM Congress, Grenoble, August 21–27, 1988). Editors: P. Germain, M. Piau and D. Caillerie. Amsterdam: North-Holland, 1989, 1–25.

[6] Arnold V. I. Contact geometry: the geometrical method of Gibbs's thermodynamics. In: Proceedings of the Gibbs Symposium (Yale University, 1989). Editors: D. G. Caldi and G. D. Mostow. Providence, RI: Amer. Math. Soc. *and* New York: American Institute of Physics, 1990, 163–179.

[7] Arnold V. I. Lectures on Partial Differential Equations, 2nd supplemented edition. Moscow: PHASIS, 1997 (in Russian).

[8] Arnold V. I., Afraĭmovich V. S., Il′yashenko Yu. S., Shil′nikov L. P. Bifurcation theory. In: Itogi Nauki i Tekhniki VINITI. Current Problems in Mathematics, Vol. 5. Moscow: VINITI, 1986, 5–218. [*The English translation in*: Bifurcation Theory and Catastrophe Theory. Editor: V. I. Arnold. Berlin: Springer, 1994, 1–205. (Encyclopædia Math. Sci., 5; Dynamical Systems, V.)]

[9] Arnold V. I., Vassiliev V. A., Goryunov V. V., Lyashko O. V. Singularities. II. Classification and Applications. Berlin: Springer, 1993, Ch. 1. (Encyclopædia Math. Sci., 39; Dynamical Systems, VIII.) [*The Russian original* 1989.]

[10] Dara L. Singularités génériques des équations différentielles multiformes. *Bol. Soc. Brasil. Mat.*, 1975, **6**(2), 95–128.

[11] Davydov A. A. Normal form of a differential equation, not solvable for the derivative, in a neighborhood of a singular point. *Funct. Anal. Appl.*, 1985, **19**(2), 81–89.

[12] Thom R. Sur les équations différentielles multiformes et leurs intégrales singulières. *Bol. Soc. Brasil. Mat.*, 1972, **3**(1), 1–11.

# 1976

## 1976-2

See the comment to problem *1978-6*.

## 1976-4 — *S. L. Tabachnikov*

A detailed discussion of this problem can be found in [1].

The question is closely related to differential geometry of pairs of transverse fields of directions in 3-space, in particular, when the distribution generated by these fields is completely non-integrable. One can construct differential invariants which are Cartan connections. See [2] for a differential-geometric study of Legendrian 2-webs.

[1] ARNOLD V. I. Geometrical Methods in the Theory of Ordinary Differential Equations, 2nd edition. New York: Springer, 1988, Sect. 1.6. (Grundlehren der Mathematischen Wissenschaften, 250.) [*The Russian original* 1978.]

[2] TABACHNIKOV S. L. Geometry of Lagrangian and Legendrian 2-web. *Differ. Geom. Appl.*, 1993, **3**(3), 265–284.

## ▽ 1976-5 — *B. A. Khesin*

ℛ The question is still open. Some related results can be found in papers [2,3] and in Section 7 of Chapter III in book [1].

[1] ARNOLD V. I., KHESIN B. A. Topological Methods in Hydrodynamics. New York: Springer, 1998. (Appl. Math. Sci., 125.)

[2] KHESIN B. A. Ergodic interpretation of integral hydrodynamic invariants. *J. Geom. Phys.*, 1992, **9**(1), 101–110.

[3] TABACHNIKOV S. L. Two remarks on the asymptotic Hopf invariant. *Funct. Anal. Appl.*, 1990, **24**(1), 74–75.

## △ 1976-5 — *S. L. Tabachnikov*

ℛ A somewhat related definition of an asymptotic Bennequin invariant was introduced in [1]. Recall that the Bennequin (or Bennequin–Thurston) invariant of a Legendrian knot $K$ in, say, a contact 3-sphere, is its linking number with the knot obtained from $K$ by a small translation in a direction, transverse to the contact structure. Likewise, one defines the Bennequin invariant of a knot, transverse to the contact structure. Similarly to the asymptotic Hopf invariant, in the definition of an asymptotic Bennequin invariant, the knot $K$ is replaced by a vector field that may be tangent to a contact distribution or transverse to it.

[1] TABACHNIKOV S. L. Two remarks on the asymptotic Hopf invariant. *Funct. Anal. Appl.*, 1990, **24**(1), 74–75.

## 1976-6 — *M. B. Mishustin*

ℛ For neighborhoods in complex surfaces the problem was solved in [1]; for the general case see the comment to problem *1989-11*.

[1] MISHUSTIN M. B. Neighborhoods of the Riemann sphere in complex surfaces. *Funct. Anal. Appl.*, 1993, **27**(3), 176–185.

## 1976-8

ℋ This is a problem in paper [1] (p. 5: Problem 3), see also the English translation [2].

[1] ARNOLD V. I. Some open problems in the theory of singularities. In: The Theory of Cubature Formulae and Applications of Functional Analysis to Problems of Mathematical Physics. Editor: S. V. Uspenskiĭ. Novosibirsk: Press of the Institute of Mathematics of the Siberian Branch of the USSR Academy of Sciences, 1976, 5–15 (in Russian). (Trudy Seminara S. L. Soboleva, 1.)

[2] ARNOLD V. I. Some open problems in the theory of singularities. In: Singularities. Part 1 (Arcata, CA, 1981). Editor: P. Orlik. Providence, RI: Amer. Math. Soc., 1983, 57–69. (Proc. Symposia Pure Math., 40.)

## 1976-9 – *A. A. Davydov*

ℛ The problem is not solved in this general formulation.

However, for generic bidynamical control systems on the plane, the existence of a regular synthesis is proved for the time optimal problem in analytic and smooth cases (see [8] and the bibliography there). The corresponding classifications of singularities are presented in [4, 6, 7].

The existence of a regular synthesis for a generic smooth system with closed quadratically convex indicatrices with phase space of dimension $n \leq 4$ follows from the classifications of singularities of wave front evolutions [1–3, 9, 10]. In this case the classification of singularities for the time optimal problem follows from [5].

[1] ARNOLD V. I. Wave front evolution and equivariant Morse lemma. *Commun. Pure Appl. Math.*, 1976, **29**(6), 557–582; correction: 1977, **30**(6), 823. [*The Russian translation in*: Vladimir Igorevich Arnold. Selecta–60. Moscow: PHASIS, 1997, 289–318.]

[2] ARNOLD V. I., GUSEIN-ZADE S. M., VARCHENKO A. N. Singularities of Differentiable Maps, Vol. I: The Classification of Critical Points, Caustics and Wave Fronts. Boston, MA: Birkhäuser, 1985. (Monographs in Math., 82.) [*The Russian original* 1982.]

[3] BOGAEVSKY I. A. Perestroikas of fronts in evolutionary families. *Proc. Steklov Inst. Math.*, 1995, **209**, 57–72.

[4] BRESSAN A., PICCOLI B. A generic classification of time-optimal planar stabilizing feedbacks. *SIAM J. Control Optim.*, 1998, **36**(1), 12–32.

[5] DAVYDOV A. A., ZAKALYUKIN V. M. The coincidence of generic singularities of solutions of extremal problems with constraints. In: Proceedings of the International Conference Dedicated to the 90th Birthday of L. S. Pontryagin (Moscow, 1998), Vol. 3: Geometic Control Theory. Itogi Nauki i Tekhniki VINITI. Contemporary Mathematics and its Applications. Thematic Surveys, Vol. 64. Moscow: VINITI, 1999, 118–143 (in Russian).

[6] PICCOLI B. Classification of generic singularities for the planar time-optimal synthesis. *SIAM J. Control Optimization*, 1996, **34**(6), 1914–1946.

[7] PICCOLI B. Regular time-optimal syntheses for smooth planar systems. *Rend. Semin. Mat. Univ. Padova*, 1996, **95**, 59–79.

[8] PICCOLI B., SUSSMANN H. J. Regular synthesis and sufficient conditions for optimality. *SIAM J. Control Optimization*, 2000, **39**(2), 359–410.

[9] ZAKALYUKIN V. M. Reconstructions of fronts and caustics depending on a parameter and versality of mappings. In: Itogi Nauki i Tekhniki VINITI. Current Problems in Mathematics, Vol. 22. Moscow: VINITI, 1983, 56–93 (in Russian) [*The English translation*: *J. Sov. Math.*, 1984, **27**, 2713–2735.]

[10] ZAKALYUKIN V. M. Envelopes of families of wave fronts and control theory. *Proc. Steklov Inst. Math.*, 1995, **209**, 114–123.

## ▽ 1976-10

H This is a problem in paper [1] (§ 3).

[1] ARNOLD V. I. Some unsolved problems in the theory of differential equations. In: Unsolved Problems in Mechanics and Applied Mathematics. Moscow: Moscow University Press, 1977, 3–9 (in Russian).

## △ 1976-10 – *A. I. Neĭshtadt*

R This problem involves systems with rotating phases, see the comment to problem *1972-9*. The question is: what is the measure of the set of initial data for which the error of description of slow variables evolution by means of the averaging method exceeds a given value? The results concerning this problem are described in the comment to problem *1972-10*.

## ▽ 1976-12

R A survey of the contemporary state of the theory of fewnomials can be found in books [1, 2].

[1] KHOVANSKIĬ A. G. Fewnomials. Providence, RI: Amer. Math. Soc., 1991. (Transl. Math. Monographs, 88.)

[2] KHOVANSKIĬ A. G. Fewnomials. Moscow: PHASIS, 1997 (in Russian). (Mathematician's Library, 2.)

### △ 1976-12 – *S. L. Tabachnikov* Also: *1968-2*

ℜ A recent survey of the subject can be found in [3]. A conjectural multivariable Descartes' rule was proposed by I. Itenberg and M. Roy in [1]; then it was disproved in [2].

[1] ITENBERG I. V., ROY M. Multivariate Descartes' rule. *Beiträge zur Algebra und Geometrie*, 1996, **37**(2), 337–346.

[2] LI T., WANG X. On multivariate Descartes' rule—a counterexample. *Beiträge zur Algebra und Geometrie*, 1998, **39**(1), 1–5.

[3] STURMFELS B. Polynomial equations and convex polytopes. *Amer. Math. Monthly*, 1998, **105**(10), 907–922.

### 1976-13

ℋ This is a problem in paper [1] (p. 5: Problem 1), see also the English translation [2].

[1] ARNOLD V. I. Some open problems in the theory of singularities. In: The Theory of Cubature Formulae and Applications of Functional Analysis to Problems of Mathematical Physics. Editor: S. V. Uspenskiĭ. Novosibirsk: Press of the Institute of Mathematics of the Siberian Branch of the USSR Academy of Sciences, 1976, 5–15 (in Russian). (Trudy Seminara S. L. Soboleva, 1.)

[2] ARNOLD V. I. Some open problems in the theory of singularities. In: Singularities. Part 1 (Arcata, CA, 1981). Editor: P. Orlik. Providence, RI: Amer. Math. Soc., 1983, 57–69. (Proc. Symposia Pure Math., 40.)

ℜ See the comment to problem *1973-7*.

### 1976-14

ℋ This is a problem in paper [1] (p. 5: Problem 2), see also the English translation [2].

[1] ARNOLD V. I. Some open problems in the theory of singularities. In: The Theory of Cubature Formulae and Applications of Functional Analysis to Problems of Mathematical Physics. Editor: S. V. Uspenskiĭ. Novosibirsk: Press of the Institute of Mathematics of the Siberian Branch of the USSR Academy of Sciences, 1976, 5–15 (in Russian). (Trudy Seminara S. L. Soboleva, 1.)

[2] ARNOLD V. I. Some open problems in the theory of singularities. In: Singularities. Part 1 (Arcata, CA, 1981). Editor: P. Orlik. Providence, RI: Amer. Math. Soc., 1983, 57–69. (Proc. Symposia Pure Math., 40.)

See the comment to problem *1979-6*.

## 1976-15

This is a problem in paper [1] (p. 5–6: Problem 4), see also the English translation [2].

[1] ARNOLD V. I. Some open problems in the theory of singularities. In: The Theory of Cubature Formulae and Applications of Functional Analysis to Problems of Mathematical Physics. Editor: S. V. Uspenskiĭ. Novosibirsk: Press of the Institute of Mathematics of the Siberian Branch of the USSR Academy of Sciences, 1976, 5–15 (in Russian). (Trudy Seminara S. L. Soboleva, 1.)

[2] ARNOLD V. I. Some open problems in the theory of singularities. In: Singularities. Part 1 (Arcata, CA, 1981). Editor: P. Orlik. Providence, RI: Amer. Math. Soc., 1983, 57–69. (Proc. Symposia Pure Math., 40.)

See the comment to problem *1973-11*.

## 1976-16

This is a problem in paper [1] (p. 6: Problem 5). The English translation [2] of that paper contains V. I. Arnold's comments of October 5, 1981, and March 3, 1982.

[1] ARNOLD V. I. Some open problems in the theory of singularities. In: The Theory of Cubature Formulae and Applications of Functional Analysis to Problems of Mathematical Physics. Editor: S. V. Uspenskiĭ. Novosibirsk: Press of the Institute of Mathematics of the Siberian Branch of the USSR Academy of Sciences, 1976, 5–15 (in Russian). (Trudy Seminara S. L. Soboleva, 1.)

[2] ARNOLD V. I. Some open problems in the theory of singularities. In: Singularities. Part 1 (Arcata, CA, 1981). Editor: P. Orlik. Providence, RI: Amer. Math. Soc., 1983, 57–69. (Proc. Symposia Pure Math., 40.)

## 1976-17

ℋ This is a problem in paper [1] (p. 6: Problem 6). The English translation [2] of that paper contains V. I. Arnold's comment of October 5, 1981.

[1] ARNOLD V. I. Some open problems in the theory of singularities. In: The Theory of Cubature Formulae and Applications of Functional Analysis to Problems of Mathematical Physics. Editor: S. V. Uspenskiĭ. Novosibirsk: Press of the Institute of Mathematics of the Siberian Branch of the USSR Academy of Sciences, 1976, 5–15 (in Russian). (Trudy Seminara S. L. Soboleva, 1.)

[2] ARNOLD V. I. Some open problems in the theory of singularities. In: Singularities. Part 1 (Arcata, CA, 1981). Editor: P. Orlik. Providence, RI: Amer. Math. Soc., 1983, 57–69. (Proc. Symposia Pure Math., 40.)

## 1976-18

ℋ This is a problem in paper [1] (p. 6: Problem 7). The English translation [2] of that paper contains V. I. Arnold's comment of October 5, 1981.

[1] ARNOLD V. I. Some open problems in the theory of singularities. In: The Theory of Cubature Formulae and Applications of Functional Analysis to Problems of Mathematical Physics. Editor: S. V. Uspenskiĭ. Novosibirsk: Press of the Institute of Mathematics of the Siberian Branch of the USSR Academy of Sciences, 1976, 5–15 (in Russian). (Trudy Seminara S. L. Soboleva, 1.)

[2] ARNOLD V. I. Some open problems in the theory of singularities. In: Singularities. Part 1 (Arcata, CA, 1981). Editor: P. Orlik. Providence, RI: Amer. Math. Soc., 1983, 57–69. (Proc. Symposia Pure Math., 40.)

## 1976-19

ℋ This is a problem in paper [1] (p. 6: Problem 8). The English translation [2] of that paper contains V. I. Arnold's comment of October 5, 1981.

[1] ARNOLD V. I. Some open problems in the theory of singularities. In: The Theory of Cubature Formulae and Applications of Functional Analysis to Problems of Mathematical Physics. Editor: S. V. Uspenskiĭ. Novosibirsk: Press of the Institute of Mathematics of the Siberian Branch of the USSR Academy of Sciences, 1976, 5–15 (in Russian). (Trudy Seminara S. L. Soboleva, 1.)

[2] ARNOLD V. I. Some open problems in the theory of singularities. In: Singularities. Part 1 (Arcata, CA, 1981). Editor: P. Orlik. Providence, RI: Amer. Math. Soc., 1983, 57–69. (Proc. Symposia Pure Math., 40.)

## 1976-20

ℋ This is a problem in paper [1] (p. 7: Problem 9). The English translation [2] of that paper contains V. I. Arnold's comment of October 5, 1981.

[1] ARNOLD V. I. Some open problems in the theory of singularities. In: The Theory of Cubature Formulae and Applications of Functional Analysis to Problems of Mathematical Physics. Editor: S. V. Uspenskiĭ. Novosibirsk: Press of the Institute of Mathematics of the Siberian Branch of the USSR Academy of Sciences, 1976, 5–15 (in Russian). (Trudy Seminara S. L. Soboleva, 1.)

[2] ARNOLD V. I. Some open problems in the theory of singularities. In: Singularities. Part 1 (Arcata, CA, 1981). Editor: P. Orlik. Providence, RI: Amer. Math. Soc., 1983, 57–69. (Proc. Symposia Pure Math., 40.)

## 1976-21

ℋ This is a problem in paper [1] (p. 7: Problem 10), see also the English translation [2].

[1] ARNOLD V. I. Some open problems in the theory of singularities. In: The Theory of Cubature Formulae and Applications of Functional Analysis to Problems of Mathematical Physics. Editor: S. V. Uspenskiĭ. Novosibirsk: Press of the Institute of Mathematics of the Siberian Branch of the USSR Academy of Sciences, 1976, 5–15 (in Russian). (Trudy Seminara S. L. Soboleva, 1.)

[2] ARNOLD V. I. Some open problems in the theory of singularities. In: Singularities. Part 1 (Arcata, CA, 1981). Editor: P. Orlik. Providence, RI: Amer. Math. Soc., 1983, 57–69. (Proc. Symposia Pure Math., 40.)

## ▽ 1976-22

ℋ This is a problem in paper [1] (p. 7–8: Problem 11). The English translation [2] of that paper contains V. I. Arnold's comment of October 5, 1981. The problem on evaluation of the indices $\beta(l)$ appears in paper [3] (XVI(E), p. 58).

[1] ARNOLD V. I. Some open problems in the theory of singularities. In: The Theory of Cubature Formulae and Applications of Functional Analysis to Problems of Mathematical Physics. Editor: S. V. Uspenskiĭ. Novosibirsk: Press of the Institute of Mathematics of the Siberian Branch of the USSR Academy of Sciences, 1976, 5–15 (in Russian). (Trudy Seminara S. L. Soboleva, 1.)

[2] ARNOLD V. I. Some open problems in the theory of singularities. In: Singularities. Part 1 (Arcata, CA, 1981). Editor: P. Orlik. Providence, RI: Amer. Math. Soc., 1983, 57–69. (Proc. Symposia Pure Math., 40.)

[3] Problems of present day mathematics. Editor: F. E. Browder. In: Mathematical Developments Arising from Hilbert Problems (Northern Illinois University, 1974). Part 1. Editor: F. E. Browder. Providence, RI: Amer. Math. Soc., 1976, 35–79. (Proc. Symposia Pure Math., 28.)

## △ 1976-22 – *V. N. Karpushkin*

ℛ The tables of indices $\beta(n,l)$ presented below follow from works [1–6] (for notations, see the formulation of the current problem).

| $l$ | 1 | 2 | 3 | 4 | 5 | 6 | 7 | 8 | 9 | 10 |
|---|---|---|---|---|---|---|---|---|---|---|
| $n=3$ | 1/6 | 1/4 | 1/3 | 3/8 | 5/12 | 1/2 | 1/2 | 13/24 | 9/16 | 7/12 |
| $n\geq 4$ | 1/6 | 1/4 | 1/3 | 3/8 | 5/12 | 1/2 | 1/2 | 13/24 | 9/16 | 2/3 |

| $l$ | 1 | 2 | 3 | 4 | 5 | 6 | 7 | 8 | 9 | 10 | 11 | 12 |
|---|---|---|---|---|---|---|---|---|---|---|---|---|
| $n=2$ | 1/6 | 1/4 | 1/3 | 3/8 | 5/12 | 4/9 | 1/2 | 1/2 | 8/15 | 11/20 | 9/16 | 3/5 |

[1] ARNOLD V. I. Remarks on the stationary phase method and Coxeter numbers. *Russian Math. Surveys*, 1973, **28**(5), 19–48.

[2] KARPUSHKIN V. N. Uniform estimates of oscillatory integrals with parabolic or hyperbolic phase. *Trudy Semin. Petrovskogo*, 1983, **9**, 1–39 (in Russian). [*The English translation*: *J. Sov. Math.*, 1986, **33**, 1159–1188.]

[3] KARPUSHKIN V. N. A theorem concerning uniform estimates of oscillatory integrals when the phase is a function of two variables. *Trudy Semin. Petrovskogo*, 1984, **10**, 150–169 (in Russian). [*The English translation*: *J. Sov. Math.*, 1986, **35**, 2809–2826.]

[4] KARPUSHKIN V. N. Dominant term in the asymptotics of oscillatory integrals with a phase of the series $T$. *Math. Notes*, 1994, **56**(6), 1304–1305.

[5] KARPUSHKIN V. N. Uniform estimates of oscillatory integrals with phase from the series $\tilde{R}_m$. *Math. Notes*, 1998, **64**(3), 404–406.

[6] VARCHENKO A. N. Newton polyhedra and estimation of oscillating integrals. *Funct. Anal. Appl.*, 1976, **10**(3), 175–196.

## 1976-23

ℋ This is a problem in paper [1] (p. 8–9: Problem 12). The English translation [2] of that paper contains V. I. Arnold's comment of October 5, 1981.

[1] ARNOLD V. I. Some open problems in the theory of singularities. In: The Theory of Cubature Formulae and Applications of Functional Analysis to Problems of Mathematical Physics. Editor: S. V. Uspenskiĭ. Novosibirsk: Press of the Institute of Mathematics of the Siberian Branch of the USSR Academy of Sciences, 1976, 5–15 (in Russian). (Trudy Seminara S. L. Soboleva, 1.)

[2] ARNOLD V. I. Some open problems in the theory of singularities. In: Singularities. Part 1 (Arcata, CA, 1981). Editor: P. Orlik. Providence, RI: Amer. Math. Soc., 1983, 57–69. (Proc. Symposia Pure Math., 40.)

## 1976-24

This is a problem in papers [1] (p. 7: Problem 9) and [2] (VIII, p. 46). The English translation [3] of the former paper contains V. I. Arnold's comment of October 5, 1981.

[1] ARNOLD V. I. Some open problems in the theory of singularities. In: The Theory of Cubature Formulae and Applications of Functional Analysis to Problems of Mathematical Physics. Editor: S. V. Uspenskiĭ. Novosibirsk: Press of the Institute of Mathematics of the Siberian Branch of the USSR Academy of Sciences, 1976, 5–15 (in Russian). (Trudy Seminara S. L. Soboleva, 1.)

[2] Problems of present day mathematics. Editor: F. E. Browder. In: Mathematical Developments Arising from Hilbert Problems (Northern Illinois University, 1974). Part 1. Editor: F. E. Browder. Providence, RI: Amer. Math. Soc., 1976, 35–79. (Proc. Symposia Pure Math., 28.)

[3] ARNOLD V. I. Some open problems in the theory of singularities. In: Singularities. Part 1 (Arcata, CA, 1981). Editor: P. Orlik. Providence, RI: Amer. Math. Soc., 1983, 57–69. (Proc. Symposia Pure Math., 40.)

## 1976-25

This is a problem in paper [1] (p. 10: Problem 14). The English translation [2] of that paper contains V. I. Arnold's comment of October 5, 1981. The more general statement of the problem is given in paper [3] (XVI(F), p. 58–59).

[1] ARNOLD V. I. Some open problems in the theory of singularities. In: The Theory of Cubature Formulae and Applications of Functional Analysis to Problems of Mathematical Physics. Editor: S. V. Uspenskiĭ. Novosibirsk: Press of the Institute of Mathematics of the Siberian Branch of the USSR Academy of Sciences, 1976, 5–15 (in Russian). (Trudy Seminara S. L. Soboleva, 1.)

[2] ARNOLD V. I. Some open problems in the theory of singularities. In: Singularities. Part 1 (Arcata, CA, 1981). Editor: P. Orlik. Providence, RI: Amer. Math. Soc., 1983, 57–69. (Proc. Symposia Pure Math., 40.)

[3] Problems of present day mathematics. Editor: F. E. Browder. In: Mathematical Developments Arising from Hilbert Problems (Northern Illinois University, 1974). Part 1. Editor: F. E. Browder. Providence, RI: Amer. Math. Soc., 1976, 35–79. (Proc. Symposia Pure Math., 28.)

## 1976-26

This is a problem in paper [1] (p. 10: Problem 15). The English translation [2] of that paper contains V. I. Arnold's comment of October 5, 1981.

[1] ARNOLD V. I. Some open problems in the theory of singularities. In: The Theory of Cubature Formulae and Applications of Functional Analysis to Problems of Mathematical Physics. Editor: S. V. Uspenskiĭ. Novosibirsk: Press of the Institute of Mathematics of the Siberian Branch of the USSR Academy of Sciences, 1976, 5–15 (in Russian). (Trudy Seminara S. L. Soboleva, 1.)

[2] ARNOLD V. I. Some open problems in the theory of singularities. In: Singularities. Part 1 (Arcata, CA, 1981). Editor: P. Orlik. Providence, RI: Amer. Math. Soc., 1983, 57–69. (Proc. Symposia Pure Math., 40.)

## 1976-27

This is a problem in paper [1] (p. 10–11: Problem 16). The English translation [2] of that paper contains V. I. Arnold's comment of October 5, 1981.

[1] ARNOLD V. I. Some open problems in the theory of singularities. In: The Theory of Cubature Formulae and Applications of Functional Analysis to Problems of Mathematical Physics. Editor: S. V. Uspenskiĭ. Novosibirsk: Press of the Institute of Mathematics of the Siberian Branch of the USSR Academy of Sciences, 1976, 5–15 (in Russian). (Trudy Seminara S. L. Soboleva, 1.)

[2] ARNOLD V. I. Some open problems in the theory of singularities. In: Singularities. Part 1 (Arcata, CA, 1981). Editor: P. Orlik. Providence, RI: Amer. Math. Soc., 1983, 57–69. (Proc. Symposia Pure Math., 40.)

## 1976-28

This is a problem in paper [1] (p. 10: Problem 15). The English translation [2] of that paper contains V. I. Arnold's comment of October 5, 1981.

[1] ARNOLD V. I. Some open problems in the theory of singularities. In: The Theory of Cubature Formulae and Applications of Functional Analysis to Problems of Mathematical Physics. Editor: S. V. Uspenskiĭ. Novosibirsk: Press of the Institute of Mathematics of the Siberian Branch of the USSR Academy of Sciences, 1976, 5–15 (in Russian). (Trudy Seminara S. L. Soboleva, 1.)

[2] ARNOLD V. I. Some open problems in the theory of singularities. In: Singularities. Part 1 (Arcata, CA, 1981). Editor: P. Orlik. Providence, RI: Amer. Math. Soc., 1983, 57–69. (Proc. Symposia Pure Math., 40.)

ℛ See also other related problems and the comments to them: *1975-19*, *1975-24*, *1980-15*, *1985-7*, *1985-22*, and *1998-8*.

## 1976-29

ℋ This is a problem in papers [1] (p. 12: Problem 18a), [2] (XVII(C), p. 59–60), and [3] (§ 4). The English translation [4] of the first paper contains V. I. Arnold's comment of October 5, 1981.

[1] ARNOLD V. I. Some open problems in the theory of singularities. In: The Theory of Cubature Formulas and Applications of Functional Analysis to Problems of Mathematical Physics. Editor: S. V. Uspenskiĭ. Novosibirsk: Press of the Institute of Mathematics of the Siberian Branch of the USSR Academy of Sciences, 1976, 5–15 (in Russian). (Trudy Seminara S. L. Soboleva, 1.)

[2] Problems of present day mathematics. Editor: F. E. Browder. In: Mathematical Developments Arising from Hilbert Problems (Northern Illinois University, 1974). Part 1. Editor: F. E. Browder. Providence, RI: Amer. Math. Soc., 1976, 35–79. (Proc. Symposia Pure Math., 28.)

[3] ARNOLD V. I. Some problems in the theory of differential equations. In: Unsolved Problems of Mechanics and Applied Mathematics. Moscow: Moscow University Press, 1977, 3–9 (in Russian).

[4] ARNOLD V. I. Some open problems in the theory of singularities. In: Singularities. Part 1 (Arcata, CA, 1981). Editor: P. Orlik. Providence, RI: Amer. Math. Soc., 1983, 57–69. (Proc. Symposia Pure Math., 40.)

ℛ See the comment to problem *1971-4*.

## 1976-30

ℋ This is a problem in paper [1] (p. 5: Problem 2), see also the English translation [2].

[1] ARNOLD V. I. Some open problems in the theory of singularities. In: The Theory of Cubature Formulae and Applications of Functional Analysis to Problems of Mathematical Physics. Editor: S. V. Uspenskiĭ. Novosibirsk: Press of the Institute of Mathematics of the Siberian Branch of the USSR Academy of Sciences, 1976, 5–15 (in Russian). (Trudy Seminara S. L. Soboleva, 1.)

[2] ARNOLD V. I. Some open problems in the theory of singularities. In: Singularities. Part 1 (Arcata, CA, 1981). Editor: P. Orlik. Providence, RI: Amer. Math. Soc., 1983, 57–69. (Proc. Symposia Pure Math., 40.)

ℛ See the comments to problem *1972-23*.

## 1976-31

ℋ This is a problem in papers [1] (p. 12: Problem 19), [2] (XVII(B), p. 59), and [3] (§ 4). The English translation [4] of the first paper contains V. I. Arnold's comment of October 5, 1981.

[1] ARNOLD V. I. Some open problems in the theory of singularities. In: The Theory of Cubature Formulas and Applications of Functional Analysis to Problems of Mathematical Physics. Editor: S. V. Uspenskiĭ. Novosibirsk: Press of the Institute of Mathematics of the Siberian Branch of the USSR Academy of Sciences, 1976, 5–15 (in Russian). (Trudy Seminara S. L. Soboleva, 1.)

[2] Problems of present day mathematics. Editor: F. E. Browder. In: Mathematical Developments Arising from Hilbert Problems (Northern Illinois University, 1974). Part 1. Editor: F. E. Browder. Providence, RI: Amer. Math. Soc., 1976, 35–79. (Proc. Symposia Pure Math., 28.)

[3] ARNOLD V. I. Some problems in the theory of differential equations. In: Unsolved Problems of Mechanics and Applied Mathematics. Moscow: Moscow University Press, 1977, 3–9 (in Russian).

[4] ARNOLD V. I. Some open problems in the theory of singularities. In: Singularities. Part 1 (Arcata, CA, 1981). Editor: P. Orlik. Providence, RI: Amer. Math. Soc., 1983, 57–69. (Proc. Symposia Pure Math., 40.)

ℛ See the comment to problem *1973-4*.

## 1976-32

ℋ This is a problem in paper [1] (p. 12–13: Problem 20). The English translation [2] of that paper contains V. I. Arnold's comments of October 5, 1981, and March 3, 1982.

[1] ARNOLD V. I. Some open problems in the theory of singularities. In: The Theory of Cubature Formulae and Applications of Functional Analysis to Problems of Mathematical Physics. Editor: S. V. Uspenskiĭ. Novosibirsk: Press of the Institute of Mathematics of the Siberian Branch of the USSR Academy of Sciences, 1976, 5–15 (in Russian). (Trudy Seminara S. L. Soboleva, 1.)

[2] ARNOLD V. I. Some open problems in the theory of singularities. In: Singularities. Part 1 (Arcata, CA, 1981). Editor: P. Orlik. Providence, RI: Amer. Math. Soc., 1983, 57–69. (Proc. Symposia Pure Math., 40.)

R See the comments to problems *1972-12* and *1981-28*.

## 1976-33

H This is a problem in paper [1] (p. 13–14: Problem 21), see also the English translation [2].

[1] ARNOLD V. I. Some open problems in the theory of singularities. In: The Theory of Cubature Formulae and Applications of Functional Analysis to Problems of Mathematical Physics. Editor: S. V. Uspenskiĭ. Novosibirsk: Press of the Institute of Mathematics of the Siberian Branch of the USSR Academy of Sciences, 1976, 5–15 (in Russian). (Trudy Seminara S. L. Soboleva, 1.)

[2] ARNOLD V. I. Some open problems in the theory of singularities. In: Singularities. Part 1 (Arcata, CA, 1981). Editor: P. Orlik. Providence, RI: Amer. Math. Soc., 1983, 57–69. (Proc. Symposia Pure Math., 40.)

## 1976-34

H This is a problem in paper [1] (VII, p. 45–46). The problem was posed jointly with G. Shimura.

[1] Problems of present day mathematics. Editor: F. E. Browder. In: Mathematical Developments Arising from Hilbert Problems (Northern Illinois University, 1974). Part 1. Editor: F. E. Browder. Providence, RI: Amer. Math. Soc., 1976, 35–79. (Proc. Symposia Pure Math., 28.)

R See the comment to problem *1972-27*.

## ▽ 1976-35

H This is a problem in paper [1] (XIII, p. 50).

[1] Problems of present day mathematics. Editor: F. E. Browder. In: Mathematical Developments Arising from Hilbert Problems (Northern Illinois University, 1974). Part 1. Editor: F. E. Browder. Providence, RI: Amer. Math. Soc., 1976, 35–79. (Proc. Symposia Pure Math., 28.)

## △▽ 1976-35 — *V. M. Kharlamov*

ℛ Rough general bounds can be easily obtained by means of the Morse theory. For example, the number in question is bounded by $(n-1)^m + (n-1)^{m-1} + \cdots + 1$. The only dimensions where we know a sharp bound is $m \leq 2$. In the case of plane curves the maximal number of connected components of the complement is $1 + \frac{n(n-1)}{2}$. This number of connected components is attained for $n$ lines in general position. The bound itself follows from the Harnack–Klein bound $b_1(\mathbb{R}A) \leq \frac{(d-1)(d-2)}{2} + 1$ for irreducible real plane curves of degree $d$ and the recurrent relation $b_1(\mathbb{R}A \cup \mathbb{R}B) - 1 \leq b_1(\mathbb{R}A) - 1 + b_1(\mathbb{R}B) - 1) + \deg A \deg B$. For surfaces in $\mathbb{R}\mathrm{P}^3$ the answer is unknown for degree 5 and higher (for lower degrees from 1 to 4 the answers are 1, 2, 4, 11).

In fact, this Arnold's problem is related to an erroneous theorem of Courant and Hermann on zeros of linear combinations of eigenfunctions of the Laplace operator. In the case of the Laplace operator the theorem would give the bound $1 + \binom{m}{n+m-2}$ in the Arnold problem. In dimension 2 this bound coincides with the above sharp bound. But in dimension 3 for degree 5 and higher it is no longer true, as it was shown by O. Viro (see [2]) who found examples of nonsingular surfaces with $(n^3 - 2n^2 + 4)/4$ components for each even $n$.

For nonsingular hypersurfaces it is easy to show, using the Viro patchworking method, that for $m$ fixed the number $c_n(m)$ in question grows like $cn^m$ with $c \in \mathbb{R}$: $\lim\limits_{m\to\infty} c_n(m)n^{-m} = c$. The precise value of $c$ is unknown. For $m = 3$, it lies between $\frac{13}{36}$ and $\frac{5}{12}$ (the lower bound follows from the Smith–Thom and Petrovskiĭ–Oleĭnik inequalities, see the comment to problem *1972-26*; the upper bound is due to F. Bihan [1]. For the class of arbitrary, not only nonsingular, hypersurfaces there are similar bounds, but I do not know how one can prove the fact that the corresponding sequence is equivalent to $cn^m$ for $n$ increasing to infinity.

Let me point out another related problem: if a subset of $\mathbb{R}^m$ is defined by an inequality $p \geq 0$ where $p$ is a polynomial in $m$ variables of given degree $n$, how many connected components can it have?

[1] BIHAN F. Asymptotics of Betti numbers of real algebraic surfaces. *Comment. Math. Helvetici*, 2003, **78**(2), 227–244.

[2] VIRO O. YA. Construction of multicomponent real algebraic surfaces. *Sov. Math. Dokl.*, 1979, **20**, 991–995.

### △ 1976-35

ℛ See the comment to problem *1972-26* by V. I. Arnold.

### ▽ 1976-36

ℋ This is a problem in paper [1] (XIII, p. 50).

[1] Problems of present day mathematics. Editor: F. E. Browder. In: Mathematical Developments Arising from Hilbert Problems (Northern Illinois University, 1974). Part 1. Editor: F. E. Browder. Providence, RI: Amer. Math. Soc., 1976, 35–79. (Proc. Symposia Pure Math., 28.)

### △▽ 1976-36 — *V. M. Kharlamov*

ℛ Similar to problems *1972-26*, *1976-35*, and many other problems of the classical real algebraic geometry, this problem is algorithmically solvable. However, to my knowledge, no nontrivial results in this direction were obtained via computer implementation of an algorithm.

A complete solution is known only up to degree 7, and in degree 8 there remain only a few uncertain arrangements (in degree 6 the classification was completed by D. A. Gudkov [3] (see also [4]) and in degree 7, by O. Viro [10]; for most recent published results on degree 8 and further references see [2]; for some information on the growth of the number of arrangements see [7].

Up to degree 5 the solution is quite easy. For example, all the prohibitions on arrangements of ovals of curves of degree $\leq 5$ can be deduced from the Bézout theorem. First spectacular specific properties of $M$-curves (that is, the curves with the maximal number of ovals) were noticed by Hilbert in the case of curves of degree 6 and included by him as a conjecture in the sixteenth problem of his famous list. This Hilbert's discovery was corrected and proved by D. A. Gudkov. Gudkov was probably the first who noticed that the Hilbert conjecture in question reflects some periodicity phenomenon (contrary to the more usual interpretations as sole bounds). Gudkov [5] conjectured, in turn, that for any even degree $d = 2k$ the number $p$ of even and the number $n$ of odd ovals of an $M$-curve satisfy the congruence $p-n=k^2 \pmod 8$ (an oval is *even* if it lies inside an even number of other ovals, otherwise it is called *odd*). [1]

This Gudkov's conjecture, proved in a weakened form by Arnold [1] and in its full generality by Rokhlin [9], influenced considerably the subsequent development of the subject and led to a bunch of other congruences in real algebraic geometry. In particular, the Gudkov congruence became a special case of a general Gudkov–Arnold–Rokhlin congruence, which is one of the extremal properties of the Smith–Thom (Harnack) bound; see the comments to problem *1972-26*. In the case of even dimensional hypersurfaces it states that $\chi(\mathbb{R}A) = \operatorname{sign}(\mathbb{C}A) \mod 16$ (sign is the signature) as soon as $\sum b_i(\mathbb{R}A;\mathbb{Z}/2) = \sum b_i(\mathbb{C}A;\mathbb{Z}/2)$.

For $M$-curves of even degree $d$ the inequality $p \le \frac{3}{8}d(d-2)+1$ (equivalent in this case to $n \ge \frac{1}{2}(\frac{d}{2}-1)(\frac{d}{2}-2)$) conjectured by V. Ragsdale in 1907 [8] is still neither proved nor disproved. It is curious to notice that among her conjectures this was the only one that she discussed in detail and that she found worthy of the name "theorem," while pointing out explicitly that she had no proof of it. As to the other laws observed by Ragsdale in examples, the bound $n \le \frac{3}{8}d(d-2)$ for $M$-curves was refuted by O. Viro, who constructed $M$-curves with $n = \frac{3}{8}d(d-2)+1$, and for non-$M$-curves both the bounds $p \le \frac{3}{8}d(d-2)+1$ and $n \le \frac{3}{8}d(d-2)$ were disproved by I. Itenberg. For $M$-curves the bound $n \le \frac{3}{8}d(d-2)+1$ is neither proved nor disproved. (For more information on the Ragsdale conjectures see [6].)


[1] Arnold V. I. Distribution of ovals of real plane algebraic curves, involutions of four-dimensional smooth manifolds, and the arithmetic of integral quadratic forms. *Funct. Anal. Appl.*, 1971, **5**(3), 169–176. [*The Russian original is reprinted in*: Vladimir Igorevich Arnold. Selecta–60. Moscow: PHASIS, 1997, 175–187.]

[2] Chevallier B. Four $M$-curves of degree 8. *Funct. Anal. Appl.*, 2002, **36**(1), 76–78.

[3] Gudkov D. A. Complete topological classification of the disposition of ovals of a sixth order curve in the projective plane. *Uchen. Zap. Gor'kov. Univ.*, 1969, **87**, 118–153 (in Russian).

[4] Gudkov D. A. Position of the circuits of a curve of sixth order. *Sov. Math. Dokl.*, 1969, **10**, 332–335.

[5] Gudkov D. A. Construction of a new series of $M$-curves. *Sov. Math. Dokl.*, 1971, **12**, 1559–1563.


[1] Gudkov verified this conjecture for many particular cases (proving both the existence of some confirming curves and the nonexistence of curves forbidden by the congruence violation) in his thesis. But he did not formulate the general conjecture, which was first formulated (and called Gudkov conjecture) in my talk at the defence of Gudkov's thesis where I was an opponent. — *V. I. Arnold.*

[6] ITENBERG I. V., VIRO O. YA. Patchworking algebraic curves disproves the Ragsdale conjecture. *Math. Intelligencer*, 1996, **18**(4), 19–28. [*Internet*: `http://www.math.uu.se/~oleg/preprints.html`]

[7] KHARLAMOV V. M., OREVKOV S. YU. Growth order of the number of classes of real plane algebraic curves as the degree grows. *Zap. Nauch. Semin. St. Peterburg. Otdel. Mat. Inst. Steklova*, 2000, **266**, 218–233 (in Russian). (Theory of representations of dynamical systems. Combinatorial and algorithmic methods, 5.)

[8] RAGSDALE V. On the arrangement of the real branches of plane algebraic curves. *Amer. J. Math.*, 1906, **28**, 377–404.

[9] ROKHLIN V. A. Congruences modulo 16 in Hilbert's sixteenth problem. *Funct. Anal. Appl.*, 1972, **6**(4), 301–306.

[10] VIRO O. YA. Curves of degree 7, curves of degree 8, and the Ragsdale conjecture. *Sov. Math. Dokl.*, 1980, **22**, 566–570.


## △ 1976-36

See the comment to problem *1972-26* by V. I. Arnold.

## ▽ 1976-37

This is a problem in papers [1] (XIII, p. 51) and [2] (§ 5).


[1] Problems of present day mathematics. Editor: F. E. Browder. In: Mathematical Developments Arising from Hilbert Problems (Northern Illinois University, 1974). Part 1. Editor: F. E. Browder. Providence, RI: Amer. Math. Soc., 1976, 35–79. (Proc. Symposia Pure Math., 28.)

[2] ARNOLD V. I. Some problems in the theory of differential equations. In: Unsolved Problems of Mechanics and Applied Mathematics. Moscow: Moscow University Press, 1977, 3–9 (in Russian).


## △ 1976-37 — *S. Yu. Yakovenko*

The question is apparently motivated by Bautin's famous result [1] asserting that in a quadratic perturbation of the linear center (the Hamiltonian linear vector field corresponding to the Hamiltonian $H(x,y) = x^2 + y^2$) not more than 3 limit cycles can be born. The original proof was obtained by somewhat mysterious calculations. Simplified proofs were obtained in [3] and [4].

It was long believed that this result implies that quadratic vector fields cannot have more than 3 limit cycles. In 1980 Shi Song Ling [2] constructed a counterexample with 4 limit cycles by explicitly perturbing a quadratic system with an ultra-ultra-weak focus at the origin (generating three small limit cycles in the perturbation) and one more "large" limit cycle far away.

[1] Bautin N. N. On the number of limit cycles which appear with the variation of coefficients from an equilibrium position of focus or center type. *AMS Transl.*, 1954, **100**, 19 pp.

[2] Ling S. S. A concrete example of the existence of four limit cycles for plane quadratic systems. *Sci. Sinica*, 1980, **23**(2), 153–158.

[3] Yakovenko S. A geometric proof of the Bautin theorem. In: Concerning the Hilbert 16th Problem. Editors: Yu. Il′yashenko and S. Yakovenko. Providence, RI: Amer. Math. Soc., 1995, 203–219. (AMS Transl., Ser. 2, 165; Adv. Math. Sci., 23.)

[4] Żołądek H. Quadratic systems with center and their perturbations. *J. Differ. Equations*, 1994, **109**(2), 223–273.

## 1976-38

This is a problem in paper [1] (XVI(D), p. 58).

[1] Problems of present day mathematics. Editor: F. E. Browder. In: Mathematical Developments Arising from Hilbert Problems (Northern Illinois University, 1974). Part 1. Editor: F. E. Browder. Providence, RI: Amer. Math. Soc., 1976, 35–79. (Proc. Symposia Pure Math., 28.)

## 1976-39

This is a problem in papers [1] (XX, p. 66) and [2] (§ 6). The conjectures on the number of fixed points of symplectomorphisms were first formulated by V. I. Arnold in paper [3a] (see also [3b]), see problems *1965-1 – 1965-3*.

[1] Problems of present day mathematics. Editor: F. E. Browder. In: Mathematical Developments Arising from Hilbert Problems (Northern Illinois University, 1974). Part 1. Editor: F. E. Browder. Providence, RI: Amer. Math. Soc., 1976, 35–79. (Proc. Symposia Pure Math., 28.)

[2] Arnold V. I. Some problems in the theory of differential equations. In: Unsolved Problems of Mechanics and Applied Mathematics. Moscow: Moscow University Press, 1977, 3–9 (in Russian).

[3a] Arnold V. I. Sur une propriété topologique des applications globalement canoniques de la mécanique classique. *C. R. Acad. Sci. Paris*, 1965, **261**(19), 3719–3722.

*The Russian translation in*:

[3b] Vladimir Igorevich Arnold. Selecta–60. Moscow: PHASIS, 1997, 81–86.

ℛ See the comment to problem *1972-33*.

### 1976-40

ℋ This is a problem in paper [1] (XXI(E), p. 67–68).

[1] Problems of present day mathematics. Editor: F. E. Browder. In: Mathematical Developments Arising from Hilbert Problems (Northern Illinois University, 1974). Part 1. Editor: F. E. Browder. Providence, RI: Amer. Math. Soc., 1976, 35–79. (Proc. Symposia Pure Math., 28.)

### 1976-41

ℋ This is a problem in paper [1] (§ 1).

[1] ARNOLD V. I. Some problems in the theory of differential equations. In: Unsolved Problems of Mechanics and Applied Mathematics. Moscow: Moscow University Press, 1977, 3–9 (in Russian).

### 1976-42

ℋ This is a problem in paper [1] (§ 2).

[1] ARNOLD V. I. Some problems in the theory of differential equations. In: Unsolved Problems of Mechanics and Applied Mathematics. Moscow: Moscow University Press, 1977, 3–9 (in Russian).

## 1977

### 1977-3 — *V. V. Goryunov*

ℛ Both uni- and bimodal boundary function singularities were classified by V. I. Matov [1,2].

[1] MATOV V. I. Singularities of the maximum function on a manifold with boundary. *Trudy Semin. Petrovskogo*, 1981, **6**, 195–222 (in Russian). [*The English translation*: *J. Sov. Math.*, 1986, **33**, 1103–1127.]

[2] MATOV V. I. Unimodal and bimodal germs of functions on a manifold with boundary. *Trudy Semin. Petrovskogo*, 1981, **7**, 174–189 (in Russian). [*The English translation*: *J. Sov. Math.*, 1985, **31**, 3193–3205.]

## 1977-4 – *S. M. Gusein-Zade*

ℛ This was done in paper [1].

[1] LYASHKO O. V. Classification of critical points of functions on a manifold with singular boundary. *Funct. Anal. Appl.*, 1983, **17**(3), 187–193.

## 1977-7 – *S. M. Gusein-Zade*

ℛ This was (partially) done in [1, 2]. Related results for indices of vector fields can be found in a number of papers by J. A. Seade, X. Gómez-Mont, A. Verjovsky, P. Mardešić, and others.

[1] EBELING W., GUSEIN-ZADE S. M. On the index of a holomorphic 1-form on an isolated complete intersection singularity. *Dokl. Math*, 2001, **64**(2), 221–224.

[2] EBELING W., GUSEIN-ZADE S. M. Indices of 1-forms on an isolated complete intersection singularity.
[*Internet*: `http://www.arXiv.org/abs/math.AG/0105242`]

## ▽ 1977-8 – *V. A. Vassiliev*

ℛ Consider a complicated function singularity $f : (\mathbb{R}^n, 0) \to (\mathbb{R}, 0)$ and two of its non-discriminant Morsifications $f_a$, $f_b$. How to prove rapidly that they belong to different components of the complement of the discriminant? An obvious invariant of this sort is the homotopy type of the local set of negative values

$$(f_a^{-1}(-\delta, 0) \cap B_\varepsilon)/(f_a^{-1}(-\delta) \cap B_\varepsilon)$$

or, more effectively, its homology group, which can be calculated via the Morse complex of critical points of $f_a$ with negative critical values. However, a more refined invariant probably can be defined in the terms of the Whitehead torsion of this complex. A similar invariant (calculated over all critical points) probably can distinguish different real forms of one and the same complex singularity.

△ **1977-8**

ℛ See the comment to problem *1973-24*.

**1977-9 — *V. I. Arnold* (1977)**

ℛ A set of weights $\{A_s, D_j\}$, $1 \le s \le m$, $1 \le j \le n$, is *nondegenerate* if there exists a quasihomogeneous mapping $f : \mathbb{C}^m \to \mathbb{C}^n$ with $\mu(f) < \infty$, weights $A_s$ in the domain, and weights $D_j$ in the range. Here $\mu = \dim_{\mathbb{C}} M/I$ is the dimension of the versal deformation base: $M$ is the free module of column vectors $\alpha = \sum_j \alpha_j \frac{\partial}{\partial y_j}$, $M \approx A^n, A = \mathbb{C}[[x_1, \ldots, x_s, \ldots, x_m]], I \subset M$ is the submodule with generators $f_i\, \partial/\partial y_j$ and $\partial f/\partial x_s$. I am unaware of whether $\mu < \infty$ is the same for all $f$ with $\mu < \infty$ in the case of $\mathbb{C}^3 \to \mathbb{C}^2$, but for $\mathbb{C}^2 \to \mathbb{C}^2$ there is a counterexample ($A_1 = A_2 = 1, D_1 = 2$, $D_2 = 3$: $\mu(x^2, y^3) = 7$, $\mu(xy, x^3 + y^3) = 6$). Therefore, a mapping $f_0$ is called *fully nondegenerate* if $\mu(f_0) = \min$ on all $f$ with given weights $\{A_s, D_j\}$. Conjecturally, in the case of $\mathbb{C}^3 \to \mathbb{C}^2$ this minimum is equal to

$$\mu_{\text{alg}} := 1 - \frac{\prod D}{\prod A}\left(\sum A - \sum D\right) \qquad \left(\text{for } \mathbb{C}^2 \to \mathbb{C}^2 \text{ — } \frac{\prod D}{\prod A}\right).$$

For $\mathbb{C}^2 \to \mathbb{C}^2$, the equality $\mu_{\min}(A, D) = \mu_{\text{alg}}(A, D)$ is disproved by the following example:

$$\begin{aligned} A_1 &= 5, \quad A_2 = 2, \quad f_1 = xy^2, \\ D_1 &= 9, \quad D_2 = 10, \quad f_2 = x^2 + y^5, \end{aligned}$$

$\mu_{\min} = 10$, $\mu_{\text{alg}} = \frac{\prod D}{\prod A} = 9$.

Conjecturally $\mu_{\min} \ge \mu_{\text{alg}}$ for all $m \ge n$ as well.

**1977-10 — *V. V. Goryunov***

ℋ These Lyashko conjectures concerning isolated complete intersection singularities were published in [3].

ℛ They were proved by A. G. Aleksandrov in [1, 2].

[1] Aleksandrov A. G. The de Rham complex of a quasihomogeneous complete intersection. *Funct. Anal. Appl.*, 1983, **17**(1), 48–49.

[2] Aleksandrov A. G. Cohomology of a quasihomogeneous complete intersection. *Math. USSR, Izv.*, 1986, **26**, 437–477.

[3] Arnold V. I. On some problems in singularity theory. In: Geometry and Analysis. Papers dedicated to the memory of V. K. Patodi. Bangalore: Indian Acad. Sci., 1980, 1–9. [*Reprinted in*: *Proc. Indian Acad. Sci. Math. Sci.*, 1981, **90**(1), 1–9.]

## 1977-12 — *A. I. Neĭshtadt*

ℛ The equation under consideration is the truncated and rescaled normal form for the following stability loss problem. A periodic solution $x = x(t)$ of a system of differential equations loses its stability when a pair of complex conjugate multipliers of this periodic solution crosses the unit circle near the points $\pm i$ (and hence there is an approximate resonance $1:4$ between the period of $x(t)$ and the period of oscillations in the system linearized near $x(t)$). The complex variable $z$ characterizes the deviation from the periodic solution, and the complex parameter $\varepsilon$ characterizes the deviation of the multiplier from the point $i$ (for $\varepsilon \neq 0$ it is possible to make $|\varepsilon| = 1$ by rescaling). After [1, 2] the problem was studied, in particular, in [4–7, 10–13] and the summary of results is contained in [3, 8, 9].

It was shown numerically that the plane of the complex parameter $A$ is divided by piecewise smooth curves (called bifurcation curves) into 48 regions. Each of these regions corresponds to a sequence of bifurcations of the phase portrait of the equation considered; these bifurcations occur when the parameter $\varepsilon$ circumscribes the unit circle (there are 12 essentially different regions, the others can be obtained by symmetries). Many of the bifurcation curves and bifurcation sequencies are obtained analytically; however, the complete analytic theory is lacking. One of the unsolved problems is the study of bifurcations for values of the parameter $A$ close to $i$ (at the point $i$, several bifurcation curves meet).


[1] ARNOLD V. I. Loss of stability of self-oscillations close to resonance and versal deformations of equivariant vector fields. *Funct. Anal. Appl.*, 1977, **11**(2), 85–92.

[2] ARNOLD V. I. Geometrical Methods in the Theory of Ordinary Differential Equations, 2nd edition. New York: Springer, 1988. (Grundlehren der Mathematischen Wissenschaften, 250.) [*The Russian original* 1978.]

[3] ARNOLD V. I., AFRAĬMOVICH V. S., IL'YASHENKO YU. S., SHIL'NIKOV L. P. Bifurcation theory. In: Itogi Nauki i Tekhniki VINITI. Current Problems in Mathematics, Vol. 5. Moscow: VINITI, 1986, 5–218 (in Russian). [*The English translation*: Bifurcation Theory and Catastrophe Theory. Editor: V. I. Arnold. Berlin: Springer, 1994, 1–205. (Encyclopædia Math. Sci., 5; Dynamical Systems, V.)]

[4] BEREZOVSKAYA F. S., KHIBNIK A. I. On the problem of bifurcations of auto-oscillations near a 1 : 4 resonance. Preprint, Research Computing Center of the USSR Academy of Sciences, Pushchino, 1979 (in Russian).

[5] BEREZOVSKAYA F. S., KHIBNIK A. I. On bifurcations of separatrices in the problem of loss of stability to auto-oscillations near a 1 : 4 resonance. *J. Appl. Math. Mech.*, 1981, **44**, 663–667.

[6] CHENG CH. -Q. Hopf bifurcations in nonautonomous systems at points of resonance. *Science in China, Ser. A*, 1990, **33**(2), 206–219.

[7] CHENG CH.-Q., SUN Y.-S. Metamorphoses of phase portraits of vector field in the case of symmetry of order 4. *J. Differ. Equations*, 1992, **95**(1), 130–139.

[8] CHOW S.-N., LI C., WANG D. Normal Forms and Bifurcation of Planar Vector Fields. Cambridge: Cambridge University Press, 1994.

[9] KRAUSKOPF B. Bifurcation sequences at 1 : 4 resonance: an inventory. *Nonlinearity*, 1994, **7**(3), 1073–1091.

[10] NEĬSHTADT A. I. Bifurcations of the phase portrait of a system of equations arising from the problem of loss stability of an auto-oscillations close to a 1 : 4 resonance. *J. Appl. Math. Mech.*, 1979, **42**, 896–907.

[11] WAN Y.-H. Bifurcation into invariant tori at points of resonance. *Arch. Rat. Mech. Anal.*, 1978, **68**(4), 343–357.

[12] WANG D. Hopf bifurcation at the nonzero foci in 1 : 4 resonance. *Acta Math. Sinica (N. S.)*, 1990, **6**(1), 10–17.

[13] ZEGELING A. Equivariant unfoldings in the case of symmetry of order 4. *Serdica Math. J.*, 1993, **19**(1), 71–79.

# 1978

## 1978-1

See the comment to problem *1972-3*.

## 1978-2 — *A. A. Davydov*

The problem has been solved completely for generic control systems on a closed surface. In that case the closure of the attainable set (= positive orbit) of a generic point (a generic initial submanifold) is a manifold with boundary having only standard singularities from the finite list up to a diffeomorphism [1,3]. Moreover, this set is asymptotically stable, and, for an orientable phase space, a generic system is structurally stable in the classical Andronov–Pontryagin sense (= for a generic system and any one being sufficiently close to it, the orbits of points for one of the systems can be carried to the orbits of points of the other by homeomorphism of the phase space which is close to identity) [3].

For a generic multidimensional control system the boundary of attainable set is a hypersurface satisfying Hölder condition [2]. The respective classification of singularities is unknown even for a three-dimensional control system with quadratically convex indicatrix. For a generic control system with quadratically convex indicatrices (for example, for simple motion with the drift $(\dot{x}-v(x))^2 \le 1$, $x \in \mathbb{R}^n$) this classification has to include all generic singularities of the boundary of the domain of the relative minimum under the equality constraints (for low dimensions the singularities of such a minimum were investigated in [4]).

[1] ARNOLD V. I. Catastrophe theory. Berlin: Springer, 1992. [*The Russian original* 1990.]

[2] DAVYDOV A. A. The quasi-Hölder nature of the boundary of attainability. *Trudy Semin. Vekt. Tenz. Anal.*, 1985, **22**, 25–30 (in Russian). [*The English translation*: *Selecta Math. Sov.*, 1990, **9**(3), 229–234.]

[3] DAVYDOV A. A. Qualitative Theory of Control Systems. Providence, RI: Amer. Math. Soc., 1994. (Transl. Math. Monographs, 141.)

[4] DAVYDOV A. A., ZAKALYUKIN V. M. The coincidence of generic singularities of solutions of extremal problems with constraints. In: Proceedings of the International Conference Dedicated to the 90th Birthday of L. S. Pontryagin (Moscow, 1998), Vol. 3: Geometic Control Theory. Itogi Nauki i Tekhniki VINITI. Contemporary Mathematics and its Applications. Thematic Surveys, Vol. 64. Moscow: VINITI, 1999, 118–143 (in Russian).

## 1978-3 – *M. B. Sevryuk*

ℛ For typical integrable Hamiltonian systems with no more than three degrees of freedom, the steepness indices were calculated by E. E. Landis (see [1]). In [1], the following definitions of the local steepness indices were used.

**Definition 1.** A number $k(f)$ is called a *steepness index* of a function $f:(\mathbb{R}^n,0)\to(\mathbb{R},0)$ if there exist constants $C>0$ and $\delta>0$ such that for all $\varepsilon$, $0<\varepsilon<\delta$, there is a sphere $S^{n-1}_\rho := \{|\mathbf{x}| = \rho < \varepsilon\}$ in the ball $\{|\mathbf{x}|<\varepsilon\}$ such that $|\operatorname{grad} f(\mathbf{x})| > C\varepsilon^{k(f)}$ for all $\mathbf{x}\in S^{n-1}_\rho$.

**Definition 2.** A number $K(F)$ is called a *uniform steepness index* of a family $F(\mathbf{x},\lambda)$, $\mathbf{x}\in\mathbb{R}^n$, $\lambda\in U\subset\mathbb{R}^m$ if, for all $\lambda\in U$, the number $K(F)$ is a steepness index of the function $F(\cdot,\lambda)$ with the same $C$ and $\delta$.

**Theorem** [1]. *Given a family $F(\mathbf{x},\lambda)$, let the origin be a simple critical point of the function $f(\mathbf{x})=F(\mathbf{x},0)$. Then, of all the possible indices $k(f)$ and $K(F)$, the* *minimal ones for $\lambda$ small enough are listed in the following table* (*therein,* $f_0$ *is the stable normal form of a function* $f$):

| *the type of f* | $A_\mu$ | $D_\mu$ | $E_6$ | $E_7$ | $E_8$ |
|---|---|---|---|---|---|
| $f_0$ | $x^{\mu+1}$ | $x^{\mu-1}+xy^2$ | $x^3+y^4$ | $x^3+xy^3$ | $x^3+y^5$ |
| $k(f)$ | $\mu$ | $\mu-2$ | 3 | 3.5 | 4 |
| $K(F)$ | $\mu$ | $\mu-1$ | ? | ? | ? |

**Conjecture** [1]. *For* $E_\mu$, *the question marks are to be replaced with* $\mu-2$ ($\mu=$ 6, 7, 8).

A brief survey of the Nekhoroshev theory is presented in the comment to problem *1966-2.*

[1] LANDIS E. E. Uniform steepness indices. *Uspekhi Mat. Nauk*, 1986, **41**(4), 179 (in Russian).

## 1978-5 — *S. M. Gusein-Zade*

This means a generalization of a number of observations like the following one. Let $R^\mu$ be the base space of the real versal deformation of the singularity $A_\mu$, and let $C$ be the set of points of $R^\mu$ for which all the critical points of the corresponding function are real. Then $C$ is a very tiny pyramid with the vertex at the origin. The ratio of its volume in the $\varepsilon$-neighborhood of the origin to the whole neighborhood volume tends to zero as $\varepsilon \to 0$.

## 1978-6 — *S. Yu. Yakovenko*

Also: *1976-2, 1979-16, 1980-1, 1983-11, 1989-17, 1990-24, 1990-25, 1994-51, 1994-52*

The problem on zeros of Poincaré integrals, known also as the infinitesimal Hilbert 16th problem, is one of the most recurring in Arnold's lists. It was published in [4], reappeared in the list [5] and rather recently the expanded formulation was again given in [3]. Two very closely related problems, *1979-26* and *1980-3*, which could well be included in the list, are singled out because they are essentially solved. Some related questions are discussed in problems *1984-10, 1985-12.*

**Origins and preliminary remarks.** The problems on zeros of the Poincaré integral

$$I = I(h; H, \omega) = \oint_{\gamma_h} M^{-1}\omega,$$
$$\gamma_h \subseteq \{H = h\}, \qquad \omega = P(x,y)\,dx + Q(x,y)\,dy, \tag{1}$$

for the polynomial perturbation

$$M\,dH + \varepsilon\omega = 0, \tag{2}$$

in particular, *complete Abelian integrals* corresponding to $M = 1$ and a polynomial $H$, appeared as an attempt to find an amenable relaxation of the Hilbert problem on limit cycles.

As a function of $h$, $I(h)$ is the first variation of the Poincaré return map with respect to the small parameter $\varepsilon$, at $\varepsilon = 0$. Thus the problem on zeros of integrals of the form (1) becomes a localized (better to say, linearized or *infinitesimal*) version of the Hilbert 16th problem on the number of limit cycles of planar polynomial vector fields, for systems infinitesimally close to integrable ones.

Probably, the question was also inspired by the works by I. G. Petrovskiĭ and E. M. Landis [33–35] who tried to reduce the Hilbert 16th problem stated in full generality, to perturbations of integrable systems.

It should be stressed that vanishing of the Poincaré integral is only a necessary condition for appearance of limit cycles, and it works only for limit cycles born out of *nonsingular* level curves of the first integral. Description of limit cycles born from *separatrix polygons* (carrying singular points of the non-perturbed vector field) is a considerably more delicate subject, which admits a satisfactory solution only in the simplest case of a separatrix loop carrying one nondegenerate saddle (R. Roussarie [37, 38]).

Further, identical vanishing of the Poincaré integral (1) does not mean in general that the family (2) consists of integrable systems only: higher variations in $\varepsilon$ may still be nonzero and it is their zeros that will determine the number and location of limit cycles born in the perturbation. However, for Abelian integrals this is impossible: in [14] Yu. Il′yashenko proved that, for a sufficiently generic polynomial Hamiltonian $H$, the integral of a form of degree $\deg\omega = \max(\deg P, \deg Q) + 1$ no greater than $\deg H$ vanishes identically if and only if $\omega$ itself is exact on $\mathbb{R}^2$. Clearly, in this case the system is Hamiltonian for all $\varepsilon$. This result, generalized by L. Gavrilov [10] for higher degree forms and by I. Pushkar′ [36] for higher dimensions, provides an effective criterion for nontriviality of the perturbation (2). Everywhere below only *isolated* zeros of the Abelian integrals are counted.

From the very beginning it should be said that *general* results for perturbations of conservative non-Hamiltonian systems are practically absent, with few exceptions concerning perturbed Lotka–Volterra systems. Therefore we will mostly discuss the problem on zeros of Abelian integrals with $H \in \mathbb{R}[x,y]$ and $M \equiv 1$.

**Brief history.** The first nontrivial case (for $H$ quadratic the Abelian integrals are rational functions of $h$) corresponds to cubic Hamiltonians. R. Bogdanov studied the complete *elliptic integral*

$$I(h) = \oint_{\{H=h\}} (a+bx)y\,dx, \qquad H(x,y) = \tfrac{1}{2}y^2 + \tfrac{1}{3}x^3 - x \tag{3}$$

and proved that it has at most one real isolated zero. His results, announced with no details of the proof in [1], were published with a proof in [6], and later were rediscovered by F. Takens [39]. This problem appeared in connection with construction of the versal deformation of what is known today as the cuspidal singularity of Bogdanov–Takens [7]. Later, in [15] Il′yashenko suggested another proof of the same result, based on the complexification of the Abelian integral as a function of $t \in \mathbb{C}$ ramified over the collection of critical values of the complexified Hamiltonian $H(x,y) \in \mathbb{C}[x,y]$. Since then, complexification became a primary tool in the investigation of complete Abelian integrals.

Shortly after that, a number of different particular cases of elliptic integrals were studied, but the major breakthrough occurred in the works by G. Petrov. He proved that for the standard elliptic Hamiltonian as in (3), integrals of *all polynomial forms* of arbitrarily high degree form a *nonoscillating*, or *Chebyshev*, family: the maximal number of real isolated zeros is by one less than the dimension of this family considered as a vector space over $\mathbb{R}$ [31]. Later Petrov proved that the same non-oscillatory property holds also for *complex* isolated zeros counted in a slit plane [32]. These proofs rely substantially on the fact that the elliptic integrals $\oint y\,dx$ and $\oint xy\,dx$ satisfy an explicitly written system of *Picard–Fuchs* linear ordinary differential equations with rational coefficients, so that their ratio satisfies a Riccati equation. On the other hand, these two integrals generate the space of all Abelian integrals over the ring of polynomial functions of $h$. The results of Petrov settle the particular question raised in problem *1979-16* and give an affirmative answer to problem *1983-11* in the part related to the elliptic integrals.

Earlier, simultaneously and independently, A. Khovanskiĭ [18] and A. Varchenko [40] proved the general finiteness result: for any combination of degrees $n$ and $d$, the number of isolated zeros of all Abelian integrals of forms of degree $\leq d$ over the level curves of Hamiltonians of degree $\leq n$ is uniformly bounded by a constant $C(n,d)$ depending only on $n$ and $d$. Their proofs, however, gave no idea of how to estimate the constant $C(n,d)$: its mere existence is ultimately derived from compactness arguments.

This result remains until nowadays the only general assertion valid for all Hamiltonians and all forms without restriction. Since it was achieved, the emphasis moved to *computability* of the bounds.

**Digression: fewnomials theory and Pfaffian manifolds.** The proof of the Khovanskiĭ–Varchenko theorem is based on a beautiful geometric theory of Pfaffian manifolds, developed by Askold Khovanskiĭ. The central idea behind this theory can be described roughly as follows: a real affine variety, defined by a mixture of algebraic and Pfaffian equations, shares many properties of real algebraic varieties provided that it "looks like an algebraic variety" topologically. A simple example is that of integral trajectories of planar polynomial vector fields. If these trajectories are not spirals (they should subdivide the real plane into two parts, in particular, being limit cycles), then the number of isolated intersections of these trajectories, say with straight lines, is explicitly bounded in terms of the degree of the planar vector field. This observation immediately allows one to solve problem *1976-2*.

The constructions in the Pfaffian manifolds theory, especially the *Pfaffian elimination*, are explicit and efficient. Geometrically they could be described as a multidimensional generalization of the Rolle theorem on alternation between roots of a smooth function of one real variable, and roots of its derivative.

One of the most spectacular achievements of this theory is an upper bound for the number of isolated solutions of a system of algebraic equations, given *not* in terms of the degrees of this equation as in the Bézout theorem, but rather through the number of different *monomial terms* occurring in the equations, uniformly over all degrees. This explains the alternative code name "*fewnomials theory*" used to designate the entire toolkit. A typical fewnomials theory result is described in problem *1979-22*.

Applications of the Pfaffian manifolds theory can sometimes be very unexpected. Thus, if the resonant Poincaré–Dulac formal normal form [2] for all singular saddle points of an analytic planar vector field is convergent, then any polycycle carrying only these points cannot accumulate near itself an infinite number of limit cycles of this field. This particular case of the finiteness theorem (see the comment to problem *1981-16*) was discovered by R. Moussu and C. Roche in [22]. Their key argument is integrability of the resonant normal form which in turn implies the fact that the Poincaré map can be described by a mixture of Pfaffian and analytic equations.

This theory, together with its numerous ramifications, is expounded in book [19]. The revised Russian edition [20] contains new applications to Hardy fields, complexity problems, Tarski problem, etc.

**Recent achievements: low degree cases.** Despite their diversity, recent results related to the infinitesimal Hilbert 16th problem can be organized into several clusters.

The most abundant group of results deals with particular cubic or quartic Hamiltonians and special choices of low degree (usually the same) perturbation forms. If the number of essential parameters is small enough, sometimes bifurcation diagrams of zeros can be constructed. Usually problems of this type appear in connection with bifurcations of limit cycles in families of vector fields exhibiting certain resonances. Though it is impossible to mention all results, probably the most spectacular single recent achievement in this direction is due to L. Gavrilov [11], see problem *1979-26.* Gavrilov proved that for a real cubic Hamiltonian with 4 distinct (complex) critical values, the number of zeros of any integral of a quadratic 1-form can be at most 2.

The advantage of cubic Hamiltonians is that their level curves are elliptic, thus the corresponding integrals can be in some sense reduced to elliptic integrals. The Picard–Fuchs system satisfied by these integrals, admits as a factor the 2-dimensional linear system reducible to a Riccati equation similar to that from [31]. Zeros of functions obtained as rational combinations of solutions of a Riccati equation, can be produced using the "fewnomials" technique introduced by Khovanskiĭ [19]. This idea after an appropriate (rather sophisticated) elaboration was used to prove that, for any cubic Hamiltonian and any polynomial form of degree $\deg\omega \le d$, the number of isolated zeros can be at most $5d+10$ (Horozov and Iliev, see [12]).

In the same paper it is shown that a generic cubic Hamiltonian admits a quartic 1-form $\omega$ yielding 5 isolated zeros to the integral (2). This gives a generally negative answer to the question raised in problem *1983-11*, whether Abelian integrals are always non-oscillating (as was the case in the standard elliptic case). Yet the conjecture from problem *1990-25* about non-oscillation of hyperelliptic integrals, remains open.

**Note added in proof.** In February 2002 Chengzhi Li and Zenghua Zhang showed that the genericity condition appearing in the Gavrilov theorem [11] is in fact obsolete. For more details see problem *1979-26.*

**Asymptotic bounds.** The roles played by the Hamiltonian $H$ and the polynomial 1-form $\omega$ are clearly unequal. Ignoring the origins of the infinitesimal Hilbert problem, one may further relax it by fixing the Hamiltonian and investigating how the bound on the number of zeros may depend on the form. This suggestion is tacitly made in the formulations of problems *1994-51* and *1994-52.*

First results in this direction were obtained by Yu. Il′yashenko, D. Novikov and S. Yakovenko. Assuming that the Hamiltonian is generic, they proved in [16, 25, 26] that, as $\deg \omega = d \to \infty$, the number of isolated zeros may grow at most as $O(\exp cd)$, where $c = c(H)$ is a constant depending only on $H$. The demonstration leaves a theoretical opportunity to compute $c(H)$ in terms of the monodromy group of $H$ and a geometry of its critical values, but the result of the computation must necessarily explode as some of the critical values of $H$ approach each other. The key idea behind the proof is to exploit the irreducibility of the monodromy group of the Picard–Fuchs equation in the complex domain.

An asymptotically accurate answer was obtained by Petrov and Khovanskiĭ in 1996. They proved that the number of isolated zeros can grow at most as $K_1(n)\,d + K_0(H)$, where $K_1(n)$ is an explicit constant depending only on the degree $n = \deg H$ while $K_0(H)$ is independent of $\omega$ but depends on $H$. Apparently, one can prove that this constant is uniformly bounded over all Hamiltonians of degree $n$ by some $K_0(n)$, but the bound $K_0(n)$ is absolutely non-efficacious exactly as the Varchenko–Khovanskiĭ bound $C(n,d)$ mentioned above. Though the proof is not yet formally published, some of its ingredients were already incorporated in other constructions [28, 41].

This result to a certain extent answers the question as it is formulated in problem *1994-52.* Though the constant $K_1(n)$ is greater than 1, still the relative excess of this upper estimate over the lower estimate guaranteed by the dimensionality arguments, is bounded uniformly over all forms of all degrees (for fixed $\deg H$), thus partially corroborating the conjecture that appeared in the earlier problem *1983-11.*

**Algorithmically constructive bounds.** The fewnomials theory applies to functions defined by planar polynomial differential equations, such as the Riccati equation mentioned above, describing their zeros in terms of the *degrees* of the defining equations.

There is no such "fewnomials theory" for polynomial vector fields in $\mathbb{R}^n$ or $\mathbb{C}^n$ with $n > 2$ [24]. However, one may compute an explicit upper bound on the number of isolated intersections between integral trajectories of a polynomial vector field and an arbitrary algebraic hypersurface in the $n$-space, not solving the equations. The answer depends (polynomially) on the *magnitude* of coefficients of the vector field, as well as on its degree and dimension (as a tower function, i. e., an iterated exponent). This result (the "*meandering theorem*"), obtained by Novikov and Yakovenko [27,29], can be applied to Picard–Fuchs systems of linear ordinary differential equations with rational coefficients, satisfied by Abelian integrals.

A precondition for such application is an explicit knowledge of the magnitude of the coefficients of the system. An explicit derivation of Picard–Fuchs equations allowing to bound their coefficients, was achieved in [30], see also [23].

The construction in the hyperelliptic case has an especially transparent form. Application of the meandering theorem in this case yields an *explicitly computable* upper bound in the form of a tower function (iterated exponent) of $n$, on the number of zeros of hyperelliptic integrals, under the additional technical assumption that all critical values of the potential are real (Novikov and Yakovenko [28]).

Actually, the result on zeros of hyperelliptic integrals is obtained as a particular case of the following general principle. A collection of (analytic multivalued) functions $f_1(t), \dots, f_n(t)$ on the Riemann sphere, satisfying a Fuchsian system of linear equations, behaves algebraically if the monodromy group of this system possesses certain spectral properties. The quasialgebraicity property mentioned above means that the question on the number of (complex isolated) zeros of any function $f$ from the differential Picard–Vessiot extension field $\mathbb{C}(f_1, \dots, f_n)$ can be explicitly answered in terms of the complexity of $f$ in this field. See [41] for the exact formulations and discussion.

**Restricted problems.** Various approaches to obtaining asymptotic or algorithmic bounds on the number of zeros of Abelian integrals are based on different properties of Abelian integrals (usually in the complex domain). For instance, the exponential asymptotic bounds from [16] are based on the irreducibility of the monodromy group of the Abelian integrals, whereas the key results from [41] are valid for any complex analytic functions satisfying Fuchsian systems of differential equations with bounded residue matrices.

These methods, though not giving a complete answer for the problem in full generality, sometimes allow for explicit upper bounds for almost all Hamiltonians, except for a proper semialgebraic subset of zero measure. As a rule, the *estimates* explode to infinity when approaching this exceptional "bad" subset, while the number of zeros remains in fact bounded by the Varchenko–Khovanskiĭ theorem. Yet the explicit nature of the estimates for a "large" portion of Hamiltonians is of obvious interest. Following Yu. Il′yashenko, we call such problems *restricted* versions of the infinitesimal Hilbert problem. Expanding the meaning of the "restrictedness," one can include in this class also the problem of majorizing the number of isolated zeros of Abelian integrals in some specific domains (e. g., on a specified distance from the set of critical values of $H$).

In this restricted sense the infinitesimal Hilbert problem is in principle solved in [30]: for any $H$ with a properly normalized principal homogeneous part and any $\varepsilon > 0$, one can place an explicit upper bound for the number of isolated zeros of all Abelian integrals, at least $\varepsilon$-distant from the critical values of $H$ (the bound depends on $H$ and $\varepsilon$). Moreover, for all Hamiltonians with the principal homogeneous part normalized as above, and pairwise distant critical values, the number of all isolated zeros of all integrals can be bounded uniformly in terms of $n$, $d$ and the (inverse) minimal distance between the critical values. The bounds are given by tower functions of height 4.

Very recently A. Glutsyuk and Yu. Il'yashenko achieved considerable progress towards solving the restricted infinitesimal problem for the particular class of Hamiltonians of the form $H(x,y) = p(x) + q(y)$ with two monic polynomials of the same degree $\deg p = \deg q = n+1$. Using different ideas partly stemming from [16, 17], in [13] they obtained an explicit upper bound for the number of isolated zeros, growing as $\exp(2435n^4)$, provided that all $n^2$ critical points of $H$ are in the disk of radius 2 but at least $(1/n^2)$-distant from each other.

**Non-Hamiltonian case.** As was already remarked, the case of general Poincaré integrals with nontrivial integrating factors is much more complicate. To begin with, merely a classification of integrable polynomial systems is very complicated. While all center conditions in the quadratic case are known since the work by Dulac [8], the analogous problem for cubic systems is not solved. Thus, as suggested in problem *1983-11*, one should begin with a certain typical (or simplest) class of integrable systems. A natural candidate is the class of Darboux integrable systems $M\,dH = 0$, where $H(x,y) = F_1^{\alpha_1} \cdots F_n^{\alpha_n}$ is the first integral equal to the product of polynomials $F_i \in \mathbb{R}[x,y]$ in real powers $\alpha_i \in \mathbb{R}$, and $M = F_1 \cdots F_n H^{-1}$ is the nontrivial integrating factor. The famous Lotka–Volterra system corresponds to three linear terms $F_1 = x$, $F_2 = y$ and $F_3 = 1 - x - y$, and seems to be one of the two simplest examples (the other one is a product of two terms with $F_1$ linear and $F_2$ quadratic).

It is much more difficult to describe the analytic continuation of the Poincaré integrals, since the "level curves" $H = h$ after complexification will not be affine Riemann surfaces continuously depending on $h$, but rather essentially noncompact leaves of the holomorphic foliation $\{M\,dH = 0\}$ with singularities on $\mathbb{C}P^2$. This makes it very difficult (if possible at all) to apply complex analytic methods that were the main tools of research in the Hamiltonian case. As a consequence, it is not possible to exhibit a finite-dimensional system of Picard–Fuchs equations (an infinite system was derived for the Darbouxian case by H. Żołądek in [9]).

Concerning the particular low-degree cases, one should mention paper [42] by Żołądek; see problem *1980-3*. In most other results concerning specific perturbations of the Lotka–Volterra system, usually monotonicity of some ratios of "monomial" Poincaré integrals is obtained by using very specific methods that do not admit generalizations for the general Darbouxian case or perturbations of higher than second degree. This monotonicity implies uniqueness of zero of the corresponding "binomial" linear combination of integrals. A useful tool for establishing such monotonicity for systems with the first integral of the form $H(x,y) = \Phi(x) + \Psi(y)$ was discovered by Chengzhi Li and Zhifen Zhang [21]: despite its seemingly artificial form, it proves to be working in many independently arising particular cases.

[1] ARNOLD V. I. Lectures on bifurcations and versal families. *Russian Math. Surveys*, 1972, **27**(5), 119–184.

[2] ARNOLD V. I. Geometrical Methods in the Theory of Ordinary Differential Equations, 2nd edition. New York: Springer, 1988. (Grundlehren der Mathematischen Wissenschaften, 250.) [*The Russian original* 1978.]

[3] ARNOLD V. I. Sur quelques problèmes de la théorie des systèmes dynamiques. *Topol. Methods Nonlinear Anal.*, 1994, **4**(2), 209–225. [*The Russian translation in*: Vladimir Igorevich Arnold. Selecta–60. Moscow: PHASIS, 1997, 533–551.]

[4] ARNOLD V. I., OLEĬNIK O. A. Topology of real algebraic varieties. *Moscow Univ. Math. Bull.*, 1979, **34**(6), 5–17.

[5] ARNOLD V. I., VISHIK M. I., IL′YASHENKO YU. S., KALASHNIKOV A. S., KONDRAT′EV V. A., KRUZHKOV S. N., LANDIS E. M., MILLIONSHCHIKOV V. M., OLEĬNIK O. A., FILIPPOV A. F., SHUBIN M. A. Some unsolved problems in the theory of differential equations and mathematical physics. *Russian Math. Surveys*, 1989, **44**(4), 157–171. [*The Russian original of the section "Problems by V. I. Arnold"* (*Uspekhi Mat. Nauk*, 1989, **44**(4), p. 191–192) *is reprinted in*: Vladimir Igorevich Arnold. Selecta–60. Moscow: PHASIS, 1997, 471–472.]

[6] BOGDANOV R. I. Bifurcations of the limit cycle of a family of plane vector fields. *Trudy Semin. Petrovskogo*, 1976, **2**, 23–35. [*The English translation*: *Selecta Math. Sov.*, 1981, **1**(4), 373–387.]

[7] BOGDANOV R. I. Versal deformation of a singularity point of a vector field on the plane in the case of zero eigenvalues. *Trudy Semin. Petrovskogo*, 1976, **2**, 37–65 (in Russian). [*The English translation*: *Selecta Math. Sov.*, 1981, **1**(4), 389–421.]

[8] DULAC H. Détermination et intégration d'une certaine classe d'équations différéntielles ayant pour point singulier un centre. *Bull. Sci. Math. France*, 1908, **32**(2), 230–252.

[9] DUMORTIER F., ROUSSARIE R., SOTOMAYOR J., ŻOLĄDEK H. Bifurcations of planar vector fields. Nilpotent Singularities and Abelian Integrals. Berlin: Springer, 1991. (Lecture Notes in Math., 1480.)

[10] GAVRILOV L. Petrov modules and zeros of Abelian integrals. *Bull. Sci. Math.*, 1998, **122**(8), 571–584.

[11] GAVRILOV L. The infinitesimal 16th Hilbert problem in the quadratic case. *Invent. Math.*, 2001, **143**(3), 449–497.

[12] GAVRILOV L. Linear estimate for the number of zeros of Abelian integrals with cubic Hamiltonians. *Nonlinearity*, 1998, **11**(6), 1521–1537.

[13] GLUTSYUK A. A., IL'YASHENKO YU. S. An estimate of the number of zeros of Abelian integrals for special Hamiltonians of arbitrary degree. [*Internet*: `http://www.arXiv.org/abs/math.DS/0112156`]

[14] IL'YASHENKO YU. S. The origin of limit cycles under perturbation of the equation $dw/dz = -R_z/R_w$, where $R(z,w)$ is a polynomial. *Math. USSR, Sb.*, 1969, **7**, 353–364.

[15] IL'YASHENKO YU. S. The multiplicity of limit cycles arising from perturbations of the form $dw/dz = P_2/Q_1$ of a Hamilton equation, in the real and complex domain. *Trudy Semin. Petrovskogo*, 1978, **3**, 49–60 (in Russian). [*The English translation*: *AMS Transl., Ser.* 2, 1982, **118**, 191–202.]

[16] IL'YASHENKO YU. S., YAKOVENKO S. YU. Double exponential estimate for the number of zeros of complete Abelian integrals and rational envelopes of linear ordinary differential equations with an irreducible monodromy group. *Invent. Math.*, 1995, **121**(3), 613–650.

[17] IL'YASHENKO YU. S., YAKOVENKO S. YU. Counting real zeros of analytic functions satisfying linear ordinary differential equations. *J. Differ. Equations*, 1996, **126**(1), 87–105.

[18] KHOVANSKIĬ A. G. Real analytic varieties with the finiteness property and complex Abelian integrals. *Funct. Anal. Appl.*, 1984, **18**(2), 119–127.

[19] KHOVANSKIĬ A. G. Fewnomials. Providence, RI: Amer. Math. Soc., 1991. (Transl. Math. Monographs, 88.)

[20] KHOVANSKIĬ A. G. Fewnomials. Moscow: PHASIS, 1997 (in Russian). (Mathematician's Library, 2.)

[21] LI CH. AND ZHANG ZH.-F. A criterion for determining the monotonicity of the ratio of two Abelian integrals. *J. Differ. Equations*, 1996, **124**(2), 407–424.

[22] MOUSSU R., ROCHE C. Théorie de Hovanskiĭ et problème de Dulac. *Invent. Math.*, 1991, **105**(2), 431–441.

[23] NOVIKOV D. I. Singular points of Picard–Fuchs systems, to appear.

[24] NOVIKOV D. I. Systems of linear ordinary differential equations with bounded coefficients may have very oscillating solutions. *Proc. Amer. Math. Soc.*, 2001, **129**(12), 3753–3755 (electronic).

[25] NOVIKOV D. I., YAKOVENKO S. YU. Simple exponential estimate for the number of real zeros of complete Abelian integrals. *C. R. Acad. Sci. Paris, Sér.* I *Math.*, 1995, **320**(7), 853–858.

[26] NOVIKOV D. I., YAKOVENKO S. YU. Simple exponential estimate for the number of real zeros of complete Abelian integrals. *Ann. Inst. Fourier* (*Grenoble*), 1995, **45**(4), 897–927.

[27] NOVIKOV D. I., YAKOVENKO S. YU. Meandering of trajectories of polynomial vector fields in the affine $n$-space. *Publ. Mat.*, 1997, **41**(1), 223–242.

[28] NOVIKOV D. I., YAKOVENKO S. YU. Tangential Hilbert problem for perturbations of hyperelliptic Hamiltonian systems. *Electron. Res. Announc. Amer. Math. Soc.*, 1999, **5**, 55–65 (electronic).

[29] NOVIKOV D. I., YAKOVENKO S. YU. Trajectories of polynomial vector fields and ascending chains of polynomial ideals. *Ann. Inst. Fourier* (*Grenoble*), 1999, **49**(2), 563–609.

[30] NOVIKOV D. I., YAKOVENKO S. YU. Redundant Picard–Fuchs system for Abelian integrals. *J. Differ. Equations*, 2001, **177**(2), 267–306.

[31] PETROV G. S. Elliptic integrals and their nonoscillation. *Funct. Anal. Appl.*, 1986, **20**(1), 37–40.

[32] PETROV G. S. The Chebyshev property of elliptic integrals. *Funct. Anal. Appl.*, 1988, **22**(1), 72–73.

[33] PETROVSKIĬ I. G., LANDIS E. M. On the number of limit cycles of the equation $dy/dx = M(x,y)/N(x,y)$, where the $M$ and $N$ are polynomials of second degree. *Dokl. Akad. Nauk SSSR* (*N. S.*), 1955, **102**, 29–32 (in Russian). [*The English translation*: *AMS Transl., Ser.* 2, 1958, **10**, 177–221.]

[34] PETROVSKIĬ I. G., LANDIS E. M. On the number of limit cycles of the equation $dy/dx = P(x,y)/Q(x,y)$, where $P$ and $Q$ are polynomials of degree $n$. *Dokl. Akad. Nauk SSSR* (*N. S.*), 1957, **113**, 748–751 (in Russian). [*The English translation*: *AMS Transl., Ser.* 2, 1960, **14**, 181–199.]

[35] PETROVSKIĬ I. G., LANDIS E. M. Corrections to the articles "On the number of limit cycles of the equations $dy/dx = P(x, y)/Q(x, y)$, where $P$ and $Q$ are polynomials of 2nd degree" and "On the number of limit cycles of the equation $dy/dx = P(x, y)/Q(x, y)$, where $P$ and $Q$ are polynomials." *Mat. Sb.* (*N. S.*), 1959 **48(90)**, 253–255 (in Russian).

[36] PUSHKAR′ I. A. Multidimensional generalization of the Il′yashenko theorem on Abelian integrals. *Funct. Anal. Appl.*, 1997, **31**(2), 100–108.

[37] ROUSSARIE R. On the number of limit cycles which appear by perturbation of separatrix loop of planar vector fields. *Bol. Soc. Brasil. Mat.*, 1986, **17**(2), 67–101.

[38] ROUSSARIE R. Cyclicité finie des lacets et des points cuspidaux. *Nonlinearity*, 1989, **2**(1), 73–117.

[39] TAKENS F. Forced oscillations and bifurcations. In: Applications of Global Analysis, I (Utrecht, 1973). Preprint, Utrecht: Math. Inst. Rijksuniv. Utrecht, 1974, 1–59. (Comm. Math. Inst. Rijksuniv. Utrecht, 3-1974.) [*Published in*: Global Analysis of Dynamical Systems. Festschrift dedicated to Floris Takens for his 60th birthday. Editors: H. W. Broer, B. Krauskopf and G. Vegter. Bristol: Inst. Phys., 2001, 1–61.]

[40] VARCHENKO A. N. Estimate of the number of zeros of an Abelian integral depending on a parameter, and limit cycles. *Funct. Anal. Appl.*, 1984, **18**(2), 98–108.

[41] YAKOVENKO S. YU. Quantitative theory of ordinary differential equations and tangential Hilbert 16th problem. Lecture notes of the course delivered at the Workshop "Asymptotic series, differential algebra and finiteness theorems" (Montreal, 2000). [*Internet*: `http://www.arXiv.org/abs/math.DS/0104140`]

[42] ŻOŁĄDEK H. Quadratic systems with center and their perturbations. *J. Differ. Equations*, 1994, **109**(2), 223–273.

## 1978-7 — *A. I. Neĭshtadt, M. B. Sevryuk*

ℛ The precise statement of the problem is probably as follows. Consider a smooth ($C^\infty$) Hamiltonian system with $n \geq 2$ degrees of freedom and equilibrium 0. Let all the eigenvalues of the system linearized at 0 be distinct (in particular, different from zero) and purely imaginary. Then, for each $N \geq 2$, the Hamilton function can be represented in a neighborhood of 0 in the form (up to an additive constant)

$$H = \tfrac{1}{2}\omega_1(p_1^2+q_1^2)+\cdots+\tfrac{1}{2}\omega_n(p_n^2+q_n^2)+H_3+H_4+\cdots+H_N \\ +O\left(|p|^{N+1}+|q|^{N+1}\right),$$

where $p=(p_1,\ldots,p_n)$ and $q=(q_1,\ldots,q_n)$ are suitable canonically conjugate variables while $H_s$ is a form of degree $s$ in $p_j$, $q_j$. The coefficients $\omega_j$ are the eigenfrequencies of the system (and $\pm i\omega_j$ are the eigenvalues of the linearized system). A resonance relation (or just a *resonance*) is by definition an equality of the form $l_1\omega_1+\cdots+l_n\omega_n=0$ with $l=(l_1,\ldots,l_n)\in\mathbb{Z}^n\setminus\{0\}$ an irreducible integer vector. The number $|l|=|l_1|+\cdots+|l_n|$ is called the *resonance order*.

Given an irreducible integer vector $k=(k_1,\ldots,k_n)\in\mathbb{Z}^n\setminus\{0\}$ with $|k|=|k_1|+\cdots+|k_n|\leq N$, assume that the frequencies $\omega_j$ satisfy resonance relations only of orders greater than $N$, with the possible exception of the relation $k_1\omega_1+\cdots+k_n\omega_n=0$. Then in some neighborhood of 0, one can reduce the Hamilton function to a $k$-resonant normal form up to the terms $O\left(|P|^{N+1}+|Q|^{N+1}\right)$ by a canonical change of variables $(p,q) \mapsto (P,Q)$ that differs from the identity transformation in the terms $O\left(|P|^2+|Q|^2\right)$, see, e. g., book [1]. This means that, in the expansion of the Hamilton function in variables $P_j$, $Q_j$, the sum of all the terms of orders $\le N$, if written in the symplectic polar coordinates $\rho_j$, $\varphi_j$:

$$P_j = \sqrt{2\rho_j}\cos\varphi_j, \quad Q_j = \sqrt{2\rho_j}\sin\varphi_j, \qquad 1 \le j \le n,$$

depends on the angles $\varphi_j$ via their combination $\psi = k_1\varphi_1 + \cdots + k_n\varphi_n$ only. After having cut off the terms $O\left(|P|^{|k|+1}+|Q|^{|k|+1}\right)$ in the Hamilton function, one obtains a truncated Hamilton function of the form

$$\mathcal{H} = \omega_1\rho_1 + \cdots + \omega_n\rho_n + F(\rho_1, \ldots, \rho_n) + B\rho_1^{|k_1|/2} \cdot \ldots \cdot \rho_n^{|k_n|/2}\cos(\psi+\psi_0).$$

Here $F$ is a polynomial in $\rho_j$ of degree $\le |k|/2$ without the constant and linear terms while $B$ and $\psi_0$ are certain constants. The system afforded by the Hamilton function $\mathcal{H}$ is integrable.

Indeed, introduce the new angular variables $\psi, \chi_1, \ldots, \chi_{n-1}$, so that the change

$$(\varphi_1, \ldots, \varphi_n) \mapsto (\psi, \chi_1, \ldots, \chi_{n-1})$$

is given by an integer unimodular matrix. Let $J, I_1, \ldots, I_{n-1}$ be the momenta canonically conjugate to the angles $\psi, \chi_1, \ldots, \chi_{n-1}$ (these momenta are linear combinations of the quantities $\rho_1, \ldots, \rho_n$ with integer coefficients). In the new variables, the Hamilton function $\mathcal{H}$ does not depend on $\chi_1, \ldots, \chi_{n-1}$. Therefore, $I_1, \ldots, I_{n-1}$ are first integrals for the system with Hamilton function $\mathcal{H}$, whereas for $J, \psi$, one obtains a one-degree-of-freedom Hamiltonian system whose Hamilton function depends on $I_1, \ldots, I_{n-1}$ as on parameters. In the studies of the motion near the resonance, the dependence on one more parameter, namely, the *resonance detuning* $\delta = k_1\omega_1 + \cdots + k_n\omega_n$, is of importance. The partition of a neighborhood of the origin in the space of parameters $\delta, I_1, \ldots, I_{n-1}$ into the subsets corresponding to different types of the phase portraits on the $(J,\psi)$-plane is called the *bifurcation diagram* of Hamilton function $\mathcal{H}$ (the rest of the parameters entering the Hamilton function—$B$ and the coefficients of the polynomial $F$—are treated as being fixed and meeting some genericity conditions [1]).

---

[1] To different values of these parameters, there may correspond different bifurcation diagrams. Consequently, each resonance is characterized, generally speaking, by a finite collection of diagrams rather than by a single diagram.

A similar analysis can be made also in the case where all the eigenvalues of the linearized Hamiltonian system are different from zero and purely imaginary but, among them, multiple eigenvalues are allowed. Within the framework of the problem in question, however, this case is important for $n = 2$ only, the corresponding resonances manifesting themselves already in the quadratic terms of the Hamilton function.

What is called the *strong resonances* in the formulation of the problem is probably the resonances of the minimal order generating topologically different bifurcation diagrams. For the case of two degrees of freedom ($n = 2$), all the bifurcation diagrams are known, and the results have been compiled in book [1]; the strong resonances are listed in the formulation of the problem. As far as the authors of the present comment know, the problem is still open for the case of three degrees of freedom ($n = 3$).

The following question is of great interest: What are the connections between the bifurcations—as the system passes through the resonance—of the phase portrait of the system with the truncated Hamilton function $\mathcal{H}$ and that of the initial system with the complete Hamilton function $H$. In general, the bifurcations of smooth one-parameter families of periodic trajectories in the system with Hamilton function $\mathcal{H}$ correspond to the same bifurcations of smooth one-parameter families of periodic trajectories in the system with Hamilton function $H$. On the other hand, the bifurcations of smooth $\nu$-parameter families of invariant $\nu$-tori in the system with Hamilton function $\mathcal{H}$ (the motion on those tori being parallel) for $\nu \geq 2$ correspond, generally speaking, to the bifurcations of *Cantor*—due to nonintegrability—$\nu$-parameter families of invariant $\nu$-tori in the system with Hamilton function $H$ (the motion on these tori being quasi-periodic).

One-parameter families of periodic trajectories in Hamiltonian systems near and at the instant of resonance are examined in a very rich body of literature. Many works were briefly surveyed in thesis [17] and in book [16]. Of the most important sources, we mention [1, 7, 12, 15]. A detailed study of the bifurcations of families of periodic trajectories in the important particular case of the three body problem is made in monograph [6].

On the other hand, the bifurcations of Cantor $\nu$-parameter families of invariant $\nu$-tori ($\nu \geq 2$) as the Hamiltonian system passes through a resonance have hardly been explored even in the case of two degrees of freedom. Such bifurcations pertain to the so-called *quasi-periodic bifurcation theory*. Up to now, the latter has been developed mainly for general (non-Hamiltonian) systems [2–5] (of the first works on the quasi-periodic bifurcation theory for Hamiltonian system, we mention paper [11]).

An extensive literature is devoted to the study of the dynamics near resonant equilibria of Hamiltonian systems as a whole. Here we point out, just as an illustration, only six recent works [8–10, 13, 14, 18]. In [8–10, 13, 18], multiple resonances ($|\omega_1| : |\omega_2| : |\omega_3| = m_1 : m_2 : m_3$) are considered as well and the relevant bibliography is given.

[1] ARNOLD V. I., KOZLOV V. V., NEĬSHTADT A. I. Mathematical Aspects of Classical and Celestial Mechanics, 2nd edition. Berlin: Springer, 1993, Ch. 7, § 3. (Encyclopædia Math. Sci., 3; Dynamical systems, III.) [*The Russian original* 1985.] [*The second, revised and supplemented, Russian edition* 2002.]

[2] BRAAKSMA B. L. J., BROER H. W. On a quasi-periodic Hopf bifurcation. *Ann. Institut Henri Poincaré, Analyse non linéaire*, 1987, **4**(2), 115–168.

[3] BRAAKSMA B. L. J., BROER H. W., HUITEMA G. B. Towards a quasi-periodic bifurcation theory. *Mem. Amer. Math. Soc.*, 1990, **83**(421), 83–170.

[4] BROER H. W. On some quasi-periodic bifurcations. *Delft Progress Report*, 1988, **12**(1), 79–96.

[5] BROER H. W., HUITEMA G. B., SEVRYUK M. B. Quasi-Periodic Motions in Families of Dynamical Systems: Order amidst Chaos. Berlin: Springer, 1996, Sect. 4.3. (Lecture Notes in Math., 1645.)

[6] BRUNO A. D. The Restricted 3-Body Problem: Plane Periodic Orbits. Berlin: Walter de Gruyter, 1994. [*The Russian original* 1990.]

[7] DUISTERMAAT J. J. Bifurcations of periodic solutions near equilibrium points of Hamiltonian systems. In: Bifurcation Theory and Applications. Editor: L. Salvadori. Berlin: Springer, 1984, 57–105. (Lecture Notes in Math., 1057.)

[8] HALLER G. Chaos near Resonance. New York: Springer, 1999, Sect. 4.1. (Appl. Math. Sci., 138.)

[9] HALLER G., WIGGINS S. Whiskered tori and chaos in resonant Hamiltonian normal forms. In: Normal Forms and Homoclinic Chaos. Editors: W. F. Langford and W. Nagata. Providence, RI: Amer. Math. Soc., 1995, 129–149. (Fields Institute Commun., 4.)

[10] HALLER G., WIGGINS S. Geometry and chaos near resonant equilibria of 3-DOF Hamiltonian systems. *Physica D*, 1996, **90**(4), 319–365.

[11] HANSSMANN H. The quasi-periodic centre-saddle bifurcation. *J. Differ. Equations*, 1998, **142**(2), 305–370.

[12] HENRARD J. Periodic orbits emanating from a resonant equilibrium. *Celest. Mech.*, 1970, **1**(3/4), 437–466.

[13] HOVEIJN I. Aspects of resonance in dynamical systems. Ph. D. Thesis, University of Utrecht, 1992.

[14] JOYEUX M. Classical dynamics of the 1 : 1, 1 : 2 and 1 : 3 resonance Hamiltonians. *Chem. Physics*, 1996, **203**(3), 281–307.

[15] SCHMIDT D. S. Periodic solutions near a resonant equilibrium of a Hamiltonian system. *Celest. Mech.*, 1974, **9**(1), 81–103.

[16] SEVRYUK M. B. Reversible Systems. Berlin: Springer, 1986. (Lecture Notes in Math., 1211.)

[17] SEVRYUK M. B. Reversible dynamical systems. Ph. D. Thesis, Moscow State University, 1987 (in Russian).

[18] WIGGINS S. Phase space geometry and dynamics associated with the 1 : 2 : 2 resonance. In: Hamiltonian Systems with Three or More Degrees of Freedom (S'Agaró, 1995). Editor: C. Simó. Dordrecht: Kluwer Acad. Publ., 1999, 254–269. (NATO Adv. Sci. Inst. Ser. C Math. Phys. Sci., 533.)

## 1978-10 – *V. V. Goryunov*

ℜ This was done by E. E. Landis [2] and O. A. Platonova [3]. See also survey [1].

[1] ARNOLD V. I. Singularities of Caustics and Wave Fronts. Dordrecht: Kluwer Academic Publ., 1990, Sect. 6.1. (Math. Appl., Sov. Ser., 62.)

[2] LANDIS E. E. Tangential singularities. *Funct. Anal. Appl.*, 1981, **15**(2), 103–114.

[3] PLATONOVA O. A. Projections of smooth surfaces. *Trudy Semin. Petrovskogo*, 1984, **10**, 135–149 (in Russian). [*The English translation*: *J. Sov. Math.*, 1986, **35**, 2796–2808.]

## 1978-17 – *I. A. Bogaevsky*

ℜ Let us consider the system of Euler–Lagrange linear partial differential equations originating from some variational principle

$$\delta \int L\, dt\, dx^1 \cdots dx^D = 0$$

with a Lagrangian $L(t,x,u_t,u_x) = T(t,x,u_t) - V(t,x,u_x)$ where $t, x^1, \ldots, x^D$ are independent variables, $u^1, \ldots, u^m$ are dependent variables, and the kinetic energy density $T$ is a positive definite quadratic form of the first derivatives of the dependent variables by time; the potential energy density $V$ is a positive semidefinite quadratic form of the first spatial derivatives of the dependent variables. In the general case, the coefficients of these forms are functions of $t$ and $x$. A good example of this situation is a perturbation spreading in an elastic medium. In this case, $u$ is a vector of displacement of a point in the medium; and the number of dependent variables is equal to the dimension of the $x$-space ($m = D$).

As known, the propagation of fronts and rays of shock and short waves is described in the framework of geometric optics by a *light hypersurface* which lies in a projective cotangent bundle over the space-time, and where the principal symbol of the original system of partial differential equations becomes degenerate. Written in coordinates $t$ and $x$, this symbol is a symmetric $(m \times m)$-matrix with coefficients

$$T_{ij}(t,x)\omega^2 - \sum_{k,l=1}^{D} V_{ij}^{kl}(t,x)p_k p_l,$$

where $(t,x)$ is a point in the space-time; a momentum $(\omega, p)$ is a cotangent vector to the space-time; $T_{ij}$ and $V_{ij}^{kl}$ are coefficients of the kinetic and potential energy densities:

$$T(t,x,u_t) = \frac{1}{2}\sum_{i,j=1}^{m} T_{ij}(t,x)u_t^i u_t^j, \quad V(t,x,u_x) = \frac{1}{2}\sum_{i,j=1}^{m}\sum_{k,l=1}^{D} V_{ij}^{kl}(t,x)u_{x^k}^i u_{x^l}^j.$$

Because the kinetic and potential energies are positive definite and positive semidefinite respectively, the system of Euler–Lagrange equations is hyperbolic. Singularities of the light hypersurface are projected into points of the space-time where the hyperbolicity is improper. If there are two or more dependent variables, then the light hypersurface can have singularities which cannot be removed by a small perturbation of the Lagrangian coefficients as functions of $t$ and $x$. The problem is to describe both the singularities themselves and the singularities of wave propagation caused by them. (More details can be found in the comments to problems *1988-3* and *1989-10*.)

It is supposed that the Lagrangian coefficients are generic functions of the space coordinate and, probably, the time. From the physical point of view it means particularly that the medium under consideration is non-homogeneous and anisotropic. A similar phenomenon occurs also in homogeneous media, and it is called the Hamilton conic refraction in crystals. However, the geometric optics of interior scattering in typical nonhomogeneous and anisotropic media differs essentially from the Hamilton conic refraction.

Typical singularities of a light hypersurface were described in [2]. It turns out that they are the same as typical singularities of the subset of degenerate matrices in the space of symmetric ones. For example, the simplest singularity of the light hypersurface (called *conic*) is the product of the usual two-dimensional cone $\xi^2 + \eta^2 = \zeta^2$ and the $(2D-2)$-dimensional vector space. The description of singularities of the light hypersurface is based on the transversality theorem for homogeneous maps proved in [9].

For a pair consisting of a generic contact structure and a $2D$-dimensional hypersurface, normal forms in a neighborhood of its conic singularity were obtained in [2] for $D = 1$; and for $D \geq 2$ in [1] (see also the problem *1988-3*). In both cases there are two normal forms: elliptic and hyperbolic ones. However, the explicit formulae are simpler for $D \geq 2$. According to [2], for $D = 1$ the elliptic normal form is not possible for light hypersurfaces due to the hyperbolicity of the system of Euler–Lagrange equations, but, as it was shown in [8], it is realized for $D \geq 2$. The conic singularities of the light hypersurface cause the Hamilton refraction in crystals, but in this case the contact structure is not generic due to homogeneity, and so it cannot be reduced either to the elliptic or hyperbolic normal form.

Singularities of typical Legendre submanifolds of the light hypersurface, going through its conic singularities for $D = 2$, are described in [3, 4], and the corresponding singularities of big fronts, ray systems on them and bifurcations of instantaneous fronts are studied in [5–7] (see problem *1989-10*). However, for $D = 3$ the singularities of typical Legendre submanifolds occuring in the case of interior scattering have not been described, not to mention the corresponding singularities of big fronts, ray systems on them and bifurcations of instantaneous fronts.


[1] ARNOLD V. I. On the interior scattering of waves, defined by hyperbolic variational principles. *J. Geom. Phys.*, 1988, **5**(3), 305–315.

[2] ARNOLD V. I. Surfaces defined by hyperbolic equations. *Math. Notes*, 1988, **44**(1), 489–497. [*The Russian original is reprinted in*: Vladimir Igorevich Arnold. Selecta–60. Moscow: PHASIS, 1997, 397–412.]

[3] ARNOLD V. I. Singularities of Caustics and Wave Fronts. Dordrecht: Kluwer Acad. Publ., 1990. (Math. Appl., Sov. Ser., 62.)

[4] ARNOLD V. I. Singularities of Caustics and Wave Fronts. Moscow: PHASIS, 1996 (in Russian). (Mathematician's Library, 1.)

[5] BOGAEVSKY I. A. Singularities of the propagation of short waves on the plane. *Sb. Math.*, 1995, **186**(11), 1581–1597.

[6] BOGAEVSKY I. A. The interior scattering of rays and wave fronts on the plane. In: ARNOLD V. I. Singularities of Caustics and Wave Fronts. Moscow: PHASIS, 1996, § 8.5, 300–316 (in Russian).

[7] BOGAEVSKY I. A. Singularities of short linear waves on the plane. In: The Arnold–Gelfand Mathematical Seminars: Geometry and Singularity Theory. Editors: V. I. Arnold, I. M. Gelfand, V. S. Retakh and M. Smirnov. Boston, MA: Birkhäuser, 1997, 107–112.

[8] BRAAM P. J., DUISTERMAAT J. J. Normal forms of real symmetric systems with multiplicity. *Indag. Math.* (*N. S.*), 1993, **4**(4), 407–421.

[9] KHESIN B. A. Singularities of light hypersurfaces and structure of hyperbolicity sets for systems of partial differential equations. In: Theory of Singularities and its Applications. Editor: V. I. Arnold. Providence, RI: Amer. Math. Soc., 1990, 119–127. (Adv. Sov. Math., 1.)

### 1978-19

R See V. I. Bakhtin's paper [1].

[1] BAKHTIN V. I. Topologically normal forms of caustic transformations of $D_\mu$-series. *Moscow Univ. Math. Bull.*, 1987, **42**(4), 63–66.

# 1979

### 1979-2

R See the comment to problem *1972-3*.

### 1979-3 — *S. V. Chmutov*

H The problem was repeated as *1980-11*.

R It was solved by A. N. Varchenko and J. Steenbrink. See the comment to problem *1980-11*.

### ▽ 1979-4 — *B. A. Khesin* Also: *1980-12*

R A. Rosly and B. Khesin proposed in [1,2] "polar homology" groups, which are holomorphic analogues of the homology groups in topology. A polar counterpart of orientation of a real manifold in this theory is the fixing of a meromorphic form on a complex manifold. The polar $k$-chains in a complex projective manifold are subvarieties of complex dimension $k$ with meromorphic forms on them, while the boundary operator is defined by taking the polar divisor and the Poincaré residue on it. Furthermore, one can define the polar analogues of the intersection index of cycles and holomorphic linking numbers for complex submanifolds (e. g., for a pair of complex curves in a complex 3-fold), see [1–3]. (The relation of this complexification of orientation to spinor structures is unknown.)

[1] KHESIN B. A., ROSLY A. A. Polar homology. [*Internet*: `http://www.arXiv.org/abs/math.AG/0009015`]

[2] KHESIN B. A., ROSLY A. A. Polar homology and holomorphic bundles. *Phil. Trans. Roy. Soc. London, Ser. A*, 2001, **359**, 1413–1427.

[3] KHESIN B. A., ROSLY A. A. Symplectic geometry on moduli spaces of holomorphic bundles over complex surfaces. In: The Arnoldfest. Proceedings of a conference in honour of V. I. Arnold for his sixtieth birthday (Toronto, 1997). Editors: E. Bierstone, B. A. Khesin, A. G. Khovanskiĭ and J. E. Marsden. Providence, RI: Amer. Math. Soc., 1999, 311–323. (Fields Inst. Commun., 24.)

### △ **1979-4** – ***V. A. Vassiliev*** Also: *1985-11*

R Orientability is the triviality of the first Stiefel–Whitney class. Therefore it is natural to expect that its "complexification" is the triviality of the first Chern class. Of course, the latter condition implies the existence of a spinor structure (i. e., the condition $w_2 = 0$) but is not equivalent to it.

### **1979-6** – ***V. A. Vassiliev*** Also: *1976-14, 1980-8, 1980-16*

R In paper [1] a $\mu =$ const stratum was presented such that the real codimension of its intersection with the space of real polynomials of a given multiplicity is strictly greater than the complex codimension of the similar complex intersection space. By a theorem of A. M. Gabrielov [2] in the complex case the modality and the proper modality (i. e., the codimension of the $\mu =$ const stratum) coincide. By the Zariski conjecture, the multiplicity of a function singularity is constant along a $\mu =$ const stratum. Therefore the example of [1] implies that at least one of the following three is true:

a) the real and complex modalities can be different;

b) the Gabrielov theorem does not hold in the real case;

c) the Zariski conjecture is false.

I believe that a) and b) are true, and c) is not.

See also the comment to problem *1975-23*.

[1] VASSILIEV V. A., SERGANOVA V. V. On the number of real and complex moduli of singularities of smooth functions and matroid realizations. *Math. Notes*, 1991, **49**(1), 15–20.

[2] GABRIELOV A. M. Bifurcations, Dynkin diagrams, and modality of isolated singularities. *Funct. Anal. Appl.*, 1974, **8**(2), 94–98.

## 1979-8 – *V. A. Vassiliev*

ℛ The caustic is the image (under a certain map) of the set of singular points of the discriminant (other than just its self-intersections); thus it is sufficient to prove the irreducibility of this set.

The space $\mathfrak{m}^2$ of all function germs with singularity at the origin can be considered as a deformation of our function $f$: to any function $\varphi \in \mathfrak{m}^2$ there corresponds the perturbation $f+\varphi$ of the function $f$. By the definition of the versal deformation, it (or, for simplicity, an arbitrarily large finite-dimensional subspace $T$ of it) is equivalent to the map induced from any versal deformation of $f$ by a certain map $\gamma\colon (T,f) \to (B,0)$, where $B$ is the base of this deformation. By a theorem of Gabrielov [1] the discriminant $\Sigma(B)$ is exactly the image of this map. The singular points of the discriminant are exactly the images under this map $\gamma$ of the points of $\mathfrak{m}^2$ corresponding to the functions with degenerate quadratic part. Therefore the irreducibility of the caustic follows from the irreducibility of the set of all degenerate quadratic forms.

The generic (of type $A_3$) singularities of the caustic are exactly the images of $\varphi \in \mathfrak{m}^2$ such that the kernel of the quadratic part of $\varphi$ is one-dimensional and belongs to the cone of zeros of the cubic part of $f$.

Therefore the number of irreducible components of the singular variety of the caustic of a non-Morse singularity can be equal to 1, 2, or 3. It is equal to the number of irreducible components of the cone of common zeros of the cubic and quadratic parts of the Taylor expansion of $f$. (In particular, there can be three components only if the corank of $f$ is equal to 2.)

[1] GABRIELOV A. M. Bifurcations, Dynkin diagrams, and modality of isolated singularities. *Funct. Anal. Appl.*, 1974, **8**(2), 94–98.

## 1979-14 – *A. A. Davydov*

ℛ The general answer is negative. The minimum of a family of functions is always continuous (see [2,3,7]) but the time optimal solution can have discontinuities even in a generic situation and in the interior of the attainable set. For example, the time optimal solution for a dynamic inequality with drift $(\dot{x}-v(x))^2 \leq 1$,

$x, v(x) \in \mathbb{R}^n$ ($v$ being some smooth vector field) can have discontinuites which are unremovable by small smooth perturbations of the inequality [4]. The classification of generic singularities of relative minimum [5, 6] provides another stable realization of discontinuities which can be realized as time optimal singularities.

The Pontryagin maximum principle and Matov theorems [7] imply the positive answer when the indicatrix is quadratically convex, smoothly depends on a point of the phase space and contains zero velocity in its interior. Conjecturally, the convexity condition is not essential.

[1] BOGAEVSKY I. A. Perestroikas of fronts in evolutionary families. *Proc. Steklov Inst. Math.*, 1995, **209**, 57–72.

[2] BRYZGALOVA L. N. Singularities of the maximum of a parametrically dependent function. *Funct. Anal. Appl.*, 1977, **11**(1), 49–51.

[3] BRYZGALOVA L. N. Maximum functions of a family of functions depending on parameters. *Funct. Anal. Appl.*, 1978, **12**(1), 50–51.

[4] DAVYDOV A. A. Local controllability of typical dynamic inequalities on surfaces. *Proc. Steklov Inst. Math.*, 1995, **209**, 73–106.

[5] DAVYDOV A. A., ZAKALYUKIN V. M. Point singularities of the conditional minimum on a three-dimensional manifold. *Proc. Steklov Inst. Math.*, 1998, **220**, 109–125.

[6] DAVYDOV A. A., ZAKALYUKIN V. M. The coincidence of generic singularities of solutions of extremal problems with constraints. In: Proceedings of the International Conference Dedicated to the 90th Birthday of L. S. Pontryagin (Moscow, 1998), Vol. 3: Geometic Control Theory. Itogi Nauki i Tekhniki VINITI. Contemporary Mathematics and its Applications. Thematic Surveys, Vol. 64. Moscow: VINITI, 1999, 118–143 (in Russian).

[7] MATOV V. I. The topological classification of germs of the maximum and minimax functions of a family of functions in general position. *Russian Math. Surveys*, 1982, **37**(4), 127–128.

## 1979-16

H This is a problem in paper [1] (p. 16: Problem 10).

[1] ARNOLD V. I., OLEĬNIK O. A. Topology of real algebraic varieties. *Vestnik Moskov. Univ. Ser. Mat. Mekh.*, 1979, № 6, 7–17 (in Russian). [*The English translation*: *Moscow Univ. Math. Bull.*, 1979, **34**(6), 5–17.]

R See the comment to problem *1978-6*.

## ▽ 1979-17

H This is a problem in paper [1] (p. 15: Problem 1).

[1] ARNOLD V. I., OLEĬNIK O. A. Topology of real algebraic varieties. *Vestnik Moskov. Univ. Ser. Mat. Mekh.*, 1979, №6, 7–17 (in Russian). [*The English translation*: *Moscow Univ. Math. Bull.*, 1979, **34**(6), 5–17.]

## △▽ 1979-17 – *V. M. Kharlamov*

R Still, very little is known. Let us denote by $I(n,d)$ the number of diffeotopy types of pairs $(P^n_{\mathbb{R}}, X_{\mathbb{R}})$, where $X_{\mathbb{R}}$ is a nonsingular hypersurface in $P^n_{\mathbb{R}}$ of degree $d$, and by $D(n,d)$ the number of deformation classes (i. e., the number of connected components of the space of nonsingular hypersurfaces). It is clear that $I(n,d) \le D(n,d) < \infty$ and it is easy to show (see [1]) that there are constants $c_n$, $c'_n$ such that $c_n d^n \le \log I(n,d) \le \log D(n,d) \le c'_n d^n \log d$ for any $d$. A bit more information is available in the case of plane curves; in particular, in paper [1] it is shown that $\log I(2,d) \asymp d^2$ ($\asymp$ states for $O$-equivalence).

A related question is when the diffeotopy type of $(P^n_{\mathbb{R}}, X_{\mathbb{R}})$ determines a connected component of the space of nonsingular hypersurfaces of given degree (cf. problem *1979-24*). The only two known general phenomena concern the elliptic and hyperbolic cases: if $X_{\mathbb{R}}$ is empty or, respectively, if it consists of $\frac{d}{2}$ standardly embedded spheres, then such hypersurfaces form a sole connected component in the space of hypersurfaces. The elliptic case is trivial: the corresponding set is convex. The connectedness in the hyperbolic case was proved by W. Nuij [5].

On the other hand, as follows from A. Nabutovsky's results [2], if $n \ge 6$, then there is no connectedness of this set even when $X_{\mathbb{R}}$ is a sphere embedded in the standard way. At least for an infinite sequence of values of $d$ the corresponding part of the space of hypersurfaces is disconnected. Moreover, there is no recursive function $\phi(d)$ such that those hypersurfaces of degree $d$ become deformation equivalent in degree $\phi(d)$; to consider degree $d$ hypersurfaces as hypersurfaces of higher degree, one multiplies the equation by a power of the sum of squares of coordinates; the proofs are based on algorithmic unsolvability of the diffeomorphism problem. According to Nabutovsky (private communication), the above nonconnectedness takes place for all sufficiently large $d$, the proof should follow the corresponding arguments from [3] and an explicit lower bound for the number of connected components can be deduced from [4].

It would be interesting to know whether the space is still disconnected if the diffeotopy type of $(\mathbb{C}P^n, X_\mathbb{C}, \mathrm{conj})$ is given (and $X_\mathbb{R}$, as above, is a sphere embedded in the standard way), cf. problem *1979-24*.

[1] KHARLAMOV V. M., OREVKOV S. YU. Growth order of the number of classes of real plane algebraic curves as the degree grows. *Zap. Nauch. Semin. St. Peterburg. Otdel. Mat. Inst. Steklova*, 2000, **266**, 218–233 (in Russian). (Theory of representations of dynamical systems. Combinatorial and algorithmic methods, 5.)

[2] NABUTOVSKY A. Nonrecursive functions in real algebraic geometry. *Bull. Amer. Math. Soc.* (*N. S.*), 1989, **20**(1), 61–65.

[3] NABUTOVSKY A. Disconnectedness of sublevel sets of some Riemannian functionals. *Geom. Funct. Anal.*, 1996, **6**(4), 703–725.

[4] NABUTOVSKY A. Geometry of the space of triangulations of a compact manifold. *Commun. Math. Phys.*, 1996, **181**(2), 303–330.

[5] NUIJ W. A note on hyperbolic polynomials. *Math. Scand.*, 1968, **23**, 69–72.

### △ 1979-17

See the comment to problem *1972-26* by V. I. Arnold.

### 1979-18

This is a problem in paper [1] (p. 15: Problem 2).

[1] ARNOLD V. I., OLEĬNIK O. A. Topology of real algebraic varieties. *Vestnik Moskov. Univ. Ser. Mat. Mekh.*, 1979, № 6, 7–17 (in Russian). [*The English translation*: *Moscow Univ. Math. Bull.*, 1979, **34**(6), 5–17.]

### ▽ 1979-19

This is a problem in paper [1] (p. 15: Problem 3).

[1] ARNOLD V. I., OLEĬNIK O. A. Topology of real algebraic varieties. *Vestnik Moskov. Univ. Ser. Mat. Mekh.*, 1979, № 6, 7–17 (in Russian). [*The English translation*: *Moscow Univ. Math. Bull.*, 1979, **34**(6), 5–17.]

### △ 1979-19 — *S. L. Tabachnikov* Also: *1985-6*

See paper [1].

[1] ITENBERG I. V., VIRO O. YA. Patchworking algebraic curves disproves the Ragsdale conjecture. *Math. Intelligencer*, 1996, **18**(4), 19–28.
[*Internet*: `http://www.math.uu.se/~oleg/preprints.html`]

## 1979-20

H This is a problem in paper [1] (p. 15–16: Problem 4).

[1] ARNOLD V. I., OLEĬNIK O. A. Topology of real algebraic varieties. *Vestnik Moskov. Univ. Ser. Mat. Mekh.*, 1979, №6, 7–17 (in Russian). [*The English translation*: *Moscow Univ. Math. Bull.*, 1979, **34**(6), 5–17.]

## 1979-21

H This is a problem in paper [1] (p. 16: Problem 5).

[1] ARNOLD V. I., OLEĬNIK O. A. Topology of real algebraic varieties. *Vestnik Moskov. Univ. Ser. Mat. Mekh.*, 1979, №6, 7–17 (in Russian). [*The English translation*: *Moscow Univ. Math. Bull.*, 1979, **34**(6), 5–17.]

## ▽ 1979-22

H This is a problem in paper [1] (p. 16: Problem 6).

[1] ARNOLD V. I., OLEĬNIK O. A. Topology of real algebraic varieties. *Vestnik Moskov. Univ. Ser. Mat. Mekh.*, 1979, №6, 7–17 (in Russian). [*The English translation*: *Moscow Univ. Math. Bull.*, 1979, **34**(6), 5–17.]

## △ 1979-22 — *S. Yu. Yakovenko*

R This problem was solved in [1] (Ch. 2, Theorems 4 and 5).

[1] KHOVANSKIĬ A. G. Fewnomials. Providence, RI: Amer. Math. Soc., 1991. (Transl. Math. Monographs, 88.)

## 1979-23

H This is a problem in paper [1] (p. 16: Problem 7).

[1] ARNOLD V. I., OLEĬNIK O. A. Topology of real algebraic varieties. *Vestnik Moskov. Univ. Ser. Mat. Mekh.*, 1979, №6, 7–17 (in Russian). [*The English translation*: *Moscow Univ. Math. Bull.*, 1979, **34**(6), 5–17.]

## ▽ 1979-24

ℋ This is a problem in paper [1] (p. 16: Problem 8).

[1] ARNOLD V. I., OLEĬNIK O. A. Topology of real algebraic varieties. *Vestnik Moskov. Univ. Ser. Mat. Mekh.*, 1979, №6, 7–17 (in Russian). [*The English translation*: *Moscow Univ. Math. Bull.*, 1979, **34**(6), 5–17.]

## △ 1979-24 — *V. M. Kharlamov*

ℛ Counterexamples were found by A. Marin (published in [2], but found at about the same time as similar counterexamples for nonmaximal curves due to T. Fiedler [1]; both use as obstruction the position of the ovals with respect to properly selected real lines intersecting the curve only at real points).

The further question, whether the topology of the triple $(\mathbb{C}P^2, \mathbb{C}A, \mathrm{conj})$ determines a connected component of the space of nonsingular curves, is still open. As is shown by J. Y. Welschinger [3], the answer is negative if $\mathbb{C}P^2$ is replaced with a ruled surface $\Sigma_a$, $a \geq 2$ (in his examples, the obstruction is originated by intersections with the exceptional section).

Here is another related problem: find all possible diffeotopy types of the triple $(\mathbb{C}P^2, \mathbb{C}A, \mathrm{conj})$ which can be realized by a plane projective real curve $A$ of given degree.

[1] FIEDLER T. New congruences in the topology of real plane algebraic curves. *Sov. Math. Dokl.*, 1983, **27**, 566–568.

[2] MARIN A. $\mathbb{C}P^2/\sigma$ ou Kuiper et Massey au pays des coniques. In: A la recherche de la topologie perdue. I. Le côté de chez Rohlin. II. Le côté de Casson. Editors: L. Guillou and A. Marin. Boston, MA – Basel – Stuttgart: Birkhäuser, 1986, 141–152. (Progr. Math., 62.)

[3] WELSCHINGER J. Y. Courbes algèbriques réelles et courbes flexibles sur les surfaces réglées de base $\mathbb{C}P^1$. *Proc. London Math. Soc., Ser.* 3, 2002, **85**(2), 367–392.

## 1979-25

ℋ This is a problem in paper [1] (p. 16: Problem 9).

[1] ARNOLD V. I., OLEĬNIK O. A. Topology of real algebraic varieties. *Vestnik Moskov. Univ. Ser. Mat. Mekh.*, 1979, №6, 7–17 (in Russian). [*The English translation*: *Moscow Univ. Math. Bull.*, 1979, **34**(6), 5–17.]

## ▽ 1979-26

*H* This is a problem in paper [1] (p. 16: Problem 11).

[1] ARNOLD V. I., OLEĬNIK O. A. Topology of real algebraic varieties. *Vestnik Moskov. Univ. Ser. Mat. Mekh.*, 1979, №6, 7–17 (in Russian). [*The English translation*: *Moscow Univ. Math. Bull.*, 1979, **34**(6), 5–17.]

## △ 1979-26 — *S. Yu. Yakovenko*

*R* For perturbations of a *generic* Hamiltonian system with a real cubic polynomial $H$, the problem was solved by L. Gavrilov [2]. He proved that, for any real cubic Hamiltonian with four distinct critical values and any cubic differential form $\omega$, the number of limit cycles born in the corresponding *quadratic* perturbation

$$M\,dH + \varepsilon\omega = 0,$$

is at most 2. This technically involved theorem incorporates the results previously obtained by E. Horozov and I. Iliev [3]. Gavrilov's theorem solves problem *1979-26* for perturbations of Hamiltonian quadratic systems, as well as numerous other results.

The central moment is a theorem on zeros of the corresponding Abelian integrals, see the comment to problem *1978-6.*

In February 2002 Chengzhi Li and Zenghua Zhang announced that *the number of isolated zeros of any Abelian integral of a real quadratic polynomial* 1*-form over closed level curves of a real cubic Hamiltonian is at most* 2. This result was achieved in a series of case study works treating the degenerate cases not covered by the Gavrilov theorem (actually, some of these results chronologically preceded [2]). Namely, when one of the critical points of $H$ escapes to infinity, the bound was obtained in [5, 6]. The case of two coinciding critical values attained at two *distinct* critical points (in this case the Hamiltonian system exhibits a heteroclinic loop) was covered in [1]. The final blow was dealt in [4] by Chengzhi Li and Zenghua Zhang who announced the solution for the case of cubic Hamiltonians exhibiting a cuspidal singularity.

It is important to emphasize that in these degenerate cases a bound on the number of zeros of Abelian integrals does not yet imply a bound on the number of limit cycles. It may happen that the integral $I(t) = \oint_{H=t} \omega$ vanishes identically, whereas the system $dH + \varepsilon\omega = 0$ for any $\varepsilon \neq 0$ is non-integrable.

P. S. (*V. I. Arnold*): The infinitesimal problem is still open for perturbations of integrable non-Hamiltonian systems, for instance, quadratic perturbations of the Lotka–Volterra system.

[1] CHOW SH.-N., LI CH., YI Y. The cyclicity of period annuli of degenerate quadratic Hamiltonian systems with elliptic segment loops. *Ergod. Theory Dynam. Systems*, 2002, **22**(2), 349–374.

[2] GAVRILOV L. The infinitesimal 16th Hilbert problem in the quadratic case. *Invent. Math.*, 2001, **143**(3), 449–497.

[3] HOROZOV E., ILIEV I. D. On the number of limit cycles in perturbations of quadratic Hamiltonian systems. *Proc. London Math. Soc., Ser.* 3, 1994, **69**(1), 198–224.

[4] LI CH., ZHANG Z. Weak Hilbert problem for $n = 2$. Preprint, February 2002.

[5] MARKOV Y. Limit cycles of perturbations of a class of quadratic Hamiltonian vector fields. *Serdica Math. J.*, 1996, **22**(2), 91–108.

[6] ZHANG ZH., LI CH. On the number of limit cycles of a class of quadratic Hamiltonian systems under quadratic perturbations. *Adv. Math.* (*China*), 1997, **26**(5), 445–460.

## ▽ 1979-27

H This is a problem in paper [1] (p. 16: Problem 12).

[1] ARNOLD V. I., OLEĬNIK O. A. Topology of real algebraic varieties. *Vestnik Moskov. Univ. Ser. Mat. Mekh.*, 1979, № 6, 7–17 (in Russian). [*The English translation*: *Moscow Univ. Math. Bull.*, 1979, **34**(6), 5–17.]

## △ 1979-27 — *S. Yu. Yakovenko* Also: *1983-16*

R These two problems can be considered as an initial step in an attempt to relax the Hilbert 16th problem by localizing it on a neighborhood of a singular point (another type of localization, with respect to parameters, leads to the problem on zeros of Poincaré integrals, see problem *1979-16* and its follow-up).

Yet it turned out that, besides limit cycles born from centers with nonzero linear part (Andronov–Hopf) and cuspidal points (Bogdanov [1]), there were only a few results, some of them incomplete. Perhaps the main reason is that all difficulties characteristic of global problems reappear in the local problem after a suitable blow-up procedure.

Considerably better is the situation with other types of polycycles (graphics, separatrix polygons) which can also generate limit cycles by small perturbations. (One singular point, degenerate or not, is a particular case of a polycycle.) *Cyclicity* of a polycycle is the number of limit cycles that can be born this way.

The known results can be arranged according to the number and types of singular points on the polycycle, that is, ultimately, according to the codimension of occurrence of polycycles in generic families of planar vector fields. For all polycycles of small codimension 1 and 2 the cyclicity is known. The list of polycycles of cyclicity 3, the *Kotova Zoo*, has been compiled and for many beasts from this zoo the cyclicity is known or at least estimated from above. These and other results can be found with appropriate references in book [2].

Results of a general nature are scarce. It is known that cyclicity of a *generic* polycycle of any finite codimension $n$, carrying only *elementary* singularities (with non-nilpotent linear parts), is finite and bounded by an algorithmically computable function $E(n)$. This result by Il′yashenko and Yakovenko [5] was recently improved by V. Kaloshin, who simplified some parts of the construction and achieved the transparency that allowed him to prove that $E(n) \leq 2^{25n^2}$ [4,6,7]. To get rid of the elementariness assumption, a parametric desingularization procedure is required. There were attempts to construct such theory (Roussarie–Denkowska, Trifonov), but all failed to reach the level of applicability required for further progress in this direction: even for bifurcations of a cuspidal loop, an upper bound for the number of limit cycles is proved only modulo an assertion on monotonicity of a certain transcendental function [3].

[1] BOGDANOV R. I. Versal deformation of a singularity point of a vector field on the plane in the case of zero eigenvalues. *Trudy Semin. Petrovskogo*, 1976, **2**, 37–65 (in Russian). [*The English translation*: *Selecta Math. Sov.*, 1981, **1**(4), 389–421.]

[2] Concerning the Hilbert 16th Problem. Editors: Yu. Il′yashenko and S. Yakovenko. Providence, RI: Amer. Math. Soc., 1995. (AMS Transl., Ser. 2, 165; Adv. Math. Sci., 23.)

[3] DUMORTIER F., ROUSSARIE R., SOTOMAYOR J. Bifurcations of cuspidal loops. *Nonlinearity*, 1997, **10**(6), 1369–1408.

[4] IL′YASHENKO YU. S., KALOSHIN V. YU. Bifurcation of planar and spatial polycycles: Arnold's program and its development. In: The Arnoldfest. Proceedings of a conference in honour of V. I. Arnold for his sixtieth birthday (Toronto, 1997). Editors: E. Bierstone, B. A. Khesin, A. G. Khovanskiĭ and J. E. Marsden. Providence, RI: Amer. Math. Soc., 1999, 241–271. (Fields Inst. Commun., 24.)

[5] IL′YASHENKO YU. S., YAKOVENKO S. YU. Finite cyclicity of elementary polycycles in generic families. In: Concerning the Hilbert 16th Problem. Editors: Yu. Il′yashenko and S. Yakovenko. Providence, RI: Amer. Math. Soc., 1995, 21–95. (AMS Transl., Ser. 2, 165; Adv. Math. Sci., 23.)

[6] KALOSHIN V. YU. Around Hilbert–Arnold problem. [*Internet*: `http://www.arXiv.org/abs/math.DS/0111053`]

[7] KALOSHIN V. YU. The Hilbert–Arnold problem and an estimate of the cyclicity of polycycles of the plane and in space. *Funct. Anal. Appl.*, 2001, **35**(2), 146–147.

# 1980

## 1980-1

ℜ See the comment to problem *1978-6*.

## 1980-2 – *S. Yu. Yakovenko* Also: *1984-16*

ℜ This is one more attempt to relax the Hilbert problem on limit cycles, this time modifying the class of admissible differential equations. The answer for the trigonometric problem as it is stated, is known only for $\deg f = 1$, where it is shown that at most 2 limit cycles can occur [3]. For $\deg f = 2$ at least 6 cycles are possible (ibid.).

The "polynomial" periodic problem *1980-2* is somewhat better understood. The differential equation of the form $\dot{x} = P(x,t)$ with the monic polynomial $P = x^d + \sum_0^{d-1} a_k(t)\, x^k$ and arbitrary dependence on $t$, the so-called *Abel equation*, was studied in [6]. For $d = \deg_x P \leq 3$ there can be at most $d$ solutions with $x(0) = x(1)$, while for $d \geq 4$ the number of cycles can be arbitrary, depending on the coefficients of the polynomial $f$ [7–9].

Very recently Yu. Il′yashenko constructed an upper bound for the number of limit cycles of the *periodic* Abel equation (as in the initial formulation) in terms of the *magnitude* of the coefficients, $C = \max_k \max_t |a_k(t)|$. In [5] he proved that the number of cycles can be majorized by an explicit expression doubly exponential in $C$. Results of similar nature were also obtained for limit cycles of the Liénard equation.

A different approach to studying the cubic Abel equation

$$y' = p(x)y^2 + q(x)y^3, \qquad p(x), q(x) \in \mathbb{R}[x], \tag{1}$$

with polynomial $p(x)$ and $q(x)$, was suggested recently by J.-P. Françoise, Y. Yomdin and coauthors. Very roughly, the idea is to solve this equation in formal series and study the algebra and geometry of coefficients of these series.

For instance, the growth of coefficients of a converging series $\sum_{k\geq 0} a_k y^k$, closely related to the growth rate of the sum of this series, is responsible for the distribution of its zeros. If the coefficients are themselves polynomials in the additional parameter(s), then *uniform* bounds on the number of zeros can be derived from analysis of the ascending chain of *Bautin ideals* $I_k = \langle a_0, a_1, \dots, a_k \rangle$ and the (infinite) descending chain of ideals $J_k = \langle a_{k+1}, a_{k+2}, \dots \rangle$ [4, 10].

In application to the cubic Abel equation (1), consider the "Green function" $G(x,y)$, defined as the value at the moment of time $x$ of the solution of this equation, defined by the initial condition $y(0) = y$. The expansion of this function $G(x,y) = y + \sum_{k\geq 2} a_k(x)\, y^k$ has polynomial coefficients $a_k \in \mathbb{R}[x]$, and the recurrent rule for $a_k$ can be easily written. The questions on zeros of the function $G$ contain in a nutshell many difficulties characteristic for the Hilbert problem. The number of isolated roots of $G(1,y)$ will be an equivalent of the problem on limit cycles. Determination of the points $x = b$ for which $G(b,y) \equiv y$ is the "Poincaré center problem" for the Abel equation. In this case we say that the points $x = 0$ and $x = b$ are *conjugated* along the equation (1).

One can easily construct examples of Abel equations with conjugated points as follows: starting from an arbitrary Abel equation, make a many-to-one polynomial change of the independent variable $x = x(t)$. The result will be a "foldable" Abel equation with all points of each preimage $t^{-1}(b)$ conjugated with each other for any choice of $b$. The conjecture is that this is the only possibility for the appearance of conjugated points. In the language of composition algebra of coefficients, this is tantamount to existence of a non-trivial compositional common factor for the primitives $P = \int p$ and $Q = \int q$ [2].

One can also formulate an infinitesimal version of this problem. If $q(x) \equiv 0$, then the equation (1) becomes integrable and the conjugate points occur only at the roots of $P(x)$. Adding a *small perturbation* $\varepsilon q(x) y^3$ to this integrable equation makes the Green function $G(x,y)$ depending on $\varepsilon$, and the first variation in $\varepsilon$, the "Poincaré integral" $F(x,y) = \frac{d}{d\varepsilon}\big|_{\varepsilon=0} G(x,y)$ (cf. the comment to problem *1978-6*), can be reduced to the integral

$$F(x,y) = \int_0^x \frac{q(t)\,dt}{1 - yp(t)} = \sum_{k=0}^{\infty} m_k(x)\, y^k, \qquad m_k(x) = \int_0^x P^k(t) q(t)\,dt. \tag{2}$$

The coefficients $m_k(x)$ are the moments of $q$ with respect to the weight $P(x)$, and their common zeros determine the "infinitesimally conjugate" points. As before, zeros of a compositional common factor of $P$ and $Q$ are common zeros of all the moments $m_k$, and the problem is to describe the other such roots.

Returning to the initial problem involving the Abel equation (1) with trigonometric polynomials $p, q \in \mathbb{C}[\exp ix]$, it is proved that this equation is a center (i. e., all trajectories are $2\pi$-periodic) when all 2-dimensional moments $\int_0^{2\pi} P^k Q^l \, dP(t)$ are zeros [1]. This assertion fails for arbitrary trigonometric functions. The reason behind this fact is that the moments can be computed as periods of polynomial 1-forms on a certain naturally arising algebraic curve, and the assertion is valid if this curve is rational.

[1] Blinov M., Yomdin Y. Center and composition conditions for Abel differential equation, and rational curves. *Qualit. Theory Dynam. Systems*, 2001, **2**(1), 111–127.

[2] Briskin M., Françoise J.-P., Yomdin Y. Center conditions, compositions of polynomials and moments on algebraic curves. *Ergod. Theory Dynam. Systems*, 1999, **19**(5), 1201–1220.

[3] Ershov E. K. On the number of cycles of some differential equations on a two-dimensional torus. *Differ. Uravneniya*, 1991, **27**(12), 2167–2169, 2207 (in Russian).

[4] Françoise J.-P., Yomdin Y. Bernstein inequalities and applications to analytic geometry and differential equations. *J. Funct. Anal.*, 1997, **146**(1), 185–205.

[5] Il'yashenko Yu. S. Hilbert-type numbers for Abel equations, growth and zeros of holomorphic functions. *Nonlinearity*, 2000, **13**(4), 1337–1342.

[6] Lins Neto A. On the number of solutions of the equation $dx/dt = \sum_{j=0}^{n} a_j(t)x^j$, $0 \le t \le 1$, for which $x(0) = x(1)$. *Invent. Math.*, 1980, **59**(1), 67–76.

[7] Panov A. A. The number of periodic solutions of polynomial differential equations. *Math. Notes*, 1998, **64**(5), 622–628.

[8] Panov A. A. On the diversity of Poincaré mappings for cubic equations with variable coefficients. *Funct. Anal. Appl.*, 1999, **33**(4), 310–312.

[9] Shahshahani S. Periodic solutions of polynomial first order differential equations. *Nonlinear Anal.*, 1981, **5**(2), 157–165.

[10] Yomdin Y. Global finiteness properties of analytic families and algebra of their Taylor coefficients. In: The Arnoldfest. Proceedings of a conference in honour of V. I. Arnold for his sixtieth birthday (Toronto, 1997). Editors: E. Bierstone, B. A. Khesin, A. G. Khovanskiĭ and J. E. Marsden. Providence, RI: Amer. Math. Soc., 1999, 527–555. (Fields Inst. Commun., 24.)

## 1980-3 — *S. Yu. Yakovenko*

This problem is a particular low-degree case of the general infinitesimal Hilbert problem, see the comment to problem *1978-6*.

The question about the number of limit cycles born from the quadratic Lotka–Volterra system was answered by H. Żołądek [1]. He proved using tremendously heavy and absolutely mysterious computations that in the quadratic perturbation of the Lotka–Volterra system the corresponding Poincaré integral may have at most 2 isolated zeros.

[1] Żołądek H. Quadratic systems with center and their perturbations. *J. Differ. Equations*, 1994, **109**(2), 223–273.

## 1980-4 — *M. B. Sevryuk*

The problem deals with E. A. Demëkhin's studies published in paper [4]. This paper discusses how $2\pi$-periodic solutions of the equation

$$s^2U^{\mathrm{IV}} + U'' + (U')^2 = Q \tag{1}$$

branch off from $(2\pi/n)$-periodic solutions (the notation of the paper differs somewhat from E. A. Demëkhin's original notation used in the formulation of the problem; in fact, the paper examines mainly the equivalent problem on periodic solutions of the equation

$$s^2H''' + H' + H^2 = Q$$

of zero mean; this equation describes stationary traveling waves in a layer of viscous liquid on an inclined plane; here $H$ is the width of the liquid film, $s$ denotes the wave number, and $Q$ is the so-called nonlinear expenditure distortion [1]).

It turns out that the values of the parameters $s$ and $Q$ for which equation (1) has $2\pi/n$-periodic solutions constitute, on the $(s,Q)$-plane, curves $\Gamma_n$ emanating from the points $(s = 1/n,\ Q = 0)$ towards smaller values of $s$. From some points on these curves, other curves $\gamma_n$ branch off that are characterized by the following property: Equation (1) admits $2\pi$-periodic solutions for $(s,Q) \in \gamma_n$.

Such bifurcations are due to the fact that equation (1) is *reversible* with respect to involution

$$G\colon (U,U',U'',U''') \mapsto (U,-U',U'',-U''')$$

of the phase space $\mathbb{R}^4$ and invariant with respect to translations

$$U \mapsto U + \mathrm{const}$$

[1] Concerning the more general equation $s^2H''' + H' - cH + H^2 = Q$ for the same physical phenomenon, see works [3, 8].

(that the background equations of [4] are reversible was first pointed out in paper [2]). Because of reversibility, $G$-invariant closed phase trajectories of equation (1) constitute, for $s$ and $Q$ fixed, one-parameter families

$$U_C = U_0 + C, \quad C = \text{const},$$

which are *structurally stable with respect to small changes of parameters s and Q* (see works [1, 5]). The condition that the period of these trajectories is equal to a prescribed number $T > 0$ determines a curve on the parameter plane $(s, Q)$. The curves $\gamma_n$ branch off from those points on the curves $\Gamma_n$ where the Floquet multipliers of the corresponding $2\pi/n$-periodic solutions are equal to 1, $e^{\pm 2\pi pi/n}$ $[1 \leq p \leq n-1, \text{GCD}(p,n) = 1]$.

Of the rich body of literature devoted to periodic solutions of reversible systems of differential equations, we mention here only the most important papers [6, 7, 9–12] where it is considered how solutions of period $nT$ branch off from solutions of period $T$ (or, equivalently, how periodic orbits of period $n$ are born from fixed points of reversible mappings).

[1] ARNOLD V. I. Reversible systems. In: Nonlinear and Turbulent Processes in Physics (Kiev, 1983), V. 3. Editor: R. Z. Sagdeev. Chur: Harwood Acad. Publ., 1984, 1161–1174. [*For the Russian translation see, e. g.*: Vladimir Igorevich Arnold. Selecta–60. Moscow: PHASIS, 1997, 355–363.]

[2] ARNOLD V. I., SEVRYUK M. B. Oscillations and bifurcations in reversible systems. In: Nonlinear Phenomena in Plasma Physics and Hydrodynamics. Editor: R. Z. Sagdeev. Moscow: Mir, 1986, 31–64.

[3] BUNOV A. V., DEMËKHIN E. A., SHKADOV V. YA. On the non-uniqueness of nonlinear wave solutions in a viscous layer. *J. Appl. Math. Mech.*, 1984, **48**(4), 495–499.

[4] DEMËKHIN E. A. Branching of a solution of the problem on stationary traveling waves in a layer of viscous liquid on an inclined plane. *Izvestiya Akad. Nauk SSSR, Ser. Mekh. Zhidkosti i Gaza* (*Mech. Liquid and Gas*), 1983, **5**, 36–44 (in Russian).

[5] DEVANEY R. L. Reversible diffeomorphisms and flows. *Trans. Amer. Math. Soc.*, 1976, **218**, 89–113.

[6] FURTER J. E. On the bifurcations of subharmonics in reversible systems. In: Singularity Theory and its Applications, Part II. Editors: M. Roberts and I. Stewart. Berlin: Springer, 1991, 167–192. (Lecture Notes in Math., 1463.)

[7] GERVAIS J.-J. Bifurcations of subharmonic solutions in reversible systems. *J. Differ. Equations*, 1988, **75**(1), 28–42; addendum: 1989, **78**(2), 400.

[8] NEPOMNYASHCHIĬ A. A. Stability of wave regimes in a film flowing down on an inclined plane. *Izvestiya Akad. Nauk SSSR, Ser. Mekh. Zhidkosti i Gaza* (*Mech. Liquid and Gas*), 1974, **3**, 28–34 (in Russian).

[9] SEVRYUK M. B. Reversible Systems. Berlin: Springer, 1986, § 5.4. (Lecture Notes in Math., 1211.)

[10] VANDERBAUWHEDE A. Bifurcation of subharmonic solutions in time-reversible systems. *Z. Angew. Math. Phys.*, 1986, **37**(4), 455–478.

[11] VANDERBAUWHEDE A. Subharmonic branching in reversible systems. *SIAM J. Math. Anal.*, 1990, **21**(4), 954–979.

[12] VANDERBAUWHEDE A. Branching of periodic solutions in time-reversible systems. In: Geometry and Analysis in Nonlinear Dynamics. Editors: H. W. Broer and F. Takens. Harlow: Longman, 1992, 97–113. (Pitman Research Notes Math. Ser., 222.)

## 1980-6 — *A. G. Khovanskiĭ*

The Jacobian problem still remains open; nobody has advanced in applying the mixed Hodge structure to it.

## 1980-8

See the comment to problem *1979-6*.

## ▽ 1980-9

This is a problem in paper [1] (§ 5); see also the comment in [2] (p. 68).

[1] ARNOLD V. I. On some problems in singularity theory. In: Geometry and Analysis. Papers dedicated to the memory of V. K. Patodi. Bangalore: Indian Acad. Sci., 1980, 1–9. [*Reprinted in*: *Proc. Indian Acad. Sci. Math. Sci.*, 1981, **90**(1), 1–9.]

[2] ARNOLD V. I. Some open problems in the theory of singularities. In: Singularities. Part 1 (Arcata, CA, 1981). Editor: P. Orlik. Providence, RI: Amer. Math. Soc., 1983, 57–69. (Proc. Symposia Pure Math., 40.)

## △ 1980-9 — *S. V. Chmutov*

First time the mixed Hodge structures were applied to the topology of the real algebraic manifolds in [1, 2].

[1] ARNOLD V. I. The index of a singular point of a vector field, the Petrovskiĭ–Oleĭnik inequalities, and mixed Hodge structures. *Funct. Anal. Appl.*, 1978, **12**(1), 1–12.

[2] KHARLAMOV V. M. A generalized Petrovskiĭ inequality, I; II. *Funct. Anal. Appl.*, 1974, **8**(2), 132–137; 1975, **9**(3), 266–268.

## ▽ 1980-11

This is a problem in paper [1] (§ 1); see also the comment in [2] (p. 68).

[1] ARNOLD V. I. On some problems in singularity theory. In: Geometry and Analysis. Papers dedicated to the memory of V. K. Patodi. Bangalore: Indian Acad. Sci., 1980, 1–9. [*Reprinted in*: *Proc. Indian Acad. Sci. Math. Sci.*, 1981, **90**(1), 1–9.]

[2] ARNOLD V. I. Some open problems in the theory of singularities. In: Singularities. Part 1 (Arcata, CA, 1981). Editor: P. Orlik. Providence, RI: Amer. Math. Soc., 1983, 57–69. (Proc. Symposia Pure Math., 40.)

## △ 1980-11 — *S. V. Chmutov* Also: *1975-8*, *1979-3*

The current problem actually coincides with *1979-3*. It was solved by A. N. Varchenko [3, 5] and J. Steenbrink [2]. Varchenko proved the semicontinuity in the case of quasihomogeneous singularities under the lower deformations. He applied this results to Bruce's problem (*1981-24*). For the semicontinuity of the spectrum and related questions see the survey [1]. The complex singularity index from problem *1979-3* equals $-(1+l_{\min})$, where $l_{\min}$ is the smallest spectral number. So this problem generalizes *1979-3*. The semicontinuity of the complex singularity index was proved by A. N. Varchenko in [4].

[1] ARNOLD V. I., VASSILIEV V. A., GORYUNOV V. V., LYASHKO O. V. Singularities. I. Local and Global Theory. Berlin: Springer, 1993. (Encyclopædia Math. Sci., 6; Dynamical Systems, VI.) [*The Russian original* 1988.]

[2] STEENBRINK J. H. M. Semicontinuity of the singularity spectrum. *Invent. Math.*, 1985, **79**(3), 557–565.

[3] VARCHENKO A. N. Asymptotic integrals and Hodge structures. In: Itogi Nauki i Tekhniki VINITI. Current Problems in Mathematics, Vol. 22. Moscow: VINITI, 1983, 130–166 (in Russian). [*The English translation*: *J. Sov. Math.*, 1984, **27**, 2760–2784.]

[4] VARCHENKO A. N. Semicontinuity of the complex singularity exponent. *Funct. Anal. Appl.*, 1983, **17**(4), 307–309.

[5] VARCHENKO A. N. On semicontinuity of the spectrum and an upper estimate for the number of singular points of a projective hypersurface. *Sov. Math. Dokl.*, 1983, **27**, 735–739.

## 1980-12

See the comment to problem *1979-4* by B. A. Khesin.

### 1980-14

ℋ This is a problem in paper [1] (§ 2); see also the comment in [2] (p. 68).

[1] ARNOLD V. I. On some problems in singularity theory. In: Geometry and Analysis. Papers dedicated to the memory of V. K. Patodi. Bangalore: Indian Acad. Sci., 1980, 1–9. [*Reprinted in*: *Proc. Indian Acad. Sci. Math. Sci.*, 1981, **90**(1), 1–9.]

[2] ARNOLD V. I. Some open problems in the theory of singularities. In: Singularities. Part 1 (Arcata, CA, 1981). Editor: P. Orlik. Providence, RI: Amer. Math. Soc., 1983, 57–69. (Proc. Symposia Pure Math., 40.)

ℛ See the comment to problem *1993-27*.

### ▽ 1980-15

ℋ This is a problem in paper [1] (§ 3); see also the comment in [2] (p. 68).

[1] ARNOLD V. I. On some problems in singularity theory. In: Geometry and Analysis. Papers dedicated to the memory of V. K. Patodi. Bangalore: Indian Acad. Sci., 1980, 1–9. [*Reprinted in*: *Proc. Indian Acad. Sci. Math. Sci.*, 1981, **90**(1), 1–9.]

[2] ARNOLD V. I. Some open problems in the theory of singularities. In: Singularities. Part 1 (Arcata, CA, 1981). Editor: P. Orlik. Providence, RI: Amer. Math. Soc., 1983, 57–69. (Proc. Symposia Pure Math., 40.)

### △ 1980-15 – *V. A. Vassiliev*

ℛ This homomorphism is canonical in the stable dimensions: if these singularities are sufficiently complicated with respect to a number $c$ and to the number $n$ of variables (namely, their $n(2c+1)$-jets are equal to zero) then this homomorphism is well-defined for all $i \leq c$. Moreover, it is an isomorphism, and these "stable" groups $H^i$ are isomorphic to the correesponding cohomology groups of the iterated loop space $\Omega^{2n}S^{2n+1}$, see [1, 2].

This is enough to define the stable cohomology ring; however, I do not know the answers for cohomology and homomorphisms in nonstable dimensions.

See also the comments to problems *1975-19*, *1975-24*, *1976-28*, *1985-7*, and *1985-22*.

[1] VASSILIEV V. A. Topology of complements to discriminants and loop spaces. In: Theory of Singularities and its Applications. Editor: V. I. Arnold. Providence, RI: Amer. Math. Soc., 1990, 9–21 (Adv. Sov. Math., 1.)

[2] Vassiliev V. A. Complements of Discriminants of Smooth Maps: Topology and Applications, revised edition. Providence, RI: Amer. Math. Soc., 1994. (Transl. Math. Monographs, 98.)

### 1980-16

ℋ This is a problem in paper [1] (§ 4); see also the comment in [2] (p. 68).

[1] Arnold V. I. On some problems in singularity theory. In: Geometry and Analysis. Papers dedicated to the memory of V. K. Patodi. Bangalore: Indian Acad. Sci., 1980, 1–9. [*Reprinted in*: *Proc. Indian Acad. Sci. Math. Sci.*, 1981, **90**(1), 1–9.]

[2] Arnold V. I. Some open problems in the theory of singularities. In: Singularities. Part 1 (Arcata, CA, 1981). Editor: P. Orlik. Providence, RI: Amer. Math. Soc., 1983, 57–69. (Proc. Symposia Pure Math., 40.)

ℛ See the comment to problem *1979-6*.

### 1980-17

ℋ This is a problem in paper [1] (§ 4); see also the comment in [2] (p. 68).

[1] Arnold V. I. On some problems in singularity theory. In: Geometry and Analysis. Papers dedicated to the memory of V. K. Patodi. Bangalore: Indian Acad. Sci., 1980, 1–9. [*Reprinted in*: *Proc. Indian Acad. Sci. Math. Sci.*, 1981, **90**(1), 1–9.]

[2] Arnold V. I. Some open problems in the theory of singularities. In: Singularities. Part 1 (Arcata, CA, 1981). Editor: P. Orlik. Providence, RI: Amer. Math. Soc., 1983, 57–69. (Proc. Symposia Pure Math., 40.)

ℛ See the comment to problem *1972-3*.

# 1981

### 1981-1 — *M. B. Sevryuk*

ℛ The discrete groups generated by reflections in the walls of simplices in the Lobachevskian space $\mathcal{H}^n$ were enumerated in paper [1] for $n \leq 3$, and in work [2] for $n = 4$. Besides that, F. Lannér [2] proved that there are no such groups for $n > 4$.

[1] COXETER H. S. M., WHITROW G. J. World-structure and non-Euclidean honeycombs. *Proc. Roy. Soc. London, Ser. A*, 1950, **201**, 417–437.

[2] LANNÉR F. On complexes with transitive groups of automorphisms. *Commun. Sém. Math. Univ. Lund*, 1950, **11**, 71 pp.

## 1981-2 — *N. N. Nekhoroshev*

ℛ The definition of steep functions $H$ used in the perturbation theory (see the comment to problem *1966-2*) is based on a more elementary principal definition of *simple* steep functions $f$. (In the comment we shall use the term "simple" in order to distinguish from steep functions $H$). The definition by E. E. Landis of the *uniform exponent k of the simple steepness* of a function $f$ at the point $x = 0$ is given in the comment to problem *1978-3*. This latter definition can also be reformulated as follows. Let us denote by $m_f(\rho)$ the minimal value of the length of the gradient of the function $f$ on the sphere of radius $\rho$: $m_f(\rho) = \min_{|x|=\rho} \left|\operatorname{grad} f|_x\right|$. For any family $F$ of functions $\{F(\cdot, \alpha), \alpha \in \mathcal{V}\}$, where $F(x, 0) \equiv f(x)$, there exist constants $C > 0$, $\delta > 0$ and a neighborhood $V \subset \mathcal{V}$ of the point $\alpha = 0$ in the space of parameters, such that

$$\max_{0 \le \rho \le \varepsilon} m_{f_\alpha}(\rho) > C\varepsilon^k \quad \text{for all } \varepsilon \in (0, \delta], \ \alpha \in V, \tag{1}$$

where $f_\alpha(x) \equiv F(x, \alpha)$. This inequality means that, for each function $f_\alpha$ of the family and each $\varepsilon \in (0, \delta]$, there exists a sphere $|x| = \rho$ in the ball $|x| \le \varepsilon$ such that $m_{f_\alpha}(\rho) > C\varepsilon^k$ where $\rho = \rho_{\alpha,\varepsilon}$. The constants $C$ and $\delta$ are called *coefficients of simple steepness* of the function $f$ at the point $x = 0$.

The original definition in [3, 4] is slightly different. It supposes that the point $x = 0$ is a critical point of every function $f_\alpha$, $\alpha \in V$. One may think that the difference is not principal. However, there is no proof that both definitions of the exponent $k(f)$ for any function $f$ at its critical point are equivalent.

Below we will explain the reason why the estimate (1) has such a complicated expression. This estimate appears in applications and must be uniform for any deformation $f_\alpha$ of the function $f$. An estimate simpler than (1), which also characterizes the growth of the minimal length of the gradient of $f$, can be readily given. It is the well known estimate of "Łojasiewicz type," $m_f(\varepsilon) > C\varepsilon^k$ for all $\varepsilon \in (0, \delta]$. However, in the degenerate, i. e., non-Morse, case, this estimate cannot be made uniform on the parameter even for the definition in [3,4]. The estimate (1) has no such defect.

Steep functions $H$ used in the theory of stability of Hamiltonian perturbations of integrable systems are defined in the following way. Let a function $H$ be defined in a domain of a Euclidean vector space $\mathbb{E}^n$ such that the point $x = 0$ of the domain is not its critical point. Then $H$ is a *steep function* at the point $x = 0$ if there are some constants $k_m$, $C_m > 0$, $\delta_m > 0$, $m = 1,\dots,n-1$, and a neighborhood $U$ of the point $x = 0$ having the following property. Let $\xi$ be a arbitrary point of $U$, and $\lambda$ be an arbitrary affine subspace of $\mathbb{E}^n$ passing through this point and lying on the hyperspace $dH|_{x=\xi} = 0$, i. e., in the tangent space at the point $x = \xi$ to the level surface $H^{-1}(C)$ of the function $H$ where $H(\xi) = C$. Then the restriction $f = f_{\xi,\lambda} := H|_\lambda$ of the function $H$ to $\lambda$ is the simple steep function at the point $x = \xi$ with the exponent $k = k_m$ and the constants $C = C_m$, $\delta = \delta_m$, where $m = \dim\lambda$. The numbers $k_1,\dots,k_{n-1}$ are called the *steepness exponents* of the steep function $H$ at the point $x = 0$. Note that this definition essentially is not invariant: it depends on the affine structure on the space of variables $x \in \mathbb{E}^n$.

Arnold's problem about worst steepness exponents $k_m(H)$ for a typical function $H$ can be divided into two parts. Let $\Sigma_m(k)$ denote the set of germ functions $f$ in $m$ variables with the uniform exponent (of the simple steepness) which is greater than or equal to $k$. We denote by $c_m(k)$ the codimension of $\Sigma_m(k)$ in the space $J_m$ of all germ functions $f$ at the critical point. The first problem is to establish the relation between the worst $k_m(H)$ of the typical $H$ and the codimension $c = c_m(k)$. The second problem is finding the value of $c = c_m(k)$.

In particular, if the above relation is, in some sense, a simplest one, then we can assume that the expression for the worst steepness exponents $k_m^w$ of typical function $H$ is given by formula (2) which can be defined in the following way. Consider the quantity $k_m(c)$ which is, in a sense, an inverted $c_m(k)$. More precisely, let $k_m(c)$ equal the maximal $k$ such that the codimension $c_m(k)$ does not exceed $c$: $k_m(c) := \max\{k | c_m(k) \le c\}$. Let us denote by $d_{m,n}$ the dimension of Grassmannian manifold of $m$-dimensional subspaces in the $n$-dimensional vector space. Then

$$k_m^w = k_m(d_{m,n-1}). \tag{2}$$

The uniform exponents $k(f)$ of simple steep functions $f$ as well as the uniform exponents $k_m(H)$ of steep functions $H$ were investigated in papers [1–3]. E. E. Landis computed the steepness exponents $k$ of typical functions $f$ in 1 and 2 variables. In particular, she obtained the worst exponents $k_m^w$ of functions $H$ in the simplest cases $n = 2, 3$.

The result of Yu. S. Il′yashenko is formulated in the comment to problem *1966-2*. Starting from this result and formula (2), we can obtain the answer to Arnold's question on the worst steepness exponents of the typical functions $H$: $k_m^w = d_{m,n-1} + 1$. However, there is no information about the relation between $k_m^w$ and $c_m(k)$. Thus, this problem of V. I. Arnold still remains unsolved.

[1] IL′YASHENKO YU. S. A steepness criterion for analytic functions. *Russian Math. Surveys*, 1986, **41**(1), 229–230.

[2] LANDIS E. E. Uniform steepness indices. *Uspekhi Mat. Nauk*, 1986, **41**(4), 179 (in Russian).

[3] NEKHOROSHEV N. N. Stable lower estimates for smooth mappings and for gradients of smooth functions. *Math. USSR, Sb.*, 1973, **19**(3), 425–467.

[4] NEKHOROSHEV N. N. An exponential estimate of the time of stability of nearly-integrable Hamiltonian systems. *Russian Math. Surveys*, 1977, **32**(6), 1–65.

## 1981-3 — *D. A. Popov*

ℛ The problem asks whether it is possible to generalize the following theorem by Colin de Verdière [1,2]: if the phases $S(x,\lambda)$ have only simple or parabolic singularities for $\lambda \in U$ then the following bound uniform on $\lambda$ holds in the parameter domain $U$:

$$|I_h(\lambda)| \equiv \left| \int_{\mathbb{R}^n} a(x,\lambda) e^{\frac{i}{h} S(x,\lambda)}\, dx \right| \le C_n(a,\lambda) h^{n/2} \sum_{x_c \in \operatorname{supp} a} |\det \partial_x^2 S(x_c,\lambda)|^{-1/2},$$

where $\partial_x^2 S(x,\lambda)$ is the Hessian, $x_c \in \mathbb{R}^n$ are critical points, $a \in C_0^\infty$ and $C_n(a,\lambda)$ is a finite constant. The fact that the theorem fails in such a form has been known for long (V. P. Palamodov, I. A. Ikromov). To convince oneself, it suffices to consider the problem on uniform estimates for Peirce's integral studied in detail in the paper [3]. This is a one-dimensional oscillating integral with the phase $S(x,\lambda) = \frac{1}{4}x^4 + \frac{\lambda_1}{2}x^2 + \lambda_2 x$, and its estimates in the circle $U$ ($\lambda_1^2 + \lambda_2^2 \le 1$) are sought. In this case, the caustic $K$ is given by the equation $A \equiv \frac{\lambda_1^3}{27} + \frac{\lambda_2^2}{4} = 0$, and it breaks the circle $U$ into two domains $U_1$ ($A > 0$) and $U_2$ ($A < 0$), containing respectively one ($x_0$) and three ($x_0$, $x_1$, $x_2$) critical points. If $\lambda \in U_1$ and $\lambda_1^2 + \lambda_2^2 \ge C$ then the value $|\det S(x_0,\lambda)|^{-1}$ is bounded ($|\det S(x_0,\lambda)| \ge C|\lambda_2|^{2/3}$ when $\lambda_2^2 > A$, and $|\det S(x_0,\lambda)| \ge C|\lambda_1|$ when $\lambda_2^2 < A$) and, therefore, Colin de Verdière's bound cannot hold in the domain $U_1$. The question about the limits of its validity arises.

In [3] it is shown that Colin de Verdière's bound is true for an arbitrary smooth family of phases $S(x,\lambda)$ (with any degeneracy) outside a narrow neighborhood of the caustic $K$. This neighborhood is given by the inequalities

$$|\det \partial_x^2 S(x,\lambda)|^{-1} \leq C_n(\varepsilon) h^{\frac{1-\varepsilon}{6}},$$
$$|g(\lambda)| \leq C_n(\varepsilon) h^{\frac{1-\varepsilon}{2}},$$

where $g(\lambda) = \min G(x_{ex})^{1/2}$, $x_{ex}$ are extremum points of the function $G = \sum_i (\partial_i S)^2$ ($\partial_i \equiv \partial/\partial x_i$) determined by the relations $\partial_i G = 0$, $x_{ex} \neq x_c$, and $\varepsilon$ is any number in the interval $0 < \varepsilon < 1$; $C_n(\varepsilon) \to \infty$ ($\varepsilon \to 0$). These inequalities, of course, do not exactly determine the boundary of the domain $V$ where Colin de Verdière's bound fails. Conjecturally, this domain is characterized by the following properties:

1) the caustic $K \subset V$, but

$$|\det \partial_x^2 S(x_c,\lambda)|^{-1} \leq C \qquad (\forall x_c,\ \lambda \in V);$$

2) the "thickness" of the domain $V$ is exponentially small, i. e.,

$$d(\lambda,K) \leq C_1 e^{-C_2 h^{-\gamma}}, \quad \gamma > 0 \qquad (\lambda \in V),$$

where $d(\lambda,K)$ is the distance between the point $\lambda$ and the caustic in any reasonable metric in the parameter space.

[1] Arnold V. I., Gusein-Zade S. M., Varchenko A. N. Singularities of Differentiable Maps, Vol. II: Monodromy and Asymptotics of Integrals. Boston, MA: Birkhäuser, 1988. (Monographs in Math., 83.) [*The Russian original* 1984.]

[2] Colin de Verdière Y. Nombre de points entiers dans une famille homothétique de domaines de $\mathbb{R}^n$. *Ann. Sci. École Norm. Sup., Sér.* 4, 1977, **10**(4), 559–575.

[3] Popov D. A. Estimates with constants for some classes of oscillatory integrals. *Russian Math. Surveys*, 1997, **52**(1), 73–145.

## 1981-4 — *E. Ferrand* Also: *1984-17*

ℜ M. L. Gromov [1] proved that there exist no embedded exact Lagrangian tori in $\mathbb{R}^4$.

[1] Gromov M. L. Pseudo holomorphic curves in symplectic manifolds. *Invent. Math.*, 1985, **82**(2), 307–347.

## 1981-6 — *V. A. Vassiliev*

ℛ In the case $\mathrm{U}(n)/\mathrm{O}(n)$, the problem was solved in [4]. Namely, the desired ring is isomorphic to $\mathbb{Z}_2[x_5, x_9, x_{11}, x_{13}, \ldots]$, i. e., to the ring of manifolds over $\mathbb{Z}_2$ in variables $x_i$, where $i = \deg x_i$ runs over all odd natural numbers except those of the form $2^k - 1$.

In the case $\mathrm{U}(n)/\mathrm{SO}(n)$, the rational homology is the same as of the stable Lagrangian Grassmannian itself: $\lim \pi_{n+k} T\lambda_n \otimes \mathbb{Q} \simeq H^*(\mathrm{U}(\infty)/\mathrm{O}(\infty), \mathbb{Q})$, see [3]. Moreover, the first 10 groups over $\mathbb{Z}$ were calculated by M. Audin [2]: $L_2 = 0$, $L_3 = L_4 = \mathbb{Z}_3$, $L_5 = L_6 = \mathbb{Z}$, $L_7 = L_8 = \mathbb{Z}_{15}$, $L_9 = \mathbb{Z}$, $L_{10} = \mathbb{Z} \oplus \mathbb{Z}_2$ (while $L_1 = \mathbb{Z}$ by [1]).


[1] ARNOLD V. I. Lagrange and Legendre cobordisms, I; II. *Funct. Anal. Appl.*, 1980, **14**(3), 167–177; **14**(4), 252–260.

[2] AUDIN M. Cobordismes d'immersions lagrangiennes et legendriennes. Paris: Hermann, 1987. (Travaux en Cours, 20.)

[3] ELIASHBERG YA. M. Cobordisme des solutions de relations différentielles. In: Séminaire Sud-Rhodanien de Géométrie. I (Lyon, 1983). Editors: P. Dazord and N. Desolneux-Moulis. Paris: Hermann, 1984, 17–31. (Travaux en Cours.)

[4] SMITH L., STONG R. E. Exotic cobordism theories associated with classical groups. *J. Math. Mech.*, 1968, **17**(12), 1087–1102.


## 1981-7 — *E. Ferrand*

ℛ This conjecture has been neither proved nor disproved yet. Weak forms were proved in papers [2, 3] which generalize partial results obtained previously by C. C. Conley and E. Zehnder, M. Chaperon, A. Weinstein and other authors. M. Damian in [1] (corrolaire 7.2) has found some conditions on the $\pi_1$ of $V$ under which the conjecture holds in its original form.


[1] DAMIAN M. Homologie de Novikov et applications en topologie symplectique. Thèse, École Polytechnique, Palaiseau, France, 1999.

[2] HOFER H. Lagrangian embeddings and critical point theory. *Ann. Institut Henri Poincaré, Analyse non linéaire*, 1985, **2**(6), 465–499.

[3] LAUDENBACH F., SIKORAV J.-C. Persistance d'intersection avec la section nulle au cours d'une isotopie hamiltonienne dans un fibré cotangent. *Invent. Math.*, 1985, **82**(2), 349–357.

## 1981-8 – *E. Ferrand*

ℛ This question was solved by P. Pushkar′ in [1].

[1] PUSHKAR′ P. E. Relative Morse theory. Preprint, 1999.

## 1981-9 – *V. L. Ginzburg*

Also: *1982-24, 1984-4, 1994-14, 1994-35, 1996-17, 1996-18*

*As I see it now, the work will consist of an introduction of about sixty pages, the translation proper (some two hundred printed pages) and over three hundred pages of various notes and commentaries.*

— Vladimir Nabokov, Letter to Henry Allen Moe, March 1953; see [32] (p. 135)

ℛ **1. The magnetic problem.** First recall the Hamiltonian description of the motion of a charge in a magnetic field on a manifold. Let $M$ be a Riemannian manifold and let $\sigma$ be a closed 2-form on $M$ (the magnetic field). Consider the twisted symplectic form $\omega = \omega_0 + \pi^*\sigma$ on $T^*M$. Here $\omega_0$ is the standard symplectic form on $T^*M$ and $\pi : T^*M \to M$ is the natural projection. The motion of a unit charge on $M$ in the magnetic field $\sigma$ is given by the Hamiltonian flow of the standard kinetic energy $H : T^*M \to \mathbb{R}$. In what follows we will refer to this flow as a *twisted geodesic flow*. If $M$ is a surface, the integral curves of the twisted geodesic flow project to the curves on $M$ whose geodesic curvature is $k = \sigma/dA$, where $dA$ is the area form. Hence, the problems in question all concern the existence of periodic orbits of twisted geodesic flows. (Note that $k$ need not be constant.)

We emphasize that, in contrast with the geodesic flow (the case when $\sigma = 0$), the dynamics of the twisted geodesic flow on a level $\{H = c\}$ depends on $c$. When $c$ is large, the twisted geodesic flow is close to the geodesic flow, although not necessarily equivalent to it. On the other hand, when $c > 0$ is small the flow is in some sense "governed" by $\sigma$.

To illustrate this point, consider the twisted geodesic flow on a closed surface of constant negative curvature $-1$ and $\sigma = dA$, see [1]. For $c \in (0; 1/2)$, the flow is periodic and the period goes to infinity as $c \to (1/2)-$. In this case, all orbits are periodic and contractible in $M$. For $c = 1/2$, the flow is the horocycle flow which is known to have no periodic orbits [17]. (This is a counterexample to the conjecture from problem *1994-35*, cf. problem *1981-9* and [13].) When $c > 1/2$, the flow is smoothly equivalent to the geodesic flow. In particular, every free nontrivial homotopy class of a map $\mathbb{S}^1 \to M$ contains the projection to $M$ of a periodic orbit and there are no periodic orbits whose projections are contractible.

Hence, when studying the existence of periodic orbits, it makes sense to treat separately the cases of high and low energy levels. For an arbitrary magnetic field $\sigma$, the existence question for periodic orbits on low energy levels is still poorly understood (see, however, [40]) and we will focus mainly on the case where $\sigma$ is symplectic. Note also that periodic orbits (contractible in $M$) may persist for all values of $H$. This is true, for example, for a flat torus and positive $k$, see [2, 23].

Below we focus exclusively on the symplectic geometry approach to the existence problem for periodic orbits, originating from [2]. A different approach based on the Morse–Novikov theory is not discussed here; see, e. g., [16, 35, 36, 41, 42].

**2. Twisted geodesic flows on surfaces.** In this section we briefly list results on the existence of periodic orbits of twisted geodesic flows on surfaces with nonvanishing magnetic field. Thus, throughout this section we assume that $k \neq 0$. As has already been pointed out, for a flat torus, every level $\{H = c\}$ carries at least three periodic orbits (four, if the orbits are non-degenerate) whose projection to the torus is contractible [2, 23]. This result, essentially solving problem *1984-4*, was obtained by Arnold in the mid-eighties as a consequence of the Conley–Zehnder theorem [7]. Since then it has served as the main motivation for applications of symplectic techniques to the study of periodic orbits of twisted geodesic flows. In a similar vein, consider an arbitrary closed orientable surface $M$ of genus $g$ with any metric. Then there are at least three periodic orbits on every low energy level (two, if $M = S^2$) and at least $2g + 2$ when the orbits are non-degenerate. See [2, 11, 13, 24] and the survey [12] for further details and references. This result is a partial solution of problem *1981-9*, cf. problems *1982-24* and *1996-18*.

The dynamics on low energy levels can also be studied by using the averaging method (cf. problem *1996-17*). We refer the reader to [3, 4, 6, 25, 43] and the references therein for a detailed discussion of this method, applications of KAM, and adiabatic invariants.

**3. Twisted geodesic flows in higher dimensions.** Before formulating a conjecture concerning the lower bounds for the number of periodic orbits of twisted geodesic flows in higher dimensions, let us discuss some of the relevant results. Throughout this section, $(M, \sigma)$ denotes a compact symplectic manifold of dimension $2m$. Note that $(M, \sigma)$ is then a symplectic submanifold of $(T^*M, \omega)$. Denote by $N$ the vector bundle over $M$ formed by symplectic orthogonals to $M$ in $T^*M$. For every $x \in M$, we have the Hamiltonian flow of $d^2H$ on $N_x$ and, as a result, a fiberwise linear flow on $N$, called the limiting flow. The Hamiltonian flow on $\{H = \varepsilon\}$ for a small $\varepsilon > 0$ is close, after suitable rescaling, to the limiting flow. When the eigenvalues of $d^2H|_{N_x}$ do not bifurcate, i. e., periodic orbits of the limiting flow form smooth manifolds in $\{d^2H = \varepsilon\}$, the results of the previous section extend to higher dimensions [14, 22].

Assume, for instance, that the metric is conformal to an almost-Kähler metric on $(M, \sigma)$, i. e., $H(X,X) = f \cdot \sigma(X, JX)$, where $f$ is a positive function and $J$ is an almost complex structure compatible with $\sigma$. (This assumption implies that all orbits of the limiting flow are closed and holds automatically when $M$ is a surface.) Then, on a low energy level, there are at least $\mathrm{CL}(M) + m$ periodic orbits [22], and at least $\mathrm{SB}(M)$, if the orbits are nondegenerate [14]. Here $\mathrm{CL}(M)$ stands for the cup-length of $M$ over $\mathbb{R}$ and $\mathrm{SB}(M)$ denotes the sum of Betti numbers. Under different non-bifurcation conditions the lower bounds can be improved: for example, when at every point of $M$ all eigenvalues are distinct, there are at least $m\mathrm{CL}(M)$ periodic orbits. The lower bounds from [22] are obtained using Moser's method [31], which can be viewed as a higher-dimensional version of the averaging over the limiting flow (cf. problem *1994-4*).

These results lead to the conjecture that in general, for a symplectic magnetic field and low energy levels, the number of periodic orbits is no less than $\mathrm{CL}(M) + 1$, or even $\mathrm{CL}(M) + m$, and $\mathrm{SB}(M)$ when the orbits are nondegenerate. Note that it is still unknown whether or not periodic orbits exist on a dense set of low energy levels. However, one can show that there are contractible (actually, "small") periodic orbits for a sequence of energy values converging to zero [15].

The main difficulty which arises in showing, by symplectic topology methods, that every low energy level carries a periodic orbit lies in the fact that it is hard to find a tractable variational principle which would pick up periodic orbits on a fixed energy level of non-contact type. Clearly, the fiberwise convexity of $H$ should be used here in an essential way; see the comment to problem *1994-13*. However, the existence of periodic orbits for a dense set of levels appears to be accessible by using symplectic or Floer homology. Finally, to obtain a lower bound on the number of periodic orbits, one needs a way to show that the action functional in question has sufficiently many critical points corresponding to geometrically distinct periodic orbits.

A different perspective on the magnetic problem in higher dimensions comes from the Weinstein–Moser theorem. Consider the following question: let $W$ be a $2m$-dimensional symplectic manifold and let $H: W \to \mathbb{R}$ be a proper smooth function which has a Morse–Bott nondegenerate minimum $H = 0$ along a compact symplectic submanifold $M$ of $W$. Does the Hamiltonian flow of $H$ have a periodic orbit on every energy level $\{H = \varepsilon\}$, where $\varepsilon > 0$ is small? We will refer to the affirmative answer to this question as the generalized Weinstein–Moser conjecture. According to Moser and Weinstein, when $W = \mathbb{R}^{2n}$ and hence $M$ is a point, every low energy level carries at least $n$ periodic orbits [31,45]. On the other hand, taking $W = T^*M$ and $H$ and $\omega$ as above, we see that the magnetic problem is just a particular case of the generalized Weinstein–Moser conjecture. In fact, the results of [14,15,22] are all proved in the context of this conjecture and apply to a broader class of Hamiltonian systems than the motion of a charge in a magnetic field.

**4. High energy levels and degenerate magnetic fields.** In this section we only mention some of the relevant results.

For any metric on $M = T^n$ and any magnetic field $\sigma$, almost all (in the sense of measure theory) levels of $H$ carry at least one periodic orbit. (See [14,20,21,26, 27].) This fact is established by showing that $T^*M$ has bounded Hofer–Zehnder capacity [19]. For any weakly exact $\sigma$ on a compact manifold $M$, there exists a sequence of positive numbers $c_k \to 0$ such that every level $\{H = c_k\}$ carries a contractible periodic orbit [29,40]. As has been shown by J. Mather, for $M = T^2$ with any $\sigma$ and a nonflat metric, there is a noncontractible periodic orbit on every high energy level; see [12]. It is a simple consequence of the Viterbo theorem [44] that, for any metric and $\sigma$ on $S^2$, there is a periodic orbit on every high energy level (cf. problem *1996-18*). Furthermore, the results of Bialy [5] on Hopf rigidity also serve as indirect evidence of existence of contractible periodic orbits when $M$ is a torus and $\sigma \neq 0$.

The dynamics of twisted geodesic flows is much better understood for exact magnetic fields. For example, in this case there is a periodic orbit on every high energy level, as is easy to see, e. g., from [18]. A sharper result can be obtained by directly applying variational methods [8]. Note in this connection that for low energy levels the action functional need not satisfy the Palais–Smale condition (see, e. g., [9]). The energy lower bound arising here is closely related to Mañe's critical value; see [8,9,30,39]. For exact magnetic fields on surfaces, the sharpest results for low energy levels come from the Morse–Novikov theory [16,41,42]. Finally, we refer the reader to the forthcoming paper [10] for a theorem closing the gap between the dynamics on low and high energy levels for exact twisted geodesic flows on surfaces.

Topological entropy of twisted geodesic flows was studied in, e. g., [28,33, 34,37,38].

**Acknowledgments.** The author is deeply grateful to Vladimir Arnold, Başak Gürel, and Debra Lewis for numerous remarks and suggestions.

[1] ARNOLD V. I. Some remarks on flows of line elements and frames. *Sov. Math. Dokl.*, 1961, **2**, 562–564.

[2] ARNOLD V. I. On some problems in symplectic topology. In: Topology and Geometry. Rohlin Seminar. Editor: O. Ya. Viro. Berlin: Springer, 1988, 1–5. (Lecture Notes in Math., 1346.) [*The Russian translation in*: Vladimir Igorevich Arnold. Selecta–60. Moscow: PHASIS, 1997, 425–429.]

[3] ARNOLD V. I. Remarks concerning the Morse theory of a divergence-free vector field, the averaging method, and the motion of a charged particle in a magnetic field. *Proc. Steklov Inst. Math.*, 1997, **216**, 3–13.

[4] BENETTIN G., SEMPIO P. Adiabatic invariants and trapping of a point charge in a strong non-uniform magnetic field. *Nonlinearity*, 1994, **7**(1), 281–303.

[5] BIALY M. Rigidity for periodic magnetic fields, *Ergod. Theory Dynam. Systems*, 2000, **20**(6), 1619–1626.

[6] CASTILHO C. The motion of a charged particle on a Riemannian surface under a non-zero magnetic field. *J. Differ. Equations*, 2001, **171**(1), 110–131.

[7] CONLEY C. C., ZEHNDER E. Birkhoff–Lewis fixed point theorem and a conjecture of V. I. Arnold. *Invent. Math.*, 1983, **73**(1), 33–49.

[8] CONTRERAS G., ITURRIAGA R., PATERNAIN G. P., PATERNAIN M. Lagrangian graphs, minimizing measures and Mañe's critical value. *Geom. Funct. Anal.*, 1998, **8**(5), 788–809.

[9] CONTRERAS G., ITURRIAGA R., PATERNAIN G. P., PATERNAIN M. The Palais–Smale condition and Mañe's critical values. *Ann. Henri Poincaré*, 2000, **1**(4), 655–684.

[10] CONTRERAS G., ITURRIAGA L., PATERNAIN G. P. Periodic orbits for exact magnetic flows on surfaces, in preparation.

[11] GINZBURG V. L. New generalizations of Poincaré's geometric theorem. *Funct. Anal. Appl.*, 1987, **21**(2), 100–106.

[12] GINZBURG V. L. On closed trajectories of a charge in a magnetic field. An application of symplectic geometry. In: Contact and Symplectic Geometry (Cambridge, 1994). Editor: C. B. Thomas. Cambridge: Cambridge University Press, 1996, 131–148. (Publ. Newton Inst., 8.)

[13] GINZBURG V. L. On the existence and non-existence of closed trajectories for some Hamiltonian flows. *Math. Z.*, 1996, **223**(3), 397–409.

[14] GINZBURG V. L., KERMAN E. Periodic orbits in magnetic fields in dimensions greater than two. In: Geometry and Topology in Dynamics (Winston-Salem, NC, 1998; San Antonio, TX, 1999). Editors: M. Barge and K. Kuperberg. Providence, RI: Amer. Math. Soc., 1999, 113–121. (Contemp. Math., 246.)

[15] GINZBURG V. L., KERMAN E. Periodic orbits of Hamiltonian flows near symplectic extrema. *Pacific J. Math.*, 2002, **206**(1), 69–91.
[*Internet*: `http://www.arXiv.org/abs/math.DG/0011011`]

[16] GRINEVICH P. G., NOVIKOV S. P. Nonselfintersecting magnetic orbits on the plane. Proof of the overthrowing of cycles principle. In: Topics in Topology and Mathematical Physics. Editor: S. P. Novikov. Providence, RI: Amer. Math. Soc., 1995, 59–82. (AMS Transl., Ser. 2, 170.)

[17] HEDLUND G. A. Geodesics on a two-dimensional Riemannian manifold with periodic coefficients. *Ann. Math., Ser.* 2, 1932, **33**, 719–739.

[18] HOFER H., VITERBO C. The Weinstein conjecture in cotangent bundles and related results. *Ann. Scuola Norm. Sup. Pisa Cl. Sci.* (4), 1988, **15**(3), 411–445.

[19] HOFER H., ZEHNDER E. Symplectic Invariants and Hamiltonian Dynamics. Basel: Birkhäuser, 1994. (Birkhäuser Adv. Texts.)

[20] JIANG M.-Y. Hofer–Zehnder symplectic capacity for 2-dimensional manifolds. *Proc. Roy. Soc. Edinburgh, Sect. A*, 1993, **123**(5), 945–950.

[21] JIANG M.-Y. Periodic solutions of Hamiltonian systems on hypersurfaces in a torus. *Manuscr. Math.*, 1994, **85**(3–4), 307–321.

[22] KERMAN E. Periodic orbits of Hamiltonian flows near symplectic critical submanifolds. *Internat. Math. Res. Notices*, 1999, **17**, 953–969.

[23] KOZLOV V. V. Variational calculus in the large and classical mechanics. *Russian Math. Surveys*, 1985, **40**(2), 37–71.

[24] KOZLOV V. V. Closed orbits and the chaotic dynamics of a charge in a periodic electromagnetic field. *Reg. Khaot. Dinamika*, 1997, **2**(1), 3–12 (in Russian).

[25] LITTLEJOHN R. G. A guiding center Hamiltonian: a new approach. *J. Math. Phys.*, 1979, **20**(12), 2445–2458.

[26] LU G. The Weinstein conjecture on some symplectic manifolds containing the holomorphic spheres. *Kyushu J. Math.*, 1998, **52**(2), 331–351.

[27] MA R. A remark on Hofer–Zehnder symplectic capacity in symplectic manifolds $M \times \mathbb{R}^{2n}$. *Chinese Ann. Math., Ser. B*, 1997, **18**(1), 89–98.

[28] MACARINI L. Entropy rigidity and harmonic fields. *Nonlinearity*, 2000, **13**(5), 1761–1774.

[29] MACARINI L. Periodic orbits for Hamiltonian flows and magnetic fields. Preprint, IMPA, 2001.

[30] MAÑE R. Lagrangian flows: the dynamics of globally minimizing orbits. In: International Conference on Dynamical Systems (Montevideo, 1995). A tribute to Ricardo Mañé. Editors: F. Ledrappier, J. Lewowicz and S. Newhouse. Harlow: Longman, 1996, 120–131. (Pitman Res. Notes Math. Ser., 362.)

[31] MOSER J. Periodic orbits near an equilibrium and a theorem by Alan Weinstein. *Commun. Pure Appl. Math.*, 1976, **29**, 727–747.

[32] NABOKOV V. Selected letters, 1940–1977. Editors: D. Nabokov and M. J. Bruccoli. San Diego, CA: Harcourt Brace Jovanovich, 1989.

[33] NICHE C. On the topological entropy of an optical Hamiltonian flow. *Nonlinearity*, 2001, **14**, 817–827.

[34] NICHE C. On the topological entropy of an optical Hamiltonian flow. [*Internet*: `http://www.arXiv.org/abs/math.DS/0010058`]

[35] NOVIKOV S. P. The Hamiltonian formalism and a many-valued analogue of Morse theory. *Russian Math. Surveys*, 1982, **37**(5), 1–56.

[36] NOVIKOV S. P., TAĬMANOV I. A. Periodic extremals of many-valued or not everywhere positive functionals. *Sov. Math. Dokl.*, 1984, **29**(1), 18–20.

[37] PATERNAIN G. P., PATERNAIN M. Anosov geodesic flows and twisted symplectic structures. In: International Conference on Dynamical Systems (Montevideo, 1995). A tribute to Ricardo Mañé. Editors: F. Ledrappier, J. Lewowicz and S. Newhouse. Harlow: Longman, 1996, 132–145. (Pitman Res. Notes Math. Ser., 362.)

[38] PATERNAIN G. P., PATERNAIN M. First derivative of topological entropy for Anosov geodesic flows in the presence of magnetic fields. *Nonlinearity*, 1997, **10**(1), 121–131.

[39] PATERNAIN G. P., PATERNAIN M. Critical values of autonomous Lagrangian systems. *Comment. Math. Helvetici*, 1997, **72**(3), 481–499.

[40] POLTEROVICH L. Geometry on the group of Hamiltonian diffeomorphisms. In: Proceedings of the International Congress of Mathematicians, Vol. II (Berlin, 1998). *Doc. Math.*, 1998, Extra Vol. II, 401–410 (electronic).

[41] TAĬMANOV I. A. Closed extremals on two-dimensional manifolds. *Russian Math. Surveys*, 1992, **47**(2), 163–211; errata: **47**(4), 247.

[42] TAĬMANOV I. A. Closed non-self-intersecting extremals of multivalued functionals. *Sib. Math. J.*, 1992, **33**(4), 686–692.

[43] TRUC F. Trajectoires bornées d'une particule soumise à un champ magnétique symétrique linéaire. *Ann. Institut Henri Poincaré, Physique théorique*, 1996, **64**(2), 127–154.

[44] VITERBO C. A proof of Weinstein's conjecture in $\mathbb{R}^{2n}$. *Ann. Institut Henri Poincaré, Analyse non linéaire*, 1987, **4**(4), 337–356.

[45] WEINSTEIN A. Normal modes for nonlinear Hamiltonian systems. *Invent. Math.*, 1973, **20**, 47–57.

## 1981-10 — *S. L. Tabachnikov* Also: *1984-3*

See paper [1].

[1] CHEKANOV YU. V. Caustics in geometrical optics. *Funct. Anal. Appl.*, 1986, **20**(3), 223–226.

## 1981-11 — *S. V. Chmutov*

Solved by O. P. Shcherbak [3]. The answer is presented in [2] (Ch. 7). See also [1].

[1] ARNOLD V. I. Singularities of Caustics and Wave Fronts. Dordrecht: Kluwer Acad. Publ., 1990. (Math. Appl., Sov. Ser., 62.)

[2] GIVENTAL A. B. Singular Lagrangian manifolds and their Lagrangian maps. In: Itogi Nauki i Tekhniki VINITI. Current Problems in Mathematics. Newest Results, Vol. 33. Moscow: VINITI, 1988, 55–112 (in Russian) [*The English translation*: *J. Sov. Math.*, 1990, **52**(4), 3246–3278.]

[3] SHCHERBAK O. P. Wave fronts and reflection groups. *Russian Math. Surveys*, 1988, **43**(3), 149–194.

## 1981-12

ℛ See the comments to problems *1993-27* and *1998-9*.

## 1981-13

ℛ See the comment to problem *1970-13*.

## 1981-14 — *I. A. Bogaevsky* Also: *1982-11, 1994-31*

ℛ Let us consider a collision-free medium of non-interacting particles. Suppose that, at the initial time moment, the particle at a point $x$ has a velocity $v_0(x)$, where $v_0$ is a smooth vector field of initial velocities. Then the movement of the medium can be described as a one-parameter family of maps; namely, the map $g_t : x \mapsto x + v_0(x)t$ sending an initial particle position into the final one is defined at each $t$. The map $g_0$ is the identity map, and for small $t$ the map $g_t$ is a one-to-one map and does not have critical points (i. e., points at which the derivative rank is smaller than the maximal possible one). However, starting from some time, faster particles outstrip slower ones, the velocity field becomes multi-valued, and the map $g_t$ acquires critical points that results in formation of particle clusters because the medium density becomes infinite at the critical values of the map $g_t$.

If the field of initial velocities is potential, the map $g_t$ is Lagrangian and the medium density is infinite at its caustics. Hence, the problem of describing the density singularities and their bifurcations in the course of time is reduced to the problem of describing Lagrangian singularities and their bifurcations, which has already been solved in small dimensions (for example, see [1]). It remains valid even if particles move in an external smooth potential field which can depend on time. The situation changes radically if we allow gravitational interaction between the particles themselves. The point is that, though the gravitational field created by the particles remains potential after formation of a caustic, it is not smooth any more, and the usual theory of Lagrangian singularities does not work in this case. However, the conjecture discussed in detail in [2] states that the singularities of the gravitational field do not destroy the topological picture of caustic bifurcations.

This conjecture was proved in [5] in the simplest case of a caustic arising in a homogeneous medium for typical real analytic initial conditions for the velocity field and density. From the mathematical point of view, in this work local analytic solutions for the so-called system of Vlasov–Poisson equations are obtained. These solutions correspond to all possible bifurcations of caustic appearance for typical analytic initial conditions. The velocity field of each solution is a surface of pleat type lying over the two-dimentional space-time. This surface is not smooth along the inverse image of the caustic, and its points can be divided into three types: a pleat point, two fold lines, two additional lines over the fold image. Slightly unexpectedly, the two fold lines together with the pleat point form a smooth curve, and the caustic is an analytic semicubic. It means that in the case under consideration the caustic bifurcation is the same as that without gravitation not only topologically but analytically as well! In [4, 5] local analytic solutions describing in detail the singularities along the two fold lines and the two additional lines over the fold image are obtained.

Thus, a completed mathematical study has been achieved only for the bifurcation corresponding to typical caustic appearance in a one-dimensional medium of non-colliding particles interacting gravitationally to each other, and only for analytic initial conditions.


[1] ARNOLD V. I. Catastrophe Theory. Berlin: Springer, 1992. [*The Russian original* 1990.]

[2] ARNOLD V. I. Mathematical problems in classical physics. In: Trends and Perspectives in Applied Mathematics. Editors: F. John, J. E. Marsden and L. Sirovich. New York: Springer, 1994, 1–20. (Appl. Math. Sci., 100.) [*The Russian translation in*: Vladimir Igorevich Arnold. Selecta–60. Moscow: PHASIS, 1997, 553–575.]

[3] ROYTVARF A. A. The motion of a continuous medium in the force field with a rooted singularity. *Moscow Univ. Mech. Bull.*, 1987, **42**(1), 24–27.

[4] ROYTVARF A. A. Two-valued velocity field with a square root singularity. *Moscow Univ. Mech. Bull.*, 1988, **43**(3), 16–19.

[5] ROYTVARF A. A. On the dynamics of a one-dimensional self-gravitating medium. *Physica D*, 1994, **73**(3), 189–204.

## 1981-16 — *S. Yu. Yakovenko*

ℜ The assertion, now commonly referred to as the *Dulac conjecture* or *Dulac problem*, was solved independently and by two completely different methods by Yu. Il′yashenko [6] and J. Ecalle [1]. In both cases the affirmative answer is derived from the *nonaccumulation theorem* asserting that limit cycles of an *analytic* vector field on the plane cannot accumulate to a *polycycle*, a separatrix polygon formed by one or more singular points of the vector field and arcs connecting these points.

Each proof is extremely involved and occupies an entire book. Il′yashenko's publication was preceded by several articles [2–5] proving the nonaccumulation theorem for special classes of polycycles and containing in a nutshell the basic ingredients of the general proof.

The finiteness theorem is widely considered as an absolute peak achievement in all the activity concerning the Hilbert 16th problem.

[1] ECALLE J. Introduction aux fonctions analysables et preuve constructive de la conjecture de Dulac. Paris: Hermann, 1992. (Actualités Math.)

[2] IL′YASHENKO YU. S. Limit cycles of polynomial vector fields with nondegenerate singular points on the real plane. *Funct. Anal. Appl.*, 1984, **18**(3), 199–209.

[3] IL′YASHENKO YU. S. Dulac's memoir "On limit cycles" and related questions of the local theory of differential equations. *Russian Math. Surveys*, 1985, **40**(6), 1–49.

[4] IL′YASHENKO YU. S. Separatrix lunes of analytic vector fields on the plane. *Moscow Univ. Math. Bull.*, 1986, **41**(4), 28–35.

[5] IL′YASHENKO YU. S. Finiteness theorems for limit cycles. *Russian Math. Surveys*, 1990, **45**(2), 129–203.

[6] IL′YASHENKO YU. S. Finiteness Theorems for Limit Cycles. Providence, RI: Amer. Math. Soc., 1991. (Transl. Math. Monographs, 94.)

## 1981-18 — *B. A. Khesin*

Also: *1993-31*, *1994-28*, *1994-29*

ℜ The construction in problem *1993-31*, apparently due to Lattès [6], is well known in the theory of dynamical systems (see, e. g., [2, 7]).

Yes, an ideal (nondissipative) $L^2$ dynamo exists on any manifold. For its existence it is sufficient for a vector field on a plane to have at least one nondegenerate saddle point. In the case of an $L^1$ dynamo and a generic magnetic field, the growth exponent is determined by the topological entropy of the velocity field; see the papers by I. Klapper and L.-S. Young [3] and by O. Kozlovskiĭ [4].

The question about a dissipative dynamo (problem *1994-28*) is much more subtle. Recently it was solved by O. Kozlovskiĭ under the strongest assumptions

(i. e., in the most general case). Namely, he constructed a steady Euler flow on a three-dimensional ball, which acts upon a divergence-free vector field such that the energy of the latter exponentially increases to infinity in spite of nonzero diffusion [5].

For a survey of the topological aspects in the problems of dissipative and nondissipative dynamo, as well as for possible modifications of these constructions, see book [1].

[1] ARNOLD V. I., KHESIN B. A. Topological Methods in Hydrodynamics. New York: Springer, 1998. (Appl. Math. Sci., 125.)

[2] KATOK A. B. Lyapunov exponents, entropy and periodic points for diffeomorphisms. *Inst. Hautes Études Sci. Publ. Math.*, 1980, **51**, 137–174.

[3] KLAPPER I., YOUNG L.-S. Rigorous bounds on the fast dynamo growth rate involving topological entropy. *Commun. Math. Phys.*, 1995, **173**(3), 623–646.

[4] KOZLOVSKIĬ O. S. An integral formula for topological entropy of $C^\infty$ maps. *Ergod. Theory Dynam. Systems*, 1998, **18**(2), 405–424.

[5] KOZLOVSKIĬ O. S. Private communication, 2000; paper in preparation.

[6] LATTÈS S. Sur l'itération des substitutions rationnelles et les fonctions de Poincaré. *C. R. Acad. Sci. Paris*, 1918, **166**(1), 26–28; erratum: 1918, **166**(2), 88.

[7] LYUBICH M. Dynamics of rational transformations: topological picture. *Russian Math. Surveys*, 1986, **41**(4), 43–117.

## 1981-21 – *A. A. Davydov*

The problem is very general. It includes investigations of singularities of local and global controllability, singularities of attainability, singularities of Bellman functions. Note that, for different types of control systems, the classification of generic singularities can be different. For example, generic singularities of the boundary of an attainable set on a torus for the system $\dot{x} = f(x,u)$ do not coincide in the cases where the set of values of the control parameter is either a circle or only two points (see [1]).

See also problems *1975-29, 1976-9, 1978-2, 1979-11, 1979-12, 1979-13, 1979-14* which are the specializations of this general question.

[1] DAVYDOV A. A. Structural stability of control systems on orientable surfaces. *Math. USSR, Sb.*, 1992, **72**(1), 1–28.

## 1981-22 — *V. P. Kostov, S. K. Lando*

ℜ The form $f(x)(dx)^\alpha$ could be thought of as an object which changes under the diffeomorphism $x \mapsto g(x)$ into $f(g(x))(g'(x))^\alpha(dx)^\alpha$ provided that $g'$ is close to 1. One can consider two situations: when $f$ is a germ of a holomorphic function at 0 (and $\alpha \in \mathbb{C}^*$), and when it is $C^\infty$-smooth (and $\alpha \in \mathbb{R}^*$). In the case where $x$ is of dimension $n > 1$, we set $dx = dx_1 \wedge \ldots \wedge dx_n$. The following cases are of particular interest: $n = 1$, $\alpha = -1$ (vector fields on a line), $\alpha = 1/2$ (semidensities), $n = 1$, $\alpha = 1$ (differential 1-forms).

Next, one can ask the question when a deformation of a given form is versal, i. e., every other deformation of the form is equivalent to one induced by the given deformation. "Equivalent" means "obtained by a diffeomorphism of the variable(s) depending smoothly or analytically on the parameters"; "induced" means "obtained by replacing the parameters by smooth or analytic functions of new parameters."

The problem was posed in [1, 2] and solved there for Poisson structures on the plane ($\alpha = -1$, $x \in \mathbb{C}^2$). A complete solution in all cases except for some "resonant" values of the power $\alpha$, can be found in the thesis [7] of the second author of the present comment. The main theorem states essentially that almost all versal deformations $F(x,\lambda)$ of the coefficient $f$ considered as a germ of a function yield versal deformations $F(x,\lambda)(dx)^\alpha$ of the differential form. The complete proof of the general case, including the resonant values of the power, can be found in [6, 9]. Earlier, A. B. Givental [3] proved the versality theorem in the case of the 1-dimensional argument $x$ and the series $\alpha = 1, 1/2, 1/3, \ldots, 1/n, \ldots$ of powers, and V. P. Kostov [4, 5] generalized this result to the 1-dimensional argument $x$ and arbitrary powers; the $C^\infty$-case is treated in [5].

Normal forms for the powers of volume forms were described by A. N. Varchenko [10] and, independently, by the second author of the present comment [8].

[1] ARNOLD V. I. Classification of the Poisson structures on the plane. *Uspekhi Mat. Nauk*, 1984, **39**(5), 256 (in Russian).

[2] ARNOLD V. I. Remarks on Poisson structures on the plane and other powers of volume forms. *Trudy Semin. Petrovskogo*, 1987, **12**, 37–46 (in Russian) [*The English translation*: *J. Sov. Math.*, 1989, **47**(3), 2509–2516.]

[3] GIVENTAL A. B. Lagrangian manifolds with singularities and irreducible $sl_2$-modules. *Russian Math. Surveys*, 1983, **38**(6), 121–122.

[4] KOSTOV V. P. Versal deformations of differential forms of degree $\alpha$ on the line. *Funct. Anal. Appl.*, 1984, **18**(4), 335–337.

[5] KOSTOV V. P. Versal deformations of differential forms of real power on a line. *Math. USSR, Izv.*, 1991, **37**(3), 525–537.

[6] KOSTOV V. P., LANDO S. K. Versal deformations of powers of volume forms. In: Computational Algebraic Geometry (Nice, 1992). Editors: F. Eyssette and A. Galligo. Boston, MA: Birkhäuser, 1993, 143–162. (Progr. Math., 109.)

[7] LANDO S. K. Deformations of differential forms. Ph. D. Thesis, Moscow State University, 1985 (in Russian).

[8] LANDO S. K. Normal forms of the degrees of a volume form. *Funct. Anal. Appl.*, 1985, **19**(2), 146–148.

[9] LANDO S. K. Deformations of differential forms. *Proc. Steklov Inst. Math.*, 1995, **209**, 145–173.

[10] VARCHENKO A. N. Local classification of volume forms in the presence of a hypersurface. *Funct. Anal. Appl.*, 1985, **19**(4), 269–276.

## 1981-23 — *V. A. Vassiliev*

ℜ This problem was solved by Val. S. Kulikov [1]. But later it turned out that the answer had already been known in the classical literature [2].

[1] KULIKOV V. S. Calculation of singularities of an imbedding of a generic algebraic surface in projective space $P^3$. *Funct. Anal. Appl.*, 1983, **17**(3), 176–186.

[2] SALMON G. Treatise on the Analytic Geometry of Three Dimensions, Vol. II, 5th edition. London: Longmans and Green, 1914; reprinted by Chelsea: New York, 1965.

## 1981-24 — *S. V. Chmutov*

ℜ The interest of Arnold's seminar to the problem about the maximal number of Morse singularities on a hypersurface of degree $d$ started with Bruce's paper [4]. His estimate was improved by A. B. Givental in [7]. However, the asymptotic (as $d \to \infty$) of the upper bound in these works remains $d^n/2$ for a hypersurface of degree $d$ in $\mathbb{C}P^n$. The first time this asymptotic was improved was by A. N. Varchenko using mixed Hodge structures and spectra of singularities (see the comment to problem *1980-11*). In the particular case $n = 3$ Varchenko's asymptotic is $\frac{23}{48}d^3$. Then Miyaoka [11] improved the asymptotic further (for the case $n = 3$ only and without using mixed Hodge structures) and obtained $\frac{4}{9}d^3$. The best known asymptotical lower bound was obtained by the author of the present comment (see [1], p. 419). For the case $n = 3$ it is $\frac{5}{12}d^3$, see [6].

For $n = 3$ the exact answer is known in the following cases.

— $d=3$: the answer is 4 (Cayley's cubic);

— $d=4$: the answer is 16 (Kummer's surface);

— $d=5$: the answer is 31 (the example is due to Togliatti [12], the upper estimate is due to Beauville [3], Givental, Varchenko);

— $d=6$: the answer is 65 (the example is due to Barth [2], the upper estimate is due to Jaffe and Ruberman [9]).

For other dimensions the situation is the following. The exact answer is known for $n=4$, $d=4$, and it is 45 (Burkhardt's quartic [5], the upper estimate is due to Varchenko). It is also known to be $\binom{[n/2]}{n+1}$ for $d=3$ and all $n$ (the upper estimate belongs to Varchenko, the examples were constructed by T. Kalker [10] and V. Goryunov [8]).

In [8] V. Goryunov obtained the best known asymptotic as $n \to \infty$ and $d=4$. It is

$$\sim \frac{3}{4} \cdot \frac{3^{n+1}}{\sqrt{\pi n}}.$$

In this case Varchenko's upper bound gives

$$\sim \frac{3}{2} \cdot \frac{3^{n+1}}{\sqrt{\pi n}}.$$

If one first takes the asymptotic as $d \to \infty$ and then as $n \to \infty$, the main term in the asymptotic of Chmutov's examples will be $\sim \sqrt{\frac{2}{\pi n}} \cdot d^n$. And in Varchenko's upper bound it will be $\sim \sqrt{\frac{6}{\pi n}} \cdot d^n$.


[1] Arnold V. I., Gusein-Zade S. M., Varchenko A. N. Singularities of Differentiable Maps, Vol. II: Monodromy and Asymptotics of Integrals. Boston, MA: Birkhäuser, 1988. (Monographs in Math., 83.) [*The Russian original* 1984.]

[2] Barth W. Two projective surfaces with many nodes, admitting the symmetries of the icosahedron. *J. Algebraic Geom.*, 1996, **5**(1), 173–186.

[3] Beauville A. Sur le nombre maximum de points doubles d'une surface dans $P^3$ ($\mu(5)=31$). *J. Géométrie Algébrique d'Angers / Algebraic Geometry* (Angers, 1979). Editor: A. Beauville. Sijthoff and Noordhoff, 1980, 207–215.

[4] Bruce J. W. An upper bound for the number of singularities on a projective hypersurface. *Bull. London Math. Soc.*, 1981, **13**(1), 47–50.

[5] Burkhardt H. Untersuchungen aus dem Gebiete der hyperelliptischen Modulfunktionen, Zweiter Teil. *Math. Ann.*, 1891, **38**, 161–224.

[6] Chmutov S. V. Examples of projective surfaces with many singularities. *J. Algebraic Geom.*, 1992, **1**(2), 191–196.

[7] GIVENTAL A. B. Maximum number of singular points on a projective hypersurface. *Funct. Anal. Appl.*, 1983, **17**(3), 223–225.

[8] GORYUNOV V. V. Symmetric quartics with many nodes. In: Singularities and Bifurcations. Editor: V. I. Arnold. Providence, RI: Amer. Math. Soc., 1994, 147–161. (Adv. Sov. Math., 21.)

[9] JAFFE D. B., RUBERMAN D. A sextic surface cannot have 66 nodes. *J. Algebraic Geom.*, 1997, **6**(1), 151–168.

[10] KALKER T. Cubic fourfolds with fifteen ordinary double points. Ph. D. Thesis, Leiden, 1986.

[11] MIYAOKA Y. The maximal number of quotient singularities on surfaces with given numerical invariants. *Math. Ann.*, 1984, **268**(2), 159–171.

[12] TOGLIATTI E. G. Sulle superficie algebriche col massimo numero di punti doppi. *Rend. Semin. Mat. Univ. Politec. Torino*, 1950, **9**, 47–59.

## ▽ 1981-26 – *D. A. Popov*

ℛ The problem of finding nontrivial estimates of the form

$$R(\lambda, D) = O(\lambda^{\beta(D)+\varepsilon}) \quad \forall \varepsilon > 0 \quad (\lambda \to \infty)$$

has a long history originating with Gauss. The method of trigonometric sums in the number theory was developed rather reasonably in order to solve this problem [4, 10]. The work [3] contains a review of known results. There the interplay between the problem of estimating $R(\lambda, D)$ and the problem of finding the number $\mathcal{N}(x, g, M)$ of eigenvalues $\lambda_n$, $|\lambda_n| < x$, for the Laplace–Beltrami operator defined by a Riemannian metric $g$ on a compact manifold $M$ without boundary is considered. In particular, the famous and yet unsolved circle problem (the problem of determining the infimum of the value $\beta(D)$ for the circle centered at the origin) is equivalent to the problem of estimating the second term of the asymptotic of $\mathcal{N}(\lambda^2, g, \mathbb{T}^2)$ for $\lambda \to \infty$ where $\mathbb{T}^2$ is the torus with planar metric. The infima $\beta_0(D)$ of $\beta(D)$ are known in exclusive cases only; and the situation with the values $\beta_0(\{D\}) = \max_{D \in \{D\}} \beta(D)$ for classes $\{D\}$ of domains is the same. It is known, for example, that if $D$ is a ball and $n \geq 4$ then $\beta_0(D) = n - 2$. The only considerably general result concerning nonconvex domains was obtained by Colin de Verdière in [1] where it was shown that for the class $\{D\}$ of domains with a smooth generic boundary and $n \leq 7$, the estimate for $R(\lambda, D)$ can be given by

$$\beta(D) = n - 2 + \frac{2}{n+1}, \quad \varepsilon = 0.$$

In papers [7–9] it is proved that with averaging on all rotations of the domain $D$ this result holds for arbitrary domains $D$ with smooth boundary and for all $n$. Recently it has been shown that Colin de Verdière's estimate is exact for $n = 3$ on some rather general class of three-dimensional bodies of revolution, and the equality $\beta_0(D) = \frac{3}{2}$ is realized for the solid torus with standardly embedded boundary [6]. The two-dimensional case has been examined most thoroughly. In Huxley's works the original estimate of $R(\lambda, D)$ with $\beta(D) = \frac{2}{3}$ for two-dimensional domains with a piecewise smooth boundary without flattening points was developed to that with $\beta(D) = \frac{2}{3}(1 - \frac{4}{73})$ [2]. The situation where flattenings of the boundary curve $\partial D$ are allowed was studied in papers [1, 5]. In [1], it is proved that $\beta_0(D) = 1 - \frac{1}{p+2}$ in this case, where $p$ is the maximal order of a curvature zero. The latter result does not survive small deformations of the boundary curve. The estimate with $\beta_0(D) = 1 - \frac{1}{p+2}$ is realized only when the tangent's slope in the point of maximal flattening is rational. If for each flattening point the corresponding slope does not have a good rational approximation (which means, for almost all slopes) the estimate of $R(\lambda, D)$ is the same as in the convex case, that is, flattening points give no contribution to the bound for $R(\lambda, D)$ [5].

[1] COLIN DE VERDIÈRE Y. Nombre de points entiers dans une famille homothétique de domaines de $\mathbb{R}^n$. *Ann. Sci. École Norm. Sup., Sér.* 4, 1977, **10**(4), 559–575.

[2] HUXLEY M. N. Exponential sums and lattice points, II. *Proc. London Math. Soc., Ser.* 3, 1993, **66**(2), 279–301.

[3] KOSYGIN D. V., MINASOV A. A., SINAĬ YA. G. Statistical properties of the spectra of Laplace–Beltrami operators on Liouville surfaces. *Russian Math. Surveys*, 1993, **48**(4), 1–142.

[4] LOO-KENG HUA. Abschätzungen von Exponentialsummen und ihre Anwendung in der Zahlentheorie. Leipzig: B. G. Teubner, 1959.

[5] MÜLLER W., NOWAK W. G. Lattice points in planar domains: applications of Huxley's "discrete Hardy–Littlewood method." In: Number-Theoretic Analysis (Vienna, 1988–1989). Editors: H. Hlawka and R. F. Tichy. Berlin: Springer, 1990, 139–164. (Lecture Notes in Math., 1452.)

[6] POPOV D. A. On the number of lattice points in three-dimensional bodies of revolution. *Izv. Math.*, 2000, **64**(2), 343–361.

[7] RANDOL B. On the Fourier transform of the indicator function of a planar set. *Trans. Amer. Math. Soc.*, 1969, **139**, 271–278.

[8] RANDOL B. On the asymptotic behavior of the Fourier transform of the indicator function of a convex set. *Trans. Amer. Math. Soc.*, 1969, **139**, 279–285.

[9] VARCHENKO A. N. On the number of integer points in families of homothetic domains in $\mathbb{R}^n$. *Funct. Anal. Appl.*, 1983, **17**(2), 79–83.

[10] VINOGRADOV I. M. Special Variants of the Method of Trigonometric Sums. Moscow: Nauka, 1976 (in Russian).

## △ 1981-26 — *M. B. Sevryuk*

ℜ Let $D \subset \mathbb{R}^n$ be a bounded domain with smooth boundary $\partial D$. Denote by $R(\lambda, D) = \lambda^n \operatorname{Vol}(D) - N(\lambda, D)$ the difference between the volume of domain $D$ stretched by a factor $\lambda$ and the number $N(\lambda, D)$ of the points with integer coordinates inside the stretched domain. In works by B. Randol [5–7], Y. Colin de Verdière [3], and A. N. Varchenko [8, 9], estimates of the difference $R(\lambda, D)$ as $\lambda \to +\infty$ are obtained as well as estimates averaged over the rotations and translations of the domain $D$. The asymptotic behavior of $R(\lambda, D)$ for $\lambda$ large depends on the singularities of the curvatures and the arithmetical properties of the boundary $\partial D$. Estimates of $R(\lambda, D)$ and related problems are discussed in detail in book [1]. In works [1, 4], an analysis of the asymptotics of oscillating integrals is used to obtain estimates of the quantities $R(\lambda, D)$.

As far as the author of the present comment knows, the effects of the singularities on the asymptotics of the numbers of integer points on submanifolds of the Euclidean space of positive codimensions and on Diophantine approximations on these submanifolds have been hardly studied.

Related questions on the number of rational points of so-called bounded height on the projective algebraic varieties are considered in paper [2].

[1] ARNOLD V. I., GUSEIN-ZADE S. M., VARCHENKO A. N. Singularities of Differentiable Maps, Vol. II: Monodromy and Asymptotics of Integrals. Boston, MA: Birkhäuser, 1988, § 6, n° 6.7. (Monographs in Math., 83.) [*The Russian original* 1984.]

[2] BATYREV V. V., MANIN YU. I. Sur le nombre des points rationnels de hauteur borné des variétés algèbriques. *Math. Ann.*, 1990, **286**(1–3), 27–43. [*Reprinted in*: Selected Papers of Yu. I. Manin. River Edge, NJ: World Scientific, 1996, 373–389.]

[3] COLIN DE VERDIÈRE Y. Nombre de points entiers dans une famille homothétique de domaines de $\mathbb{R}^n$. *Ann. Sci. École Norm. Sup., Sér.* 4, 1977, **10**(4), 559–575.

[4] POPOV D. A. Estimates with constants for some classes of oscillating integrals. *Russian Math. Surveys*, 1997, **52**(1), 73–145.

[5] RANDOL B. A lattice-point problem. I; II. *Trans. Amer. Math. Soc.*, 1966, **121**, 257–268; **125**, 101–113.

[6] RANDOL B. On the Fourier transform of the indicator function of a planar set. *Trans. Amer. Math. Soc.*, 1969, **139**, 271–278.

[7] RANDOL B. On the asymptotic behavior of the Fourier transform of the indicator function of a convex set. *Trans. Amer. Math. Soc.*, 1969, **139**, 279–285.

[8] VARCHENKO A. N. On the number of integer points in a domain. *Russian Math. Surveys*, 1982, **37**(3), 223–224.

[9] VARCHENKO A. N. On the number of integer points in families of homothetic domains in $\mathbb{R}^n$. *Funct. Anal. Appl.*, 1983, **17**(2), 79–83.

## 1981-28 — *I. A. Bogaevsky* Also: *1973-2*

The *convex hull* of a subset of an affine space is the intersection of all semispaces containing the subset. The boundary of the convex hull of a smooth compact hypersurface can have singularities (even in the case of a hypersurface without boundary). For example, the singularities of the boundary of the convex hull of a generic closed curve in the plane are exhausted by the second derivative discontinuities.

The problem is to study the singularities of the convex hulls of generic compact smooth (of the type $C^\infty$) hypersurfaces without boundary embedded into a four-dimensional affine space with respect to diffeomorphisms. The case of 3-dimensional space was investigated in [5] (see below).

A *singularity* of a convex hull is called its germ at a point where the boundary is not smooth. As usual, *generic* hypersurfaces are embeddings forming an open everywhere dense subset in the space of all considered embeddings equipped with the $C^\infty$-topology. In other words, we are interested only in the singularities of the convex hull which are not removed by a small perturbation of the original hypersurface.

At present this problem is "almost" solved; namely, in the generic case all singularities are enumerated: $\mathcal{R}_1$, $\mathcal{R}_2^0$, $\mathcal{R}_2^\pm$ [3], $\mathcal{R}_3$ and $\mathcal{V}_3$ [1]. Normal forms containing not more than one number *modulus* (a continuous invariant with respect to diffeomorphisms) are obtained for all of them except for $\mathcal{R}_3$. The singularity $\mathcal{R}_3$ does not have function moduli but contains at least nine number moduli; the exact quantity of the number moduli is unknown. In a neighborhood of a typical boundary point the convex hull of the generic compact hypersurface is diffeomorphic to a closed semispace. The boundary points where the convex hull has singularities $\mathcal{R}_1$ form a two-dimensional smooth surface. Singularities $\mathcal{R}_2^0$ appear along smooth curves, while singularities $\mathcal{R}_2^\pm$, $\mathcal{R}_3$ and $\mathcal{V}_3$ occur at isolated points of the boundary of the convex hull.

As for the singularities of the convex hulls of smooth submanifolds of an affine space of higher codimensions (for example, space curves), see the comment to problem *1972-12*.

**Three-dimensional space.** According to [5], the convex hull of a generic compact surface without boundary in three-dimensional space can have singularities of two types only. We call them "simplest" and "angular." *Simplest* singularities are diffeomorphic to the subgraph of the square of the distance from a point in the plane to a semiplane lying in it. *Angular* singularities are diffeomorphic to the subgraph of the square of the distance from a point in the plane to an angle $\beta$ lying in it, where $0 < \beta < \pi$ is a number modulus. In a neighborhood of a typical boundary point the convex hull of a generic compact surface without boundary is diffeomorphic to a closed semispace. The simplest singularities appear along smooth curves and angular singularities occur at isolated points of the convex hull boundary.

**Four-dimensional space.** Singularities $\mathcal{R}_1$, $\mathcal{R}_2^0$, and $\mathcal{R}_2^{\pm}$ are stabilizations of simplest and angular singularities in the four-dimensional space, and they are studied in [3]. Singularities $\mathcal{R}_1$ are diffeomorphic to the subgraph of the square of the distance from a point in the three-dimensional space to a semispace lying in it. Singularities $\mathcal{R}_2^0$, $\mathcal{R}_2^+$, and $\mathcal{R}_2^-$ are diffeomorphic to the square of the distance from a point in the three-dimensional space to a dihedral angle $\{x \geq y \cot\beta(z),\ y \geq 0\}$ lying in it, where $\beta(z) = \beta_0 + z$, $\beta(z) = \beta_0 + z^2$, and $\beta(z) = \beta_0 - z^2$, respectively, and $0 < \beta_0 < \pi$ is the only number modulus in each of the three normal forms.

Singularities $\mathcal{R}_3$ and $\mathcal{V}_3$ are described in [1] where it is proved that they complete the list of singularities of the convex hull of the generic compact hypersurfaces without boundary embedded into four-dimensional space. The singularity $\mathcal{R}_3$ does not have function moduli but its normal form has not been found, nor has the exact quantity of the number moduli. It is only known that this quantity is not less than nine though it is apparently much more. By definition, the singularity $\mathcal{R}_3$ is diffeomorphic to the germ at 0 of the set

$$\Big\{(x,y,z,t) \in \mathbb{R}^3 \times \mathbb{R} : \min_{p,q,r \geq 0} F(p,q,r;x,y,z,t) \leq 0\Big\}$$

where $F$ is a polynomial which, for $\deg p = \deg q = \deg r = 1$, $\deg x = \deg y = \deg z = 1$ and $\deg t = 2$, is expanded into quasihomogeneous terms $F = F_2 + F_3 + \cdots$ where

$$F_2 = p^2 + q^2 + r^2 + 2apq + 2bpr + 2cqr + 2px + 2qy + 2rz + t,$$

and numbers $a$, $b$ and $c$ are such that the square form $p^2+q^2+r^2+2apq+2bpr+2cqr$ is positive definite. For some $F$ (for example, $F=F_2$), the singularity $\mathcal{R}_3$ is diffeomorphic to the subgraph of the square of the distance from a point in the three-dinensional space to a curvelinear trihedral angle lying in it, though probably it is not correct in the general case.

The singularity $\mathcal{V}_3$ does not have moduli and, by definition, is diffeomorphic to the subgraph of the square of the distance to the closure $V_3$ of the component of the swallowtail complement which consists of polynomials without real roots:

$$V_3=\left\{(x,y,z)\in\mathbb{R}^3 : \forall\tau\in\mathbb{R}\ \tau^4+x\tau^2+y\tau+z\geq 0\right\}.$$

Locally the boundary of the convex hull is the subgraph of a continuously differentiable function (see, for instance, [5]), and its typical singularities are second derivative discontinuities. It is shown in [4] that the set of such discontinuities in a neighborhood of a singularity $\mathcal{V}_3$ is a sail-boat. A *sail-boat* is the union of the cut swallowtail (it is the boundary of the set $V_3$) and a half of the Whitney umbrella; their lines of self-intersections coincide and the tangent cones are transversal. According to [4], all sail-boats are locally diffeomorphic to each other.

**Higher dimensions.** According to [2], starting from five-dimensional space, a convex hull can have function moduli which are not removed by a small perturbation of the original compact hypersurface without boundary. For example, they appear as relations between number moduli (which exist in four-dimesional space) along lines formed by the singularities $\mathcal{R}_3$.

[1] BOGAEVSKY I. A. Singularities of convex hulls of three-dimensional hypersurfaces. *Proc. Steklov Inst. Math.*, 1998, **221**, 71–90.

[2] SEDYKH V. D. Functional moduli of singularities of convex hulls of manifolds of codimension 1 and 2. *Math. USSR, Sb.*, 1984, **47**, 223–236.

[3] SEDYKH V. D. Stabilization of singularities of convex hulls. *Math. USSR, Sb.*, 1989, **63**(2), 499–505.

[4] SEDYKH V. D. The sewing of a swallowtail and a Whitney umbrella in a four-dimensional controlled system. In: Proceedings of Gubkin State Oil and Gas Academy. Moscow: Neft′ i Gaz, 1997, 58–68 (in Russian).

[5] ZAKALYUKIN V. M. Singularities of convex hulls of smooth manifolds. *Funct. Anal. Appl.*, 1977, **11**(3), 225–227.

# 1982

## 1982-2 — *B. A. Khesin*

See A. Vaintrob's paper [1].


[1] VAINTROB A. YU. Darboux theorem and equivariant Morse lemma. *J. Geom. Phys.*, 1996, **18**(1), 59–75.


## 1982-5 — *M. B. Sevryuk*

The most important results in this direction (for torus mappings as well as for vector fields on the torus) were obtained by V. I. Arnold in paper [1] and by O. G. Galkin in the series of works [2–7].

See also problem *1984-16*.


[1] ARNOLD V. I. Remarks on the perturbation theory for Mathieu type problems. *Russian Math. Surveys*, 1983, **38**(4), 215–233.

[2] GALKIN O. G. Resonance regions for Mathieu type dynamical systems. *Russian Math. Surveys*, 1989, **44**(3), 191–192.

[3] GALKIN O. G. Resonance regions for Mathieu type dynamical systems on a torus. *Physica D*, 1989, **39**(2–3), 287–298.

[4] GALKIN O. G. Phase-locking for Mathieu type vector fields on the torus. *Funct. Anal. Appl.*, 1992, **26**(1), 1–6.

[5] GALKIN O. G. Phase-locking for maps of a torus: a computer assisted study. *Chaos*, 1993, **3**(1), 73–82.

[6] GALKIN O. G. Phase-locking for Mathieu type mappings of the torus. *Funct. Anal. Appl.*, 1993, **27**(1), 1–9.

[7] GALKIN O. G. Phase-locking for dynamical systems on the torus and perturbation theory for Mathieu-type problems. *J. Nonlinear Sci.*, 1994, **4**(2), 127–156.


## 1982-6 — *B. A. Khesin*

See paper [1] and the survey in book [2].


[1] ARNOLD V. I. Some remarks on the antidynamo theorem. *Moscow Univ. Math. Bull.*, 1982, **37**(6), 57–66.

[2] ARNOLD V. I., KHESIN B. A. Topological Methods in Hydrodynamics. New York: Springer, 1998. (Appl. Math. Sci., 125.)

## ▽ **1982-7** – *B. Z. Shapiro* Also: *1983-15, 1985-10*

ℛ For elliptic polynomials the question is considered in [1], and for hyperbolic polynomials in [2, 3]. In particular, it is shown that the list of singularities of elliptic and hyperbolic polynomials, as well as the list of corresponding singularities of the boundaries, stabilize when the number of parameters is fixed while either both the degree and the number of variables or just the degree grow. In [3] the list of simple (multi)singularities is presented and the relation between the singularities of elliptic and hyperbolic polynomials is clarified.


[1] MATOV V. I. Elliptic domains of general families of homogeneous polynomials and extreme functions. *Funct. Anal. Appl.*, 1985, **19**(2), 102–111.

[2] VAINSHTEIN A. D., SHAPIRO B. Z. Singularities of hyperbolic polynomials and of the boundary of the hyperbolicity domain. *Uspekhi Mat. Nauk*, 1985, **40**(6), 305 (in Russian)

[3] VAINSHTEIN A. D., SHAPIRO B. Z. Singularities of the boundary of the hyperbolicity domain. In: Itogi Nauki i Tekhniki VINITI. Current Problems in Mathematics. Newest Results, Vol. 33. Moscow: VINITI, 1988, 193–214 (in Russian). [*The English translation*: *J. Sov. Math.*, 1990, **52**(4), 3326–3337.]


## △ **1982-7** – *V. A. Vassiliev*

ℛ For a review of results on this problem and related ones, see Chapter III in book [1].

Elliptic polynomials were studied by V. I. Matov, see [2–4]. Hyperbolic manifolds were studied by A. D. Vainshtein and B. Z. Shapiro, see the comment by B. Z. Shapiro.


[1] ARNOLD V. I., VASSILIEV V. A., GORYUNOV V. V., LYASHKO O. V. Singularity Theory. II. Classification and Applications. Berlin: Springer, 1993. (Encyclopædia Math. Sci., 39; Dynamical Systems, VIII.) [*The Russian original* 1989.]

[2] MATOV V. I. The topological classification of germs of the maximum and minimax functions of a family of functions in general position. *Russian Math. Surveys*, 1982, **37**(4), 127–128.

[3] MATOV V. I. Elliptic domains of general families of homogeneous polynomials and extreme functions. *Funct. Anal. Appl.*, 1985, **19**(2), 102–111.

[4] MATOV V. I. Extremum functions of finite families of convex homogeneous functions. *Funct. Anal. Appl.*, 1987, **21**(1), 42–52.

## 1982-11

ℛ See the comment to problem *1981-14*.

## 1982-12

ℛ See the comment to problem *1975-15*.

## 1982-16 – *S. K. Lando*

ℛ A proof of a stronger statement, the semicontinuity of the spectrum of a singularity, exploiting mixed Hodge structures, was given by A. N. Varchenko in [2] and by J. Steenbrink in [1]. In the case $n = 2$ an elementary proof of the semicontinuity was given by the author of the present comment in 1981 (unpublished). I do not know whether an elementary proof in arbitrary dimension has ever been written.

[1] STEENBRINK J. Semicontinuity of the singularity spectrum. *Invent. Math.*, 1985, **79**(3), 557–565.

[2] VARCHENKO A. N. Asymptotic integrals and Hodge structures. In: Itogi Nauki i Tekhniki VINITI. Current Problems in Mathematics, Vol. 22. Moscow: VINITI, 1983, 130–166 (in Russian). [*The English translation*: *J. Sov. Math.*, 1984, **27**, 2760–2784.]

## 1982-17

ℋ The case of quadrics mentioned in the statement of this problem is considered in paper [1].

[1] VAINSHTEIN A. D., SHAPIRO B. Z. Higher-dimensional analogs of the theorems of Newton and Ivory. *Funct. Anal. Appl.*, 1985, **19**(1), 17–20.

## 1982-24

ℛ See the comment to problem *1981-9*.

# 1983

## 1983-1 — *Yu. M. Baryshnikov, M. Garay*

Also: *1984-20, 1990-19*

ℜ The following result was proved by Arnold [1]. Consider a smooth hypersurface $M$ in the projective space $\mathbb{C}P^n$. Fix a point $p \in M$ and denote by $T_pM$ the hyperplane of $\mathbb{C}P^n$ tangent to $M$ at $p$. Denote by $\Sigma \subset \mathbb{C}^n$ the set of values of the parameter $\lambda = (\lambda_1, \ldots, \lambda_n)$ for which the polynomial

$$x^{n+1} + \lambda_1 x^{n-1} + \lambda_2 x^{n-2} + \cdots + \lambda_n$$

has a double root. A swallowtail singular point of a hypersurface $N \subset \mathbb{C}P^n$ is a point $q$ such that the germ of $N$ at $q$ is biholomorphically equivalent to the germ at the origin of $\Sigma$. A *special parabolic point* (Arnold calls them multidimensional inflections) of the hypersurface $M$ is a smooth point $p \in M$ such that the projectively dual hypersurface to $M$ has a swallowtail singular point at $(T_pM) \in (\mathbb{C}P^n)^\vee$. Here $(\mathbb{C}P^n)^\vee$ denotes the dual space to $\mathbb{C}P^n$. Let $f : U \to \mathbb{C}$ be a function with a generic isolated Morse critical point at the origin $0 \in U$. Consider the hypersurfaces $V_\varepsilon = \{p \in U : f(p) = \varepsilon\}$. Then there are $(n+1)!$ special parabolic points of the hypersurfaces $V_\varepsilon$ that converge to $p_0$ when $\varepsilon \to f(0)$. If $n = 2$, we recover Plücker's result [3] asserting that there are 6 inflection points vanishing at a generic Morse double point. [1]

These $(n+1)!$ multidimensional inflections in fact split into $n!$ clusters, with $n+1$ inflections in each.

In the real situation, the number of real inflections in each group may be 1 (for $n$ even), or 0 or 2 (for $n$ odd), like the number of real roots of $x^{n+1} = \varepsilon$. The real clusters correspond to the real roots in $\mathbb{R}P^{n-1}$ of $n-1$ real homogeneous polynomials of degrees $2, 3, \ldots, n$ (that is, the number is even and ranges between 0 and $n!$), see [2].

[1] ARNOLD V. I. Vanishing inflections. *Funct. Anal. Appl.*, 1984, **18**(2), 128–130.

[1] A. Cayley, G. Salmon [4] and H.-G. Zeuthen investigated thoroughly the numerology of special points on the dual pairs of hypersurfaces in three-dimensional projective space. Yet it seems they never considered a Morse singular point on the surface. The reason for this is perhaps the very nonlocal structure of the surface dual to such surface singularity.

[2] BARYSHNIKOV Y. Real vanishing inflections and boundary singularities. In: Theory of Singularities and its Applications. Editor: V. I. Arnold. Providence, RI: Amer. Math. Soc., 1990, 129–136. (Adv. Sov. Math., 1.)

[3] PLÜCKER J. System Der Analytischen Geometrie (1834). In: Gesammelte Wissenschatfliche Abhandlungen, Band 1, Vol. 2. Leipzig: B. G. Teubner, 1898.

[4] SALMON G. A Treatise on the Analytic Geometry of Three Dimensions. Dublin: Hodges, Smith & Co., 1865. [*The German extended translation*: SALMON G., FIEDLER R. Analytische Geometrie des Raumes, II Teil. Leipzig: B. G. Teubner, 1880.]

## 1983-3 – *M. B. Sevryuk*

ℜ "The Conley–Zehnder theory" is another name for the theory of fixed points of symplectomorphisms. In a series of works [1] (see problems *1965-1–1965-3*), [2] (see problems *1966-4* and *1966-5*), [3] (see problems *1972-17*, *1972-18* and *1972-33* as well as problem *1970-10*), [4] (see problem *1976-39*), [5] (see problem *1976-39*), and [7], V. I. Arnold formulated several conjectures on fixed points of symplectomorphisms preserving the center-of-mass (see the comment to problem *1972-33*) and proved them for the symplectomorphisms which are not too far from the identity map. The best known conjecture is as follows: a symplectomorphism $F$ of a closed (i. e., compact and without boundary) symplectic manifold $M$ onto itself possesses at least as many fixed points as a smooth function on $M$ must have critical points, whenever $F$ preserves the center-of-mass. It is this statement that is usually referred to as the Arnold conjecture. In the rich history of the attempts to prove the conjecture, one of the milestones was paper [9] by C. C. Conley and E. Zehnder who verified the Arnold conjecture for tori $\mathbb{T}^{2n}$ of all the even dimensions with the standard symplectic structure (the comment to problem *1972-33* provides other references).

On the other hand, it is well known that many theorems concerning autonomous Hamiltonian flows or symplectomorphisms preserving the center-of-mass can be carried over respectively to reversible flows and reversible diffeomorphisms (note that up to now, reversible analogues of autonomous locally Hamiltonian flows and those of symplectomorphisms that do not preserve the center-of-mass are unknown). The most striking example of the Hamiltonian-reversible parallelism is the existence of the reversible KAM theory which resembles the Hamiltonian one in many aspects. Paper [18] lists 7 pairs of parallel Hamiltonian and reversible theories. Works [12, 13, 15, 16] present detailed surveys of the modern state of the theory of reversible systems, and an extensive bibliography is given therein. V. I. Arnold raised the problem (still unsolved) of constructing a unified "supertheory" whose even component corresponds to reversible systems, and odd component to Hamiltonian ones, see [17] (p. 297) and problem *1984-23*.

At the same time, the author of the present comment is unaware of reversible analogues of the Conley–Zehnder theory (the question on such analogues was raised by V. I. Arnold in paper [6], see also [17], p. 294–295). Now, we shall give some heuristic arguments suggesting that such analogues do not exist at all (and simultaneously explaining why the situation with the KAM theory is much more favorable). Consider the symplectic manifold

$$M = \mathbb{T}^{2n}_{\varphi} \times \mathbb{T}^{s}_{\psi} \times \mathbb{R}^{2m}_{x} \times \mathbb{R}^{s}_{y}$$

(the subscript at each factor indicates the letter denoting the coordinate in this factor) equipped with the standard symplectic structure

$$\omega^2 = \sum_{i=1}^{n} d\varphi_{2i-1} \wedge d\varphi_{2i} + \sum_{j=1}^{m} dx_{2j-1} \wedge dx_{2j} + \sum_{k=1}^{s} dy_k \wedge d\psi_k$$

(here $\mathbb{T}^N = \mathbb{R}^N/2\pi\mathbb{Z}^N$). One easily verifies that the shift symplectomorphism

$$F: M \to M, \qquad F: (\varphi, \psi, x, y) \mapsto (\varphi + a, \psi + \alpha, x + b, y + \beta)$$

preserves the center-of-mass if and only if $a \equiv 0 \pmod{2\pi}$ and $\beta = 0$.

The condition $a \equiv 0 \pmod{2\pi}$ is a *conditio sine qua non* of the Conley–Zehnder theory: A non-identical translation of the torus $\mathbb{T}^{2n}$ has no fixed points. The condition $\beta = 0$ is a heuristic "prerequisite" of the KAM theory for symplectomorphisms preserving the center-of-mass, since this condition excludes a systematic drift along the action variables $y_k$.

Now, consider the manifold

$$\widetilde{M} = \mathbb{T}^{N}_{\chi} \times \mathbb{T}^{n}_{\varphi} \times \mathbb{T}^{\nu}_{\psi} \times \mathbb{R}^{m}_{x} \times \mathbb{R}^{\mu}_{y}$$

where the involution

$$G: \widetilde{M} \to \widetilde{M}, \qquad G: (\chi, \varphi, \psi, x, y) \mapsto (\chi + \Pi_N, -\varphi, \psi, -x, y)$$

acts, where $\Pi_N = (\pi, \ldots, \pi) \in \mathbb{R}^N$. It is easy to verify that the shift diffeomorphism

$$\widetilde{F}: \widetilde{M} \to \widetilde{M}, \qquad \widetilde{F}: (\chi, \varphi, \psi, x, y) \mapsto (\chi + A, \varphi + a, \psi + \alpha, x + b, y + \beta)$$

is reversible with respect to the involution $G$ if and only if $A \equiv 0 \pmod{\pi}$, $\alpha \equiv 0 \pmod{\pi}$, and $\beta = 0$.

As before, the condition $\beta = 0$ is a heuristic "prerequisite" of the KAM theory (for reversible diffeomorphisms). But the reversibility of $\widetilde{F}$ imposes *no restrictions* on the vector $a$ whereas the vectors $A$ and $\alpha$ vanish, for $\widetilde{F}$ reversible, modd $\pi$ *only* (rather than modd $2\pi$). Thus, a torus diffeomorphism reversible with respect to an involution of this torus can well possess no fixed points at all.

In the reversible context, one can develop a theory of fixed points of *involutions themselves* rather than a theory of fixed points of diffeomorphisms reversible with respect to some non-trivial involution. (Of course, involutions are just mappings reversible with respect to the identity transformation but, in the general theory of reversible systems, such an example of reversible mappings is regarded as trivial.) Numerous results on the sets of fixed points of involutions (of compact as well as non-compact manifolds) and an extensive bibliography are presented in books [8, 10, 11]. In paper [14], the following theorem on fixed points of torus involutions is proved: *If all the fixed points of a continuous involution of the $N$-dimensional torus are isolated then the number of those fixed points is equal to either zero or* $2^N$. This theorem strongly resembles the Conley–Zehnder theorem concerning fixed points of torus symplectomorphisms preserving the center-of-mass.

Involutions of the torus $\mathbb{T}^N_\varphi$ without fixed points are exemplified by the involution $\varphi \mapsto \varphi + \Pi_N$, and those with $2^N$ isolated fixed points by the involution $\varphi \mapsto -\varphi$.

An analogous theorem holds for fixed points of sphere involutions: *If all the fixed points of a continuous involution of the $N$-dimensional sphere are isolated then the number of those fixed points is equal to either zero or* 2 (this is a very particular case of Theorem III.7.11 in book [8]).

Involutions of the sphere $S^N$ without fixed points are exemplified by the antipodal involution, and those with 2 isolated fixed points by the reflection $(x_1, \ldots, x_N, x_{N+1}) \mapsto (-x_1, \ldots, -x_N, x_{N+1})$ for the standard embedding of $S^N$ in $\mathbb{R}^{N+1}_x$.

However, the statement that *if all the fixed points of a continuous involution of a compact manifold $M$ are isolated then the number of those fixed points either is zero or is equal to (or no less than) $B(M)$, where $B(M)$ denotes the sum of the Betti numbers of $M$ over* $\mathbb{Z}$, is in general not valid. For instance, for each $g \geq 0$, one can easily construct an involution of the surface $M_g$ of genus $g$ with 2 isolated fixed points for $g$ even and with 4 isolated fixed points for $g$ odd. It suffices to consider an embedding of $M_g$ in $\mathbb{R}^3$ symmetric with respect to some straight line $l$ that intersects $M_g$ at 2 (respectively at 4) points. Then the desired involution is the reflection in $l$.

Note that for each $g \geq 0$, there is also an involution of surface $M_g$ with $2g+2 = B(M_g)$ isolated fixed points: It suffices to consider an embedding of $M_g$

in $\mathbb{R}^3$ symmetric with respect to some straight line $l$ that intersects $M_g$ at $2g+2$ points.


[1] ARNOLD V. I. Sur une propriété topologique des applications globalement canoniques de la mécanique classique. *C. R. Acad. Sci. Paris*, 1965, **261**(19), 3719–3722. [*The Russian translation in*: Vladimir Igorevich Arnold. Selecta–60. Moscow: PHASIS, 1997, 81–86.]

[2] ARNOLD V. I. The stability problem and ergodic properties of classical dynamical systems. In: Proceedings of the International Congress of Mathematicians (Moscow, 1966). Moscow: Mir, 1968, 387–392 (in Russian). [The English translation: *AMS Transl., Ser.* 2, 1968, **70**, 5–11.] [*The original is reprinted in*: Vladimir Igorevich Arnold. Selecta–60. Moscow: PHASIS, 1997, 95–101.]

[3] ARNOLD V. I. A comment to H. Poincaré's paper "Sur un théorème de géométrie." In: POINCARÉ H. Selected Works in Three Volumes (in Russian). Editors: N. N. Bogolyubov, V. I. Arnold and I. B. Pogrebysskiĭ. Vol. II. New methods of celestial mechanics. Topology. Number theory. Moscow: Nauka, 1972, 987–989 (in Russian).

[4] ARNOLD V. I. Fixed points of symplectic diffeomorphisms. In: Problems of present day mathematics. Editor: F. E. Browder. In: Mathematical Developments Arising from Hilbert Problems (Northern Illinois University, 1974). Part 1. Editor: F. E. Browder. Providence, RI: Amer. Math. Soc., 1976, 35–79; see XX, p. 66. (Proc. Symposia Pure Math., 28.)

[5] ARNOLD V. I. Some problems in the theory of differential equations. In: Unsolved Problems of Mechanics and Applied Mathematics. Moscow: Moscow University Press, 1977, 3–9 (in Russian).

[6] ARNOLD V. I. Reversible systems. In: Nonlinear and Turbulent Processes in Physics (Kiev, 1983), V. 3. Editor: R. Z. Sagdeev. Chur: Harwood Acad. Publ., 1984, 1161–1174. [*For the Russian translation see, e. g.*: Vladimir Igorevich Arnold. Selecta–60. Moscow: PHASIS, 1997, 355–363.]

[7] ARNOLD V. I. Mathematical Methods of Classical Mechanics, 2nd edition. New York: Springer, 1989, Appendix 9. (Graduate Texts in Math., 60.) [*The Russian original* 1974.]

[8] BREDON G. E. Introduction to Compact Transformation Groups. New York: Academic Press, 1972.

[9] CONLEY C. C., ZEHNDER E. The Birkhoff–Lewis fixed point theorem and a conjecture of V. I. Arnold. *Invent. Math.*, 1983, **73**(1), 33–49.

[10] CONNER P. E. Differentiable Periodic Maps. Berlin: Springer, 1979. (Lecture Notes in Math., 738.)

[11] CONNER P. E., FLOYD E. E. Differentiable Periodic Maps. Berlin: Springer, 1964.

[12] LAMB J. S. W. Reversing symmetries in dynamical systems. Ph. D. Thesis, University of Amsterdam, 1994.

[13] LAMB J. S. W., ROBERTS J. A. G. Time-reversal symmetry in dynamical systems: a survey. *Physica D*, 1998, **112**(1–2), 1–39.

[14] MONTALDI J. Caustics in time reversible Hamiltonian systems. In: Singularity Theory and its Applications, Part II. Editors: M. Roberts and I. Stewart. Berlin: Springer, 1991, 266–277. (Lecture Notes in Math., 1463.)

[15] ROBERTS J. A. G. Some characterisations of low-dimensional dynamical systems with time-reversal symmetry. In: Control and Chaos. Editors: K. Judd, A. Mees, K. L. Teo and T. L. Vincent. Boston, MA: Birkhäuser, 1997, 106–133. (Math. Model., 8.)

[16] ROBERTS J. A. G., QUISPEL G. R. W. Chaos and time-reversal symmetry. Order and chaos in reversible dynamical systems. *Phys. Rep.*, 1992, **216**(2–3), 63–177.

[17] SEVRYUK M. B. Reversible Systems. Berlin: Springer, 1986. (Lecture Notes in Math., 1211.)

[18] SEVRYUK M. B. Lower-dimensional tori in reversible systems. *Chaos*, 1991, **1**(2), 160–167.

## 1983-4 – *S. L. Tabachnikov*

ℛ See paper [1].

[1] CHEKANOV YU. V. Asymptotic behavior of the number of maxima of the product of linear functions of two variables. *Vestnik Moskov. Univ. Ser. Mat. Mekh.*, 1986, № 3, 93–94 (in Russian).

## 1983-5 – *S. V. Chmutov*

ℛ About this problem see paper [1]. For a nontrivial upper bound and an example of a polynomial with only maxima see [2]. The case of all $(d-1)^2$ critical points being real is dealt in [3].

[1] DURFEE A., KRONENFELD N., MUNSON H., ROY J., WESTBY I. Counting critical points of real polynomials in two variables. *Amer. Math. Monthly*, 1993, **100**(3), 255–271.

[2] DURFEE A. The index of grad $f(x,y)$. *Topology*, 1998, **37**(6), 1339–1361.

[3] SHUSTIN E. Critical points of real polynomials, subdivisions of Newton polyhedra and topology of real algebraic hypersurfaces. In: Topology of Real Algebraic Varieties and Related Topics. Dedicated to the memory of Dmitriĭ Andreevich Gudkov. Editors: V. Kharlamov, A. Korchagin, G. Polotovskiĭ and O. Viro. Providence, RI: Amer. Math. Soc., 1996, 203–223. (AMS Transl., Ser. 2, 173; Adv. Math. Sci., 29.)

## 1983-7 — *A. A. Glutsyuk*

ℛ The problem was solved independently by M. V. Yakobson [2] and M. R. Herman [1].

It is proved in [1, 2] that a circle diffeomorphism given by a trigonometric polynomial of degree $n > 0$ can have at most $2n$ periodic trajectories, and this estimate is sharp. This statement is proved by considering the extension of the diffeomorphism up to a holomorphic mapping $\mathbb{C}^* \to \mathbb{C}^*$ and using the theory of iterations of holomorphic mappings. This proof is generalized and yields an analogous result for trigonometrically-rational diffeomorphisms. (This general result is not stated explicitly in [2]. At the same time the proof presented in [2] for trigonometric polynomials remains valid without changes in this general case as well.) It is proved analogously that if $z \mapsto R(z)$ is a rational mapping of degree $d$ whose restriction to a circle (line) is a diffeomorphism, then the number of cycles in this circle (line) is not greater than $2d - 2$ if the diffeomorphism preserves the orientation of the circle, and is not greater than $2d - 1$ if it reverses the orientation [2]. It is not known, whether or not these estimates are sharp. If a smooth mapping of a circle onto itself is not a homeomorphism, then the number of its periodic orbits of the length $n$ grows as $e^{nh}$, where $h$ is a topological entropy of the mapping [2].

[1] HERMAN M. R. Majoration du nombre de cycles périodiques pour certaines familles de difféomorphismes du cercle. *An. Acad. Brasil. Ciênc.*, 1985, **57**(3), 261–263.

[2] YAKOBSON M. V. On the number of periodic trajectories for analytic diffeomorphisms of the circle. *Funct. Anal. Appl.*, 1985, **19**(1), 79–80.

## 1983-11

ℛ See the comment to problem *1978-6*.

## 1983-14 — *B. A. Khesin*

Also: *1985-4, 1987-1, 1990-10*

ℛ A discussion of V. V. Fock's results on the asymptotics of the corresponding eigenvalues and eigenfunctions, as well as comments by M. A. Shubin and C. King, can be found in book [1] (Chapter V, Remark 3.14).

[1] ARNOLD V. I., KHESIN B. A. Topological Methods in Hydrodynamics. New York: Springer, 1998. (Appl. Math. Sci., 125.)

## 1983-15

ℛ See the comment to problem *1982-7* by B. Z. Shapiro.

## 1983-16

ℛ See the comment to problem *1979-27*.

# 1984

## 1984-1 – *B. Z. Shapiro* Also: *1985-2*

ℛ Singularities of the boundary of the space of all Chebyshev systems (i. e., disconjugate linear ordinary differential equations) defined on the interval [0; 1] within the space of all linear ordinary differential equations of a given order was studied in [1, 2].

It is shown that the list of these singularities occuring in typical families with a fixed number of parameters coincides with the list of singularities of generic sections with the same number of parameters of the so-called train. A *train* is the hypersurface in the space of complete real flags consisting of all flags nontransversal to a given one. For more information about trains which are closely related to Schubert calculus on flag varieties, see the comment to problem *1987-7*. Singularities of the boundary of the space of all Chebyshev systems are also closely related to singularities of the boundary of the space of fundamental systems, see problem *1985-1*.


[1] Shapiro B. Z. Linear differential equations and real flag manifolds. *Funct. Anal. Appl.*, 1989, **23**(1), 82–83.

[2] Shapiro B. Z. Space of linear differential equations, and flag manifolds. *Math. USSR, Izv.*, 1991, **36**(1), 183–197.


## 1984-3

ℛ See the comment to problem *1981-10*.

## 1984-4

ℛ See the comment to problem *1981-9*.

## 1984-5

ℛ It is easily verified that

$$a_0 = \frac{(2^{2/3}-1)^{3/2}}{2} = 0.225098\ldots$$

And, herein, the point of cubic tangency of the parabola with the circle is $\left(\sqrt{1-2^{-2/3}}, 2^{-1/3}\right)$.

## 1984-6 — *B. S. Kruglikov*

ℛ According to Weinstein's theorem [11], the germ of a Poisson structure can be decomposed into the direct sum of a nondegenerate Poisson structure and a degenerate one. Since the first is integrable by the Darboux theorem, one should classify only degenerate Poisson structures.

The linearization of a Poisson structure $\nabla$ vanishing at $x \in M$ gives a Lie algebra structure on $T_x^*M$. If this Lie algebra $\mathfrak{g}$ is semisimple then the structure is formally linearizable [11]. Weinstein conjectured that this sufficient condition is necessary as well. In fact this hypothesis contradicted an example of stable Poisson structure on the plane considered by Arnold much earlier [1] (see also Appendix 13 in [2]). Other counterexamples to the Weinstein conjecture are contained in [8].

In general, the obstructions for linearization belong to the group $H^2(\mathfrak{g}; S^*\mathfrak{g})$ and are calculated via a spectral sequence associated with the $\mu$-filtration in the Poisson cohomology complex with differential $\partial_\nabla$ [8] (note that those important tools were already introduced in paper [7]). Moreover, these obstructions form a complete set of invariants giving the formal normal forms of Poisson structures.

In dimension 3, the Bianchi classification of Lie algebras gives different types of linearizations. For all non-Abelian Lie groups the technique of Lychagina allows the derivation of formal normal forms, as done in [9]. One result had been previously obtained by Dufour [5].

There are some other results on formal normal forms [6]. It is likely that some Poisson structures from the formal list allows also analytic/smooth conjugation to the corresponding normal form, as was done by Conn [3,4] for semisimple algebras $\mathfrak{g}$ (of compact type in the smooth case). A particular success in this direction is a result of [10]. But it seems that an important first step for the realization of this problem is a formal classification of Poisson structures degenerate along subvarieties.

If the algebra $\mathfrak{g}$ is commutative (the 1-jet of $\nabla$ vanishes, $[\nabla]_x^1 = 0$), then the 2-jet $[\nabla]_x^2$ defines a new structure on $T_x^*$ called the Lie–Sklyanin algebra, see [8]. Normal forms are given then in terms of the second cohomology group of this algebra, but applying the technique requires first the classification of Lie–Sklyanin algebras in small dimensions.

[1] ARNOLD V. I. Small denominators and problems of stability of motion in classical and celestial mechanics. *Russian Math. Surveys*, 1963, **18**(6), 85–191.

[2] ARNOLD V. I. Mathematical Methods of Classical Mechanics, 2nd edition. New York: Springer, 1989. (Graduate Texts in Math., 60.) [*The Russian original* 1974.]

[3] CONN J. F. Linearization of analytic Poisson structures. *Ann. Math., Ser.* 2, 1984, **119**(3), 577–601.

[4] CONN J. F. Normal forms for smooth Poisson structures. *Ann. Math., Ser.* 2, 1985, **121**(3), 565–593.

[5] DUFOUR J.-P. Linéarisation de certaines structures de Poisson. *J. Differ. Geom.*, 1990, **32**(2), 415–428.

[6] DUFOUR J.-P. Une nouvelle famille d'algèbres de Lie non dégénérées. *Indag. Math.* (*N. S.*), 1995, **6**(1), 67–82.

[7] LYCHAGIN V. V. Spectral sequences and normal forms of Lie algebras of vector fields. *Russian Math. Surveys*, 1983, **37**(5), 152–153.

[8] LYCHAGINA O. V. Normal forms of Poisson structures. *Math. Notes*, 1997, **61**(2), 180–192.

[9] LYCHAGINA O. V. Degenerate Poisson structures in dimension 3. *Math. Notes*, 1998, **63**(4), 509–521.

[10] WADE A. Normalisation formelle de structures de Poisson. *C. R. Acad. Sci. Paris, Sér.* I *Math.*, 1997, **324**(5), 531–536.

[11] WEINSTEIN A. The local structure of Poisson manifolds. *J. Differ. Geom.*, 1983, **18**(3), 523–557.

## ▽ **1984-7** – *A. A. Bolibruch*

For the first question (description of versal deformations of Fuchsian systems) see [3].

The answer to the second question is positive. More precisely, the following holds (see [1] for details):

**Theorem.** *Every system with regular singular points on the Riemann sphere can be presented as an isomonodromic confluence of singular points of some family of Fuchsian systems.*

The situation with the third question is the following. Every irregular singular point can be presented as a non isomonodromic confluence of Fuchsian ones. At every such Fuchsian singularity consider a special Levelt's basis (i. e., basis of solutions with all possible rates of growth). It turns out that under some generic assumptions connection matrices between these bases tend to Stokes matrices of the initial system, see [2].

J.-P. Ramis proved that monodromy matrices and Stokes matrices belong to the Galois group of the system of linear differential equations, see [4].


[1] BOLIBRUCH A. A. Regular singular points as isomonodromic confluences of Fuchsian singularities. *Russian Math. Surveys*, 2001, **56**(4), 745–746.

[2] GLUTSYUK A. Stokes operators via limit monodromy of generic perturbation. *J. Dynam. Control Systems*, 1999, **5**(1), 101–135.

[3] KOSTOV V. P. Normal forms for systems of linear differential equations with bifurcating Fuchsian singularities. *Differ. Uravneniya*, 1990, **26**(2), 353–356 (in Russian, for the English translation see *Differ. Equations*).

[4] RAMIS J. -P. Phénomène de Stokes et resommation. *C. R. Acad. Sci. Paris, Sér.* I *Math.*, 1985, **301**(4), 99–102.


## △ 1984-7 — *A. A. Glutsyuk, V. P. Kostov*

**The first question: find versal deformations.** A Fuchsian system is a linear system of ordinary differential equations meromorphically depending on a complex time and with poles of the first order. This is a particular case of regular systems, i. e., meromorphic linear systems whose solutions have a moderate growth rate (not faster than some real power of the distance to a given pole provided that the time remains within some sector with vertex at the pole). Versal deformations of Fuchsian systems are considered in [9, 10]. Unfoldings of non-Fuchsian systems are dealt with in [11, 12]. In both cases families of systems with bifurcation of the poles are considered. These families are polynomials in the time.

There is a related result by A. A. Bolibruch who described isomonodromic deformations of Fuchsian systems [2].

**The second question: regular singularities as limits of isomonodromic confluences of Fuchsian ones.** Partially solved by A. A. Bolibruch who showed that the limit of each confluenting isomonodromic family of Fuchsian systems on the Riemann sphere is a system with only regular singular points (i. e., a system with polynomial growth of solutions at the singular points).

**The third question: Stokes operators via limit monodromy.** A slightly different version of the third question was solved in [6]. This paper deals with an irregular nonresonant singularity of the linear equation:

$$\dot{z} = \frac{A(t)}{t^{k+1}}, \quad z \in \mathbb{C}^n, \quad |t| < 1, \quad n, k \in \mathbb{N}, \quad n \geq 2, \tag{1}$$

where $A(t)$ is a holomorphic $(n \times n)$-matrix valued function in the unit disk such that $A(0)$ has distinct eigenvalues. It is shown that

(2) for a generic deformation of (1) that splits its irregular singularity 0 into $k+1$ Fuchsian singularities of the perturbed equation, appropriate branches of appropriately normalized eigenbases of the monodromy operators around singularities of the perturbed equation converge to appropriate canonical sectorial solution bases of (1). In particular, the transition matrices between appropriately normalized eigenbases converge to appropriate Stokes matrices of the nonperturbed equation.

**Remark 1.** In the case where $k = 1$ and $n = 2$, both Stokes matrices are limits of transition matrices. In the general case, for any given generic deformation of (1), each Stokes matrix can be expressed in terms of the limit transition matrices.

**Remark 2** (from an unpublished paper by A. A. Glutsyuk)**.** In the simplest case where $k = 1$ and $n = 2$ (then the perturbed equation has two singularities), statement (2) implies that for a generic equation (1) and its generic deformation *each operator from the monodromy group of the perturbed equation (except for those corresponding to circuits along a fixed circle centered at 0) tends to infinity*, and *no operator from the monodromy group converges to a Stokes operator*. On the other hand, *commutators of appropriate (noninteger) powers (e. g., cubic roots) of appropriate monodromy operators around the singularities of the perturbed equation tend to the Stokes operators*.

Thus, the statement from the third question as it was formulated is false in general. On the other hand, to express all the Stokes operators as limit products of appropriate (*non-integer*) powers of the monodromy operators is an open question in the general case.

**History of the third question.** It is well known that each equation (1) is formally analytically equivalent to a direct sum of one-dimensional equations with right-hand sides polynomial in $t$ of degrees at most $k$. This direct sum is called the *formal normal form*. Generically, (1) is not analytically equivalent to its formal normal form: the normalizing series diverge.

In 1919 R. Garnier [5] studied some particular deformations of equations (1) under the additional condition that the eigenvalues of $A(0)$ form a convex polygon. He obtained analytic classification invariants for such equations by studying these deformations. For the general equation (1) the complete system of additional analytic invariants (Stokes operators) complementing the formal normal form to a complete system of analytic classification invariants was obtained later in the 1970s in the papers by Jurkat, Lutz, Peyerimhoff [8], Sibuya [15] and Balser, Jurkat, Lutz [1]. These invariants are linear operators acting in the space of solutions of (1). In 1984 V. I. Arnold proposed the conjecture that Stokes operators of equation (1) can be expressed as limits of some operators from the monodromy group of its deformation with Fuchsian singularities. A little later, a similar but slightly different conjecture was independently stated by J.-P. Ramis (1988), who proved the particular case of statement (2) for classical confluenting family of hypergeometric equations by direct calculation in 1989 [14]. In the late 1980s B. A. Khesin also proved a particular case of (2), but his result was not published. In 1991 A. Duval [4] proved (2) for biconfluenting family of hypergeometric equations (where the nonperturbed equation is equivalent to the Bessel equation) by direct calculation. In 1994 C. Zhang [16] found the expression of Garnier's invariants via Stokes operators. The results of Garnier and Zhang imply (2) for the deformations they considered, which are more general than those treated in [4, 14]. The conjecture that Stokes operators of the nonperturbed equation can be expressed via limits of transition operators between appropriate monodromy eigenbases of the perturbed equation was first stated by A. A. Bolibruch in 1994 and proved in [6]. An analogous result for the generic resonant case was obtained in an unpublished paper of the first author of the comment. The analogues of (2) for the nonlinear Stokes phenomena were proved in [7] (on one-dimensional conformal mappings tangent to the identity and their Écalle–Voronin moduli, two-dimensional saddle-node holomorphic vector fields and their Martinet–Ramis moduli, arbitrary-dimensional saddle-node fields and their sectorial central manifolds). For example, Écalle–Voronin moduli of a one-dimensional conformal mapping quadratically-tangent to the identity are expressed in terms of transitions between linearizing charts of its perturbation at the fixed points. This statement is extended to the case of arbitrary order of tangency.

A particular case of the latter result when the tangency is quadratic was obtained earlier by J. Martinet [13]. The first author of the present comment has also learned that in the case of tangency of arbitrary order for analytic deformations this result was known to specialists (e. g., to A. Douady), but the proof communicated to him by Douady was found to be based on quite different ideas. Very recently the first author became aware that in 2001 X. Buff, Tan Lei and some students of Douady had extended this result to a larger class of deformations (unpublished).


[1] BALSER W., JURKAT W. B., LUTZ D. A. Birkhoff invariants and Stokes' multipliers for meromorphic linear differential equations. *J. Math. Anal. Appl.*, 1979, **71**(1), 48–94.

[2] BOLIBRUCH A. A. On isomonodromic deformations of Fuchsian systems. *J. Dynam. Control Systems*, 1997, **3**(4), 589–604.

[3] BOLIBRUCH A. A. On isomonodromic confluences of Fuchsian singularities. *Proc. Steklov Inst. Math.*, 1998, **221**, 117–132.

[4] DUVAL A. Biconfluence et groupe de Galois. *J. Fac. Sci. Tokyo, Sect. IA Math.*, 1991, **38**(2), 211–223.

[5] GARNIER R. Sur les singularités irrégulières des équations différentielles linéaires. *J. Math. Pures Appl., Sér.* 8, 1919, **2**, 99–198.

[6] GLUTSYUK A. A. Stokes operators via limit monodromy of generic perturbation. *J. Dynam. Control Systems*, 1999, **5**(1), 101–135.

[7] GLUTSYUK A. A. Confluence of singular points and the nonlinear Stokes phenomena. *Trudy Moskov. Mat. Obshch.*, 2000, **62**, 54–104 (in Russian, for the English translation see *Trans. Moscow Math. Soc.*).

[8] JURKAT W. B., LUTZ D. A., PEYERIMHOFF A. Birkhoff invariants and effective calculations for meromorphic linear differential equations. *J. Math. Anal. Appl.*, 1976, **53**(2), 438–470.

[9] KOSTOV V. P. Normal forms for systems of linear differential equations with bifurcating Fuchsian singularities. *Differ. Uravneniya*, 1990, **26**(2), 353–356 (in Russian, for the English translation see *Differ. Equations*).

[10] KOSTOV V. P. Miniversal and isomonodromic deformations of Fuchsian systems with bifurcation poles. *Differ. Equations*, 1990, **26**(5), 561–571.

[11] KOSTOV V. P. Normal forms of unfoldings of non-Fuchsian systems. In: Aspects of Complex Analysis, Differential Geometry, Mathematical Physics and Applications (St. Konstantin, 1998). Editors: S. Dimiev and K. Sekigawa. River Edge, NJ: World Scientific, 1999, 1–18. [*Prépublication* Université de Nice, no. 348, juin 1993.]

[12] KOSTOV V. P. Normal forms of unfoldings of non-Fuchsian systems. *C. R. Acad. Sci. Paris, Sér.* I *Math.*, 1994, **318**(7), 623–628.

[13] MARTINET J. Remarques sur la bifurcation nœud-col dans le domaine complexe. In: Singularités d'équations différentielles (Dijon, 1985). *Astérisque*, 1987, **150–151**, 131–149.

[14] RAMIS J.-P. Confluence et résurgence. *J. Fac. Sci. Tokyo, Sect. IA Math.*, 1989, **36**(3), 703–716.

[15] SIBUYA Y. Stokes phenomena. *Bull. Amer. Math. Soc.*, 1977, **83**(5), 1075–1077.

[16] ZHANG C. Quelques études en théorie des équations fonctionnelles et en analyse combinatoire. Thèse, Institut de Recherche Mathematique Avancée, Université Louis Pasteur et CNRS (URA 01), 1994.

## 1984-8 — *S. V. Chmutov*

ℛ The problem is still open. However, there is a classification of groups generated by skew-symmetric transvections which includes the monodromy groups of all singularities. This classification was given by W. A. M. Janssen in [2] based on preceding works [1, 6].

The theory from papers [1, 2, 6] was applied to the problem about counting components of the intersection of two Schubert cells [3, 4]. Probably subsequent works [4, 5] of these authors contains an approach to the required axiomatic definition of the skew-symmetric monodromy groups.

[1] CHMUTOV S. V. Monodromy groups of critical points of functions, I; II. *Invent. Math.*, 1982, **67**(1), 123–131; 1983, **73**(3), 491–510.

[2] JANSSEN W. A. M. Skew-symmetric vanishing lattices and their monodromy groups, I; II. *Math. Ann.*, 1983, **266**(1), 115–133; 1985, **272**(1), 17–22.

[3] SHAPIRO B., SHAPIRO M., VAINSHTEIN A. Connected components in the intersection of two open opposite Schubert cells in $SL_n(\mathbb{R})/\mathcal{B}$. *Internat. Math. Res. Notices*, 1997, **10**, 469–493.

[4] SHAPIRO B., SHAPIRO M., VAINSHTEIN A. Skew-symmetric vanishing lattices and intersections of Schubert cells. *Internat. Math. Res. Notices*, 1998, **11**, 563–588.

[5] SHAPIRO B., SHAPIRO M., VAINSHTEIN A., ZELEVINSKY A. Simply laced Coxeter groups and groups generated by symplectic transvections. *Michigan Math. J.*, 2000, **48**, 531–551.

[6] WAJNRYB B. On the monodromy group of plane curve singularities. *Math. Ann.*, 1979/80, **246**(2), 141–154.

## 1984-9 — *V. A. Vassiliev*

ℛ The answer is negative. Indeed, suppose that the intersection index of two vanishing cycles of some distinguished basis is equal to 2, and the sum of these two cycles has non-zero intersection index with some third vanishing cycle. (These conditions hold, e. g., for any hyperbolic singularity, and also for any singularity more complicated than some hyperbolic one.) Then the consecutive reflections in the first two cycles provide diagrams with arbitraly large multilicities of edges. If I remember correctly, this or nearly this solution was given by A. M. Gabrielov at the same session of the seminar, at which the problem was posed.

## 1984-10 – *S. Yu. Yakovenko* Also: *1985-12*

H The problem reappeared recently in [2].

The question was motivated by Arnold's "Lagrangian Sturm theorem" [1] describing moments of non-transversality between Lagrangian planes moving by virtue of a linear Hamiltonian system with quadratic (nonautonomous) Hamiltonian, and a fixed Lagrangian plane.

R In [3] A. Givental proved that the system of Picard–Fuchs equations for *hyperelliptic* Abelian integrals are Hamiltonian. More precisely, integrals of the forms $(x^{2g-k}/y)\,dx$, $k = 1, \ldots, 2g$, over the level curves $\{H(x,y) = t\}$ of the Hamiltonian $H(x,y) = y^2 + x^{2g+1} + \lambda_1 x^{2g-1} + \cdots + \lambda_{2g-1}x$ satisfy a linear system which is Hamiltonian with respect to the symplectic form obtained from the intersection form.

The corresponding Hamiltonian is positive for those values of $t$ for which the real level curve $\{H = t\}$ possesses the maximal number $g+1$ of components. By the above Arnold theorem, this means that certain determinants involving the hyperelliptic integrals are non-oscillating on such intervals of $t$. This does not imply, however, any information on zeros of the integrals themselves.

In [4] it is shown that the Picard–Fuchs system for hyperelliptic integrals has a *hypergeometric* form, $(tE + A)\dot{I} = BI$ with $A$, $B$ constant matrices depending on $H$ (and $I$ is the column vector of hyperelliptic integrals). This is true also for Abelian integrals associated with a generic bivariate Hamiltonian $H$, if $I$ consists of integrals of all cohomologically independent monomial 1-forms of degree $\leq 2\deg H$ (ibid.).

[1] ARNOLD V. I. The Sturm theorems and symplectic geometry. *Funct. Anal. Appl.*, 1985, **19**(4), 251–259.

[2] ARNOLD V. I. Sur quelques problèmes de la théorie des systèmes dynamiques. *Topol. Methods Nonlinear Anal.*, 1994, **4**(2), 209–225. [*The Russian translation in*: Vladimir Igorevich Arnold. Selecta–60. Moscow: PHASIS, 1997, 533–551.]

[3] GIVENTAL A. B. Sturm's theorem for hyperelliptic integrals. *Leningrad Math. J.*, 1990, **1**(5), 1157–1163.

[4] NOVIKOV D. I., YAKOVENKO S. YU. Redundant Picard–Fuchs system for Abelian integrals. *J. Differ. Equations*, 2001, **177**(2), 267–306.

### 1984-11

ℛ See the comment to problem *1973-15.*

### 1984-12 – *B. A. Khesin*

ℛ Certain partial results are obtained by B. A. Khesin in paper [3], see also book [1]. Recently, T. Riviére [4] extended the ergodic approach to a certain restricted class of two-dimensional foliations, following the idea of [2]: to take the average over the leaves of two dimensional foliation one can consider a time average of a Brownian motion along the leaves.

See also the comment to problem *1973-23.*


[1] ARNOLD V. I., KHESIN B. A. Topological Methods in Hydrodynamics. New York: Springer, 1998. (Appl. Math. Sci., 125.)

[2] GARNETT L. Foliations, the ergodic theorem and Brownian motion. *J. Funct. Anal.*, 1983, **51**(3), 285–311.

[3] KHESIN B. A. Ergodic interpretation of integral hydrodynamic invariants. *J. Geom. Phys.*, 1992, **9**(1), 101–110.

[4] RIVIÉRE T. High-dimensional helicities and rigidity of linked foliations. Preprint, Courant Institute, New York, 1999.


### 1984-15

ℛ See the comment to problem *1993-27.*

### ▽ 1984-16 – *M. B. Sevryuk*

ℛ The most important results in this direction (in the multidimensional case as well; moreover, not only for differential equations on the torus but also for torus diffeomorphisms) were obtained by O. G. Galkin in the series of works [2–7] pointed out in the comment to problem *1982-5* and (for circle diffeomorphisms) by O. S. Kozlovskiĭ in paper [1].


[1] KOZLOVSKIĬ O. S. Resonance zone boundaries for families of circle diffeomorphisms. *Physica D*, 1991, **54**(1–2), 1–4.


### △ 1984-16

ℛ See the comment to problem *1980-2.*

## 1984-17

ℜ See the comment to problem *1981-4*.

## 1984-20

ℜ See the comment to problem *1983-1*.

## 1984-21 – *M. L. Kontsevich*

ℜ I am sure that other types of behavior are possible, the problem is that in this area it is very hard to prove results about a non-Gaussian behavior. For example, predictions of physicists on universality and conformal invariance seem to be yet out of reach by rigorous methods.

## 1984-22 – *B. A. Khesin*

ℜ It is still unknown whether the number of different typical local bifurcations of gradient dynamical systems depending on 4 parameters is finite or infinite. The 3-parameter case, where the number of different typical bifurcations is equal to 7, was described in papers [2,4], see also [3]. The corresponding global problem for 2 parameters was solved in [1].

[1] CARNEIRO M. J. D., PALIS J. Bifurcations and global stability of families of gradients. *Inst. Hautes Études Sci. Publ. Math.*, 1989, **70**, 103–168.

[2] KHESIN B. A. Bifurcations of gradient dynamical systems. In: Itogi Nauki i Tekhniki VINITI. Current Problems in Mathematics. Newest Results, Vol. 33. Moscow: VINITI, 1988, 113–155, 236 (in Russian). [*The English translation*: *J. Sov. Math.*, 1990, **52**(4), 3279–3305.]

[3] VEGTER G. Bifurcations of gradient vectorfields. In: Bifurcations, Theorie Ergodique et Applications (Dijon, 1981). *Astérisque*, 1982, **98–99**, 39–73.

[4] VEGTER G. Bifurcations of gradient vector fields. Ph. D. Thesis, University of Groningen, 1983.

## 1984-23

ℜ See the comment to problem *1983-3*.

# 1985

## 1985-2

ℜ See the comment to problem *1984-1*.

## 1985-4

ℜ See the comment to problem *1983-14*.

## 1985-5 – *E. Ferrand*

ℜ Klaus Mohnke proved in [1] that there exists at least one chord when the contact structure is the standard one (but when the contact form realizing it is arbitrary).

[1] MOHNKE K. Holomorphic disks and the chord conjecture. [*Internet*: `http://xxx.arXiv.org/abs/math.SG/0008014`]

## 1985-6

ℜ Cf. problem *1979-19*.

## 1985-7 – *V. A. Vassiliev*

ℜ 1. For any $n$ the cohomology rings of complements of bifurcation sets of levels (= the discriminants) of function singularities in $\mathbb{C}^n$ stabilize to some stable rings (for $n = 1$ see [1], for arbitrary $n$ see [4]).

2. Namely, this stable ring is isomorphic to $H^*(\Omega^{2n}S^{2n+1})$, where $\Omega^i$ denotes the $i$-fold loop space, see [5, 6]. For $n = 1$ this is the May–Segal formula for the cohomology of the stable braid group.

3. The similar stabilization holds for the cohomology rings of the complements of caustics; the stable ring is naturally isomorphic to $H^*(\Omega^{2n}\Sigma^{2n}\Lambda(n))$, where $\Sigma^i$ denotes the $i$-fold suspension and $\Lambda(n)$ is the $n$-th Lagrangian Grassmannian. All these isomorphisms are induced by the jet embeddings (sending a

function $f_\lambda$ to its 1-jet extension in the case of discriminants and to the 2-jet extension in the case of caustics). Thus, these facts provide a homological version of the Smale–Hirsh "*h*-principle."

4. An analog of the first statement is true for complements of the sets of complex functions having singularities of any fixed type. If the real codimension of this set is greater than 1 then it is true also for real singularities. (For $n = 1$ see [2]; for arbitrary $n$ the proof is essentially the same as in the complex case.)

5. For singular sets defined in terms of multisingularities (like the bifurcation diagrams of functions and the closures of the Maxwell sets) the stabilization in the $\mathbb{C}^1$-case was proved in [3]. For the case of functions $\mathbb{C}^n \to \mathbb{C}$ with arbitrary $n$ the proof is essentially the same (although probably is not published yet). Moreover, if the set of multisingular functions has real codimension $\geq 2$ then the same is true for real function multisingularities. For the complement of the entire bifurcation diagram (of zeros or functions) in $\mathbb{R}^n$ the similar statement formally fails: already the quantity of components of complements of such diagrams grows to infinity together with the complexity of singularities.

6. There are *stable irreducibility theorems* for the adjacency of strata of singularities and multisingularities of functions in $\mathbb{C}^n$. Namely, for any class of singularities or multisingularities (of finite codimension) there exists a sufficiently complicated singularity such that the corresponding stratum in the space of its versal deformation consists of a single component; see [4, 5]. Almost the same proof holds in the real situation. (We assume, of course, that all classes of (multi)singularities are determined by appropriate irreducible subsets in the spaces of (multi)jets of some finite order.)

See also problems *1975-19, 1975-24, 1976-28, 1980-15, 1985-22,* and *1998-8.*


[1] ARNOLD V. I. On some topological invariants of algebraic functions. *Trans. Moscow Math. Soc.*, 1970, **21**, 30–52.

[2] ARNOLD V. I. Spaces of functions with moderate singularities. *Funct. Anal. Appl.*, 1989, **23**(3), 169–177. [*The Russian original is reprinted in*: Vladimir Igorevich Arnold. Selecta–60. Moscow: PHASIS, 1997, 455–469].

[3] NEKRASOV N. A. On the cohomology of the complement of the bifurcation diagram of the singularity $A_\mu$. *Funct. Anal. Appl.*, 1993, **27**(4), 245–250.

[4] VASSILIEV V. A. Stable cohomologies of the complements of the discriminants of deformations of singularities of smooth functions. In: Itogi Nauki i Tekhniki VINITI. Current Problems in Mathematics. Newest Results, Vol. 33. Moscow: VINITI, 1988, 3–29 (in Russian). [*The English translation*: *J. Sov. Math.*, 1990, **52**(4), 3217–3230.]

[5] VASSILIEV V. A. Topology of complements to discriminants and loop spaces. In: Theory of Singularities and its Applications. Editor: V. I. Arnold. Providence, RI: Amer. Math. Soc., 1990, 9–21. (Adv. Sov. Math., 1.)

[6] VASSILIEV V. A. Topology of Complements to Discriminants. Moscow: PHASIS, 1997 (in Russian).

## 1985-8 — *Yu. M. Baryshnikov, M. Garay*

ℜ This problem is related to problems *1983-1, 1984-20, 1990-19.*

For smooth curves in affine or projective spaces, a local theory was constructed by M. E. Kazarian [4–6]. Kazarian mainly studies the bifurcation diagrams associated with the vanishing flattening points (i. e., the points where the osculating hyperplane has a higher order of tangency than generically). The complete list of bifurcation diagrams appearing in two or three parameter families is given. Then a classification of degenerate flattening points with respect to the Young diagram is introduced. A relation with the stratification of grassmannians by Schubert cells is found.

The first steps of the contactification of Kazarian's theory were made in [3]. They allow the generalization of Kazarian's theory for higher dimensional manifolds than curves, for example for bifurcations of parabolic points on surfaces. This gives generalizations of some of the results found in [1] and [2].

For complete intersections with isolated singular points, the first steps were made in [3]. The case of plane curves is already non-trivial and far from being understood. For instance, consider a generic $k$-parameter family of holomorphic Morse functions $f_\lambda : U \to \mathbb{C}$ with an only critical point at the origin. Here $U$ denotes an open neighborhood in the affine space $\mathbb{C}^2$ and $\lambda \in \mathbb{C}^k$. The following results were proved in [3].

1) there exist generic families $f_\lambda$ for which the number of vanishing inflection points of $V_{0,\mathbf{e}}$ at the critical point of $f_0$ is equal to $6+k$.

2) if 1) holds then for any $m \leq k$ there is a variety $V_m$ of dimension $k-m$ in the space of the parameter $\lambda \in \mathbb{C}^k$ for which the number of vansihing flattening point is equal to $6+m$.

3) if 1) holds then, for any other generic family $g_\lambda$ such that $g_0 = f_0$, the stratifications from 2) associated with $f_\lambda$ and $g_\lambda$ are biholomorphically equivalent.

Similar results hold in the real case.

To prove these results, a special versal deformation theory is introduced and a classification analogous to Arnold's $A$, $D$, $E$ classification is started. The monodromy of the vanishing inflection points as well as other invariants related to the projective structure are also studied in [3].

[1] ARNOLD V. I. Wave front evolution and equivariant Morse lemma. *Commun. Pure Appl. Math.*, 1976, **29**(6), 557–582; correction: 1977, **30**(6), 823. [*The Russian translation in*: Vladimir Igorevich Arnold. Selecta–60. Moscow: PHASIS, 1997, 289–318.]

[2] BRUCE J. W., GIBLIN P. J., TARI F. Families of surfaces: height functions and projections to planes. *Math. Scand.*, 1998, **82**(2), 165–185.

[3] GARAY M. The classical and Legendrian theory of vanishing flattening points of plane and spatial curves. Ph. D. Thesis, Université Paris 7, 2001.

[4] KAZARIAN M. E. Singularities of the boundary of fundamental systems, flattenings of projective curves, and Schubert cells. In: Itogi Nauki i Tekhniki VINITI. Current Problems in Mathematics. Newest Results, Vol. 33. Moscow: VINITI, 1988, 215–234 (in Russian). [*The English translation*: *J. Sov. Math.*, 1990, **52**, 3338–3349.]

[5] KAZARIAN M. E. Bifurcations of flattenings and Schubert cells. In: Theory of Singularities and its Applications. Editor: V. I. Arnold. Providence, RI: Amer. Math. Soc., 1990, 145–156. (Adv. Sov. Math., 1.)

[6] KAZARIAN M. E. Flattening of projective curves, singularities of Schubert stratifications of Grassmannians and flag varieties, and bifurcations of Weierstrass points of algebraic curves. *Russian Math. Surveys*, 1991, **46**(5), 91–136.

## 1985-10

See the comment to problem *1982-7.*

## 1985-11

See the comments to problem *1979-4.*

## 1985-12

See the comment to problem *1984-10.*

## 1985-13 – *M. L. Kontsevich*

Asymptotically it is clear by "geometric quantization" philosophy that dimensions, Klebsch–Gordan coefficients, $6j$-symbols, etc. (for representations with a large highest weight) can be expressed in terms of volumes of convex polytopes. Some results were obtained by A. Klyachko and others, see [1]. For explicit values of representation-theoretic numbers, there is no explicit geometric-combinatorial formula yet.

The $6j$-symbol is nonvanishing only if all triangles satisfy the triangle inequality. Asymptotically it is proportional to the $(-1/2)$-power of the volume of the tetrahedron if it is Euclidean, and exponentially small if the tetrahedron is Minkowskian [2].

[1] FULTON W. Eigenvalues, invariant factors, highest weights, and Schubert calculus. [*Internet*: `http://www.arXiv.org/abs/math.AG/9908012`]

[2] ROBERTS J. Classical $6j$-symbols and the tetrahedron. *Geometry & Topology*, 1999, **3**, 21–66 (electronic).
[*Internet*: `http://www.arXiv.org/abs/math-ph/9812013`]

## 1985-14 — *V. N. Karpushkin*

$\mathfrak{R}$ There is an analog of the theorem on uniform estimates with individual singularity index in the theory of exponential integrals in $\mathbb{R}^n$, $n = 1, 2$. Let $n = 2$, and let $f_1, \ldots, f_Q$ be real polynomials in $\mathbb{R}^2$, $Q \geq 1$, and $f_1(0) = \cdots = f_Q(0) = 0$.

Suppose that $\alpha$ is the singularity index of $\sum_{j=1}^{Q} f_j^2$ in $O \in \mathbb{R}^2$ and $p$ is the multiplicity of $\alpha$. Let $D$ be a neighborhood of the point $O \in \mathbb{C}^2$.

**Theorem 1.** *For arbitrary $f_1, \ldots, f_N$ there exist $C > 0$, $\varepsilon > 0$ and a neighborhood $A$ of the point $O \in \mathbb{R}^2$ such that*

$$\int_A \exp\left(-\lambda \sum_{j=1}^{Q} (f_j + F_j)^2\right) dx \leq C\lambda^{\alpha-1} (\ln \lambda)^p$$

*for all $\lambda \geq 2$ and real polynomials $F_j$ such that $\sup\limits_{z \in D} |F_j(z)| < \varepsilon$.*

This result follows from Theorem 2.3 in [1]. There is a similar result for $\mathbb{R}^1$.

The analog of the conjecture on the semicontinuity of the singularity index for exponential integrals fails for $\mathbb{R}^n$, $n \geq 3$.

**Theorem 2.** *Let $B_{\varepsilon,q} = -(x_1^4 - \varepsilon x_1^2 + x_2^2 + x_3^2)^2 - x_1^{2q} - x_2^{2q} - x_3^{2q}$, $q \geq 5$. Then for every $\varepsilon > 0$ and any neighborhood $A$ of the point $O \in \mathbb{R}^3$*

$$\int_A e^{\lambda B_{\varepsilon,q}}\, dx \Big/ C_{\varepsilon,q} \lambda^{-1/2+\gamma(q)} \to 1, \quad \lambda \to +\infty,$$

*for some $C_{\varepsilon,q} > 0$, $\gamma(q)$.*

Here $\gamma(q) \to 0$ when $q \to +\infty$, see [2].

In addition, for any neighborhood $A$ of the point $O \in \mathbb{R}^3$,

$$\int_A e^{\lambda B_{0,q}}\, dx \Big/ C_q \lambda^{-5/8} \to 1, \quad \lambda \to +\infty, \qquad C_q > 0.$$

This result follows from the counterexample by A. N. Varchenko [2]. For exponential integrals in $\mathbb{R}^n$, $n \geq 1$, without perturbations of the phase, see [3].


[1] KARPUSHKIN V. N. Uniform estimates of volumes. *Proc. Steklov Inst. Math.*, 1998, **221**, 214–220.

[2] VARCHENKO A. N. Newton polyhedra and estimation of oscillating integrals. *Funct. Anal. Appl.*, 1976, **10**(3), 175–196.

[3] VASSILIEV V. A. The asymptotic of exponential integrals, Newton diagrams, and classification of minimum points. *Funct. Anal. Appl.*, 1977, **11**(3), 163–172.


### 1985-15 – *V. I. Arnold*

Done by A. B. Givental in [1].


[1] GIVENTAL A. B. Singular Lagrangian manifolds and their Lagrangian mappings. In: Itogi Nauki i Tekhniki VINITI. Current Problems in Mathematics. Newest Results, Vol. 33. Moscow: VINITI, 1988, 55–112 (in Russian). [*The English translation*: *J. Sov. Math.*, 1990, **52**, 3246–3278.]


### 1985-17 – *S. M. Gusein-Zade*

It is not known yet.

### 1985-19 – *V. P. Kostov* Also: *1985-26*

The problem of proving the existence of the constant $C$ from problem *1985-26* was solved in [3] where the author used earlier results of V. I. Arnold and A. B. Givental, see [1,2]; a positive answer to problem *1985-19* was also given in [3]. The pyramid $\Pi$ is the hyperbolicity domain of the polynomial $x^{n+1} + a_1 x^{n-1} + \cdots + a_n$, i. e., the set of values of its coefficients for which it has only real roots. The pyramid $\Pi$ has a stratified structure which was studied in [1,3,4,7–9]. Its projection in the space $Oa_1 \ldots a_k$ is the set of points between and on the graphs of two Lipschitz functions defined on its projection in $Oa_1 \ldots a_{k-1}$, see [3]. The program to study hyperbolic polynomials (arrangements of their roots and those of their derivatives) is pursued in [5,6,10].

[1] ARNOLD V. I. Hyperbolic polynomials and Vandermonde mappings. *Funct. Anal. Appl.*, 1986, **20**(2), 52–53.

[2] GIVENTAL A. B. Moments of random variables and the equivariant Morse lemma. *Russian Math. Surveys*, 1987, **42**(2), 275–276.

[3] KOSTOV V. P. On the geometric properties of Vandermonde's mapping and on the problem of moments. *Proc. Roy. Soc. Edinburgh, Sect. A*, 1989, **112**(3–4), 203–211.

[4] KOSTOV V. P. On the hyperbolicity domain of the polynomial $x^n + a_1x^{n-1} + \cdots + a_n$, *Serdica Math. J.*, 1999, **25**(1), 47–70.

[5] KOSTOV V. P. Discriminant sets of families of hyperbolic polynomials of degree 4 and 5. *Serdica Math. J.*, 2002, **28**(2), 117–152. [*Prépublication* Université de Nice, no. 620, juin 2001.]

[6] KOSTOV V. P., SHAPIRO B. Z. On arrangements of roots for a real hyperbolic polynomial and its derivatives. *Bull. Sci. Math.*, 2002, **126**(1), 45–60. [*Prépublication* Université de Nice, no. 619, juin 2001.]

[7] MEGUERDITCHIAN I. A theorem on the escape from the space of hyperbolic polynomials. *Math. Z.*, 1992, **211**(3), 449–460.

[8] MEGUERDITCHIAN I. Géométrie du discriminant réel et des polynômes hyperboliques. Thèse de doctorat, Université de Rennes I, soutenue le 24.01.1991.

[9] MEGUERDITCHIAN I. Géométrie locale des polynômes hyperboliques. *C. R. Acad. Sci. Paris, Sér.* I *Math.*, 1991, **312**(11), 849–852.

[10] SHAPIRO B. Z., SHAPIRO M. Z. This strange and mysterious Rolle's theorem. *Amer. Math. Monthly*, submitted.


## 1985-20 — *B. A. Khesin*

A homogeneous vector field $f\colon \mathbb{R}^3 \to \mathbb{R}^3$ is called nondegenerate if the field vanishes only at the origin. Let $d$ be the homogeneity degree of the field. Note that the $\mathbb{Z}$-valued or $\mathbb{Z}_2$-valued index is a homotopy invariant of the fields with, respectively, odd or even degrees $d$.

V. M. Kharlamov classified the homogeneous fields in $\mathbb{R}^3$ for polynomials of lower degrees [1]. For each of the cases $d = 1$ and $d = 2$, there are 2 connected components in the space of the nondegenerate homogeneous fields. For $d = 3$ there are 8 classes. For $d = 4$ there are at least two classes, but a more precise answer is unknown. For higher $d$ the question is open. (The index $\operatorname{ind}(f)$ for even $d$ is subject to Arnold's inequality $|\operatorname{ind}(f)| \leq 3(d^2-1)/4+1$ and the comparison $\operatorname{ind}(f) \equiv d \pmod 2$.)

Singularities of the boundary of nondegenerate homogeneous vector fields are the Whitney umbrellas, see [2].

[1] KHARLAMOV V. M. Homotopy of real homogeneous vector fields. *Uspekhi Mat. Nauk*, 1983, **38**(5), 173 (in Russian).

[2] KHESIN B. A. Homogeneous vector fields and the Whitney umbrellas. *Russian Math. Surveys*, 1987, **42**(5), 171–172.

### ▽ 1985-22 — *F. Napolitano* Also: *1998-8*

ℛ Consider the miniversal unfolding $F_\lambda(x) = x^{k+1} + \lambda_1 x^{k-1} + \cdots + \lambda_{k-1}x$ of the singularity $A_k$ where $x \in \mathbb{C}$, $\lambda_i \in \mathbb{C}$. The cohomology ring of the complement of the Maxwell stratum (respectively, of the bifurcation diagram) in the space of parameters $\lambda$ stabilizes as $k \to \infty$ (N. A. Nekrasov [5]). Analogous stabilization occurs for the cohomology of the complement of the Maxwell stratum (respectively of the bifurcation diagram) of complex *bipolynomials* as defined by V. I. Arnold [2] (F. Napolitano [3]). Even for ordinary polynomials, the cohomology groups of the complement of the Maxwell stratum are not known. The cohomology groups of complements of any stratum of the discriminant of singularities $A_k$ are described in [1,4]: the cohomology of the complement of any stratum stabilizes as $\mu \to \infty$.

[1] ARNOLD V. I. On some topological invariants of algebraic functions. *Trans. Moscow Math. Soc.*, 1970, **21**, 30–52.

[2] ARNOLD V. I. Topological classification of complex trigonometric polynomials and combinatorics of graphs with an equal number of vertices and edges. *Funct. Anal. Appl.*, 1996, **30**(1), 1–14.

[3] NAPOLITANO F. Discriminant and bifurcation diagram of complex trigonometric polynomials. *C. R. Acad. Sci. Paris, Sér.* I *Math.*, 1998, **327**(8), 771–776.

[4] NAPOLITANO F. Topology of complements of strata of the discriminant of polynomials. *C. R. Acad. Sci. Paris, Sér.* I *Math.*, 1998, **327**(7), 665–670.

[5] NEKRASOV N. A. On the cohomology of the complement of the bifurcation diagram of the singularity $A_\mu$. *Funct. Anal. Appl.*, 1993, **27**(4), 245–250.

### △ 1985-22 — *V. A. Vassiliev* Also: *1998-8*

ℛ In the complex case the stabilization of cohomology rings of the complements of Maxwell strata holds in the following sense:

There exist "stable" graded rings $M(n)$ and $M = M(\infty)$ such that

a) for any $n$, any isolated singularity $f$ in $n$ variables, and any number $k$ there exists a singularity $g$ more complicated than $f$ (i. e., the singularities equivalent to $f$ appear in any neighborhood of $g$) such that the cohomology ring of the complement of the Maxwell set of the versal deformation of $g$ is isomorphic to $M(n)$ in dimensions not exceeding $k$;

b) for any isolated singularity $f$ and any $k$ there exists a singularity $g$ in the same or greater number of variables than $f$ and more complicated than an appropriate stabilization of $f$, such that the cohomology ring of the complement of the Maxwell set of $g$ is isomorphic to $M$ up to dimension $k$.

There is no problem with the abstract definition of stable rings $M(n)$ and $M$ (as inductive limits of similar rings for individual singularities); however, statements a) and b) imply that they are finitely generated in any fixed dimension.

The proof for $n = 1$ was given in [1]. For an arbitrary $n$ the proof is essentially the same, with additional use of the stable irreducibility theorems for multisingularities from [2] and Section 4 of [3]. However, the calculation (and the answer) seems to be quite complicated.

We do not assert that, for *any* monotone sequence of adjacent singularities with infinitely growing Milnor numbers, the limit of cohomology classes of complements of Maxwell sets over this sequence coincides with $M(n)$ or $M$: moreover, conjecturally this is false.

The literal analog of the problem for real function singularities has a negative answer: already the numbers of components of complements of Maxwell sets grow infinitely). However, if we remove not the entire Maxwell set but some natural subset of it of positive codimension (e. g., the closure of the set of functions with $m \geq 3$ critical points on the same level) then exactly the same stabilization theorems would hold.


[1] NEKRASOV N. A. On the cohomology of the complement of the bifurcation diagram of the singularity $A_\mu$. *Funct. Anal. Appl.*, 1993, **27**(4), 245–250.

[2] VASSILIEV V. A. Stable cohomologies of the complements of the discriminants of deformations of singularities of smooth functions. In: Itogi Nauki i Tekhniki VINITI. Current Problems in Mathematics. Newest Results, Vol. 33. Moscow: VINITI, 1988, 3–29 (in Russian). [*The English translation*: *J. Sov. Math.*, 1990, **52**(4), 3217–3230.]

[3] VASSILIEV V. A. Complements of Discriminants of Smooth Maps: Topology and Applications, revised edition. Providence, RI: Amer. Math. Soc., 1994. (Transl. Math. Monographs, 98.)


**1985-24**

See the comment to problem *1993-24*. Cf. also problem *1996-4*.

## 1985-26

See the comment to problem *1985-19*.

# 1986

## 1986-1 – *E. Ferrand*

The problem is still open, the known results are surveyed in paper [1].

[1] VITERBO C. Properties of embedded Lagrange manifolds. In: First European Congress of Mathematics (Paris, July 6–10, 1992), Vol. II. Invited lectures, Part 2. Editors: A. Joseph, F. Mignot, F. Murat, B. Prum and R. Rentschler. Basel: Birkhäuser, 1994, 463–474. (Progr. Math., 120.)

## 1986-4 – *M. B. Sevryuk*

Let $X$ be a Hamiltonian or reversible autonomous system of ordinary differential equations in $\mathbb{R}^{2N}$ ($N \geq 1$) with equilibrium 0 (in the reversible context, it is also assumed that the point 0 is invariant under the reversing involution $G$).

**Theorem 1.** *Suppose that the linearization $D_0X$ of system $X$ at* 0 *possesses a pair of simple purely imaginary eigenvalues $\pm i\omega$* ($\omega > 0$) *whereas all the remaining eigenvalues $\pm\lambda_2, \ldots, \pm\lambda_N$ satisfy the following* nonresonance condition:

$$\frac{\lambda_j}{i\omega} \notin \mathbb{Z}, \quad 2 \leq j \leq N.$$

*Then, through the equilibrium* 0, *there passes a two-dimensional surface $M$ possessing the following properties.*

1) *The surface $M$ has the same smoothness as the system $X$ itself* (*to be more precise, the surface $M$ is analytic, of smoothness class $C^\infty$, or finitely smooth whenever the system $X$ is respectively analytic, infinitely smooth, or finitely smooth*).

2) *The tangent plane to $M$ at* 0 *is an invariant plane of the linear system $D_0X$ corresponding to the eigenvalues $\pm i\omega$.*

3) *The surface* $M \setminus \{0\}$ *is analytically,* $C^\infty$, *or finitely smoothly (depending on the smoothness class of the system* $X$*) foliated into closed trajectories of the system* $X$, *the periods of these trajectories being close to* $2\pi/\omega$ *and tending to* $2\pi/\omega$ *as the trajectories shrink down to* 0. *In the reversible context, these trajectories are invariant under the reversing involution* $G$ *as well.*

4) *If the system* $X$ *depends on parameters then the surface* $M$ *also depends on these parameters with the same smoothness.*

In the Hamiltonian context, this theorem is the classical *Lyapunov center theorem* of 1892 [13] (Ch. II, n° 45), one of the main results in the theory of small oscillations in nonlinear Hamiltonian systems. As far as the author of the present comment knows, the Lyapunov theorem was first carried over to reversible systems by R. L. Devaney [8] in 1976, and in the reversible context, Theorem 1 is usually called the *Lyapunov–Devaney theorem.* [1] Note, however, that not later than in 1893 A. M. Lyapunov himself pointed out [12] (n° 15) that equilibria of the center type are characteristic for two classes of planar autonomous systems, namely, conservative ones and reversible ones (the trivial case of Theorem 1 for $N = 1$). The surface $M$ in Theorem 1 is called the *Lyapunov invariant surface*.

To the Lyapunov center theorem and, to a lesser extent, the Lyapunov–Devaney theorem, there is devoted a rich body of literature. By now, rather many different proofs of these results have been obtained. The Lyapunov center theorem can be generalized to systems which possess a non-trivial first integral but are not necessarily Hamiltonian. In the present comment, we would confine ourselves to a reference to thesis [19] containing an extensive bibliography.

Now consider a one-parameter family $X_\varepsilon$ of Hamiltonian or reversible autonomous systems of ordinary differential equations in $\mathbb{R}^{2N}$ ($N \geq 2$) with common equilibrium 0 (in the reversible context, it is again assumed that the point 0 is invariant under the reversing involution $G$), $\varepsilon \in \mathbb{R}$. Suppose that, for $\varepsilon = 0$, the linearization $D_0X_0$ of the system $X_0$ at 0 possesses two resonant pairs of purely imaginary eigenvalues $\pm pi\omega$, $\pm qi\omega$, where $1 \leq p \leq q$ are mutually prime positive integers and $\omega > 0$, while the remaining eigenvalues $\pm\lambda_3, \ldots, \pm\lambda_N$ satisfy the following condition:

$$\frac{\lambda_j}{i\omega} \notin \mathbb{Z}, \quad 3 \leq j \leq N$$

---

[1] One also sometimes calls the reversible Theorem 1 the Lyapunov–Bruno–Devaney theorem, having in view A. D. Bruno's general results [5] on analytic invariant manifolds and normal forms of analytic systems of differential equations.

(the $p:q$ resonance). For $p=q=1$, one additionally assumes that, to each of the eigenvalues $i\omega$ and $-i\omega$ of multiplicity two, there corresponds a Jordan block of order 2 (this is a genericity condition).

For small $\varepsilon$, the system $X_\varepsilon$ can admit, near 0, an invariant two-dimensional surface $M_\varepsilon$ foliated into closed trajectories (invariant under the reversing involution $G$ in the reversible context) with periods close to $2\pi/\omega$. One has to study the perestroikas of this surface as the parameter $\varepsilon$ passes through the critical value 0.

**Remark 1.** The $p:q$ resonance is said to be *integral* (or *anti-Lyapunov*) for $p=1$ and *subharmonic* for $q>p\geq 2$.

**Remark 2.** According to Theorem 1, in the case of integral resonances $1:q$ with $q\geq 2$, the system $X_\varepsilon$ possesses, for all small $\varepsilon$, an invariant two-dimensional surface $M'_\varepsilon$ passing through 0 and foliated into closed trajectories (invariant under the involution $G$ in the reversible context) with periods close to $2\pi/(q\omega)$. This surface depends on $\varepsilon$ smoothly or analytically, and it undergoes no bifurcation as the parameter $\varepsilon$ passes through the critical value 0. The closed trajectories of the system $X_\varepsilon$ with periods close to $2\pi/\omega$ (into these trajectories, the surface $M_\varepsilon$ is foliated) and those with periods close to $2\pi/(q\omega)$ are respectively called *long period cycles* and *short period cycles*.

Again according to Theorem 1, in the case of subharmonic resonances $p:q$, the system $X_\varepsilon$ possesses, for all small $\varepsilon$, invariant two-dimensional surfaces $M'_\varepsilon$ and $M''_\varepsilon$ passing through 0 and foliated into closed trajectories (invariant under the involution $G$ in the reversible context) with periods close respectively to $2\pi/(p\omega)$ and $2\pi/(q\omega)$. These surfaces depend on $\varepsilon$ smoothly or analytically, and they undergo no bifurcations as the parameter $\varepsilon$ passes through the critical value 0. The closed trajectories of the system $X_\varepsilon$ with periods close to $2\pi/\omega$ (into these trajectories, the surface $M_\varepsilon$ is foliated), those with periods close to $2\pi/(p\omega)$, and those with periods close to $2\pi/(q\omega)$ are respectively called *very long period cycles*, *long period cycles*, and *short period cycles*.

**Remark 3.** The perestroika of the surface $M_\varepsilon$ in the case of the $1:1$ resonance is called the *Hamiltonian Hopf bifurcation* in the Hamiltonian context and the *reversible Hopf bifurcation* in the reversible one.[2]

Up to now, the perestroikas of the surface $M_\varepsilon$ have been studied in detail (and the stability type of the closed trajectories has been determined) for all the resonances in the Hamiltonian as well as reversible contexts. In the Hamiltonian

[2] It would be more correct to speak of the Hamiltonian or reversible Poincaré–Andronov bifurcation—by analogy with the usual "Hopf bifurcation"; see [2] (§ 33, A–C).

context, the normal forms and versal unfoldings of the surface $M_0$ have been found for all the resonances $p:q$ except the $1:1$ resonance [9]. The dimension of the base of the versal unfolding is equal to 1 for $p=1$, $q=2$, to 2 for $p+q \geq 5$, and to 3 for $p=1$, $q=3$. In the reversible context, the normal forms of the family of curves $\Gamma_\varepsilon$ have been obtained for all the resonances; here $\Gamma_\varepsilon$ are the lines along which the surfaces $M_\varepsilon$ intersect the $N$-dimensional manifold $\operatorname{Fix} G$ of fixed points of the reversing involution $G$.[3] The corresponding references for both the Hamiltonian and reversible contexts are given in, e. g., thesis [19], and a more extensive bibliography for the $1:1$ resonance is presented in, e. g., paper [11].

The most important references will be given here. The most thorough information on the perestroikas of the surface $M_\varepsilon$ in the Hamiltonian context for all the resonances $p:q$ except the $1:1$ resonance is contained in article [9]. Of the previous works, we mention papers [10, 17]. The Hamiltonian Hopf bifurcation is explored in detail in, e. g., works [15, 16].[4] All the resonances in the Hamiltonian context are considered from a unified viewpoint in book [3] (Ch. 7, § 3). In the reversible context, the main sources on the perestroikas of the surface $M_\varepsilon$ for all the resonances are paper [4] (§§ 2.14–2.16), book [18] (Ch. 6), and thesis [19]. V. I. Arnold and the author of the present comment started studying the resonances in question in reversible systems in 1984, trying to explain certain bifurcations of periodic solutions in distributed kinetic systems discovered (via exploring formal solutions of the corresponding equations) by the physicists S. I. Anisimov, S. M. Gol′berg, A. B. Konstantinov, B. A. Malomed, and M. I. Tribel′skiĭ in 1982–84 (actually, they observed the $1:1$ resonance); see paper [14] and references therein.

It turns out that for each resonance, the general picture of the perestroikas of the surface $M_\varepsilon$ is almost the same in the Hamiltonian context and in the reversible one. This is one of the manifestations of the general phenomenon of parallelism between the Hamiltonian dynamics and the reversible dynamics (see the comment to problem *1983-3*). Under certain nondegeneracy conditions, there occur only two essentially different types of perestroika of the surface $M_\varepsilon$ for each resonance. These types are called the *elliptic regime* and *hyperbolic regime*. The nondegeneracy conditions are imposed on the derivative $\frac{d}{d\varepsilon}D_0X_\varepsilon\big|_{\varepsilon=0}$ and on the

[3] Any closed $G$-invariant trajectory of a reversible system contains just two points fixed under the reversing involution $G$.

[4] In the Hamiltonian context, the $1:1$ resonance has been also studied in the degenerate set-up where the eigenvalues $\pm i\omega$ of the linear system $D_0X_0$ of multiplicity two are semisimple (to them, there corresponds the diagonal matrix $\operatorname{diag}\{i\omega, i\omega, -i\omega, -i\omega,\}$) [6] (see also announcement [7] which appeared after thesis [6] itself).

$k$-jet of the system $X_0$ at the point 0, where $k = 3$ for the $1:1$ resonance and the subharmonic resonances and $k = q$ for the $1:q$ resonances with $q \geq 2$.[5] The regime is determined by the 3-jet of the system $X_0$ at the point 0 for all the resonances except the $1:2$ resonance. For the $1:2$ resonance, the regime is determined by the $s$-jet of the system $X_0$ at the point 0 where $s = 1$ in the Hamiltonian context and $s = 2$ in the reversible one.

**Remark 4.** One of the main characteristics of a resonant *Hamiltonian* system for any resonance (except the $1:1$ resonance) is also the set of signs of the resonant eigenfrequencies, i. e., the signature of the quadratic part of the Hamilton function at equilibrium 0: the behavior patterns of Hamiltonian systems in the definite case (the resonant eigenfrequencies are of like sign) and in the indefinite case (the resonant eigenfrequencies are of opposite sign) are different. For the $p:q$ resonances with $q \geq 3$ in the Hamiltonian systems, both the regimes occur generically in the definite as well as indefinite cases. For the $1:2$ resonance, the regime is completely determined by just the quadratic part of the Hamilton function: the regime is hyperbolic in the definite case and elliptic in the indefinite case.

Below we shall give, for the reversible context, the normal forms of the families of curves $\Gamma_\varepsilon = M_\varepsilon \cap \operatorname{Fix} G$, $\Gamma'_\varepsilon = M'_\varepsilon \cap \operatorname{Fix} G$ (for all the resonances except $1:1$), and $\Gamma''_\varepsilon = M''_\varepsilon \cap \operatorname{Fix} G$ (for the subharmonic resonances). For definiteness, assume that the systems $X_\varepsilon$ are smooth ($C^\infty$) and depend on the parameter $\varepsilon$ smoothly. Under the appropriate nondegeneracy conditions, in the $N$-dimensional manifold $\operatorname{Fix} G$ of fixed points of the reversing involution $G$, there is a smooth two-dimensional surface $\Sigma_\varepsilon \ni 0$ (for each $|\varepsilon|$ small enough) depending on $\varepsilon$ smoothly and containing the curve $\Gamma_\varepsilon$ (respectively the curves $\Gamma_\varepsilon$, $\Gamma'_\varepsilon$ or the curves $\Gamma_\varepsilon$, $\Gamma'_\varepsilon$, $\Gamma''_\varepsilon$). By choosing a suitable coordinate system $(\xi, \eta)$ on $\Sigma_\varepsilon$ (this coordinate system depending on $\varepsilon$ smoothly) and multiplying, if necessary, $\varepsilon$ by a suitable non-zero constant, one can reduce the equations of the curves in question on the surface $\Sigma_\varepsilon$ to the following forms.

A) The $1:1$ resonance:

$$\Gamma_\varepsilon = \{ (\varepsilon \pm \xi^2)\xi^2 = \eta^2 \}$$

where the plus sign corresponds to the hyperbolic regime, and the minus sign to the elliptic regime. In the elliptic regime, the surface $M_0$ is just the origin. In the

[5] To determine the stability type of the closed trajectories in the case of subharmonic resonances $p:q$, one needs also some nondegeneracy condition imposed on the $(p+q-1)$-jet of the system $X_0$ at the point 0. In the reversible context only, the author of the present comment has checked carefully all the pointed out orders of the jets of the system $X_0$ at the point 0 on which the nondegeneracy conditions are imposed.

hyperbolic regime, the surface $M_0$ consists of two leaves which are $C^1$-smooth at 0 but *not* $C^2$-smooth.

B) The 1 : 2 resonance:

$$\Gamma_\varepsilon = \{\,\eta(\varepsilon-\eta)\pm\xi^2=0\,\}, \quad \Gamma'_\varepsilon=\{\xi=0\}$$

where the plus sign corresponds to the hyperbolic regime, and the minus sign to the elliptic regime. In the elliptic regime, the surface $M_0$ is just the origin. In the hyperbolic regime, the surface $M_0$ consists of two leaves which are even not $C^1$-smooth at 0.

C) The 1 : 3 resonance:

$$\Gamma_\varepsilon = \{\,\eta(\varepsilon+\eta\xi\cos\theta+\eta^2\sin\theta)+\xi^3=0\,\}, \quad \Gamma'_\varepsilon=\{\,\xi=C\eta+\eta^2\Theta(\eta,\varepsilon)\,\}$$

where $\theta$ is a real constant defined $\operatorname{mod} 2\pi$ such that

$$\Delta(\theta) := 4\cos^3\theta+27\sin^2\theta\neq 0,$$

$C$ is a real constant, and $\Theta$ is a smooth function. The inequality $\Delta(\theta)<0$ corresponds to the hyperbolic regime, and the inequality $\Delta(\theta)>0$ to the elliptic regime. In the elliptic regime, the surface $M_0$ consists of a single leaf and in the hyperbolic regime, of three leaves. All these leaves are even not $C^1$-smooth at 0.

D) The 1 : $q$ resonances, $q\geq 4$:

$$\Gamma_\varepsilon = \{\,\eta(\varepsilon\pm\eta^2-\xi^2)+\xi^q=0\,\}, \quad \Gamma'_\varepsilon=\{\,\xi=\eta^2\Theta(\eta,\varepsilon)\,\}$$

where $\Theta$ is a smooth function, the plus sign corresponds to the hyperbolic regime, and the minus sign to the elliptic regime. In the elliptic regime, the surface $M_0$ consists of a single leaf which is $C^{q-3}$-smooth at 0 but *not* $C^{q-2}$-smooth. In the hyperbolic regime, the surface $M_0$ consists of three leaves. One of them is $C^{q-3}$-smooth at 0 but *not* $C^{q-2}$-smooth. Two other leaves are even not $C^1$-smooth at 0.

E) The subharmonic resonances:

$$\Gamma_\varepsilon = \{\,\xi^2\pm\eta^2=\varepsilon\,\}, \quad \Gamma'_\varepsilon=\{\eta=0\}, \quad \Gamma''_\varepsilon=\{\xi=0\}$$

where the minus sign corresponds to the hyperbolic regime, and the plus sign to the elliptic regime. In the elliptic regime, the surface $M_0$ is just the origin. In the hyperbolic regime, the surface $M_0$ consists of two leaves which are even not $C^1$-smooth at 0.

For the proofs and further information, see [4, 18, 19]. For the general properties of closed trajectories in reversible systems, see [1,4,8,18,19].

[1] ARNOLD V. I. Reversible systems. In: Nonlinear and Turbulent Processes in Physics (Kiev, 1983), V. 3. Editor: R. Z. Sagdeev. Chur: Harwood Acad. Publ., 1984, 1161–1174. [*For the Russian translation see, e. g.*: Vladimir Igorevich Arnold. Selecta–60. Moscow: PHASIS, 1997, 355–363.]

[2] ARNOLD V. I. Geometrical Methods in the Theory of Ordinary Differential Equations, 2nd edition. New York: Springer, 1988. (Grundlehren der Mathematischen Wissenschaften, 250.) [*The Russian original* 1978.]

[3] ARNOLD V. I., KOZLOV V. V., NEĬSHTADT A. I. Mathematical Aspects of Classical and Celestial Mechanics, 2nd edition. Berlin: Springer, 1993. (Encyclopædia Math. Sci., 3; Dynamical systems, III.) [*The Russian original* 1985.] [*The second, revised and supplemented, Russian edition* 2002.]

[4] ARNOLD V. I., SEVRYUK M. B. Oscillations and bifurcations in reversible systems. In: Nonlinear Phenomena in Plasma Physics and Hydrodynamics. Editor: R. Z. Sagdeev. Moscow: Mir, 1986, 31–64.

[5] BRUNO A. D. Analytical form of differential equations. I; II. *Trans. Moscow Math. Soc.*, 1971, **25**, 131–288; 1972, **26**, 199–239.

[6] COTTER C. The 1 : 1 semisimple resonance. Ph. D. thesis, University of California at Santa Cruz, 1986.

[7] COTTER C. Bifurcations of Hamiltonian periodic orbits near an equilibrium in 1 : 1 resonance. *Abstracts of Papers Presented to the AMS*, 1987 (issue 53), **8**(6), 394–395.

[8] DEVANEY R. L. Reversible diffeomorphisms and flows. *Trans. Amer. Math. Soc.*, 1976, **218**, 89–113.

[9] DUISTERMAAT J. J. Bifurcations of periodic solutions near equilibrium points of Hamiltonian systems. In: Bifurcation Theory and Applications. Editor: L. Salvadori. Berlin: Springer, 1984, 57–105. (Lecture Notes in Math., 1057.)

[10] HENRARD J. Periodic orbits emanating from a resonant equilibrium. *Celest. Mech.*, 1970, **1**(3/4), 437–466.

[11] LAHIRI A., BHOWAL A., ROY T. K., SEVRYUK M. B. Stability of invariant curves in four-dimensional reversible mappings near 1 : 1 resonance. *Physica D*, 1993, **63**(1–2), 99–116.

[12] LYAPUNOV A. M. An analysis of one of the special cases in the problem of the stability of motion. In: Collected Papers, Vol. II. Moscow–Leningrad: the USSR Academy of Sciences Press, 1956, 272–331 (in Russian). [*The original publication* 1893.]

[13] LYAPUNOV A. M. The General Problem of the Stability of Motion. London: Taylor & Francis, 1992. [*The Russian original* 1892.] [*The first French translation* 1907.]

[14] MALOMED B. A., TRIBELSKY M. I. Bifurcations in distributed kinetic systems with aperiodic instability. *Physica D*, 1984, **14**(1), 67–87.

[15] VAN DER MEER J.-C. The Hamiltonian Hopf Bifurcation. Berlin: Springer, 1985. (Lecture Notes in Math., 1160.)

[16] VAN DER MEER J.-C. Bifurcation at nonsemisimple 1 : −1 resonance. *Z. Angew. Math. Phys.*, 1986, **37**(3), 425–437.

[17] SCHMIDT D. S. Periodic solutions near a resonant equilibrium of a Hamiltonian system. *Celest. Mech.*, 1974, **9**(1), 81–103.

[18] SEVRYUK M. B. Reversible Systems. Berlin: Springer, 1986. (Lecture Notes in Math., 1211.)

[19] SEVRYUK M. B. Reversible dynamical systems. Ph. D. Thesis, Moscow State University, 1987 (in Russian).

## 1986-6 — *B. A. Khesin*

ℛ Ya. M. Eliashberg and T. Ratiu proved that the diameter of the symplectomorphism group of an even-dimensional ball is infinite (see [2, 3]). See also A. I. Schnirelmann's results in [4, 5] and the survey of other results in book [1].

[1] ARNOLD V. I., KHESIN B. A. Topological Methods in Hydrodynamics. New York: Springer, 1998. (Appl. Math. Sci., 125.)

[2] ELIASHBERG YA. M., RATIU T. On the diameter of the symplectomorphism group of the ball. In: Symplectic Geometry, Groupoids, and Integrable Systems (Berkeley, 1989). Editors: P. Dazord and A. Weinstein. New York: Springer, 1991, 169–172. (Math. Sci. Res. Inst. Publ., 20.)

[3] ELIASHBERG YA. M., RATIU T. The diameter of the symplectomorphism group is infinite. *Invent. Math.*, 1991, **103**(2), 327–340.

[4] SCHNIRELMANN A. I. The geometry of the group of diffeomorphisms and the dynamics of an ideal incompressible fluid. *Math. USSR, Sb.*, 1987, **56**(1), 79–105.

[5] SCHNIRELMANN A. I. Attainable diffeomorphisms. *Geom. Funct. Anal.*, 1993, **3**(3), 279–294.

## ▽ 1986-7 — *M. L. Kontsevich* Also: *1989-18*

ℛ A physicist guesses from conformal field theory techniques the exact critical exponent for the meander number:

$$\alpha = \sqrt{29}(\sqrt{29}+\sqrt{5})/12 = 3.4206\ldots,$$

see [1].

The meander number is the number of closed roads crossing an infinite river at $2n$ bridges, and its asymptotic form as $n$ goes to $\infty$ is expected to be

$$\text{const} \times R^{2n}/n^{\alpha} \times (1+o(1)).$$

Physics says nothing about the values of the constant and $R$, but numerically $R = 12.262874\ldots$

For semi-meander numbers (embeddings of an interval intersecting a line at $n$ points) the critical exponent is

$$\alpha' = 1 + \sqrt{11}(\sqrt{29} + \sqrt{5})/24.$$


[1] DI FRANCESCO P., GOLINELLI O., GUITTER E. Meanders: exact asymptotic. *Nucl. Phys. B*, 2000, **570**(3), 699–712.
[*Internet*: `http://www.arXiv.org/abs/cond-mat/9910453`]


### △ 1986-7 – *S. K. Lando* Also: *1989-18*

For the explicit statement of the problem see problem *1989-18*. It was formulated in [1]. Up to now, even the leading term of the asymptotics is not known. Meandric numbers $M_n$ for even values of the index are known up to $n = 86$ [5]. In [6, 7] the asymptotic upper bound

$$M_n < cn^{-\frac{5}{2}} \left( \frac{4 - \pi}{\pi} \right)^n$$

was given. Conjecturally, the leading term of the asymptotics behaves as $cR^n n^\alpha$ for some constants $c$, $R$, and $\alpha$. Conjectural values of the constants are $R \approx 3.501837$ and $\alpha = (29 + \sqrt{145})/12 \approx 3.4201328$ [4]. For the statistics of meanders and related problems see [3]. In [6,7] the relation between meanders and matrix models in field theory is discussed, and in [2] meanders are related to the Temperley–Lieb algebra.


[1] ARNOLD V. I. A branched covering $\mathbb{C}P^2 \to S^4$, hyperbolicity and projectivity topology. *Sib. Math. J.*, 1988, **29**(5), 717–726. [*The Russian original is reprinted in*: Vladimir Igorevich Arnold. Selecta–60. Moscow: PHASIS, 1997, 431–448.]

[2] DI FRANCESCO P., GOLINELLI O., GUITTER E. Meanders and the Temperley–Lieb algebra. *Commun. Math. Phys.*, 1997, **186**(1), 1–59.

[3] DI FRANCESCO P., GOLINELLI O., GUITTER E. Meander, folding and arch statistics. *Math. Comp. Modelling*, 1997, **26**(8–10), 97–147.

[4] DI FRANCESCO P., GOLINELLI O., GUITTER E. Meanders: exact asymptotics. *Nucl. Phys. B*, 2000, **570**(3), 699–712.
[*Internet*: `http://www.arXiv.org/abs/cond-mat/9910453`]

[5] JENSEN I., GUTTMAN A. J. Critical exponents of plane meanders. Preprint, April 2000.

[6] Lando S. K., Zvonkin A. K. Meanders. *Selecta Math. Sov.*, 1992, **11**(2), 117–144.

[7] Lando S. K., Zvonkin A. K. Plane and projective meanders. *Theor. Comp. Sci.*, 1993, **117**(1–2), 227–241.

## 1986-8 – *I. A. Bogaevsky, V. D. Sedykh*

Let $M$ be a smooth closed convex hypersurface in $\mathbb{R}^{n+1}$ and let $\pi : \mathbb{R}^{n+1} \to \mathbb{R}^n$ be a linear projection. The image $\pi(M)$ of a hypersurface $M$ is called its *shadow*. The boundary of the shadow is called the *apparent contour* (of the hypersurface $M$).

If the second quadratic form of the hypersurface $M$ is positive definite, then its apparent contour is smooth. Otherwise, the apparent contour can be nonsmooth (with the exception of the trivial case $n = 1$).

**Theorem** [3]. *Let $M$ be a smooth closed convex surface in $\mathbb{R}^3$. Then*

1) *the apparent contour of the surface $M$ belongs to the class $D^2$ (is a twice differentiable curve);*

2) *there exists a surface $M$ such that its apparent contour does not belong to the class $C^2$ (the second derivative is discontinuous).*

*If the surface $M$ is real analytic, then*

1) *its apparent contour belongs to the class $C^{2+\varepsilon}$ for some $\varepsilon > 0$ (the second derivative satisfies the Hölder condition with index $\varepsilon$);*

2) *for any odd integer $q \geq 3$ there exists a surface $M$ such that its apparent contour belongs exactly to the class $C^{2+2/q}$.*

**Theorem** [4]. *Let $M$ be a smooth closed convex hypersurface in $\mathbb{R}^{n+1}$ where $n \geq 3$. Then*

1) *the apparent contour of the hypersurface $M$ belongs to the class $C^{1+1}$ (the first derivative is Lipschitz continuous);*

2) *there exists a hypersurface $M$ such that its apparent contour does not belong to the class $D^2$.*

We think that in the analytic case the following conjecture is true:

**Conjecture.** *The apparent contour of a real analytic smooth closed convex hypersurface in $\mathbb{R}^{n+1}$, where $n \geq 3$, belongs to the class $D^2$.*

This conjecture cannot be improved:

**Theorem** [1]. *There exists a real analytic smooth closed convex hypersurface in* $\mathbb{R}^{n+1}$*, where* $n \geq 3$*, such that its apparent contour does not belong to the class* $C^2$.

The set of hypersurfaces having a positive definite second quadratic form is an open subset $\mathcal{U}$ (in the $C^\infty$-topology) in the functional space of all closed hypersurfaces which are diffeomorphic to a given hypersurface. The apparent contours of such hypersurfaces are smooth. The boundary of the set $\mathcal{U}$ consists of convex hypersurfaces whose apparent contours can be nonsmooth.

Now consider a *generic* $m$-parameter family of smooth hypersurfaces in $\mathbb{R}^{n+1}$ (which belong to some open everywhere dense subset in the fine Whitney $C^\infty$-topology in the space of all smooth $m$-parameter families of hypersurfaces which are diffeomorphic to a given hypersurface). If $m = 0$, then the convex hypersurfaces in such a family belong to the set $\mathcal{U}$. Hypersurfaces lying on the boundary of the set $\mathcal{U}$ can be in such a family if $m \geq 1$.

**Theorem** [1]. *For any* $\pi$ *the apparent contour of a smooth convex hypersurface in* $\mathbb{R}^{n+1}$ *lying in a generic m-parameter family of hypersurfaces belongs to the class* $C^{2+2/q(m)}$ *if* $n = 2$ *and* $D^2$ *if* $n \geq 3$ *(*$C^\infty$ *for* $m = 0$ *and any n). Here* $q(m) = 2[(m+2)/3] + 1$ *and* $[\cdot]$ *denotes the integral part.*

These estimates cannot be improved, as was proved in [1]. For example, the apparent contour of a typical convex hypersurface lying on the boundary of the set $\mathcal{U}$ is not smooth for a suitable projection $\pi$. Normal forms of germs of such projections were found in [2].

The last theorem shows that the counterexamples from the theorems of [3] and [4] have infinite codimension, and the smoothnesses of apparent contours in the analytic and the finite-parametric cases are the same.

[1] BOGAEVSKY I. A. Degree of smoothness for visible contours of convex hypersurfaces. In: Theory of Singularities and its Applications. Editor: V. I. Arnold. Providence, RI: Amer. Math. Soc., 1990, 119–127. (Adv. Sov. Math., 1.)

[2] BOGAEVSKY I. A. Singularities of projective hypersurfaces. *Funct. Anal. Appl.*, 1990, **24**(2), 97–103.

[3] KISELMAN C. O. How smooth is the shadow of a smooth convex body? *J. London Math. Soc., Ser.* 2, 1986, **33**(1), 101–109.

[4] SEDYKH V. D. An infinitely smooth compact convex hypersurface with a shadow whose boundary is not twice-differentiable. *Funct. Anal. Appl.*, 1989, **23**(3), 246–248.

### **1986-12** — ***B. A. Khesin*** Also: *1987-11, 1994-30*

ℛ For numerical studies of the singularities in this variational problem we refer to the papers of H. K. Moffatt and his school (see, e. g., [2, 5–8]). Certain restrictions on the topology of stationary solutions of the Euler equation are given in papers [4] (for even-dimensional fluids) and [3] (for odd-dimensional fluids); see also the survey in book [1]. Cf. also a seemingly related problem *1973-20.*


[1] ARNOLD V. I., KHESIN B. A. Topological Methods in Hydrodynamics. New York: Springer, 1998. (Appl. Math. Sci., 125.)

[2] BAJER K. Flow kinematics and magnetic equilibria. Ph. D. Thesis, Cambridge University, 1989.

[3] ETNYRE J., GHRIST R. Contact topology and hydrodynamics: I. Beltrami fields and the Seifert conjecture. *Nonlinearity*, 2000, **13**(2), 441–458.

[4] GINZBURG V. L., KHESIN B. A. Steady fluid flows and symplectic geometry. *J. Geom. Phys.*, 1994, **14**(2), 195–210.

[5] MOFFATT H. K. Some developments in the theory of turbulence. *J. Fluid Mech.*, 1981, **106**, 27–47.

[6] MOFFATT H. K. Magnetostatic equilibria and analogous Euler flows of arbitrarily complex topology. I. Fundamentals. *J. Fluid Mech.*, 1985, **159**, 359–378.

[7] MOFFATT H. K. Magnetostatic equilibria and analogous Euler flows of arbitrarily complex topology. II. Stability considerations. *J. Fluid Mech.*, 1986, **166**, 359–378.

[8] MOFFATT H. K. Structure and stability of solutions of the Euler equations: a Lagrangian approach. *Phil. Trans. Roy. Soc. London, Ser. A*, 1990, **333**(1631), 321–342.


# 1987

### **1987-1**

ℛ See the comment to problem *1983-14.*

### ▽ **1987-3** — ***S. L. Tabachnikov***

ℛ A similar relation between the oscillator and the Coulomb system on the sphere or pseudosphere is discussed in [3]. The problem concerns the Bohlin theorem; see V. I. Arnold's book [1] and also [2].

[1] ARNOLD V. I. Huygens and Barrow, Newton and Hooke. Pioneers in Mathematical Analysis and Catastrophe Theory from Evolvents to Quasicrystals. Basel: Birkhäuser, 1990. [*The Russian original* 1989.]

[2] KASNER E. Differential Geometric Aspects of Dynamics. AMS Colloq. Publ., 1913.

[3] NERSESSIAN A., POGOSYAN G. Relation of the oscillator and Coulomb systems on spheres and pseudospheres.
[*Internet*: `http://www.arXiv.org/abs/quant-ph/0006118`]

## △ 1987-3 – *V. I. Arnold*

ℛ For the quantum version of Bohlin's theorem, see notes [1,2].

[1] FAURE R. Méthodes d'intégration communes à la mécanique classique et à la mécanique ondulatoire. *C. R. Acad. Sci. Paris*, 1950, **230**(4), 364–365.

[2] FAURE R. Transformations conformes en mécanique ondulatoire. *C. R. Acad. Sci. Paris*, 1953, **237**, 603–605.

## 1987-4 – *A. G. Khovanskiĭ, D. I. Novikov*

ℋ This problem appears in paper [1]. See also problems *1990-4, 1993-37.*

ℛ The condition that a hypersurface has constant signature means that, in some suitable affine chart of each point, the hypersurface has a local equation $z = x_1^2 + \cdots + x_k^2 - x_{k+1}^2 - \cdots - x_{k+l}^2 + o(|x|)$, and $k$ and $l$ are the same for all points. Nonsingular quadrics are a classical example of such surfaces.

Here we give a short description of the results of [2–4] (we do not know about any other results in this direction).

In [4] various affine versions of the conjecture are considered: the hypersurface is embedded into the affine space $\mathbb{R}^{k+l+1}$, but we assume that its behavior at infinity is simple. The first result says that, if a hypersurface approaches the standard cone $x_1^2 + \cdots + x_{k+1}^2 = y_1^2 + \cdots + y_l^2$ at infinity, then it does not intersect this cone. One can obtain a proof of the fourth part of the problem above by a slight modification of the proof of this result.

Next, the authors prove that in the case $k = l = 1$ the projection of this surface to a sphere centered at the origin (= the vertex of the cone) is one-to-one.

Then the authors consider the case $k = l = 1$ and a slightly different asymptotic behavior: the surface approaches the union of two halves of the cone $x^2 + y^2 = z^2$ moved apart one from the other (so they do not intersect). An example of such a surface not separating any two lines in $\mathbb{R}^3$ is constructed. This surface bounds a convex-concave set—a set whose horizontal sections are all nonempty, convex, and convexly depend on the height. Unlike domains that are bounded by hyperbolic surfaces, these sets are easy to manipulate (add, glue together, etc.), and their boundaries are hyperbolic if smooth.

A projective analogue of the class of convex-concave sets is the main object of [2, 3]. Namely, a set $S \subset \mathbb{RP}^n$ is called $L$-convex-concave (where $L \subset \mathbb{RP}^n$ is a $k$-dimensional subspace not intersecting $S$) if, first, sections of $S$ by $(k+1)$-dimensional subspaces containing $L$ are closed, convex, and nonempty; second, projections of $S$ from $(k-1)$-dimensional subspaces of $L$ have convex (open) complement. The relation of $L$-convex-concave sets to hypersurfaces of constant signature is similar to the relation of compact convex sets to smooth connected locally convex hypersurfaces. An analogue of the second part of the problem claims that an $L$-convex-concave set in $\mathbb{RP}^n$ contains a projective subspace of dimension $n - \dim L - 1$ (analogues of the third and the first parts of the problem then easily follow).

In [3] the authors define $L$-duality, such that the $L$-dual of an $L$-convex-concave set is an $L^*$-convex-concave set in the dual space, and $L^*$-dual of $L$-dual is the set itself. This $L$-duality becomes the standard duality of convex bodies for $k = 0$. The authors prove that the analogue of the second part of the problem is simultaneously true (or wrong) for an $L$-convex-concave set and for its $L$-dual, and they define certain ($L$-dual) surgeries on $L$-convex-concave sets.

In [2] the authors prove the analogue of the second part of the problem for $k = l = 1$, i. e., that an $L$-convex-concave subset of $\mathbb{RP}^3$ with $\dim L = 1$ contains a line. Apart from the results of [3], the proof uses classical tools of convex geometry: the Helly theorem, Chebyshev approximation, and the Browder theorem.

[1] Arnold V. I. A branched covering $\mathbb{C}P^2 \to S^4$, hyperbolicity and projectivity topology. *Sib. Math. J.*, 1988, **29**(5), 717–726. [*The Russian original is reprinted in*: Vladimir Igorevich Arnold. Selecta–60. Moscow: PHASIS, 1997, 431–448.]

[2] Khovanskiĭ A. G., Novikov D. I. Convex-concave body in $\mathbb{RP}^3$ contains a line. [*Internet*: `http://www.arXiv.org/abs/math.DG/0203200`]

[3] Khovanskiĭ A. G., Novikov D. I. $L$-convex-concave sets in real projective space and $L$-duality. [*Internet*: `http://www.arXiv.org/abs/math.DG/0203203`]

[4] Khovanskiĭ A. G., Novikov D. I. On affine hypersurfaces with everywhere nondegenerate second quadratic form. [*Internet*: `http://www.arXiv.org/abs/math.DG/0203202`]

## 1987-6 – *V. A. Vassiliev*

**1. The complex problem.** Here it is assumed that $\mathbb{C}^n$ is the space of all complex polynomials of the form $z^n + \lambda_1 z^{n-1} + \cdots + \lambda_{n-1} z + \lambda_n$.

The initial problem concerning the group $\pi_3(\mathbb{C}^n \setminus \Sigma^{n-2})$ was essentially solved by V. I. Arnold [1]. Indeed, it is proved there that $H_3 = \mathbb{Z}$ and $H_2 = 0$. Since the space $\mathbb{C}^n \setminus \Sigma^{n-2}$ obviously is simply-connected, we have also $\pi_3 = \mathbb{Z}$ by the Hurewicz theorem.

Further, let $\Sigma^{n-k} \subset \mathbb{C}^n$ be the set of complex polynomials having at least one root of multiplicity $k+1$. Then, according to [1], for any $k < n$ we have $H_{2k-1}(\mathbb{C}^n \setminus \Sigma^{n-k}) \simeq \mathbb{Z}$, and all groups $H_i$, $i < 2k-1$, are trivial. Hence also the equality $\pi_{2k-1}(\mathbb{C}^n \setminus \Sigma^{n-k})$ follows.

For other groups $\pi_i(\mathbb{C}^n \setminus \Sigma^{n-k})$ the problem is comparatively simple only if $n$ is either sufficiently large with respect to $i$ or sufficiently small with respect to $k$. Namely, for any $n > k \geq 1$ there is a more or less canonical embedding of the space $\mathbb{C}^n \setminus \Sigma^{n-k}$ into the space $\Omega^2 S^{2k+1}$ of double loops of the $(2k+1)$-dimensional sphere (see [4,5] and the comments to problems *1988-9* and *1988-10*). Supposing that $k > 1$, this embedding induces an isomorphism of the groups $\pi_i$ of these spaces if $i < (2k-1)([n/(k+1)]+1)$. In particular, if $n$, $k$ and $i$ satisfy this relation then $\pi_i(\mathbb{C}^n \setminus \Sigma^{n-k}) \simeq \pi_{i+2}(S^{2k+1})$. On the other hand, if $k \geq n/2$ then $\mathbb{C}^n \setminus \Sigma^{n-k}$ is homotopy equivalent to $S^{2k-1}$, cf. [2].

**2. The real problem** with $k \geq 3$ is completely similar. If $k \geq n/2$ then the space $\mathbb{R}^n \setminus \Sigma^{n-k}$ is homotopy equivalent to $S^{k-1}$, see [2]. Homology groups of these spaces with all $n$ and $k$ also were calculated in [2]. Their homotopy groups in stable dimensions can be calculated using the following statement (see [3,5]). The space $\mathbb{R}^n \setminus \Sigma^{n-k}$ is canonically embedded into the loop space $\Omega S^k$. This embedding induces isomorphisms of all groups $\pi_i$ with $i < (k-1)([n/(k+1)]+1)-1$. Therefore if $n$, $k$ and $i$ satisfy this relation then we have $\pi_i(\mathbb{R}^n \setminus \Sigma^{n-k}) \simeq \pi_{i+1}(\Omega S^k)$.

[1] ARNOLD V. I. Some topological invariants of algebraic functions. *Trans. Moscow Math. Soc.*, 1970, **21**, 30–52.

[2] ARNOLD V. I. Spaces of functions with moderate singularities. *Funct. Anal. Appl.*, 1989, **23**(3), 169–177.

[3] VASSILIEV V. A. Topology of spaces of functions withoup compound singularities. *Funct. Anal. Appl.*, 1989, **23**(4), 277–286.

[4] VASSILIEV V. A. Topology of complements to discriminants and loop spaces, In: Theory of Singularities and its Applications. Editor: V. I. Arnold. Providence, RI: Amer. Math. Soc., 1990, 9–21. (Adv. Sov. Math., 1.)

[5] VASSILIEV V. A. Complements of Discriminats of Smooth Maps: Topology and Applications. Providence, RI: Amer. Math. Soc., 1994. (Transl. Math. Monographs, 98.)

## **1987-7 – *B. Z. Shapiro*** Also: *1990-26*

The topology of the complement of a trail of a complete flag in $\mathbb{R}^n$ was studied in a series of papers [2, 5–10, 12]. The initial question was finally solved in [8, 10]. The answer is as follows:

| $n$ | 2 | 3 | 4 | 5 | 6 | 7 | ... |
|---|---|---|---|---|---|---|---|
| $\#_{\text{components}}(n)$ | 2 | 6 | 20 | 52 | $3\cdot 2^5$ | $3\cdot 2^6$ | ... |

The stable answer $\#_{\text{components}}(n) = 3\cdot 2^{n-1}$ is valid starting from $n = 6$.

There exists an intriguing interpretation of the above problem of counting the connected components as counting orbits for the certain groups acting by reflections in vector spaces over the field $\mathbb{F}_2$. There also exists a (so far) mystical relation between the reflection groups occuring in the above problem and the groups appearing in the papers by Zuber and Varchenko with Gusein-Zade in the connection with the Verlinde algebra; see [1]. An attempt to construct an informal complexification (à la V. Arnold) of the problem of counting connected components and the appropriate group over $\mathbb{Z}$ can be found in [11]. In [3, 4, 13] the authors address a more general question of calculating the number of connected components in the intersection of open real Schubert cells for a variety of complete flags for all Weyl groups as well as in the so-called real double Schubert cells $G^{u,v}(\mathbb{R})$ defined as $G^{u,v}(\mathbb{C}) = BuB \cap B_-vB_-$ in a simply-connected, connected complex semisimple group $G$; here $B$ and $B_-$ are opposite Borel subgroups, and $u$ and $v$ are elements of the Weyl group $W$.

In [2, 5] one finds examples of flag varieties of the form $\mathbb{P}T^*\mathbb{P}^n$ for which the intersection of any arrangement of open real Schubert cells enjoys the so-called $M$-property, i. e., the sum of Betti numbers with $\mathbb{F}_2$-coefficients of such an intersection coincides with that of its complexification.

Papers [6, 7, 12] deal with the study of the $E_{p,q}$-characteristics (after Khovanskiĭ–Danilov) for the complement of a trail of a complete flag in $\mathbb{R}^n$ and $\mathbb{C}^n$ and other intersections of Schubert cells. The papers contain a construction of a certan decomposition of the intersection of any pair of Schubert cells which allows in principle to calculate the $E_{p,q}$-characteristics for any chosen intersection. But the problem of getting the answer in the closed form is apparently hopeless. On the other hand, it is quite plausible that the methods developed in [8, 10] can be used not only to count the number of connected components but also to get information about the low-dimensional homology groups for different intersections of Schubert cells.

Let us finally mention that the $E_{p,q}$-characteristics are closely related to the so-called $R$-polynomials in Kazhdan–Lusztig theory of the structure constants in Hecke algebra.


[1] GUSEIN-ZADE S. M., VARCHENKO A. N. Verlinde algebras and the intersection form on vanishing cycles. *Selecta Math (N. S.)*, 1997, **3**(1), 79–97.

[2] KOSTOV V. P., SHAPIRO B. Z. Flags in $\mathbb{R}^3$ transversal to a given set of flags form an $M$-manifold. *Moscow Univ. Math. Bull.*, 1989, **44**(5), 31–36.

[3] RIETSCH K. Intersections of Bruhat cells in real flag varieties. With an appendix by G. Lusztig. *Internat. Math. Res. Notices*, 1997, **13**, 623–640.

[4] RIETSCH K. The intersection of opposed big cells in real flag varieties. *Proc. Roy. Soc. London, Ser. A*, 1997, **453**(1959), 785–791.

[5] SHAPIRO B. Z., SHAPIRO M. Z. The $M$-property of flag varieties. *Topology Appl.*, 1992, **43**(1), 65–81.

[6] SHAPIRO B. Z., SHAPIRO M. Z., VAINSHTEIN A. D. Topology of intersections of Schubert cells and Hecke algebra. *Discrete Math.*, 1996, **153**(1–3), 305–318.

[7] SHAPIRO B. Z., SHAPIRO M. Z., VAINSHTEIN A. D. On combinatorics and topology of pairwise intersections of Schubert cells in $\mathrm{SL}_n/\mathcal{B}$. In: The Arnold–Gelfand Mathematical Seminars: Geometry and Singularity Theory. Editors: V. I. Arnold, I. M. Gelfand, V. S. Retakh and M. Smirnov. Boston, MA: Birkhäuser, 1997, 397–437.

[8] SHAPIRO B. Z., SHAPIRO M. Z., VAINSHTEIN A. D. Connected components in the intersection of two open opposite Schubert cells in $\mathrm{SL}_n(\mathbb{R})/\mathcal{B}$. *Internat. Math. Res. Notices*, 1997, **10**, 469–493.

[9] SHAPIRO B. Z., SHAPIRO M. Z., VAINSHTEIN A. D. Kazhdan–Lusztig polynomials for certain varieties of incomplete flags. *Discrete Math.*, 1998, **180**(1–3), 345–355.

[10] SHAPIRO B. Z., SHAPIRO M. Z., VAINSHTEIN A. D. Skew-symmetric vanishing lattices and intersections of Schubert cells. *Internat. Math. Res. Notices*, 1998, **11**, 563–588.

[11] SHAPIRO B. Z., SHAPIRO M. Z., VAINSHTEIN A. D., ZELEVINSKY A. Simply laced Coxeter groups and groups generated by symplectic transvections. *Michigan J. Math.*, 2000, **48**, 531–552.

[12] SHAPIRO B. Z., VAINSHTEIN A. D. Euler characteristics for links of Schubert cells in the space of complete flags. In: Theory of Singularities and its Applications. Editor: V. I. Arnold. Providence, RI: Amer. Math. Soc., 1990, 273–286. (Adv. Sov. Math., 1.)

[13] ZELEVINSKY A. Connected components of real double Bruhat cells. *Internat. Math. Res. Notices*, 2000, **21**, 1131–1154.

## 1987-10 — *V. N. Karpushkin*

ℛ A stronger conjecture seems to be true: the number of critical points of the $N$-th eigenfunction of the Laplacian in an $n$-dimensional domain increases at least as $a(D)N^{1/n}$ and no faster than $b(D)N$ (provided that the number of critical points is finite), where $a(D), b(D) > 0$.

From the Bézout theorem it follows that the number of critical points (should their number be finite) for the $N$-th eigenfunction of the Laplace operator in $\mathbb{R}\mathrm{P}^n$ and $S^n$ grows no faster than $n!\,N$ as $N \to \infty$.

The explicit form of spherical harmonics from the canonical basis of $S^2$ or $\mathbb{R}\mathrm{P}^2$ shows that the number of their critical points increases no slower than $4\sqrt{N}$ on $S^2$ and $2\sqrt{2N}$ on $\mathbb{R}\mathrm{P}^2$ and no faster than $N$ both on $S^2$ and $\mathbb{R}\mathrm{P}^2$. (The number of critical points is finite if the eigenfunction from the canonical basis of $S^2$ or $\mathbb{R}\mathrm{P}^2$ is not zonal.)

The upper bound for the number of critical points (should their number be finite) on the line of level 0 of the $N$-th eigenfunction $\psi_n$ of the Laplacian ($\Delta\psi_N = \lambda_n\psi_N$) is given in [2], and is applicable for two-dimensional compact Riemannian manifolds without boundary. Namely, it is proved in [2] that

$$L_N = \sum_{a\in\Gamma}(k(a)-1) \leq N - \chi(M^2),$$

where $\Gamma$ is the line of level 0 for $\psi_N$, $k(a)$ is the degree of the main homogeneous part of the function $\psi_N$ at the point $a \in \Gamma$, and $\chi(M^2)$ is the Euler characteristic of the manifold $M^2$.

A more precise estimate of $L_N$ for the case where $M^2$ is $\mathbb{R}\mathrm{P}^2$ or $S^2$ was obtained in [1].


[1] KARPUSHKIN V. N. Topology of the zeros of eigenfunctions. *Funct. Anal. Appl.*, 1989, **23**(3), 218–220.

[2] KARPUSHKIN V. N. Multiplicities of singularities of eigenfunctions for the Laplace–Beltrami operator. *Funct. Anal. Appl.*, 1995, **29**(1), 62–64.


## 1987-11

ℛ See the comment to problem *1986-12*.

## 1987-12 — *A. A. Bolibruch*

ℛ It is proved that under some generic conditions an isomonodromic confluence of Fuchsian singularities leads to regular singular points only, see [1]. But in the resonant case this statement has not been proved yet. (See also my comment to problem *1984*-7.)

In the case of irregular singular points B. Malgrange stated the similar problem: *Prove that the index of irregularity of a system cannot increase under an isomonodromic deformation of singularities* (an isomonodromic deformation in this case is a deformation preserving monodromy matrices, Stokes matrices, and connection matrices; the index of irregularity is the sum of degrees of polynomials in the exponential part of a formal solution to the system).

V. P. Kostov in [4] studied an algebraic structure of the set of linear complex differential equations: different strata, their adjacency.

The problem of describing the isomonodromic classes of such systems is closely connected with investigation of singular points of the Schlesinger equation of isomonodromic deformatrions (see [2,5]) and with Painlevé equations (see [3]).

[1] BOLIBRUCH A. A. On isomonodromic confluences of Fuchsian singularities. *Proc. Steklov Inst. Math.*, 1998, **221**, 117–132.

[2] BOLIBRUCH A. A. On orders of movable poles of the Schlesinger equation. *J. Dynam. Control Systems*, 2000, **6**(1), 57–73.

[3] ITS A. R., NOVOKSHENOV V. YU. The Isomonodromic Deformation Method in the Theory of Painlevé Equations. Berlin: Springer, 1986. (Lecture Notes in Math., 1191.)

[4] KOSTOV V. P. On the existence of monodromy groups of Fuchsian systems on Riemann's sphere with unipotent generators. *J. Dynam. Control Systems*, 1996, **2**(1), 125–155. [*Prépublication* Université de Nice, no. 401, septembre 1994.]

[5] MALGRANGE B. Sur les déformations isomonodromiques: I. Singularités régulières. In: Mathématique et physique. Lectures presented at the seminar held at the École Normale Supérieure, Paris, 1979/1982. Editors: L. Boutet de Monvel, A. Douady and J.-L. Verdier. Boston, MA – Basel – Stuttgart: Birkhäuser, 1983, 401–426. (Progr. Math., 37.)

## 1987-14 — *V. A. Vassiliev* Also: *1988-13*, *1990-27*

ℜ The volume cut from a compact *convex* domain bounded by a smooth algebraic hypersurface in an *even-dimensional* space is never algebraic. In an odd-dimensional space smooth hypersurfaces of degree $\geq 3$ such that this volume function is algebraic, conjecturally do not exist. Even if they exist then they form a very thin subset in the space of all algebraic hypersurfaces of a given degree: for instance, the projective closure of their complexifications in $\mathbb{C}P^n$ cannot be nonsingular. Moreover, the complexification of a hypersurface with algebraic volume function cannot have parabolic points of finite multiplicity (i. e., points of non-Morse discrete tangency with the tangent hyperplane).

The proofs are based on the *Picard–Lefschetz theory*, i. e., on the study of the monodromy of the (relative) homology class of the piece cut from the domain defined by the loops in the complexified space of hyperplanes. Generically, the orbit of this homology class is infinite, as well as the set of values of integrals of the volume form along the elements of this orbit. Therefore the analytic continuation of the volume function should have logarithmic ramification.

For these and related results see [1–3].

[1] ARNOLD V. I. Huygens and Barrow, Newton and Hooke. Pioneers in Mathematical Analysis and Catastrophe Theory from Evolvents to Quasicrystals. Basel: Birkhäuser, 1990. [*The Russian original* 1989.]

[2] ARNOLD V. I., VASSILIEV V. A. Newton's *Principia* read 300 years later. *Notices Amer. Math. Soc.*, 1989, **36**(9), 1148–1154; addendum: 1990, **37**(2), 144.

[3] VASSILIEV V. A. Ramified Integrals, Singularities and Lacunas. Dordrecht: Kluwer Acad. Publ., 1995. (Math. Appl., 315.)

# 1988

### 1988-3 — *I. A. Bogaevsky*

ℛ Let us consider a hypersurface in a $(2D+1)$-dimensional contact space being diffeomorphic to the product of a two-dimensional cone $x^2+y^2=z^2$ and a $(2D-2)$-dimensional vector space. The question is: what are local normal forms of this hypersurface with respect to diffeomorphisms preserving the contact structure in the generic case? This question is related to the theory of symmetric hyperbolic systems of partial differential equations (see problem *1978-17*), interior wave scattering (see problem *1989-10*), relaxation oscillations, and the theory of ordinary differential equations unsolved with respect to the derivative [2].

Normal forms of the hypersurface at its typical singular point with respect to formal diffeomorphisms preserving the contact structure are obtained for $D=1$ in [2] and for $D\geq 2$ in [1]. In both cases there are two normal forms: elliptic and hyperbolic, though the explicit formulae are simpler for $D\geq 2$. The point is that in both forms for $D=1$ there is a continuous number invariant (a *modulus*) disappearing for $D\geq 2$.

In the analytic case the series defining the formal normalizing diffeomorphism diverges, as a rule, for any $D$. Probably, there exist infinitly smooth diffeomorphisms reducing the original hypersurface to the normal forms mentioned above. One can find an outline of the proof for $D = 1$ in [2–4], however its details have not been published. This proof is based on an interesting connection between reducing a contact structure to a normal form in a neighborhood of the vertex of a quadratic cone and the theory of normal forms of equivariant plane vector fields at singular points. The question about the existence of a infinitely smooth normalizing diffeomorphism for $D \geq 2$ is open.

For $D \geq 2$ the hypersurface can stably acquire degenerations with respect to the contact structure which is naturally called *parabolic*. Indeed, on the manifold of the singular points of this hypersurface, the elliptic and hyperbolic domains are naturally selected in accordance with the normal form of the hypersurface in a neighborhood of a given point. In the generic case the elliptic and hyperbolic domains are separated by hypersurfaces with parabolic degenerations. Most likely, the hypersurface has function moduli at a typical parabolic degeneration even with respect to formal diffeomorphisms.

[1] ARNOLD V. I. On the interior scattering of waves, defined by hyperbolic variational principles. *J. Geom. Phys.*, 1988, **5**(3), 305–315.

[2] ARNOLD V. I. Surfaces defined by hyperbolic equations. *Math. Notes*, 1988, **44**(1–2), 489–497. [*The Russian original is reprinted in*: Vladimir Igorevich Arnold. Selecta–60. Moscow: PHASIS, 1997, 397–412.]

[3] ARNOLD V. I. Singularities of Caustics and Wave Fronts. Dordrecht: Kluwer Acad. Publ., 1990. (Math. Appl., Sov. Ser., 62.)

[4] ARNOLD V. I. Singularities of Caustics and Wave Fronts. Moscow: PHASIS, 1996 (in Russian). (Mathematician's Library, 1.)

## 1988-4 – *O. S. Kozlovskiĭ*

This problem can be solved by standard methods of complex dynamics, and this was done by O. S. Kozlovskiĭ when he was a university student (1990).

There must be an attracting (or degenerate) periodic point between any two repelling periodic points of a circle diffeomorphism. So it is sufficient to estimate the number of attracting and degenerate periodic points. For this purpose, it is necessary to holomorphically extend the action of a diffeomorphism on the fundamental parallelogram of an elliptic function and observe that any immediate attraction domain of an attracting or degenerate periodic trajectory contains a critical point of the holomorphic extension of the considered diffeomorphism. Furthermore, due to the symmetry with respect to compex conjugation, the immediate attraction domain of a real periodic trajectory contains two critical points.

Thus, the number of periodic orbits of the diffeomorphism of $\mathbb{S}^1$ of the type $x \mapsto x + a + f(x)$, where $f(x)$ is an elliptic function, is bounded above by the number of roots of the equation $f'(z) = -1$, where $z$ belongs to the fundamental parallelogram of the function $f$.

## ▽ 1988-5 — *Yu. M. Baryshnikov*

ℛ The box-counting argument implies that, for any mapping taking the $m$-dimensional cube *onto* the $n$-dimensional cube and satisfying the Hölder condition of order $\alpha$, the inequality $\alpha \le m/n$ holds.

Many of the examples of Peano curves filling the $n$-dimensional cube clearly satisfy the Hölder condition of order $1/n$ (if this needs a proof, there is one in print, see [1]).

Taking products, one obtains (best possible) Hölder coefficients for Peano mappings from the $k$- onto the $kn$-dimensional cube. The case of dimensions $2 \to 3$ is the first non-trivial one.

One can construct, quite easily and for any $m, n$ with $\gcd(m,n) = 1$, an *onto* mapping $f : I^m \to I^n$ such that

$$\lim_{x \neq y,\ |x-y| \le \varepsilon} \frac{\log(|f(x) - f(y)|)}{\log(|x-y|)} = \frac{m}{n};$$

this mapping clearly satisfies the Hölder condition of any order $\alpha < m/n$. I do not know whether such a mapping exists for $\alpha = m/n$. [1]

**Remark.** I remember that V. I. Arnold cited this problem in connection with cortex packing in skull, and attributed it to A. N. Kolmogorov.

[1] BUCKLEY S. Space-filling curves and related functions. *Irish Math. Soc. Bull.*, 1996, **36**, 9–18.

## △ 1988-5 — *E. V. Shchepin*

ℛ A few days after V. I. Arnold had posed this question to me (in 1988), I proposed a way to construct the desired mapping. To define a mapping of $I^2$ onto $I^3$,

[1] I could not find a reference to any relevant work by Shchepin.

we subdivide $I^2$ into rectangles, and $I^3$ into parallelepipeds, step by step, and establish a one-to-one correspondence between the constructed partitions. Such correspondence generates a (set-valued) mapping $f: I^2 \to I^3$ by the following rule: $f(x)$ is defined as the intersection of all parallelepipeds corresponding to the rectangles containing the point $x$. This construction defines a single-valued mapping provided that the diameters of the parallelepipeds tend to zero. To ensure the continuity of $f$, the following condition must be satisfied.

**Adjacency condition.** *The parallelepipeds that correspond to adjacent rectangles are themselves adjacent.*

If this condition is satisfied, then the Hölder exponent $2/3$ would be provided by the boundedness of the ratio of the maximal diameter and the minimal thickness for the rectangles and the parallelepipeds of those partitions. Since there is apparently no obstruction to construct a Peano-type continuous mapping of $I^2$ onto $I^3$, the proposed method *seems* to solve the discussed problem. I think, under this impression Arnold attributed its solution to me. But I had not tried to write any article on this subject or even to check the construction.

When I was asked to write a comment on this problem, I tried to check the proof. I found out that the above-mentioned algorithm of subsequent subdivisions in general fails to satisfy the adjacency condition, and therefore fails to generate even a continuous mapping of $I^2$ onto $I^3$.

To correct the proof, one could apply a more sophisticated correspondence between subdivisions of the square and the cube. This led me to the following

**Problem.** *Does there exist a bijective correspondence $f$ between $n^6$ unit squares constituting a square with side $n^3$, and $n^6$ unit cubes constituting a cube with side $n^2$, such that $P \cap Q \neq \emptyset$ implies $f(P) \cap f(Q) \neq \emptyset$?*

The question becomes very sharp for 64-element partitions (i. e., the case $n = 2$). Its positive solution was obtained with the help of a computer program written by my son Nikita, and the computation lasted 8 hours. But this result does not provide the positive solution of Arnold's problem. More details on the connection between Arnold's problem and its discrete version can be found in my paper "On Hölder mappings of cubes" available via my homepage `http://genesis.mi.ras.ru/~scepin`.

## ▽ **1988-6** — *O. S. Kozlovskiĭ* Also: *1988-7, 1989-2, 1994-45*

The problems on the rate of increase of the number of periodic trajectories of a diffeomorphism and the number of intersection points of two manifolds of complementary dimension, one of which is fixed, and the other is iterated by a diffeomorphism action, are closely related to each other. Indeed, let $F: M \to M$ be a diffeomorphism. Then the number of periodic points of the diffeomorphism $F$ of period $n$ (which is not necessarily minimal) is equal to the number of intersection points of manifolds $\widetilde{F}^n(X) \cap X$, where $\widetilde{F} = (\mathrm{id}, F): M \times M \to M \times M$, and $X$ is the diagonal in $M \times M$.

It turns out that the rate of increase of the number of periodic points and the number of intersection points depends on the smoothness class of the diffeomorphism $F$.

If $F$ is an *algebraic* diffeomorphism, then it is evident that the rate of increase is bounded by an exponent. In particular, because algebraic diffeomorphisms are dense in a smooth diffeomorphism space, there is a dense diffeomorphism set in this space, for which the rate of increase of the number of periodic points is majorized by an exponent [2]. It seems likely that the exponent index is equal to the topological entropy $h_{\mathrm{top}}$ of the diffeomorphism. This conjecture is supported by the the fact that, if $f: [0;1] \to [0;1]$ is a $C^3$-map of a segment with nonplanar critical points, then $h_{\mathrm{top}} = \lim_{n\to\infty} \frac{1}{n} \ln \#\mathrm{Per}_n$, where $\#\mathrm{Per}_n$ is the number of periodic points of period $n$ for the map $f$. (This formula follows from [5,6].)

For any numerical sequence $a_n$ one can construct an *analytic* diffeomorphism $F$ of the two-dimensional torus such that the number of intersection points $F^{n_i}(\mathbb{S}^1) \cap \mathbb{S}^1$ is greater than $a_{n_i}$, where $\mathbb{S}^1$ is a fixed circle on the torus, and $n_i$ is a subsequence of positive integers [4]. A similar example of the rate of increase of the number of periodic points for an analytic diffeomorphism is unknown.

In the case where $F$ is a *smooth* diffeomorphism, there is an example where the number of periodic trajectories increases arbitrarily fast [3]. Moreover, it turns out that there are sufficiently many diffeomorphisms with a superexponential growth of the number of periodic trajectories; the set of these diffeomorphisms does not belong to the first category. An example of arbitrarily fast growth of the number of intersection points was constructed in [1].

[1] Arnold V. I. Dynamics of intersections. In: Analysis, et cetera. Research papers published in honor of Jürgen Moser's 60th birthday. Editors: P. H. Rabinowitz and E. Zehnder. Boston, MA: Academic Press, 1990, 77–84.

[2] Artin M., Mazur B. On periodic points. *Ann. Math., Ser.* 2, 1965, **81**(1), 82–99.

[3] Kaloshin V. Yu. Generic diffeomorphisms with superexponential growth of number of periodic orbits. *Commun. Math. Phys.*, 2000, **211**(1), 253–271.

[4] Kozlovskiĭ O. S. The dynamics of intersections of analytic manifolds. *Dokl. Math.*, 1992, **45**(2), 425–427.

[5] DE MELO W., VAN STRIEN S. One-dimensional Dynamics. Berlin: Springer, 1993.

[6] MISIUREWICZ M., SZLENK W. Entropy of piecewise monotone mappings. *Studia Math.*, 1980, **67**(1), 45–63.

△ **1988-6** – ***M. B. Sevryuk*** Also: *1988-7, 1988-8, 1989-2, 1990-1, 1990-20, 1990-21, 1992-12–1992-14, 1994-45–1994-50*

Problems *1988-6, 1988-7, 1989-2, 1990-21, 1992-12–1992-14, 1994-45–1994-48* and the first part of problem *1988-8* pertain to the question on the asymptotics of topological complexity of intersections. In its most general form (see problems *1988-8* and *1994-48*), this question is formulated as follows. Let $M^m$ be a compact manifold of dimension $m$, and $X^k$ and $Y^l$ its compact submanifolds of dimensions $k$ and $l$, respectively. Let also $A: M \to M$ be a certain mapping. One has to estimate the asymptotics of one or another topological invariant (in the simplest case, that of the number of connected components) of the intersection $(A^n X) \cap Y$ (where $A^n: M \to M$ denotes the $n$-th iteration of the mapping $A$) as $n \to +\infty$.

One of the particular cases of this question consists in estimating the asymptotics of the number of periodic trajectories of period $n$ for a mapping $B: N \to N$ of a certain manifold $N$ (cf. problems *1988-7, 1989-2, 1992-13, 1994-45*, and *1994-47*).

Indeed, if $M = N \times N$, $X = Y = \{(x,x) \mid x \in N\}$ is the diagonal in $M$ and a mapping $A: M \to M$ is defined by the formula $A(x,y) = (Bx, y)$, then the intersection points of $A^n X$ with $Y$ are just the points of the form $(x,x)$, for which $B^n x = x$. As a rule, one assumes in this case that all the periodic points are nondegenerate.

The question on the asymptotics of topological complexity of intersections can be raised in the algebraic, smooth ($C^r$ with $r \leq +\infty$), and analytic categories.

The main results here are as follows. In paper [6], M. Artin and B. Mazur obtained an exponential estimate $\nu(n) \leq Ce^{\lambda n}$ for the number $\nu(n)$ of periodic trajectories of period $n$ for *algebraic* manifolds and mappings. In the same paper, they showed that, in the functional space of smooth (of class $C^r$ for each positive integer $r$) mappings $A: M \to M$, there exists an everywhere dense subset of mappings satisfying the same exponential estimate. In works [1,2] ([1] is the note cited in the formulation of problem *1992-12*), V. I. Arnold proved an exponential estimate for the values of various topological and integro-topological invariants [such as the $(k+l-m)$-dimensional volume or the Morse and Betti numbers, in particular, the number of connected components] of the intersection $(A^n X^k) \cap Y^l$ in the case of diffeomorphisms $A: M^m \to M^m$. However, this estimate holds "almost always" ("with probability 1") rather than always, to be more precise, for almost all (in the sense of the Lebesgue measure) the values of the parameter $t \in \mathbb{R}^p$ in typical families of fixed manifolds $Y_t^l$ provided that the dimension $p$ of the parameter space is large enough. In work [2], a $C^\infty$-example was also constructed of an *arbitrarily fast growth* of the number of connected components of $(A^n X^k) \cap Y^l$ (in this example, $M$ is a two-torus while $X = Y$ is a circle on $M$). An analogous analytic example was constructed by O. S. Kozlovskiĭ in paper [9], so that the answer to the question of problem *1988-6* is affirmative. In works [10, 11], examples were constructed of an arbitrarily fast growth of both the number of periodic trajectories of period $n$ in a diffeomorphism $A: M \to M$ (as $n \to +\infty$) and the number of periodic trajectories of period no greater than $T$ in smooth vector field $V$ on $N$ (as $T \to +\infty$) for an *arbitrary* compact manifold $M$ of dimension $\geq 2$ and an *arbitrary* compact manifold $N$ of dimension $\geq 3$. Moreover, in these examples, all the periodic points or closed trajectories are nondegenerate.

As far as the author of the present comment knows, the question of problem *1992-12* is still open.

The second part of problem *1988-8* and problems *1990-1, 1990-20, 1994-49, 1994-50* are devoted to the local analogue of the question on the asymptotics of topological complexity of intersections. This local theory is considered in paper [3].

The whole circle of questions touched upon in problems *1988-6, 1988-7, 1988-8, 1989-2, 1990-1, 1990-20, 1990-21, 1992-12–1992-14, 1994-45–1994-50* is discussed in detail in V. I. Arnold's survey works [4, 5].

For further results, see the papers by E. Rosales-González [12, 13] and V. Yu. Kaloshin [7, 8]. Problems *1958-3, 1971-9, 1991-3, 1992-5, 1993-35*, and *1993-36* are devoted to topics close to those considered in the present comment.

[1] ARNOLD V. I. Dynamics of complexity of intersections. *Bol. Soc. Brasil. Mat.* (*N. S.*), 1990, **21**(1), 1–10. [*The Russian translation in*: Vladimir Igorevich Arnold. Selecta–60. Moscow: PHASIS, 1997, 489–499.]

[2] ARNOLD V. I. Dynamics of intersections. In: Analysis, et cetera. Research papers published in honor of Jürgen Moser's 60th birthday. Editors: P. H. Rabinowitz and E. Zehnder. Boston, MA: Academic Press, 1990, 77–84.

[3] ARNOLD V. I. Bounds for Milnor numbers of intersections in holomorphic dynamical systems. In: Topological Methods in Modern Mathematics. Proceedings of the symposium in honor of John Milnor's sixtieth birthday (Stony Brook, NY, 1991). Editors: L. R. Goldberg and A. V. Phillips. Houston, TX: Publish or Perish, 1993, 379–390.

[4] ARNOLD V. I. Problems on singularities and dynamical systems. In: Developments in Mathematics: the Moscow School. Editors: V. I. Arnold and M. Monastyrsky. London: Chapman & Hall, 1993, 251–274.

[5] ARNOLD V. I. Sur quelques problèmes de la théorie des systèmes dynamiques. *Topol. Methods Nonlinear Anal.*, 1994, **4**(2), 209–225. [*The Russian translation in*: Vladimir Igorevich Arnold. Selecta–60. Moscow: PHASIS, 1997, 533–551.]

[6] ARTIN M., MAZUR B. On periodic points. *Ann. Math., Ser.* 2, 1965, **81**(1), 82–99.

[7] KALOSHIN V. YU. An extension of the Artin–Mazur theorem. *Ann. Math., Ser.* 2, 1999, **150**(2), 729–741.

[8] KALOSHIN V. YU. Generic diffeomorphisms with superexponential growth of number of periodic orbits. *Commun. Math. Phys.*, 2000, **211**(1), 253–271.

[9] KOZLOVSKIĬ O. S. The dynamics of intersections of analytic manifolds. *Dokl. Math.*, 1992, **45**(2), 425–427.

[10] ROSALES-GONZÁLEZ E. On the growth of the numbers of periodic orbits of dynamical systems. *Funct. Anal. Appl.*, 1991, **25**(4), 254–262.

[11] ROSALES-GONZÁLEZ E. On the growth of the number of long period solutions of differential equations. *Funct. Anal. Appl.*, 1992, **26**(2), 99–105.

[12] ROSALES-GONZÁLEZ E. Milnor numbers in dynamical systems. In: Singularity Theory (Trieste, 1991). Editors: D. T. Lê, K. Saito and B. Teissier. River Edge, NJ: World Scientific, 1995, 627–634.

[13] ROSALES-GONZÁLEZ E. Intersection dynamics on Grassmann manifolds. *Bol. Soc. Mat. Mexicana, Ser.* 3, 1996, **2**(2), 129–138.

## 1988-7

See the comments to problem *1988-6*.

## 1988-8

See the comment to problem *1988-6* by M. B. Sevryuk.

## 1988-9 – *V. A. Vassiliev*

Homology groups of these complements were calculated in [1]. Homotopy types (in particular, the cohomology rings) of spaces $\mathbb{R}^\mu \setminus \bar{A}_k$, where $k \geq 2$ is fixed and $\mu \to \infty$, stabilize to those of the loop space $\Omega S^k$ of the $k$-dimensional sphere. This stabilization is realized by the jet embeddings $\mathbb{R}^\mu \to \Omega\mathbb{R}^{k+1}$ sending any polynomial $f$ to the (suitably improved at infinity) map $(f, f', \ldots, f^{(k)}) : \mathbb{R}^1 \to \mathbb{R}^{k+1}$: this embedding sends $\mathbb{R}^\mu \setminus \bar{A}_k$ to $\Omega(\mathbb{R}^{k+1} \setminus 0)$ and induces an isomorphism of all homotopy groups of these spaces up to dimension $\mu/(k+1) - 1$, see [4].

Also, the spaces $\mathbb{R}^\mu \setminus \bar{A}_k$ homotopically converge to the space of all smooth functions $\mathbb{R}^1 \to \mathbb{R}^1$ with some fixed behavior at infinity (say, equal to 1 outside some compact set) and having no $(k+1)$-fold zeros (i. e., points where $f = f' = \cdots = f^{(k)} = 0$. The jet embedding of the latter space into the space $\Omega S^k$ is a homotopy equivalence for any $k \geq 2$.

For an arbitrary closed singularity class $A \subset J^m(\mathbb{R}^n, \mathbb{R})$ of codimension $\geq 2$ the cohomology rings of complements of the corresponding strata of discriminants of sufficiently complicated isolated singularities in $\mathbb{R}^n$ stabilize to the cohomology ring of the iterated loop space $\Omega^n(J^m(\mathbb{R}^n, \mathbb{R}) \setminus A)$ when the complexity of these singularities grows. Moreover, if the codimension of the class $A$ is greater than 3 then the homotopy types also stabilize. This is proved in [5, 6] in the case of complex singularities but is true also in the real situation by the same arguments.

Probably in the "homotopy" statement the condition codim $\geq 3$ can be replaced by $\geq 2$ (unlike the complex situation). At least, in the similar "non-local" problem concerning the space of functions $M^n \to \mathbb{R}$ without singularities of type $A_3$ such an "$h$"-principle holds: this space is homotopy equivalent to the corresponding space of admissible sections of the jet bundle $J^k(M^n, \mathbb{R}) \to M^n$. (K. Igusa [3] proved the homotopy equivalence up to dimension $n-1$, and Ya. M. Eliashberg and N. M. Mishachev [2] in all dimensions.)


[1] ARNOLD V. I. Spaces of functions with moderate singularities. *Funct. Anal. Appl.*, 1989, **23**(3), 169–177. [*The Russian original is reprinted in*: Vladimir Igorevich Arnold. Selecta–60. Moscow: PHASIS, 1997, 455–469].

[2] ELIASHBERG YA. M., MISHACHEV N. M. Wrinkling of smooth mappings and its applications. *Invent. Math.*, 1997, **130**(2), 345–369.

[3] IGUSA K. Higher singularities of smooth functions are unnecessary. *Ann. Math., Ser.* 2, 1984, **119**(1), 1–58.

[4] VASSILIEV V. A. Topology of spaces of functions without compound singularities. *Funct. Anal. Appl.*, 1989, **23**(4), 277–286.

[5] VASSILIEV V. A. Topology of complements to discriminants and loop spaces. In: Theory of Singularities and its Applications. Editor: V. I. Arnold. Providence, RI: Amer. Math. Soc., 1990, 9–21. (Adv. Sov. Math., 1.).

[6] VASSILIEV V. A. Complements of Discriminants of Smooth Maps: Topology and Applications, revised edition. Providence, RI: Amer. Math. Soc., 1994. (Transl. Math. Monographs, 98.)

## 1988-10 — *V. A. Vassiliev*

The stabilization of homology groups was proved by V. I. Arnold [1].

If $k \geq 2$ then the homotopy types of these complements converge to that of the double loop space $\Omega^2 S^{2k+1}$. More precisely, for any $\mu$ and $k$ we have the (more or less canonical) embedding $\mathbb{C}^\mu \setminus A_k \to \Omega^2 S^{2k+1}$ inducing an isomorphism of all homotopy groups of dimensions $\leq (2k-1)[\mu/(k+1)] - 1$. Namely, this is the jet embedding sending any polynomial $f \in \mathbb{C}^\mu \setminus A_k$ to the map $(f, f', \ldots, f^{(k)}) : \mathbb{C}^1 \to \mathbb{C}^{k+1} \setminus 0 \sim S^{2k+1}$ improved in some uniform way close to the infinity in $\mathbb{C}^1$.

If $k = 1$ (i. e., $A_k$ is the entire discriminant) then only the *stable* homotopy types of spaces $\mathbb{C}^\mu \setminus A_k$ converge to that of $\Omega^2 S^3$. In particular, all homology groups, including the extraordinary ones, of these spaces also converge to to similar groups of $\Omega^2 S^3$. The usual homotopy types in this case converge to that of the classifying space of the stable braid group (which is stably homotopy equivalent to $\Omega^2 S^3$ by a theorem of J. P. May and G. Segal).

In general, for any $n$ and any singularity class $A \subset J^k(\mathbb{C}^n, \mathbb{C})$ the stable homotopy type of complements of the stratum $\{A\}$ in the spaces of deformations of function singularities $(\mathbb{C}^n, 0) \to (\mathbb{C}, 0)$ converges (when the complexity of these singularities grows) to the stable homotopy type of the iterated loop space $\Omega^{2n}(J^k(\mathbb{C}^n, \mathbb{C}) \setminus A)$. If $\operatorname{codim}_{\mathbb{C}} A \geq 2$, then the same stabilization holds for the usual homotopy type, see [2, 3].

[1] ARNOLD V. I. On some topological invariants of algebraic functions. *Trans. Moscow Math. Soc.*, 1970, **21**, 30–52.

[2] VASSILIEV V. A. Topology of complements to discriminants and loop spaces. In: Theory of Singularities and its Applications. Editor: V. I. Arnold. Providence, RI: Amer. Math. Soc., 1990, 9–21. (Adv. Sov. Math., 1.).

[3] VASSILIEV V. A. Complements of Discriminants of Smooth Maps: Topology and Applications, revised edition. Providence, RI: Amer. Math. Soc., 1994. (Transl. Math. Monographs, 98.)

## ▽ 1988-11 — *V. D. Sedykh*

There are many works devoted to generalizations of the classical theorem [15] on four vertices of a smooth closed simple curve in Euclidean plane.

Kneser [13] in 1912 remarked that vertices of a plane curve pass into flattening points of its image in the space under the stereographic projection (it was shown in [22] that this is a manifestation of a general relation between singularities of the envelope of the family of normals to a manifold and singularities of the front of tangent hyperplanes to its stereographic image). Therefore, any smooth closed simple curve in the sphere $S^2$ in the three-dimensional space has at least four flattening points.

Later on, the theorem on four flattening points was proved for wider classes of spatial curves. Namely: for curves lying on surfaces that intersect any straight line at most at two points [14]; for curves any two points of which lie on a plane that does not intersect the curve anymore [7]; for curves lying on the boundary of its convex hull (we call such curves *weakly convex*) and intersecting any plane at finitely many points and any straight line at most at two points [8]; for weakly convex generic curves [16].

In 1992 the theorem on four flattening points of weakly convex space curves was proved under the weakest assumption:

**Theorem** [19]. *Any closed $C^3$-embedded weakly convex curve in the three-dimensional space has at least four flattening points if its curvature nowhere vanishes.*

The theorems of [15, 19] have discrete versions (for polygonal lines on a plane and in a space; see paper [25] and references therein). Some estimates of the number of flattening points were obtained in [17] for weakly convex space curves with zero-curvature points and in [18] for curves which are not weakly convex.

Numerous works are connected with generalizations of the 4-vertex theorem in terms of symplectic and contact topology (see, for example, [2,5,6,27]). In the framework of projective topology, V. I. Arnold proved the tennis ball theorem:

**Theorem** [1, 3]. *If a smooth closed curve embedded into the standard sphere $S^2$ bisects its area, then this curve has at least four points of a spherical inflection.*

Some multidimensional versions of the 4-vertex theorem were given by Barner [7]. He proved, in particular, that a smooth closed curve in the $n$-dimensional projective space has at least $n+1$ flattening points if it satisfies the following property: for any $n-1$ points of a curve there is a plane which passes through these points and does not intersect the curve at other points (curves with such property are called *Barner-convex*). In the partial case of curves having a convex projection onto a hyperplane, this fact was also obtained in [4]. In [31], the results of [7] were generalized to curves in Lobachevskian spaces.

Note that, in paper [16], the theorem on four flattening points of a smooth closed weakly convex generic curve was deduced from the relation

$$\chi(A_3) - \chi(3A_1) = 4$$

which binds the number $\chi(A_3)$ of supporting osculating planes to a curve and the number $\chi(3A_1)$ of supporting planes tangent to the curve at three points. This relation is a three-dimensional version of the Bose–Haupt formula [9, 10] which claims that the number of supporting curvature circles to a smooth simple closed generic curve in the Euclidean plane exceeds the number of supporting circles tangent to the curve at three points by 4.

The Bose–Haupt formula was generalized for supporting hyperplanes to smooth closed weakly convex generic submanifolds in multidimensional spaces in [20] (a proof is given in [21, 24]; it is based on the calculation of the homology groups of the set of singular supporting hyperplanes to a strictly convex submanifold) and for supporting hyperspheres to smooth closed generic submanifolds in Euclidean spaces in [23] (see also [26]). In [11, 12] some formulae of this type were obtained for singularities of the global minimum (maximum) function of a generic family of smooth functions on a circle (to get the Bose–Haupt formula for a plane curve, one should just consider the square of the distance from a point of the curve to a given point in the plane).

Finally, the recent results [28] on the topology of wave fronts led to formulae of Bose–Haupt type for isolated singularities of corank 1 on generic wave fronts satisfying some topological conditions. As a corollary, we proved, for example, that if a smooth closed connected generic curve in $\mathbb{R}^{11}$ is Barner-convex, then the numbers $\chi(A_{\mu_1} + \cdots + A_{\mu_m})$ of its supporting $A_{\mu_1} + \cdots + A_{\mu_m}$-planes having the eleventh multiplicity of tangency with it are bound by the equality

$$\begin{aligned}
42\chi(A_{11}) - 14\chi(A_9 + 2A_1) - 5\chi(A_7 + A_3 + A_1) + 5\chi(A_7 + 4A_1) - 4\chi(2A_5 + A_1) - \\
- 2\chi(A_5 + 2A_3) + 2\chi(A_5 + A_3 + 3A_1) - 2\chi(A_5 + 6A_1) + \chi(3A_3 + 2A_1) - \\
- \chi(2A_3 + 5A_1) + \chi(A_3 + 8A_1) - \chi(11A_1) = 504.
\end{aligned}$$

Here a supporting $A_{\mu_1} + \cdots + A_{\mu_m}$-plane is a supporting hyperplane which is tangent to a curve at $m$ points with multiplicities $\mu_1, \ldots, \mu_m$ respectively; $kA_\mu = A_\mu + \cdots + A_\mu$ ($m$ times).

There are many other works which are connected in either case with different generalizations of the 4-vertex theorem, see, e. g., [29,30]. However, necessary and sufficient conditions under which a curve in the three-dimensional space has at least four flattening points have not been found so far.

[1] ARNOLD V. I. Topological Invariants of Plane Curves and Caustics. Dean Jacqueline B. Lewis Memorial Lectures, Rutgers University. Providence, RI: Amer. Math. Soc., 1994. (University Lecture Series, 5.)

[2] ARNOLD V. I. The geometry of spherical curves and the algebra of quaternions. *Russian Math. Surveys*, 1995, **50**(1), 1–68.

[3] ARNOLD V. I. Topological properties of Legendre projections in contact geometry of wave fronts. *St. Petersburg Math. J.*, 1995, **6**(3), 439–452.

[4] ARNOLD V. I. On the number of flattening points of space curves. In: Sinaĭ's Moscow Seminar on Dynamical Systems. Editors: L. A. Bunimovich, B. M. Gurevich and Ya. B. Pesin. Providence, RI: Amer. Math. Soc., 1996, 11–22. (AMS Transl., Ser. 2, 171; Adv. Math. Sci., 28.)

[5] ARNOLD V. I. Topological problems of the theory of wave propagation. *Russian Math. Surveys*, 1996, **51**(1), 1–47.

[6] ARNOLD V. I. Towards the Legendre Sturm theory of space curves. *Funct. Anal. Appl.*, 1998, **32**(2), 75–80.

[7] BARNER M. Über die Mindestanzahl stationärer Schmiegebenen bei geschlossenen streng-konvexen Raumkurven. *Abh. Math. Semin. Univ. Hamburg*, 1956, **20**, 196–215.

[8] BISZTRICZKY T. Inflectional convex space curves. *Canad. J. Math.*, 1984, **36**(3), 537–549.

[9] BOSE R. C. On the number of circles of curvature perfectly enclosing or perfectly enclosed by a closed convex oval. *Math. Z.*, 1932, **35**, 16–24.

[10] HAUPT O. Verallgemeinerung eines Satzes von R. C. Bose über die Anzahl der Schmiegkreise eines Ovals, die vom Oval umschlossen werden oder das Oval umschließen. *J. Reine Angew. Math.*, 1969, **239/240**, 339–352.

[11] KAZARIAN M. E. The Chern–Euler number of circle bundle via singularity theory. *Math. Scand.*, 1998, **82**(2), 207–236.

[12] KAZARIAN M. E. Relative Morse theory of one-dimensional bundles and cyclic homology. *Funct. Anal. Appl.*, 1997, **31**(1), 16–24.

[13] KNESER A. Bemerkungen über die Anzahl der Extreme der Krümmung geschlossenen Kurven und über verwandte Fragen in einer nicht-euclidischen Geometrie. In: Festschrift H. Weber 70. Leipzig, 1912, 170–180.

[14] MOHRMANN H. Die Minimalzahl der stationären Ebenen eines räumlichen Ovals. *Sitz. Ber. kgl. Bayerichen Akad. Wiss., Math.-Phys. Kl.*, 1917, 1–3.

[15] MUKHOPADHAYA S. New methods in the geometry of a plane arc. I. Cyclic and sextactic points. *Bull. Calcutta Math. Soc.*, 1909, **1**, 31–37.

[16] ROMERO-FUSTER M. C. Convexly generic curves in $\mathbb{R}^3$. *Geom. Dedicata*, 1988, **28**(1), 7–29.

[17] ROMERO-FUSTER M. C., SEDYKH V. D. On the number of singularities, zero curvature points and vertices of a simple convex space curve. *J. Geometry*, 1995, **52**(1–2), 168–172.

[18] ROMERO-FUSTER M. C., SEDYKH V. D. A lower estimate for the number of zero-torsion points of a space curve. *Beiträge zur Algebra und Geometrie*, 1997, **38**(1), 183–192.

[19] SEDYKH V. D. A theorem on four vertices of a convex space curve. *Funct. Anal. Appl.*, 1992, **26**(1), 28–32.

[20] SEDYKH V. D. Invariants of convex manifolds. *Dokl. Math.*, 1993, **46**(2), 392–396.

[21] SEDYKH V. D. Invariants of strictly convex manifolds. *Funct. Anal. Appl.*, 1993, **27**(3), 205–210.

[22] SEDYKH V. D. Connection of Lagrangian singularities with Legendre singularities under stereographic projection. *Sb. Math.*, 1995, **83**(2), 533–540.

[23] SEDYKH V. D. Invariants of nonflat manifolds. *Funct. Anal. Appl.*, 1995, **29**(3), 180–187.

[24] SEDYKH V. D. Strict convexity of a generic convex manifold. *Proc. Steklov Inst. Math.*, 1995, **209**, 174–190.

[25] SEDYKH V. D. Discrete versions of the four-vertex theorem. In: Topics in Singularity Theory. V. I. Arnold's 60th Anniversary Collection. Editors: A. Khovanskiĭ, A. Varchenko and V. Vassiliev. Providence, RI: Amer. Math. Soc., 1997, 197–207. (AMS Transl., Ser. 2, 180; Adv. Math. Sci., 34.)

[26] SEDYKH V. D. Invariants of submanifolds in Euclidean space. In: The Arnold–Gelfand Mathematical Seminars: Geometry and Singularity Theory. Editors: V. I. Arnold, I. M. Gelfand, V. S. Retakh and M. Smirnov. Boston, MA: Birkhäuser, 1997, 389–395.

[27] SEDYKH V. D. Some invariants of admissible homotopies of space curves. *Funct. Anal. Appl.*, 2001, **35**(4), 284–293.

[28] SEDYKH V. D. Relations between Euler numbers of manifolds of corank 1 singularities of a generic front. *Dokl. Math.*, 2002, **65**(2), 276–279.

[29] THORBERGSSON G., UMEHARA M. A unified approach to the four vertex theorems, II. In: Differential and Symplectic Topology of Knots and Curves. Editor: S. Tabachnikov. Providence, RI: Amer. Math. Soc., 1999, 229–252. (AMS Transl., Ser. 2, 190; Adv. Math. Sci, 42.)

[30] UMEHARA M. A unified approach to the four vertex theorems, I. In: Differential and Symplectic Topology of Knots and Curves. Editor: S. Tabachnikov. Providence, RI: Amer. Math. Soc., 1999, 185–228. (AMS Transl., Ser. 2, 190; Adv. Math. Sci, 42.)

[31] URIBE-VARGAS R. On the $(2k+2)$-vertex and $(2k+2)$-flattening theorems in higher dimensional Lobatchevskian space. *C. R. Acad. Sci. Paris, Sér.* I *Math.*, 1997, **325**(5), 505–510.

## △ 1988-11 — *R. Uribe-Vargas*

ℛ This is a complement to the comment by V. D. Sedykh (mainly for curves in $\mathbb{R}^3$).

*A closed curve in* $\mathbb{R}^3$ (*in* $\mathbb{RP}^3$) *which can be projected, from a point exterior to it, onto a convex curve of* $\mathbb{R}^2$ (*of* $\mathbb{RP}^2$) *has at least* 4 *flattenings* [2].

A closed curve in $\mathbb{R}^3$ (in $\mathbb{RP}^3$) is called a *Barner curve* if for every 2 (not necessarily geometrically different) points of the curve there exists a plane through these points that does not intersect the curve elsewhere. *A Barner curve has at least* 4 *flattenings* [3]. *A closed curve in* $\mathbb{R}^3$ (*in* $\mathbb{RP}^3$) *having a convex projection is a Barner curve* [10] and *there is a non-empty open set of embedded closed curves in* $\mathbb{R}^3$ (*in* $\mathbb{RP}^3$) *which are Barner curves and have no convex projection* [8].

A closed $C^3$-smooth curve with non-vanishing curvature in $\mathbb{R}^3$ lying on the boundary of its convex hull is called (here) a *Carathéodory curve*. *A generic Carathéodory curve has at least* 4 *flattenings* [6]; in [4], Blaschke attributes an equivalent result to Carathéodory.

**Sedykh's Theorem** [7]. *Any Carathéodory curve* (without genericity conditions) *has at least* 4 *flattenings.*

A closed curve in $\mathbb{R}^3$ with non-vanishing curvature and no parallel tangents with the same orientation is called a *Segre curve* [11].

**Segre's Theorem** [9]. *Any Segre curve has at least* 4 *flattenings.*

*Any Barner curve in* $\mathbb{R}^3$ *is a Carathéodory curve and it is also a Segre curve* [11].

Sedykh's theorem is considered, in a major part of the concerned literature, as the most general 4-flattening theorem for curves in $\mathbb{R}^3$. However, in [11] it is proved that *there is a non-empty open set of Segre curves in* $\mathbb{R}^3$ *which are not Carathéodory curves* and also that *there is a non-empty open set of Carathéodory curves in* $\mathbb{R}^3$ *which are not Segre curves*. See also the comment to problem *1994-6*.

For $n$-dimensional spaces, a class of curves generalizing the convex ones was introduced in [12], where the $n$-dimensional Lobachevskian space is denoted by $\mathbb{L}^n$:

**Definition.** A closed curve embedded in $\mathbb{R}^n$ (in $S^n$ or in $\mathbb{L}^n$) is called *spherically convex* if for each $k$-tuple of different points of the curve, $k \leq n$, with positive multiplicities satisfying $m_1 + \cdots + m_k = n$, there exists at least one hypersphere of $\mathbb{R}^n$ (or hypersphere of $S^n$ or hyperbolic hypersphere of $\mathbb{L}^n$, respectively) intersecting the curve at these points, with corresponding multiplicities, that does not intersect the curve elsewhere. The hyperspheres of infinite radius are not excluded.

**Remark.** Affine transformations of $\mathbb{R}^n$ preserve convex curves but do not preserve vertices. Conformal transformations of $\mathbb{R}^n$ (respectively of $S^n$ or of $\mathbb{L}^n$) preserve both vertices and spherically convex curves, but do not preserve convex curves. So, in the study of problems about vertices it seems to be more natural to consider spherically convex curves instead of convex ones.

**Theorem** [12]. *A spherically convex curve in a 2k-dimensional space of constant curvature (Euclidean, spherical or hyperbolic) has not less than $2k+2$ vertices.*

In [5], M. Kazarian gave a nonlinear (symplectic) version of Arnold's theorem [2] on the number of flattenings of a small perturbation of a hyperplane convex curve in $\mathbb{R}^{2k+1}$. This generalization justifies the assertion that Arnold's theorem belongs to symplectic topology.


[1] ARNOLD V. I. The geometry of spherical curves and the algebra of quaternions. *Russian Math. Surveys*, 1995, **50**(1), 1–68.

[2] ARNOLD V. I. On the number of flattening points of space curves. In: Sinaĭ's Moscow Seminar on Dynamical Systems. Editors: L. A. Bunimovich, B. M. Gurevich and Ya. B. Pesin. Providence, RI: Amer. Math. Soc., 1996, 11–22. (AMS Transl., Ser. 2, 171; Adv. Math. Sci., 28.)

[3] BARNER M. Über die Mindestanzahl stationärer Schmiegeebenen bei geschlossenen streng-konvexen Raumkurven. *Abh. Math. Semin. Univ. Hamburg*, 1956, **20**, 196–215.

[4] BLASCHKE W. Vorlesungen über Differential-Geometrie, Vol. I, 3te Edition. Berlin: Springer, 1930.

[5] KAZARIAN M. E. Nonlinear version of Arnold's theorem on flattening points. *C. R. Acad. Sci. Paris, Sér.* I *Math.*, 1996, **323**(1), 63–68.

[6] ROMERO-FUSTER M. C. Convexly generic curves in $\mathbb{R}^3$. *Geom. Dedicata*, 1988, **28**(1), 7–29.

[7] SEDYKH V. D. A theorem on four vertices of a convex space curve. *Funct. Anal. Appl.*, 1992, **26**(1), 28–32.

[8] SEDYKH V. D. On some classes of curves in a projective space. In: Geometry and Topology of Caustics—CAUSTICS'98 (Warsaw). Editors: S. Janeczko and V. M. Zakalyukin. Warsaw: Polish Academy of Sciences, Institute of Mathematics, 1999, 237–266. (Banach Center Publ., 50.)

[9] SEGRE B. Alcune proprietà differenziali in grande delle curve chiuse sghembe. *Rend. Mat., Ser.* 6, 1968, **1**, 237–297.

[10] TABACHNIKOV S. L. On zeros of the Schwarzian derivative. In: Topics in Singularity Theory. V. I. Arnold's 60th Anniversary Collection. Editors: A. Khovanskiĭ, A. Varchenko and V. Vassiliev. Providence, RI: Amer. Math. Soc., 1997, 229–239. (AMS Transl., Ser. 2, 180; Adv. Math. Sci., 34.)

[11] URIBE-VARGAS R. Symplectic and contact singularities in the differential geometry of curves and surfaces. Ph. D. Thesis, Université Paris 7, 2001, Ch. 3, § 7.

[12] URIBE-VARGAS R. Four-vertex theorems in higher-dimensional spaces for a larger class of curves than the convex ones. *C. R. Acad. Sci. Paris, Sér.* I *Math.*, 2000, **330**(12), 1085–1090.

## 1988-13

ℛ See the comment to problem *1987-14*.

## 1988-15 — *R. Uribe-Vargas*

ℛ A Lagrangian manifold $L \subset T^*M$ is called *pseudo-optical* if there exists a vector field on $L$ (its *framing*) whose image on $M$ under the natural projection $T^*M \to M$ does not vanish at any point of $L$. Two pseudo-optical immersed Lagrangian submanifolds of $T^*M$ are said to be *isotopic* if they may be connected by a continuous family of pseudo-optical immersed Lagrangian submanifolds $L_t$ with framing continuously depending on the parameter and such that the caustic (the set of critical values of the map $L_t \subset T^*M \to M$) of each submanifold of the family is compact.

For the singularities $D_4^{\pm}$ the kernel of the tangent map $T_mL \to T_{f(m)}M$, $m \in L$, is two-dimensional. The one-dimensional image of this map is called the *tangent line* to the caustic singularity $D_4^{\pm}$. The tangent line to the purse ($D_4^+$) is that to its cuspidal edge, and the tangent line to the pyramid ($D_4^-$) is the common one to the three cuspidal edges. One of the two branches of the cuspidal edge of the purse consists of points of type $A_3^+$ and the other of points of type $A_3^-$. For the pyramid there are three branches of cuspidal edges of type $A_3^+$ on one side of the singularity $D_4^-$ and three ones of type $A_3^-$ on the other side. The tangent line to the caustic $D_4^{\pm}$ carries thus a *natural orientation*. This orientation is chosen [4, 5] in such a way that the orienting vector is directed towards the cuspidal edges of type $A_3^-$.

For a Lagrangian pseudo-optical manifold $L$, the image under the tangent map $T_mL \to T_{f(m)}M$ of the framing vector, at a point of Lagrangian singularity $D_4^{\pm}$, does not vanish and lies on the tangent line to the caustic. There is a natural way to define a *sign* of each singularity $D_4^\pm$. Both $D_4^+$ and $D_4^-$ singularities may have positive or negative signs.

**Definition.** A Lagrangian singularity $D_4^\pm$ on a pseudo-optical Lagrangian 3-fold is called *positive* or *negative* depending on the coincidence or not of the natural orientation of the tangent line to the caustic with the given by the framing vector. The notation $D_4^{++}$, $D_4^{+-}$ and $D_4^{-+}$, $D_4^{--}$ is used for these singularities.

**Theorem** [4, 5]. *The algebraic number of singularities $D_4^\pm$ of a pseudo-optical 3-fold L counted with their signs,*

$$d(L) = \#D_4^{++} - \#D_4^{+-} + \#D_4^{-+} - \#D_4^{--},$$

*is an invariant of isotopy class of immersed pseudo-optical Lagrangian submanifolds of $T^*M$.*

The *standard Lagrangian cylinder* [2] is the Lagrangian submanifold $L_0$ of $\mathbb{R}^6 = T^*\mathbb{R}^3$, given by

$$L_0 = \{(p,q) \in \mathbb{R}^6 : \|p\| = 1,\ q = sp,\ s \in \mathbb{R}\}.$$

The standard Lagrangian cylinder is pseudo-optical—with framing $\partial/\partial s$—and its caustic (one point) is compact.

**Proposition** [4, 5]. *For any pseudo-optical generic Lagrangian submanifold L isotopic to $L_0$, $d(L) = 4$. In particular, the caustic of L has at least* 4 *points of singularity $D_4^\pm$.*

According to [4,5], the definition of the invariant $d(L)$ (for the optical case) is implicitly contained in [3]. In [3] there are interesting results relating the Euler characteristic of a compact critical surface of an optical Lagrangian projection (having only singularities $A_\mu$ and $D_4^\pm$) with the number of singularities $D_4^\pm$. The results of [3] are well described in [1].

[1] ARNOLD V. I. Singularities of Caustics and Wave Fronts. Dordrecht: Kluwer Acad. Publ., 1990, § 2.4. (Math. Appl., Sov. Ser., 62.)

[2] ARNOLD V. I. Sur les propriétés des projections Lagrangiennes en géométrie symplectique des caustiques. CEREMADE (UMR 7534), Université Paris-Dauphine, № 9320, 14/06/1993; *Rev. Mat. Univ. Complut. Madrid*, 1995, **8**(1), 109–119. [*The Russian translation in*: Vladimir Igorevich Arnold. Selecta–60. Moscow: PHASIS, 1997, 525–532.]

[3] CHEKANOV YU. V. Caustics in geometrical optics. *Funct. Anal. Appl.*, 1986, **20**(3), 223–226.

[4] KAZARIAN M. E. Umbilical characteristic number of Lagrangian mappings of a 3-dimensional pseudo-optical manifold. *J. Math. Sci.*, 1996, **32**(5), 3672–3679.

[5] KAZARIAN M. E. Umbilical characteristic number of Lagrangian mappings of 3-dimensional pseudo-optical manifolds. In: Singularities and Differential Equations (Warsaw, 1993). Editors: S. Janeczko, W. M. Zajączkowski and B. Ziemian. Warsaw: Polish Academy of Sciences, Institute of Mathematics, 1996, 161–170. (Banach Center Publ., 33.)

## 1988-16

See the comments to problems *1993-27* and *1998-9*.

## 1988-23 – *P. M. Akhmet′ev*

The investigations started in 1970, see [3]. The idea of replacing fibrations with algebraic functions and studying their classifying spaces homotopy theory had been explicitly stated in [2], and had immediately led to the cobordisms and to new versions of the Pontryagin–Thom theories. The most important result was obtained by V. I. Arnold in [4] (this paper is mentioned in the problem).

In 1976 Andras Szűcs, following the idea by M. Gromov, constructed the classified space for immersions with prescribed multiplicity of multiple self-intersection points up to cobordism [8]. In 1980 Szűcs constructed a classified space for the maps with simplest singularities and described topological applications by means of calculation of homotopy of this space, the most interesting application being in [9]. Similar problems were investigated by U. Koschorke by a different method, see [6].

R. Rimányi developed ideas by A. Szűcs and constructed the classified space for singularities of the type $\Sigma^2$ and more complicated, see [7].

The space of functions with prescribed singularities and the space of plane curves with no horizontal inflection points generalize the Pontryagin–Thom construction in a different way, see [4]. The generalizations developed by P. M. Akhmet′ev, M. E. Kazarian and V. A. Vassiliev are related with higher singularities of functions and involve immersed submanifolds in codimension 1 (the Cerf diagrams), see [1, 5, 10].

[1] AKHMET′EV P. M. Embedding of compacta, stable homotopy groups of spheres, and singularity theory. *Russian Math. Surveys*, 2000, **55**(3), 405–462.

[2] ARNOLD V. I. Cohomology classes of algebraic functions invariant under Tschirnhausen transformations. *Funct. Anal. Appl.*, 1970, **4**(1), 74–75. [*The Russian original is reprinted in*: Vladimir Igorevich Arnold. Selecta–60. Moscow: PHASIS, 1997, 151–154.]

[3] ARNOLD V. I. Topological invariants of algebraic functions, II. *Funct. Anal. Appl.*, 1970, **4**(2), 91–98.

[4] ARNOLD V. I. Spaces of functions with moderate singularities. *Funct. Anal. Appl.*, 1989, **23**(3), 169–177. [*The Russian original is reprinted in*: Vladimir Igorevich Arnold. Selecta–60. Moscow: PHASIS, 1997, 455–469].

[5] KAZARIAN M. E. Characteristic classes of Lagrangian and Legendre singularities. *Russian Math. Surveys*, 1995, **50**(4), 701–726.

[6] KOSCHORKE U. Vector Fields and Other Vector Bundle Morphisms—a Singularity Approach. Berlin: Springer, 1981. (Lecture Notes in Math., 847.)

[7] RIMÁNYI R., SZŰCS A. Pontrjagin-Thom-type construction for maps with singularities. *Topology*, 1998, **37**(6), 1177–1191.

[8] SZŰCS A. Cobordism groups of *l*-immersions, I; II. *Acta Math. Acad. Sci. Hungar.*, 1976, **27**(3–4), 343–358; **28**(1–2), 93–102 (in Russian).

[9] SZŰCS A. Cobordism groups of immersions of oriented manifolds. *Acta Math. Hungar.*, 1994, **64**(2), 191–230.

[10] VASSILIEV V. A. Topology of Complements to Discriminants. Moscow: PHASIS, 1997 (in Russian).

## ▽ 1988-24 – *S. V. Duzhin*

ℛ N. V. Ilyushechkin [2] found an explicit expression for the discriminant of the characteristic polynomial of a normal complex matrix as a sum of (a large number) of squares of modules of certain polynomials depending on the coefficients of the matrix. In the particular case of real symmetric matrices a similar result was obtained by the same author earlier [1].

[1] ILYUSHECHKIN N. V. On a certain class of smooth matrix-valued functions. *Russian Math. Surveys*, 1985, **40**(1), 223–224.

[2] ILYUSHECHKIN N. V. The discriminant of the characteristic polynomial of a normal matrix. *Math. Notes*, 1992, **51**(3), 230–235.

## △ 1988-24 – *M. B. Sevryuk*

ℛ The stratification of the spaces of quadratic, Hermitian, and hyper-Hermitian forms according to the multiplicities of the eigenvalues was studied in works

[1] (the real case), [2] (the Hermitian, i. e., complex case, studied in [1] with less details, missing the remark that it provides the integer quantum Hall effect explanation), and [13] (the hyper-Hermitian, i. e., quaternionic case). For a unified viewpoint see [4].

The parallelism of the theories of quadratic, Hermitian, and hyper-Hermitian forms is a particular manifestation of the general phenomenon of the $\mathbb{R}$–$\mathbb{C}$–$\mathbb{H}$-ternarity in mathematics. The lists of $\mathbb{R}$–$\mathbb{C}$–$\mathbb{H}$-triads appearing in many very different branches of the mathematical science are given in works [3, 5–7]; see also the comments to problems *1997-9* and *1998-16* and papers [8–12].

[1] ARNOLD V. I. Modes and quasimodes. *Funct. Anal. Appl.*, 1972, **6**(2), 94–101. [*The Russian original is reprinted in*: Vladimir Igorevich Arnold. Selecta–60. Moscow: PHASIS, 1997, 189–202.]

[2] ARNOLD V. I. Remarks on eigenvalues and eigenvectors of Hermitian matrices, Berry phase, adiabatic connections and quantum Hall effect. *Selecta Math.* (*N. S.*), 1995, **1**(1), 1–19. [*The Russian translation in*: Vladimir Igorevich Arnold. Selecta–60. Moscow: PHASIS, 1997, 583–604.]

[3] ARNOLD V. I. Mysterious Mathematical Trinities. Topological Economy Principle in Algebraic Geometry. Moscow: Moscow Center for Continuous Mathematical Education Press, 1997 (in Russian).

[4] ARNOLD V. I. Relatives of the quotient of the complex projective plane by complex conjugation. *Proc. Steklov Inst. Math.*, 1999, **224**, 46–56.
[*Internet*: `http://www.pdmi.ras.ru/~arnsem/Arnold/arn-papers.html`]

[5] ARNOLD V. I. Symplectization, complexification and mathematical trinities. In: The Arnoldfest. Proceedings of a conference in honour of V. I. Arnold for his sixtieth birthday (Toronto, 1997). Editors: E. Bierstone, B. A. Khesin, A. G. Khovanskiĭ and J. E. Marsden. Providence, RI: Amer. Math. Soc., 1999, 23–37. (Fields Institute Commun., 24.); CEREMADE (UMR 7534), Université Paris-Dauphine, № 9815, 04/03/1998.
[*Internet*: `http://www.pdmi.ras.ru/~arnsem/Arnold/arn-papers.html`]

[6] ARNOLD V. I. Mysterious mathematical trinities. In: Students' Readings in the Mathematical College of the Independent University of Moscow, Vol. 1. Editor: V. V. Prasolov. Moscow: Moscow Center for Continuous Mathematical Education Press, 2000, 4–16 (in Russian).

[7] ARNOLD V. I. Polymathematics: is mathematics a single science or a set of arts? In: Mathematics: Frontiers and Perspectives. Editors: V. I. Arnold, M. Atiyah, P. Lax and B. Mazur. Providence, RI: Amer. Math. Soc., 2000, 403–416; CEREMADE (UMR 7534), Université Paris-Dauphine, № 9911, 10/03/1999.
[*Internet*: `http://www.pdmi.ras.ru/~arnsem/Arnold/arn-papers.html`]

[8] ARNOLD V. I. The complex Lagrangian Grassmannian. *Funct. Anal. Appl.*, 2000, **34**(3), 208–210.

[9] ARNOLD V. I. The Lagrangian Grassmannian of a quaternionic hypersymplectic space. *Funct. Anal. Appl.*, 2001, **35**(1), 61–63.

[10] ARNOLD V. I. Complexification of tetrahedron and pseudoprojective transformations. *Funct. Anal. Appl.*, 2001, **35**(4), 241–246.

[11] ARNOLD V. I. Pseudoquaternion geometry. *Funct. Anal. Appl.*, 2002, **36**(1), 1–12.

[12] ARNOLD V. I. The Geometry of Complex Numbers, Quaternions, and Spins. Moscow: Moscow Center for Continuous Mathematical Education Press, 2002 (in Russian).

[13] KAZARIAN M. E. A remark on the eigenvectors and eigenvalues of hyper-Hermitian matrices. Preprint, 1998 (in Russian).
[*Internet*: http://www.pdmi.ras.ru/~arnsem/papers/]

## 1988-25 — *V. I. Arnold*

ℜ For 3 generators, the simplicity (absence of moduli) has been proved by E. I. Korkina, but starting from 4 the simplicity need not take place: relevant counterexamples starting with $(1,3,4,7)$ have been constructed by D. Eisenbud and B. Sturmfels. The classification of simple quadruples is still unknown. See [1, 2] for more details.

[1] ARNOLD V. I. Continued Fractions. Moscow: Moscow Center for Continuous Mathematical Education Press, 2001 (in Russian). ("Mathematical Education" Library, 14.)

[2] STURMFELS B. Gröbner Bases and Convex Polytopes. Providence, RI: Amer. Math. Soc., 1996, 85–98. (University Lecture Series, 8.)

## 1988-26 — *A. M. Leontovich*

ℜ The value $\Delta(n) = \lim\limits_{N\to\infty} N(r(N))^n$ is the density of the most compact *packing* of the $n$-dimensional space $\mathbb{R}^n$ by disjoint balls of the same radius, and the value $\Theta(n) = \lim\limits_{N\to\infty} N(R(N))^n$ is the density of the most loose *covering* of the $n$-dimensional space by overlapping balls of the same radius. Therefore, the eccentricity $\rho_n = \left(\frac{\Theta(n)}{\Delta(n)}\right)^{1/n}$. Numerous books and papers are devoted to the problem on the best packings and coverings; this problem is closely related to various topics in mathematics and physics: lattice theory, coding theory, the geometry of numbers, quadratic forms, prime groups, crystallography, the theory of superstrings, ... This subject is covered in a very comprehensive and detailed book [1] (see especially Chapters 1, 2). A more popular presentation is given in book [2], but much less material is included therein.

Let us recollect some results on the best packings and coverings.

By $\Delta^P(n)$, $\Theta^P(n)$ we shall denote the densities of the best *lattice* packings and coverings, i. e., those having the property that the centers of their balls form a lattice in $\mathbb{R}^n$. It is natural to try to find the best packings and coverings among the lattice ones. For $n \geq 2$,

$$\Delta^P(n) \leq \Delta(n) < 1 < \Theta(n) \leq \Theta^P(n).$$

(Obviuously, for $n = 1$ we have $\Delta^P(1) = \Delta(1) = \Theta^P(1) = \Theta(1) = 1$.)

1) The exact values of $\Delta(n)$, $\Theta(n)$ have been found only for $n = 1, 2$. In this case, $\Delta^P(n) = \Delta(n)$, $\Theta^P(n) = \Theta(n)$, and the best packings and coverings are really lattice ones. For $n = 2$, the lattice for the best packing and the best covering is the same, namely, the so-called *hexagonal* lattice whose nodes are the vertices of regular triangles that fill the plane. For this case,

$$\Delta(2) = \frac{\pi}{\sqrt{12}} \approx 0.9069, \qquad \Theta(2) = \frac{2\pi}{3\sqrt{3}} \approx 1.2092.$$

2) The best *lattice* packings for $n \leq 8$ and the best *lattice* coverings for $n \leq 5$ have been found; thus, the exact values of $\Delta^P(n)$, $\Theta^P(n)$ are known for these $n$. Apparently (not proved though), for these $n$ the values $\Delta^P(n)$, $\Theta^P(n)$ coincide with the correspomding values $\Delta(n)$, $\Theta(n)$, that is, lattice packings and coverings can really serve as the best (although there exist non-lattice packings and coverings with these optimal densities $\Delta(n)$, $\Theta(n)$). Let us point out the best lattices for $n = 3$. It was already Gauss who proved that the most compact three-dimensional lattice packing is determined by the so-called *face-centered* lattice (which is the natural generalization of the hexagonal lattice to the three-dimensional case). The corresponding density value is $\Delta^P(3) = \frac{\pi}{\sqrt{18}} \approx 0.7405$. R. P. Bambah proved that the most loose three-dimensional lattice covering is given by the so-called *volume-centered* lattice (this is dual to the face-centered one); the corresponding density value is $\Theta^P(3) \approx 1.4635$. Note that, for $3 \leq n \leq 8$, the lattices providing the best lattice packing and covering differ.

3) It seems that the best lattice packings and coverings have been found for $n = 12, 16, 24$ (most probably they are the best among all packings and coverings, not only the lattice ones). All these lattices are interesting enough. The so-called J. Leech's lattice in dimension $n = 24$ is of special importance; conjecturally, it gives both the most compact packing and the most loose covering. This lattice appears in a variety of mathematical problems.

One should not think, however, that for all dimensions $n$ lattice packings and coverings can be taken as the best ones. Thus, probably it is not the case for $n = 10, 11, 13$.

4) Minkowski obtained the following lower bound for the densities $\Delta(n)$, $\Delta^P(n)$ of the most compact packings:

$$\Delta(n) \geq \Delta^P(n) \geq \frac{\zeta(n)}{2^{n-1}}, \tag{1}$$

where $\zeta(n)$ is the Riemann $\zeta$-function. Therefore,

$$\log_2 \Delta(n) \geq -n + 1 + \varepsilon_n, \quad \text{where} \quad \varepsilon_n \to 0 \quad \text{as} \quad n \to \infty. \tag{2}$$

Minkowski's proof is not constructive. It involves averaging by lattices and calculating the density average for lattice packings. For large $n$, all known packing constructions give results worse than Minkowski's bound. For example, in the case of lattices naturally generalizing the hexagonal and the face-centered lattice to arbitrary $n$, the packing density decreases superexponentially as $n$ grows. (This is moreover true for cubic lattices.) However, there exist constructively defined lattices in dimension $n$ for which the packing density decreases exponentially as $n \to \infty$, but still the exponent index for them is less than that in Minkowski's bound (which is equal to $1/2$).

5) As for upper bounds for $\Delta(n)$, one having a clear meaning was obtained by C. A. Rogers. Namely, he proved that

$$\Delta(n) \leq \sigma_n, \tag{3}$$

where $\sigma_n$ is the ratio of the volume of the intersection of a regular $n$-simplex whose edge has length 2 with balls of radius 1 centered at the vertices of the simplex, to the volume of this $n$-simplex. For large $n$, it then follows that

$$\log_2 \Delta(n) \leq -\frac{1}{2}n,$$

i. e., the right-hand side in Rogers' estimate (3) decreases exponentially as $n$ grows, with the exponent index $\frac{1}{\sqrt{2}}$. G. A. Kabatyanskiĭ and V. I. Levenshtein obtained a better estimate from which it follows that, for large $n$,

$$\log_2 \Delta(n) \leq -0.5990\, n.$$

Hence, if $n$ is large then

$$-1 \leq \frac{1}{n} \log_2 \Delta(n) \leq -0.5990. \tag{4}$$

From the latter it follows that

$$\frac{1}{2} \leq \sqrt[n]{\Delta(n)} \leq 2^{-0.5990}. \tag{4'}$$

Results analogous to 4) and 5) are true for coverings as well.

6) Rogers proved that there exist lattice coverings with density not exceeding $cn(\ln n)^{\frac{1}{2}\log_2(2\pi e)}$ for some constant $c$, i. e.,

$$\Theta^P(n) \leq cn(\ln n)^{\frac{1}{2}\log_2(2\pi e)} \approx cn(\ln n)^{2.0471}. \tag{5}$$

(This bound is an analog of (1).) Also, Rogers obtained a better estimate for non-lattice coverings:

$$\Theta(n) \leq n\ln n + n\ln\ln n + 5n. \tag{6}$$

The estimates (5), (6) are not constructive (like (1)).

7) On the other hand, H. S. M. Coxeter, L. Few and Rogers proved the following lower bound for $\Theta(n)$ (analogous to (3)):

$$\Theta(n) \geq \tau_n, \tag{7}$$

where $\tau_n$ is the ratio of the volume of a regular $n$-simplex whose edge has length 2 with balls of radius $\left(\frac{2n}{n+1}\right)^{1/2}$ centered at the vertices of the simplex, to the volume of this $n$-simplex (balls of this radius just cover the simplex). Therefore

$$\tau_n = \left(\frac{2n}{n+1}\right)^{n/2} \sigma_n \sim \frac{n}{e\sqrt{e}} \quad \text{as} \quad n \to \infty.$$

From the latter, together with (6) and (7), it follows that

$$\frac{n}{e\sqrt{e}} \leq \Theta(n) \leq n\ln n + n\ln\ln n + 5n. \tag{8}$$

Note that the upper and the lower bounds for $\Theta(n)$ in (8) are closer to each other than the corresponding bounds for $\Delta(n)$ implied by (4). From (8) it follows that

$$\sqrt[n]{\Theta(n)} \to 1 \quad \text{as} \quad n \to \infty. \tag{8'}$$

8) From $(4')$ and $(8')$ we find that for the eccentricity $\rho_n$ asymptotically

$$2^{0.5990} \leq \rho_n \leq 2.$$

This is consistent with the assertion that $\rho_n \geq \sqrt{\frac{2n}{n+1}}$, but opposes the conjecture that $\lim_{n\to\infty} \rho_n = \sqrt{2}$. Apparently, this conjecture was proposed when the result of Kabatyanskiĭ and Levenshtein (which is a work of 1978) was not yet known.

[1] CONWAY J. H, SLOANE N. J. A. Sphere Packings, Lattices and Groups, Vol. 1, 2. New York: Springer, 1988. (Grundlehren der Mathematischen Wissenschaften, 290.)

[2] ROGERS C. A. Packing and Covering. New York: Cambridge University Press, 1964. (Cambridge Tracts in Math. and Math. Phys., 54.)

## 1988-27

This is a problem in paper [1a] (3°, I.A; see also [1b], p. 428).

[1a] ARNOLD V. I. On some problems in symplectic topology. In: Topology and Geometry. Rohlin Seminar. Editor: O. Ya. Viro. Berlin: Springer, 1988, 1–5. (Lecture Notes in Math., 1346.)

*The Russian translation in*:

[1b] Vladimir Igorevich Arnold. Selecta–60. Moscow: PHASIS, 1997, 425–429.

## 1988-28

This is a problem in paper [1a] (3°, I.B; see also [1b], p. 428).

[1a] ARNOLD V. I. On some problems in symplectic topology. In: Topology and Geometry. Rohlin Seminar. Editor: O. Ya. Viro. Berlin: Springer, 1988, 1–5. (Lecture Notes in Math., 1346.)

*The Russian translation in*:

[1b] Vladimir Igorevich Arnold. Selecta–60. Moscow: PHASIS, 1997, 425–429.

## 1988-29

This is a problem in paper [1a] (3°, I.C; see also [1b], p. 428).

[1a] ARNOLD V. I. On some problems in symplectic topology. In: Topology and Geometry. Rohlin Seminar. Editor: O. Ya. Viro. Berlin: Springer, 1988, 1–5. (Lecture Notes in Math., 1346.)

*The Russian translation in*:

[1b] Vladimir Igorevich Arnold. Selecta–60. Moscow: PHASIS, 1997, 425–429.

## 1988-30

This is a problem in paper [1a] (3°, II.A; see also [1b], p. 428).

[1a] ARNOLD V. I. On some problems in symplectic topology. In: Topology and Geometry. Rohlin Seminar. Editor: O. Ya. Viro. Berlin: Springer, 1988, 1–5. (Lecture Notes in Math., 1346.)

*The Russian translation in*:

[1b] Vladimir Igorevich Arnold. Selecta–60. Moscow: PHASIS, 1997, 425–429.

## 1988-31

This is a problem in paper [1a] (3°, II.B; see also [1b], p. 428).

[1a] ARNOLD V. I. On some problems in symplectic topology. In: Topology and Geometry. Rohlin Seminar. Editor: O. Ya. Viro. Berlin: Springer, 1988, 1–5. (Lecture Notes in Math., 1346.)

*The Russian translation in*:

[1b] Vladimir Igorevich Arnold. Selecta–60. Moscow: PHASIS, 1997, 425–429.

## 1988-32

This is a problem in paper [1a] (3°, II.C; see also [1b], p. 428).

[1a] ARNOLD V. I. On some problems in symplectic topology. In: Topology and Geometry. Rohlin Seminar. Editor: O. Ya. Viro. Berlin: Springer, 1988, 1–5. (Lecture Notes in Math., 1346.)

*The Russian translation in*:

[1b] Vladimir Igorevich Arnold. Selecta–60. Moscow: PHASIS, 1997, 425–429.

# 1989

## 1989-2

See the comments to problems *1988-6* and *1992-13*.

## 1989-3 – *V. A. Vassiliev* Also: *1990-18*

ℛ If $\mathbb{R}^n$ is the space of real manifolds of the form $x^n + \lambda_1 x^{n-1} + \cdots + \lambda_{n-1}x + \lambda_n$, then for $n < 6$ the space $\mathbb{R}^n \setminus \bar{A}_3$ is homotopy equivalent to a circle, and for $n = 6$, 7 or 8 it is homotopy equivalent to the wedge (bouquet) of $S^1$ and $S^2$ (Yu. M. Makhlin, 1990, see [1]). Therefore $\pi_2(\mathbb{R}^n \setminus \bar{A}_3)$ is trivial for $n < 6$ and is equal to $\mathbb{Z}^{\mathbb{Z}}$ if $n = 6, 7, 8$. The stabilization appears at $n = 9$, so for all $n \geq 9$ we have $\pi_2 = \pi_2(\Omega S^2) = \pi_3(S^2) = \mathbb{Z}$.

[1] VASSILIEV V. A. Complements of Discriminants of Smooth Maps: Topology and Applications, revised edition. Providence, RI: Amer. Math. Soc., 1994. (Transl. Math. Monographs, 98.)

## 1989-7 – *S. M. Gusein-Zade*

ℛ The existence of such densities (i. e., the corresponding limits) for finite sums of the specified type and for almost all pseudoperiodic functions was proved in [2–4]. The bounds were published in [1].

[1] ARNOLD V. I. The longest curves of given degree and the quasicrystallic Harnack theorem in pseudoperiodic topology. *Funct. Anal. Appl.*, 2002, **36**(3), 165–171.

[2] ESTEROV A. I. Densities of the Betti numbers of pre-level sets of quasi-periodic functions. *Russian Math. Surveys*, 2000, **55**(2), 338–339.

[3] GUSEIN-ZADE S. M. Number of critical points for a quasiperiodic potential. *Funct. Anal. Appl.*, 1989, **23**(2), 129–130.

[4] GUSEIN-ZADE S. M. On the topology of quasiperiodic functions. In: Pseudoperiodic Topology. Editors: V. Arnold, M. Kontsevich and A. Zorich. Providence, RI: Amer. Math. Soc., 1999, 1–7. (AMS Transl., Ser. 2, 197; Adv. Math. Sci., 46.)

## ▽ 1989-10 – *V. I. Arnold* (1989)

ℛ These systems are considered in the phase space in paper [1] (and in book [2]); though what would result from Legendrian manifolds and their characteristics upon projection on the physical space-time has not been investigated completely.

[1] ARNOLD V. I. On the interior scattering of waves, defined by hyperbolic variational principles. *J. Geom. Phys.*, 1988, **5**(3), 305–315.

[2] ARNOLD V. I. Singularities of Caustics and Wave Fronts. Dordrecht: Kluwer Acad. Publ., 1990, § 2.4. (Math. Appl., Sov. Ser., 62.)

## △ 1989-10 — *I. A. Bogaevsky*

ℛ Consider the system of Euler–Lagrange linear partial differential equations arising from some variational principle

$$\delta \int L\,dt\,dx^1 \cdots dx^D = 0$$

with the Lagrangian $L(t,x,u_t,u_x) = T(t,x,u_t) - V(t,x,u_x)$ where $t, x^1, \ldots, x^D$ are independent variables, $u^1, \ldots, u^m$ are dependent variables, the kinetic energy density $T$ is a positive definite quadratic form of the first derivatives of the dependent variables by time, and the potential energy density $V$ is a positive semidefinite quadratic form of the first spatial derivatives of the dependent variables. Generically, the coefficients of both forms are functions of $t$ and $x$. The propagation of perturbation in a resilient medium is a good model example of the described situation, $u$ being the shift vector of a point of the medium, and the number of dependent variables hence being equal to the dimension of the $x$-space ($m = D$).

It is known that the propagation of fronts and rays of shock and short waves is described in the framework of geometric optics by a *light hypersurface*, which lies in the projectivized cotangent bundle over the space-time, and where the main matrix symbol of the initial system of partial differential equations degenerates (for its explicit formulae see the comment to problem *1978-17*). Namely, the *big front* of a shock wave is the hypersurface in the space-time where the solution is discontinuous, and the *big front* of a short wave asymptotic is the hypersurface in the space-time where the phase of the asymptotic is constant, big fronts are stratified into rays, and their sections by isochrones $t = \text{const}$ are the *instantaneous fronts* developing in the course of time. Legendrian submanifolds of the light hypersurface project into big fronts, and its characteristics into rays. A big front is not uniquely determined by the initial front; physically, this means that the initial front splits into several instantaneous fronts developing independently of each other.

In the situation considered, the initial front generically splits into $2m$ instantaneous fronts, the main matrix symbol is a symmetric matrix, and the Euler–Lagrange system is hyperbolic. Singularities of the light hypersurface project into points of the space-time where this hyperbolicity is improper.

If there are at least two dependent variables then a light hypersurface can have singularities which are not removable by a small perturbation of the Lagrangian coefficients as functions of $t$ and $x$. The problem is to describe the singularities of a big front, the rays system on it, and the perestroikas of an instantaneous front under *interior scattering*, i. e., as the corresponding Legendrian manifold passes through these singulatities. (In the case where the light hypersurface is smooth, this problem is solved for $D \leq 5$, see, e. g., the comment to problem *1974-8*.) It is assumed that the initial front is smooth and typical, and the Lagrangian coefficients depend generically on the point of the space and possibly on time. The last assertion physically means, in particular, that the medium is non-homogeneous, anisotropic and possibly non-autonomous. A similar phenomenon is observed also in homogeneous media and is called the Hamilton conic refraction in crystals. Nevertheless, the geometric optics of the interior scattering in typical non-homogeneous and anisotropic media differs substantially from the Hamilton conic refraction.

The problem posed above is trivial for the line ($D = 1$), solved for the plane ($D = 2$), and open for the space ($D = 3$).

Typical singularities of a light hypersurface were described in [2]. These singularities turned out to be the same as the singularities of the set of degenerate matrices in the space of symmetric matrices. For example, the simplest singularity of a light hypersurface (called *conic*) is the product of the usual two-dimensional cone $\xi^2 + \eta^2 = \zeta^2$ with the $(2D-2)$-dimensional vector space. The description of singularities of a light hypersurface is based on the transversality theorem for homogeneous mappings proved in [9].

The normal forms of the pair consisting of a generic contact structure and a $2D$-dimensional hypersurface in a neighborhood of its conic singularity were found in [2] for $D = 1$ and in [1] for $D \geq 2$ (see also problem *1988-3*). Two normal forms, an elliptic and a hyperbolic one, are present in both cases, but explicit formulae are simpler for $D \geq 2$. According to [2], for $D = 1$ the elliptic normal form for light hypersurfaces is impossible because of the hyperbolicity of the Euler–Lagrange system itself, but, as shown in [8], it is realized if $D \geq 2$. Conic singularities of a light hypersurface are responsible for the Hamilton refraction in crystals, but the contact structure in this case is not generic due to homogeneity and hence cannot be reduced to either an elliptic normal form or a hyperbolic one.

For $D = 1$, an instantaneous front is a ponit on the $x$-line, a big front consists of a single (curvilinear) ray on the $(t, x)$-plane, a light hypersurface is a surface in the three-dimensional contact space, and a Legendrian submanifold consists of a single characteristic. From a typical initial point $2m$ rays go out, and neither of the corresponding $2m$ characteristics passes through the singularities of the light hypersurface which include only conic ones. Thus, the case $D = 1$ is trivial.

For $D = 2$, interior scattering occurs in the generic case only in hyperbolic conic singularities of the light hypersurface. Indeed, an instantaneous front is a curve on the $x$-plane, a big front is a surface in the three-dimensional $(t, x)$-space, a Legendrian submanifold is two-dimensional, and a light hypersurface is four-dimensional and, as always, stratified into characteristics. Normal forms show that, unlike the elliptic case, a hyperbolic conic singularity of the light hypersurface has two characteristics passing through it. All such characteristics form a 3-submanifold $S$ on the light hypersurface. Next, the initial front determines a Legendrian curve in the projectivized cotangent bundle over the $x$-space; under the natural pullback, this curve determines $2m$ isotropic curves on the light hypersurface which are called *initial conditions* and are continued by means of the characteristics to Legendrian surfaces projecting into $2m$ big fronts. These, in turn, determine evolutions of $2m$ instantaneous fronts. If the initial front is typical, then, by dimensional reasons, all initial conditions lie on the smooth part of the light hypersurface and can intersect the manifold $S$ at individual points. The characteristic leaving each of such intersection points gets into a hyperbolic conic singularity in a neighborhood of which the Legendrian surface whose initial condition intersects the submanifold $S$ has a singularity. These singularities for a typical initial condition were described in [3,4], and the corresponding singularities of big fronts, rays systems on them and perestroikas of instantaneous fronts were studied in [5–7].

For $D = 3$, even the singularities of Legendrian submanifolds resulting from interior scattering have not been described, to say nothing of the corresponding singularities of big fronts, rays systems and perestroikas of instantaneous fronts. Let us point out two new phenomena arising in this case. Firstly, initial conditions can stably pass through elliptic and hyperbolic conic singularities of the light hypersurface. Secondly, a Legendrian manifold can stably pass through conic singularities of the light hypersurface which are naturally called *parabolic*. They are unremovable already for $D \geq 2$. Indeed, the manifold of conic singularities of a light hypersurface naturally involves elliptic and hyperbolic domains, respective to the normal form to which the contact structure in a neighborhood of the given point reduces. Elliptic and hyperbolic domains are separated by parabolic singularities. Apparently, in a neighborhood of a typical parabolic singularity, the pair consising of the contact structure and the light hypersurface has function moduli.

All above-mentioned results are featured in detail in book [4] (see Chapter 8 “Wave transformations determined by hyperbolic variational principles”). Some open topological question related to the problem under discussion are formulated there on page 282. This book, unlike its original English edition [3], contains a description of singularities of big fronts, rays systems on them and perestroikas of instantaneous fronts for $D = 2$ first published in paper [5]. However, the reference to this paper in book [4] has a misprint: the note on page 300 should read [201] instead of [20].


[1] ARNOLD V. I. On the interior scattering of waves, defined by hyperbolic variational principles. *J. Geom. Phys.*, 1988, **5**(3), 305–315.

[2] ARNOLD V. I. Surfaces defined by hyperbolic equations. *Math. Notes*, 1988, **44**(1), 489–497. [*The Russian original is reprinted in*: Vladimir Igorevich Arnold. Selecta–60. Moscow: PHASIS, 1997, 397–412.]

[3] ARNOLD V. I. Singularities of Caustics and Wave Fronts. Dordrecht: Kluwer Acad. Publ., 1990. (Math. Appl., Sov. Ser., 62.)

[4] ARNOLD V. I. Singularities of Caustics and Wave Fronts. Moscow: PHASIS, 1996 (in Russian). (Mathematician's Library, 1.)

[5] BOGAEVSKY I. A. Singularities of the propagation of short waves on the plane. *Sb. Math.*, 1995, **186**(11), 1581–1597.

[6] BOGAEVSKY I. A. The interior scattering of rays and wave fronts on the plane. In: ARNOLD V. I. Singularities of Caustics and Wave Fronts. Moscow: PHASIS, 1996, § 8.5, 300–316 (in Russian). (Mathematician's Library, 1.)

[7] BOGAEVSKY I. A. Singularities of short linear waves on the plane. In: The Arnold–Gelfand Mathematical Seminars: Geometry and Singularity Theory. Editors: V. I. Arnold, I. M. Gelfand, V. S. Retakh and M. Smirnov. Boston, MA: Birkhäuser, 1997, 107–112.

[8] BRAAM P. J., DUISTERMAAT J. J. Normal forms of real symmetric systems with multiplicity. *Indag. Math. (N. S.)*, 1993, **4**(4), 407–421.

[9] KHESIN B. A. Singularities of light hypersurfaces and structure of hyperbolicity sets for systems of partial differential equations. In: Theory of Singularities and its Applications. Editor: V. I. Arnold. Providence, RI: Amer. Math. Soc., 1990, 119–127. (Adv. Sov. Math., 1.)


## **1989-11** — *M. B. Mishustin*

**1. Formal moduli of neighborhoods.** One can cover a neighborhood of a submanifold by small balls and try to choose coordinates in these balls to put collating mappings in some normal form. The collating mappings can be presented by series in variables transverse to submanifold, so one can sequentially construct normal forms of jets of collating mappings.

In abstract terms these normal forms present first cohomologies of a sheaf of filtered groups over submainfold. These cohomologies can be obtained as the limit of a (non-Abelian) spectral sequence. This technique allows us to define formal moduli of neighborhoods of complex subspaces in spaces with singularities, see [7].

The calculation of the first term of this sequence gives the following answer. Let $E_p = \bigoplus_{q=1}^{\infty}(H^p(N^{*q}) \oplus H^p(N^{*q} \otimes TM))$ , where $H^p(N^{*q})$ denotes $p$-th cohomology group of a manifold $M$ with coefficients in $q$-th symmetric degree of a vector bundle $N$, similar to $H^p(N^{*q} \otimes TM)$. Then $E_1$ is the space of formal moduli of neighborhoods of embeddings of manifold $M$ with normal bundle $N$, i. e., to every such embedding $\iota$ we can assign an element $\mathfrak{E}(\iota) \in E_1$ so that:

— the correspondence $\mathfrak{E}$ epimorphically maps embeddings to a subset of a series in $E_1$ that converges in the natural sense and lies in the kernel of the differential $E_1 \to E_2$ defined by the spectral sequence;

— neighborhoods of the embeddings $\iota$ and $\iota'$ can be collated by mappings with coinciding $q$-jets if and only if the $q$-jets of $\mathfrak{E}(\iota)$ and $\mathfrak{E}(\iota')$ belong to the same orbit of the action of the group with Lie algebra $E_0$ defined by spectral sequence;

— the correspondence $\mathfrak{E}$ maps analytic families to analytic families.

In order to include the moduli of deformations of the submanifold and the normal bundle in the modulus of a neighborhood, one should include in $E_p$ the zero term of filtration by starting the sum from $q = 0$. For further purposes this is more convenient.

The calculation of the first term of the sequence in many cases delivers the final formal answer, because Kodaira's theorem of triviality, semicontinuity of dimensions of cohomology groups and other reasons often provide triviality of cohomology groups. In particular, one can obtain the following description of formal moduli of neighborhoods of embeddings of Riemann curves in complex surfaces:

— if the index $s$ of self-intersection of the trivial section of the normal bundle is negative, then the moduli space is finite-dimensional; moreover, it is trivial, when $s < 4 - 4g$;

— if $s > 0$, then the moduli space is infinite-dimensional, namely, is the sum of spaces of formal series of mappings $\mathbb{C}^2 \to \mathbb{C}^{2s}$ and $\mathbb{C} \to \mathbb{C}^{4g-4}$;

— if $s = 0$, then the structure of moduli space is typical for entities with dense resonances: the dimension of a modulus of a $q$-jet depends on the choice of a $(q-1)$-jet. From the genericity viewpoint the answers are as follows:

— for almost any normal bundle the formal modulus is presented by a series of mappings $\mathbb{C} \to \mathbb{C}^{4g-4}$. Exceptional bundles are also dense;

— in general $k$-dimensional families there are neighborhoods for which the base of formal versal deformation increases by a $2k$-dimensional modulus;

— in exceptional cases of infinite codimension the modulus can increase by a series of mappings $\mathbb{C} \to \mathbb{C}^2$.

**2. Convergence of formal series, stability of moduli and pseudoconvexity of neighborhoods.** The following properties of topological type of neighborhood seem to be highly correlated:

— formal equivalence of neighborhoods implies analytic equivalence;

— the spectral sequence converges after finitely many steps;

— the normal bundle admits metrics of negative or positive curvature.

The last property is essential for the following reason: it provides some kind of pseudoconvexity/concavity of systems of small neighborhoods, and so the $\bar{\partial}$-operator is closed in them and "small denominators" do not appear in the process of transformation to normal form.

With respect to the last property we have the following classification:

1) Negative neighborhoods. In this case, as G. Grauert proved in [3], a submanifold can be modified to a point in a manifold. Its neighborhood then is modified to a germ of a complex space at this point, and so the classification of negative neighborhoods can be obtained from the classification of germs of fixed topology.

Formal modulus of negative neighborhoods is finite-dimensional.

Formal equivalence always implies analytic equivalence.

2) Positive neighborhoods. This class is studied much less, though, I suppose, admits complete classification. Positive neighborhoods of elliptic curves were classified by Yu. S. Il′yashenko in [4]. Positive neighborhoods of Riemann curves of any genre in the case when $s > 4g-3$ were classified in [8]. Both papers construct universal infinite-dimensional moduli that coincide with the formal answer. Both authors avoid studying the convergence of series and exploit properties of deformations of curves in a surface.

I suppose that moduli spaces of positive neighborhoods are always infinite-dimensional and asymptotically are spaces of series of mappings $\mathbb{C}^m \to \mathbb{C}^n$ for some $m$ and $n$.

I also think that formal equivalence of positive neighborhoods always implies analytic equivalence.

3) Other neighborhoods. In this case, with rare exceptions, formal moduli spaces are spaces with dense resonances, and formally isomorphic neighborhoods can be analytically non-isomorphic. In fact, the only studied neighborhoods are those of a multi-dimensional torus with topologically trivial normal bundle. Neighborhoods of elliptic curves in surfaces were studied by V. I. Arnold in [2], Section 27, a research inspired by a close relation of these neighborhoods to bifurcations of invariant submanifolds that was descibed in [1]. In [2] V. I. Arnold

showed that, for a subset of full measure in the set of normal bundles, a neighborhood is isomorphic to a neighborhood in a normal bundle. This result was extended to neighborhoods of a multi-dimensional torus by Yu. S. Il′yashenko and A. S. Pyartli in [5].

Examples of neighborhoods of a torus that are not formally isomorphic to a neighborhood in the linear bundle can be obtained from the formal theory. Examples of neighborhoods of a torus that are isomorphic to the linear version formally, but not analytically, can be constructed with the help of the theory of germs of automorphisms of $\mathbb{C}$.

The technique developed by Arnold seems to be applicable to zero-type neighborhoods of Riemann curves of higher genera in surfaces.

Still, for some manifolds neighborhoods of the third class have simple moduli due to the topology of the manifold itself. For example, a neighborhood of the Riemann sphere in a surface with zero self-intersection index is isomorphic to a neighborhood in a trivial bundle, which was proved by V. I. Savel′ev in [9]. This fact is conjectured for all neighborhoods of projective spaces with topologically trivial normal bundle.

**3. Obstacles to convergence of formal series.** Following the concepts of materialization of resonances, Arnold also supposed that obstacles to analytic isomorphism of formally isomorphic neighborhoods are compact complex curves in any neighborhood of an elliptic curve. Indeed, in some cases such curves were found by Yu. S. Il′yashenko and A. S. Pyartli in [6]. Further advances of the theory of germs of automorphisms of $\mathbb{C}$ permit construction of neighborhoods which are not isomorphic to their formal normal forms but have no such obstacles.

These cases might be exceptional phenomena, similar to the absence of periodic points at the boundary of the linearizability domain in the theory of normal forms of invariant circles or fixed point neighborhoods of analytic mappings of the complex line.

The exceptional character of both events seems not to have been proved yet.

Studying these obstacles leads, in particular, to the following idea: these compact curves, as well as the central curve, are compact layers of a holomorphic foliation in some neighborhood of the curve. The existence of such foliation for any zero-type neighborhood of a Riemann curve would reduce the classification of neighborhoods to the classification of local objects with the help of monodromy operators. The existence of a foliation can be proved in most cases and there are no known obstacles to it. Still the question is open, and the comment to problem *1993-25* is too optimistic.


[1] ARNOLD V. I. Bifurcations of invariant manifolds of differential equations and normal forms of neighborhoods of elliptic curves. *Funct. Anal. Appl.*, 1976, **10**(4), 249–259.

[2] ARNOLD V. I. Geometrical Methods in the Theory of Ordinary Differential Equations, 2nd edition. New York: Springer, 1988. (Grundlehren der Mathematischen Wissenschaften, 250.) [*The Russian original* 1978.]

[3] GRAUERT H. Über modifikationen und exzeptionelle analytische Mengen. *Math. Ann.*, 1962, **146**(4), 331–368.

[4] IL′YASHENKO YU. S. Positive type embeddings of elliptic curves into complex surfaces. *Trudy Moskov. Mat. Obshch.*, 1982, **45**, 37–67 (in Russian, for the English translation see *Trans. Moscow Math. Soc.*).

[5] IL′YASHENKO YU. S., PYARTLI A. S. Neighborhoods of zero type in embedded complex tori. *Trudy Semin. Petrovskogo*, 1979, **5**, 85–90 (in Russian). [*The English translation in*: Topics in Modern Mathematics. Editor: O. A. Oleĭnik. New York: Consultant Bureau, 1985, 107–121. (Petrovskiĭ Semin., 5.)]

[6] IL′YASHENKO YU. S., PYARTLI A. S. Materialization of Poincaré resonances and divergence of normalizing series. *Trudy Semin. Petrovskogo*, 1981, **7**, 3–49 (in Russian). [*The English translation*: *J. Sov. Math.*, 1985, **31**, 3053–3092.]

[7] KOSAREW S. Modulräume holomorpher Abbildungen auf konkaven komplexen Räumen. *Ann. Sci. École Norm. Super., Ser.* 4, 1987, **20**(3), 285–310.

[8] MISHUSTIN M. B. Neighborhoods of Riemann curves in complex spaces. *Funct. Anal. Appl.*, 1995, **29**(1), 20–31.

[9] SAVEL′EV V. I. Zero-type embeddings of the sphere into complex surfaces. *Moscow Univ. Math. Bull.*, 1982, **37**(4), 34–39.


## 1989-14 — *V. A. Vassiliev*

A closely related problem was discussed in Section IV.4.2 of [1].

In [2] it was proved that the trefoil knot can be realized by a polynomial mapping of degree 5. In [3] the homology groups of spaces of all polynomial knots of degree $\leq 4$ in spaces $\mathbb{R}^n$, $n \geq 3$, were calculated. In particular, it was shown that the space of degree 4 polynomial knots in $\mathbb{R}^3$ is path-connected and contains no non-trivial knots. In the study of this space the method of conic resolutions was developed, solving some other problems mentioned in this book; see the comments to problems *1970-13* and *1990-12*.

[1] Arnold V. I., Vassiliev V. A., Goryunov V. V., Lyashko O. V. Singularities. I. Local and Global Theory. Berlin: Springer, 1993, Sect. 3.3. (Encyclopædia Math. Sci., 6; Dynamical Systems, VI.) [*The Russian original* 1988.]

[2] Shastri A. R. Polynomial representations of knots. *Tôhoku Math. J., Ser.* 2, 1992, **44**(1), 11–17.

[3] Vassiliev V. A. On spaces of polynomial knots. *Sb. Math.*, 1996, **187**(2), 193–213.

## 1989-15

This is a problem in paper [1a] (p. 191; see also [1b], p. 471).

[1a] Arnold V. I., Vishik M. I., Il′yashenko Yu. S., Kalashnikov A. S., Kondrat′ev V. A., Kruzhkov S. N., Landis E. M., Millionshchikov V. M., Oleĭnik O. A., Filippov A. F., Shubin M. A. Some unsolved problems in the theory of differential equations and mathematical physics. *Uspekhi Mat. Nauk*, 1989, **44**(4), 191–202 (in Russian). [*The English translation*: *Russian Math. Surveys*, 1989, **44**(4), 157–171.]

*The Russian original of the section "Problems by V. I. Arnold"* (*Uspekhi Mat. Nauk*, 1989, **44**(4), *p.* 191–192) *is reprinted in*:

[1b] Vladimir Igorevich Arnold. Selecta–60. Moscow: PHASIS, 1997, 471–472.

## 1989-16

This is a problem in paper [1a] (p. 191; see also [1b], p. 471).

[1a] Arnold V. I., Vishik M. I., Il′yashenko Yu. S., Kalashnikov A. S., Kondrat′ev V. A., Kruzhkov S. N., Landis E. M., Millionshchikov V. M., Oleĭnik O. A., Filippov A. F., Shubin M. A. Some unsolved problems in the theory of differential equations and mathematical physics. *Uspekhi Mat. Nauk*, 1989, **44**(4), 191–202 (in Russian). [*The English translation*: *Russian Math. Surveys*, 1989, **44**(4), 157–171.]

*The Russian original of the section "Problems by V. I. Arnold"* (*Uspekhi Mat. Nauk*, 1989, **44**(4), *p.* 191–192) *is reprinted in*:

[1b] Vladimir Igorevich Arnold. Selecta–60. Moscow: PHASIS, 1997, 471–472.

## 1989-17

This is a problem in paper [1a] (p. 191; see also [1b], p. 471–472).

[1a] ARNOLD V. I., VISHIK M. I., IL′YASHENKO YU. S., KALASHNIKOV A. S., KONDRAT′EV V. A., KRUZHKOV S. N., LANDIS E. M., MILLIONSHCHIKOV V. M., OLEĬNIK O. A., FILIPPOV A. F., SHUBIN M. A. Some unsolved problems in the theory of differential equations and mathematical physics. *Uspekhi Mat. Nauk*, 1989, **44**(4), 191–202 (in Russian). [*The English translation*: *Russian Math. Surveys*, 1989, **44**(4), 157–171.]

*The Russian original of the section "Problems by V. I. Arnold"* (*Uspekhi Mat. Nauk*, 1989, **44**(4), *p.* 191–192) *is reprinted in*:

[1b] Vladimir Igorevich Arnold. Selecta–60. Moscow: PHASIS, 1997, 471–472.

ℛ See the comment to problem *1978-6.*

## 1989-18

ℋ This is a problem in paper [1a] (p. 191–192; see also [1b], p. 472).

[1a] ARNOLD V. I., VISHIK M. I., IL′YASHENKO YU. S., KALASHNIKOV A. S., KONDRAT′EV V. A., KRUZHKOV S. N., LANDIS E. M., MILLIONSHCHIKOV V. M., OLEĬNIK O. A., FILIPPOV A. F., SHUBIN M. A. Some unsolved problems in the theory of differential equations and mathematical physics. *Uspekhi Mat. Nauk*, 1989, **44**(4), 191–202 (in Russian). [*The English translation*: *Russian Math. Surveys*, 1989, **44**(4), 157–171.]

*The Russian original of the section "Problems by V. I. Arnold"* (*Uspekhi Mat. Nauk*, 1989, **44**(4), *p.* 191–192) *is reprinted in*:

[1b] Vladimir Igorevich Arnold. Selecta–60. Moscow: PHASIS, 1997, 471–472.

ℛ See the comments to problem *1986-7*.

## 1989-19

ℋ This is a problem in paper [1a] (p. 192; see also [1b], p. 472).

[1a] ARNOLD V. I., VISHIK M. I., IL′YASHENKO YU. S., KALASHNIKOV A. S., KONDRAT′EV V. A., KRUZHKOV S. N., LANDIS E. M., MILLIONSHCHIKOV V. M., OLEĬNIK O. A., FILIPPOV A. F., SHUBIN M. A. Some unsolved problems in the theory of differential equations and mathematical physics. *Uspekhi Mat. Nauk*, 1989, **44**(4), 191–202 (in Russian). [*The English translation*: *Russian Math. Surveys*, 1989, **44**(4), 157–171.]

*The Russian original of the section "Problems by V. I. Arnold"* (*Uspekhi Mat. Nauk*, 1989, **44**(4), *p.* 191–192) *is reprinted in*:

[1b] Vladimir Igorevich Arnold. Selecta–60. Moscow: PHASIS, 1997, 471–472.

R See the comment to problem *1971-11*.

### 1989-20 — *V. P. Kostov*

R For typical one- and two-parameter families, the singularities of the pseudo-Stokes hypersurfaces are described in paper [1].

[1] Kostov V. P. On the stratification and singularities of the Stokes hypersurface of one- and two-parameter families of polynomials. In: Theory of Singularities and its Applications. Editor: V. I. Arnold. Providence, RI: Amer. Math. Soc., 1990, 251–271. (Adv. Sov. Math., 1.)

# 1990

### 1990-1

H This is a problem in paper [1] (p. 259).

[1] Arnold V. I. Problems on singularities and dynamical systems. In: Developments in Mathematics: the Moscow School. Editors: V. I. Arnold and M. Monastyrsky. London: Chapman & Hall, 1993, 251–274.

R See the comments to problem *1988-6*.

### 1990-4

H See also problems *1987-4* and *1993-37*.

### 1990-7 — *O. S. Kozlovskiĭ*

R The problem of *multiplicity* of periodic points has not been solved yet.

It is possible to change this question: is the *number* of periodic points arising at infinitely small $b$ bounded (uniformly with respect to $a$)? In this case the answer is negative. For example, one can take a lacunary Fourier series as the function $f$:

$$f(x) = \sum_{k=1}^{\infty} c_k \sin(p_k^2 x) \tag{1}$$

where $p_k$ are prime numbers ordered according to their growth; $c_k = \exp(-p_k^2)$.

Because the coefficients of the Fourier series (1) decrease exponentially, the function $f$ is analytic. In this case the $p_k$-th iteration of the diffeomorphism $x \mapsto x + 2\pi/p_k + bf(x)$ is equal to $x \mapsto x + bp_k c_k \sin(p_k^2 x) + o(b)$, so that at small $b$ this diffeomorphism has at least $p_k$ periodic orbits.

In the same way, it can be shown that there is an analytic function $f(x)$ such that *in the first approximation* with respect to small $b$ the multiplicity of periodic points of the diffeomorphism $x \mapsto x + a + bf(x)$ is not bounded uniformly with respect to $a$.

### 1990-10

See the comment to problem *1983-14*.

### ▽ 1990-11 — *F. Aicardi*

In [1] a proof of the Sakharov statement is given and a generalization to the $N$-dimensional case is conjectured. The average numbers of $m$-dimensional faces of the pieces obtained by cutting an $N$-dimensional bounded domain with randomly selected hyperplanes are in fact calculated to be equal to the numbers of the faces of this dimension on a cubical piece (neglecting the contributions to these numbers of the pieces at the boundary of the domain), but conditions for the existence of such averages are not provided.

[1] AICARDI F. Sur le découpage de domaines bornés de dimension $N$. *C. R. Acad. Sci. Paris, Sér.* I *Math.*, 1993, **316**(2), 175–181.

### △▽ 1990-11 — *A. M. Leontovich*

This problem is one of the so-called "amateur problems" of A. D. Sakharov (see [1]).

R Three proofs of Sakharov's statement are suggested.

**Proof 1.** Let $n$ lines of generic position be given on the plane, i. e., of these, no three lines intersect at one point and no two lines are parallel. Then every pair of lines intersect at exactly one point, and different pairs of lines have different intersection points. Hence such a configuration has the following properties:

1) The number $N_0$ of vertices (intersection points of lines) equals

$$C_n^2 = \frac{n(n-1)}{2}.$$

2) Each of the $n$ lines has $n-1$ vertices, and these vertices break the line into $n$ intervals (sides, or edges). Thus, the total number of edges is

$$n \cdot n = n^2.$$

3) By induction on $n$ it is proved that the number $N_2$ of regions (areas into which the plane is divided by $n$ lines) equals

$$1+(1+2+\cdots+n) = 1+\frac{n(n+1)}{2}.$$

(The same follows from the Euler formula:

$$N_0 - N_1 + N_2 = 1,$$

and hence

$$N_2 = 1+N_1-N_0 = 1+n^2-\frac{n(n-1)}{2} = \frac{n(n+1)}{2}+1.)$$

Since each edge belongs to two regions, the average number of edges per region is equal to

$$\frac{2N_1}{N_2} = \frac{2n^2}{\frac{n(n+1)}{2}+1} \to 4 \quad \text{as } n \to \infty,$$

as was to be proved. [1]

[1] Evidently, the average number of vertices per region asymptotically equals the average number of edges, because a bounded region (polygon) has as many vertices as edges, while the number of vertices for an unbounded region is the number of edges minus one, and for large $n$ almost all regions are bounded. The direct calculation shows the same: since every vertex belongs to four regions, the average number of vertices per region equals $4N_0/N_2 = 2n(n-1)[n(n+1)/2+1]^{-1} \to 4$ as $n \to \infty$. By the way, it is easy to find the exact number of unbounded regions: it is equal to $2N_1 - 4N_0 = 2n^2 - 2n(n-1) = 2n$.

**Proof 2.** Obviously, any two configurations of lines on the plane can be "homotopized" into each other by a contunuous motion of the lines, and so that at each moment only one line moves and, if the configurations are generic (nondegenerate), then they become degenerate only in finitely many moments during the homotopy, the degeneracies being only elementary: either (at most) two lines become parallel or (at most) three lines meet at one point. It is clear that passing through such an elementary degeneracy preserves the number $N_0$ of vertices, the number $N_1$ of edges, and the number $N_2$ of regions; hence the average number of edges per region does not change. Therefore, this average is the same as that for the configuration obtained by a small perturbation of the configuration consisting of two families of parallel lines for which the average number of edges is asymptotically equal to 4 (as for the square).

**Proof 3.** This proof requires a modification of the problem statement. Namely, let us consider configurations of $n$ projective lines on the projective plane.

In this case, each edge connects two vertices (unlike configurations on the usual plane), and four edges start from each vertex (for the usual plane this was also true). Hence $4N_0 = 2N_1, N_1 = 2N_0$. Next, from the Euler formula $N_0 - N_1 + N_2 = 1$ it follows that $N_2 = 1 + N_1 - N_0 = 1 + N_0$. Therefore, the average number of edges per region equals $\frac{2N_1}{N_2} = \frac{4N_0}{N_0+1} \to 4$ as $N_0 \to \infty$, as was to be proved.

**Remark 1.** Proof 3 does not require that a configuration consists of straight lines, arbitrary closed curves can be taken instead. For a configuration of *n lines* on the projective plane we have $N_0 = \frac{n(n-1)}{2}$, $N_1 = n(n-1)$, $N_2 = \frac{n(n-1)}{2} + 1$.

**Remark 2.** Instead of the problem on configurations of lines on the plane, one could study the analogous problem on configurations of great circles on the sphere. In this case, the average number of edges per region is also asymptotically equal to 4, the numbers of vertices, edges, and regions respectively being twice those for lines on the projective plane, i. e., $N_0 = n(n-1), N_1 = 2n(n-1), N_2 = n(n-1)+2$, and $N_0 - N_1 + N_2 = 2$.

**Remark 3.** Analogously, the problem on the average number of edges per region can be solved for a two-dimensional net (such that four edges meet at each vertex) on a surface of arbitrary genus $g$. In this case $N_1 = 2N_0$, $N_0 - N_1 + N_2 = 2 - 2g$, $\frac{2N_1}{N_2} = \frac{4N_0}{2-2g+N_0} \to 4$ as $N_0 \to \infty$.

**Remark 4.** An analogous solution can be found for the problem on the average number of edges per region in a generic two-dimensional net, i. e., a net such that exactly three edges meet at each vertex (such a net can be obtained from any two-dimensional net by a small deformation slightly shifting its vertices). In this case, $3N_0 = 2N_1$, $N_1 = \frac{3}{2}N_0$, and hence the average number of edges per region equals

$$\frac{2N_1}{N_2} = \frac{\frac{3}{2}N_0}{2-2g+N_1-N_0} = \frac{\frac{3}{2}N_0}{2-2g+\frac{1}{2}N_0} \to 6 \quad \text{as } N_0 \to \infty.$$

**Remark 5.** Thus, the average number of edges per region is precisely calculated for a configuration of lines (i. e., for a two-dimensional net generated by lines on the plane, or on the projective plane, or on the sphere). A natural question on the distribution of the number of edges for regions of such two-dimensional nets arises. In contrast to the average value of this number, the distribution can depend on the net. A thorough study of this problem was carried out by Lewis with coauthors in a series of works; see paper [2] where an extensive bibliography is provided. This paper contains also an exposition of some nonstrict mathematical models, based on the chaos conjecture, from which the distribution function of the number of edges for regions is found.

All the above is, as a whole, generalized to the many-dimensional case.

Proof 2 is generalized in an obvious way and results in Aicardi's statement.

Proof 1 can also be generalized. For this, induction on the dimension must be used.

For example, consider $n$ planes of generic position in the three-dimensional space. Let us introduce the following notations: $N_0$ is the number of vertices, $N_1$ is the number of edges, $N_2$ is the number of 2-faces, $N_3$ is the number of 3-cells. Next, let $\pi$ be one of the $n$ planes. Denote: by $N_0(\pi)$ the number of vertices lying in the plane $\pi$; by $N_1(\pi)$ the number of edges in $\pi$; by $N_2(\pi)$ the number of 2-faces in $\pi$. We have a generic configuration of $n-1$ lines on the plane $\pi$. Hence $N_0(\pi) = \frac{(n-1)(n-2)}{2}$, $N_1(\pi) = (n-1)^2$, $N_2(\pi) = \frac{n(n-1)}{2} + 1$. Since every vertex belongs to three planes,

$$N_0 = \frac{1}{3}\sum_{\pi} N_0(\pi) = \frac{n(n-1)(n-2)}{6}.$$

Since every edge belongs to two planes,

$$N_1 = \frac{1}{2}\sum_{\pi} N_1(\pi) = \frac{n(n-1)^2}{2}.$$

And every 2-face belongs to one plane, so

$$N_2 = \sum_{\pi} N_2(\pi) = \frac{n^2(n-1)}{2} + n.$$

Because for $n$ lines on the plane the number of regions equals $\frac{n(n+1)}{2}+1$, by induction on $n$ it is proved that

$$N_3 = 1+1+2+4+\cdots+\left(\frac{n(n-1)}{2}+1\right) = \frac{n^3+5n+6}{6}.$$

(Note that the Euler formula holds:

$$N_0 - N_1 + N_2 - N_3 = \\ = \frac{n(n-1)(n-2)}{6} - \frac{n(n-1)^2}{2} + \frac{n^2(n-1)}{2} + n - \frac{n^3+5n+6}{6} = -1.)$$

Eight 3-cells meet at every vertex, and hence the average number of vertices per 3-cell is equal to

$$\frac{8N_0}{N_3} = \frac{8\frac{n(n-1)(n-2)}{6}}{\frac{n^3+5n+6}{6}} \to 8$$

as $n \to \infty$. Since four 3-cells border on each edge, the average number of edges per 3-cell is

$$\frac{4N_1}{N_3} = \frac{4\frac{n(n-1)^2}{2}}{\frac{n^3+5n+6}{6}} \to 12$$

as $n \to \infty$. Since two 3-cells border on each 2-face, the average number of 2-faces per 3-cell equals

$$\frac{2N_2}{N_3} = \frac{n^2(n-1)+2n}{\frac{n^3+5n+6}{6}} \to 6$$

as $n \to \infty$, the same as Aicardi asserts.

And Proof 3 is also generalized analogously. In this case, one should consider a configuration of $(d-1)$-dimensional projective planes (of generic position) in the $d$-dimensional projective space.

For example, let $d = 3$, i. e., we have a configuration of projective planes in the projective 3-space. Then in each projective plane $\pi$ we have (see Proof 3 above): $N_1(\pi) = 2N_0(\pi)$, $N_2(\pi) = 1 + N_0(\pi)$. Therefore

$$N_0 = \frac{1}{3}\sum_{\pi} N_0(\pi), \qquad N_1 = \frac{1}{2}\sum_{\pi} N_1(\pi) = \sum_{\pi} N_0(\pi) = 3N_0,$$

$$N_2 = \sum_{\pi} N_2(\pi) = \sum_{\pi} N_0(\pi) + \text{the number of planes} = \\ = 3N_0 + \text{the number of planes}.$$

Finally, apply the Euler formula: $N_0 - N_1 + N_2 - N_3 = 0$, then

$$N_3 = N_0 + N_2 - N_1 = 4N_0 + \text{the number of planes} - 3N_0 =$$
$$= N_0 + \text{the number of planes}.$$

Hence the average number of vertices is

$$\frac{8N_0}{N_3} = \frac{8N_0}{N_0 + \text{the number of planes}} \to 8$$

if the average number of vertices in a plane $\to \infty$ (this is true if the number of planes $\to \infty$); the average number of edges equals

$$\frac{4N_1}{N_3} = \frac{12N_0}{N_0 + \text{the number of planes}} \to 12;$$

the average number of 2-faces equals

$$\frac{2N_2}{N_3} = \frac{2(3N_0 + \text{the number of planes})}{N_0 + \text{the number of planes}} \to 6,$$

as was to be proved.

**Remark 6.** It might seem that the latter reasoning did not require that the configuration consists just of planes. But this is not all the case. Indeed, if arbitrary (closed) surfaces are taken instead of planes, then the domains into which they break the space need not be cells. For instance, one can get a domain having two 2-faces whose intersection is a circle (not an edge connecting two vertices) and contains no vertices, and there may be arbitrarily many such domains. Therefore, the problem on the average number of vertices, edges, and 2-faces becomes unreasonable and, at least, needs a more accurate statement.

**Remark 7.** The last proof was based on the three-dimensional Euler formula: $N_0 - N_1 + N_2 - N_3 = 0$. From the two-dimensional Euler formula one can deduce another relation between the numbers $N_0$, $N_1$, $N_2$, $N_3$ for configurations of (pojective) planes in the projective 3-space. It is obtained as follows.

Let us denote by $N_0$, $N_1$, $N_2$, $N_3$ the number of vertices, edges, 2-cells, 3-cells respectively. Let $l(i_0, i_1, i_2)$ be the number of 3-cells (i. e., domains into which the projective space is broken by planes) having $i_0$ vertices, $i_1$ edges, and $i_2$ faces (2-cells). Since eight 3-cells border on each vertex,

$$\sum l(i_0, i_1, i_2) i_0 = 8N_0$$

(from now on, sums are taken on $i_0$, $i_1$, $i_2$). Because four 3-cells border on every edge, we have

$$\sum l(i_0,i_1,i_2)i_1 = 4N_1.$$

Since two 3-cells border on every 2-face,

$$\sum l(i_0,i_1,i_2)i_2 = 2N_2.$$

Further, obviously,

$$\sum l(i_0,i_1,i_2) = N_3.$$

By virtue of the two-dimensional Euler formula, $i_2 = 2+i_1-i_0$. Hence

$$2N_2 = 2N_3 + 4N_1 - 8N_0, \quad \text{i. e.} \quad 4N_0 - 2N_1 + N_2 - N_3 = 0. \tag{1}$$

Note that, from this formula and the relation $6N_0 = 2N_1$ (implied by the facts that six edges start from every vertex and every edge connects two vertices), the three-dimensional Euler formula $N_0 - N_1 + N_2 - N_3 = 0$ easily follows.

Formulas analogous to (1) can be derived for higher dimension of the space. But they do not imply the Euler formula.

**Remark 8.** An analog of the previous remark is true for "generic" three-dimensional nets in the three-dimensional projective space, i. e., for cellular partitions of the projective space such that four edges, six 2-faces, and four 3-cells meet at each vertex. In this case, three 3-cells border on every edge. Hence, for such three-dimensional nets,

$$\sum l(i_0,i_1,i_2)i_0 = 4N_0, \qquad \sum l(i_0,i_1,i_2)i_1 = 3N_1,$$
$$\sum l(i_0,i_1,i_2)i_2 = 2N_2, \qquad \sum l(i_0,i_1,i_2) = N_3.$$

From the latter and the two-dimensional Euler formula $i_0 - i_1 + i_2 = 2$ we get:

$$4N_0 - 3N_1 + 2N_2 - 2N_3 = 0. \tag{2}$$

This formula is an analog of (1).

Note that, since four edges meet at each vertex in this case, $4N_0 = 2N_1$. From the latter and (2) one can again deduce the three-dimensional Euler formula: $N_0 - N_1 + N_2 - N_3 = 0$.

**Remark 9.** Remarks 7 and 8 are very similar. Nevertheless, unlike configurations formed by planes in $\mathbb{RP}^3$, for "generic" three-dimensional nets no values for average numbers of vertices, edges and 2-faces could be found; only two linear relations between these averages have been obtained. In my opinion, such averages cannot be evaluated at all. Probably this is due to the fact that there are, in a sense, "much more" "generic" three-dimensional nets than the nets obtained by intersections of families of closed 2-surfaces; the "manifold" of "generic" nets has higher dimension.

[1] From A. D. Sakharov's "amateur problems". *Kvant*, 1991, № 5, 11–12 (in Russian).

[2] Leontovich A. M., Maresin V. M., Ogaryshev V. F., Philippov V. B. Models of formation of a one-layer tissue. In: Theoretical and Mathematical Aspects of Morphogenesis. Moscow: Nauka, 1987, 182–198 (in Russian).

### △ 1990-11 — *B. T. Polyak*

H The problem was formulated and solved (in arbitrary dimension) by L. Schläfli in 1852; his work "Theorie der vielfachen Kontinuität" is reprinted on pages 209–212 in collection [2]. The result has since been rediscovered many times, see, for example, [3]. It is also discussed in Pólya's book [1].

[1] Pólya G. Mathematics and Plausible Reasoning. Vol. I. Induction and Analogy in Mathematics. Princeton, NJ: Princeton University Press, 1954. [*Reprinted* 1990.]

[2] Schläfli L. Gesammelte mathematische Abhandlungen, Band 1. Basel: Birkhäuser, 1950.

[3] Winder R. O. Partitions of $N$-space by hyperplanes. *SIAM J. Appl. Math.*, 1966, **14**(4), 811–818.

### 1990-12 — *V. A. Vassiliev*

R **Algebraic aspect.** This manifold is algebraically very complicated. Indeed, let us consider the space of polynomials of a fixed degree $d \geq 4$ on the spaces $\mathbb{R}^n$. Then the problem of deciding on which side of this hypersurface a given polynomial lies (i. e., whether it is everywhere positive or not) is an NP-complete problem over $n \to \infty$, see [2].

**Topological aspect.** The study of this boundary provides many results in the following situation. Consider some important finite-dimensional space $\mathcal{F}$ of functions on the manifold $M$. Then the intersection of this boundary with the unit sphere in $\mathcal{F}$ is homeomorphic to a sphere, as well as the similar intersection of the dual cone in the conjugate space $\mathcal{F}^*$. On the other hand, the natural stratification of this boundary allows one to construct these boundaries from (quite complicated) strata. In particular, this construction provides strange facts about the order complexes of sets in $M$ which can serve as the singular sets of functions $f \in \mathcal{F}$, i. e., as the sets at which such functions take global minimum with value 0.

**Example 1.** If $M = S^1$ and $\mathcal{F}$ consists of all Fourier polynomials of degree $\leq k$, then we obtain the following Carathéodory's theorem: the union of all $(k-1)$-dimensional simplices in $\mathbb{R}^N$, $N \geq 2k$, whose vertices lie on a generically embedded circle, is homeomorphic to $S^{2k-1}$.

**Example 2.** If $M = \mathbb{R}^n$ and $\mathcal{F}$ consists of all quadratic forms on $\mathbb{R}^n$, then we obtain the following theorem: the (suitably topologized) order complex of all proper subspaces in $\mathbb{R}^n$ is homeomorphic to $S^{(n^2+n-4)/2}$, see [1,3].


[1] ARNOLD V. I. A branched covering $\mathbb{C}P^2 \to S^4$, hyperbolicity and projectivity topology. *Sib. Math. J.*, 1988, **29**(5), 717–726. [*The Russian original is reprinted in*: Vladimir Igorevich Arnold. Selecta–60. Moscow: PHASIS, 1997, 431–448.]

[2] BLUM L., SHUB M., SMALE S. On a theory of computation and complexity over the real numbers: NP-completeness, recursive functions and universal mashines. *Bull. Amer. Math. Soc. (N. S.)*, 1989, **21**(1), 1–46.

[3] VASSILIEV V. A. A geometric realization of the homology of classical Lie groups, and complexes *S*-dual to flag manifolds. *St. Petersburg Math. J.*, 1991, **3**(4), 809–815.


## 1990-14 — *B. A. Khesin*

For related definitions of the Hopf invariant see [1,2]. One can also regard the symplectic field theory of Ya. Eliashberg, A. Givental, and H. Hofer [3] as a construction of a Floer-type complex in the contact case.


[1] ARNOLD V. I. The asymptotic Hopf invariant and its applications. In: Proceedings of the All-Union School on Differential Equations with Infinitely Many Independent Variables and on Dynamical Systems with Infinitely Many Degrees of Freedom (Dilizhan, May 21 – June 3, 1973). Yerevan: AS of Armenian SSR, 1974, 229–256 (in Russian). [*The English translation*: *Selecta Math. Sov.*, 1986, **5**(4), 327–345.] [*The Russian original is reprinted and supplemented in*: Vladimir Igorevich Arnold. Selecta–60. Moscow: PHASIS, 1997, 215–236.]

[2] ARNOLD V. I., KHESIN B. A. Topological Methods in Hydrodynamics. New York: Springer, 1998. (Appl. Math. Sci., 125.)

[3] ELIASHBERG YA. M., GIVENTAL A. B., HOFER H. Introduction to symplectic field theory. *Geom. Funct. Anal.*, 2000, Special Volume, Part II, 560–673.

## 1990-16 – *B. A. Khesin*

R The "diffused" linking number of curves is the asymptotic linking number (or helicity) of a divergence-free vector field; see paper [2], as well as the survey in [1]. See also the comments to problem *1976-5* and references therein.


[1] ARNOLD V. I., KHESIN B. A. Topological Methods in Hydrodynamics. New York: Springer, 1998. (Appl. Math. Sci., 125.)

[2] VERJOVSKY A., VILA FREYER R. F. The Jones–Witten invariant for flows on a 3-dimensional manifold. *Commun. Math. Phys.*, 1994, **163**(1), 73–88.


## 1990-17

H This is a problem in paper [1] (§ 1, p. 1).


[1] ARNOLD V. I. Ten problems. In: Theory of Singularities and its Applications. Editor: V. I. Arnold. Providence, RI: Amer. Math. Soc., 1990, 1–8. (Adv. Sov. Math., 1.)


## 1990-18

H This is a problem in paper [1] (§ 2, p. 2).


[1] ARNOLD V. I. Ten problems. In: Theory of Singularities and its Applications. Editor: V. I. Arnold. Providence, RI: Amer. Math. Soc., 1990, 1–8. (Adv. Sov. Math., 1.)


R See the comment to problem *1989-3*.

## 1990-19

H This is a problem in paper [1] (§ 3, p. 2).


[1] ARNOLD V. I. Ten problems. In: Theory of Singularities and its Applications. Editor: V. I. Arnold. Providence, RI: Amer. Math. Soc., 1990, 1–8. (Adv. Sov. Math., 1.)


R See the comment to problem *1983-1*.

## 1990-20

$\mathcal{H}$ This is a problem in paper [1] (§ 4, p. 3).

[1] ARNOLD V. I. Ten problems. In: Theory of Singularities and its Applications. Editor: V. I. Arnold. Providence, RI: Amer. Math. Soc., 1990, 1–8. (Adv. Sov. Math., 1.)

$\mathcal{R}$ See the comment to problem *1988-6* by M. B. Sevryuk.

## 1990-21

$\mathcal{H}$ This is a problem in paper [1] (§ 4, p. 3).

[1] ARNOLD V. I. Ten problems. In: Theory of Singularities and its Applications. Editor: V. I. Arnold. Providence, RI: Amer. Math. Soc., 1990, 1–8. (Adv. Sov. Math., 1.)

$\mathcal{R}$ See the comment to problem *1988-6* by M. B. Sevryuk.

## ▽ 1990-22

$\mathcal{H}$ This is a problem in paper [1] (§ 5, p. 4).

[1] ARNOLD V. I. Ten problems. In: Theory of Singularities and its Applications. Editor: V. I. Arnold. Providence, RI: Amer. Math. Soc., 1990, 1–8. (Adv. Sov. Math., 1.)

## △ 1990-22 — *M. B. Mishustin*

$\mathcal{R}$ The problem was solved in [1]; for generalizations see the comment to problem *1989-11*.

[1] MISHUSTIN M. B. Neighborhoods of the Riemann sphere in complex surfaces. *Funct. Anal. Appl.*, 1993, **27**(3), 176–185.

## 1990-23

$\mathcal{H}$ This is a problem in paper [1] (§ 6, p. 4).

[1] ARNOLD V. I. Ten problems. In: Theory of Singularities and its Applications. Editor: V. I. Arnold. Providence, RI: Amer. Math. Soc., 1990, 1–8. (Adv. Sov. Math., 1.)

## 1990-24

H This is a problem in paper [1] (§ 7, p. 5).

[1] ARNOLD V. I. Ten problems. In: Theory of Singularities and its Applications. Editor: V. I. Arnold. Providence, RI: Amer. Math. Soc., 1990, 1–8. (Adv. Sov. Math., 1.)

R See the comment to problem *1978-6.*

## 1990-25

H This is a problem in paper [1] (§ 7, p. 5).

[1] ARNOLD V. I. Ten problems. In: Theory of Singularities and its Applications. Editor: V. I. Arnold. Providence, RI: Amer. Math. Soc., 1990, 1–8. (Adv. Sov. Math., 1.)

R See the comment to problem *1978-6.*

## 1990-26

H This is a problem in paper [1] (§ 8, p. 5).

[1] ARNOLD V. I. Ten problems. In: Theory of Singularities and its Applications. Editor: V. I. Arnold. Providence, RI: Amer. Math. Soc., 1990, 1–8. (Adv. Sov. Math., 1.)

R See the comment to problem *1987-7.*

## 1990-27

H This is a problem in paper [1] (§ 9, p. 6).

[1] ARNOLD V. I. Ten problems. In: Theory of Singularities and its Applications. Editor: V. I. Arnold. Providence, RI: Amer. Math. Soc., 1990, 1–8. (Adv. Sov. Math., 1.)

R See the comment to problem *1987-14.*

## 1990-28

H This is a problem in paper [1] (§ 10, p. 7).

[1] ARNOLD V. I. Ten problems. In: Theory of Singularities and its Applications. Editor: V. I. Arnold. Providence, RI: Amer. Math. Soc., 1990, 1–8. (Adv. Sov. Math., 1.)

# 1991

## 1991-1

R See the comment to problem *1973-25*.

## 1991-2 – *D. A. Zvonkine*

R We suggest the following simplest complexification of the Euler–Bernoulli numbers $E_n$. Note that $E_n$ is the number of real monic polynomials of degree $n+1$ (the sum of whose roots equals 0) with given (real) simple critical values. Then the complexification of $E_n$ is the number of monic *complex* polynomials with a zero sum of roots and given simple critical values. This number equals $(n+1)^{n-1}$, i. e., the multiplicity of the Lyashko–Looijenga map (cf. problems *1995-1*, *1995-2*, *1996-8*, and *1996-13*).

The multiplicity of the Lyashko–Looijenga map in the complex case is also known when the critical values are not necessarily simple. In the real case, this multiplicity depends on the order (on the real line) of the critical values. For some strata, I know the sum of the multiplicities over all possible orders. In codimension 1, the answer is $E_n$ for the caustic and $(n-2)E_n/2$ for the Maxwell stratum. I do not know how the multiplicities are distributed among the possible orders of the critical values.

## ▽ 1991-3

H This is a problem in paper [1] (p. 261).

[1] ARNOLD V. I. Problems on singularities and dynamical systems. In: Developments in Mathematics: the Moscow School. Editors: V. I. Arnold and M. Monastyrsky. London: Chapman & Hall, 1993, 251–274.

## △▽ 1991-3 – *V. I. Arnold* (1991)

R Intervals between the numbers $i$ of zero terms form a curious rapidly growing sequence already for $n=3$. For example, the sequence $i = 0, 1, 3, 8$ is attainable but cannot be continued. According to Yu. V. Matiyasevich, computer experiments (see *Math. Rev.* **51** #5479, **54** #2576, **56** #8480, **58** #187, 80b:10013, 10015,

83k:10020) have discovered only 6 zeros in this case. I do not know whether the boundedness has been proved. The question is related to the Skolem theorem and, therefore, to the problems on the asymptotics of intersections, and those on the number of cycles in dynamical systems generalizing the Hilbert 16th problem on limit cycles to the many-dimensional case, see [1].

[1] ARNOLD V. I. Bounds for Milnor numbers of intersections in holomorphic dynamical systems. In: Topological Methods in Modern Mathematics. Proceedings of the symposium in honor of John Milnor's sixtieth birthday (Stony Brook, NY, 1991). Editors: L. R. Goldberg and A. V. Phillips. Houston, TX: Publish or Perish, 1993, 379–390.

### △ 1991-3 – *S. V. Duzhin*

B. Deshommes [2] proved that a degree 3 recurrence sequence of rational numbers with two adjacent zeros has at most 6 zeros, and there are only 3 explicitly described exceptional cases where the number of zeros is 5 or 6. F. Beukers [1] proved that any recurrence sequence of degree 3 has at most 6 zeros.

[1] BEUKERS F. The zero-multiplicity of ternary recurrences. *Compos. Math.*, 1991, **77**(2), 165–177.

[2] DESHOMMES B. Puissances binomiales dans un corps cubique. *Dissertationes Math. (Rozprawy Mat.)*, 1991, **312**, 57 pp.

### 1991-8 – *M. B. Sevryuk*

V. I. Arnold's conjecture of 1991 about a beautiful answer in this problem is cited in paper [1] (see also *Math. Reviews* 97j:58144) where, in particular, the following result is obtained: If the set of integers $n$ such that $\dim[(A^nX)\cap Y]\geq 1$ is infinite then there exist integers $q$ and $r$, $0\leq r<q$, such that $\dim[(A^{r+qk}X)\cap Y]\geq 1$ for each integer $k$.

[1] ROSALES-GONZÁLEZ E. Intersection dynamics on Grassmann manifolds. *Bol. Soc. Mat. Mexicana, Ser.* 3, 1996, **2**(2), 129–138.

### 1991-10

This is a problem in paper [1] (p. 254).

[1] ARNOLD V. I. Problems on singularities and dynamical systems. In: Developments in Mathematics: the Moscow School. Editors: V. I. Arnold and M. Monastyrsky. London: Chapman & Hall, 1993, 251–274.

## ▽ 1991-11 — *V. I. Arnold*

ℛ The problem arose in connection with the study of graded algebras with the simplest Poincaré series $1+t+t^2+\cdots$; a generalization of the Lagrange theorem was published by E. I. Korkina in paper [2]. The existence of the averages (for typical cones with three faces) was proved by M. L. Kontsevich and Yu. M. Sukhov, see [1].

[1] ARNOLD V. I. Higher dimensional continued fractions. *Reg. Chaot. Dynamics*, 1998, **3**(3), 10–17.

[2] KORKINA E. I. La périodicité des fractions continues multidimensionnelles. *C. R. Acad. Sci. Paris, Sér.* I *Math.*, 1994, **319**(8), 777–780.

## △ 1991-11 — *J.-O. Moussafir*

ℛ Problem *1991-11* is a geometric formulation of the problem of finding a generalization of the Gauss–Kuz′min theorem for ordinary continued fractions, see [1, 2, 8]. Let us introduce some notations. A simplicial cone $C \subset \mathbb{R}^d$ is a cone spanned by $d$ independent vectors. The Klein polyhedron $K$ associated with $C$ is the convex hull of all nonzero points of $C \cap \mathbb{Z}^d$. The sail $V$ is the border of $K$. When $d = 2$ and $C$ is generated by $(1,0)$ and $(1,\alpha)$, the integral lengths of edges (number of integral points on an edge minus 1) of $V$ coincide with half of the coefficients that appear in the continued fraction of $\alpha$—the other half corresponds to the sail of another quadrant. The integral invariants associated with $V$ when $d = 2$ are very few, but as $d$ increases the integral invariants are more numerous. For instance, when $d = 3$, one can associate with every face of $V$ its integral area, its integral distance to the origin, the number of its vertices, and the number of edges connected to each vertex. One may also introduce integral angles, etc. One major result in this theory is due to Yu. M. Sukhov and M. L. Kontsevich [3] (see especially Preface on pages IX–XII of that book). They proved that for almost all simplicial cones the asymptotic distribution of some of these invariants does exist and is universal (cone independent). They get a more accurate result for the number of edge per face. The empirical distributions of integral area and integral distance to the origin, as shown in J.-O. Moussafir's thesis [6], reveal that no smooth curve fits the data.

The generalization of the Lagrange theorem has been investigated by Tsuchihashi [7] and Korkina [4, 5]. They get an equivalence: the periodicity of the sail is equivalent to the fact that $C$ is the eigencone of an hyperbolic operator of $\mathrm{SL}(\mathbb{Z})$. Tsuchihashi discovered also some partial generalizations to the case of complex eigenvalues, proving the periodicity of the sail topological structure (but missing the inverse result that the periodicity implies the provenance from $SL(\mathbb{Z})$).

[1] ARNOLD V. I. Continued Fractions. Moscow: Moscow Center for Continuous Mathematical Education Press, 2001 (in Russian). ("Mathematical Education" Library, 14.)

[2] BRODÉN T. Wahrscheinlichkeits bestimmungen bei der gewöhnlichen Kettenbruchentwicklung reeller Zahlen. *Akad. Föhr. Stockholm*, 1900, **57**, 239–266.

[3] KONTSEVICH M. L., SUKHOV YU. M. Statistics of Klein polyhedra and multidimensional continued fractions. In: Pseudoperiodic Topology. Editors: V. Arnold, M. Kontsevich and A. Zorich. Providence, RI: Amer. Math. Soc., 1999, 9–27. (AMS Transl., Ser. 2, 197; Adv. Math. Sci., 46.)

[4] KORKINA E. I. La périodicité des fractions continues multidimensionnelles. *C. R. Acad. Sci. Paris, Sér.* I *Math.*, 1994, **319**(8), 777–780.

[5] KORKINA E. I. Two-dimensional continued fractions. The simplest examples. *Proc. Steklov Inst. Math.*, 1995, **209**, 124–144.

[6] MOUSSAFIR J.-O. Voiles et polyédres de Klein. Géométrie, algorithmes et statistiques. Thèse, Université Paris-Dauphine, 2000.

[7] TSUCHIHASHI H. Higher-dimensional analogues of periodic continued fractions and cusp singularities. *Tôhoku Math. J., Ser.* 2, 1983, **35**(4), 607–639.

[8] WIMAN A. Über eine-wahrscheinlichkeits Auflage bei Kettenbruchentwicklungen. *Akad. Föhr. Stockholm*, 1900, **57**, 589–841.

## ▽ 1991-14

H This is a problem in paper [1] (p. 255).

[1] ARNOLD V. I. Problems on singularities and dynamical systems. In: Developments in Mathematics: the Moscow School. Editors: V. I. Arnold and M. Monastyrsky. London: Chapman & Hall, 1993, 251–274.

## △ 1991-14 – *A. V. Zorich*

R The initial progress in the problem was obtained by S. Tsarev (1982) who constructed a rather special example (using parameters with particular Diophantine properties). In his example, the intersection line has an asymptotic direction but the deviation from the asymptotic direction is unbounded. However, almost at the same time the conjecture was proved for an open dense set of directions of planes (A. Zorich [10]). More precisely, it was proved for planes close to those having a rational direction. The proof is based on the following elementary argument.

Passing to the quotient $\mathbb{R}^3/\mathbb{Z}^3$ over the lattice, one transforms a Fermi-surface to a closed surface embedded into the three-dimensional torus; the plane sections of the Fermi-surface project to the leaves of a foliation on this compact surface. It is very easy to show that, when the direction of the initial plane is sufficiently close to a rational one, the minimal components of the foliation are just tori with holes. In particular, being unfolded in $\mathbb{R}^3$, such components can be restricted between two parallel rational planes. Thus, an intersection of an unfolded minimal component with the initial irrational plane is restricted by a pair of parallel straight lines in this plane.

Actually, in the initial formulation [6] S. P. Novikov stated only the existence of an asymptotic direction; the observation [10] motivated the stronger version of the conjecture (see [1, 7, 8]) claiming that the deviation of the trajectory from the straight line is generically uniformly bounded.

The questions, whether the set of "good" directions of the planes has the full measure, what is the structure of the set of "bad" directions, what is the Hausdorff dimension of this latter set, what is the structure of abnormal intersection lines, etc, were open for quite a while; some aspects are still open. However, many answers are already found; I would distinguish several beautiful results of I. Dynnikov.

There are two natural approaches to this problem. One can fix a periodic surface, and perturb the direction of the plane, or one can fix a direction of planes and consider a family of perturbations of a periodic surface. Using the second approach I. Dynnikov proved in 1993 (see [2]) the following statement. Let the periodic surface be a level surface of a periodic Morse function in $\mathbb{R}^3$; let $a$ and $b$ be the minimum and the maximum of this function. Fix a generic (in measure-theoretic sense) direction of a family of parallel planes. There is an interval $[c;d]$, $a < c \leq d < b$, such that for any level surface corresponding to the value outside $[c;d]$ all connected components of the plane sections are closed. If $c < d$, then all unbounded components of plane sections of the remaining level surfaces go along straight lines with bounded deviations from them. This implies that for almost all level surfaces the conjecture is valid for almost all directions of the plane.

The situation when $c = d$ is, however, possible, and for this particular level surface the behavior of the plane sections might be quite complicated; it is still unclear whether the set of "good" directions for this level surface has the full measure. I. Dynnikov elaborated a highly nontrivial construction producing numerous examples of nontypical behavior of trajectories, see [3], when trajectories do not even have an asymptotic direction (see, however, [11] for numerical simulations of Dynnikov's examples, showing that in many cases some preferred direction still exists). The structure of the space of "nice" directions was studied by I. Dynnikov in paper [5].

The developments in this area permitted S. P. Novikov and A. Ya. Mal′tsev to obtain interesting applications in solid state physics, see [9], and, in particular, to correct some common prejudices based on experimental studies of Fermi-surfaces of metals. (As a matter of fact, Fermi-surfaces of metals may have quite complicated structure, in particular they may give surfaces of rather high genera after passing to a quotient over the lattice.)

[1] ARNOLD V. I. Problems on singularities and dynamical systems. In: Developments in Mathematics: the Moscow School. Editors: V. I. Arnold and M. Monastyrsky. London: Chapman & Hall, 1993, 251–274.

[2] DYNNIKOV I. A. Proof of S. P. Novikov's conjecture on the semiclassical motion of an electron. *Math. Notes*, 1993, **53**(5), 495–501.

[3] DYNNIKOV I. A. Semi-classical motion of the electron. Proof of Novikov's conjecture in general position and counterexamples. In: Solitons, Geometry, and Topology: on the Crossroad. Editors: V. M. Buchstaber and S. P. Novikov. Providence, RI: Amer. Math. Soc. 1997, 45–73. (AMS Transl., Ser. 2, 179; Adv. Math. Sci., 33.)

[4] DYNNIKOV I. A. Surfaces in 3-torus: geometry of plane sections. In: Papers from the 2nd European Congress of Mathematics (Budapest, 1996), Vol. 1. Editors: A. Balog, G. O. H. Katona, A. Recski and D. Szász. Basel: Birkhäuser, 1998, 162–177. (Progr. Math., 168.)

[5] DYNNIKOV I. A. The geometry of stability zones in Novikov's problem on the semi-classical motion of an electron. *Russian Math. Surveys*, 1999, **54**(1), 21–59.

[6] NOVIKOV S. P. The Hamiltonian formalism and a multi-valued analogue of Morse theory. *Russian Math. Surveys*, 1982, **37**(5), 1–56.

[7] NOVIKOV S. P. Quasiperiodic structures in topology. In: Topological Methods in Modern Mathematics. Proceedings of the symposium in honor of John Milnor's sixtieth birthday (Stony Brook, NY, 1991). Editors: L. R. Goldberg and A. V. Phillips. Houston, TX: Publish or Perish, 1993, 223–233.

[8] NOVIKOV S. P. The semiclassical electron in a magnetic field and lattice. Some problems of low dimensional "periodic" topology. *Geom. Funct. Anal.*, 1995, **5**(2), 434–444.

[9] NOVIKOV S. P., MAL′TSEV A. YA. Topological phenomena in normal metals. *Letters to JETP*, 1996, **63**(10), 809–813; *Physics–Uspekhi*, 1998, **41**(3), 231–239. [*Internet*: `http://www.arXiv.org/abs/cond-mat/9709007`]

[10] ZORICH A. V. S. P. Novikov's problem on the semiclassical motion of an electron in a homogeneous magnetic field that is close to rational. *Russian Math. Surveys*, 1984, **39**(5), 287–288.

[11] ZORICH A. V. Asymptotic flag of an orientable measured foliation on a surface. In: Geometric Study of Foliations (Tokyo, 1993). Editors: T. Mizutani, K. Masuda, S. Matsumoto, T. Inaba, T. Tsuboi and Y. Mitsumatsu. River Edge, NJ: World Scientific, 1994, 479–498.

# 1992

## ▽ 1992-1

ℋ The second question is a problem in paper [1] (p. 272).

[1] ARNOLD V. I. Problems on singularities and dynamical systems. In: Developments in Mathematics: the Moscow School. Editors: V. I. Arnold and M. Monastyrsky. London: Chapman & Hall, 1993, 251–274.

## △ 1992-1 – *S. M. Gusein-Zade*

ℛ For the given formulation the answer is negative. For functions of three or more variables it is not sufficient to require that the index is equal to zero. One should demand the contractibility of the set $\{f \le 0\}$ where $f\colon (\mathbb{R}^n, 0) \to (\mathbb{R}, 0)$ is the function under consideration. For functions of two variables of finite multiplicity (when contractibility of the set $\{f \le 0\}$ is equivalent to the fact that the index is equal to zero), the affirmative answer was given by S. M. Gusein-Zade [1,2].

[1] GUSEIN-ZADE S. M. On a problem of B. Teissier. In: Topics in Singularity Theory. V. I. Arnold's 60th Anniversary Collection. Editors: A. Khovanskiĭ, A. Varchenko and V. Vassiliev. Providence, RI: Amer. Math. Soc., 1997, 117–125. (AMS Transl., Ser. 2, 180; Adv. Math. Sci., 34.)

[2] GUSEIN-ZADE S. M. On the existence of deformations without critical points (the Teissier problem for functions of two variables). *Funct. Anal. Appl.*, 1997, **31**(1), 58–60.

## 1992-3

ℋ This is a problem in paper [1] (p. 252).

[1] ARNOLD V. I. Problems on singularities and dynamical systems. In: Developments in Mathematics: the Moscow School. Editors: V. I. Arnold and M. Monastyrsky. London: Chapman & Hall, 1993, 251–274.

## 1992-7

H This is a problem in paper [1] (p. 251).

[1] ARNOLD V. I. Problems on singularities and dynamical systems. In: Developments in Mathematics: the Moscow School. Editors: V. I. Arnold and M. Monastyrsky. London: Chapman & Hall, 1993, 251–274.

## 1992-8 — *B. A. Khesin*

R The linear functional on divergence-free vector fields corresponding to a curve is the flux of the field across an oriented (Seifert) surface bounded by the curve. See the discussion of this symplectic structure in books [1, 2].

[1] ARNOLD V. I., KHESIN B. A. Topological Methods in Hydrodynamics. New York: Springer, 1998. (Appl. Math. Sci., 125.)

[2] BRYLINSKI J. -L. Loop Spaces, Characteristic Classes and Geometric Quantization. Boston, MA: Birkhäuser, 1993. (Progr. Math., 107.)

## 1992-11

R See the comment to problem *1971-11*.

## 1992-12

R See the comments to problems *1988-6* (by M. B. Sevryuk) and *1992-13*.

## 1992-13 — *V. Yu. Kaloshin*

Also: *1989-2, 1992-12, 1994-47, 1994-48*

R Let $M$ be a compact manifold of dimension at least 2. Let $\mathrm{Diff}^r(M,M)$ be the space of $C^r$ invertible mappings (diffeomorphisms) of $M$ into itself with the uniform $C^r$-topology.

For a diffeomorphism $f \in C^r(M)$, consider the number of *isolated* periodic points of period $n$ (i. e., the number of isolated fixed points of $f^n$)

$$P_n(f) = \#\{\text{isolated } x \in M : x = f^n(x)\}.$$

Artin and Mazur [1] proved that there exists a dense set $\mathcal{D}$ in $\mathrm{Diff}^r(M,M)$ of diffeomorphisms $f$ such that, for any diffeomorphism $f \in \mathcal{D}$, the number $P_n(f)$ grows at most exponentially with $n$, i. e., for some number $C = C(f) > 0$

$$P_n(f) \leq \exp(Cn) \quad \text{for all } n \in \mathbb{Z}_+. \tag{1}$$

We call a diffeomorphism, satisfying (1), a *diffeomorphism of Artin–Mazur* or an *A–M diffeomorphism.*

Artin and Mazur [1] posed the following problem: *What can be said about the set of A–M mappings with only transversal periodic orbits*? Recall that a periodic orbit of period $n$ is called *transversal* if the linearization $df^n$ at this point does not have an $n$-th root of unity as an eigenvalue. Notice that a hyperbolic periodic point is always transversal, but not vise versa.

The first result is the following.

**Theorem 1** [7]. *Let* $0 \leq r < \infty$. *Then the set of A–M diffeomorphisms with only hyperbolic periodic orbits is dense in the space* $\mathrm{Diff}^r(M)$.

This theorem says that diffeomorphisms which satisfy a stronger condition than that of Artin–Mazur are dense, because any hyperbolic point is transversal, but not vise versa.

Finite smoothness ($r < \infty$) is essential in Theorem 1.

S. Smale [17] and R. Bowen [3] posed questions about relations between the rate of growth of the number of periodic points on one hand and dynamical $\zeta_f$-function or topological entropy on the other hand for (Baire) generic diffeomorphisms. In particular, these questions ask whether A–M diffeomorphisms are generic.

Recall that, according to the standard terminology, a subset of a topological space is called *residual or Baire residual* if it contains a countable intersection of open dense sets. Elements of such a set are called *generic or Baire generic.*

The second main result is

**Theorem 2** [8]. *Let* $2 \leq r < \infty$. *Then the set of A–M diffeomorphisms is not* $C^r$*-residual in the space* $\mathrm{Diff}^r(M)$ *with the uniform* $C^r$*-topology.*

*Moreover, there is an open set* $\mathcal{N} \subset \mathrm{Diff}^r(M)$ *in the space of* $C^r$*-diffeomorphisms such that for an arbitrary sequence of positive integers* $\{a_n\}_{n=1}^{\infty}$ *there exists* *a residual set* $\mathcal{R}_a \subset \mathcal{N}$, *depending on the sequence* $\{a_n\}_{n=1}^{\infty}$, *with the property that* $f \in \mathcal{R}_a$ *implies that*

$$\overline{\lim_{n\to\infty}} P_n(f)/a_n = \infty. \tag{2}$$

A proof of this theorem is based on a theorem of Gonchenko–Shil′nikov–Turaev [4]. Paper [6] contains a detailed proof of this theorem of Gonchenko–Shil′nikov–Turaev following the strategy proposed in [4].

The open set appearing in Theorem 2 for $\dim M = 2$ is usually called a *Newhouse domain.* The existence of these domains was proved by Newhouse [12]. Moreover, using a theorem of Pujals–Sambarino [14] and Theorem 2, one can prove that, for surface diffeomorphisms, $\dim M = 2$ outside uniformly hyperbolic or Axiom A diffeomorphisms, $\mathrm{Diff}^r(M) \setminus$ Axiom A diffeomorphisms with an arbitrarily fast growth of the number of periodic points (2) form a $C^1$-dense $C^r$-residual set, i. e., a set containing a countable intersection of $C^1$-dense $C^r$-open sets. Note that $2 \le r$ is essential in Theorem 2.

Surprisingly, even for a $C^r$-unimodal map of an interval $f\colon [-1,1] \to [-1,1]$ with any $r \ge 1$, an arbitrarily fast growth of the number of periodic points (2) is also possible [11].

It looks plausible that Theorem 2 is true for $r = 1$ and $\dim M \ge 3$ using the construction from [2], where a $C^1$-version of a Newhouse domain is constructed.

In [15], an example of a diffeomorphism with an arbitrarily fast growth of the number of periodic points is constructed.

According to the second part of Theorem 2, there is no upper estimate for the growth of the number of periodic points of Baire generic diffeomorphisms. However, Baire generic sets in Euclidean spaces can have zero Lebesgue measure (see [5, 6, 13, 16] where various examples of such phenomena are discussed).

Problems *1992-12* and *1992-13* are about at most exponential growth of the number of periodic points "with probability one." In particular, problem *1992-13* can be rephrased as follows: *Prove that "with probability one" a diffeomorphism is A–M.*

Our third result nearly solves this problem.

First of all we define precisely "with probability one." Let $B^N$ be the closed unit ball in $\mathbb{R}^N$. Denote by $\mathrm{Diff}^r(B^N)$ the space of $C^r$-diffeomorphisms from $B^N$ into the interior of $B^N$. Fix a coordinate system in $\mathbb{R}^N$. Let $\alpha = (\alpha_1, \dots, \alpha_N)$ be a multi-index from $\mathbb{Z}_+^N$ and $|\alpha| = \sum_i \alpha_i$. For a point $x = (x_1, \dots, x_N) \in \mathbb{R}^N$ we write $x^\alpha = \prod_{i=1}^N x_i^{\alpha_i}$. Associate with a real analytic function $\phi : B^N \to \mathbb{R}^N$ the set of coefficients of its expansion:

$$\phi_{\vec{\varepsilon}}(x) = \sum_{\alpha \in \mathbb{Z}_+^N} \vec{\varepsilon}_\alpha x^\alpha. \tag{3}$$

Denote by $W_{k,N}$ the space of $N$-component homogeneous vector-polynomials of degree $k$ in $N$ variables and by $\nu(k,N) = \dim W_{k,N}$ the dimension of $W_{k,N}$. According to the notation of the expansion (3), denote coordinates in $W_{k,N}$ by

$$\vec{\varepsilon}_k = \left(\{\vec{\varepsilon}_\alpha\}_{|\alpha|=k}\right) \in W_{k,N}. \tag{4}$$

In $W_{k,N}$ we use a scalar product that is invariant with respect to orthogonal transformations of $\mathbb{R}^N \supset B^N$, defined as follows:

$$\langle \vec{\varepsilon}_k, \vec{\zeta}_k \rangle_k = \sum_{|\alpha|=k} \binom{k}{\alpha}^{-1} \langle \vec{\varepsilon}_\alpha, \vec{\zeta}_\alpha \rangle, \quad \|\vec{\varepsilon}_k\|_k = \left(\langle \vec{\varepsilon}_k, \vec{\varepsilon}_k \rangle_k\right)^{1/2}. \tag{5}$$

Denote by

$$B_k^N(r) = \{\vec{\varepsilon}_k \in W_{k,N} : \ \|\vec{\varepsilon}_k\|_k \le r\} \tag{6}$$

the closed $r$-ball in $W_{k,N}$ centered at the origin. Let $\mathrm{Leb}_{k,N}$ be Lebesgue measure on $W_{k,N}$ induced by the scalar product (5) and normalized by a constant so that the volume of the unit ball is one: $\mathrm{Leb}_{k,N}(B_k^N(1)) = 1$.

Fix a non-increasing sequence of positive numbers $\mathbf{r} = \left(\{r_k\}_{k=0}^\infty\right)$ such that $r_k \to 0$ as $k \to \infty$ and define a Hilbert brick of size $\mathbf{r}$

$$\begin{aligned} HB^N(\mathbf{r}) &= \{\vec{\varepsilon} = \{\vec{\varepsilon}_\alpha\}_{\alpha \in \mathbb{Z}_+^N} : \text{for all } k \in \mathbb{Z}_+, \ \|\vec{\varepsilon}_k\|_k \le r_k\} \\ &= B_0^N(r_0) \times B_1^N(r_1) \times \cdots \times B_k^N(r_k) \times \cdots \subset W_{0,N} \times W_{1,N} \times \cdots \times W_{k,N} \times \cdots. \end{aligned}$$

Define a *product Lebesgue probability measure* $\mu_{\mathbf{r}}^N$ *associated with the Hilbert brick* $HB^N(\mathbf{r})$ *of size* $\mathbf{r}$ by normalizing for each $k \in \mathbb{Z}_+$ the corresponding Lebesgue measure $\mathrm{Leb}_{k,N}$ on $W_{k,N}$ to the Lebesgue probability measure on the $r_k$-ball $B_k^N(r_k)$:

$$\mu_{k,r}^N = r^{-\nu(k,N)}\, Leb_{k,N} \quad \text{and} \quad \mu_{\mathbf{r}}^N = \times_{k=0}^\infty \mu_{k,r_k}^N.$$

Let $f \in \mathrm{Diff}^r(B^N)$ be a $C^r$ diffeomorphism of $B^N$ into its interior. We call $HB^N(\mathbf{r})$ a Hilbert brick of *an admissible size* $\mathbf{r} = \left(\{r_k\}_{k=0}^\infty\right)$ with respect to $f$ if:

A) for each $\vec{\varepsilon} \in HB^N(\mathbf{r})$, the corresponding function $\phi_{\vec{\varepsilon}}(x) = \sum_{\alpha \in \mathbb{Z}_+^N} \vec{\varepsilon}_\alpha x^\alpha$ is analytic on $B^N$;

B) for each $\vec{\varepsilon} \in HB^N(\mathbf{r})$, the corresponding map $f_{\vec{\varepsilon}}(x) = f(x) + \phi_{\vec{\varepsilon}}(x)$ is a diffeomorphism from $B^N$ into its interior, i. e., $\{f_{\vec{\varepsilon}}\}_{\vec{\varepsilon} \in HB^N(\mathbf{r})} \subset \mathrm{Diff}^r(B^N)$;

C) for all $\delta > 0$ and all $C > 0$, the sequence $r_k \exp(Ck^{1+\delta}) \to \infty$ as $k \to \infty$.

An example of an admissible sequence $\mathbf{r} = (\{r_k\}_{k=0}^{\infty})$ is $r_k = \tau/k!$, where $\tau$ depends on $f$ and is chosen sufficiently small to ensure that condition B holds.

The third main result is the following

**Theorem 3** [10]. *For any $1 < \rho \leq \infty$ or $\omega$ and any $C^\rho$ diffeomorphism $f \in \mathrm{Diff}^\rho(B^N)$, consider a Hilbert brick $HB^N(\mathbf{r})$ of an admissible size $\mathbf{r}$ with respect to $f$ and the family of analytic perturbations of $f$*

$$\{f_{\vec{\varepsilon}}(x) = f(x) + \phi_{\vec{\varepsilon}}(x)\}_{\vec{\varepsilon} \in HB^N(\mathbf{r})}$$

*with the Lebesgue product probability measure $\mu_{\mathbf{r}}^N$ associated with $HB^N(\mathbf{r})$. Then for every $\delta > 0$ and for $\mu_{\mathbf{r}}^N$-a. e. $\vec{\varepsilon}$ there is $C = C(\vec{\varepsilon}, \delta) > 0$ such that for all $n \in \mathbb{Z}_+$*

$$P_n(f_{\vec{\varepsilon}}) < \exp(Cn^{1+\delta}).$$

Similar definitions of "with probability one" or prevalence were proposed in [5, 18]. Recall that $\omega$ denotes real analytic. Theorem 3 is a considerable improvement of the theorem of Yomdin [19].

In fact, a stronger statement has been proved. Define for each point $x \in B^N$ of period $n \in \mathbb{Z}_+$ of a diffeomorphism $f \in \mathrm{Diff}^r(B^N)$, $f^n(x) = x$, its *hyperbolicity* as the closest distance of the linearization $df^n(x)$ to the unit circle $\{|z| = 1\}$:

$$\gamma_n(x, f) = \inf_{\phi \in [0,1)} \left\| (\exp(2\pi i \phi) I - df^n(x))^{-1} \right\|^{-1}.$$

In particular, this means that all the eigenvalues of $df^n(x)$ are at least $\gamma_n(x, f)$-away from the unit circle $\{|z| = 1\}$. Hyperbolicity of a hyperbolic periodic point is positive. Let

$$\gamma_n(f) = \min_x \gamma_n(x, f),$$

where the minimum is taken over all periodic points of period $n$.

**Theorem 4** [10]. *With the notation of Theorem 3, for every $\delta > 0$ and $\mu_{\mathbf{r}}^N$-a. e. $\vec{\varepsilon}$, there is $C = C(\vec{\varepsilon}, \delta) > 0$ such that for all $n \in \mathbb{Z}_+$*

$$\gamma_n(f_{\vec{\varepsilon}}) > \exp(-Cn^{1+\delta}).$$

By virtue of regularity of the first derivative of $f$, the equation $f^n(x) = x$ cannot have two solutions which are too close together. Therefore, this theorem implies Theorem 3. The proof of this Theorem develops a new technique of perturbation of diffeomorphisms by Newton interpolation polynomials. This technique is outlined in [10]. The complete proof is in [9] and is in preparation for publication.

See also the comment to problem *1988-6* by M. B. Sevryuk.

[1] ARTIN M., MAZUR B. On periodic points. *Ann. Math., Ser.* 2, 1965, **81**(1), 82–99.

[2] BONATTI C., DÍAZ L. Connexions hétéroclines et généricité d'une infinité de puits et de sources. *Ann. Sci. École Norm. Sup.* (4), 1999, **32**(1), 135–150.

[3] BOWEN R. On Axiom A Diffeomorphisms. Providence, RI: Amer. Math. Soc., 1978. (Regional Conf. Ser. in Math., 35.)

[4] GONCHENKO S. V., SHIL'NIKOV L. P., TURAEV D. V. On models with nonrough Poincaré homoclinic curves. *Physica D*, 1993, **62**(1–4), 1–14.

[5] HUNT B. R., SAUER T., YORKE J. A. Prevalence: a translation-invariant "almost every" for infinite-dimensional spaces. *Bull. Amer. Math. Soc.* (*N. S.*), 1992, **27**(2), 217–238; addendum: 1993, **28**(2), 306–307.

[6] KALOSHIN V. YU. Some prevalent properties of smooth dynamical systems. *Proc. Steklov Inst. Math.*, 1996, **213**, 115–140.

[7] KALOSHIN V. YU. An extension of the Artin–Mazur theorem. *Ann. Math., Ser.* 2, 1999, **150**(2), 729–741.

[8] KALOSHIN V. YU. Generic diffeomorphisms with superexponential growth of number of periodic orbits. *Commun. Math. Phys.*, 2000, **211**(1), 253–271.

[9] KALOSHIN V. YU. A stretched exponential estimate on growth of number of periodic points for diffeomorphisms with probability one. Ph. D. Thesis, Princeton University, 2001.

[10] KALOSHIN V. YU., HUNT B. R. A stretched exponential estimate on growth of number of periodic points for diffeomorphisms with probability one, I; II. *Electron. Res. Announc. Amer. Math. Soc.*, 2001, **7**, 17–27; 28–36 (electronic).

[11] KALOSHIN V. YU., KOZLOVSKIĬ O. S. A $C^r$-unimodal map with a superexponential growth of the number of periodic points. Preprint.

[12] NEWHOUSE S. Diffeomorphisms with infinitely many sinks. *Topology*, 1974, **13**, 9–18.

[13] OXTOBY J. Measure and Category. A Survey of the Analogies Between Topological and Measure Spaces, 2nd edition. New York – Berlin: Springer, 1980. (Graduate Texts in Math., 2.)

[14] PUJALS E. R., SAMBARINO M. Homoclinic tangencies and hyperbolicity for surface diffeomorphisms. *Ann. Math., Ser.* 2, 2000, **151**(3), 961–1023.

[15] ROSALES-GONZÁLEZ E. Growth of periodic orbits of dynamical systems. *Funct. Anal. Appl.*, 1991, **25**(4), 254–262.

[16] SIMON B. Operators with singular continuous spectrum. I. General operators. *Ann. Math., Ser.* 2, 1995, **141**(1), 131–145.

[17] SMALE S. Differentiable dynamical systems. *Bull. Amer. Math. Soc.*, 1967, **73**, 747–817.

[18] VISHIK M. I., KUKSIN S. B. Quasilinear elliptic equations and Fredholm manifolds. *Moscow Univ. Math. Bull.*, 1985, **40**(6), 26–34.

[19] YOMDIN Y. A quantitative version of the Kupka–Smale theorem. *Ergod. Theory Dynam. Systems*, 1985, **5**(3), 449–472.

## 1992-14

H This is a problem in paper [1] (p. 266).

[1] ARNOLD V. I. Problems on singularities and dynamical systems. In: Developments in Mathematics: the Moscow School. Editors: V. I. Arnold and M. Monastyrsky. London: Chapman & Hall, 1993, 251–274.

R See the comment to problem *1988-6* by M. B. Sevryuk.

## 1992-15

H This is a problem in paper [1] (p. 252).

[1] ARNOLD V. I. Problems on singularities and dynamical systems. In: Developments in Mathematics: the Moscow School. Editors: V. I. Arnold and M. Monastyrsky. London: Chapman & Hall, 1993, 251–274.

# 1993

## 1993-1

R See the comment to problem *1993-25*.

## 1993-3 — *R. Uribe-Vargas*

ℛ This problem can be transferred to a problem of families of functions on the circle: On one hand, the curve $f_{a,b}(x,y) = c > 0$, for $c$ sufficiently small, is a small convex curve $\gamma_c$. This curve can be parametrized by the angle, $\theta \in \mathbb{S}^1$, formed by its external unit normal vector, $\nu(\theta)$, with a fixed oriented line, at each point of the curve, $\gamma_c : \theta \mapsto \gamma_c(\theta)$. Thus, for $c$ sufficiently small, the *support function* of the curve $\gamma_c$, $h_c : \mathbb{S}^1 \to \mathbb{R}$, $h_c(\theta) = \langle \nu(\theta), \gamma_c(\theta) \rangle$, is well defined and the vertices of $\gamma_c$ correspond to the points $\theta_0 \in \mathbb{S}^1$ for which $h'_c(\theta_0) + h'''_c(\theta_0) = 0$. On the other hand, given the family of circles centered at the origin, $x^2 + y^2 = r^2$, on the plane with coordinates $x$, $y$, the graph of a function $(x,y) \mapsto f(x,y)$, restricted to a circle $\mathbb{S}^1$ of the family, is a curve in $\mathbb{R}^3$ whose *flattenings* (points at which the torsion vanishes) correspond to the points $\theta_0 \in \mathbb{S}^1$ for which $h'(\theta_0) + h'''(\theta_0) = 0$, where $x = r\cos\theta$, $y = r\sin\theta$ and $h = f_{|\mathbb{S}^1}$.

So in [1] small curves in the real projective 3-space are considered as the intersections of a smooth surface (depending on parameters) in the 3-space, given as the graph of a smooth function $f : \mathbb{R}^2 \to \mathbb{R}$, with the family of cylinders, $x^2 + y^2 = r^2$, whose radius $r$ shrinks to zero. These curves are denoted by $\gamma(f,r) : \mathbb{S}^1 \to \mathbb{R}^3$. The germs near the degenerate cylinder ($r = 0$) are considered. The hypersurface in this space of curves—all $f$ and all $r \geq 0$—separating the domains of small curves having non-equal numbers of flattenings, is called the *bifurcation diagram* of flattenings.

The vanishing curves belonging to the bifurcation diagram of flattenings form a set of codimension 2 in this diagram and of codimension 3 in this space of curves. The singularity of the bifurcation diagram at a vanishing curve was studied in [1] where it was proved that its intersection with a 3-dimensional transversal space is a parabolic cup, based on a hypocycloid with 6 cusps (see Fig. 1).

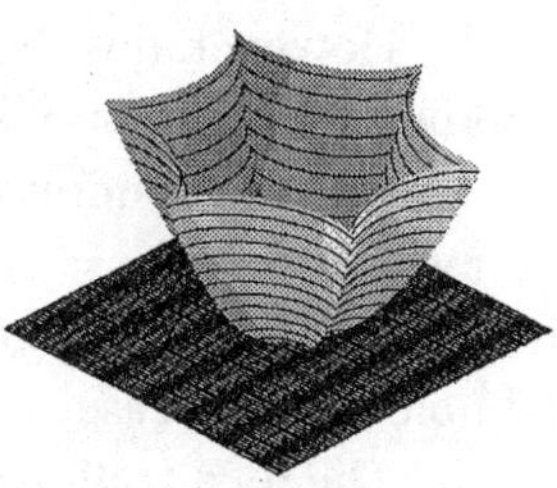

**Fig. 1:** The cupocycloid

More precisely, let $F : \mathbb{R}^2 \times \Lambda \to \mathbb{R}$ be a germ at $\lambda = 0$ of a function family $\{f_\lambda : (\mathbb{R}^2, 0) \to \mathbb{R}\}$ depending on a parameter $\lambda \in \Lambda = \mathbb{R}^l$.

**Definition.** The *bifurcation diagram* of $F$ is the subset of $\Lambda \times \mathbb{R}$ formed by the pairs $(\lambda, c)$, where $c = r^2$, such that the curve $\gamma(f_\lambda, r)$ has a degenerate flattening or the curve itself is not smooth. The family $F$ is said to be the *generating family* of its bifurcation diagram.

**Theorem** [1]. *The bifurcation diagram* $\Sigma_\Gamma$ *has sections* ($c = \text{const}$) *diffeomorphic to hypocycloids with* 6 *cusps whose radius tends to* 0 *as* $c \to 0$, *more precisely: when* $\varepsilon \to 0$,

$$\frac{\Sigma_\Gamma \cap (c = \varepsilon)}{\sqrt{\varepsilon}} \longrightarrow \textit{standard hypocycloid.}$$

Denote by $x$, $y$ the Euclidean coordinates on $\mathbb{R}^2$. Let $h(x,y) = x^2 + y^2$.

**Definition.** Two function family germs $F_1, F_2 : \mathbb{R}^2 \times \mathbb{R}^l \to \mathbb{R}$, at $\lambda = 0$, are called *polar-equivalent* if there exist a diffeomorphism $\Phi_1$, preserving the origin in $\mathbb{R}^2 \times \mathbb{R}^l$ and the foliation by circles $S_r$: $h(x,y) = r^2$, in $\mathbb{R}^2$ for all the values of the parameters, and a diffeomorphism $\Phi_2$, making commutative the following diagram:

$$\begin{array}{ccc} \mathbb{R}^2 \times \mathbb{R}^l & \xrightarrow{\Phi_1} & \mathbb{R}^2 \times \mathbb{R}^l \\ {\scriptstyle h\times \mathrm{Id}}\big\downarrow & & {\scriptstyle h\times \mathrm{Id}}\big\downarrow \\ \mathbb{R}_+ \times \mathbb{R}^l & \xrightarrow{\Phi_2} & \mathbb{R}_+ \times \mathbb{R}^l, \end{array} \qquad \Phi_2 \circ (h \times \mathrm{Id}) = (h \times \mathrm{Id}) \circ \Phi_1,$$

and a non-zero function family $P : \mathbb{R}^2 \times \mathbb{R}^l \to \mathbb{R}$ such that $F_2 = P \cdot (F_1 \circ \Phi_1)$.

**Remark.** Polar equivalent families have diffeomorphic bifurcation diagrams.

**Theorem** [1]. *The generating family* (*of the germ at the origin*) *of the bifurcation diagram has exactly one modulus* (*a continuous invariant of the polar orbit—or polar-equivalence class—of the generating family*) *in the formal and* $C^\infty$ *categories.*

However, this theorem does not imply that the bifurcation diagram has one modulus. One can only say that the bifurcation diagram of flattenings has at most one modulus (a continuous invariant of its differentiable type) but it is unknown if the modality $= 1$ or $= 0$. So the absence of functional moduli is proved in [1] for the formal and $C^\infty$ categories, but this provides no information on the analytic and the holomorphic case!

The localization technique described in [1] permits the problem to be presented as a stability property of a germ of deformation.

[1] URIBE-VARGAS R. Symplectic and contact singularities in the differential geometry of curves and surfaces. Ph. D. Thesis, Université Paris 7, 2001, Ch. 6.

## 1993-5 — *J.-O. Moussafir*

This question goes back to Arnold's paper [1].

ℜ Sturmfels' examples can be found in [3, 4] along with many examples and explanations about the connection between sails and effective algebraic geometry. See also [2] for an introduction to toric varieties.

[1] ARNOLD V. I. *A*-graded algebras and continued fractions. *Commun. Pure Appl. Math.*, 1989, **42**(7), 993–1000. [*The Russian translation in*: Vladimir Igorevich Arnold. Selecta–60. Moscow: PHASIS, 1997, 473–482.]

[2] FULTON W. Introduction to Toric Varieties. Princeton, NJ: Princeton University Press, 1993. (Ann. Math. Studies, 131.)

[3] STURMFELS B. Gröbner bases of toric varieties. *Tôhoku Math. J., Ser.* 2, 1991, **43**(2), 249–261.

[4] STURMFELS B. Gröbner Bases and Convex Polytopes. Providence, RI: Amer. Math. Soc., 1996. (University Lecture Series, 8.)

## 1993-6 – *S. V. Chmutov*

ℜ In addition to the paper cited in the problem, see also [1].

[1] FINTUSHEL R., STERN R. J. Instanton homology of Seifert fibred homology three-spheres. *Proc. London Math. Soc., Ser.* 3, 1990, **61**(1), 109–137.

## ▽ 1993-10 – *F. Aicardi*

ℜ Here a *curve* is the image of a smooth immersion of the circle into the plane.

**Definition.** A curve with *n* double points is called minimal if all curves in its $J^+$ class have at least *n* double points.

**Conjecture 1.** *Two minimal curves with n double points belonging to the same $J^+$ class are connectable by a generic path, avoiding the $J^+$ discriminant, in the space of curves having no more than $n+4$ double points.*

**Conjecture 2.** *There exists a positive number c such that two minimal curves with n double points belonging to the same $J^+$ class are connectable by a generic path crossing the safe discriminant ($J^-$ and* St*) not more than $cn^2$ times.*

(From a letter to V. I. Arnold dated October 25, 1995.)

### △ 1993-10 — *S. V. Chmutov*

ℛ As far as I know the problem is still open for $J^+$. For the St classification the problem was solved by S. L. Tabachnikov and S. V. Chmutov (1995, unpublished). The solution is algorithmical and the growth of the function is clear from it. The function grows no faster than a linear function in $n$.

### ▽ 1993-11 — *M. B. Sevryuk*

ℛ In 1997–98, very strong results in this direction (moreover, for the multidimensional case) were obtained by M. L. Kontsevich and Yu. M. Sukhov, see [1–3]. The whole topic arises from paper [4].

[1] ARNOLD V. I. Higher dimensional continued fractions. *Reg. Chaot. Dynamics*, 1998, **3**(3), 10–17.

[2] KONTSEVICH M. L., SUKHOV YU. M. Statistics of Klein polyhedra and multidimensional continued fractions. In: Pseudoperiodic Topology. Editors: V. Arnold, M. Kontsevich and A. Zorich. Providence, RI: Amer. Math. Soc., 1999, 9–27. (AMS Transl., Ser. 2, 197; Adv. Math. Sci., 46.)

[3] ARNOLD V. I. Continued Fractions. Moscow: Moscow Center for Continuous Mathematical Education Press, 2001 (in Russian). ("Mathematical Education" Library, 14.)

[4] ARNOLD V. I. *A*-graded algebras and continued fractions. *Commun. Pure Appl. Math.*, 1989, **42**(7), 993–1000. [*The Russian translation in*: Vladimir Igorevich Arnold. Selecta–60. Moscow: PHASIS, 1997, 473–482.]

### △ 1993-11 — *V. I. Arnold*

ℛ The solution of problem C published in [1] seems to be robust.

[1] AVDEEVA M. O., BYKOVSKIĬ V. A. A solution of Arnold's problem on the Gauss–Kuz′min statistics. Preprint № 8, Far Eastern Branch of Russian Academy of Sciences, Institute for Applied Mathematics, Khabarovsk Division. Vladivostok: Dal′nauka, 2002, 12 pp. (in Russian).

### 1993-12 — *V. I. Arnold* (1993)

ℛ D. Siersma's constructions probably give (at least in the simplest cases where $\dim_{\mathbb{C}} T^*F_{n+1} = 4$) a definition of natural "monodromy at ∞" corresponding to a non-isolated singularity of the preimage of a polynomial with multiple roots. In other words, a set of concepts (possibly, partial) may be concealed in these constructions, and we should *calculate* these objects in our case.

## ▽ 1993-13 — *V. I. Arnold* (1993)

ℛ Let us point out an adjacent problem. In paper [1] for $A_\mu$ and $D_\mu$ in $\mathbb{C}^2$ it is proved that the mapping $\pi_1(\mathbb{C}^\mu \setminus \Sigma) \to \pi_0(\mathrm{Diff}(V, \partial V))$ has trivial kernel. Is this true for suspensions in $\mathbb{C}^n$? For $E_k$? For nonsimple singularities? (I always thought it was true.)

[1] PERRON B., VANNIER J.-P. Groupe de monodromie géométrique des singularités simples. *C. R. Acad. Sci. Paris, Sér.* I *Math.*, 1992, **315**(10), 1067–1070.

## △ 1993-13 — *S. V. Chmutov*

ℛ B. Wajnryb [1] proved that for the singularities $E_6$, $E_7$, $E_8$ the map $\pi_1(\mathbb{C}^\mu \setminus \Sigma) \to \pi_0(\mathrm{Diff}(V, \partial V))$ has a nontrivial kernel.

[1] WAJNRYB B. Artin groups and geometric monodromy. *Invent. Math.*, 1999, **138**(3), 563–571.

## 1993-17 — *V. I. Arnold*

ℛ If no, this is a good problem for "Kvant" ("Quantum"). Here is its solution:

**Lemma.** *$\forall$ integer $x$, $0 < x < p$, $\exists$ integer $Y(x)$ such that $xY \equiv 1 \mod p^p$.*

**Proof.** As $p$ is prime, $\exists$ integer $y$ such that $xy = 1 + pz$ ($z$ is integer). Denote $pz = Z$. Since $p$ is odd, $Z^p + 1 = (1+Z)(Z^{p-1} - Z^{p-2} + \cdots + 1)$. Therefore, $xy(Z^{p-1} - Z^{p-2} + \cdots + 1) \equiv 1 \mod p^p$, which proves the lemma with $Y = y(Z^{p-1} - Z^{p-2} + \cdots + 1)$.

**Proof of the statement.** $i(i-1)\cdots(i-x+1)$ is a degree $x$ polynomial with integer coefficients. And the denominator in $C_i^x$ is invertible modulo $p^p$ due to the following lemma: $\frac{1}{x!} \equiv \Pi := \prod_{1\le\xi\le x} Y(\xi) \mod p^p$, where the $Y(\xi)$ are integer for $\xi < p$; thus, $C_i^x \equiv \Pi \cdot i(i-1)\cdots(i-x+1) \mod p^p$.

Another problem for "Kvant": ($\forall a, b$ the tori $T_{a\times b}$ are covered by rectangles $m \times 1$ and $1 \times n$ if and only if $a$ is a multiple of $m$ or $b$ is a multiple of $n$) $\Longleftrightarrow$ ($m = p^k$, $n = p^l$, $p$ is prime), see paper [2].

Related new extensions of the small Fermat theorem: 1) $(\mathrm{tr} A)^p - \mathrm{tr}(A^p)$ is divisible by the prime number $p$ for any matrix $A \in \mathrm{SL}(n, \mathbb{Z})$ with integer elements and determinant 1; 2) $(\lambda_1 + \cdots + \lambda_n)^p - (\lambda_1^p + \cdots + \lambda_n^p) = pF(\sigma_1, \ldots, \sigma_p)$ for some polynomial $F$ in the variables $\sigma_1,\dots,\sigma_p$ with integer coefficients, where $\sigma_i$ is the $i$-th basic symmetric function of $\lambda_1,\dots,\lambda_n$. The proofs are contained in paper [1].

[1] ARNOLD V. I. Fermat dynamics, matrix arithmetics, finite circles and finite Lobachevsky planes. *Funct. Anal. Appl.*, 2004, **38**(1), 20 pp.

[2] REMILA É. Sur le pavage de tore $T_{a\times b}$ par $h_m$ et $v_n$. *C. R. Acad. Sci. Paris, Sér.* I *Math.*, 1993, **316**(9), 949–952.

## 1993-20 — *V. I. Arnold* (1993)

R. Fintushel and R. Stern [1] define Floer homology with *integer* (not mod 8) calibration, but the answers anyway fail to "fit" into any Newton polyhedra (therefore, apparently, something very nontrivial should be done to the polyhedron). For example, they found the following Poincaré polynomials for the sums of degrees:

| | | |
|---|---|---|
| 2 3 5 | $t+t^5$ | 1 0 1 |
| 2 3 11 | $t+t^3+t^5+t^7$ | 1 1 1 1 |
| 2 3 17 | $t+t^3+2t^5+t^7+t^9$ | 1 1 2 1 1 |
| 2 3 23 | $t+2t^3+2t^5+2t^7+t^9$ | 1 2 2 2 1 |
| 2 3 29 | $t+t^3+3t^5+2t^7+2t^9+t^{11}$ | 1 1 3 2 2 1 |
| 2 3 35 | $t+2t^3+3t^5+3t^7+2t^9+t^{11}$ | 1 2 3 3 2 1 |
| 2 3 41 | $t+t^3+4t^5+3t^7+3t^9+2t^{11}$ | 1 1 4 3 3 2 |
| 2 3 47 | $t+2t^3+3t^5+4t^7+3t^9+2t^{11}+t^{13}$ | 1 2 3 4 3 2 1 |
| 2 3 53 | | 1 1 4 4 4 3 1 |
| 2 3 59 | | 1 2 3 5 4 3 2 |
| 2 3 65 | | 1 1 4 5 5 4 2 |
| 2 3 71 | | 1 2 3 5 5 4 3 1 |

| | |
|---|---|
| 2 3 7 | $1t^{-1}+0t+1t^3$ |
| 13 | $1t^{-1}+1t+1t^3+1t^5$ |
| 19 | 2 1 2 1 |
| 25 | 2 2 2 2 |
| 31 | 2 2 3 2 1 |

| | |
|---|---|
| 37 | 1 3 3 2 1 |
| 43 | 1 4 4 4 3 |
| 49 | 2 3 5 4 3 1 |
| 3 4 13 | $t^{-5}+2t^{-3}+3t^{-1}+2t+2t^3$ |
| 4 5 21 | $4t^{-9}+5t^{-7}+8t^{-5}+6t^{-3}+5t^{-1}+t+t^3$ |
| 5 6 31 | $2t^{-15}+10t^{-13}+11t^{-11}+15t^{-9}+3t^{-7}+8t^{-5}+$ $+5t^{-3}+4t^{-1}+t+t^3$ |
| 2 3 39 | $2t^{-1}+3t+4t^3+4t^5+3t^7$ |

What are these numbers? Can one really guess (I would try to prove something about the asymptotic form in the series $[2,3,6k-1]$, $k\to\infty$)?

But, whereas in the work of Fintushel–Stern the idea of calculation is more or less clarified, it seems that paper [2] still does not permit the answers to be found.

[1] FINTUSHEL R., STERN R. J. Integer graded instanton homology groups for homology three-spheres. *Topology*, 1992, **31**(3), 589–604.

[2] LESCOP C. Sur l'invariant de Casson–Walker: formule de chirurgie globale et généralisation aux variétés de dimension 3 fermées orientées. *C. R. Acad. Sci. Paris, Sér.* I *Math.*, 1992, **315**(4), 437–440.

## 1993-24 — *V. I. Arnold*

**Observation 1** ([1], p. 39, 81 and Fig. 28)**.** The Maxwell stratum $A_4$, like the caustic, is the swallowtail (Fig. 1).

A. B. Givental (in paper [3] on the mapping of periods written jointly with A. N. Varchenko) generalized this observation to the many-dimensional case of singularities $A_{2k}$ for arbitrary $k$; the manifolds of polynomials with many double roots and polynomials with a root of large multiplicity are diffeomorphic and even Lagrange equivalent:

$$\{(x^a+p_1x^{a-1}+\cdots+p_a)^2(x-2p_1)\}\sim$$
$$\sim\{(x-u)^{a+1}(x^a+(a+1)ux^{a-1}+v_1x^{a-2}+\cdots+v_{a-1})\}.$$

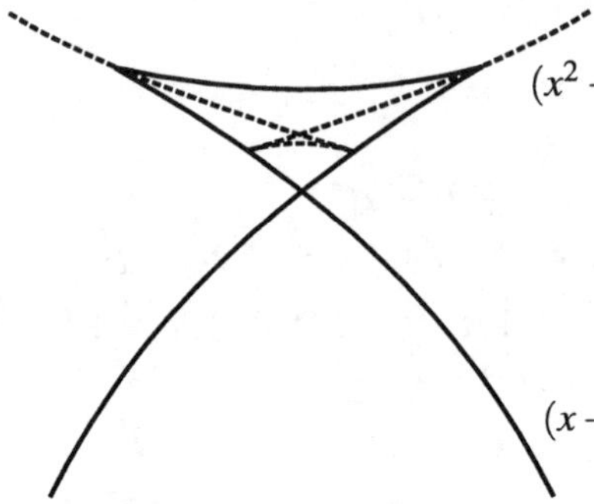


$$\{x^5+ax^3+bx^2+cx+d=(x^2+px+q)^2(x-2p)=(x-u)^3(x^2+3ux+v)\}$$

**Fig. 1:** The section by the plane $a=-1$ of the caustic (continuous line) and the Maxwell strata (dotted line) of the family $x^5+ax^3+bx^2+cx$ (the constant term $d$ is discarded)

**Observation 2** ([1], p. 141–142 and Fig. 91). For $B_4$ (or $C_4$) the caustic and the Maxwell stratum are also diffeomorphic (the similar assertion for the singularities $B_{2k}$ and $C_{2k}$ with arbitrary $k$ was proved by F. Napolitano [2]). In the case of $C_4$ the family of functions

$$C_4 \sim \{f+ax^3+by^2+cy,\ f=x^4+xy, \text{ the boundary } x=0\},$$

Fig. 2 depicts

the caustic: $\{y^4+ay^3+by^2+cy$ with a non-Morse critical point $(A^2)$ or zero $(C_2)\}$,

the Maxwell stratum: $\begin{cases} A_1A^1, & \text{coincidence of the boundary and internal critical values,} \\ 2A^1, & \text{coincidence of the two boundary critical values.} \end{cases}$

We have:

$$\frac{A^2}{C_2}=\frac{A^1A_1}{2A^1}. \quad \text{In the section } a=\text{const.}$$

A generalization is available near medium dimension, where the manifolds can be Lagrangian or Legendrian.

Cf. also problems *1985-24* and *1996-4*.

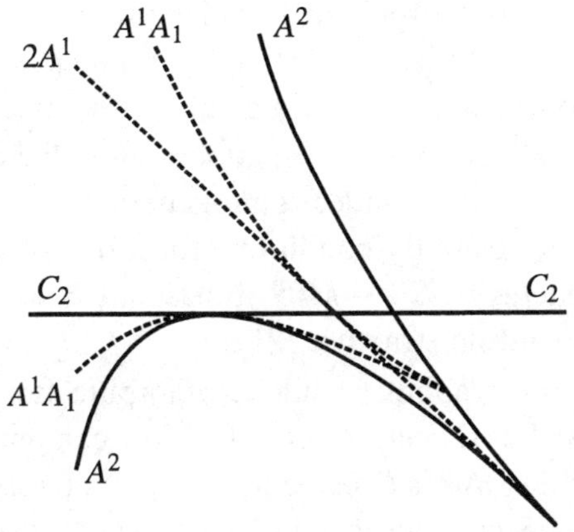


**Fig. 2:** The bifurcation diagram of the boundary singularity $C_4$. Components of the caustic are shown by continuous lines, and Maxwell strata by dotted lines


[1] ARNOLD V. I. Singularities of Caustics and Wave Fronts. Dordrecht: Kluwer Acad. Publ., 1990. (Math. Appl., Sov. Ser., 62.)

[2] NAPOLITANO F. Duality between the generalized caustic and Maxwell stratum for the singularities $B_{2k}$ and $C_{2k}$. *C. R. Acad. Sci. Paris, Sér.* I *Math.*, 1997, **325**(3), 313–317.

[3] VARCHENKO A. N., GIVENTAL A. B. Mapping of periods and intersection form. *Funct. Anal. Appl.*, 1982, **16**(2), 83–93.


## 1993-25 — *B. S. Kruglikov* Also: *1993-1*

ℜ A pair $(M^{2n},J)$ is called an *almost complex manifold* if the operator field $J \in T^*M \otimes TM$ satisfies the condition $J^2 = -\mathbf{1}$. A submanifold $N \subset M$ is called *pseudoholomorphic* if its tangent bundle $TN \subset TM$ is invariant under the action of $J$. In general, an almost complex manifold has no pseudoholomorphic submanifolds $N$ of dimension $\dim N > 2$. An obstruction for this is the Nijenhuis tensor of the structure, which is not zero on subspaces of the tangent space with dimension $\geq 4$. But pseudoholomorphic curves $N^2 \subset M^{2n}$ do exist and they play an important role in investigation of the almost complex manifolds.

Such curves appeared for the first time in paper [10]. It was proved there that there exists a small pseudoholomorphic disk $u : (D^2_\varepsilon, J_0) \to (M^{2n}, J)$, $u(0) = p$, $u_*(0)e = v \in T_pM$, through every point $p \in M$ in any direction $v \in T_pM$, where $e \in T_0D^2_\varepsilon \simeq \mathbb{C}$ is the unit vector. Note that, on almost complex manifolds, not necessarily small pseudoholomorphic disks appear in families. Every big pseudoholomorphic disk $u: D_R \to M^{2n}$ has in its neighborhood pseudoholomorphic disks of almost the same size $\tilde{u}: D_{R-\varepsilon} \to M^{2n}$ in nearby directions [6].

The first results about compact pseudoholomorphic disks appeared in paper [3]. It was shown there that, under some genericity conditions for an almost complex structure $J$ and positivity conditions for a homology class $A \in H^2(M)$, pseudoholomorphic spheres $u: S^2 \to M^{2n}$ of class $[u(S^2)] = A$ appear in families which have a smooth manifold structure [8].

Arnold's question is about pseudoholomorphic tori $T^2 \subset M^{2n}$. Even in the complex case [1] such tori occur discretely. How can one recognize whether a given pseudoholomorphic torus is isolated in the set of all tori of a given homology class or belongs to some parametric family? In [1] (§ 27) Arnold associates a pair of numbers $(\lambda, \omega)$ with every holomorphic torus (elliptic curve) in a complex surface. In the case when this pair is nonresonant there are no other elliptic curves of the same homology class in a neighborhood. It is important to note that such a pair no longer appears in the almost complex case. Actually, the number $\omega$ is an invariant of the structure $J$ restricted to the torus $T^2 = T^2(2\pi, \omega)$ and so it is defined canonically. But $\lambda$ in the complex case fixes the rule for the gluing of charts of the normal bundle. In the complex situation the normal bundle of an elliptic curve considered in [1] is given by the 1-jet of the structure on the curve (here the structure of complex multiplication is $J_0$ or $i$). In the almost complex situation the 1-jet of an almost complex structure $J$ on the submanifold $T^2 \subset M^{2n}$ is given by the Nijenhuis tensor [5] along the submanifold. So, firstly, the notion of the normal bundle is different and $\lambda$ is not defined and, secondly, it would not be in any sense sufficient for the characterization of the 1-jet of $J$.

Thus, the study of the 1-jet of an almost complex structure $J$ leads to the investigation of Nijenhuis tensors field $N_J$ along the pseudoholomorphic torus $T^2 \subset (M^4, J)$. This field can be an arbitrary field of $J$-antilinear skew-symmetric $(2,1)$-tensors [5]. So one can expect no canonical normal form of the germ of the structure $J$ in coordinates, similar to the form of a complex structure obtained in [1]. This is especially well-illustrated if we consider the 2-jet of an almost complex operator on the torus $T^2$. Actually, the investigation of Nijenhuis tensor structure in dimension 4 shows that, with a nondegenerate field of tensors $N_J$ on $M^4$, a distribution $\Pi^3 \subset TM$ with transversal measure is associated [5]. Intersection of this distribution with the tangent space to the torus $T^2$ gives a foliation of the torus with transversal measure. In particular, the rotation number and a pair of invariant functions are canonically defined. So germs of neighborhoods of pseudoholomorphic tori have natural moduli. Obstructions for pseudoholomorphic tori deformations should be sought in terms of the invariants obtained with this approach.

The cited paper of Moser [9] considers KAM-theory for pseudoholomorphic foliations of an almost complex torus $T^{2n}$. It is clear that there can be no pseudoholomorphic foliations by tori (note however Kuksin's results [7] on the existence of pseudoholomorphic tori in a fixed homology class for special types of structures $J$). Therefore, Moser considers foliations of the torus $(T^4, J)$ by pseudoholomorphic lines $u: \mathbb{C} \to T^4$. It is stated in his main result that, for a small perturbation $J$ of the standard complex structure $J_0$ on the $T^4$, most leaves of the prescribed foliation $u_\alpha: \mathbb{C} \to T^4$ persist, i. e., deform to pseudoholomorphic lines for the structure $J$.

Moser poses a question to what extent his theory remains true. A partial answer to the question is a theorem of Bangert [2], which states that some leaves persist if the perturbation is such that the new structure $J$ is still tamed by the standard symplectic form $\omega_0$, i. e., $\omega_0(X, JX) > 0$ for any vector $X \neq 0$. In other words, the existence of nontrivial maps of complex lines $u: \mathbb{C} \to (T^{2n}, J)$ for subordinate structures $J$ is proved. Bangert uses Brody's reparametrization lemma. In complex analysis this lemma is applied to obtain a criterion of hyperbolicity for manifolds. The corresponding theory for almost complex manifolds was constructed in paper [4]. In the journal version of the paper there is another proof of Bangert's theorem which, by means of an analog of Brody's complex criterion, can be reformulated as the non-hyperbolicity of $(T^{2n}, J)$. The construction permits to "see" the direction of the constructed curve $u(\mathbb{C})$. Moreover, there are as many such directions as holomorphic tori $T^2$ in the standard complex torus $(T_0^{2n}, J_0) = \mathbb{C}^n/\mathbb{Z}^{2n}$. In other words, we get a more general theorem, which states the existence of an entire pseudoholomorphic curve $u: \mathbb{C} \to (T^{2n}, J)$ through every point in the dierection given by an arbitrary $J_0$-invariant lattice $\mathbb{Z}^2 \subset H_2(T_0^{2n}; \mathbb{Z})$, and different lines correspond to different directions.

Applying an analogous technique for investigations of neighborhoods of pseudoholomorphic tori $T^2 \subset (M^4, J)$, we can show that this neighborhood is "foliated" into pseudoholomorphic cylinders. Moreover, this foliation in some cases is equivalent to the holomorphic foliation of an elliptic curve neighborhood, which is given in the canonical Arnold coordinates $(r, \varphi)$ by the equation $r = \text{const}$. This is one of the generalizations of Floquet-type theory [1] for the almost complex case. Note however, that attempts to construct complex coordinates even on a single pseudoholomorphic transversal to the torus lead to some interesting questions of monodromy of the given pseudoholomorphic foliation, which are alien for the complex situation.

Everything in the above discussion mainly concerns neighborhoods of the torus with zero self-intersection $T^2 \cdot T^2 = 0$. In the case of negative self-intersection, the impossibility of the curve deformation follows from the positivity of intersections of pseudoholomorphic curves [3, 8]. Grauert's theorem in its original form is wrong in the almost complex category; it requires a modification similar to what was discussed above. For positive neighborhoods ($T^2 \cdot T^2 > 0$) the Riemann–Roch theorem predicts the existence of deformations of pseudoholomorphic tori, but this is still a conjecture.

[1] ARNOLD V. I. Geometrical Methods in the Theory of Ordinary Differential Equations, 2nd edition. New York: Springer, 1988. (Grundlehren der Mathematischen Wissenschaften, 250.) [*The Russian original* 1978.]

[2] BANGERT V. Existence of a complex line in tame almost complex tori. *Duke Math. J.*, 1998, **94**(1), 29–40.

[3] GROMOV M. L. Pseudoholomorphic curves in symplectic manifolds. *Invent. Math.*, 1985, **82**(2), 307–347.

[4] KRUGLIKOV B. S., OVERHOLT M. The Kobayashi pseudodistance on almost complex manifolds. *Differ. Geom. Appl.*, 1999, **11**, 265–277. [*Internet*: `http://www.arXiv.org/abs/math.DG/9703005`]

[5] KRUGLIKOV B. S. Nijenhuis tensors and obstructions to constructing pseudoholomorphic mappings. *Math. Notes*, 1998, **63**(4), 476–493.

[6] KRUGLIKOV B. S. Existence of close pseudoholomorphic disks for almost complex manifolds and their application to the Kobayashi–Royden pseudonorm. *Funct. Anal. Appl.*, 1999, **33**(1), 38–48.

[7] KUKSIN S. B. Pseudoholomorphic 2-tori in $T^4$. *Differ. Geom. Appl.*, 1996, **6**(2), 109–119.

[8] MCDUFF D., SALAMON D. *J*-Holomorphic Curves and Quantum Cohomology. Providence, RI: Amer. Math. Soc., 1994. (University Lecture Series, 6.)

[9] MOSER J. On the persistence of pseudoholomorphic curves on an almost complex torus (with an appendix by PÖSCHEL J.). *Invent. Math.*, 1995, **119**(3), 401–442.

[10] NIJENHUIS A., WOOLF W. B. Some integration problems in almost-complex and complex manifolds. *Ann. Math., Ser.* 2, 1963, **77**(3), 424–489.

## 1993-27 — *F. Napolitano*

Also: *1980-14, 1981-12, 1984-15, 1988-16*

ℜ The *pseudo-homology theory* suggested in this problem was defined in [1].

To each loop $\gamma \in F_{i+1}$ there is associated the subgroup $\partial_{i+1}\gamma$ of $F_i$ generated by the elements $(A_\gamma\phi)\phi^{-1}$, $\phi \in F_i$. The union of these subgroups over all $\gamma$ in $F_{i+1}$

is equal to the normal subgroup $R_i$ of relations in $\Gamma_i$. The "kernel" of the boundary operator $\partial_{i+1}$ is the subgroup of $F_{i+1}$ of all elements whose image under $\partial_{i+1}$ is the trivial subgroup. By Arnold's "Poincaré lemma," the kernel of $\partial_i$ contains the subgroup $R_i$. Moreover, its image under $A$ is a subgroup of $\widehat{R}_i$.

Taking the quotient of the kernel by the image, as in the usual homology theory, we define the pseudo-homology groups of $\Sigma_0$: $\mathcal{H}_i(\Sigma_0) = \ker\partial_i/R_i$. The pseudo-homology groups of $\Sigma_0$ depend only on the isotopy class of $\Sigma_0$. In particular, they depend neither on the choice of generic projections nor on the base points of loop groups considered. The first pseudo-homology group $\mathcal{H}_0(\Sigma_0)$ is the Poincaré group of the complement to $\Sigma_0$ (near 0). Examples of hypersurfaces with nontrivial $n$-th pseudo-homology groups, $n > 0$, can easily be constructed (for instance, consider the "complex $n$-sphere" $\{(z_1, \ldots, z_{n+1}) \in \mathbb{C}^{n+1} : z_1^2 + \cdots + z_{n+1}^2 = 1\}$).

[1] NAPOLITANO F. Pseudo-homology of complex hypersurfaces. *C. R. Acad. Sci. Paris, Sér.* I *Math.*, 1999, **328**(11), 1025–1030.

## ▽ 1993-28 – *V. I. Arnold* (1993)

ℜ The difficulty of Cartan's activity in this area is reasonably clarified by the fact that he does not precisely formulate the conditions under which "the method works" but affirms instead that it works "always" (likewise, in *algebraic* geometry Puiseux series, the theorem on zeros, etc., always work as well).

An attempt to reduce Cartan's theory to a form similar to algebraic geometry was made by Griffiths and his disciples [1].

But, it seems to me, one should[1] clear up Cartan's theory in the framework of singularity theory, i. e., dropping first the cases of codimension 1, then 2, ..., then all cases of infinite codimension; for the latter there must exist a certain hierarchy in every class of degeneracy (or class of systems distinguished by special conditions like those distinguishing Hamiltonian equations).

As far as I know, *nothing* has been accomplished in this field, and hence a researcher smart enough to explore it would probably find here many new things at once.

[1] GRIFFITHS P. Exterior Differential Systems and the Calculus of Variations. Boston, MA: Birkhäuser, 1983. (Progr. Math., 25.)

[1] especially for a transition from analytic coefficients of equations to (generic) smooth ones

## △ 1993-28 – *B. S. Kruglikov*

ℛ Cartan's theory of differential systems has a dual counterpart—the theory of jets [7]. While the first approach is still popular, many results have formulations for differential equations considered as submanifolds $\mathcal{E} \subset J^k(\mathbb{R}^n, \mathbb{R}^m)$. In this approach, formal integrability theory can be extended to a general scheme of singularities of solutions [4]. The number of prolongations to resolve the singularity is determined by the corresponding Spencer $\delta$-complex, as in the theory of formal integrability.

The first problem in the classification of singularities of type $\Sigma_l$ is the description of orbits of the action of the group $\mathrm{GL}_l(\mathbb{R}) \oplus \mathrm{GL}_m(\mathbb{R})$ on $S^k(\mathbb{R}^l) \otimes \mathbb{R}^m$ [1]. Since the difference of dimensions $l^2 + m^2 - m\binom{k+l-1}{k}$ is negative for $l > 1$ and $k \geq k_0(l,m)$, the problem certainly has moduli. These moduli are of algebraic nature and lead to Jordan algebras, Clifford algebras etc., see [5].

In contrast, singularities of type $\Sigma_1$ can be studied with much success. For example, a (nonlinear) elliptic equation cannot have a solution with singularity $\Sigma_1$ (a classical statement in the linear theory [1]). Also, a wave front of the solution propagates along generalized bicharacteristics [2], and any singularity of type $\Sigma_1$ is realized for some differential equation [3] etc.

But, for $l > 1$ and a small order $k \leq k_0$ of the PDE's, there is a hope of obtaining a classification, see, for instance, [6] (note that though the results seem to be true, the main technical tool there—the transversality theorem—has a gap in proof).


[1] KRASIL'SHCHIK I. S., LYCHAGIN V. V., VINOGRADOV A. M. Geometry of Jet Spaces and Nonlinear Partial Differential Equations. New York: Gordon and Breach, 1986. (Adv. Studies in Contemp. Math., 1.)

[2] KRISHCHENKO A. P. The structure of the singularities of the solutions of quasilinear equations. *Uspekhi Mat. Nauk*, 1976, **31**(3), 219–220 (in Russian).

[3] KRISHCHENKO A. P. On the bends of *R*-manifolds. *Moscow Univ. Math. Bull.*, 1977, **32**(1), 13–16.

[4] LYCHAGIN V. V. Singularities of multivalued solutions of nonlinear differential equations, and nonlinear phenomena. *Acta Appl. Math.*, 1985, **3**(2), 135–173.

[5] LYCHAGIN V. V. Geometric theory of singularities of solutions of nonlinear differential equations. In: Itogi Nauki i Tekhniki VINITI. Problems in Geometry, Vol. 20. Moscow: VINITI, 1988, 207–247 (in Russian). [*The English translation*: *J. Sov. Math.*, 1990, **51**(6), 2735–2757.]

[6] RAKHIMOV A. KH. Singularities of Riemannian invariants. *Funct. Anal. Appl.*, 1993, **27**(1), 39–50.

[7] SPENCER D. C. Overdetermined systems of linear partial differential equations. *Bull. Amer. Math. Soc.*, 1969, **75**, 179–239.

## 1993-29 – *A. A. Glutsyuk*

ℜ The statement of the problem is true on the plane ($n = 2$) and is wrong in higher dimensions ($n \geq 3$).

The problem was stated by Lawrence Markus and Hidehiko Yamabe [16] as a conjecture in 1960. The two-dimensional case was investigated by C. Olech [20, 21], G. Meisters [20], P. Hartman [15], S. S. Anisov and others under some additional assumptions on the vector field. Olech [21] proved the global stability subject to the conditions of the problem under the assumption that the norm $\|v\|$ of the field $v$ is bounded away from zero outside some compact set. Later Olech and Meisters in their joint paper [20] proved the global stability of a two-dimensional polynomial vector field that satisfies the conditions of the problem. Hartman [15] proved the global stability of a two-dimensional vector field subject to the conditions of the problem under the assumption that the integral

$$\int_0^\infty \left( \min_{x^2+y^2=r^2} \|v(x,y)\| \right) dr$$

diverges.

The result of Olech [21] reduces the two-dimensional Markus–Yamabe problem to the following one: Is it true that subject to the conditions of the problem the mapping given by the coordinate representation of the vector field is injective? The positive answer to this question was obtained by S. S. Anisov in 1993 in an unpublished paper under the assumption that this mapping is a covering and under weakened conditions of the problem: one requires only that the Jacobian matrix has no real positive eigenvalues. The results of Anisov are reproduced in Section 3 of paper [10].

Two partial results for $n = 2$ were obtained in [23]. One of them is the global stability subject to the conditions of the two-dimensional problem under the assumption that at least one component of the field is a rational function.

The positive solutions of the two-dimensional problem in the general case were obtained in 1993 simultaneously and independently by C. Gutiérrez (presented at the conference on dynamical systems in Rio-de-Janeiro (August 1993) and at a conference in Trento [13] (September 1993), published in [14]), R. Fessler (presented at the conference in Trento [7], published in [8]) and a little later in the same year by the author of the present comment (announced with a brief proof in [9], the complete text of the proof being published in [10]). The proof in [10] is shorter than the two previous ones. On the other hand, Gutiérrez and Fessler proved more general theorems that imply the following statement. *Let a mapping* $\mathbb{R}^2 \to \mathbb{R}^2$ *be a local diffeomorphism everywhere whose Jacobian matrix does not have positive eigenvalues outside some compact set in the plane. Then this mapping is injective.*

A little later a positive solution of the two-dimensional problem was obtained by the Chinese mathematicians P. Chen, J. He, H. Qin [3,4].

The Markus–Yamabe problem in higher dimensions ($n \geq 3$) was investigated by N. E. Barabanov [1], J. Bernat and J. Llibre [2] and others. In 1988 Barabanov [1] made an attempt to construct a four-dimensional counterexample to the Markus–Yamabe conjecture. Later Bernat and Llibre found mistakes in his paper. In 1994 Bernat and Llibre constructed a valid four-dimensional counterexample using Barabanov's ideas (a four-dimensional counterexample generates counterexamples in higher dimensions). A three-dimensional counterexample to the Markus–Yamabe conjecture was constructed by the author of the present comment in 1955. It was announced at the conference on dynamical systems in Trieste (May 1995) and at the conference "Differential equations and related questions" (XVIII joint meeting of Petrovskiĭ seminar and Moscow Mathematical Society) in Moscow (April 1996) [11]. The complete text of the proof is presented in preprint [12].

A little later the same year (1995) A. Cima, A. van den Essen, A. Gasull, E. Hubbers and F. Mañosas [11] constructed a polynomial three-dimensional counterexample to the Markus–Yamabe conjecture (published in [6]). This counterexample is written explicitly in [5,6,11,12]:

$$\dot{x} = -x + z(x+yz)^2, \quad \dot{y} = -y - (x+yz)^2, \quad \dot{z} = -z;$$

the eigenvalues of the Jacobian matrix are constant and equal to $-1$; the origin is the unique singular point; there is an unbounded trajectory

$$x(t) = 18e^t, \quad y(t) = -12e^{2t}, \quad z(t) = e^{-t}.$$

In papers [5,6] a polynomial counterexample to the discrete version of the Markus–Yamabe conjecture was also constructed: a polynomial authomorphism $\mathbb{R}^3 \to \mathbb{R}^3$ whose Jacobian matrix has eigenvalues of module less than one everywhere (more precisely, these eigenvalues are constant and equal to $\frac{1}{2}$) with a fixed point that is not globally attractive (there is an orbit that goes to infinity).

For other results related to the Markus–Yamabe problem, see papers [6, 14] and the bibliography to them; see also reviews [22] (this is the review for papers [8, 10, 14]), [17] (this is the review for paper [2]), [18] (this is the review for paper [23]) and [19] (this is the review for paper [6]) in *Math. Reviews*.

The following version of the Markus–Yamabe problem remains open: is it true that every vector field of the type $\dot{x} = c - x + H(x)$, $x \in \mathbb{R}^n$, where $H$ is a homogeneous vector polynomial of degree 3 with a nilpotent Jacobian matrix and $c = \text{const}$, has at most one singular point?

This problem is equivalent to the well-known Jacobian problem (see [17]).


[1] BARABANOV N. E. On the Kalman problem. *Sib. Math. J.*, 1988, **29**(3), 333–341.

[2] BERNAT J., LLIBRE J. Counterexample to Kalman and Markus–Yamabe conjecture in dimension larger than 3. *Dynam. Contin. Discrete Impuls. Systems*, 1996, **2**(3), 337–379.

[3] CHEN P. N., HE J. X., QIN H. S. A proof of the Jacobian conjecture on global asymptotic stability. *Chinese Sci. Bull.*, 1996, **41**(15), 1233–1237.

[4] CHEN P. N., HE J. X., QIN H. S. A proof of the Jacobian conjecture on global asymptotic stability. *Acta Math. Sinica* (*Engl. Ser.*), 2001, **17**(1), 119–132.

[5] CIMA A., VAN DEN ESSEN A., GASULL A., HUBBERS E., MAÑOSAS F. A polynomial counterexample to the Markus–Yamabe conjecture. Rep. 9551, Dept. Math. Univ. Nijmegen, Nijmegen, 1995.

[6] CIMA A., VAN DEN ESSEN A., GASULL A., HUBBERS E., MAÑOSAS F. A polynomial counterexample to the Markus–Yamabe conjecture. *Adv. Math.*, 1997, **131**(2), 453–457.

[7] FESSLER R. A solution to the global asymptotic stability Jacobian conjecture and a generalization. In: Recent Results on the Global Asymptotic Stability Jacobian Conjecture (Editor: M. Sabatini), Matematica 429, Università di Trento, 1994. Workshop, I-38050 POVO (TN) Italy, September 14–17, 1993.

[8] FESSLER R. A proof of the two-dimensional Markus–Yamabe stability conjecture and a generalization. *Ann. Polon. Math.*, 1995, **62**(1), 45–74.

[9] GLUTSYUK A. A. A complete solution of the Jacobian problem for vector field on the plane. *Russian Math. Surveys*, 1994, **49**(3), 185–186.

[10] GLUTSYUK A. A. Asymptotic stability of linearizations of a planar vector field with a singular point implies global stability. *Funct. Anal. Appl.*, 1995, **29**(4), 238–247.

[11] GLUTSYUK A. A. A three-dimensional counterexample to the Markus–Yamabe global stability problem. *Uspekhi Mat. Nauk*, 1996, **51**(5), 169 (in Russian).

[12] GLUTSYUK A. A. Asymptotic stability of linearizations of a vector field in $\mathbb{R}^3$ with a singular point does not imply global stability. Preprint, Comunicaciónes del CIMAT, Guanajuato, Mexico, 1996.

[13] GUTIÉRREZ C. A solution to the bidimensional global asymptotic stability conjecture. In: Recent Results on the Global Asymptotic Stability Jacobian Conjecture (Editor: M. Sabatini), Matematica 429, Università di Trento, 1994. Workshop, I-38050 POVO (TN) Italy, September 14–17, 1993.

[14] GUTIÉRREZ C. A solution to the bidimensional global asymptotic stability conjecture. *Ann. Institut Henri Poincaré, Analyse non linéaire*, 1995, **12**(6), 627–671.

[15] HARTMAN P. Ordinary Differential Equations. New York – London – Sydney: John Wiley, 1964, Ch. XIV, Part III.

[16] MARKUS L., YAMABE H. Global stability criteria for differential systems. *Osaka Math. J.*, 1960, **12**, 305–317.

[17] MEISTERS G. H. Featured Review 98c:34079. *Math. Reviews*, 1998.

[18] MEISTERS G. H. Featured Review 98c:34080. *Math. Reviews*, 1998.

[19] MEISTERS G. H. Featured Review 98k:34084. *Math. Reviews*, 1998.

[20] MEISTERS G. H., OLECH C. Solution of the global asymptotic stability Jacobian conjecture for the polynomial case. In: Analyse Mathématique et Applications (Contributions en l'honneur de J.-L. Lions). Paris: Gauthier-Villars, 1988, 373–381.

[21] OLECH C. On the global stability of an autonomous system on the plane. *Contrib. Differ. Equations*, 1963, **1**(3), 389–400.

[22] OLECH C. Featured Review 96k:34099a,b,c. *Math. Reviews*, 1996.

[23] PARTHASARATHY T., SABATINI M. Some new results on the global asymptotic stability Jacobian conjecture. *Bull. Polish Acad. Sci. Math.*, 1993, **41**(3), 221–228 (1994).

## 1993-30 — *Yu. M. Baryshnikov*

The Stokes phenomenon is the bifurcation of a saddle-saddle connection, i. e., trajectory connecting in a non-generic way two zeros of a vector field. The Stokes phenomenon is important in many areas, such as the asymptotic expansion of integrals, asymptotics of singular differential equations, Morse theory, ...

A Stokes set is the corresponding bifurcation diagram. More precisely, consider a family of vector fields $v_\lambda$ on a smooth manifold $M^m$ which has only a finite number of isolated critical points of the same index $k$ (such that both $k$ and $m-k$ are positive), and which, generically, do not have any trajectories tending to different critical points, as time goes to $\pm\infty$ ("instantons").

As the parameter $\lambda$ varies, instantons appear at a subset $\Sigma \subset \Lambda$ of real codimension 1, which is referred to as the *Stokes set*. Most studies of the Stokes set deal with the families of gradient vector fields of the real part of the analytic function $f_\lambda$ in one complex variable depending on the parameter $\lambda$.

Wright, and then Berry and his students, published several papers (see, for example, [4, 8]) studying systematically the situation where $f_\lambda$ is a family depending on a small number of real parameters. Berry also investigated the asymptotics near the Stokes lines, see [3].

Kostov described the singularities of the Maxwell stratum for the *imaginary part* of $f$, a singular variety of real codimension 1 which contains the Stokes set, see [6]; this variety has been also studied by Lando.

A "combinatorial model" of the Stokes sets for the monic (complex) polynomes in one variable of a fixed degree $n+1$ and with vanishing sum of roots was studied in [2]. It turned out that the number of components of the complement $\mathbb{C}^{n-1}-\Sigma$ of the Stokes set is $\frac{1}{2n+1}\binom{3n}{n}$ and that these components are contractible.

The links of the Stokes sets outside the bifurcation diagram (parameters corresponding to multiple critical values) are homeomorphic to fans dual to certain convex polyhedra. These polyhedra form a finite family (for each $n$) including the famous Stasheff polyhedron (convex polyhedron whose vertices correspond to triangulations of convex $n+2$-gon and, more generally, whose faces correspond to subdivisions of convex $n+2$-gon into convex polygons).

Other instances where Stasheff polyhedra appear hidden in Stokes-like bifurcation diagrams are described in [1, 5].


[1] Baryshnikov Yu. M. Bifurcation diagrams of quadratic differentials. *C. R. Acad. Sci. Paris, Sér.* I *Math.*, 1997, **325**(1), 71–76.

[2] Baryshnikov Yu. M. On Stokes sets. In: New Developments in Singularity Theory (Cambridge, 2000). Editors: D. Siersma, C. T. C. Wall and V. M. Zakalyukin. Dordrecht: Kluwer Acad. Publ., 2001, 65–86 (NATO Sci. Ser. II Math. Phys. Chem., 21.)

[3] Berry M. V. Stokes' phenomenon; smoothing a Victorian discontinuity. *Inst. Hautes Études Sci. Publ. Math.*, 1988, **68**, 211–221.

[4] Berry M. V., Howls C. J. Stokes surfaces of difraction catastrophes with codimension three. *Nonlinearity*, 1990, **3**(2), 281–291.

[5] Kapranov M. M., Saito M. Hidden Stasheff polytopes in algebraic $K$-theory and in the space of Morse functions. In: Higher Homotopy Structures in Topology and Mathematical Physics. Proceedings of the International Conference held in honor of the 60th birthday of Jim Stasheff (Poughkeepsie, NY, 1996). Editor: J. McCleary. Providence, RI: Amer. Math. Soc., 1999, 191–225. (Contemp. Math., 227.)

[6] Kostov V. P. On the stratification and singularities of the Stokes hypersurface of one- and two-parameter families of polynomials. In: Theory of Singularities and its Applications. Editor: V. I. Arnold. Providence, RI: Amer. Math. Soc., 1990, 251–271. (Adv. Sov. Math., 1.)

[7] Lando S. K. Geometry of the Stokes sets of families of functions of one variable. *J. Math. Sci.*, 1997, **83**(4), 534–538.

[8] WRIGHT F. J. The Stokes set of the cusp diffraction catastrophe. *J. Phys. A*, 1980, **13**(9), 2913–2928.

### 1993-31

ℛ See the comment to problem *1981-18*.

### ▽ 1993-33 – *V. I. Arnold* (1993)

ℛ For example, the Gauss statistics can be (I think) obtained in the following way: choose a point $(x_1, x_2) \in \mathbb{Z}_+^2$ and build a continued fraction (the envelope of the convex hull of integer points of the quadrant without 0 strictly below the line joining our point to the origin).

The elements of our continued fraction are the numbers of integer points on the sides of this hull (and its upper counterpart). For the statistic, any of these two can be taken. We claim that, for $N \to \infty$, the average distribution of 1 anywhere in the fraction, or 2, or any intervals $[a_1, a_2, \ldots, a_k]$, on (indivisible, i. e., having relatively prime coordinates $x_1$ and $x_2$) points satisfying $x_1^2 + x_2^2 \leq N^2$ (or $x_1 \leq N$, $x_2 \leq N$, or $x_1 + x_2 \leq N$—it does not matter), converges to that evaluated with the Gauss measure. We recall that the Gauss measure is invariant under the endomorphism $z \mapsto \frac{1}{z-[z]}$ of the semiaxis $z > 1$, and the invariance = (the measure of the *pre*image equals the measure of the image). Gauss found the density of such measure, $\rho(z)\,dz$; here $z = x_2/x_1$.

This result needs be generalized to $\mathbb{R}^n$ instead of $\mathbb{R}^2$. The measure is taken on a piece of the projective space (maybe on higher Grassmannians?). The mappings (generalizations of the endomorphism) are constructed with the aid of $\mathrm{SL}(n, \mathbb{Z})$ instead of $\mathrm{SL}(2, \mathbb{Z})$. One may start, e. g., with sails—hulls of integer points in $n$ simplicial cones with direction edges

$$(e_1, \ldots, e_{\hat{\imath}}, \ldots, e_n; X),$$

where $X = (x_1, \ldots, x_n) \in \mathbb{Z}_+^n$ is a random (indivisible?) vector. Then one can count, say, the numbers of integer points on faces, or the numbers of vertices of a face, or other invariants: What is their statistic for a random $X$ with norm $\leq N \to \infty$?

Probably a (computer?) experiment worth performing. It is suitable to construct the sail by means of Gröbner bases of integer linear programming, but the key point is to choose the right object for the statistic. Perhaps the "integer-valued area" of faces has the best behavior?

## △ 1993-33 – *M. B. Sevryuk*

R The statistics of multidimensional continued fractions was recently studied by M. L. Kontsevich and Yu. M. Sukhov, see [2–4]. The whole topic arose in paper [1].

[1] ARNOLD V. I. *A*-graded algebras and continued fractions. *Comm. Pure Appl. Math.*, 1989, **42**(7), 993–1000. [*The Russian translation in*: Vladimir Igorevich Arnold. Selecta–60. Moscow: PHASIS, 1997, 473–482.]

[2] ARNOLD V. I. Higher dimensional continued fractions. *Reg. Chaot. Dynamics*, 1998, **3**(3), 10–17.

[3] ARNOLD V. I. Continued Fractions. Moscow: Moscow Center for Continuous Mathematical Education Press, 2001 (in Russian). ("Mathematical Education" Library, 14.)

[4] KONTSEVICH M. L., SUKHOV YU. M. Statistics of Klein polyhedra and multidimensional continued fractions. In: Pseudoperiodic Topology. Editors: V. Arnold, M. Kontsevich and A. Zorich. Providence, RI: Amer. Math. Soc., 1999, 9–27. (AMS Transl., Ser. 2, 197; Adv. Math. Sci., 46.)

## 1993-35

H This is a problem in paper [1] (p. 265).

[1] ARNOLD V. I. Problems on singularities and dynamical systems. In: Developments in Mathematics: the Moscow School. Editors: V. I. Arnold and M. Monastyrsky. London: Chapman & Hall, 1993, 251–274.

## 1993-36

H This is a problem in paper [1] (p. 265).

[1] ARNOLD V. I. Problems on singularities and dynamical systems. In: Developments in Mathematics: the Moscow School. Editors: V. I. Arnold and M. Monastyrsky. London: Chapman & Hall, 1993, 251–274.

## 1993-37

H This is a problem in paper [1] (p. 267). See also problems *1987-4* and *1990-4*.

[1] ARNOLD V. I. Problems on singularities and dynamical systems. In: Developments in Mathematics: the Moscow School. Editors: V. I. Arnold and M. Monastyrsky. London: Chapman & Hall, 1993, 251–274.

### 1993-38

$\mathcal{H}$ This is a problem in paper [1] (p. 267).

[1] ARNOLD V. I. Problems on singularities and dynamical systems. In: Developments in Mathematics: the Moscow School. Editors: V. I. Arnold and M. Monastyrsky. London: Chapman & Hall, 1993, 251–274.

### 1993-39

$\mathcal{H}$ This is a problem in paper [1] (p. 269).

[1] ARNOLD V. I. Problems on singularities and dynamical systems. In: Developments in Mathematics: the Moscow School. Editors: V. I. Arnold and M. Monastyrsky. London: Chapman & Hall, 1993, 251–274.

$\mathcal{R}$ See the comments to problem *1995-13*.

### 1993-40

$\mathcal{H}$ This is a problem in paper [1] (p. 269).

[1] ARNOLD V. I. Problems on singularities and dynamical systems. In: Developments in Mathematics: the Moscow School. Editors: V. I. Arnold and M. Monastyrsky. London: Chapman & Hall, 1993, 251–274.

$\mathcal{R}$ See the comments to problem *1995-13*.

### ▽ 1993-41

$\mathcal{H}$ This is a problem in paper [1] (p. 270).

[1] ARNOLD V. I. Problems on singularities and dynamical systems. In: Developments in Mathematics: the Moscow School. Editors: V. I. Arnold and M. Monastyrsky. London: Chapman & Hall, 1993, 251–274.

### △ 1993-41 – *M. R. Entov* Also: *1993-42*

$\mathcal{R}$ The problem was solved in [1], and the answer is "yes" (which is, of course, also the answer for problem *1993-42*).

[1] ENTOV M. R. Surgery on Lagrangian and Legendrian singularities. *Geom. Funct. Anal.*, 1999, **9**(2), 298–352.

## 1993-42

ℋ This is a problem in paper [1] (p. 271).

[1] ARNOLD V. I. Problems on singularities and dynamical systems. In: Developments in Mathematics: the Moscow School. Editors: V. I. Arnold and M. Monastyrsky. London: Chapman & Hall, 1993, 251–274.

ℛ See the comment to problem *1993-41*.

## 1993-43

ℋ This is a problem in paper [1] (p. 271).

[1] ARNOLD V. I. Problems on singularities and dynamical systems. In: Developments in Mathematics: the Moscow School. Editors: V. I. Arnold and M. Monastyrsky. London: Chapman & Hall, 1993, 251–274.

## 1993-44

ℋ This is a problem in paper [1] (p. 271).

[1] ARNOLD V. I. Problems on singularities and dynamical systems. In: Developments in Mathematics: the Moscow School. Editors: V. I. Arnold and M. Monastyrsky. London: Chapman & Hall, 1993, 251–274.

## 1993-45

ℋ This is a problem in paper [1] (p. 272).

[1] ARNOLD V. I. Problems on singularities and dynamical systems. In: Developments in Mathematics: the Moscow School. Editors: V. I. Arnold and M. Monastyrsky. London: Chapman & Hall, 1993, 251–274.

## 1993-46

ℋ This is a problem in paper [1] (p. 272).

[1] ARNOLD V. I. Problems on singularities and dynamical systems. In: Developments in Mathematics: the Moscow School. Editors: V. I. Arnold and M. Monastyrsky. London: Chapman & Hall, 1993, 251–274.

ℛ Cf. problem *1993-34*.

## 1993-47

ℋ This is a problem in paper [1] (p. 273).

[1] ARNOLD V. I. Problems on singularities and dynamical systems. In: Developments in Mathematics: the Moscow School. Editors: V. I. Arnold and M. Monastyrsky. London: Chapman & Hall, 1993, 251–274.

ℛ Cf. problem *1992-1*.

## 1993-48 — *M. B. Sevryuk*

ℛ Note that if all the fixed points of a continuous involution of the $n$-torus are isolated then the number of these points is equal to either zero or $2^n$ [1] (see the comment to problem *1983-3*).

One restriction on the numbers $p_i$ was obtained in paper [2]. Namely, one can treat the set of fixed points of the involution $G|_L$ as a ring $(\mathbb{Z}_2)^n$ with respect to the operations of componentwise addition and multiplication (provided that the angular coordinate $\varphi$ on the torus $L$ is defined modd 2). It turns out that for any pair of fixed points $a_i$ and $a_j$ of the involution $G|_L$, the sum of the numbers $p$ corresponding to the four fixed points 0, $a_i$, $a_j$, and $a_i + a_j$ is even. It is unknown whether other restrictions exist.

Note that if there are numbers of different parity among $p_i$, then the ambient manifold $M$ is nonorientable [2].

For $n = 1$ and $N = 2$, all the possibilities do occur. One can easily construct the type collections $(1,1)$, $(1,1)$ and $(2,0)$, $(2,0)$ by considering involutions of the cylinder $M = \mathbb{S}^1 \times \mathbb{R}$ leaving the circle $L = \mathbb{S}^1 \times \{0\}$ invariant. The collection $(1,1)$, $(2,0)$ occurs for some involutions of the Möbius band which leave the central circle invariant.

[1] MONTALDI J. Caustics in time reversible Hamiltonian systems. In: Singularity Theory and its Applications, Part II. Editors: M. Roberts and I. Stewart. Berlin: Springer, 1991, 266–277. (Lecture Notes in Math., 1463.)

[2] QUISPEL G. R. W., SEVRYUK M. B. KAM theorems for the product of two involutions of different types. *Chaos*, 1993, **3**(4), 757–769.

# 1994

## 1994-2 – *S. L. Tabachnikov*

ℜ The "tennis ball" theorem states that an embedded spherical curve bisecting the area has at least 4 inflection points. This theorem is a special case of a theorem in [1] that has the same conclusion with the weaker assumption that the center of the sphere is inside the convex hull of the curve. A simpler proof is given in [2].


[1] SEGRE B. Alcune proprietà differenziali in grande delle curve chiuse sghembe. *Rend. Mat., Ser.* 6, 1968, **1**, 237–297.

[2] WEINER J. Global properties of spherical curves. *J. Differ. Geom.*, 1977, **12**(3), 425–434.


## 1994-5 – *V. D. Sedykh*

ℜ The problem was solved in paper [1] (the answers are "yes").


[1] ANISOV S. S. Convex curves in $\mathbb{RP}^n$. *Proc. Steklov Inst. Math.*, 1998, **221**, 3–39.


## ▽ 1994-6 – *V. D. Sedykh*

ℜ The desired class of deformations (*admissible homotopies*) and one of their invariants (*Sturmianity*) was described by V. I. Arnold in [1]. Some other invariants of these deformations can be found in [2] (see the comment to problem *1998-6*).


[1] ARNOLD V. I. Towards the Legendre Sturm theory of space curves. *Funct. Anal. Appl.*, 1998, **32**(2), 75–80.

[2] SEDYKH V. D. Some invariants of admissible homotopies of space curves. *Funct. Anal. Appl.*, 2001, **35**(4), 284–293.


## △ 1994-6 – *R. Uribe-Vargas*

ℜ This is a complement to the comment by V. D. Sedykh.

In [1], V. I. Arnold gave the first step towards a Legendrian Sturm theory of space curves. He imposed some conditions on curves in terms of a 2-dimensional Legendrian knot of the space $PT^*\mathbb{R}^3$ of contact elements of $\mathbb{R}^3$ associated with each curve in $\mathbb{R}^3$ (or in $\mathbb{R}P^3$). This Legendrian 2-dimensional knot consists of the contact elements of $\mathbb{R}^3$ (or of $\mathbb{R}P^3$) tangent to the curve.

The space of oriented orthonormal frames in $\mathbb{R}^3$ is isomorphic to the contact manifold $ST^*S^2 \simeq \mathbb{R}P^3$. The Frenet frames at the points of a closed curve in $\mathbb{R}^3$ determine a Legendrian knot in $\mathbb{R}P^3$.

**Conjecture** [2]. *Any smooth closed curve in* $\mathbb{R}^3$ *obtained as a deformation of a plane convex curve has at least* 4 *flattenings if, during the deformation process, the associated Legendrian knot in* $\mathbb{R}P^3$ *does not change its knot type.*

This conjecture is in the spirit of Arnold's and Chekanov's philosophy, and the class of curves under consideration is more general than the class dealt with in [1]. But instead of the 2-dimensional Legendrian knot in $PT^*\mathbb{R}^3$ (or in $PT^*\mathbb{R}P^3$) associated with a curve in $\mathbb{R}^3$ (or in $\mathbb{R}P^3$), the 1-dimensional Legendrian knot in $ST^*S^2 = \mathbb{R}P^3$ associated with the Frenet frames of the curve in $\mathbb{R}^3$ is considered.


[1] ARNOLD V. I. Towards the Legendre Sturm theory of space curves. *Funct. Anal. Appl.*, 1998, **32**(2), 75–80.

[2] URIBE-VARGAS R. Symplectic and contact singularities in the differential geometry of curves and surfaces. Ph. D. Thesis, Université Paris 7, 2001, Ch. 3, § 7.


## 1994-8 — *S. L. Tabachnikov*

Upper bounds for the Bennequin invariant of Legendrian links in the space of contact elements in the plane are given in [3, 4] in terms of the HOMFLY and Kauffman polynomials of the link. See also [1, 2].


[1] CHMUTOV S. V., GORYUNOV V. V. Polynomial invariants of Legendrian links and plane fronts. In: Topics in Singularity Theory. V. I. Arnold's 60th Anniversary Collection. Editors: A. Khovanskiĭ, A. Varchenko and V. Vassiliev. Providence, RI: Amer. Math. Soc., 1997, 25–43. (AMS Transl., Ser. 2, 180; Adv. Math. Sci., 34.)

[2] CHMUTOV S. V., GORYUNOV V. V., MURAKAMI H. Regular Legendrian knots and the HOMFLY polynomial of immersed plane curves. *Math. Ann.*, 2000, **317**(3), 389–413.

[3] TABACHNIKOV S. L. Estimates for the Bennequin number of Legendrian links from state models for knot polynomials. *Math. Res. Lett.*, 1997, **4**(1), 143–156.

[4] TABACHNIKOV S. L., FUCHS D. B. Invariants of Legendrian and transverse knots in the standard contact space. *Topology*, 1997, **36**(5), 1025–1053.

## 1994-9 – *V. I. Arnold*

ℛ Here is a tentative plan to carry over this construction to $\mathbb{C}^3$. Let us fix a value $\lambda_1 \neq 0$. Then identify close lines parallel to the $\lambda_2$-axis by deforming the point $\lambda_2$ on such line in a fixed way. On the initial line and the final line, identify the points $\infty$ and the pairs of intersection points with the discriminant. This is done by the same construction that is used for identifying close elliptic curves. Now build a twofold covering of our line $\lambda_1 = \text{const}$ ramified at 4 points: the chosen point $\lambda_2$, the two degeneracy points on the discriminant, and a point at infinity. Identify the two resulting elliptic curves as it was done in $\mathbb{C}^2$. Since the point at infinity is preserved under the deformation of $\lambda_2$, the real identification of the deformed covering elliptic curve and the initial one commutes with the involution, i. e., drops into the base. This determines a homeomorphism of our line $\lambda_1 = \text{const}$ onto the deformed line which "respects" $\infty$ and the discriminant points and moves the point $\lambda_2$ in the desired way. Now one needs to extend the local trivialization of the fibering of ramification curves $x^3+\lambda_1 x+\lambda_2+y^2=0$ on the plane $\mathbb{C}^2$ to the similar trivialization of the fibering of twofold coverings of $\mathbb{C}^2$ with branching along these curves. To do this, we again use the identification of elliptic level curves of the function $\lambda_2 = -x^3 - \lambda_1 x - y^2$ (we already know which of these curves must be identified with each deformed curve from the construction of the identification of lines $\lambda_1 = \text{const}$). It should be verified there that the whole construction of the identification of adjacent elliptic curves has a continuous extension up to the discriminant, in particular, up to $\lambda_1 = \lambda_2 = 0$ (using the asymptotics of elliptic integrals).

## ▽ 1994-10 – *S. V. Duzhin*

ℛ The number of equivalence classes of plane curves with $n \leq 10$ self-intersections was found by S. M. Gusein-Zade and F. S. Duzhin [1] with the help of a computer. They consider one version of the problem in the case of long curves (no orientations) and four versions in the case of closed curves: oriented and non-oriented curves in the oriented and non-oriented plane.

The number of long curves up to 20 crossings was found in 2001 by J. L. Jacobsen and P. Zinn-Justin [2].

No general formulae or exact asymptotics are known (December 2001).

[1] Gusein-Zade S. M., Duzhin F. S. On the number of topological types of plane curves. *Russian Math. Surveys*, 1998, **53**(3), 626–627.

[2] JACOBSEN J. L., ZINN-JUSTIN P. A transfer matrix approach to the enumeration of knots.
[*Internet*: `http://www.arXiv.org/abs/math-ph/0102015`]

## △ 1994-10 – *S. K. Lando*

ℋ The problem was formulated in [1].

ℛ Enumeration algorithms and data for small values of the number of double points can be found in [2, 3]. The methods of Hermitian matrix integration provide the exponential upper bound $c \cdot 12^n n^{-3/2}$ for the number of plane curves with $n$ double points [4].

[1] ARNOLD V. I. Plane curves, their invariants, perestroikas and classifications. In: Singularities and Bifurcations. Editor: V. I. Arnold. Providence, RI: Amer. Math. Soc., 1994, 33–91. (Adv. Sov. Math., 21.)

[2] GUSEIN-ZADE S. M. On the enumeration of curves from infinity to infinity. In: Singularities and Bifurcations. Editor: V. I. Arnold. Providence, RI: Amer. Math. Soc., 1994, 189–198. (Adv. Sov. Math., 21.)

[3] GUSEIN-ZADE S. M., DUZHIN F. S. On the number of topological types of plane curves. *Russian Math. Surveys*, 1998, **53**(3), 626–627.

[4] LANDO S. K. On enumeration of unicursal curves. In: Differential and Symplectic Topology of Knots and Curves. Editor: S. Tabachnikov. Providence, RI: Amer. Math. Soc., 1999, 77–81. (AMS Transl., Ser. 2, 190; Adv. Math. Sci., 42.)

## 1994-11 – *V. I. Arnold*

ℛ The simplest singularity—on the stratum $\lambda_1 = \lambda_2$ of codimension 3—is characterized by a closed 2-form on a three-dimensional transversal. Interpreting a 2-form in $\mathbb{R}^3$ as a divergence-free vector field, we obtain a field with Newton singularity.

The next singularity—on the stratum $\lambda_1 = \lambda_2 = \lambda_3$ of codimension 8—is characterized by a closed 2-form on $\mathbb{R}^8$. Its restriction to $\mathbb{S}^7$ has Newton-type singularities on two four-dimensional "poles." Thus, we get an unexplored generalization of the Newton field (M. Berry's comment on paper [1]).

[1] ARNOLD V. I. Remarks on eigenvalues and eigenvectors of Hermitian matrices, Berry phase, adiabatic connections and quantum Hall effect. *Selecta Math. (N. S.)*, 1995, **1**(1), 1–19. [*The Russian translation in*: Vladimir Igorevich Arnold. Selecta–60. Moscow: PHASIS, 1997, 583–604.]

## 1994-13 – *V. L. Ginzburg*

R The Seifert conjecture is the question posed by Seifert [16] whether or not every smooth nonvanishing vector field on the 3-dimensional sphere has a periodic orbit. Of course, a similar question can be asked for other manifolds or some special classes of vector fields (e. g., of a certain smoothness class, volume-preserving, Hamiltonian, Legendrian, etc.). Traditionally, nontrivial examples of nonvanishing vector fields without periodic orbits are referred to as *counterexamples* to the Seifert conjecture.

A $C^1$-smooth counterexample to the Seifert conjecture on $\mathbb{S}^3$ was constructed by Schweitzer [15]. The $C^1$-smoothness constraint was later improved to $C^{2+\alpha}$ by Harrison [7]. Finally, a $C^\infty$-counterexample was found by K. Kuperberg [12]; see also [11]. A construction of a $C^1$-smooth volume-preserving counterexample on $\mathbb{S}^3$ was carried out by G. Kuperberg [10]. To the date, it is not known whether or not there exists such a $C^\infty$-smooth counterexample.

The vector field arising in the magnetic problem is not only Legendrian and volume-preserving but also Hamiltonian. The Hamiltonian Seifert conjecture is the question whether or not there exists a proper function on $\mathbb{R}^{2n}$ whose Hamiltonian flow has no periodic orbits on at least one regular level set; such a $C^2$-smooth function on $\mathbb{R}^4$ was constructed in [5, 6]. As a consequence, it is not hard to see that in the twisted cotangent bundle of $S^2$ there exists a compact $C^2$-hypersurface enclosing the zero section and having no closed characteristics.

Counterexamples to the Seifert conjecture are much easier to find in higher dimensions. (Hence, this is perhaps where one should start the investigation of the Legendrian Seifert conjecture.) For $2n+1 \geq 5$, a $C^\infty$-smooth counterexample to the Seifert conjecture on $S^{2n+1}$ was found by Wilson [17]; volume-preserving and Hamiltonian $C^\infty$-smooth counterexamples have also been constructed (see [1, 2, 8, 9]). We refer the reader to surveys [3, 4, 13, 14] for further references and details.

It is *not* known whether any of these counterexamples can be made *Legendrian.*

[1] GINZBURG V. L. An embedding $S^{2n-1} \to \mathbb{R}^{2n}$, $2n-1 \geq 7$, whose Hamiltonian flow has no periodic trajectories. *Internat. Math. Res. Notices*, 1995, **2**, 83–97 (electronic).

[2] GINZBURG V. L. A smooth counterexample to the Hamiltonian Seifert conjecture in $\mathbb{R}^6$. *Internat. Math. Res. Notices*, 1997, **13**, 641–650.

[3] GINZBURG V. L. Hamiltonian dynamical systems without periodic orbits. In: Northern California Symplectic Geometry Seminar. Editors: Ya. Eliashberg, D. Fuchs, T. Ratiu and A. Weinstein. Providence, RI: Amer. Math. Soc., 1999, 35–48. (AMS Transl., Ser. 2, 196; Adv. Math. Sci., 45.)

[4] GINZBURG V. L. The Hamiltonian Seifert conjecture: examples and open problems. In: European Congress of Mathematics (Barcelona, 2000), Vol. II. Editors: C. Casacuberta, R. M. Miró-Roig, J. Verdera and S. Xambó-Descamps. Basel: Birkhäuser, 2001, 547–555. (Progr. Math., 202.)
[*Internet*: `http://www.arXiv.org/abs/math.DG/0004020`]

[5] GINZBURG V. L., GÜREL B. Z. On the construction of a $C^2$-counterexample to the Hamiltonian Seifert Conjecture in $\mathbb{R}^4$.
[*Internet*: `http://www.arXiv.org/abs/math.DG/0109153`]

[6] GINZBURG V. L., GÜREL B. Z. A $C^2$-smooth counterexample to the Hamiltonian Seifert Conjecture in $\mathbb{R}^4$. *Electron. Res. Announc. Amer. Math. Soc.*, 2002, **8**, 11–19 (electronic).
[*Internet*: `http://www.arXiv.org/abs/math.DG/0110047`]

[7] HARRISON J. $C^2$ counterexamples to the Seifert conjecture. *Topology*, 1988, **27**(3), 249–278.

[8] HERMAN M. -R. Examples of compact hypersurfaces in $\mathbb{R}^{2p}$, $2p \geq 6$, with no periodic orbits. In: Hamiltonian Systems with Three or More Degrees of Freedom (S'Agaró, 1995). Editor: C. Simó. Dordrecht: Kluwer Acad. Publ., 1999, 126. (NATO Adv. Sci. Inst. Ser. C Math. Phys. Sci., 533.)

[9] KERMAN E. New smooth counterexamples to the Hamiltonian Seifert conjecture. *J. Symplectic Geom.*, 2002, **1**(2), 253–267.
[*Internet*: `http://www.arXiv.org/abs/math.DG/0101185`]

[10] KUPERBERG G. A volume–preserving counterexample to the Seifert conjecture. *Comment. Math. Helvetici*, 1996, **71**(1), 70–97.

[11] KUPERBERG G., KUPERBERG K. Generalized counterexamples to the Seifert conjecture. *Ann. Math., Ser.* 2, 1996, **143**(3), 547–576; **144**(2), 239–268.

[12] KUPERBERG K. A smooth counterexample to the Seifert conjecture in dimension three. *Ann. Math., Ser.* 2, 1994, **140**(3), 723–732.

[13] KUPERBERG K. Counterexamples to the Seifert conjecture. In: Proceedings of the International Congress of Mathematicians, Vol. II (Berlin, 1998). *Doc. Math.*, 1998, Extra Vol. II, 831–840 (electronic).

[14] KUPERBERG K. Aperiodic dynamical systems. *Notices Amer. Math. Soc.*, 1999, **46**(9), 1035–1040.

[15] SCHWEITZER P. A. Counterexamples to the Seifert conjecture and opening closed leaves of foliations. *Ann. Math., Ser.* 2, 1974, **100**, 386–400.

[16] SEIFERT H. Closed integral curves in the 3-space and isotopic two-dimensional deformations. *Proc. Amer. Math. Soc.*, 1950, **1**, 287–302.

[17] WILSON F. W., JR. On the minimal sets of non-singular vector fields. *Ann. Math., Ser.* 2, 1966, **84**(3), 529–536.

## 1994-14

See the comment to problem *1981-9*.

## ▽ 1994-15 – *V. D. Sedykh*

The number of connected components of the set of convex curves in $\mathbb{RP}^n$ was calculated by M. Z. Shapiro (see [1]).

The negative answer to the first question of the problem (for $n = 3$) was given by S. S. Anisov in [2].


[1] SHAPIRO M. Z. Topology of the space of non-degenerate curves. *Izv. Math.*, 1994, **43**(2), 291–310.

[2] ANISOV S. S. Convex curves in $\mathbb{RP}^n$. *Proc. Steklov Inst. Math.*, 1998, **221**, 3–39.


## △▽ 1994-15 – *B. Z. Shapiro*

The negative answer to the first question of the problem in the case of nonclosed curves follows from the main result of [2] claiming that the boundary curves in the space of convex (nonclosed) curves enjoy the property that their initial and final osculating flags are nontransversal. What is essential is that any type of nontransversality can be realized by boundary curves, while the problem discusses only nontransversality of points and hyperplanes. The homotopy type of the space of convex curves in $\mathbb{RP}^n$ was determined in [1]. In particular, the space of parameterized closed convex curves $(\mathbb{S}^1, 0) \to \mathbb{RP}^n$ with a given Frenet frame at 0 is contractible. Note that any Chebyshev system consisting of $n+1$ functions (see problem *1984-1*) produces a convex curve in $\mathbb{RP}^n$ by taking the usual projectivization map $\mathbb{R}^{n+1} \to \mathbb{RP}^n$. Therefore, the stratification of the boundary of the space of convex curves in $\mathbb{RP}^n$ and the occurring singularities are essentially the same as for the space of Chebyshev systems of $n+1$ functions.


[1] ANISOV S. S. Convex curves in $\mathbb{RP}^n$. *Proc. Steklov Inst. Math.*, 1998, **221**, 3–39.

[2] SHAPIRO B. Z. Spaces of linear differential equations, and flag manifolds. *Math. USSR, Izv.*, 1991, **36**(1), 183–197.


## △ 1994-15 – *R. Uribe-Vargas*

A closed curve without flattenings and whose osculating hyperplane intersects it only at the point of osculation is said to be *weakly convex* [2]. A convex curve in $\mathbb{RP}^n$ is always weakly convex. A curve in $\mathbb{RP}^2$ (in $\mathbb{R}^2$) is convex if and only if it is weakly convex. However, it was proved—in [1] for $n = 3$ and in [2] for any $n > 2$—that the answer to the first problem is negative, i. e., for $n > 2$ there exist weakly convex curves which are not convex.

**Examples** [2]. 1) The curve in $\mathbb{RP}^{2k}$, $k \geq 2$, given in affine coordinates by

$$\omega \mapsto (\cos\omega,\ \sin\omega,\ \cos 2\omega,\ \sin 2\omega,\ \ldots,\ \cos(k-1)\omega,\ \sin(k-1)\omega,\ \cos(k+1)\omega,\ \sin(k+1)\omega),$$

is weakly convex but not convex.

2) The curve $\gamma$ in $\mathbb{RP}^{2k-1}$, $k \geq 2$, given in homogeneous coordinates by

$$\omega \mapsto [\cos\omega : \sin\omega : \cos 3\omega : \cdots : \cos(2k-3)\omega : \sin(2k-3)\omega : \cos(2k+1)\omega : \sin(2k+1)\omega],$$

is weakly convex but not convex.

[1] Anisov S. S. Convex curves in $\mathbb{RP}^n$. *Proc. Steklov Inst. Math.*, 1998, **221**, 3–39.

[2] Uribe-Vargas R. Symplectic and contact singularities in the differential geometry of curves and surfaces. Ph. D. Thesis, Université Paris 7, 2001, Ch. 3, § 5.

## 1994-16 – *B. A. Khesin*

See paper [1].

[1] Ovsienko V. Yu., Khesin B. A. Symplectic leaves of the Gelfand–Dikiĭ brackets and homotopy classes of nondegenerate curves. *Funct. Anal. Appl.*, 1990, **24**(1), 33–40.

## ▽ 1994-17 – *B. Z. Shapiro*

The analogous problem for the Euclidean duality is directly connected with the isoperimetric inequalities for Euclidean curves, see [1].

[1] Sedykh V. D., Shapiro B. Z. On Young hulls of convex curves in $\mathbb{R}^{2n}$ *J. Geometry*, 1998, **63**(1–2), 168–182.

## △ 1994-17 – *S. L. Tabachnikov*

One may consider a more restrictive problem. The projective duality assigns a point $x^*$ of the dual curve $\gamma^*$ to a point $x$ of the original curve $\gamma$ ($x^*$ is the tangent line to $\gamma$ at $x$), and one may request that this correspondence $x \to x^*$ is a projective equivalence between $\gamma$ and $\gamma^*$. In the classical book [1] by E. Wilczyński such curves are called *identically self-dual.* If, in addition, the curve is convex then identical self-duality implies that the curve is a conic—this is proved in Section 3.1 of [1].

[1] WILCZYŃSKI E. Projective Differential Geometry of Curves and Ruled Surfaces. Leipzig: B. G. Teubner, 1906; reprinted by Chelsea: New York, 1962.

## 1994-20 – *F. Aicardi*

In [1], the caustic of the ellipsoid in $\mathbb{R}^4$ is analyzed and some general properties of singularities of caustics of ellipsoids in $\mathbb{R}^n$, $n > 4$, are discussed.

[1] JOETS A., RIBOTTA R. Caustique de la surface ellipsoîdale à trois dimensions. *Experim. Math.*, 1999, **8**(1), 49–55.

## 1994-21 – *S. V. Chmutov, E. Ferrand*

The problem was solved by A. Shumakovich (1995) and independently by V. V. Goryunov [1]. The statement remains true if $\mathbb{R}^2$ is replaced by an arbitrary two-dimensional surface [2].

[1] CHMUTOV S. V., GORYUNOV V. V., MURAKAMI H. Regular Legendrian knots and the HOMFLY polynomial of immersed plane curves. *Math. Ann.*, 2000, **317**(3), 389–413.

[2] FERRAND E. Singularities cancellation on wave fronts. *Topology Appl.*, 1999, **95**(2), 155–163.

## 1994-22 – *V. D. Sedykh*

A proof is contained in paper [1] by S. S. Anisov.

[1] ANISOV S. S. Projective convex curves. In: The Arnold–Gelfand Mathematical Seminars: Geometry and Singularity Theory. Editors: V. I. Arnold, I. M. Gelfand, V. S. Retakh and M. Smirnov. Boston, MA: Birkhäuser, 1997, 93–99.

## 1994-23 – *B. Z. Shapiro*

ℛ This problem has been solved. It was shown in [1] that the fronts of any two closed convex curves in $\mathbb{RP}^n$ are homeomorphic, and even diffeomorphic (the latter had been proved by V.M.Zakalyukin but still remains unpublished). Worth mentioning also is the following conjecture about the fronts of closed convex curves in $\mathbb{RP}^n$, the conjecture being wide open.

*For any convex curve* $\gamma\colon \mathbb{S}^1 \to \mathbb{RP}^n$ *the "degree" of its front* $D_\gamma \subset \mathbb{RP}^n$ *equals* $2(n-1)$.

(Here the "degree" of a real projective hypersurface is the maximal total multiplicity of its intersection with projective lines.)

The only case settled in [2] is $n = 3$. This paper contains an even more general conjecture about the degrees of the higher fronts associated with a convex curve $\gamma$ in the Grassmannians of $k$-dimensional subspaces in $\mathbb{RP}^n$.

[1] SHAPIRO B. Z. Discriminants of convex curves are homeomorphic. *Proc. Amer. Math. Soc.*, 1998, **126**(7), 1923–1930.

[2] SHAPIRO B. Z., SHAPIRO M. Z. Projective convexity in $\mathbb{P}^3$ implies Grassmann convexity. *Internat. J. Math.*, 2000, **11**(4), 579–588.

## 1994-24

ℋ This is a problem in paper [1a] (§ 1.1; see also [1b], p. 555).

[1a] ARNOLD V. I. Mathematical problems in classical physics. In: Trends and Perspectives in Applied Mathematics. Editors: F. John, J. E. Marsden and L. Sirovich. New York: Springer, 1994, 1–20. (Appl. Math. Sci., 100.)

*The Russian translation in*:

[1b] Vladimir Igorevich Arnold. Selecta–60. Moscow: PHASIS, 1997, 553–575.

## 1994-25

ℋ This is a problem in paper [1a] (§ 1.2; see also [1b], p. 556).

[1a] ARNOLD V. I. Mathematical problems in classical physics. In: Trends and Perspectives in Applied Mathematics. Editors: F. John, J. E. Marsden and L. Sirovich. New York: Springer, 1994, 1–20. (Appl. Math. Sci., 100.)

*The Russian translation in*:

[1b] Vladimir Igorevich Arnold. Selecta–60. Moscow: PHASIS, 1997, 553–575.

## 1994-26

This is a problem in paper [1a] (§ 1.3; see also [1b], p. 558).

[1a] ARNOLD V. I. Mathematical problems in classical physics. In: Trends and Perspectives in Applied Mathematics. Editors: F. John, J. E. Marsden and L. Sirovich. New York: Springer, 1994, 1–20. (Appl. Math. Sci., 100.)

*The Russian translation in*:

[1b] Vladimir Igorevich Arnold. Selecta–60. Moscow: PHASIS, 1997, 553–575.

See the comment to problem *1971-11*.

## 1994-27

This is a problem in paper [1a] (§ 1.3; see also [1b], p. 558).

[1a] ARNOLD V. I. Mathematical problems in classical physics. In: Trends and Perspectives in Applied Mathematics. Editors: F. John, J. E. Marsden and L. Sirovich. New York: Springer, 1994, 1–20. (Appl. Math. Sci., 100.)

*The Russian translation in*:

[1b] Vladimir Igorevich Arnold. Selecta–60. Moscow: PHASIS, 1997, 553–575.

See the comment to problem *1971-11*.

## 1994-28

This is a problem in paper [1a] (§ 1.4; see also [1b], p. 559–560).

[1a] ARNOLD V. I. Mathematical problems in classical physics. In: Trends and Perspectives in Applied Mathematics. Editors: F. John, J. E. Marsden and L. Sirovich. New York: Springer, 1994, 1–20. (Appl. Math. Sci., 100.)

*The Russian translation in*:

[1b] Vladimir Igorevich Arnold. Selecta–60. Moscow: PHASIS, 1997, 553–575.

See the comment to problem *1981-18*.

## 1994-29

𝓗 This is a problem in paper [1a] (§ 1.4; see also [1b], p. 561).

[1a] ARNOLD V. I. Mathematical problems in classical physics. In: Trends and Perspectives in Applied Mathematics. Editors: F. John, J. E. Marsden and L. Sirovich. New York: Springer, 1994, 1–20. (Appl. Math. Sci., 100.)

*The Russian translation in*:

[1b] Vladimir Igorevich Arnold. Selecta–60. Moscow: PHASIS, 1997, 553–575.

𝓡 See the comment to problem *1981-18.*

## 1994-30

𝓗 This is a problem in paper [1a] (§ 1.5; see also [1b], p. 563).

[1a] ARNOLD V. I. Mathematical problems in classical physics. In: Trends and Perspectives in Applied Mathematics. Editors: F. John, J. E. Marsden and L. Sirovich. New York: Springer, 1994, 1–20. (Appl. Math. Sci., 100.)

*The Russian translation in*:

[1b] Vladimir Igorevich Arnold. Selecta–60. Moscow: PHASIS, 1997, 553–575.

𝓡 See the comment to problem *1986-12.*

## 1994-31

𝓗 This is a problem in paper [1a] (§ 1.6; see also [1b], p. 567–569).

[1a] ARNOLD V. I. Mathematical problems in classical physics. In: Trends and Perspectives in Applied Mathematics. Editors: F. John, J. E. Marsden and L. Sirovich. New York: Springer, 1994, 1–20. (Appl. Math. Sci., 100.)

*The Russian translation in*:

[1b] Vladimir Igorevich Arnold. Selecta–60. Moscow: PHASIS, 1997, 553–575.

𝓡 See the comment to problem *1981-14.*

## 1994-32

ℋ This is a problem in paper [1a] (§ 1.7; see also [1b], p. 570).

[1a] ARNOLD V. I. Mathematical problems in classical physics. In: Trends and Perspectives in Applied Mathematics. Editors: F. John, J. E. Marsden and L. Sirovich. New York: Springer, 1994, 1–20. (Appl. Math. Sci., 100.)

*The Russian translation in*:

[1b] Vladimir Igorevich Arnold. Selecta–60. Moscow: PHASIS, 1997, 553–575.

## 1994-33

ℋ This is a problem in paper [1a] (§ 1.8; see also [1b], p. 571).

[1a] ARNOLD V. I. Mathematical problems in classical physics. In: Trends and Perspectives in Applied Mathematics. Editors: F. John, J. E. Marsden and L. Sirovich. New York: Springer, 1994, 1–20. (Appl. Math. Sci., 100.)

*The Russian translation in*:

[1b] Vladimir Igorevich Arnold. Selecta–60. Moscow: PHASIS, 1997, 553–575.

ℛ See the comment to problem *1963-1*.

## 1994-34

ℋ This is a problem in paper [1a] (§ 1.8; see also [1b], p. 571).

[1a] ARNOLD V. I. Mathematical problems in classical physics. In: Trends and Perspectives in Applied Mathematics. Editors: F. John, J. E. Marsden and L. Sirovich. New York: Springer, 1994, 1–20. (Appl. Math. Sci., 100.)

*The Russian translation in*:

[1b] Vladimir Igorevich Arnold. Selecta–60. Moscow: PHASIS, 1997, 553–575.

## 1994-35

ℋ This is a problem in paper [1a] (§ 1.8; see also [1b], p. 572).

[1a] ARNOLD V. I. Mathematical problems in classical physics. In: Trends and Perspectives in Applied Mathematics. Editors: F. John, J. E. Marsden and L. Sirovich. New York: Springer, 1994, 1–20. (Appl. Math. Sci., 100.)

*The Russian translation in*:

[1b] Vladimir Igorevich Arnold. Selecta–60. Moscow: PHASIS, 1997, 553–575.

ℛ See the comment to problem *1981-9*.

## 1994-36

ℋ This is a problem in paper [1a] (§ 1.8; see also [1b], p. 573).

[1a] ARNOLD V. I. Mathematical problems in classical physics. In: Trends and Perspectives in Applied Mathematics. Editors: F. John, J. E. Marsden and L. Sirovich. New York: Springer, 1994, 1–20. (Appl. Math. Sci., 100.)

*The Russian translation in*:

[1b] Vladimir Igorevich Arnold. Selecta–60. Moscow: PHASIS, 1997, 553–575.

## 1994-37

ℋ This is a problem in paper [1a] (§ 1.8; see also [1b], p. 573).

[1a] ARNOLD V. I. Mathematical problems in classical physics. In: Trends and Perspectives in Applied Mathematics. Editors: F. John, J. E. Marsden and L. Sirovich. New York: Springer, 1994, 1–20. (Appl. Math. Sci., 100.)

*The Russian translation in*:

[1b] Vladimir Igorevich Arnold. Selecta–60. Moscow: PHASIS, 1997, 553–575.

ℛ See the comment to problem *1973-4*.

## 1994-38

ℋ This is a problem in paper [1a] (§ 3; see also [1b], p. 580).

A similar definition of an analytically solvable problem can be found in paper [2a] (§ 8; see also [2b], p. 547–549).

[1a] ARNOLD V. I. Problèmes résolubles et problèmes irrésolubles analytiques et géométriques. In: Passion des Formes. Dynamique Qualitative Sémiophysique et Intelligibilité. Dédié à R. Thom. Fontenay-St Cloud: ENS Éditions, 1994, 411–417; In: Formes et Dynamique, Renaissance d'un Paradigme. Hommage à René Thom. Paris: Eshel, 1995.

*The Russian translation in*:

[1b] Vladimir Igorevich Arnold. Selecta–60. Moscow: PHASIS, 1997, 577–582.

[2a] ARNOLD V. I. Sur quelques problèmes de la théorie des systèmes dynamiques. *Topol. Methods Nonlinear Anal.*, 1994, **4**(2), 209–225.

*The Russian translation in*:

[2b] Vladimir Igorevich Arnold. Selecta–60. Moscow: PHASIS, 1997, 533–551.

## 1994-39

This is a problem in paper [1a] (§ 3; see also [1b], p. 581).

[1a] ARNOLD V. I. Problèmes résolubles et problèmes irrésolubles analytiques et géométriques. In: Passion des Formes. Dynamique Qualitative Sémiophysique et Intelligibilité. Dédié à R. Thom. Fontenay-St Cloud: ENS Éditions, 1994, 411–417; In: Formes et Dynamique, Renaissance d'un Paradigme. Hommage à René Thom. Paris: Eshel, 1995.

*The Russian translation in*:

[1b] Vladimir Igorevich Arnold. Selecta–60. Moscow: PHASIS, 1997, 577–582.

## 1994-40

This is a problem in paper [1a] (§ 3; see also [1b], p. 581).

[1a] ARNOLD V. I. Problèmes résolubles et problèmes irrésolubles analytiques et géométriques. In: Passion des Formes. Dynamique Qualitative Sémiophysique et Intelligibilité. Dédié à R. Thom. Fontenay-St Cloud: ENS Éditions, 1994, 411–417; In: Formes et Dynamique, Renaissance d'un Paradigme. Hommage à René Thom. Paris: Eshel, 1995.

*The Russian translation in*:

[1b] Vladimir Igorevich Arnold. Selecta–60. Moscow: PHASIS, 1997, 577–582.

## 1994-41

This is a problem in paper [1a] (§ 3; see also [1b], p. 581).

The question about the geometric solvability of the problem on the stability of an equilibrium of a vector field in $\mathbb{R}^n$ (cf. problem *1994-38*) is also contained in paper [2a] (§ 8; see also [2b], p. 549).

[1a] ARNOLD V. I. Problèmes résolubles et problèmes irrésolubles analytiques et géométriques. In: Passion des Formes. Dynamique Qualitative Sémiophysique et Intelligibilité. Dédié à R. Thom. Fontenay-St Cloud: ENS Éditions, 1994, 411–417; In: Formes et Dynamique, Renaissance d'un Paradigme. Hommage à René Thom. Paris: Eshel, 1995.

*The Russian translation in*:

[1b] Vladimir Igorevich Arnold. Selecta–60. Moscow: PHASIS, 1997, 577–582.

[2a] ARNOLD V. I. Sur quelques problèmes de la théorie des systèmes dynamiques. *Topol. Methods Nonlinear Anal.*, 1994, **4**(2), 209–225.

*The Russian translation in*:

[2b] Vladimir Igorevich Arnold. Selecta–60. Moscow: PHASIS, 1997, 533–551.

## 1994-42

This is a problem in paper [1a] (§ 3; see also [1b], p. 582).

[1a] ARNOLD V. I. Problèmes résolubles et problèmes irrésolubles analytiques et géométriques. In: Passion des Formes. Dynamique Qualitative Sémiophysique et Intelligibilité. Dédié à R. Thom. Fontenay-St Cloud: ENS Éditions, 1994, 411–417; In: Formes et Dynamique, Renaissance d'un Paradigme. Hommage à René Thom. Paris: Eshel, 1995.

*The Russian translation in*:

[1b] Vladimir Igorevich Arnold. Selecta–60. Moscow: PHASIS, 1997, 577–582.

## 1994-43

This is a problem in paper [1a] (§ 3; see also [1b], p. 533).

[1a] ARNOLD V. I. Sur quelques problèmes de la théorie des systèmes dynamiques. *Topol. Methods Nonlinear Anal.*, 1994, **4**(2), 209–225.

*The Russian translation in*:

[1b] Vladimir Igorevich Arnold. Selecta–60. Moscow: PHASIS, 1997, 533–551.

## 1994-44

This is a problem in paper [1a] (§ 3; see also [1b], p. 534).

[1a] ARNOLD V. I. Sur quelques problèmes de la théorie des systèmes dynamiques. *Topol. Methods Nonlinear Anal.*, 1994, **4**(2), 209–225.

*The Russian translation in*:

[1b] Vladimir Igorevich Arnold. Selecta–60. Moscow: PHASIS, 1997, 533–551.

## 1994-45

ℋ This is a problem in paper [1a] (§ 3; see also [1b], p. 536).

[1a] ARNOLD V. I. Sur quelques problèmes de la théorie des systèmes dynamiques. *Topol. Methods Nonlinear Anal.*, 1994, **4**(2), 209–225.

*The Russian translation in*:

[1b] Vladimir Igorevich Arnold. Selecta–60. Moscow: PHASIS, 1997, 533–551.

ℛ See the comments to problem *1988-6*.

## 1994-46

ℋ This is a problem in paper [1a] (§ 3; see also [1b], p. 536).

[1a] ARNOLD V. I. Sur quelques problèmes de la théorie des systèmes dynamiques. *Topol. Methods Nonlinear Anal.*, 1994, **4**(2), 209–225.

*The Russian translation in*:

[1b] Vladimir Igorevich Arnold. Selecta–60. Moscow: PHASIS, 1997, 533–551.

ℛ See the comment to problem *1988-6* by M. B. Sevryuk.

## 1994-47

ℋ This is a problem in paper [1a] (§ 3; see also [1b], p. 536).

[1a] ARNOLD V. I. Sur quelques problèmes de la théorie des systèmes dynamiques. *Topol. Methods Nonlinear Anal.*, 1994, **4**(2), 209–225.

*The Russian translation in*:

[1b] Vladimir Igorevich Arnold. Selecta–60. Moscow: PHASIS, 1997, 533–551.

ℛ See the comments to problems *1988-6* (by M. B. Sevryuk) and *1992-13*.

## 1994-48

This is a problem in paper [1a] (§ 4; see also [1b], p. 537–538).

[1a] Arnold V. I. Sur quelques problèmes de la théorie des systèmes dynamiques. *Topol. Methods Nonlinear Anal.*, 1994, **4**(2), 209–225.

*The Russian translation in*:

[1b] Vladimir Igorevich Arnold. Selecta–60. Moscow: PHASIS, 1997, 533–551.

See the comments to problems *1988-6* (by M. B. Sevryuk) and *1992-13*.

## 1994-49

This is a problem in paper [1a] (§ 5; see also [1b], p. 539).

[1a] Arnold V. I. Sur quelques problèmes de la théorie des systèmes dynamiques. *Topol. Methods Nonlinear Anal.*, 1994, **4**(2), 209–225.

*The Russian translation in*:

[1b] Vladimir Igorevich Arnold. Selecta–60. Moscow: PHASIS, 1997, 533–551.

See the comment to problem *1988-6* by M. B. Sevryuk.

## 1994-50

This is a problem in paper [1a] (§ 5; see also [1b], p. 540).

[1a] Arnold V. I. Sur quelques problèmes de la théorie des systèmes dynamiques. *Topol. Methods Nonlinear Anal.*, 1994, **4**(2), 209–225.

*The Russian translation in*:

[1b] Vladimir Igorevich Arnold. Selecta–60. Moscow: PHASIS, 1997, 533–551.

See the comment to problem *1988-6* by M. B. Sevryuk.

## 1994-51

This is a problem in paper [1a] (§ 6; see also [1b], p. 541).

[1a] Arnold V. I. Sur quelques problèmes de la théorie des systèmes dynamiques. *Topol. Methods Nonlinear Anal.*, 1994, **4**(2), 209–225.

*The Russian translation in*:

[1b] Vladimir Igorevich Arnold. Selecta–60. Moscow: PHASIS, 1997, 533–551.

See the comment to problem *1978-6*.

### 1994-52

ℋ This is a problem in paper [1a] (§ 6; see also [1b], p. 542).

[1a] ARNOLD V. I. Sur quelques problèmes de la théorie des systèmes dynamiques. *Topol. Methods Nonlinear Anal.*, 1994, **4**(2), 209–225.

*The Russian translation in*:

[1b] Vladimir Igorevich Arnold. Selecta–60. Moscow: PHASIS, 1997, 533–551.

ℛ See the comment to problem *1978-6.*

### 1994-53

ℋ This is a problem in paper [1a] (§ 7; see also [1b], p. 545).

[1a] ARNOLD V. I. Sur quelques problèmes de la théorie des systèmes dynamiques. *Topol. Methods Nonlinear Anal.*, 1994, **4**(2), 209–225.

*The Russian translation in*:

[1b] Vladimir Igorevich Arnold. Selecta–60. Moscow: PHASIS, 1997, 533–551.

ℛ See the comment to problem *1972-20.*

# 1995

### 1995-1 — *B. Z. Shapiro*

ℛ It was shown in [2, 3] that the number of connected components in the space of all real trigonometric polynomials having the maximal number of real and distinct critical values coincides with the Euler–Bernoulli number, i. e., with the number of all cyclic up-down permutations of a certain length. Note that the same question in the case where not all critical values are real is still open, see problem *1997-1*. The motivation for this problem comes from [1].

The case of polynomials in the hyperbolic functions $\sinh x$ and $\cosh x$ was recently considered in [4]. The number of connected components in the space of polynomials in hyperbolic functions having the maximal number of real distinct critical values can be expressed as a sum with certain binomial coefficients of the classical Entringer numbers $e_{p,q}$ counting all up-down permutations of length $q$ whose first entry equals $p$.

See also the comments to problems *1970-15*, *1973-27*, and *1991-2*.

[1] ARNOLD V. I. Topological classification of trigonometric polynomials and combinatorics of graphs with an equal number of vertices and edges. *Funct. Anal. Appl.*, 1996, **30**(1), 1–14.

[2] ARNOLD V. I. Topological classification of real trigonometric polynomials and cyclic serpents polyhedron. In: The Arnold-Gelfand Mathematical Seminars: Geometry and Singularity Theory. Editors: V. I. Arnold, I. M. Gelfand, V. S. Retakh and M. Smirnov. Boston, MA: Birkhäuser, 1997, 101–106. [*The Russian translation in*: Vladimir Igorevich Arnold. Selecta–60. Moscow: PHASIS, 1997, 619–625.]

[3] SHAPIRO B. Z. On the number of components of the space of trigonometric polynomials of degree $n$ with $2n$ different critical values. *Math. Notes*, 1997, **62**(4), 529–534.

[4] SHAPIRO B. Z., VAINSHTEIN A. D. On the number of connected components in the space of $M$-polynomials in hyperbolic functions. *Adv. Appl. Math.*, to appear.

## 1995-2

See the comments to problem *1970-15*.

## 1995-3 — *F. Aicardi*

This conjecture follows from the *economy principle* in algebraic geometry formulated by Arnold [2]: *If a geometric or topological phenomenon can be realized by algebraic objects, then the simplest algebraic realizations are topologically as simple as possible.*

In this sense the "simplest" algebraic object realizing the deformation of the projective plane is a cubic equation.

This conjecture is a generalization to the projective real three-dimensional space of the dual version of the Möbius theorem, stating that the dual curve to a small perturbation of the projective real line in the projective real plane has at least three cusps.

In 1997, D. Panov proved a theorem giving conditions to reduce the number of elliptic domains of a surface [3]. The surface is perturbed (smoothly) only over an arbitrarily small neighborhood of a curve, sited in the hyperbolic domain, joining two parabolic points belonging to the boundaries of two elliptic domains. This curve must be transversal to the asymptotic directions. The perturbation makes the surface elliptic on a narrow strip containing this curve.

To apply this theorem to the cubic surface, where the repartitions of special parabolic points on the boundaries of the four elliptic ovals is $\{(0),(1),(2),(3)\}$, one has to perturb it to have the repartition $\{(0),(2),(2),(2)\}$. Applying the Panov theorem three times, the elliptic domain with zero special parabolic points on its boundary is joined with the other domains, so obtaining a single elliptic domain bordered by a connected parabolic curve.

This procedure, however, creates two new special parabolic points on the boundary of each elliptic strip. Therefore, the resulting surface has a parabolic curve with a single component and 12 special points.

The figure (from [1]) shows the cusp edge of the dual surface before and after application of the Panov procedure (for a definition of elliptic and hyperbolic swallowtails, dual to elliptic and hyperbolic special points, see problem *1997-6*).

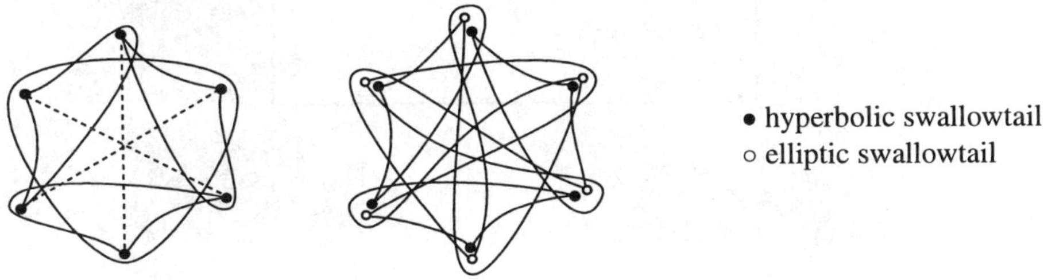


The self-intersection curve (dotted line) of the dual surface is not drawn in the case of a single cusp-edge: it is even more complicated than the cusp-edge.

This figure indicates that the number of connected components of the parabolic curve is not a good parameter of the geometric "complexity."

The conjecture that the cubic surface attains the minimal complexity holds. The conjecture generalizing the Möbius theorem should be reformulated in this way: *there are no smooth perturbations of the real projective plane whose parabolic curve has less than six special parabolic points* (see problem *1998-2*).

[1] AICARDI F. Letter to D. Panov, November 1997.

[2] ARNOLD V. I. Topological problems in wave propagation theory and topological economy principle in algebraic geometry. In: The Arnoldfest. Proceedings of a conference in honour of V. I. Arnold for his sixtieth birthday (Toronto, 1997). Editors: E. Bierstone, B. A. Khesin, A. G. Khovanskiĭ and J. E. Marsden. Providence, RI: Amer. Math. Soc., 1999, 39–54. (Fields Inst. Commun., 24.)

[3] PANOV D. A. Parabolic curves and gradient mappings. *Proc. Steklov Inst. Math.*, 1998, **221**, 261–278.

## 1995-8 — *F. Aicardi*

ℛ Here a *curve* is the image of a smooth immersion of the circle into the plane. A cooriented *front* is the projection to the plane of the image of a Legendrian immersion of the circle into the space of cooriented contact elements of the plane. A cooriented curve is a front without cusps.

All cooriented curves with at most four double points with equal Legendrian knot (i. e., connectable by a path in the space of fronts, avoiding the equally directed self-tangencies) are connectable without equally directed self-tangencies in the space of curves.

The first example of curves, with equal Legendrian knot, that I am unable to connect in the space of curves without equally directed self-tangencies, consists in the following pair of curves with five double points.

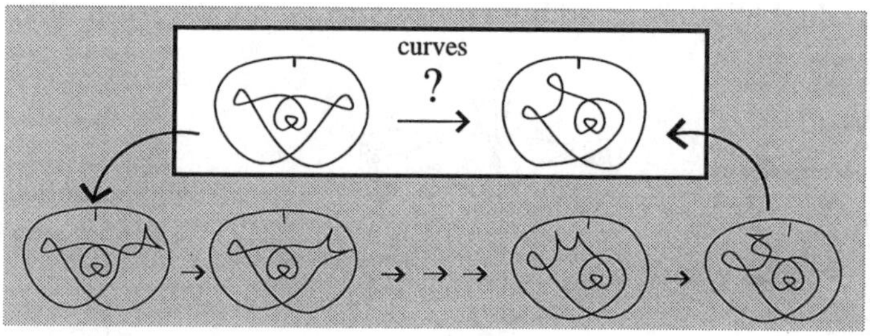


## ▽ 1995-9 — *F. Aicardi*

ℛ All pairs of Legendrian knots in $ST^*\mathbb{R}^2$, having smooth fronts with at most four double points, and isotopic as equipped knots, are Legendrian isotopic. The first pair of smooth curves, lifted to Legendrian knots isotopic as framed knots, that I am unable to connect by a path avoiding equally directed self-tangencies in the space of fronts, are shown in the following figure (the curves have the same Bennequin invariant, because both of them have $J^+ = 0$; a knot in the solid torus isotopic to their Legendrian lifting is shown at the right).

Yu. V. Chekanov and P. E. Pushkar′ [1] proved that there exist pairs of Legendrian knots which are Legendrian not isotopic but isotopic as framed knots in

$\mathbb{R}^3$ with standard contact form. In their example the corresponding fronts have six double points and ten cusps. The topological type of these Legendrian knots, according to [2], is the simplest one.

In the same paper they also give a proof of the Arnold conjecture (see problem *1993-9*): two circles with opposite coorientations are not connectable, avoiding equally directed self-tangencies, in the space of fronts with less than four cusps.

A new natural question is the following: let $c$ be a smooth cooriented curve in the plane (or a front with zero Maslov index), and $\bar{c}$ the curve obtained from $c$ by reversing the coorientation. Are there pairs of curves $c$, $\bar{c}$ such that their liftings to Legendrian knots are not Legendrian isotopic? I conjectured that they exist, and that the simplest examples are the following:

Recent results related to this question are presented in [5], section 4.

**Remarks.** Examples of pairs of isotopic Legendrian links in $ST^*\mathbf{R}^2$ which are not equivalent via an isotopy of contact diffeomorphisms are given in [6] (see also [3,4]).


[1] CHEKANOV YU. V., PUSHKAR′ P. E. The combinatorics of fronts of Legendrian knots. Arnold's conjecture and invariants of Legendrian knots. Preprint (in Russian).

[2] ETNYRE J. B., HONDA K. Knots and contact geometry. [*Internet*: `http://www.arXiv.org/abs/math.DG/0006112`]

[3] FERRAND E. Familles génératrices et nœuds legendriens. Thèse, École Polytechnique, 1997, Chapitre II, section 4.2.6.

[4] FRASER M. Example of nonisotopic Legendrian curves not distinguished by the invariants *tb* and *r*. *Internat. Math. Res. Notices*, 1996, **19**, 923–928.

[5] LENNY NY L. Computable Legendrian invariants. Preprint, 2000.

[6] TRAYNOR L. Legendrian circular helix links. *Math. Proc. Cambridge Phil. Soc.*, 1997, **122**(2), 301–314.


### △ 1995-9 – *E. Ferrand*

The problem was solved in 1995 by Yu. V. Chekanov (still unpublished). The solution was also announced by Ya. M. Eliashberg and H. Hofer at the conference in Toronto in honor of V. I. Arnold for his sixtieth birthday (Arnoldfest, 1997).

### ▽ 1995-10 – *E. Ferrand*

The problem was solved independently in papers [1, 2].


[1] ENTOV M. R. On the necessity of Legendrian fold singularities. *Internat. Math. Res. Notices*, 1998, **20**, 1055–1077.

[2] FERRAND E., PUSHKAR′ P. E. Non-cancellation of cusps on wave fronts. *C. R. Acad. Sci. Paris, Sér.* I *Math.*, 1998, **327**(9), 827–831.


### △ 1995-10 – *F. Aicardi*

In fact, it is proved in [1, 2] (see the comment by E. Ferrand), that there exist Legendrian isotopy classes with zero Maslov index containing no Legendrian knots with immersed fronts. An example of a cooriented front, for which at least two of the cusps cannot be eliminated by means of Legendrian isotopy, is given in [2]. This front is shown in the figure (left), together with a simpler front (right), for which the non-cancellation of cusps is implicitly proved in [3]. This last one seems the simplest example of a front with the requested property.

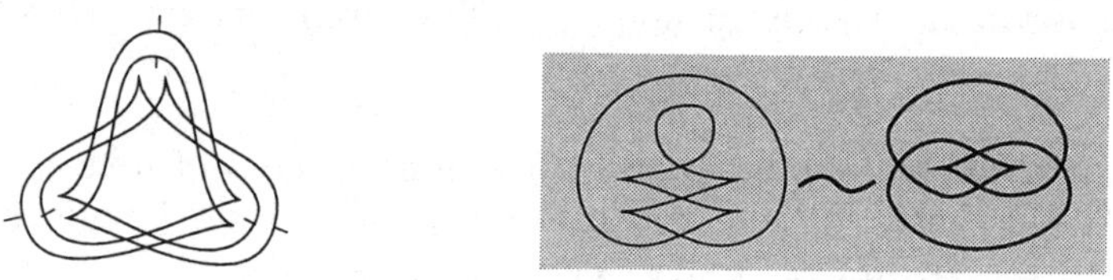


[3] FERRAND E., PUSHKAR′ P. E. Non-cancellation of singularities on wave fronts. Preprint, CEREMADE 9842, Université Paris-Dauphine, 1998.


### ▽ 1995-11 – *E. Ferrand*

The problem is still open in this form.

### △ 1995-11 – *B. Z. Shapiro*

Paper [1] contains a criterion when a planar tree-like curve can be realized without inflections. (The treelikeness of a plane curve is the property that the curve splits into two connected components upon the removal of its arbitrary double point.) The problem on the calculation of the minimal number of inflections for the class of tree-like curves is reduced in [1] to a discrete optimization problem. The latter apparently has no solution in a closed form.

[1] SHAPIRO B. Z. Tree-like curves and their number of inflection points. In: Differential and Symplectic Topology of Knots and Curves. Editor: S. Tabachnikov. Providence, RI: Amer. Math. Soc., 1999, 113–129. (AMS Transl., Ser. 2, 190; Adv. Math. Sci., 42.)

## 1995-12

See the comments to problem *1998-17*.

## ▽ 1995-13 – *V. I. Arnold* Also: *1993-39, 1993-40*

That the number of cusps is not less than four, is true for an arbitrary convex surface. Conjecturally, not only the first, but the second, third, ..., caustic of a generic "pole" on a generic convex surface has at least four cusps, but this has not been proved even for surfaces $C^\infty$-close to a sphere. Nevertheless, for each $N$ there are at least four cusps on the $N$-th caustic provided that the distance from a sphere is less than some sufficiently small $\varepsilon(N)$.

Jacobi integrated the equation for geodesics on the ellipsoid in θ-functions, and hence the question on the singular points of caustics on ellipsoids belongs to algebraic geometry.

However, this problem is apparently beyond the scope of contemporary algebraic geometrists that are interested in number theory and complex geometry rather than *real* problems of *real* geometry.

## △ 1995-13 – *V. M. Zakalyukin* Also: *1993-39, 1993-40*

The result of V. I. Arnold on estimating the number of cusp points of caustics of exponential mappings on surfaces close to a sphere was published in [1].

This problem has generated lots of others. For example, M. E. Kazarian in [2] found a topological invariant counting the number of singularities of type $D_4$ of the caustics of families of normals to hypersurfaces in Riemannian 3-dimensional manifolds.

In [3] it was proved that the Maxwell stratum of the system of normals to a many-dimensional ellipsoid has minimal complexity with respect to an arbitrary closed hypersurface of the Euclidean space.

[1] ARNOLD V. I. Topological Invariants of Plane Curves and Caustics. Dean Jacqueline B. Lewis Memorial Lectures, Rutgers University. Providence, RI: Amer. Math. Soc., 1994. (University Lecture Series, 5.)

[2] KAZARIAN M. E. Umbilical characteristic numbers of Lagrangian mappings of 3-dimensional pseudo-optical manifolds. In: Singularities and Differential Equations (Warsaw, 1993). Editors: S. Janeczko, W. M. Zając̨zkowski and B. Ziemian. Warsaw: Polish Academy of Sciences, Institute of Mathematics, 1996, 161–170. (Banach Center Publ., 33.)

[3] ZAKALYUKIN V. M. Maxwell stratum of Lagrangian collapse. *Proc. Steklov Inst. Math.*, 1998, **221**, 187–201.

# 1996

## 1996-2 — *V. A. Vassiliev*

ℛ The theory of such cohomology groups is similar to the theory of finite type invariants of generic immersions, in particular for all of them the natural filtration (by "orders") is well defined. The calculation of these groups is based on the natural resolution of the discriminant variety of all immersions with forbidden singularities, and leads to the study of complexes of connected hypergraphs in the same way as the homological study of knot spaces is related to complexes of connected graphs; see [2]. However, the answers look more exotic; see [3]. For instance, the 1-dimensional cohomology group of the space of immersions $\mathbb{S}^1 \to \mathbb{R}^2$ without 4-fold self-intersections has the following components $\mathcal{F}_i$ of low orders: $\mathcal{F}_1 = \mathcal{F}_2 = 0$, $\mathcal{F}_3 = \mathbb{Z}_2$, $\mathcal{F}_4/\mathcal{F}_3 = \mathbb{Z}_5$, $\mathcal{F}_5/\mathcal{F}_4 = \mathbb{Z}^2$. The absence of a free cohomology class of the lowest possible order (equal to 3) is related to the fact that the configuration space of quadruples of points in $S^1$ is non-orientable (while the existence of the "strangeness" invariant from [1] is ensured by the orientability of the configuration space of triples). The first free group $\mathcal{F}_5/\mathcal{F}_4$ in this list is the homology group of the complex of connected 3-hypergraphs with 6 vertices, *anti-invariant* under the cyclic permutation of these vertices.

[1] ARNOLD V. I. Plane curves, their invariants, perestroikas and classifications. In: Singularities and Bifurcations. Editor: V. I. Arnold. Providence, RI: Amer. Math. Soc., 1994, 33–91. (Adv. Sov. Math., 21.)

[2] VASSILIEV V. A. Complexes of connected graphs. In: The Gelfand Mathematical Seminars 1990–1992. Editors: L. Corwin, I. Gelfand and J. Lepowsky. Boston, MA – Basel: Birkhäuser, 1993, 223–235.

[3] VASSILIEV V. A. On finite order invariants of triple point free plane curves. In: Differential Topology, Infinite-Dimensional Lie Algebras, and Applications. D. B. Fuchs' 60th Anniversary Collection. Editors: A. Astashkevich and S. Tabachnikov. Providence, RI: Amer. Math. Soc., 1999, 275–300. (AMS Transl., Ser. 2, 194; Adv. Math. Sci., 44.)

## 1996-3 — *S. V. Duzhin, Ya. G. Mostovoy*

The fact that $SP^n(\mathbb{R}P^2)$ (the $n$-th symmetric power of $\mathbb{R}P^2$) is homeomorphic to $\mathbb{R}P^{2n}$ was established in [3]. In the appendix of this paper the authors discuss the symmetric powers of arbitrary non-orientable surfaces.

Speaking more generally, there are two approaches to the proof of this equality:

(a) through the "topological Maxwell theorem" (see, e. g., [1, 2]);

(b) as a quaternionic version of the fact that $SP^n(S^2) = \mathbb{C}P^n$. Consider the vector space of all meromorphic functions on the Riemannian sphere such that: (1) the antipodal map takes the function into its complex conjugate, (2) the poles are only at 0 and $\infty$ and have order no higher than $n$. The projectivization of this space, on one hand, coincides with $\mathbb{R}P^{2n}$ and, on the other hand, with $SP^n(\mathbb{R}P^2)$ (thinking in terms of the zeros of these functions). The quaternions enter because the antipodal map of $S^2$ can be realized as the multiplication by the quaternionic unit $j$.

[1] ARNOLD V. I. Topological content of the Maxwell theorem on multipole representation of spherical functions. *Topol. Methods Nonlinear Anal.*, 1996, **7**(2), 205–217.

[2] ARNOLD V. I. Lectures on Partial Differential Equations, 2nd supplemented edition. Moscow: PHASIS, 1997 (in Russian).

[3] DUPONT J. L., LUSZTIG G. On manifolds satisfying $w_1^2 = 0$. *Topology*, 1971, **10**, 81–92.

## 1996-4

See the comment to problem *1993-24*. Cf. also problem *1985-24*.

### ▽ 1996-5 — *A. E. Eremenko, D. I. Novikov*

ℛ The answer is positive under some mild assumptions, e. g., if $f$ is a locally integrable temperate distribution. With more general interpretations of Fourier transform, counterexamples exist; see preprint [1].

[1] EREMENKO A. E., NOVIKOV D. I. Oscillation of Fourier integrals with a spectral gap.
[*Internet*: `http://www.arXiv.org/abs/math.CA/0301060`]

### △ 1996-5 — *S. B. Kuksin*

ℛ Sturm's oscillation theorem [8] states that the $k$-th ($k \geq 1$) eigenfunction of the Sturm–Liouville equation

$$-(f(x)y'(x))' + g(x)y(x) = \lambda_k y(x), \quad x \in [0, \pi], \tag{1}$$

supplemented by general boundary conditions, has as many zeros as the $k$-th eigenfunction of the trivial equation $-\tilde{y}'' = \tilde{\lambda}_k \tilde{y}$ has under the same boundary conditions; see, e. g., [4, 6]. This classical result was generalised and developed in the last century (some generalisations are given in [6], Chapter X). For example, in 1916 Kellogg obtained the following result which we formulate according to [3]: *Let $\varphi_k(x)$ be the k-th eigenfunction of equation* (1), *supplemented by "sufficiently general" boundary conditions. Then for all integers k and m* ($k < m$), *linear combinations $\sum_{j=k}^m c_j \varphi_j(x)$ (where $\sum_{j=k}^m c_j^2 > 0$) have not less than $k-1$ sign changes and not more than $m-1$ zeros in the interval* $(0; \pi)$. For example, a non-zero function $\sum_{j=k}^m c_j \sin jx$, $x \in (0; \pi)$, has no less than $k-1$ changes of sign since $\sin jx$ is the $j$-th eigenfunction of the operator $-\partial^2/\partial x^2$ with zero boundary conditions. Moreover, the Sturm–Hurwitz theorem holds for convex curves (i. e., Chebyshev systems) regardless of eigenfunctions [5].

The theorem stated above has been rediscovered more than once. In particular, in 1990 it was reproved by S. L. Tabachnikov [9] for the case of the simplest equation $-y'' = \lambda_k y$ under $2\pi$-periodic boundary conditions: *a non-zero real periodic function $\sum_{j=k}^\infty (c_j \sin jx + d_j \cos jx)$ has at least 2k changes of sign within its period.*

Problem *1996-5* is a continuous (in $k$) analogue of this theorem, since a function $f(x)$ under the above condition equals $2\int_\omega^\infty (\operatorname{Re} F(l) \cos lx - \operatorname{Im} F(l) \times \sin lx)\, dl$, i. e., there are no frequencies $\sin lx$ and $\cos lx$ with $l < \omega$ in its Fourier integral.

This problem (and Kellogg's theorem) has an important analogue in the theory of random processes. Let $\xi(t)$ be a real stationary *Gaussian* process such that that $\mathsf{M}\xi(0)=0$ and $\mathsf{M}\xi(0)^2=1$ (M stands for the mathematical expectation). Let $r(t)$ be its correlation function: $r(t)=\mathsf{M}\xi(0)\xi(t)$. Suppose that $r$ is integrable and denote by $f(\lambda)$ the spectral density of the process. Then $r(t)=\int e^{it\lambda}f(\lambda)\,d\lambda$. Physically, $f(\lambda)$ is the spectral density of the energy of the process (see [2]). Suppose also that $f(\lambda)=0$ for $|\lambda|\le\omega$, which means that the energy of the process $\xi(x)$ stays out of the "low frequency" interval $(-\omega;\omega)$ (now the Fourier transform $\hat{\xi}(\lambda)$ of the random function $\xi(x)$, understood as in the theory of generalized functions of moderate growth, almost certainly vanishes for $|\lambda|<\omega$). What can be said concerning the density of zeros of the random function $\xi(t)$?

Let $\mathcal{E}_{(T_0;T_0+T]}$ be a random variable, equal to the number of zeros of the random function $\xi(t)$ in an interval $(T_0;T_0+T]$. Due to the stationarity, its mean value $E_T$ does not depend on $T_0$ and is proportional to $T$: $E_T=TE_1$. For $E_1$, in the Gaussian case which we discuss, the following Reiss's formula holds (see [2]):

$$E_1=\frac{1}{\pi}\lambda_2^{1/2},$$

where $\lambda_2$ is the second spectral moment: $\lambda_2=\int\lambda^2 f(\lambda)\,d\lambda$. Since $f=0$ for $|\lambda|\le\omega$, then $\lambda_2\ge\omega^2\int f(\lambda)\,d\lambda=\omega^2$, and hence

$$E_1\ge\frac{\omega}{\pi}.$$

For a fixed trajectory $\xi(t)$ we have:

$$\frac{1}{T}\mathcal{E}_{(0;T]}=\frac{1}{T}\left(\mathcal{E}_{(0;1]}+\cdots+\mathcal{E}_{(T-1;T]}\right). \tag{2}$$

Since the correlation $r(t)$ is integrable, the identically distributed random variables $\mathcal{E}_{(j;j+1]}$ and $\mathcal{E}_{[(l;l+1]}$ are weakly correlated for large $|l-j|$. Therefore, the Strong Law of Large Numbers applies to the right-hand side of (2), and

$$\frac{1}{T}\mathcal{E}_{(0;T]}\to E_1\ge\frac{\omega}{\pi}\quad\text{as}\quad T\to\infty,$$

for almost every trajectory $\xi(t)$. This proves the claim of problem *1996-5* for stationary Gaussian processes.

In the non-Gaussian case, very little is known concerning the random variables $\mathcal{E}_{(0;T]}$.

For a real Fourier integral $f(x) = \int F(\lambda)e^{i\lambda x}\, d\lambda$ the following analogue of the second claim of Kellogg's theorem is true: if $F(\lambda) = 0$ for $|\lambda| \geq m$ and $n(l)$ is the number of zeros of $f(x)$ in the interval $[-l; l]$, then $\underline{\lim}\, n(l)/2l \leq m/\pi$ as $l \to \infty$. This result follows immediately from Jensen's formula (see [7], § 5), applied to the function $f$, extended to an integer function $f(z)$ of the complex variable $z$.

In the multidimensional case, the following weaker form of Sturm's theorem is true: the number of domains of the sign changes for $k$-th eigenfunction of a self-adjoint elliptic differential operator of the second order is less then $k$, notwithstanding the number of independent variables (see [1]). Now there is no analogue of the second claim of the Kellogg's theorem: the above is not true for the linear combination of the first $k$ eigenfunctions.

See also the comment to problem *2000-9* and references therein.


[1] COURANT R., HILBERT D. Methods of Mathematical Physics, Vol. I. New York: Wiley Interscience, 1953; reprinted 1989. [*The German original* 1943.]

[2] CRAMÉR H., LEADBETTER M. R. Stationary and Related Stochastic Processes. Sample Function Properties and their Applications. New York – London – Sydney: John Wiley, 1967.

[3] GANTMACHER F. R., KREĬN M. G. Oscillating Matrices and Small Oscillations of Mechanical Systems Moscow–Leningrad: OGIZ, 1941 (in Russian). [*The German translation*: Oscillationsmatrizen, Oscillationskerne und kleine Schwingungen mechanischer Systeme. Berlin: Akademie Verlag, 1960.]

[4] HARTMAN P. Ordinary Differential Equations. New York – London – Sydney: John Wiley, 1964.

[5] HURWITZ A. Über die Fourierschen Konstanten integrierbarer Funktionen. *Math. Ann.*, 1903, **57**(4), 425–446.

[6] INCE E. L. Ordinary Differential Equations. New York: Dover, 1944.

[7] LEVIN B. YA. Distribution of Zeros of Entire Functions. Providence, RI: Amer. Math. Soc., 1964. [*The Russian original* 1956.]

[8] STURM J. C. F. Mémoire sur les équations différentielles du second ordre. *J. Math. Pures Appl.*, 1836, **1**, 106–186.

[9] TABACHNIKOV S. L. Around four vertices. *Russian Math. Surveys*, 1990, **45**(1), 229–230.


## 1996-6 — *F. Aicardi*

This problem arises from the attempt to generalize to higher dimensions theorems and conjectures related to the eversion of the circle in the plane [1, 2] (see also problem *1993-9*). The starting point is to consider families of surfaces equidistant (called fronts) to some convex surfaces (in a certain class of smooth perturbations of the sphere) in the Euclidean tridimensional space and to observe their singularities. Hence to formulate conjectures on the set of those singularities that are unavoidable for every family of fronts realizing the eversion of a surface in that class and in larger ones.

The economy principle (see the comment to problem *1995-3*) suggests that the simplest scenario of eversion of a compact surface diffeomorphic to a sphere is realized by the equidistant fronts from surfaces given by the simplest equations. This is what happens in the plane: the equidistant fronts from an ellipse have at most four cusps. In fact, four cusps are necessary for the eversion, not only of a convex curve diffeomorphic to a circle by equidistant fronts, but also for a more general class of curves and a more general definition of fronts (see [1] and problem *1993-9*). For this reason Arnold proposed the study of what happens in the case of ellipsoids and quadraticoids.

Some surprising facts came from the observation.

In the space of ellipsoids, we consider equivalently two ellipsoids having similar scenarios, i. e., whose fronts experience the same ordered sequence of singularities. The number of such events (called basic events) is 10 for all generic scenarios. For the most part, these events consist of multiple singularities, i. e., up to four codimension one singularities occurring at the same time. So, the single (codimension one) events are in all 19: they should occur at different times in a family of equidistant fronts from a small perturbation of the ellipsoid breaking its symmetries.

Among these 19 events, 16 are typical singularities of wave fronts and 3 are self-tangencies of the front.

Two equivalence classes are separated by degenerate ellipsoids, which form the discriminant. For these ellipsoids two basic events happen at the same time. Crossing the discriminant, the two events reverse the order of occurrence. It turn out that the discriminant is itself degenerate, because on it up to three pairs of basic events become pairs of simultaneous events (at different times but along the same scenario).

The non-evident fact is that for the the quadraticoids the situation is the same: there are no quadraticoids realizing scenarios different from those realized by equidistant fronts from ellipsoids and vice-versa.

Notice that a front equidistant from a quadraticoid with semiaxes $A$, $B$ and $C$ is itself a quadraticoid (with semiaxes $A+t$, $B+t$ and $C+t$).

A quadraticoid is convex if the ratios between any two semiaxes is inside $(\frac{1}{2}, 2)$, i. e., a quadraticoid cannot be thin or flat as an ellipsoid can be.

Choosing the semiaxes $a$, $b$, $c$ of the ellipsoids and $A$, $B$, $C$ of the quadraticoids (with the convexity restrictions) as coordinates in the spaces of these objects, the diffeomorphism between the discriminants is given by: $a = e^A$, $b = e^B$, $c = e^C$.

In the figure, sections of the spaces of ellipsoids and quadraticoids are shown (respectively, $c = \text{const}$, $a < c$, $b < c$, and $C = \text{const}$, $A < C$, $B < C$). The discriminant (black lines) separates six classes of ellipsoid ($a < b < c$) and quadraticoids ($A < B < C$). The diffeomorphism acts on discriminants inside the grey squares.

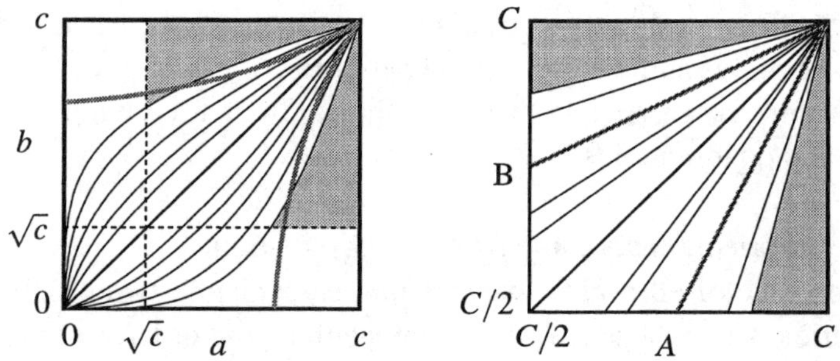


**Remark.** The property of ellipsoids and quadraticoids of having the same images under the Gauss map, discussed by Darboux [4], was uselessly rediscovered in order to understand such diffeomorphism. Indeed, an ellipsoid (with semiaxes $a \le b \le c$) and a quadraticoid (with semiaxes $A \le B \le C$) have the same image if their semiaxes satisfy: $a = 1/\sqrt{C}$, $b = 1/\sqrt{B}$ and $c = 1/\sqrt{A}$. So, their equidistant fronts may have different scenarios.

The caustics of ellipsoids are all diffeomorphic, as well as those of the quadraticoids, except for three cones in the space of ellipsoids and three planes in the space of quadraticoids. On these surfaces the two closed cusp-edges of the caustic intersect in two opposite points. Their sections $c = \text{const}$ and $C = \text{const}$ respectively, are shown in the figure (grey lines). Such sorts of discriminant, evidently, are not mapped one into another by the above diffeomorphism.

There is no evidence to disprove the conjecture by Zakalyukin that the caustics of the ellipsoids are the simplest ones.

Observing the families of equidistant fronts from ellipsoids and quadraticoids, one may formulate the following conjectures on the minimal number of unavoidable singularities attained in a family of fronts realizing the eversion of a deformed sphere:

**Conjecture 1.** *There exists a front with at least* 4 *swallowtails.*

**Conjecture 2.** *There exists a front whose cuspidal edge has at least four components.*

**Remark.** There are families containing fronts with 8 swallowtails, but not more than four components of the cuspidal edge.

However, the following conjecture is against the economy principle.

**Conjecture 3.** *The minimal number of singularities necessary to the eversion by equidistant fronts of a convex surface diffeomorphic to a sphere is* 12 (*not counting self-tangencies*).

**Remark.** Conjecture 3 is equivalent to the following generalization of the four vertices theorem [3], stating that the curvature radius of a convex curve on the plane has at least four extremes.

Let $r$ and $R$ ($r \leq R$) be the principal curvature radii of a generic convex surface diffeomorphic to a sphere.

**Conjecture 4.** *On the surface, $r$ attains at least a minimum, a saddle, and two maxima; $R$ attains at least two minima, a saddle and a maximum.*

**Remark.** For ellipsoids and quadraticoids both $r$ and $R$ attain two maxima, two minima and two saddles.

[1] Arnold V. I. Topological Invariants of Plane Curves and Caustics. Dean Jacqueline B. Lewis Memorial Lectures, Rutgers University. Providence, RI: Amer. Math. Soc., 1994. (University Lecture Series, 5.)

[2] Arnold V. I. Invariants and perestroikas of plane fronts. *Proc. Steklov Inst. Math.*, 1995, **209**, 11–56.

[3] Barner M., Flohr F. Der Vierscheitelsatz und seine Verallgemeinerung. *Der Mathematikunterricht*, 1958, **4**, 43–73.

[4] Darboux J. G. Leçons sur la théorie générale des surfaces, Vol. 1, 2$^{e}$ édition. Paris, 1914, § VII, 291–293.

## 1996-8

See the comments to problem *1970-15*.

## 1996-9 — *V. D. Sedykh* Also: *1998-7*

ℜ I know only two facts (see [1]) related to these two problems.

**Theorem 1.** *There is an open set of smooth closed curves in* $\mathbb{R}\mathrm{P}^n$ *that are Barner-convex and have no convex projections into any hyperplane.*

**Theorem 2.** *Let* $\gamma$ *be a smooth closed curve in* $\mathbb{R}\mathrm{P}^n$*. Then the set of points in* $\mathbb{R}P^n \setminus \gamma$ *such that* $\gamma$ *is projected from them into a convex curve is either empty or a connected component of the complement in* $\mathbb{R}\mathrm{P}^n$ *of the union of the osculating hyperplanes at flattening points of the curve* $\gamma$.

[1] SEDYKH V. D. On some classes of curves in a projective space. In: Geometry and Topology of Caustics—CAUSTICS'98 (Warsaw). Editors: S. Janeczko and V. M. Zakalyukin. Warsaw: Polish Academy of Sciences, Institute of Mathematics, 1999, 237–266. (Banach Center Publ., 50.)

## 1996-13

ℜ See the comment to problem *1970-15.*

## 1996-14 — *M. L. Kontsevich*

ℜ The same question can be asked for an arbitrary odd-dimensional manifold with a closed 2-form of the maximal possible rank everywhere. This is a big business now; Ya. Eliashberg and H. Hofer defined cohomology in this case similarly to Floer cohomology. See, e. g., Eliashberg's talk [1] at ICM Berlin, 1998.

[1] ELIASHBERG YA. M. Invariants in contact topology. In: Proceedings of the International Congress of Mathematicians, Vol. II (Berlin, 1998). *Doc. Math.*, 1998, Extra Vol. II, 327–338 (electronic).

## 1996-15 — *V. I. Arnold*

ℜ The group of linear transformations preserving the cone $x^2 + y^2 = z^2$ acts on the two-sheeted hyperboloid and respects the Lobachevskian metric determined by the intersection of a tangent plane to the hyperboloid with another infinitely close hyperboloid. On the hyperboloid of one sheet, such a metric is Lorentzian. Projecting the hyperboloids into the projective plane of lines passing through the origin, we obtain the metric of the Klein model of the Lobachevskian plane inside the circle and the de Sitter metric on the complementary plane.

This construction is discussed in detail in [1]. See also problem *2002-16.*

[1] ARNOLD V. I. Arithmetics of Binary Quadratic Forms, Symmetry of Their Continued Fractions, and Geometry of Their de Sitter World. Moscow: Moscow Center for Continuous Mathematical Education Press, 2002.

## 1996-17

ℜ See the comment to problem *1981-9.*

## 1996-18

ℜ See the comment to problem *1981-9.*

## 1996-20 — *M. B. Sevryuk*

ℜ The source of this problem is Herman's theorem [2, 3]. According to this theorem, an invariant torus of a Hamiltonian system (or symplectomorphism) is automatically *isotropic* provided that this torus carries quasi-periodic motions and that the symplectic structure is exact. In particular, if the symplectic structure is exact then all the invariant tori of the KAM theory are isotropic and, consequently, their dimensions do not exceed the number of degrees of freedom. These questions are discussed in detail in book [1]; see also the earlier paper [4].

[1] BROER H. W., HUITEMA G. B., SEVRYUK M. B. Quasi-Periodic Motions in Families of Dynamical Systems: Order amidst Chaos. Berlin: Springer, 1996. (Lecture Notes in Math., 1645.)

[2] HERMAN M. R. Existence et non existence de tores invariants par des difféomorphismes symplectiques. Palaiseau: École Polytech., Centre de Math., 1988. (Séminaire sur les Équations aux Dérivées Partielles 1987–1988, Exp. № 14.)

[3] HERMAN M. R. Inégalités "a priori" pour des tores lagrangiens invariants par des difféomorphismes symplectiques. *Inst. Hautes Études Sci. Publ. Math.*, 1989, **70**, 47–101.

[4] QUISPEL G. R. W., SEVRYUK M. B. KAM theorems for the product of two involutions of different types. *Chaos*, 1993, **3**(4), 757–769.

## 1996-21 — *M. B. Sevryuk*

ℛ Give an example of a field $V$ with the phase space $M$ different from $\mathbb{R}^n$ such that $V$ is irreversible but the time 1 flow map of $V$ is reversible.

Let $M$ be the disjoint union of two copies $\mathbb{R}_1^2$ and $\mathbb{R}_2^2$ of the plane $\mathbb{R}^2$ with polar coordinates $(r_1, \varphi_1)$ and $(r_2, \varphi_2)$, respectively. Introduce the notation $f(r) := r(r^2-1)$ and consider the vector field $V$ on $M$ governing the system of equations

$$\begin{cases} \dot\varphi_1 = 2\pi \\ \dot r_1 = f(r_1), \end{cases} \qquad \begin{cases} \dot\varphi_2 = 4\pi \\ \dot r_2 = -f(r_2). \end{cases}$$

This field is not reversible with respect to any involution of the phase space $M$.

Indeed, suppose that some involution $G: M \to M$ reverses the field $V$. Since $\gamma_1 = \{r_1 = 1\} \subset \mathbb{R}_1^2$ is the only periodic trajectory of the field $V$ of period 1 while $\gamma_2 = \{r_2 = 1\} \subset \mathbb{R}_2^2$ is the only periodic trajectory of the field $V$ of period $1/2$, then $G(\gamma_1) = \gamma_1$, $G(\gamma_2) = \gamma_2$ and, consequently, $G(\mathbb{R}_1^2) = \mathbb{R}_1^2$, $G(\mathbb{R}_2^2) = \mathbb{R}_2^2$. Thus, one concludes that each of the two systems on $\mathbb{R}^2$

$$\begin{cases} \dot\varphi = 2\pi \\ \dot r = f(r) \end{cases} \quad \text{and} \quad \begin{cases} \dot\varphi = 4\pi \\ \dot r = -f(r) \end{cases}$$

taken separately is reversible, which is not true: the only equilibrium $r = 0$ of the first system is an attractor and that of the second system is a repeller.

On the other hand, the time 1 flow map of the field $V$ coincides with the time 1 flow map of the field governing the system

$$\begin{cases} \dot\varphi_1 = 0 \\ \dot r_1 = f(r_1), \end{cases} \qquad \begin{cases} \dot\varphi_2 = 0 \\ \dot r_2 = -f(r_2). \end{cases}$$

The latter system is reversible with respect to the involution

$$(\varphi_1, r_1) \mapsto (\varphi_2, r_2), \qquad (\varphi_2, r_2) \mapsto (\varphi_1, r_1)$$

interchanging $\mathbb{R}_1^2$ and $\mathbb{R}_2^2$.

I am unaware of an example of a vector field with the properties pointed out in the second part of the problem—on any phase space.

# 1997

## 1997-2 – *F. Aicardi*

The number of regions of $\mathbb{R}^n$ where the coordinates are all different is equal to $n!$. Since a selector chooses one of the $n$ coordinates $x_i$ in each region, the total number of selectors is $n^{n!}$.

The image of a Matov selector is the entire set of coordinates. I call a selector having this property *proper*. The number of proper selectors is $n^{n!} - (n-1)^{n!+1}$.

Calculations indicate that the number of Matov selectors in $\mathbb{R}^n$ is less than $(n-1)^n$. The fraction of Matov selectors among proper selectors diminishes very rapidly: it is of exponent $10^{-2}$ for $n=3$, $10^{-557}$ for $n=6$ and $10^{-346279}$ for $n=9$.

## 1997-6 – *F. Aicardi*

A generic function $f$ on the plane, elliptic in a domain which is bounded by a smooth connected parabolic curve and contains two elliptic and one hyperbolic special points, is constructed as follows [1].

A cubic equation, giving the perturbation of the real projective plane in the projective space has the form $z(x^2+y^2+t^2) = p(x,y,t)$, where $p(x,y,t)$ is a homogeneous polynomial of degree 3 in $x$, $y$ and $t$.

The affine equation $z = p(x,y,1)/(1+x^2+y^2)$ defines the graph of a function $f$ on the plane having, for appropriate values of the coefficients of the polynomial $p$, four bounded elliptic domains. In the generic case, there are six singular parabolic points whose repartition on the boundaries of the four elliptic domains is $\{(0)(1)(2)(3)\}$. We request also that, in the elliptic domains whose boundaries contain zero and one special point respectively, the Hessian of $f$ has the same signature. In this case we can apply the Panov procedure (see the comment to problem *1995-3*) to join these two elliptic domains by an elliptic strip. This procedure creates two new special points of elliptic type on the boundary of the strip. So we obtain a parabolic curve with two elliptic special points and one hyperbolic.

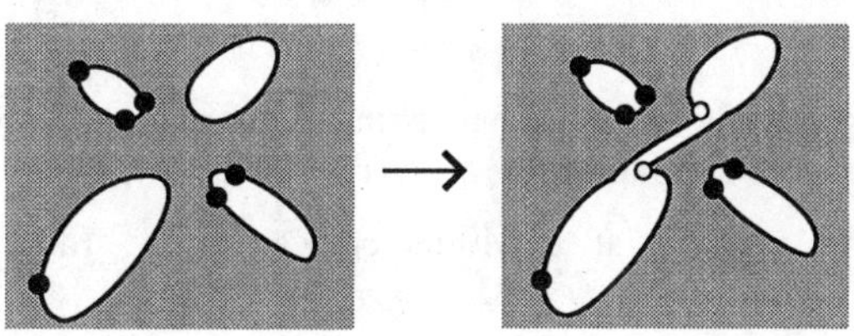

• hyperbolic special points
∘ elliptic special points

[1] AICARDI F. Letter to D. Panov, November 1997.

## ▽ 1997-8 — *V. I. Arnold*

ℛ Probably, the point to which a "conic neighborhood" of the pyramid contracts has moduli and even functional moduli. The idea for the proof of stability in a neighborhood of the main part of the caustic (say, for $A \neq 0$ in Example 3) is the following.

Fix values of the parameters $A_0$, $a_0$, $b_0$, $c_0$ under which the function of the family has four real critical points with distinct critical values. A nearby family also has four real critical points. The values of the four parameters of this nearby family may be chosen in such a way that the critical values of the nearby family coincide with those of our trigonometric polynomial. The corresponding function on the circle can be obtained from this trigonometric polynomial by a (uniquely determined) small diffeomorphism of the circle.

This construction is then extended on our four parameters up to the caustic (the boundary for the existence of four critical points) and even somewhat further (in the analytic case when one can monitor critical values at complex points). Thus the holomorphic stability in a "conic neighborhood" of the caustic can be proved. In the smooth case, the justification of the possibility of extending the normalizing diffeomorphism beyond the caustic is analogous to the use of the preliminary Malgrange theorem in the similar local problem. However, this construction must be applied here to functions on the circle but not to function germs at a point.

## △▽ 1997-8 — *Yu. M. Baryshnikov, M. Garay*

ℛ The stability of the bifurcation diagram from Example 4 was proved using the methods of [1]. The proof will be published in an extended version of the thesis.

Partial results were obtained by R. Uribe-Vargas who also found other appearances of the same bifurcation diagram in related problems [2].

[1] GARAY M. The classical and Legendrian theory of vanishing flattening points of plane and spatial curves. Ph. D. Thesis, Université Paris 7, 2001.

[2] URIBE-VARGAS R. On vanishing vertices at a Morse critical point. Talk at V. I. Arnold's seminar, 1995.

## △ 1997-8 – *R. Uribe-Vargas*

ℛ Up to now, there is no general solution (or theory) for these problems. Some of them have been solved (or partially solved) by different people.

In [1] it is proved that Agrachev's caustic (Example 2) is, in the simplest case, not stable at the origin, in the formal and $C^\infty$ categories: it has one modulus whose geometric meaning is given in [1].

In [2] it is proved that the bifurcation diagram (front) of Example 4 has at most one modulus at the origin, in the formal and $C^\infty$ categories. It is discussed in the comment to problem *1993-3*.

[1] AGRACHEV A. A., CHARLOT G., GAUTHIER J. P., ZAKALYUKIN V. M. On stability of generic sub-Riemannian caustics in the three-space. *C. R. Acad. Sci. Paris, Sér.* I *Math.*, 2000, **330**(6), 465–470.

[2] URIBE-VARGAS R. Symplectic and contact singularities in the differential geometry of curves and surfaces. Ph. D. Thesis, Université Paris 7, 2001, Ch. 6.

## ▽ 1997-9 – *V. I. Arnold*

ℛ The first line in the table shows that moving to the right in each triple is a sort of complexification. Thus, the octahedron should appear, in a mysterious sense, as the complexification of the tetrahedron, and the icosahedron as that of the octahedron. The number of edges of each of the regular polyhedra equals $k(k+1)$ where $k = 1+1$ for the real case (tetrahedron), $k = 1+2$ for the complex case (octahedron), and $k = 1+4$ for the quaternionic case (icosahedron).

The parallelism of the lines $\mathbb{R}$, $\mathbb{C}$, $\mathbb{H}$ and $E_6$, $E_7$, $E_8$ was noticed by D. A. Kazhdan in his lecture at I. M. Gelfand's jubilee in Rutgers (1993). He used number-theoretic arguments.

Parabolic unimodal singularities $P_8$, $X_9$, $J_{10}$ adjoin to simple singularities $E_6$, $E_7$, $E_8$ respectively and border them; the relationship between these two trinities is duly studied. It is interesting, however, that, considering boundary singularities (or singularities of meromorphic functions), one encounters two other related bordering triples of unimodal singularities. One of these triples consists of three families of codimension 5:

$$P_8^{\#} = \frac{x}{xy+y^3+ay^2z+z^3},$$

$$X_9^{\#} = \frac{x}{x^2+axy^2+y^4+z^2},$$

$$J_{10}^{\#} = \frac{x}{x^3+ax^3y+y^3+z^2}.$$

The connection of this triple with the rest of our triples had been undiscovered for a long time (so, the notations of the corresponding singularities in the 1978 lists are in no way related to $P_8$, $X_9$, $J_{10}$).

$A_3$ is the symmetry group of the tetrahedron, $B_3$ of the octahedron, and $H_3$ of the icosahedron. The way to obtain the trinity $E_6$, $E_7$, $E_8$ from the trinity $A_3$, $B_3$, $H_3$ is well known. For example, one may take the syzygies of the three invariants of the corresponding binary group acting in $\mathbb{C}^2$ and get a surface with a singularity $E_6$, $E_7$, or $E_8$ in $\mathbb{C}^3$ as the quotient of $\mathbb{C}^2$ by this binary group, and then the group $E_6$ ($E_7$, $E_8$) is reconstructed as the monodromy group of this singularity.

Another means of transition from the line $A_3$ to the line $E_6$ is suggested by the McCay correspondence: the extended Dynkin diagram of $E_6$ ($E_7$, $E_8$) describes the factorization of the tensor product of an irreducible representation of the desired group $A_3$ ($B_3$, $H_3$) with the standard representation onto irreducibles.

The transition from the line $A_3$ to the line $D_4$ is suggested by the decomposition of Springer cones into Weyl chambers. The numbers of chambers in the eight simplicial cones into which $\mathbb{R}^3$ is divided by the three mirror planes bounding a Weyl chamber are, respectively,

$$\begin{aligned} A_3 &: \ 24 = 2(1+3+3+5), \\ B_3 &: \ 48 = 2(1+5+7+11), \\ H_3 &: \ 120 = 2(1+11+19+29). \end{aligned}$$

Increasing the numbers in brackets by 1, we obtain the weights of invariants for

$$D_4: \ 2,4,4,6; \quad F_4: \ 2,6,8,12; \quad H_4: \ 2,12,20,30.$$

By the way, these weights provide the numbers of vertices, faces and edges for a tetrahedron, an octahedron and an icosahedron (deuces probably correspond to 3-faces).

The relation between the three drawing triangles (with angles $\frac{\pi}{a}, \frac{\pi}{b}, \frac{\pi}{c}$) and parabolic singularities $P_8, X_9, J_{10}$ is well known. For instance, the quasihomogeneity indices of the parabolic singularities

$$P_8: \ \left(\frac{1}{3}+\frac{1}{3}+\frac{1}{3}=1\right); \quad X_9: \ \left(\frac{1}{4}+\frac{1}{4}+\frac{1}{2}=1\right); \quad J_{10}: \ \left(\frac{1}{6}+\frac{1}{3}+\frac{1}{2}=1\right)$$

provide all the solutions of the equation $\frac{\pi}{a}+\frac{\pi}{b}+\frac{\pi}{c}=\pi$ for the angle sum of a planar triangle. Recall that regular polyhedra are associated with the solutions

$$A_3:\ \left(\frac{1}{3}+\frac{1}{3}+\frac{1}{2}=1+\frac{1}{6}\right),$$
$$B_3:\ \left(\frac{1}{4}+\frac{1}{3}+\frac{1}{2}=1+\frac{1}{12}\right),$$
$$H_3:\ \left(\frac{1}{5}+\frac{1}{3}+\frac{1}{2}=1+\frac{1}{30}\right)$$

of the formula for the angle sum of a spherical triangle $\frac{\pi}{a}+\frac{\pi}{b}+\frac{\pi}{c}=\pi+\frac{\pi}{h}$. Note that the Coxeter numbers $h$ of the corresponding groups ($D_4$, $F_4$, $H_4$) are equal to the numbers of edges for a tetrahedron, an octahedron, and an icosahedron, respectively.

The two-fold Möbius covering $\mathbb{S}^1 \xrightarrow{\mathbb{S}^0} S^1$ (mapping the edge of a Möbius band onto its base) turns under complexification into the Hopf bundle $\mathbb{S}^3 \xrightarrow{\mathbb{S}^1} S^2$. Here, the first circle of the Möbius map should be regarded as $\mathrm{SO}(2)$ which, after complexification, becomes $\mathrm{SU}(2) \approx \mathbb{S}^3$. The second circle of the Möbius map should be regarded as $\mathbb{R}\mathrm{P}^1$, after complexification it becomes $\mathbb{C}\mathrm{P}^1 \approx S^2$. A pair of points of $\mathbb{S}^0$ after complexification becomes a point of $\mathbb{S}^1$ (e. g., because $\mathrm{O}/\mathrm{SO}$ turns into $\mathrm{U}/\mathrm{SU}$). Hence, comnplexification of the Möbius covering yields the Hopf bundle $\mathbb{S}^3 \to S^2$. The appearance of the second Hopf bundle $\mathbb{S}^7 \to S^4$ under quaternionization of the Möbius bundle is equally natural.

The complexification of the stratification of the bundle of eigenvectors of symmetric matrices without multiple eigenvalues (considered, together with the monodromy of the corresponding covering, in paper [1]) leads to the natural connection in the bundle of eigenvectors of Hermitian matrices without multiple eigenvalues (considered in paper [2]). We conclude that, firstly, the connection in a complex vector bundle and its curvature give complexification to the covering and its monodromy and, secondly, the theories of Berry phase and quantum Hall effect must have a further analog (where the real codimension of the singularity corresponding to the "multiplicity of roots" equals 5).

What plays the role of vector bundle and connection in this quaternionic case, is to be clarified, but in place of the curvature there should be some 4-form (A. M. Gabrielov told me that, in his investigations of expressions of the Pontryagin form in polylogarithms performed jointly with I. M. Gelfand and M. V. Losik, functions with three poles on $S^2$ appeared where two poles had occurred in the case of Chern classes).

Elliptic numbers are described in paper [3]. The line of the table starting from cohomology was suggested by A. B. Givental, and I would not comment it.

See also the comment to problem *1998-16*.


[1] ARNOLD V. I. Modes and quasimodes. *Funct. Anal. Appl.*, 1972, **6**(2), 94–101. [*The Russian original is reprinted in*: Vladimir Igorevich Arnold. Selecta–60. Moscow: PHASIS, 1997, 189–202.]

[2] ARNOLD V. I. Remarks on eigenvalues and eigenvectors of Hermitian matrices, Berry phase, adiabatic connections and quantum Hall effect. *Selecta Math. (N. S.)*, 1995, **1**(1), 1–19. [*The Russian translation in*: Vladimir Igorevich Arnold. Selecta–60. Moscow: PHASIS, 1997, 583–604.]

[3] FRENKEL I. B., TURAEV V. G. Elliptic solutions of the Yang–Baxter equation and modular hypergeometric functions. In: The Arnold–Gelfand Mathematical Seminars: Geometry and Singularity Theory. Editors: V. I. Arnold, I. M. Gelfand, V. S. Retakh and M. Smirnov. Boston, MA: Birkhäuser, 1997, 171–204.


## △ 1997-9 – *B. A. Khesin*

ℛ A partial answer to the question on the complexification of a connection and its curvature can be found in [4,5]: this is a $(0,1)$-connection and its curvature wedged with a meromorphic differential form on the manifold. This complexification also has a finite-dimensional part, the theory of "polar homology." The latter is a complex analogue of the (real) theory of singular homology in topology. The Cauchy residue formula becomes a complex counterpart of the Stokes formula (see its application to the complex gauge groups in [2]). The corresponding gauge theory on complex manifolds is discussed in [1,3].

See also the comment to problem *1979-4*.


[1] DONALDSON S. K., THOMAS R. P. Gauge theory in higher dimensions. In: The Geometric Universe. Science, Geometry, and the Work of Roger Penrose. Papers from Symp. on geometric issues in the foundations of science held in honor of the 65th birthday of Sir Roger Penrose (Oxford, June 1996). Editors: S. A. Huggett, L. J. Mason, K. P. Tod, S. T. Tsou and N. M. J. Woodhouse. Oxford: Oxford University Press, 1998, 31–47.

[2] FRENKEL I. B., KHESIN B. A. Four-dimensional realization of two-dimensional current groups. *Commun. Math. Phys.*, 1996, **178**(3), 541–562.

[3] KHESIN B. A. Informal complexification and Poisson structures on moduli spaces. In: Topics in Singularity Theory. V. I. Arnold's 60th Anniversary Collection. Editors: A. Khovanskiĭ, A. Varchenko and V. Vassiliev. Providence, RI: Amer. Math. Soc., 1997, 147–155. (AMS Transl., Ser. 2, 180; Adv. Math. Sci., 34.)

[4] KHESIN B. A., ROSLY A. A. Symplectic geometry on moduli spaces of holomorphic bundles over complex surfaces. In: The Arnoldfest. Proceedings of a conference in honour of V. I. Arnold for his sixtieth birthday (Toronto, 1997). Editors: E. Bierstone, B. A. Khesin, A. G. Khovanskiĭ and J. E. Marsden. Providence, RI: Amer. Math. Soc., 1999, 311–323. (Fields Inst. Commun., 24.)

[5] KHESIN B. A., ROSLY A. A. Polar homology and holomorphic bundles. *Phil. Trans. Roy. Soc. London, Ser. A*, 2001, **359**, 1413–1427.

# 1998

## 1998-1 – *V. I. Arnold*

ℛ Cartan's theory is similar to algebraic geometry: proved statements are valid for arbitrarily deep degeneracies but the cases of finite (and even small) codimension of the considered degeneracy in the space of all possible systems are not yet clarified. This is also the reason for the analytic case restriction (in the Cartan–Kähler theorem and analogous situations): infinitely degenerate systems can be investigated only due to the analyticity of initial data, the smooth case remains unexplored even when the codimension of the degeneracy is small.

When one approaches the problem from the viewpoint of singularity theory instead of algebraic geometry, the principal question is about the codimensions of various degeneracies and about versal deformations, i. e., the degeneracies appearing in typical families of systems depending on a finite (or even small) number of parameters. In particular, of interest is, as always, the problem on simple singularities (having no moduli) and on finitely determined (having no functional moduli) singularities of differential systems (both smooth and analytic).

See also the comments to problem *1993-28*.

## ▽ 1998-2 – *F. Aicardi*

ℛ In the case of cubic surfaces (see comment to problem *1995-3*), the six special parabolic points are hyperbolic (see their definition in problem *1997-6*). If there exists a surface with six special parabolic points and less than four elliptic ovals, then the special points cannot all be hyperbolic.

**Proof.** Let $S$ be a generic surface, a perturbation of the projective plane, having $n$ elliptic domains diffeomorphic to a disc. Consider the compact surface $W$

obtained by taking two copies of the hyperbolic domain of $S$ and gluing them along the parabolic curve. The Euler characteristic of $W$ is equal to $2-2n$. The two fields of asymptotic directions, defined on the non-elliptic domain of $S$, coincide on the parabolic curve. They define a field of directions on $W$ (one field on each sheet), which is continuous on $W$. The singular points of this field are the special parabolic points, having indices equal to 1 (elliptic) or $-1$ (hyperbolic). By the Poincaré theorem, the sum of indices of the special points is equal to $2-2n$. This implies that if $n<4$, then the sum of the indices of the special parabolic points is not less than $-4$.

## △ 1998-2 — *V. I. Arnold*

A cubic surface has four parabolic curves and six special points, as was proved by B. Segre [3].

In 1997, D. A. Panov constructed a surface with a single parabolic curve, though having 12 special points [2]. See also a series of relevant results and conjectures in the earlier paper [1].

[1] Arnold V. I. Remarks on the parabolic curves on surfaces and on the higher-dimensional Möbius–Sturm theory. *Funct. Anal. Appl.*, 1997, **31**(4), 227–239.

[2] Panov D. A. Parabolic curves and gradient mapping. *Proc. Steklov Inst. Math.*, 1998, **221**, 261–278.

[3] Segre B. The Non-singular Cubic Surfaces. Oxford: Oxford University Press, 1942.

## 1998-3 — *V. I. Arnold*

The existence of three (respectively two) parabolic curves was proved for small values of the parameter in an arbitrary one-parameter family of generic deformations of the given surface [1].

[1] Arnold V. I. Remarks on the parabolic curves on surfaces and on the higher-dimensional Möbius–Sturm theory. *Funct. Anal. Appl.*, 1997, **31**(4), 227–239.

## 1998-4 — *V. I. Arnold*

Examples of hyperbolic functions are suggested by generalized first degree spherical functions (satisfying the equation $\Delta u + 2u = 0$ almost everywhere, $\Delta$ being the spherical Laplacian). Odd generalized first degree spherical functions have at least six logarithmic poles (see [1]).

The question on parabolic curves is open even for rational generalized first degree spherical functions.

[1] ARNOLD V. I. Remarks on the parabolic curves on surfaces and on the higher-dimensional Möbius–Sturm theory. *Funct. Anal. Appl.*, 1997, **31**(4), 227–239.

## **1998-5** – ***V. I. Arnold*** Also: *1969-2*

For analytic $g$ or $h$, the existence of a surface with the given Hessian $h$ or the given Gaussian curvature $g$ is implied by the Cauchy–Kovalevskaya theorem. If $g(0) \neq 0 \neq h(0)$, then such a surface exists without the analyticity assumption.

If the critical point 0 of the function $g$ or $h$ with zero critical value has finite multiplicity, then a surface with Gaussian curvature or Hessian right-equivalent to this function does exist. But it is unknown whether one can get rid of the equivalence, i. e., of the change of variable identifying a point on a surface with a point on a plane (see [1]). In paper [1], it is also proved that a Morse critical point of the Hessian (or Gaussian curvature) at a flattening point cannot be a minimum and that a parabolic curve cannot have singularity of type $E_6$ (diffeomorpic to the singularity of the curve $x^3 = y^4$ at zero) at such a point.

[1] ARNOLD V. I. On the problem of realization of a given Gaussian curvature function. CEREMADE (UMR 7534), Université Paris-Dauphine, № 9809, 12/02/1998; *Topol. Methods Nonlinear Anal.*, 1998, **11**(2), 199–206.
[*Internet*: `http://www.pdmi.ras.ru/~arnsem/Arnold/arn-papers.html`]

## ▽ **1998-6** – ***V. I. Arnold***

The number of flattening points of a generic curve is bounded below by its "Sturmianity", which is invariant under admissible regular homotopies (see [1]).

All curves close to our curve have not less than four flattening points, because it has a convex projection. Some of these nearby curves have zero Sturmianity. It is however unclear at all whether their flattening points can be eliminated by admissible perestroikas: besides the Sturmianity, other invariants including those not yet discovered can obstruct it.

[1] ARNOLD V. I. Towards the Legendre Sturm theory of space curves. *Funct. Anal. Appl.*, 1998, **32**(2), 75–80.

### △ 1998-6 — *V. D. Sedykh*

ℛ This problem was solved in paper [2]. Namely, it was proved that the curve $x = \cos t$, $y = \sin t$, $z = \cos 3t$ cannot be deformed in the class of admissible homotopies into a curve without flattening points.

For the proof, we use an invariant of admissible homotopies of space curves, which is given by the number of closed double lines of the front of tangent planes to a curve. These lines are defined as follows. Let us consider the closure of the set of self-intersection points of a front. It is the union of curves having only cusps, generic double or triple self-intersections and end-points at vertices of swallowtails. These curves are called double lines of a front. We take only those of them that are closed.

In paper [2], we also construct an invariant generalizing the Sturmianity of a curve defined by V. I. Arnold in [1]. It is a chord diagram in which the number of chords intersecting an odd number of other chords coincides with the Sturmianity of the curve.

[1] ARNOLD V. I. Towards the Legendre Sturm theory of space curves. *Funct. Anal. Appl.*, 1998, **32**(2), 75–80.

[2] SEDYKH V. D. Some invariants of admissible homotopies of space curves. *Funct. Anal. Appl.*, 2001, **35**(4), 284–293.

### 1998-7

ℛ See the comment to problem *1996-9*.

### 1998-8

ℛ See the comments to problem *1985-22*.

### 1998-9 — *V. I. Arnold*

Also: *1980-14, 1981-12, 1984-15, 1988-16*

ℛ The first discriminant is the manifold of polynomials with multiple roots in the $a$-space; the fundamental group of its complement is the braid group.

The second discriminant is the bifurcation diagram of functions in the space with coordinates $(a_1, \ldots, a_{n-1})$. This is dealt with in the previous problem.

Starting from the third discriminant, a generic projection and the projection omitting a coordinate do not coincide.

Cf. [1] (§ 2), see also problem *1993-27.*

[1] ARNOLD V. I. On some problems in singularity theory. In: Geometry and Analysis. Papers dedicated to the memory of V. K. Patodi. Bangalore: Indian Acad. Sci., 1980, 1–9. [*Reprinted in*: *Proc. Indian Acad. Sci. Math. Sci.*, 1981, **90**(1), 1–9.]

## ▽ 1998-10 – *V. I. Arnold*

ℜ Although the dyed braid group is the complexification $\pi_1(\mathbb{C}^n \setminus \Delta)$ of the symmetric group $\pi_0(\mathbb{R}^n \setminus \Delta)$, it is unlikely that complexification of dyed braids reduces to the study of the commutative group $\pi_3(\mathbb{H}^n \setminus \Delta)$ (here $\Delta$ are the diagonals). Probably $\pi_1(\Omega^2)$ should be considered instead of $\pi_3$, where $\Omega^2$ is the second loop space with some condition of holomorphy type.

Concerning two-fiber braids, see problems *1998-14* and *1999-12.*

Upon complexification of Maslov index, the mapping $\det^2 : \mathrm{U}/\mathrm{O} \to \mathbb{S}^1$ should turn into some mapping $\mathrm{Sp}/\mathrm{U} \to S^2$. Then "index" becomes a *two-dimensional* cohomology class just as it must be (the real codimension of a caustic equals two). But what "quaternion determinant" of quaternion matrices from Sp (a determinant taking complex values on U) is involved in this construction?

Complexification of the signature definition of index also leads to interesting topological constructions related to the manifold of degenerate complex symmetric matrices and the eigenvector bundle over its complement.

More details on the complexification of $\mathbb{Z}_2 = \mathbb{Z}/2\mathbb{Z}$ (making it $\mathbb{Z}$) and on the complexifications of braids and dyed braids can be found in paper [1].

[1] ARNOLD V. I. Polymathematics: is mathematics a single science or a set of arts? In: Mathematics: Frontiers and Perspectives. Editors: V. I. Arnold, M. Atiyah, P. Lax and B. Mazur. Providence, RI: Amer. Math. Soc., 2000, 403–416; CEREMADE (UMR 7534), Université Paris-Dauphine, № 9911, 10/03/1999.
[*Internet*: `http://www.pdmi.ras.ru/~arnsem/Arnold/arn-papers.html`]

## △▽ 1998-10 – *B. A. Khesin*

ℜ The problem of complexification of the Vassiliev knot and link invariants is discussed in the book [1] (see also [2] for a complexification of linking number).

[1] ARNOLD V. I., KHESIN B. A. Topological Methods in Hydrodynamics. New York: Springer, 1998. (Appl. Math. Sci., 125.)

[2] KHESIN B. A., ROSLY A. A. Polar homology and holomorphic bundles. *Phil. Trans. Roy. Soc. London, Ser. A*, 2001, **359**, 1413–1427.

### △ 1998-10 — *M. L. Kontsevich*

ℛ A possible complexification for Vassiliev invariants is obtained by replacement of knots by algebraic curves in a 3-dimensional complex Calabi–Yau manifold. The idea is to study the perturbation theory of quantum field theory which is a complex version of the Chern–Simons theory. Up to now, I tried only the case of the theory without observables, i. e., for a Calabi–Yau manifold itself, without holomorphic curves in it. In the topological case it corresponds to the theory of finite type invariants of homology 3-spheres, see, e. g., [1].

Holomorphic formulae are morally the same as those in the topological case. The integrals are divergent but then can be canonically regularized.

[1] BOTT R., TAUBES C. On the self-linking of knots. Topology and physics. *J. Math. Phys.*, 1994, **35**(10), 5247–5287.

### ▽ 1998-11 — *V. I. Arnold*

ℛ A. M. Gabrielov and A. G. Khovanskiĭ have found out that, firstly, this question is not easy at all (even the essential term of the asymptotic expression could not be guessed) and, secondly, the answer can have important applications.

### △ 1998-11 — *A. M. Gabrielov*

ℛ An upper bound on the Milnor number in a generic $m$-parameter family of mappings $\mathbb{C}^n \to \mathbb{C}^n$ is found in Theorem 7 of [2]. This upper bound is applied in the paper to calculate the multiplicity of a zero-dimensional ideal generated by Noetherian functions.

According to the $h$-principle [1,3], the same upper upper bound is valid for a generic $m$-parameter family of functions in $n$ variables.

[1] ARNOLD V. I., VASSILIEV V. A., GORYUNOV V. V., LYASHKO O. V. Singularities. I. Local and Global Theory. Berlin: Springer, 1993. (Encyclopædia Math. Sci., 6; Dynamical Systems, VI.) [*The Russian original* 1988.]

[2] GABRIELOV A. M., KHOVANSKIĬ A. G. Multiplicity of a Noetherian intersection. In: Geometry of Differential Equations. Editors: A. G. Khovanskiĭ, A. N. Varchenko and V. A. Vassiliev. Providence, RI: Amer. Math. Soc., 1998, 119–130. (AMS Transl., Ser. 2, 186; Adv. Math. Sci., 39.)

[3] GROMOV M. L. Partial Differential Relations. Berlin: Springer, 1986.

## 1998-12

ℜ This question has been solved (affirmatively) [1].

[1] ARNOLD V. I. Topological problems in the theory of asymptotic curves. *Proc. Steklov Inst. Math.*, 1999, **225**, 5–15.
[*Internet*: `http://www.pdmi.ras.ru/~arnsem/Arnold/arn-papers.html`]

## 1998-13 — *V. I. Arnold, B. A. Khesin*

ℜ A *local* density is a function depending on a *finite* number of partial derivatives of the velocity field at a point of the flow domain. (And the conserved quantity is supposed to be the integral of the local density over this domain.) The classical invariants for the Euler equation are the energy and generalised enstrophies (i. e., the integrals of arbitrary functions of the vorticity) in the two-dimensional case, and the energy and helicity (i. e., the asymptotic Hopf invariant, which measures the average linking of the trajectories) in the three-dimensional case. Similar conservation laws exists for higher-dimensional Riemannian manifolds (these laws were established by L. Tartar and D. Serre in the Euclidean case, and by V. Yu. Ovsienko, Yu. V. Chekanov, and B. A. Khesin in the Riemmanian case). The even-dimensional case is similar to the two-dimensional one, while the odd-dimensional case mimics the three-dimensional one.

With the exception of the energy, the above integrals are conserved not only along the trajectories of the Euler equation, but also along the corresponding coadjoint orbits. In other words, these integrals have the same values on isovorticed fields (i. e., for the fields whose vorticities are related by a diffeomorphism action). See the details in book [1].

Conjecturally, the classical conservation laws exhaust all invariants that are integrals of local densities for the Euler equation on almost any Riemannian manifold. (Additional invariants, certainly, exist for the manifolds with special symmetry, e. g., for the standard sphere or the standard torus.) Apparently, to prove the non-existence of nonclassical invariants for coadjoint orbits could be easier than for trajectories, since their number should be fewer. As was suggested by J.-P. Serre, it might be even easier is to prove the non-existence of integrals of universal densities (where universality is understood as independence on the flow domain, cf. the classical invariants).

D. Serre [2, 3] proved that there are no new conservation quantities among local densities depending on the velocity field and its first partial derivaties in the Euclidean case.

[1] ARNOLD V. I., KHESIN B. A. Topological Methods in Hydrodynamics. New York: Springer, 1998. (Appl. Math. Sci., 125.)

[2] SERRE D. Les invariants du premier ordre de l'équation d'Euler en dimension 3. *C. R. Acad. Sci. Paris, Sér. A–B*, 1979, **289**(4), A267–A270.

[3] SERRE D. Les invariants du premier ordre de l'équation d'Euler en dimension trois. *Physica D*, 1984, **13**(1–2), 105–136.

## 1998-14 – *V. I. Arnold*

ℛ The "multiplication" is distributive from the right with respect to the "addition": $(a+b)c = ac+bc$. A. Dold has pointed out that our "addition" is really noncommutative already in the case of the group $\mathbb{S}^3 \times \mathbb{S}^3$: this follows from the homotopic noncommutativity (proved by I. M. James) of the operation $\mathbb{S}^3 \times \mathbb{S}^3 \to \mathbb{S}^3$ assigning the product $ab$ to each pair $(a,b)$ of quaternions.

ℋ The problem originated from an attempt to complexify the two-thread braid group (cf. problems *1998-10* and *1999-12*).

## ▽ 1998-15 – *V. I. Arnold*

ℛ Of course, one does not discuss here the homomorphisms of the mutliplicative structure. The manifold of quaternionic matrices determining degenerate operators $\mathbb{H}^n \to \mathbb{H}^n$ has real codimension four in the space of all quaternionic matrices. Is it (in the real sense) a complete intersection of four hypersurfaces?

The determinant of the corresponding *complex* matrix of order $2n$ is a *real* polynomial vanishing on a submanifold of codimension four. Is it a sum of (four?) squares of real polynomials? If there are more squares then the question on "syzygies," i. e., relations between them, arises.

A similar question would also be interesting for the discriminants of characteristic equations of symmetric (Hermitian, hyper-Hermitian, ...) matrices, see papers [1, 2].

[1] ARNOLD V. I. Relatives of the quotient of the complex projective plane by complex conjugation. *Proc. Steklov Inst. Math.*, 1999, **224**, 46–56. [*Internet*: `http://www.pdmi.ras.ru/~arnsem/Arnold/arn-papers.html`]

[2] ILYUSHECHKIN N. V. The discriminant of the characteristic polynomial of a normal matrix. *Math. Notes*, 1992, **51**(3), 230–235.

## △▽ 1998-15 — *S. V. Duzhin*

ℛ Let $A : \mathbb{H}^n \to \mathbb{H}^n$ be a quaternionic linear operator (to be definite, suppose that it is linear on the left, i. e., satisfies the identity $A(qv) = qA(v)$ for all $q \in \mathbb{H}$, $v \in \mathbb{H}^n$). Identifying $\mathbb{H}$ with $\mathbb{C}^2$, we obtain a complex-linear operator ${}^{\mathbb{C}}A : \mathbb{C}^{2n} \to \mathbb{C}^{2n}$. Let

$$H(A) := \det {}^{\mathbb{C}}A.$$

It is easy to see that $H(A)$ is a real-valued polynomial, $H(A) \geq 0$ for any $A$, and the equation $H(A) = 0$ determines a cone $\Sigma$ of real codimension 4. The problem reduces to the study of the algebraic variety $\Sigma$.

**Theorem.** *The polynomial $H(A)$ can be represented as a sum of squares of rational functions, but for any $n \geq 2$ it cannot be represented as a sum of squares of real polynomials.*

Details are available in [1].

[1] DUZHIN S. V. A remark on Arnold's problem concerning quaternionic determinants. Research Announcements of Arnold's Seminar, 12 October 1998. [*Internet*: `http://www.pdmi.ras.ru/~arnsem/papers/resann.html`]

## △ 1998-15 — *M. B. Sevryuk*

ℛ A rich body of literature is devoted to the quaternionic determinant problem. Of the latest works, we mention, e. g., [1, 4], and of earlier articles, [2, 3]. In papers [1, 3], the history of the question is expounded in detail and an extensive bibliography is presented.

[1] ASLAKSEN H. Quaternionic determinants. *Math. Intelligencer*, 1996, **18**(3), 57–65.

[2] GELFAND I. M., RETAKH V. S. The determinants of matrices over non-commutative rings. *Funct. Anal. Appl.*, 1991, **25**(2), 91–102.

[3] GELFAND I. M., RETAKH V. S. The theory of non-commutative determinants and characteristic functions of graphs. *Funct. Anal. Appl.*, 1992, **26**(4), 231–246.

[4] KAZARIAN M. E. A remark on the eigenvectors and eigenvalues of hyper-Hermitian matrices. Preprint, 1998 (in Russian). [*Internet*: `http://www.pdmi.ras.ru/~arnsem/papers/`]

## 1998-16 — *V. I. Arnold*

The concepts of a one-dimensional holomorphic bundle, and its connection and curvature form, are themselves complexifications of the concepts of a covering, the monodromy and the first Whitney class.

After the second complexification, the complex numbers should become the quaternions, the Cauchy–Riemann equations should probably become nonlinear, the curvature 2-form should turn into a "hypercurvature" 4-form, and a Chern class into a Pontryagin class.

The theory of quantum Hall effect and the theory of Berry phase are complexifications of the Wiegner – von Neumann theory of repelling the eigenvalues of quadratic forms. After the second complexification, the forms must become hyper-Hermitian (invariant under the action of the compact group Sp by real quadratic forms on the quaternion space).

These questions are discussed in works [1–7]; see also the comments to problem *1997-9*.

[1] ARNOLD V. I. Modes and quasimodes. *Funct. Anal. Appl.*, 1972, **6**(2), 94–101. [*The Russian original is reprinted in*: Vladimir Igorevich Arnold. Selecta–60. Moscow: PHASIS, 1997, 189–202.]

[2] ARNOLD V. I. Remarks on eigenvalues and eigenvectors of Hermitian matrices, Berry phase, adiabatic connections and quantum Hall effect. *Selecta Math.* (*N. S.*), 1995, **1**(1), 1–19. [*The Russian translation in*: Vladimir Igorevich Arnold. Selecta–60. Moscow: PHASIS, 1997, 583–604.]

[3] ARNOLD V. I. Mysterious Mathematical Trinities. Topological Economy Principle in Algebraic Geometry. Moscow: Moscow Center for Continuous Mathematical Education Press, 1997 (in Russian).

[4] ARNOLD V. I. Relatives of the quotient of the complex projective plane by complex conjugation. *Proc. Steklov Inst. Math.*, 1999, **224**, 46–56.
[*Internet*: `http://www.pdmi.ras.ru/~arnsem/Arnold/arn-papers.html`]

[5] ARNOLD V. I. Symplectization, complexification and mathematical trinities. In: The Arnoldfest. Proceedings of a conference in honour of V. I. Arnold for his sixtieth birthday (Toronto, 1997). Editors: E. Bierstone, B. A. Khesin, A. G. Khovanskiĭ and J. E. Marsden. Providence, RI: Amer. Math. Soc., 1999, 23–37. (Fields Institute Commun., 24.); CEREMADE (UMR 7534), Université Paris-Dauphine, № 9815, 04/03/1998.
[*Internet*: `http://www.pdmi.ras.ru/~arnsem/Arnold/arn-papers.html`]

[6] ARNOLD V. I. Polymathematics: is mathematics a single science or a set of arts? In: Mathematics: Frontiers and Perspectives. Editors: V. I. Arnold, M. Atiyah, P. Lax and B. Mazur. Providence, RI: Amer. Math. Soc., 2000, 403–416; CEREMADE (UMR 7534), Université Paris-Dauphine, № 9911, 10/03/1999.
[*Internet*: `http://www.pdmi.ras.ru/~arnsem/Arnold/arn-papers.html`]

[7] KAZARIAN M. E. A remark on the eigenvectors and eigenvalues of hyper-Hermitian matrices. Preprint, 1998 (in Russian).
[*Internet*: `http://www.pdmi.ras.ru/~arnsem/papers/`]

### ▽ 1998-17 – *V. I. Arnold* Also: *1995-12*

ℛ The base for the Liouville theorem is the natural affine structure on the fibers of the Lagrangian fibration of a symplectic manifold. The fibers of the Legendrian fibration of a contact manifold are endowed by a natural projective structure. Undoubtedly, the latter can be used for the construction of a projective generalization of the theory of fully integrable systems in contact geometry (which may even be fruitful in the theory of partial differential equations). Strangely, nobody seems to have done it so far.

ℋ The problem appears already in book [1] (see p. 22–23).

[1] ARNOLD V. I. Lectures on Partial Differential Equations, 2nd supplemented edition. Moscow: PHASIS, 1997 (in Russian).

### △ 1998-17 – *B. A. Khesin* Also: *1995-12*

ℛ Apparently, the notion of cosymmetry introduced by V. I. Yudovich (see [3–6]), plays the same role for contact systems (i. e., for dynamical systems subordinated to a given contact structure) as the notion of an additional first integral does for Hamiltonian systems.

Then a contact analogue of a completely integrable system is a vector field belonging to the intersection of $n+1$ contact structures (i. e., it is a contact system in a $(2n+1)$-dimensional contact manifold with $n$ cosymmetries). For such a system the Frobenius theorem on integrability of an $n$-dimensional distribution, which is the intersection of all the cosymmtries, becomes a counterpart of the involutivity of the first integrals in the Hamiltonian case. The latter observation gives, apparently, a partial answer to the question on what could be an analogue of the Liouville integrability theorem for contact systems. The fibers of this $n$-dimensional distribution (the intersection of all $n+1$ distributions of hyperplanes) are Legendrian for each cosymmetry.

The parallelism of the triples (Hamiltonian system, first integrals, involutivity of the integrals) $\sim$ (contact systems, cosymmetry, Frobenius integrability theorem) becomes even more explicit in the following context. Rather than describing a completely integrable system by the set of $n$ first integrals in involution for one and the same symplectic structure, we can define it by specifying a single Hamiltonian function and $n$ compatible symplectic structures. The latter is very similar to the "contact integrable" case where we have one vector field and $n+1$ contact compatible structures (of 1-forms) containing it.

Also, for some results related to an extension of the Liouville theorem to a contact manifold (via classical Hamiltonian approach) see paper [1] by A. Banyaga, as well as [2].


[1] BANYAGA A. The geometry surrounding the Arnold–Liouville theorem. In: Advances in Geometry. Boston, MA: Birkhäuser, 1999, 53–69. (Progr. Math., 172.)

[2] BANYAGA A., MOLINO P. Complete integrability in contact geometry. Preprint, 1996; book in preparation.

[3] YUDOVICH V. I. Cosymmetry, degeneration of solutions of operator equations, and the onset of filtration convection. *Math. Notes*, 1991, **49**(5-6), 540–545.

[4] YUDOVICH V. I. Secondary cycle of equilibria in a system with cosymmetry, its creation by bifurcation and impossibility of symmetric treatment of it. *Chaos*, 1995, **5**(2), 402–411.

[5] YUDOVICH V. I. Cosymmetry and dynamical systems. In: Proceedings of the Third International Congress on Industrial and Applied Mathematics (Hamburg, July 1995). Editors: E. Kreuzer and O. Mahrenholtz. *Z. Angew. Math. Mech.*, 1996, **76**, suppl. 4, 556–559.

[6] YUDOVICH V. I. The implicit function theorem for cosymmetric equations. *Math. Notes*, 1996, **60**(2), 313–317.


## 1998-18 – *V. I. Arnold*

G. Hofer proved the existence of closed orbits for Hofer fields on certain manifolds (for example, on $\mathbb{S}^3$). Thus, the solution of this problem would yield new results on the existence of closed tarjectories in the magnetic problem.

Note that the vector field posing the magnetic problem has an auxiliary property: the vectors of the field lie in planes of some (other) contact structure of the phase space (this is the tautological contact structure of the space of contact elements).

It would be interesting to examine to what extent each of the two conditions—the Hofer property and the subordination to the second contact structure—limits the possibility of constructing counterexamples to the Seifert conjecture on the existence of periodic orbits of a vector field on the 3-sphere (by the way, for divergence-free vector fields even smooth counterexapmles are not known).

## ▽ 1998-19 — *V. I. Arnold*

ℛ M. L. Kontsevich and E. I. Korkina, who had heard from me about this problem, proved the following closely related assertions. Let us assign two numbers to a lattice in a Euclidean space: the greatest radius $r$ of open balls centered at points of the lattice that do not overlap, and the least radius $R$ of closed balls centered at points of the lattice that cover the whole of the space.

Then, the product $r$(the lattice) $\cdot$ $R$(the dual lattice) is bounded below and above by constants $c$ and $C$ respectively depending only on the dimension $n$.

Korkina noticed that $c = 1/4$, the estimate being exact, and $C \leq (\sqrt{3}/4) \times \sqrt{(4/3)^n - 1}$, this estimate being non-exact.

She also proved that the product of the lengths of the shortest nonzero vectors in the initial and the dual lattice does not exceed $(2/\sqrt{3})^{n-1}$.

All these results should have been known already to Minkowski, but seemingly have not been observed by his successors, see [1].

[1] ARNOLD V. I. Higher dimensional continued fractions. *Reg. Chaot. Dynamics*, 1998, **3**(3), 10–17.

## △ 1998-19 — *N. P. Dolbilin*

ℛ Let $L$ be a lattice of rank $n$ in the Euclidean space $\mathbb{R}^n$, $r(L)$ its *packing radius* (the greatest radius of open equal balls that are centered at points of the lattice and do not overlap) and $R(L)$ the *covering radius* of the lattice (the least radius of closed equal balls that are centered at lattice points and cover the whole of the space). The length $\lambda_1(L)$ of the shortest vector of the lattice $L$ is obviously equal to $2r(L)$. The lattice $L^*$ *dual* to $L$ (also called the *polar* or *reciprocal* lattice) is defined as

$$L^* : = \{\mathbf{a}^* \in \mathbb{R}^n : (\mathbf{a}^*, \mathbf{a}) \in \mathbb{Z} \text{ for all } \mathbf{a} \in L\}.$$

Similarly, the following estimate was achieved for the first time in [3]:

$$\frac{1}{4} \leq \lambda_1^2(L) \cdot R^2(L^*) \leq \frac{1}{4} \sum_{i=1}^{n} \gamma_i^{*2},$$

where $\gamma_n^* = \max\{\gamma_i : 1 \le i \le n\}$, $\gamma_i$ is the Hermite constant:

$$\gamma_i = \sup_{\Lambda} \left\{ \frac{r^2(\Lambda)}{\det^{2/i}(\Lambda)},\ \Lambda \text{ is an } i\text{-dimensional lattice} \right\},$$

here $\det(\Lambda)$ denotes the $i$-dimensional volume of a fundamental parallelepiped of the lattice $\Lambda$. The lower bound cannot be refined. Since $\gamma_i^* \le 2i/3$ for all $i \ge 2$ and $\lambda_1(L) = 2r(L)$ the upper bound can be rewritten as follows:

$$1/4 \le r(L) \cdot R(L^*) \le \frac{1}{4} n^{3/2}.$$

W. Banaszczyk conjectured ([2], p. 751; see also [1], p. 43–44) that the power $3/2$ can be replaced by 1. The conjecture was proved by analytic methods in [4]:

$$r(\Lambda) \cdot R(\Lambda^*) \le \frac{n}{4\pi}(1 + o(1)).$$

[1] BANASZCZYK W. Additive Subgroups of Topological Vector Spaces. Berlin: Springer, 1991. (Lecture Notes in Math., 1466.)

[2] GRUBER P. M. Geometry of numbers. In: Handbook of Convex Geometry, Vol. B. Editors: P. M. Gruber and J. M. Wills. Amsterdam: North-Holland, 1993, 739–763.

[3] LAGARIAS J. C., LENSTRA H. W., JR., SCHNORR C. -P. Korkin–Zolotarëv bases and successive minima of a lattice and its reciprocal lattice. *Combinatorica*, 1990, **10**(4), 333–348.

[4] YUDIN V. A. Two extremal problems for trigonometric polynomials. *Sb. Math.*, 1996, **187**(11), 1721–1736.

## 1998-20 — *V. I. Arnold*

In the symplectic case the answer seems to be obtained by an extension of a part of the list *A*, *D*, *E* (some classes of which are divided into subclasses) [1–3].

For these cases, the simplest is the classification of *stably simple* singularities (that remain simple, i. e., having no moduli, when the contact or the symplectic space is embedded into a space of greater dimension).

By the way, such stabilizations simplify the classification of simple singularities of curves in the usual space as well [1, 4, 5].

[1] ARNOLD V. I. Simple singularities of curves. CEREMADE (UMR 7534), Université Paris-Dauphine, № 9906, 09/02/1999; *Proc. Steklov Inst. Math.*, 1999, **226**, 20–28. [*Internet*: `http://www.pdmi.ras.ru/~arnsem/Arnold/arn-papers.html`]

[2] ARNOLD V. I. First steps of local contact algebra. CEREMADE (UMR 7534), Université Paris-Dauphine, № 9909, 10/02/1999; *Canad. J. Math.*, 1999, **51**(6), 1123–1134.
[*Internet*: `http://www.pdmi.ras.ru/~arnsem/Arnold/arn-papers.html`]

[3] ARNOLD V. I. First steps of local symplectic algebra. CEREMADE (UMR 7534), Université Paris-Dauphine, № 9902, 20/01/1999; In: Differential Topology, Infinite-Dimensional Lie Algebras, and Applications. D. B. Fuchs' 60th Anniversary Collection. Editors: A. Astashkevich and S. Tabachnikov. Providence, RI: Amer. Math. Soc., 1999, 1–8. (AMS Transl., Ser. 2, 194; Adv. Math. Sci., 44.)
[*Internet*: `http://www.pdmi.ras.ru/~arnsem/Arnold/arn-papers.html`]

[4] BRUCE J. W., GAFFNEY T. J. Simple singularities of mappings $(\mathbb{C},0) \to (\mathbb{C}^2,0)$. *J. London Math. Soc., Ser.* 2, 1982, **26**(3), 465–474.

[5] GIBSON C. G., HOBBS C. A. Simple singularities of space curves. *Math. Proc. Cambridge Phil. Soc.*, 1993, **113**(2), 297–310.

## 1998-21 – *V. I. Arnold*

ℛ Penrose's conjecture claims that even if the homology linking coefficient vanishes then linking still takes place (with the meaning that the Legendrian submanifolds cannot be taken apart by a contact or even by a usual isotopy).

The question becomes nontrivial after paasing through conjugate points when swallowtails appear on the "cone" in the space-time bounded by light geodesics issuing from a point.

The problem is discussed in papers [1, 2].

[1] LOW R. J. Twistor linking and causal relations. *Classical Quantum Gravity*, 1990, **7**(2), 177–187.

[2] LOW R. J. Twistor linking and causal relations in exterior Schwarzschild space. *Classical Quantum Gravity*, 1994, **11**(2), 453–456.

## ▽ 1998-22 – *V. I. Arnold*

ℛ Clearly, the number of simplices is not less than $cn$. J. Hass and J. Lagarias proved that this number is not greater than $Cn^2$. This lemma forms the basis for their proof of the fact that an unknotted curve in $\mathbb{R}^3$ with $n$ double projection points can be transformed into a circle by at most $Ke^{Ln}$ Reidemeister moves.

It is natural to ask the question how the minimal number of moves grows in similar problems such as 1) recognition of the equivalence of two knots; 2) recognition of the Legendrian equivalence of two Legendrian knots (or at least their triviality); 3) recognition of the $J^+$-equivalence of curves immersed into the plane.

However, in cases 2) and 3) even algorithmical solvability (i. e., the existence of *any* recursive function bounding the required number of moves above) has not been proved.

The work by Hass and Lagarias was reported at the International Congress of Mathematicians in Berlin, 1998 [1].

[1] HASS J., LAGARIAS J. The number of Reidemeister moves needed for unknoting. In: ICM 1998, International Congress of Mathematicians. Abstracts of short communications and poster sessions. Berlin–Bielefeld: University of Bielefeld Press, 1998, 89.

### △ 1998-22 — *J. C. Lagarias*

ℛ For the triangulation problem in $\mathbb{R}^3$, we interpret the problem to be that of triangulating a convex polyhedron $Q$ containing the polygon in its 1-skeleton, where one is free to choose $Q$; for $\mathbb{S}^3$ it is a triangulation in which one vertex is the point "at infinity." The problem has two forms, which may be termed straight line triangulation or curvilinear triangulation. In the straight line triangulation the simplices are Euclidean simplices in $\mathbb{R}^3$ (tetrahedra), and in the curvilinear case the simplices are topological simplices embedded in $\mathbb{R}^3$.

We first consider the curvilinear triangulation case. For an unknotted polygon one can attain a $Cn$ upper bound as follows. Since the polygon is unknotted, there is a homeomorphism $\phi: \mathbb{R}^3 \to \mathbb{R}^3$ which is the identity outside a large ball, and takes the unknotted polygon to a convex $n$-gon lying in a plane. We can easily triangulate the resulting configuration using $Cn$ tetrahedra, to obtain either a triangulation of a convex polyhedron $Q$ whose boundary is located entirely outside the large ball in $\mathbb{R}^3$ where $\phi$ is not the identity, or else a triangulation of $\mathbb{S}^3$. Applying $\phi^{-1}$ then gives a curvilinear triangulation of $\phi^{-1}(Q) = Q$ in $\mathbb{R}^3$, respectively $\mathbb{S}^3$. For a general knotted polygon no better upper bound is known for a curvilinear triangulation than $Cn^2$; such a bound is already attained by a straight line triangulation.

We next consider the straight line triangulation case, that of bounding above the number of Euclidean tetrahedra in a minimal triangulation of some convex polyhedron which contains the polygonal knot in the 1-skeleton of the triangulation. One can find a direct construction using $Cn^2$ tetrahedra; see Lemma 2.3 in [7]. This bound applies to $\mathbb{S}^3$ as well since a convex polyhedron can be coned to the point at infinity to get a triangulation of $\mathbb{S}^3$. It seems likely that no better upper bound than $Cn^2$ will hold for this problem. About this possibility we add the following remarks.

1) The Euclidean simplex problem can be viewed as a problem in computational geometry in $\mathbb{R}^3$. One can potentially make use of ideas involving convexity and volume, in attempting to establish a $Cn^2$ lower bound. In this direction, Chazelle [3] (Sect. 3) showed that there exists a sequence of $n \to \infty$, where there exists a (non-convex) polyhedron in $\mathbb{R}^3$ whose surface can be triangulated with $n$ triangles, such that any partition of this polyhedron into convex sets requires at least $Cn^2$ such sets.

2) Avis and ElGindy [2] (Theorem 2.3 (c)) showed that there exist sets of $2n$ points in $\mathbb{R}^3$ such that any triangulation of their convex hull $P_{2n}$ that includes only these points as vertices requires $(n-1)^2$ tetrahedra. However, for "generic" points such a triangulation exists using $O(n)$ tetrahedra. Their bad non-generic example consists of $n$ points on each of two skew lines; their convex hull is a tetrahedron. However they do not address the question whether adding extra vertices inside $P_{2n}$ might reduce the number of tetrahedra needed to triangulate.

3) A candidate family of knots to study, which may require a large number of tetrahedra in any (linear or curvilinear) triangulation of $\mathbb{R}^3$ containing the knot in its 1-skeleton, are suitable representatives of the $(n, n-1)$ torus knot. This knot has a polygonal embedding $K_{2n}$ using $2n$ line segments, given in [1], Lemma 8.1. Any projection of this knot has at least $Cn^2$ crossings, and it is known that any orientable triangulated embedded surface having $K_{2n}$ as boundary requires at least $2n^2 - 6n + 5$ triangles, see Theorem 1.2 in [5].

**Remark.** In the application to unknotting complexity, given in [4,6], there are two possible measures of "input size" of the knot. The first is the number of edges $n$ in a polygonal knot, and the second is the number $m$ of crossings in a knot diagram, a projection of that knot into the plane. The number $m$ can be as large as $Cn^2$, and there exist examples where all projections of a polygon have at least $Cn^2$ crossings. In the unknotting questions considered in [4, 6], one takes the knot as represented in terms of its knot diagram, arising from a projection on the plane, and measures the complexity $m$ of this graph in terms of its number of vertices, which is at least as large as the number of knot crossings. The knot diagram $\mathcal{G}$ is a labeled planar graph, and it is possible to find a nice embedding of this graph in the plane (straight line embedding with vertices at small integer lattice points) so that one can construct a triangulated convex polyhedron using $Cm$ tetrahedra, which has a polygonal knot in its 1-skeleton that projects to $\mathcal{G}$; see, for example, Lemma 7.1 in [6] or Theorem 8.1 in [4]. This construction made use of the freedom to pick a suitable embedding of the graph (in its isomorphism class) to reduce the number of tetrahedra needed in the polyhedron. In problem *1998-22*, there is no freedom to move the polygonal knot.

[1] ADAMS C., BRENNAN B. M., GREILSHEIMER D. L., WOO A. K. Stick numbers and composition of knots and links. *J. Knot Theory Ramifications*, 1997, **6**, 149–161.

[2] AVIS D., ELGINDY H. Triangulating point sets in space. *Discr. & Comp. Geom.*, 1987, **2**, 99–111.

[3] CHAZELLE B. Convex partitions of polyhedra: a lower bound and worst case optimal algorithm. *SIAM J. Comput.*, 1984, **13**, 488–507.

[4] HASS J., LAGARIAS J. C. The number of Reidemeister moves needed for unknotting. *J. Amer. Math. Soc.*, 2001, **14**, 399–428.
[*Internet*: `http://www.arXiv.org/abs/math.GT/9807012`]

[5] HASS J., LAGARIAS J. C. Affine isoperimetric inequalities for piecewise linear surfaces.
[*Internet*: `http://www.arXiv.org/abs/math.GT/0202179`]

[6] HASS J., LAGARIAS J. C., PIPPENGER N. The computational complexity of knot and link problems. *J. Assoc. Comp. Mach.*, 1999, **46**(2), 185–211.
[*Internet*: `http://www.arXiv.org/abs/math.GT/9807016`]

[7] HASS J., LAGARIAS J. C., THURSTON W. P. Area inequalities for embedded disks bounding unknotted curves, in preparation.

## ▽ 1998-24 – *V. I. Arnold*

ℜ The equation means the vanishing of the curvature of level curves, which are therefore straight lines for the solution.

## △ 1998-24 – *S. V. Duzhin*

ℜ Using the standard methods of group analysis [1], one can describe all equations having this property. For example, in Monge's notation $p = u_x$, $q = u_y$, $r = u_{xx}$, $s = u_{xy}$, $t = u_{yy}$, the general form of a second order equation resolved with respect to $s$, and such that any function of an arbitrary solution is again a solution, is

$$s = \frac{1}{2}\left(\frac{pt}{q} + \frac{qr}{p}\right) + \phi\left(x, y, p, q, \frac{pt}{q} - \frac{qr}{p}\right),$$

where $\phi$ is an arbitrary function homogeneous with respect to the last 3 arguments of degree 1.

[1] Symmetries and Conservation Laws for Differential Equations of Mathematical Physics. Editors: I. S. Krasil′shchik and A. M. Vinogradov. Providence, RI: Amer. Math. Soc., 1999. (Transl. Math. Monographs, 182.) [*The Russian original* 1997.]

### ▽ 1998-25

$\mathfrak{R}$ Note that, in the case of matrices of dimension two, the union of the plane of all diagonal matrices $\left(\begin{smallmatrix} a & 0 \\ 0 & b \end{smallmatrix}\right)$ with the line of Jordan blocks $\left(\begin{smallmatrix} a & 1 \\ 0 & a \end{smallmatrix}\right)$ *does not satisfy* the problem's condition because the plane $\left\{\left(\begin{smallmatrix} a & 0 \\ 0 & b \end{smallmatrix}\right)\right\}$ contains *two* representatives $\left(\begin{smallmatrix} \lambda & 0 \\ 0 & \mu \end{smallmatrix}\right)$ and $\left(\begin{smallmatrix} \mu & 0 \\ 0 & \lambda \end{smallmatrix}\right)$ of each class of conjugate matrices with a simple spectrum $(\lambda, \mu)$.

### △▽ 1998-25 – *V. I. Arnold*

$\mathfrak{R}$ For matrices of dimension two, the appropriate choice is given by the union of the plane of Sylvester matrices $\left(\begin{smallmatrix} 0 & 1 \\ a & b \end{smallmatrix}\right)$ with the line of scalar matrices $\left(\begin{smallmatrix} a & 0 \\ 0 & a \end{smallmatrix}\right)$. For matrices of a not too high dimension, such a choice is also possible.

The computation of numbers of equivalence classes of matrices over finite fields apparently suggests that such a choice is possible for matrices of any dimension, but the description of the corresponding invariant of Jordan normal form requires nontrivial combinatorial analysis.

### △ 1998-25 – *M. L. Kontsevich*

$\mathfrak{R}$ I think that the problem with one matrix is easy, but I cannot remember the solution now. Operators $T$ in the $n$-dimensional space such that there is only one Jordan block for every eigenvalue can be canonically reduced to the canonical form in the basis $v, Tv, T^2v, \ldots, T^{n-1}v$, where $v$ is any cyclic vector of $T$.

One can ask the question for a pair (or, more generally, a $k$-tuple) of matrices. Calculation over finite fields shows that the number of orbits is a universal polynomial $P(q)$ with integral coefficients, where $q$ is the number of elements in the finite field. First of all, using the Pólya formula one can show easily that there exists a polynomial with rational coefficients. That the coefficients are integers follows from the fact that they should have interpretations as ranks of cohomology groups of algebraic varieties. It is not known whether the coefficients of this polynomial are non-negative; experimentally it is so for two matrices of sizes up to $10 \times 10$. Analogous questions can be asked for collections of commuting matrices.

I learned from Lieven le Bruyn that he had also made an attempt to classify conjugacy classes of pairs of noncommutative matrices using the same numerics over finite fields.

# 1999

## 1999-8 — *M. B. Sevryuk* Also: *1999-9*

ℜ In the present comment, we shall give complete proofs of the (simple and very well known) properties of the semigroup $S(a)$ for $n = 2$ mentioned in the formulations of problems *1999-8* and *1999-9*, as well as a proof of the existence of the number $K(a)$ for each $n \geq 2$ (the latter fact is also easy and very well known).

Let $1 \leq u < v$ be positive integers such that $\gcd(u, v) = 1$. Denote the number $(u-1)(v-1)$ by $K(u, v)$. For brevity, we shall write $K$ in place of $K(u, v)$ and $S$ in place of $S(u, v)$.

**Lemma 1.** *For each integer N, there is a unique representation of the form*

$$N = ru + sv, \qquad r \in \mathbb{Z},\ 0 \leq r \leq v-1, \quad s \in \mathbb{Z}. \tag{1}$$

**Proof.** Since $\gcd(u, v) = 1$, all the $v$ numbers $N$, $N-u$, $\ldots$, $N-(v-1)u$ give different remainders when divided by $v$. In particular, exactly one of these numbers is a multiple of $v$.

**Lemma 2.** *An integer N belongs to the semigroup S if and only if $s \geq 0$ in the representation* (1).

**Proof.** If $s \geq 0$ then $N \in S$ by definition. Now suppose that $s < 0$. If $N = \tilde{r}u + \tilde{s}v$ with some integers $\tilde{r} \geq 0$ and $\tilde{s}$ then $\tilde{r} \geq r$, whence $\tilde{s} \leq s < 0$. Hence $N \notin S$.

**Theorem 3** (The Sylvester duality [6]; see also [1, 2, 4, 5]). *An integer N belongs to the semigroup S if and only if the integer $K-1-N$ does not.*

**Proof.** If $N = ru + sv$ is the representation (1) of $N$ then

$$K-1-N = uv-u-v-(ru+sv) = (v-1-r)u + (-1-s)v$$

is the analogous representation of $K-1-N$. Of the two integers $s$ and $-1-s$, exactly one is non-negative. Consequently, of the two integers $N$ and $K-1-N$, exactly one lies in the semigroup $S$ according to Lemma 2.

Such a symmetry of $S$ and its complement is called the *Gorenstein property* (of the plane curve $x = t^u$, $y = t^v$), cf. [5].

**Corollary 4.** $K-1 \notin S$.

Indeed, $K-1-(K-1)=0 \in S$.

**Corollary 5.** $N \in S \ \forall N \geq K$.

Indeed, $\forall N \geq K$ one has $K-1-N<0$, and hence $K-1-N \notin S$.

Now let $n \geq 2$ and let $1 \leq a_1 < a_2 < \cdots < a_n$ be positive integers such that $\gcd(a_1, a_2, \ldots, a_n) = 1$.

**Theorem 6.** *There exists an integer $K = K(a) \in \mathbb{Z}_+$ such that $N \in S(a)$ for all $N \geq K$.*

**Proof** (by induction over $n$). For $n=2$, the statement of the theorem does hold with $K(a) = (a_1-1)(a_2-1)$ according to Corollary 5. Let $n \geq 3$, $\gcd(a_1, \ldots, a_{n-1}) = d \in \mathbb{N}$, and $N \in S(a_1/d, \ldots, a_{n-1}/d)$ for all $N \geq \widehat{K}$ with some $\widehat{K} \in \mathbb{Z}_+$. Set

$$K = d\widehat{K} + (d-1)a_n,$$

and let $N \geq K$. Consider the $d$ numbers $N, N-a_n, \ldots, N-(d-1)a_n$. All of them are no less than $d\widehat{K}$, and exactly one of them is a multiple of $d$ since $\gcd(d, a_n) = 1$ (cf. the proof of Lemma 1). Let this number be $N - ra_n = ds$, $s \geq \widehat{K}$. Then $s \in S(a_1/d, \ldots, a_{n-1}/d)$, and therefore $N = ds + ra_n \in S(a_1, a_2, \ldots, a_n)$ as well.

The statistics of the numbers $K(a)$ is connected with the so-called weak asymptotics (the asymptotics averaged over the translations and rotations of the space) for the numbers of integer points in the domains and on the surfaces in $\mathbb{R}^l$ (cf. [2]); see the comments to problems *1981-26* and *1999-10*. It is also related to the statistics of the group representations [3].


[1] ARNOLD V. I. Simple singularities of curves. *Proc. Steklov Inst. Math.*, 1999, **226**, 20–28; CEREMADE (UMR 7534), Université Paris-Dauphine, № 9906, 09/02/1999. [*Internet*: `http://www.pdmi.ras.ru/~arnsem/Arnold/arn-papers.html`]

[2] ARNOLD V. I. Weak asymptotics for the numbers of solutions of Diophantine problems. *Funct. Anal. Appl.*, 1999, **33**(4), 292–293.

[3] ARNOLD V. I. Frequent representations. *Moscow Math. J.*, 2003, **3**(4), 14 pp.

[4] HERZOG J. Generators and relations of Abelian semigroups and semigroup rings. *Manuscripta Math.*, 1970, **3**(2), 175–193.

[5] KUNZ E. The value-semigroup of a one-dimensional Gorenstein ring. *Proc. Amer. Math. Soc.*, 1970, **25**(4), 748–751.

[6] SYLVESTER J. J. Mathematical questions with their solutions. *Educational Times*, 1884, **41**, 21.

### 1999-9

ℛ See the comment to problem *1999-8*.

### 1999-10 — *V. I. Arnold* Also: *1999-11*

ℛ The asymptotic densities $p(N)$ and $P(N)$ should be treated as follows. Fix a (Jordan) domain $U$, e. g., a ball or a cube, in the space $\mathbb{R}^n \ni a$. Fix a large number $M$ and consider all integer points $a \in MU$ such that $\gcd(a_1, a_2, \ldots, a_n) = 1$. For every such $a$, let us consider the semigroup $S(a)$ and evaluate $K(a)$.

The images of integer points in $\mathbb{Z}^n_+$ under the normalized projection

$$\mathbb{R}^n \to \mathbb{R}, \quad k \mapsto \frac{\langle k, a\rangle}{K(a)}$$

form a locally finite set. Take an atomic measure on this set so that the measure of a point is either always equal to 1 or equal to the number of integer points $k \in \mathbb{Z}^n_+$ projected into that point.

The measure obtained in this way is a functional on, say, continuous or smooth functions; let us average it on $a \in MU$ (maybe even with some density $\rho(a)$).

**The main conjecture.** *For any "good" function $f(x)$, the ratio of the integral by this averaged measure and the integral by the measure with density $Cx^{n-1}$ converges to 1 as $M \to \infty$, where $C = M^{n-1}c$ and $c$ is the $(n-1)$-volume of the simplex*

$$\{k \in \mathbb{R}^n \mid \langle k, a\rangle = 1,\ k_1 > 0,\ k_2 > 0,\ \ldots,\ k_n > 0\}$$

*averaged on $a \in U$.*

### 1999-11

ℛ See the comment to problem *1999-10*.

### 1999-12

ℛ See the comments to problems *1998-10* and *1998-14*.

## 1999-15

ℜ On ℝ–ℂ–ℍ-trinity, see problem *1997-9* and the comments to it, as well as problems *1988-24*, *1998-15*, *1998-16*, and the comments thereto.

## 1999-17 — *F. Aicardi*

ℜ In the real Euclidean $n$-dimensional space with Cartesian coordinates $x_i$, the *antisphere* is defined by the equation

$$\sum_i x_i^{-2} = 1.$$

In the dual space with coordinates $p_i$, the hypersurface dual to the antisphere is the $n$-astroid:

$$\sum_i p_i^{2/3} = 1.$$

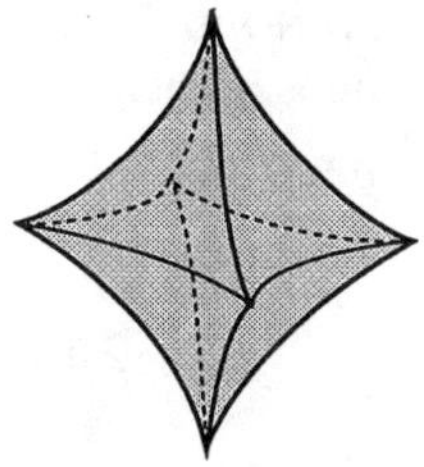

**Fig. 1:** The 3-astroid

The envelope of the normal lines to an ellipsoid in $\mathbb{R}^n$ (the caustic) is *not* the $n$-astroid.

# 2000

## 2000-7 — *V. I. Arnold*

ℜ Concerning the formula "the number of the species $\sim S^{1/4}$" and "the Arrhenius law," see [1,2].

[1] KRASILOV V. A. Macroevolution and evolutionary synthesis. In: Evolution, Ecology, Biodiversity. Proceedings of a conference in memory of N. N. Vorontsov (1934–2000) held on December 26–27, 2000. Editor: E. A. Lyapunova. Moscow: Research Center for Education Preceding the High School, 2001, 27–47 (in Russian).

[2] MALYSHEV L. I. The quantitative analysis of flora: spatial diversity, the level of specific richness, and representativity of sampling areas. *Botanicheskiĭ Zh.*, 1975, **60**(11), 1537–1550 (in Russian).

## 2000-8 — *M. B. Sevryuk*

$\mathfrak{R}$ Let $A$ be an arbitrary smooth (in the real sense) mapping of $\mathbb{C}\mathrm{P}^2$ onto itself taking complex projective straight lines to complex projective straight lines. That $A$ is necessarily either a complex projective map or the product of a complex projective map and the complex conjugation was proved in paper [1] (describing the octahedron as the complexification of the tetrahedron). The quaternionic version is conjectured to be the icosahedron (see [2]).

[1] ARNOLD V. I. Complexification of tetrahedron and pseudoprojective transformations. *Funct. Anal. Appl.*, 2001, **35**(4), 241–246.

[2] ARNOLD V. I. Pseudoquaternion geometry. *Funct. Anal. Appl.*, 2002, **36**(1), 1–12.

## 2000-9 — *M. B. Sevryuk* Also: *1996-5, 2002-13*

$\mathfrak{R}$ The Sturm–Hurwitz theorem was proved by A. Hurwitz [17] (J. C. F. Sturm [21] had obtained it for trigonometric polynomials only). This theorem has been rediscovered more than once (see, e. g., book [16] for the first nontrivial case $k = 1$ or papers [18, 22] for the general case). The Sturm–Hurwitz theorem (and especially its particular case $k = 1$) has many important applications in topology and singularity theory, see, e. g., V. I. Arnold's works [1–15]. Several quite different proofs of the theorem are known. For instance, three proofs are sketched together in works [7, 10, 13–15]. These proofs admit various generalizations.

The Sturm–Hurwitz theorem was carried over to algebraic curves of genera $g \geq 1$ by S. M. Natanzon [19] (see also survey [20]). In fact, in [19, 20], analogues of this theorem for the so-called tensors of weight $\lambda$ (for arbitrary $2\lambda \in \mathbb{Z}$) on algebraic curves of arbitrary genus $g \geq 0$ were obtained. The classical Sturm–Hurwitz theorem corresponds to the case $g = \lambda = 0$.

See also the comment to problem *1996-5*.

[1] ARNOLD V. I. Topological Invariants of Plane Curves and Caustics. Dean Jacqueline B. Lewis Memorial Lectures, Rutgers University. Providence, RI: Amer. Math. Soc., 1994. (University Lecture Series, 5.)

[2] ARNOLD V. I. Invariants and perestroikas of plane fronts. *Proc. Steklov Inst. Math.*, 1995, **209**, 11–56.

[3] ARNOLD V. I. On the topological properties of Legendrian projections in contact geometry of wave fronts. *St. Petersburg Math. J.*, 1995, **6**(3), 439–452.

[4] ARNOLD V. I. Sur les propriétés topologiques des projections lagrangiennes en géométrie symplectique des caustiques. CEREMADE (UMR 7534), Université Paris-Dauphine, № 9320, 14/06/1993; *Rev. Mat. Univ. Complut. Madrid*, 1995, **8**(1), 109–119. [*The Russian translation in*: Vladimir Igorevich Arnold. Selecta–60. Moscow: PHASIS, 1997, 525–532.]

[5] ARNOLD V. I. The geometry of spherical curves and quaternion algebra. *Russian Math. Surveys*, 1995, **50**(1), 1–68.

[6] ARNOLD V. I. On the number of flattening points of space curves. In: Sinaĭ's Moscow Seminar on Dynamical Systems. Editors: L. A. Bunimovich, B. M. Gurevich and Ya. B. Pesin. Providence, RI: Amer. Math. Soc., 1996, 11–22. (AMS Transl., Ser. 2, 171; Adv. Math. Sci., 28.)

[7] ARNOLD V. I. Topological problems in the theory of wave propagation. *Russian Math. Surveys*, 1996, **51**(1), 1–47.

[8] ARNOLD V. I. Remarks on parabolic curves on surfaces and on higher dimensional Möbius–Sturm theory. *Funct. Anal. Appl.*, 1997, **31**(4), 227–239.

[9] ARNOLD V. I. Mysterious Mathematical Trinities. Topological Economy Principle in Algebraic Geometry. Moscow: Moscow Center for Continuous Mathematical Education Press, 1997 (in Russian).

[10] ARNOLD V. I. Topological problems in wave propagation theory and topological economy principle in algebraic geometry. In: The Arnoldfest. Proceedings of a conference in honour of V. I. Arnold for his sixtieth birthday (Toronto, 1997). Editors: E. Bierstone, B. A. Khesin, A. G. Khovanskiĭ and J. E. Marsden. Providence, RI: Amer. Math. Soc., 1999, 39–54. (Fields Institute Commun., 24.);

[11] ARNOLD V. I. Topologically necessary singularities on moving wavefronts and caustics. In: Hamiltonian Systems with Three or More Degrees of Freedom (S'Agaró, 1995). Editor: C. Simó. Dordrecht: Kluwer Acad. Publ., 1999, 11–12. (NATO Adv. Sci. Inst. Ser. C Math. Phys. Sci., 533.)

[12] ARNOLD V. I. Symplectic geometry and topology. *J. Math. Phys.*, 2000, **41**(6), 3307–3343.

[13] ARNOLD V. I. Astroidal geometry of hypocycloids and the Hessian topology of hyperbolic polynomials. *Russian Math. Surveys*, 2001, **56**(6), 1019–1083.

[14] ARNOLD V. I. Astroidal Geometry of Hypocycloids and the Hessian Topology of Hyperbolic Polynomials. Moscow: Moscow Center for Continuous Mathematical Education Press, 2001 (in Russian).

[15] ARNOLD V. I. Wave Fronts and Topology of Curves. Moscow: PHASIS, 2002 (in Russian). (Young Mathematician's Library, 9.)

[16] BLASCHKE W. Kreis und Kugel. Leipzig: Veit, 1916. [*2te Auflage*: Berlin: Walter de Gruyter, 1956.]

[17] HURWITZ A. Über die Fourierschen Konstanten integrierbarer Funktionen. *Math. Ann.*, 1903, **57**(4), 425–446.

[18] KELLOGG O. D. The oscillation of functions of an orthogonal set. *Amer. J. Math.*, 1916, **38**(1), 1–5.

[19] NATANZON S. M. Trigonometric tensors on algebraic curves of arbitrary genus. An analogue of the Sturm–Hurwitz theorem. *Russian Math. Surveys*, 1995, **50**(6), 1286–1287.

[20] NATANZON S. M. Moduli of real algebraic curves and their superanalogues. Spinors and Jacobians of real curves. *Russian Math. Surveys*, 1999, **54**(6), 1091–1147.

[21] STURM J. C. F. Mémoire sur les équations différentielles du second ordre. *J. Math. Pures Appl.*, 1836, **1**, 106–186.

[22] TABACHNIKOV S. L. Around four vertices. *Russian Math. Surveys*, 1990, **45**(1), 229–230.

## 2000-10 — *M. B. Sevryuk* Also: *2000-11*

H Controls optimal on the average and phase transitions in controlled dynamical systems are studied in paper [2]. The starting point of the work was an example in chemical engineering [1].

[1] ARNOLD V. I. Convex hulls and the increase of efficiency of systems under pulsating loading. *Sib. Math. J.*, 1987, **28**(4), 540–542.

[2] ARNOLD V. I. Optimization in mean and phase transitions in controlled dynamical systems. *Funct. Anal. Appl.*, 2002, **36**(2), 83–92.

## 2000-11

R See the comment to problem *2000-10*, and also paper [1].

[1] ARNOLD V. I. On a variational problem connected with phase transitions of means in controllable dynamical systems. In: Nonlinear Problems in Mathematical Physics and Related Topics I. In honour of Professor O. A. Ladyzhenskaya. Editors: M. Sh. Birman, S. Hildebrandt, V. A. Solonnikov and N. N. Ural′tseva. Dordrecht: Kluwer Acad. Publ., 2002, 23–34. (Internat. Math. Ser., 1.)

## ▽ 2000-12 — *V. I. Arnold*

R Concerning experimental statistics and their comparison with random ones, see [1].

[1] SCHÜTT C., WERNER E. Random polytopes with vertices on the boundary of a convex body. *C. R. Acad. Sci. Paris, Sér.* I *Math.*, 2000, **331**(9), 697–701.

## △ 2000-12 – *M. B. Sevryuk*

R The sails are two-dimensional analogues of continued fractions. Concerning the periodicity of the sails, see [5–8]. The statistics of multidimensional continued fractions was recently explored by M. L. Kontsevich and Yu. M. Sukhov, see [2–4]. The modern stage of the studies in the statistics of continued fractions (conventional ones as well as higher dimensional ones) has its origin in V. I. Arnold's paper [1].

See also problems *1993-11* and *1993-33*.


[1] ARNOLD V. I. *A*-graded algebras and continued fractions. *Comm. Pure Appl. Math.*, 1989, **42**(7), 993–1000. [*The Russian translation in*: Vladimir Igorevich Arnold. Selecta–60. Moscow: PHASIS, 1997, 473–482.]

[2] ARNOLD V. I. Higher dimensional continued fractions. *Reg. Chaot. Dynamics*, 1998, **3**(3), 10–17.

[3] ARNOLD V. I. Continued Fractions. Moscow: Moscow Center for Continuous Mathematical Education Press, 2001 (in Russian). ("Mathematical Education" Library, 14.)

[4] KONTSEVICH M. L., SUKHOV YU. M. Statistics of Klein polyhedra and multidimensional continued fractions. In: Pseudoperiodic Topology. Editors: V. Arnold, M. Kontsevich and A. Zorich. Providence, RI: Amer. Math. Soc., 1999, 9–27. (AMS Transl., Ser. 2, 197; Adv. Math. Sci., 46.)

[5] KORKINA E. I. La périodicité des fractions continues multidimensionnelles. *C. R. Acad. Sci. Paris, Sér.* I *Math.*, 1994, **319**(8), 777–780.

[6] KORKINA E. I. Two-dimensional continued fractions. The simplest examples. *Proc. Steklov Inst. Math.*, 1995, **209**, 124–144.

[7] TSUCHIHASHI H. 2-dimensional periodic continued fractions and 3-dimensional cusp singularities. *Proc. Jap. Acad., Ser. A Math. Sci.*, 1982, **58**(6), 262–264.

[8] TSUCHIHASHI H. Higher-dimensional analogues of periodic continued fractions and cusp singularities. *Tôhoku Math. J., Ser.* 2, 1983, **35**(4), 607–639.


# 2001

## 2001-1

H This is a problem in works [1, 2].

[1] ARNOLD V. I. Astroidal geometry of hypocycloids and the Hessian topology of hyperbolic polynomials. *Russian Math. Surveys*, 2001, **56**(6), 1019–1083.

[2] ARNOLD V. I. Astroidal Geometry of Hypocycloids and the Hessian Topology of Hyperbolic Polynomials. Moscow: Moscow Center for Continuous Mathematical Education Press, 2001 (in Russian).

## 2001-2

ℋ This is a problem in works [1, 2].

[1] ARNOLD V. I. Astroidal geometry of hypocycloids and the Hessian topology of hyperbolic polynomials. *Russian Math. Surveys*, 2001, **56**(6), 1019–1083.

[2] ARNOLD V. I. Astroidal Geometry of Hypocycloids and the Hessian Topology of Hyperbolic Polynomials. Moscow: Moscow Center for Continuous Mathematical Education Press, 2001 (in Russian).

## 2001-4 – *V. I. Arnold*

ℛ The four-cusp theorem for the caustic is an extension of the Sturm–Hurwitz theorem on the zeros of the Fourier series omitting the first harmonics. The problem is a generalization of this Sturm–Hurwitz–Morse theory to the multivalued functions (to the Lagrangian manifolds that are not the sections of the cotangent bundle). One of the generalizations of this kind was proved by Chekanov and Pushkar′ in their recent work on an application of the contact homology to the Arnold conjecture on the wave fronts reversal. Hence the problem deals with a new extension of this contact homology and hence with continuations of the Arnold conjectures in symplectic topology (problems *1965-1–1965-3*, *1993-9*), see [1, 2].

[1] ARNOLD V. I. Topological Invariants of Plane Curves and Caustics. Dean Jacqueline B. Lewis Memorial Lectures, Rutgers University. Providence, RI: Amer. Math. Soc., 1994. (University Lecture Series, 5.)

[2] ARNOLD V. I. Topological problems in the theory of wave propagation. *Russian Math. Surveys*, 1996, **51**(1), 1–47.

## 2001-6

ℋ For more details on the problem, see works [1, 2].

[1] ARNOLD V. I. Astroidal geometry of hypocycloids and the Hessian topology of hyperbolic polynomials. *Russian Math. Surveys*, 2001, **56**(6), 1019–1083.

[2] ARNOLD V. I. Astroidal Geometry of Hypocycloids and the Hessian Topology of Hyperbolic Polynomials. Moscow: Moscow Center for Continuous Mathematical Education Press, 2001 (in Russian).

# 2002

## ▽ 2002-1

H This is Problem 1 in paper [1] which is also included in book [2].

[1] ARNOLD V. I. Problems to the Seminar: 15 January 2002. CEREMADE (UMR 7534), Université Paris-Dauphine, № 0216, 16/05/2002.

[2] ARNOLD V. I. What Is Mathematics? Moscow: Moscow Center for Continuous Mathematical Education Press, 2002, 104 pp.

## △ 2002-1 – *V. I. Arnold* Also: *2002-2, 2002-3, 2002-4*

R More details on problems *2002-1–2002-4* are described in my recent works [1, 2].

[1] ARNOLD V. I. Astroidal geometry of hypocycloids and the Hessian topology of hyperbolic polynomials. *Russian Math. Surveys*, 2001, **56**(6), 1019–1083.

[2] ARNOLD V. I. Astroidal Geometry of Hypocycloids and the Hessian Topology of Hyperbolic Polynomials. Moscow: Moscow Center for Continuous Mathematical Education Press, 2001 (in Russian).

## 2002-2

H This is Problem 2 in paper [1] which is also included in book [2].

[1] ARNOLD V. I. Problems to the Seminar: 15 January 2002. CEREMADE (UMR 7534), Université Paris-Dauphine, № 0216, 16/05/2002.

[2] ARNOLD V. I. What Is Mathematics? Moscow: Moscow Center for Continuous Mathematical Education Press, 2002, 104 pp.

R See the comment to problem *2002-1*.

## 2002-3

*H* This is Problem 3 in paper [1] which is also included in book [2].

[1] Arnold V. I. Problems to the Seminar: 15 January 2002. CEREMADE (UMR 7534), Université Paris-Dauphine, № 0216, 16/05/2002.

[2] Arnold V. I. What Is Mathematics? Moscow: Moscow Center for Continuous Mathematical Education Press, 2002, 104 pp.

*R* See the comment to problem *2002-1*.

## 2002-4

*H* This is Problem 4 in paper [1] which is also included in book [2].

[1] Arnold V. I. Problems to the Seminar: 15 January 2002. CEREMADE (UMR 7534), Université Paris-Dauphine, № 0216, 16/05/2002.

[2] Arnold V. I. What Is Mathematics? Moscow: Moscow Center for Continuous Mathematical Education Press, 2002, 104 pp.

*R* See the comment to problem *2002-1*.

## 2002-5

*H* This is Problem 5 in paper [1] which is also included in book [2].

[1] Arnold V. I. Problems to the Seminar: 15 January 2002. CEREMADE (UMR 7534), Université Paris-Dauphine, № 0216, 16/05/2002.

[2] Arnold V. I. What Is Mathematics? Moscow: Moscow Center for Continuous Mathematical Education Press, 2002, 104 pp.

## 2002-6

*H* This is Problem 6 in paper [1] which is also included in book [2].

[1] Arnold V. I. Problems to the Seminar: 15 January 2002. CEREMADE (UMR 7534), Université Paris-Dauphine, № 0216, 16/05/2002.

[2] Arnold V. I. What Is Mathematics? Moscow: Moscow Center for Continuous Mathematical Education Press, 2002, 104 pp.

## ▽ 2002-7

ℋ This is Problem 7 in paper [1] which is also included in book [2].

[1] ARNOLD V. I. Problems to the Seminar: 15 January 2002. CEREMADE (UMR 7534), Université Paris-Dauphine, № 0216, 16/05/2002.

[2] ARNOLD V. I. What Is Mathematics? Moscow: Moscow Center for Continuous Mathematical Education Press, 2002, 104 pp.

## △ 2002-7 – *V. I. Arnold*

ℋ This problem is related to the *Euler equations for stationary states on two-dimensional incompressible hydrodynamics and magnetohydrodynamics*. The smooth version of this equation is a functional dependence between some function $f$ on $M^2$ and its Laplacian (their Poisson bracket should vanish identically).

And the problem is to extend this equation to the case of "generalized functions." This problem is also a simplified (2-dimensional) version of Sakharov's 3-dimensional problem on a *minimal energy magnetic field of a star* (in the 3-dimensional case, the Euler equation means the vanishing of the Poisson bracket of a divergence-free vector field with its curl field). More details (and references to my initial 1974 paper [1] on this subject) are contained in book [2], pages 69–81 and 112–193.

In magnetohydrodynamics, one is searching for the vector field of minimal energy obtainable from a given one (from the "initial magnetic field") by a volume-preserving diffeomorphism (representing the star particles motion). In the hydrodynamical problem, one considers the (velocity) field of stationary energy among those whose curl is obtainable from a given "initial curl" by a volume-preserving diffeomorphism (in the two-dimensional case, the preserved object is the Laplacian value of the stream function at every particle of the fluid). The Euler equations of these two different variational problems are the same.

[1] ARNOLD V. I. The asymptotic Hopf invariant and its applications. In: Proceedings of the All-Union School on Differential Equations with Infinitely Many Independent Variables and on Dynamical Systems with Infinitely Many Degrees of Freedom (Dilizhan, May 21 – June 3, 1973). Yerevan: AS of Armenian SSR, 1974, 229–256 (in Russian). [*The English translation*: *Selecta Math. Sov.*, 1986, **5**(4), 327–345.] [*The Russian original is reprinted and supplemented in*: Vladimir Igorevich Arnold. Selecta–60. Moscow: PHASIS, 1997, 215–236.]

[2] ARNOLD V. I., KHESIN B. A. Topological Methods in Hydrodynamics. New York: Springer, 1998. (Appl. Math. Sci., 125.)

## ▽ 2002-8

ℋ This is Problem 8 in paper [1] which is also included in book [2].

[1] ARNOLD V. I. Problems to the Seminar: 15 January 2002. CEREMADE (UMR 7534), Université Paris-Dauphine, № 0216, 16/05/2002.

[2] ARNOLD V. I. What Is Mathematics? Moscow: Moscow Center for Continuous Mathematical Education Press, 2002, 104 pp.

## △ 2002-8 – *V. I. Arnold*

ℛ Among the 45 $(C,B,A)$-permutations for $n = 11$, there are 22 cyclic permutations (which are all transitive), and for the preceding values of $n$ the numbers of cyclic and of transitive cyclic permutations form two strange nonmonotone sequences.

Among all the $n!$ permutations of $n$ elements, those which are transitively cyclic form a small part (their number is $(n-1)!$), and the number of all cyclic ones seems to be asymptotically equivalent to the same $1/n$th part of the total number of permutations as that of transitively cyclic ones.

Thus, it seems that *the statistic of the* $(C,B,A)$*-permutations is essentially different from that of the generic ones* (and one might study the same difference for other permutations of several letters acting on a large set $\{1,\ldots,n\}$).

ℋ All these questions had been formulated in 1958 (see problem *1958-1*) at my Moscow seminar as simplified models for the segments interchange mapping problem in the theory of dynamical systems, which was later published in my 1963 long paper [1] on Hamiltonian dynamics. The present day state of this problem (studied especially by M. L. Kontsevich and A. V. Zorich) is described in Zorich's recent paper [2].

Kontsevich has recently attempted to claim that the interchange of three segments is always rotation equivalent. Problem *2002-8* requires the study of the *frequency* of the counterexamples to this "equivalence" (the existence of such counterexamples being, of course, known to me in the 1950s while I was inventing the segments interchange problem).

[1] ARNOLD V. I. Small denominators and problems of stability of motion in classical and celestial mechanics. *Russian Math. Surveys*, 1963, **18**(6), 85–191.

[2] ZORICH A. How do the leaves of a closed 1-form wind around a surface? In: Pseudoperiodic Topology. Editors: V. Arnold, M. Kontsevich and A. Zorich. Providence, RI: Amer. Math. Soc., 1999, 135–178. (AMS Transl., Ser. 2, 197; Adv. Math. Sci., 46.)

## ▽ 2002-9

*H* This is Problem 9 in paper [1] which is also included in book [2].

[1] ARNOLD V. I. Problems to the Seminar: 15 January 2002. CEREMADE (UMR 7534), Université Paris-Dauphine, № 0216, 16/05/2002.

[2] ARNOLD V. I. What Is Mathematics? Moscow: Moscow Center for Continuous Mathematical Education Press, 2002, 104 pp.

## △ 2002-9 — *V. I. Arnold* Also: *2002-10*

*R* More details on the real smooth versions of problems *2002-9* and *2002-10* can be found in my papers [1, 2].

[1] ARNOLD V. I. Complexification of tetrahedron and pseudoprojective transformations. *Funct. Anal. Appl.*, 2001, **35**(4), 241–246.

[2] ARNOLD V. I. Pseudoquaternion geometry. *Funct. Anal. Appl.*, 2002, **36**(1), 1–12.

## 2002-10

*H* This is Problem 10 in paper [1] which is also included in book [2].

[1] ARNOLD V. I. Problems to the Seminar: 15 January 2002. CEREMADE (UMR 7534), Université Paris-Dauphine, № 0216, 16/05/2002.

[2] ARNOLD V. I. What Is Mathematics? Moscow: Moscow Center for Continuous Mathematical Education Press, 2002, 104 pp.

*R* See the comment to problem *2002-9*.

## 2002-11

*H* This is Problem 11 in paper [1] which is also included in book [2].

[1] ARNOLD V. I. Problems to the Seminar: 15 January 2002. CEREMADE (UMR 7534), Université Paris-Dauphine, № 0216, 16/05/2002.

[2] ARNOLD V. I. What Is Mathematics? Moscow: Moscow Center for Continuous Mathematical Education Press, 2002, 104 pp.

### ▽ 2002-12

ℋ This is Problem 12 in paper [1] which is also included in book [2].

[1] ARNOLD V. I. Problems to the Seminar: 15 January 2002. CEREMADE (UMR 7534), Université Paris-Dauphine, № 0216, 16/05/2002.

[2] ARNOLD V. I. What Is Mathematics? Moscow: Moscow Center for Continuous Mathematical Education Press, 2002, 104 pp.

### △ 2002-12 — *V. I. Arnold*

ℛ Past experience shows that one should extend the Morse theory of ordinary functions to the theory of intersections of Lagrangian manifolds in symplectic topology (the "Arnold conjectures" of 1965, extending the "Poincaré last theorem" and being the starting point of the Floer homology and of many other things in symplectic and contact topology; see problems *1965-1–1965-3*, *1966-4*, *1966-5*, *1972-17*, *1972-33*, *1976-39* and the corresponding comments).

Recently Yu. V. Chekanov and P. E. Pushkar′ used this extension to prove my 1993 conjecture on the necessity of 4 cusps at any wave front eversion from [1]. However, it is unclear to me whether their version of the Chekanov contact cohomology theory suffices to understand the caustics of exact Lagrangian submanifolds (i. e., of Legendrian knots in the manifold of 1-jets of periodic functions).

[1] ARNOLD V. I. Topological Invariants of Plane Curves and Caustics. Dean Jacqueline B. Lewis Memorial Lectures, Rutgers University. Providence, RI: Amer. Math. Soc., 1994. (University Lecture Series, 5.)

### ▽ 2002-13

ℋ This is Problem 13 in paper [1] which is also included in book [2].

[1] ARNOLD V. I. Problems to the Seminar: 15 January 2002. CEREMADE (UMR 7534), Université Paris-Dauphine, № 0216, 16/05/2002.

[2] ARNOLD V. I. What Is Mathematics? Moscow: Moscow Center for Continuous Mathematical Education Press, 2002, 104 pp.

### △ 2002-13 — *V. I. Arnold*

ℛ The $C$-caustics are algebraic curves (of the same genus as $C$). In the case of problem *2002-12* they were rational curves, since the circle is a rational curve, and thus the caustics of trigonometric polynomials are unicursal curves and their Riemann surfaces are spheres.

By the way, with each periodic function in problem *2002-12* and with each $P$ in the present case, one can associate a 1-parameter family of "*wavefront curves*" each consisting of those points of the plane $\{(A,B)\}$ for which the restriction of $P+Ax+By$ to $C$ has a fixed critical value $p$ (which is the parameter of the family).

These fronts are algebraic curves (of the same genus as $C$), rational in the case of trigonometric polynomials in problem *2002-12*.

The problems on cusps, alternative lengths, etc. for $C$-caustics and $C$-fronts for algebraic curves $C$ of higher genera are especially interesting in the complex case, since the real four-cusp property is known, e. g., for the case of smooth convex curves (and with more cusps for some generalizations of convexity). But even in the rational (genus zero) case, and even for the degenerate elliptic curve $y^2 = x^2 + x^3$, there are interesting problems relating the cusps of the caustic and the double points of the degenerate curve. There have been published some extensions of the Sturm–Hurwitz theorem on Fourier series to the case of higher genera (by S. M. Natanzon, see the comment to problem *2000-9*). But I am not able to use them to obtain any useful knowledge on $C$-caustics and $C$-fronts for algebraic curves $C$ of higher genera.

For more details on these caustics, see paper [1] (or its monograph version [2]) where the present problem first appeared.

[1] ARNOLD V. I. Astroidal geometry of hypocycloids and the Hessian topology of hyperbolic polynomials. *Russian Math. Surveys*, 2001, **56**(6), 1019–1083.

[2] ARNOLD V. I. Astroidal Geometry of Hypocycloids and the Hessian Topology of Hyperbolic Polynomials. Moscow: Moscow Center for Continuous Mathematical Education Press, 2001 (in Russian).

## ▽ 2002-14

This is Problem 14 in paper [1] which is also included in book [2].

[1] ARNOLD V. I. Problems to the Seminar: 15 January 2002. CEREMADE (UMR 7534), Université Paris-Dauphine, № 0216, 16/05/2002.

[2] ARNOLD V. I. What Is Mathematics? Moscow: Moscow Center for Continuous Mathematical Education Press, 2002, 104 pp.

## △ 2002-14 — *V. I. Arnold*

ℜ The simplest example is provided by the "*cubic golden ratio*" matrix

$$A = \begin{pmatrix} 3 & 2 & 1 \\ 2 & 2 & 1 \\ 1 & 1 & 1 \end{pmatrix};$$

the ordinary golden ratio, $(\sqrt{5}-1)/2$, after addition of 2 makes an eigenvalue of the "quadratic golden ratio" matrix

$$A = \begin{pmatrix} 2 & 1 \\ 1 & 1 \end{pmatrix}.$$

The torus triangulation corresponding to the case of the cubic golden ratio can be described as the decomposition of the square into two triangles by a diagonal (no interior distinguished points). The simplest triangulations were studied in paper [1] by E. I. Korkina.

Studying other $\mathrm{SL}(3,\mathbb{R})$ matrices, starting with those not too large, one might obtain some hints *what are all the quotient tori triangulations and the distinguished points sets*. One might also study the statistics of these triangulations: the proportion of triangles, of quadrangles, of pentagons, and so on; also the statistics of the number of faces meeting at a vertex, that of the number of sides of a polygon of the decomposition, that of the numbers of distinguished points on faces and on edges, etc. It would be interesting to compare these statistics with that of the sails of a randomly chosen triangular cone (whose vertex is the origin). First examples of such triangulations have been recently studied by O. Karpenkov (Moscow State University), 2003.

One might even compare the results with the similar statistics for the boundaries of the convex hulls of the sets of integer points in bodies bounded by large smooth surfaces. Another interesting object to compare with would be plane decompositions by *Voronoĭ domains of random distributions of the domain centers* (a Voronoĭ domain of a discrete set of centers in the Euclidean plane consists of the points of the plane for which the given center is the closest one among all the centers).

Studying these statistics, one should compare the distribution of the areas of the domains, that of their boundary perimeters, that of the lengths of their edges, that of the number of the edges, and especially their simultaneous distributions, since they are far from being independent.

I would even suggest studying the dimensionless characteristics like the ratio "area/perimeter squared" and the number of vertices of a polygonal domain whose correlation is also an interesting characteristic of the decomposition.

While averaging I would rather study the more stable *averages per unit of area* than the *averages per domain* (which overestimate the contribution of small domains due to their larger number).

[1] KORKINA E. I. Two-dimensional continued fractions. The simplest examples. *Proc. Steklov Inst. Math.*, 1995, **209**, 124–144.

## ▽ 2002-15

ℋ This is Problem 15 in paper [1] which is also included in book [2].

[1] ARNOLD V. I. Problems to the Seminar: 15 January 2002. CEREMADE (UMR 7534), Université Paris-Dauphine, № 0216, 16/05/2002.

[2] ARNOLD V. I. What Is Mathematics? Moscow: Moscow Center for Continuous Mathematical Education Press, 2002, 104 pp.

## △ 2002-15 – *V. I. Arnold*

ℛ The guess is that these two statistics are the same, in both cases being the statistic of the elements of the continued fraction for a random real number.

The last statistic (frequency of $k$) $= \ln(1+\frac{1}{k(k+2)})/\ln 2$ was discovered by Gauss and proved by R. O. Kuz′min (1928) using the Birkhoff ergodic theorem (for every real number except those belonging to some set of Lebesgue measure zero).

However, two papers [2, 4] whose titles mention the same problem appeared already in 1900.

Unfortunately, I have not checked if these papers contained Kuz′min's theorem of 1928, and hence to read and to evaluate them from the modern viewpoint is also a problem.

The theorems by Gauss, Wiman and Kuz′min and their relation to the dynamical system $x \mapsto \{1/x\} = 1/x - [1/x]$ with invariant measure $\mu(A) = \int_A \rho(x)\,dx$, $\rho = 1/(1+x)$, are discussed in book [1].

The statistics for the sails of random triangular pyramids in $\mathbb{R}^3$ are unknown, but M. Kontsevich and Yu. Sukhov [3] have proved my conjectures on their *existence* and *universality* (independence of a pyramid).

Unfortunately, this mathematical existence proof did not answer the natural "physical" problems like: *Are there more or less integer points on the sail's edges than on a random segment of similar length between two integer points*? What are the proportions between *the numbers of triangles and quadrangles* (or triangles and pentagons) among the faces of the sail of a random pyramid? What is the distribution of the number of integer points on a face: is it more or less than in a random planar domain of the same integral area? Are the ratios "area/perimeter squared" for faces of the sail distributed in the same way as those for random planar polygons or differently?

All these questions are also problems in *experimental mathematics*, since it would be interesting to have the empirical data for, say, the "pseudorandom" cones generated by three integer vectors $(a,b,c)$ of total norm $|a|^2+|b|^2+|c|^2<N$. The average statistics over the set of all such triples depend on $N$, and the empirical mean values for, say, $N=10$ and 100, might suggest whether there is a limit for $N$ tending to infinity.

These empirical statistics might also be compared with those obtained from cubic algebraic number fields (problem *2002-14*): *Is there a significant difference between the statistic of cubic field sails and that of random pyramid sails*?

[1] ARNOLD V. I. Continued Fractions. Moscow: Moscow Center for Continuous Mathematical Education Press, 2001 (in Russian). ("Mathematical Education" Library, 14.)

[2] BRODÉN T. Wahrscheinlichkeits Bestimmungen bei der gewöhnlichen Kettenbruchentwicklung reeller Zahlen. *Akad. Föhr. Stockholm*, 1900, **57**, 239–266.

[3] KONTSEVICH M. L., SUKHOV YU. M. Statistics of Klein polyhedra and multidimensional continued fractions. In: Pseudoperiodic Topology. Editors: V. Arnold, M. Kontsevich and A. Zorich. Providence, RI: Amer. Math. Soc., 1999, 9–27. (AMS Transl., Ser. 2, 197; Adv. Math. Sci., 46.)

[4] WIMAN A. Über eine-wahrscheinlichkeits Auflage bei Kettenbruchentwicklungen. *Akad. Föhr. Stockholm*, 1900, **57**, 589–841.

## ▽ 2002-16

H Ergodic properties of dense orbits of actions of noncommutative discrete groups on manifolds were first studied in paper [1], cf. problems *1963-6–1963–12.*

[1] ARNOLD V. I., KRYLOV A. L. Uniform distribution of points on a sphere and some ergodic properties of solutions of linear ordinary differential equations in a complex region. *Sov. Math. Dokl.*, 1963, **4**(1), 1–5. [*The Russian original is reprinted in*: Vladimir Igorevich Arnold. Selecta–60. Moscow: PHASIS, 1997, 47–53.]

## △ 2002-16 — *V. I. Arnold*

These equidistribution results have been extended to the rotations of Euclidean spaces by D. Kazhdan, but the products of motions of the Lobachevskian plane or space behave differently, the orbit being non-equidistributed along the plane, preferring to escape to some infinity.

It is unknown what would happen for the de Sitter world action of $\mathrm{SL}(2,\mathbb{Z})$, which can be described as the group of the products of the Lobachevskiĭ reflections in the sides of the equilateral triangle inscribed into the absolute circle of the Klein model. This group of projective transformations of $\mathbb{R}\mathrm{P}^2$ acts on the Möbius band, complementary to the Lobachevskian plane represented by the Klein disk bounded by the absolute circle. Inside the disk, these projective transformations preserve the Lobachevskian metric, while outside they preserve the Lorentzian relativistic metric of the de Sitter world.

The orbit of a point inside the disk is discrete, but for the de Sitter world the orbit of a point is everywhere dense in the Möbius band. The problem is to understand whether it is equipartitioned there or behaves in a different way (which one?).

## 2002-22

Table 1 below presents, for the odd integers $1 < n < 128$, the following characteristics of the periodicity of the Fermat–Euler dynamical system $\{x \mapsto 2x \pmod{n}\}$. The periods are provided by the signed numbers: $(N+)$ and $(M-)$ represent the greatest solutions of the congruences $(N+)$: $2^{\varphi(n)/N} \equiv 1 \pmod{n}$; $(M-)$: $2^{\varphi(n)/M} \equiv -1 \pmod{n}$, and the next numbers $(T, \varphi)$ in the table represent the minimal period $T = \varphi(n)/N$ and the Euler function value $\varphi(n)$. The bold numbers $n$ are prime, in this case $\varphi(n) = n - 1$. The period length number $\varphi(n)/N$ is not listed for the $(2-)$ class where $M = 2$ and hence $N = 1$. For $n = 101$ the period is 50.

Numbers $n$, belonging to the classes $(N = kL+)$ and $(M = kL-)$, belong to the class $(L+)$ if $k$ is even, whereas if $k$ is odd they belong to $(L+)$ and $(L-)$ respectively.

For example, $(6-) \subset (3+)$, $(6-) \subset (2-)$, $(2-) \subset (1+)$. Thus, the minimal periods $T = \varphi(n)/N$ in these cases are the numbers $\varphi(n)/3$ and $\varphi(n)/2$; $\varphi$ is presented in the table. For $n = \mathbf{5}$ the period $T$ is $2 = 4/2$, for $n = \mathbf{43}$ it is $14 = 42/3$.

**Table 1:** The periods of the Fermat–Euler dynamical system

| | $n\,(M-)\,(\varphi)$ | | $n\,(N+)\,(T,\varphi)$ |
|---|---|---|---|
| | **3** (2−) (2) | **5** (2−) (4) | **7** (2+) (3,6) |
| 9 (2−) (6) | **11** (2−) (10) | **13** (2−) (12) | 15 (2+) (4,8) |
| **17** (4−) (8,16) | **19** (2−) (18) | 21 (2+) (6,12) | **23** (2+) (11,22) |
| 25 (2−) (20) | 27 (2−) (18) | **29** (2−) (28) | **31** (6+) (5,30) |
| 33 (4−) (10,20) | 35 (2+) (12,24) | **37** (2−) (36) | 39 (2+) (12,24) |
| **41** (4−) (20,40) | **43** (6−) (14,42) | 45 (2+) (12,24) | **47** (2+) (23,46) |
| 49 (2+) (21,42) | 51 (4+) (8,32) | **53** (2−) (52) | 55 (2+) (20,40) |
| 57 (4−) (18,36) | **59** (2−) (58) | **61** (2−) (60) | 63 (6+) (6,36) |
| 65 (8−) (12,48) | **67** (2−) (66) | 69 (2+) (22,44) | **71** (2+) (35,70) |
| **73** (8+) (9,72) | 75 (2+) (20,40) | 77 (2+) (30,60) | **79** (2+) (39,78) |
| 81 (2−) (54) | **83** (2−) (82) | 85 (8+) (8,64) | 87 (2+) (28,56) |
| **89** (8+) (11,88) | 91 (6+) (12,72) | 93 (6+) (10,60) | 95 (2+) (36,72) |
| **97** (4−) (48,96) | 99 (4−) (30,60) | **101** (2−) (100) | **103** (2+) (51,102) |
| 105 (4+) (12,48) | **107** (2−) (106) | **109** (6−) (36,108) | 111 (2+) (36,72) |
| **113** (8−) (28,112) | 115 (2+) (44,88) | 117 (6+) (12,72) | 119 (4+) (24,96) |
| 121 (2−) (110) | 123 (4+) (20,80) | 125 (2−) (100) | 127 (18+) (7,126) |

To study the behavior of the number of orbits $N(n)$ and of the orbits period $T = \varphi(n)/N(n)$, I have computed the sums of the numbers of orbits $N$ and of the period lengths $\varphi/N$ for the successive dozens of odd numbers $n$ less than 512,

$$24c < n < 24(c+1), \qquad c = 0, 1, \ldots, 20.$$

The resulting two sequences of sums are first

$$\begin{array}{l} 16,\ 25,\ 28,\ 48,\ 37,\ 46,\ 43, \\ 32,\ 38,\ 58,\ 70,\ 62,\ 32,\ 57, \\ 79,\ 49,\ 52,\ 44,\ 95,\ 50,\ 52 \end{array}$$

for the sums of the successive dozens of orbit numbers $N(n)$, and second

$$\begin{array}{rl} S = & 84,\ 210,\ 378,\ 339,\ 527,\ 682,\ 679, \\ & 1071,\ 1037,\ 804,\ 823,\ 891,\ 1375,\ 1165, \\ & 1411,\ 2026,\ 1474,\ 1927,\ 1333,\ 2088,\ 1849 \end{array}$$

for the sums of the successive dozens of period lengths $T = \varphi(n)/N(n)$. Each sum represents the contribution of 12 successive odd numbers $n$ and should be divided by 12 to provide an approximation for the averaged dependence of the orbits number $N$ and of the period $T = \varphi/n$ on $n$: $T \approx S/12$.

**Example.** The greatest number in my tables, $n = 511$, belongs to the classes (48+), (24+), 16+), (12+), (8+), (6+), (4+), (2+), having $\varphi(511) = 432$ residues, relatively prime to it. Hence the corresponding period is $T(511) = \varphi(511)/N(511) = 9$, that is, $2^9 \equiv 1 \pmod{511}$, the number of orbits $N(511)$ being 48 (the number $\varphi(2^a - 1)$ is always divisible by $a$). The value $T = 9$ is much less than $S/12 \approx 1851/12$, but $n = 511$ is exceptional.

The empirical growth of the averaged periods with $n$ in these tables looks approximately like $T = \varphi(n)/N(n) \approx S/12 \approx n/3$ (since $S$ is of order of 2000 for $n \approx 500$). For a random sequence it would grow rather like the square root of $n$.

The *linear* growth of the period $T$ with $n$ seems to show some kind of asymptotic nonrandomness of the sequence of Fermat residues $\{2^i \pmod{n}\}$, which I would consider as a mutual *repulsion* of these residues and which merits to be studied.

The tables show a strange distribution of the values of the ratio $N/T$, where $T = \varphi(n)/N(n)$ is the period, $N$ being the number of orbits.

The empirical large $n$ approximations discussed above,

$$T \approx n/3, \qquad \varphi \approx 6n/\pi^2,$$

would imply relative smallness of the orbits number

$$N \approx 18/\pi^2 \approx 2, \qquad N/T \approx 6/n,$$

confirmed by the tables for $n < 512$.

In these tables, there are many values of $n$ for which $N = 1$ or 2, contributing to make the average $N \approx 2$ in spite of the occurrences of greater $N$. To the opposite case of the greatest values of the ratio $N/T$ belong, say, the values

$$n = 2^T - 1, \qquad N \approx \frac{6}{\pi^2} n/T \approx \frac{6}{\pi^2} n/\log n,$$

where $N/T \approx (6/\pi^2)n/\log^2 n$ becomes large. Such "large $N/T$" values of $n$ are exceptions forming a small part of the totality.

There exist, however, "middle cases" in the tables, where the ratio $N/T$ is neither small nor large, like, say, the case $T \approx N$. Since $NT = \varphi(n) \approx 6n/\pi^2$, we get, for these middle cases, the approximation $N^2 \approx 6n/\pi^2$, the numbers $N$ and $T$

being both proportional to the square root of $n$, as for random $n$-element sequences of length $T$ without repetitions.

It would be interesting to prove that such "middle type" numbers $n$ (with bounded ratio $N/T$) are not exceptional but form a substantial part of the totality of the odd integers (which does not become a small part of the totality of the segment $0 \leq n \leq K$ when $K$ is growing).

Both the distribution of the values of $N(n)/T(n)$ and its asymptotic behavior for $K \to \infty$ remain mysterious. The latest numerical experiments (for $n \sim 10^9$) performed by F. Aicardi (2003) suggest the averaged behavior of the period $T$ of the type $T \sim Cn/\log n$.

It would be interesting to explore the probable behavior ($T \stackrel{?}{\sim} Cn \log n$) of the average growth of the period $T$ using the statistics of the distances between resudues of different powers $2^t$ modulo $n$. To study these distances, one might replace the geometric progression $\{2^t\}$ by the arithmetic progression $\{t \log 2\}$, whose Diophantine problems might be investigated using the (Gauss–Kuz′min) statistics of the elements of continued fractions of random real numbers.

To do this, one needs to know whether the continued fractions of numbers of type $\log_a b$ behave the same as those of random real numbers, which is also an interesting conjecture.

The survey article [1] contains more references and tables, including a discussion of the relation of these statistics to the discovery of the Kolmogorov turbulence laws and to the criteria for randomness of sequences.

[1] ARNOLD V. I. Topology and statistics of formulae of arithmetics. *Russian Math. Surveys*, 2003, **58**(4), 637–664.

# *Author Index for Comments*

(featuring problem numbers)

Printing and Binding: Strauss GmbH, Mörlenbach